Continuous Symmetries and Integrability of Discrete Equations

Volume 38

CRM MONOGRAPH SERIES

Centre de Recherches Mathématiques
Montréal

Continuous Symmetries and Integrability of Discrete Equations

Decio Levi
Pavel Winternitz
Ravil I. Yamilov

The Centre de Recherches Mathématiques (CRM) was created in 1968 to promote research in pure and applied mathematics and related disciplines. Among its activities are thematic programs, summer schools, workshops, postdoctoral programs, and publishing. The CRM receives funding from the Natural Sciences and Engineering Research Council (Canada), the FRQNT (Quebec), the Simons Foundation (USA), the NSF (USA), and its partner universities (Université de Montréal, McGill, UQAM, Concordia, Université Laval, Université de Sherbrooke and University of Ottawa). It collaborates with the Institut des Sciences Mathématiques (ISM). For more information visit www.crm.math.ca.

AMERICAN MATHEMATICAL SOCIETY
Providence, Rhode Island

2020 *Mathematics Subject Classification.* Primary 34-XX, 35-XX, 35Cxx, 35Pxx, 37Kxx, 39-XX, 39Axx; Secondary 17B67, 22E65, 34M55, 34C14, 34K04, 34K08, 34K17, 34L25, 35A22, 35B06, 35Q53, 37J35, 37K40, 37K06, 37K10, 37K15, 37K30, 37K35, 39A06, 39A14, 39A36.

For additional information and updates on this book, visit
www.ams.org/bookpages/crmm-38

Library of Congress Cataloging-in-Publication Data

Names: Levi, D. (Decio), author. | Winternitz, Pavel, author. | Yamilov, Ravil I., 1957-2020, author.
Title: Continuous symmetries and integrability of discrete equations / Decio Levi, Pavel Winternitz, Ravil I. Yamilov.
Description: Providence, Rhode Island : American Mathematical Society, [2022] | Series: CRM monograph series / Centre de Recherches Mathématiques, Montréal, 1065-8599 ; volume 38 | Includes bibliographical references and index.
Identifiers: LCCN 2022037569 | ISBN 9780821843543 (hardcover) | ISBN 9781470472382 (ebook)
Subjects: LCSH: Differential equations. | Symmetry (Mathematics) | Integral equations. | Difference equations. | Discrete mathematics. | AMS: Ordinary differential equations. | Partial differential equations. | Dynamical systems and ergodic theory. | Difference and functional equations. | Nonassociative rings and algebras – Lie algebras and Lie superalgebras – Kac-Moody (super)algebras; extended affine Lie algebras; toroidal Lie algebras. | Topological groups, Lie groups – Lie groups – Infinite-dimensional Lie groups and their Lie algebras: general properties.
Classification: LCC QA371 .L394 2022 | DDC 515/.38–dc23/eng20221024
LC record available at https://lccn.loc.gov/2022037569

∞ The paper used in this book is acid-free and falls within the guidelines
established to ensure permanence and durability.
Visit the AMS home page at `https://www.ams.org/`

10 9 8 7 6 5 4 3 2 1   27 26 25 24 23 22

Contents

Foreword

Pavel Winternitz joined the Centre de Recherches Mathématiques (CRM) in 1972. He has had an illustrious career, has been one of the founders of Montreal school in mathematical physics and for six decades until his saddening departure has kept writing many important new chapters of mathematics and physics. Of all his collaborators, the one with whom he has authored the most papers is Decio Levi, himself a distinguished scholar from the great Italian research tradition in integrable models. With common connections in Mexico, they began collaborating in the mid 80s and have been prolific. Decio Levi was appointed as associate member of the CRM and has been a frequent visitor. They have also regularly collaborated with the third author of this book Ravil Yamilov who trained with Alexei Shabat another pioneer in the modern study of integrability.

Their collaborative work is monumental, it has provided a thorough and profound understanding of the role and applications of symmetry methods in the study of the numerous systems described by non linear differential or difference equations. The present book offers a cogent synthesis of this impactful field they have established; the results presented in this monograph will keep influencing methodological developments and will bear on the solutions of various practical problems.

In the early 90s, at a time when I was director of the CRM, I told Pavel that we should organize a meeting on the symmetries of difference equations. He suggested inviting Decio to join us as organizers. This led to a colloquium entitled *Symmetry and Integrability of Difference Equations*. It is a testimony to the depth and importance of the topic of this workshop that it became the first of an ongoing and lively conference series known under the acronym SIDE that takes place every two years and which is bringing together a large international research community. This book entitled *Continuous Symmetries and Integrability of Discrete Equations* relates to the queries of all these investigators and offers a timely and much welcome broad exposition of a subject that continues to be actively explored.

In one of our many conversations, I recall Pavel telling me that he felt he had waited too long to start writing books. He had in fact put himself to task in this respect lately and he really had this book project at heart. Decio valliantly brought it to completion. I know the publication of this volume would have made Pavel very happy. We, the readers, should be thankful for all the knowledge that it contains and is passed on thus offering a legacy of passionate, inspired and seminal research. I am sure many, younger and seasoned, scientists will delight studying the contents of this book and will find in it a precious source of learned information.

Montreal, February 6 2022

Luc Vinet, CM, OQ, FRSC
Aisenstadt Professor of Physics
Université de Montréal,
Chief Executive Officer,
IVADO
and
Former Director,
Centre de Recherches Mathématiques (CRM)

List of Figures

List of Tables

Preface

The main motivation for writing a new book on integrable systems and symmetries is to present the results of the researches of the authors on the application of symmetries and integrability techniques to the case of equations defined on the lattice. This is a relatively new field which has many applications both in themselves, for example, in the description of the evolution of crystals and molecular systems defined on lattices and as the numerical approximation of a differential equation preserving its symmetries.

Few books already exist in this field [384, 408, 763, 777] which, even if dealing with integrability and symmetries of difference equations, are complementary to the present one as different materials are considered and, if the same, in a different way, using different techniques.

This book is aimed as a tool to PhD students and early researchers, both in Theoretical Physics and in Applied Mathematics interested in the study of symmetries and integrability of differential and difference equations. It contains three Chapters and five Appendices.

The first Chapter is an Introduction where we present the general ideas about symmetries, lattices, differential difference and partial difference equations and Lie point symmetries defined on them. The book should have been a two volume set but, due to Pavel Winternitz death, only a few parts of the first volume have been written and are contained in the Introduction. Here we give a definition of Lie symmetry group, Lie point symmetries, contact symmetries and generalized symmetries, show how to find them and present their main applications both to differential equations and difference equations. In particular we describe the lattices on which difference equations could live and the non obvious behavior of partial differences which one can evince from the discrete Clairaut-Schwarz-Young theorem proved here. Among the applications we present results on symmetry preserving discretization of ordinary differential equations and the extension to the lattice of Lie classification of second order ordinary differential equations.

At the end of the Introduction one can find a list of the subjects, possibly with references, which were meant to be in the first volume and have been left out of this volume.

In Chapter 2 we deal with integrable and linearizable systems in two dimensions. We start from the prototype of integrable and linearizable partial differential equations, the Korteweg de Vries and the Burgers and present all their integrability properties. Then we consider the best known integrable differential difference and partial difference equations.

For all equations we show the integrability properties by presenting their Lax pair, i.e. the overdetermined system of linear equations for a complex function whose compatibility implies the nonlinear partial differential, differential difference or partial difference equation. In the case of integrable equations the Lax pair depends essentially on a complex parameter, often called spectral parameter, which is absent in the case of linearizable equations. In such a case we say the Lax pair is fake. In all cases we construct the corresponding hierarchies of equations, introduce the Bäcklund transformations, Bianchi identities, nonlinear superposition formulas, and infinite sequence of generalized symmetries. In correspondence with the symmetries by symmetry reduction we can find an infinite sequence of exact solutions, the solitons and cnoidal waves.

Throughout the text we present the role of the Bäcklund transformations in the discretization as was presented in [480] and recently re-proposed in [605]. Moreover let us mention the non obvious result, shown in [471], that any discrete equation can be interpreted as a Bäcklund transformation, Bianchi identity or nonlinear superposition formula for a differential equation.

Integrable equations possess also an infinity of non equivalent conserved quantities. This is not the case of the linearizable ones. The proof of this result is carried out in the case of the differential difference Burgers equation in Chapter 3.

For Boll extension of the Adler, Bobenko and Suris integrable quad-graph equations of Volterra type we prove that they are all Darboux integrable and using this property we can solve them.

We extend the well known results by Bluman and Kumei on the use of symmetries for testing the linearizability of the nonlinear partial differential equations to the case of nonlinear differential difference equations.

In Chapter 3 we treat with all details the theory of symmetries as integrability criteria. This theory has been introduced in the case of partial differential equations by Shabat and collaborators in the '80 of last century for the classification of integrable equations. Here we present in detail, as an introduction, the case of partial differential equations and then we go over to the case of differential difference equations and partial difference equations. The theory is used to test for integrability classes of nonlinear equations[110], construct recursive operators, calculate Lax pair, generalized symmetries and conservation laws. The results are then used to classify classes of nonlinear evolutionary differential difference equations of the Toda, relativistic Toda and Volterra type. The proof of the integrability of the obtained autonomous differential difference equations is then given by constructing the Miura transformation with known integrable equations or by constructing the master symmetry which, starting from a symmetry, constructs recursively a denumerable number of them. Results are also presented in the case of simple non autonomous equations, i.e. equations depending explicitly on the lattice index or on time, scalar evolutionary equation of an arbitrary order or having multiple components satisfying a Jordan algebra.

The generalized symmetry method is extended to the case of partial difference equations living on a quad-graph. There the theory is more complex as the integrability conditions are defined up to the equation itself. We carry out the classification fixing the order of the symmetry. At the end we apply the theory to some simple classification problems. The linearizability of partial difference equations defined on a three point lattice or on a quad-graph is dealt with the same techniques of the generalized symmetry method.

In the Appendices we give details which are non essential but may help the reader to deepen part of the subjects presented in Chapter 2 and 3.

This book came out from the experience of writing the review articles [850] and [544] for the Journal of Physics A: Math. Gen.. It is based on material that was previously dispersed in journal articles or Conference Proceedings, all of them written by one or two and few times by the three authors of this book together with collaborators. Occasionally the text is drawn from previous publications however, also unpublished findings are included.

Within each Chapter, Section and Appendix all equations are numbered progressively and are referred accordingly with their complete address. To facilitate retrieving equations, the indication of the Chapter, Section, subSection, subsubSection, subsubsubSection or Appendix where they are situated is indicated. The references are written down in alphabetic order.

This volume is concluded by a subject index including a list of the principal mathematical symbols, when not evident.

The research of Decio Levi has been supported by INFN IS-CSN4 *Mathematical Methods of Nonlinear Physics*, by PRIN projects of the Italian Minister for Education and Scientific Research *Metodi geometrici nella teoria delle onde non lineari ed applicazioni*, 2006, *Nonlinear waves: integrable finite dimensional reductions and discretizations*, from 2007 to 2009, *Continuous and discrete nonlinear integrable evolutions: from water waves to symplectic maps*, from 2010 and from the Project of GNFM–INdAM *Sistemi dinamici nonlineari discreti: simmetrie ed integrabilità e riduzione di equazioni differenziali di interesse fisico-matematico*.

The research of Ravil Yamilov has been partially supported by numerous grants of the Russian Foundation for Basic Research, Grants number 06-01-92051-KE-a, 07-01-00081-a, 10-01-00088-a, 11-01-97005-r-povolzhie-a and 14-01- 97008-r-povolzhie-a.

The research of Pavel Winternitz was partly supported by research grants from NSERC of Canada.

It is disturbing to me that when reading and rereading the text one still find errors. My hope is that any remaining errors will not lead the reader to confusion.

Rome, June 28th, 2022

Decio Levi

Acknowledgment

Decio Levi would like to thank the Centre de Recherches Mathématiques Montréal and Pavel Winternitz for their hospitality during numerous visits there while working on the manuscript. Ravil I. Yamilov and Pavel Winternitz would like to thank the Mathematical and Physics Department of Roma Tre University for its hospitality. Moreover Decio Levi would like to thank O. Ragnisco and M. A. Rodríguez for reading this text and suggesting corrections. All of the authors thank all of their colleagues and students with whom they collaborated on the results that are present in this book. Decio Levi thanks in particular G. Gubbiotti, M. Petrera and C. Scimiterna. Ravil I. Yamilov thanks R. Garifullin. All of the authors thank R. Hernandez Heredero for providing a translation of [841] which presents part of the proof of Theorem 49.

Decio Levi thanks V. Dorodnitsyn, R. Garifullin, W. Miller Jr., A. K. Pogrebkov and L. Šnobl for helping with relevant information of a recent bibliography on the subject of the book. Moreover the authors thank the referees for many suggestions which have been implemented in the present text to improve it.

The authors are indebted to Jonathan Godin for transforming the manuscript into a publishable book and Veronique Hussin for helping with the publication.

Decio Levi dedicates this book to his wife Irene and his children Giorgina and Eugenio. Ravil I. Yamilov dedicates this book to his wife Svetlana and his daughter Ekaterina. Pavel Winternitz dedicates this book to his wife Milada and his sons Michael and Peter. The authors thank them for their support and encouragements.

Permissions

Figures 1.1, 1.2, 1.5, 1.6 are reprinted from [511]. Used with permission. Permission conveyed through Copyright Clearance Center, Inc.

Figures 1.3 and 1.4 are reprinted from [530]. ©IOP Publishing. Reproduced with permission. All rights reserved.

Figures 1.8 is reprinted from [116]. ©IOP Publishing. Reproduced with permission. All rights reserved.

Figure 3.1 is reprinted from [745]. Used with Permission. Permission conveyed through Copyright Clearance Center, Inc.

Appendix D, Translation from the Russian of R I Yamilov, *On the classification of discrete equations*, is reprinted with permission from [841].

CHAPTER 1

Introduction

Integrable systems play an important role in modern mathematics and physics. Its theory for Partial Differential Equations (PDEs) can be found in many books. For example see [3, 12, 68, 69, 76, 80, 83, 106, 133, 147, 154, 171, 185, 193, 211, 215, 216, 233, 234, 237, 247, 250, 268, 274, 317, 335, 355, 358, 395, 401, 439, 446, 447, 450, 452, 461, 467, 584, 585, 588, 619, 623, 630, 648, 649, 668, 671, 674, 738, 748, 763, 854].

The study of integrable discrete equations defined on a lattice goes back to the ‘80 of last century and not so much is present on books [9, 104, 318, 328, 384, 460, 777].

The symmetry theory of differential equations is well understood. It goes back to the classical work of Lie and is reviewed in numerous modern books and articles. Among them let us mention [36, 37, 43, 45, 97, 97–100, 111, 153, 153, 162, 281, 288, 405, 412, 413, 419, 442, 479, 604, 631, 659, 660, 666, 712, 770, 771, 773, 820, 830]. As a matter of fact, Lie group theory is the most general and useful tool we have for obtaining exact analytic solutions of large classes of differential equations, specially non linear ones. All exact analytic solutions of a differential equation are related in a way or another to Lie group theory or its extensions.

The application of Lie group theory to discrete equations is much more recent and a vigorous development of the theory only started in the 1990-ties [27, 71, 139, 162, 187, 217–224, 230–232, 258, 259, 322, 351, 367, 371–373, 376–378, 384, 424, 427, 464, 465, 468, 476–479, 481, 485–489, 498, 503, 505, 508, 510, 514–516, 526, 527, 529, 536, 537, 539–541, 544–546, 550–553, 555, 572, 575, 576, 578, 673, 711, 715, 723–726, 804, 831, 832, 834], [506, 507, 525, 528, 530–532, 535, 538, 542, 543, 547, 556–558, 563, 574, 577, 594, 612, 643, 653, 654, 681, 683, 698–703, 733, 799, 800, 809, 833, 850, 866]. In this whole field of research one uses group theory to do for difference equations what has been done for differential ones. This includes generating new solutions from old ones, identifying equations that can be transformed into each other, performing symmetry reduction, and identifying integrable equations. Moreover Lie group theory in the discrete setting can be used, see [71, 89–91, 93–96, 152, 220–223, 232, 483, 484, 514–516, 545, 701–703, 808, 809, 833], to discretize a differential equation preserving its symmetries.

When adapting the group theoretical approach from differential equations to difference ones, we must answer four basic questions:

(1) What do we mean by symmetries and what we do with them?
(2) What we know about lattices?
(3) What is a difference equation?
(4) How do we find the symmetries of a difference equation?

Let us first discuss briefly the first point for differential equations and then we pass to define the lattice, the difference equations and symmetries for them.

1. Lie point symmetries of differential equations, their extensions and applications

Let us here briefly review the situation for differential equations both Ordinary Differ<ential Equations (ODEs) and PDEs. In this exposition we will mainly follow the book by Olver[659] and will concentrate on results that will be needed for discrete equations. For more details we advice the reader to consult [659].
Let us consider a general system of N differential equations

$$E_a(x, u, u_x, u_{2x}, \ldots, u_{nx}), \quad x \in \mathbb{R}^p, u \in \mathbb{R}^q, a = 1, \ldots, N, \quad p \in \mathbb{Z}^+, q \in \mathbb{Z}^+, \tag{1.1.1}$$

for p independent variables and q dependent ones. The functions E_a are smooth but otherwise arbitrary.
The general x and u are vectors and (1.1.1) is a system of PDEs, x are independent variables, u dependent ones and u_{kx} denotes the set of all partial derivatives of order k of all components of u. For $n = 0$ (1.1.1) is a system of algebraic equation, for $n \neq 0$ and $N > 1$ it is a set of differential equations. For $p = 1$ it is a system of ODEs, for $p \geq 1$, $q = 1$, $N = 1$ it is a single PDE (or ODE).
Let us first consider the Lie point symmetry group of system (1.1.1). This is a local Lie group G of local point transformation taking solutions of the system (1.1.1) into solutions of the same system. Thus the symmetry group transformations leave the solution set invariant (but not necessarily individual solutions).
A one-parameter set of Lie point transformations has the form

$$\tilde{x} = \Lambda_\epsilon(x, u), \quad \tilde{u} = \Omega_\epsilon(x, u), \tag{1.1.2}$$

where $\Lambda_\epsilon(x, u)$ and $\Omega_\epsilon(x, u)$ are differentiable functions of x and u and analytic in ϵ. ϵ is the group parameter such that $\epsilon = 0$ corresponds to the identity transformation

$$\tilde{x} = x = \Lambda_0(x, u), \quad \tilde{u} = u = \Omega_0(x, u)$$

It is assumed that the inverse transformation, which we will indicate by $-\epsilon$, exists and it is such that

$$x = \Lambda_{-\epsilon}(\tilde{x}, \tilde{u}), \quad u = \Omega_{-\epsilon}(\tilde{x}, \tilde{u}),$$

at least locally (for $|\epsilon| \ll 1$, $|\tilde{x} - x| \ll 1$).

The closure is probably the most important condition for a Lie group of transformations. The combination of two transformations, one of parameter ϵ given by (1.1.2) and one of a different parameter, say $\tilde{\epsilon}$, given by

$$\tilde{\tilde{x}} = \Lambda_{\tilde{\epsilon}}(\tilde{x}, \tilde{u}), \quad \tilde{\tilde{u}} = \Omega_{\tilde{\epsilon}}(\tilde{x}, \tilde{u}), \tag{1.1.3}$$

gives a transformation of the same form

$$\tilde{\tilde{x}} = \Lambda_\mu(x, u), \quad \tilde{\tilde{u}} = \Omega_\mu(x, u), \tag{1.1.4}$$

where $\mu = \psi(\epsilon, \tilde{\epsilon})$, the combination law of the group parameters, is an analytic function of both parameters, ϵ and $\tilde{\epsilon}$.

The transformations (1.1.2) of local coordinates also determine the transformations of functions $u = f(x)$ and of derivatives of functions. For a differential equation the transformation (1.1.2), the set of transformations which leave the equation invariant and transform solutions into solutions, determine the transformation of the derivatives. For more details see [659].
How does one find the symmetry group G? Instead of looking for 'global' transformations

as in (1.1.2) one looks for infinitesimal ones, i.e. one looks for the Lie algebra $\mathfrak{g}$ that corresponds to G. This exists due to the analyticity and differentiability property of the functions Λ_ϵ and Ω_ϵ. A one-parameter group of infinitesimal point transformations will have the form

$$\tilde{x}_i = x_i + \epsilon\xi_i(x,u) + \mathcal{O}(\epsilon^2), \quad \tilde{u}_\alpha = u_\alpha + \epsilon\phi_\alpha(x,u) + \mathcal{O}(\epsilon^2),$$

$$|\epsilon| \ll 1, |\tilde{x} - x| \ll 1, \tag{1.1.5}$$

where

$$\xi_i(x,u) = \left.\frac{\partial \Lambda_{\epsilon i}}{\partial \epsilon}\right|_{\epsilon=0} \qquad \phi_\alpha(x,u) = \left.\frac{\partial \Omega_{\epsilon \alpha}}{\partial \epsilon}\right|_{\epsilon=0}.$$

The search for the symmetry algebra $\mathfrak{g}$ of a system of differential equations is best formulated in terms of vector fields acting on the space $X \times U$ of independent and dependent variables. Indeed, consider the vector field

$$\hat{X} = \sum_{i=1}^{p} \xi_i(x,u)\partial_{x_i} + \sum_{\alpha=1}^{q} \phi_\alpha(x,u)\partial_{u_\alpha}, \tag{1.1.6}$$

where the coefficients ξ_i and ϕ_α are defined in (1.1.5). If the functions ξ_i and ϕ_α are known, the vector field (1.1.6) can be integrated to obtain the finite transformations (1.1.2). This is the content of the First Lie theorem. Indeed to get the function Λ and Ω, all we have to do is to integrate the equations

$$\frac{d\tilde{x}_i}{d\epsilon} = \xi_i(\tilde{x},\tilde{u}), \quad \frac{d\tilde{u}_\alpha}{d\epsilon} = \phi_\alpha(\tilde{x},\tilde{u}), \tag{1.1.7}$$

subject to initial conditions

$$\tilde{x}_i|_{\epsilon=0} = x_i, \quad \tilde{u}_\alpha|_{\epsilon=0} = u_\alpha. \tag{1.1.8}$$

This provides us with a one-parameter group of local Lie point transformations of the form (1.1.2) where ϵ is the group parameter.
The vector field (1.1.6) tells us how the variables x and u transform. When dealing with a differential equation we also need to know how derivatives such as $u_x, u_{xx}, \dots, u_{nx}$ transform. This is given by the prolongation of the vector field $\hat{X}$. We have

$$\mathrm{pr}\hat{X} = \hat{X} + \sum_\alpha \Big\{ \sum_i \phi_\alpha^{x_i}\partial_{u_{\alpha,x_i}} + \sum_{i,k} \phi_\alpha^{x_i x_k}\partial_{u_{\alpha,x_i x_k}} + \sum_{i,k,l} \phi_\alpha^{x_i x_k x_l}\partial_{u_{\alpha,x_i x_k x_l}} + \dots \Big\}, \tag{1.1.9}$$

where the coefficients in the prolongation can be calculated recursively, using the total derivative operator,

$$D_{x_i} = \partial_{x_i} + u_{\alpha,x_i}\partial_{u_\alpha} + u_{\alpha,x_a x_i}\partial_{u_\alpha, x_a} + u_{\alpha,x_a x_b x_i}\partial_{u_\alpha, x_a x_b} + \dots, \tag{1.1.10}$$

(a summation over repeated indexes is to be understood). The recursive formula are

$$\phi_\alpha^{x_i} = D_{x_i}\phi_\alpha - (D_{x_i}\xi_a)u_{\alpha,x_a}, \quad \phi_\alpha^{x_i x_k} = D_{x_k}\phi_\alpha^{x_i} - (D_{x_k}\xi_a)u_{\alpha,x_i x_a},$$

$$\phi_\alpha^{x_i x_k x_l} = D_{x_l}\phi_\alpha^{x_i x_k} - (D_{x_l}\xi_a)u_{\alpha,x_i x_k x_a}, \tag{1.1.11}$$

etc.
The invariance condition for system (1.1.1), i.e. the condition that ξ_i and ϕ_α provide a transformation Λ_ϵ and Ω_ϵ which leave (1.1.1) invariant, is expressed in terms of the operator (1.1.9) as

$$\mathrm{pr}^{(n)}\,\hat{X} E_a = 0, \quad a = 1, \dots, N, \tag{1.1.12}$$

when (1.1.1) is satisfied, i.e. $E_1 = \dots = E_N = 0$. In (1.1.12) $\mathrm{pr}^{(n)}\,\hat{X}$ is the prolongation (1.1.9) calculated up to order n (where n is the order of system (1.1.1)).
Eq. (1.1.12) is a system of linear partial differential equations for the functions $\xi_i(x,u)$ and

$\phi_\alpha(x,u)$, in which the variables x and u figure as independent variables. By definition of point transformations the coefficients ξ_i and ϕ_α depend only on $(x_1,\dots,x_p,u_1,\dots,u_q)$, not on any derivatives of u_α. The action of $pr^{(n)}\hat{X}$ in (1.1.12) will, on the other hand, introduce terms in (1.1.12), involving the derivatives $\frac{\partial^k u}{\partial x_1^{k_1}\dots\partial x_p^{k_p}}$, $k = k_1 + \dots + k_p$, $1 \le k \le n$. We use the N equations (1.1.1) to eliminate N of the derivatives. We then collect all linearly independent remaining expressions in the derivatives and set the coefficients of these expressions equal to zero. This provides the *determining equations*: a set of linear partial differential equations for the functions $\xi_i(x,u)$ and $\phi_\alpha(x,u)$. The order of the system of determining equations is the same as the order of the studied system (1.1.1); however, the determining system is linear, even if the system (1.1.1) is non linear. It is usually overdetermined and not difficult to solve. Computer programs using various symbolic languages exist that construct the determining system and solve it, or at least partially solve it [75, 153, 163, 164, 166, 213, 362, 366, 705, 706, 709, 734, 740, 770, 836]. The solution of the determining system may be trivial, i.e. $\xi_i = 0, \phi_\alpha = 0$. Then the only symmetry available is the identity and symmetry approach is of no avail. Alternatively, the general solution may depend on a finite number K_0 of integration constants. In this case the Lie algebra $\mathfrak{g}$ of the symmetry group, the 'symmetry algebra', for short, is then K_0-dimensional and must be identified as an abstract Lie algebra [422, 697, 761].

The symmetry algebra $\mathfrak{g}$ obtained by solving the determining equations is usually obtained in a nonstandard form and must be transformed to some 'canonical basis'. The Lie algebra may be decomposable into a direct sum of Lie algebras

$$\mathfrak{g} \sim \mathfrak{g}_1 \oplus \mathfrak{g}_2 \oplus \dots \oplus \mathfrak{g}_j.$$

It is then advantageous to transform to a basis in which the decomposition is explicit and then consider each indecomposable component $\mathfrak{g}_i, i = 1,\dots,j$ separately.

The possibilities are:

1. $\mathfrak{g}_i$ is simple. Complete classifications of all complex and real simple Lie algebras exist and can be found in many books [155, 298, 444, 677, 697, 761].
2. $\mathfrak{g}_i$ is solvable. Then one needs to determine its (unique) nilradical (maximal nilpotent ideal) and other invariants (basis independent quantities) like dimensions of subalgebras in the derived series, upper central series and lower central series.
3. $\mathfrak{g}_i$ has a nontrivial Levi decomposition [559] into a semidirect sum

$$\mathfrak{g}_i = P \ni R(\mathfrak{g}_i)$$

where P is the Levi factor, i.e. a uniquely defined semisimple subalgebra of $\mathfrak{g}_i$ and $R(\mathfrak{g}_i)$ is the radical (the unique maximal solvable ideal).

For algorithms performing the above decomposition tasks we refer to the article [697], the book [326] and various computer programs [188, 189, 325, 326, 694, 696].

Finally, the general solution of the determining equations may involve arbitrary functions and the symmetry algebra is infinite-dimensional. For instance, for a linear PDE the linear superposition principle is reflected by the presence in the Lie algebra of an operator depending on the general solution of the studied equation. In turn, this general solution depends on arbitrary functions, e.g. the Cauchy boundary data.

Contact symmetries. So far we have considered only point transformations, as in (1.1.2), in which the new variables $\tilde{x}$ and $\tilde{u}$ depend only on the old ones, x and u. More general transformations are "contact transformations", where ϕ_α and ξ_i and $\phi_\alpha^{x_i}$ also depend on first derivatives of u [45, 100, 153, 405, 412, 659, 773] and nothing else. This turns out to happen if

$$\phi_{\alpha,u_x} = u_x \xi_{i,u_x}, \tag{1.1.13}$$

the contact condition, which implies that the coefficient of the prolongation at order n is such that $\phi_\alpha^{(n)} = \phi_\alpha^{(n)}(x, u, u_x, \dots, u_{nx})$. Contact transformations can still generate a symmetry group given by .

$$\tilde{x} = \Lambda_\epsilon(x, u, u_x), \quad \tilde{u} = \Omega_\epsilon(x, u, u_x), \quad \tilde{u}_{\tilde{x}} = \Phi_\epsilon(x, u, u_x). \tag{1.1.14}$$

On contact transformations we quote the following theorem which can be found in [72,412]:

Theorem 1. *Every group of nth order contact transformations is either*

(1) *a group of pointwise transformations* (1.1.2) *extended to derivatives up to order n if $q > 1$ or*
(2) *a Lie group of contact transformations extended to derivatives up to order n if $q = 1$.*

Generalized symmetries. A still more general class of transformations are generalized transformations, also called *Lie-Bäcklund* transformations [98, 100, 153, 412, 647, 658, 659, 771]. These involve derivatives of arbitrary orders.

When studying generalized symmetries, and sometimes also point symmetries, it is convenient to use a different formalism, namely that of evolutionary vector fields. Let us first consider the case of Lie point symmetries, i.e. vector fields of the form (1.1.6) and their prolongations (1.1.9). With each vector field (1.1.6) we can associate its evolutionary counterpart $\hat{X}_e$, defined as

$$\hat{X}_e = Q_\alpha(x, u, u_x)\partial_{u_\alpha}, \; Q_\alpha = \phi_\alpha - \xi_j u_{\alpha,x_j}. \tag{1.1.15}$$

The prolongation of the evolutionary vector field (1.1.15) is defined as

$$\begin{aligned} pr\hat{X}_e &= Q_\alpha \partial_{u_\alpha} + Q_\alpha^{x_j} \partial u_{\alpha,x_j} + Q_\alpha^{x_j x_k} \partial_{u_\alpha, x_j x_k} + \dots \\ Q_\alpha^{x_j} &= D_{x_j} Q_\alpha, \quad Q_\alpha^{x_j x_k} = D_{x_j} D_{x_k} Q_\alpha, \dots . \end{aligned} \tag{1.1.16}$$

The functions Q_α are called the characteristics of the vector field. In this formalism the operators $\hat{X}_e$ and $pr\hat{X}_e$ do not act on the independent variables x_j. For Lie point symmetries evolutionary and ordinary vector fields are entirely equivalent and it is easy to pass from one to the other. Indeed, (1.1.15) gives the connection between the two. The symmetry algorithms for calculating the symmetry algebra $\mathfrak{g}$ in terms of ordinary, or evolutionary vector fields, are also equivalent. Equation (1.1.12) is simply replaced by

$$\mathrm{pr}^{(n)} \hat{X}_e E_a = 0, \quad a = 1, \dots, N, \tag{1.1.17}$$

when $E_1 = \dots = E_N = 0$ and its differential consequences. The reason that equations (1.1.12) and (1.1.17) are equivalent is the following:

$$\mathrm{pr}^{(n)} \hat{X}_e = \mathrm{pr}^{(n)} X - \xi_i D_i. \tag{1.1.18}$$

The total derivative D_i acts like a generalized symmetry of (1.1.1), i.e.,

$$D_i E_a = 0 \quad i = 1, \dots, p, \; a = 1, \dots, N, \tag{1.1.19}$$

when $E_1 = \dots = E_N = 0$. Eqs. (1.1.18) and (1.1.19) prove that systems (1.1.12) and (1.1.17) are equivalent. Eq. (1.1.19) itself follows from the fact that $D_i E_a = 0$ is a differential consequence of equation (1.1.1); hence, every solution of (1.1.1) is also a solution of (1.1.19) (i.e. the action of D_i on solutions is trivial).

To find generalized symmetries of order k, we use (1.1.15) but allow the characteristics Q_α to depend on all derivatives of u up to order k

$$\hat{X}_e^k = Q_\alpha(x, u, u_x, \cdots, u_{kx})\partial_{u_\alpha}. \tag{1.1.20}$$

The prolongation is calculated using (1.1.16). The symmetry algorithm is again (1.1.17) when $E_1 = \cdots = E_N = 0$ together with its k derivatives.

A very useful property of evolutionary symmetries is that the functions Q_α provide compatible flows. This means that the system of equations

$$\frac{\partial u_\alpha}{\partial \epsilon} = Q_\alpha \tag{1.1.21}$$

is compatible with system (1.1.1) when $u_\alpha = u_\alpha(x, \epsilon)$. In particular, group-invariant solutions [458], i.e., solutions invariant under a subgroup of the K_0 dimensional group G, are obtained as fixed points of

$$Q_\alpha = 0 \tag{1.1.22}$$

If Q_α is the characteristic of a point and contact transformation, then (1.1.22) is a system of quasilinear first-order PDEs. They can be solved and their solutions can be substituted into (1.1.1), yielding the invariant solutions explicitly. If Q_α is the characteristic of a generalized symmetry (1.1.20), (1.1.21) does not provide a group transformation. Eq. (1.1.22) however can still be used to find invariant solutions . Usually generalized symmetries exist only for integrable systems, which have an infinite number of them and thus will form an infinite algebra, as we will see in Chapter 2. We mention that there is no guarantee that (1.1.21) or even (1.1.22) will provide physically meaningful solutions.
If Q_α is the characteristic of a generalized symmetry, then (1.1.22) is a system of generally non linear PDEs which rarely can be solved and, for integrable equations, often provide soliton solutions. In this case the Lie algebra $\mathfrak{g}$ is infinite [see Chapter 2 for examples associated to integrable PDEs].

Formal symmetries. Assuming for simplicity that (1.1.1) is just an evolutionary system of PDEs, which we can write as

$$u_{\alpha,t} = f_\alpha(x, u, u_y, \cdots, u_{ny}), \tag{1.1.23}$$

where t is one of the p independent variables $x \equiv \{t, y\}$. Then the symmetries are defined by the compatibility between (1.1.23) and (1.1.21):

$$\frac{\partial^2 u_\alpha}{\partial t \partial \epsilon} = \frac{\partial^2 u_\alpha}{\partial \epsilon \partial t}, \qquad \text{i.e.} \tag{1.1.24}$$

$$D_\epsilon f_\alpha - D_t Q_\alpha = 0, \tag{1.1.25}$$

where the total derivatives D_ϵ and D_t are given as in (1.1.10).

A formal series of order k, A_k is given by

$$A_k = a_k D^k + a_{k-1} D^{k-1} + \cdots + a_0 + a_{-1} D^{-1} + \cdots, \tag{1.1.26}$$

where a_j are function of the form

$$a_j = a_j(u, u_y, \cdots, u_{\ell y}) \qquad 0 \le \ell \le \infty. \tag{1.1.27}$$

An approximate series solution of order k of (1.1.25), i.e. containing k terms, will be called a formal symmetry of order k. The detailed complete definition of formal symmetry will be given in Section 3.1.

Applications. A further important aspect of the group analysis of differential equations is the construction of equations invariant under a given Lie group. The way to do this for an ODE or a PDE is to represent the Lie algebra $\mathfrak{g}$ of the given Lie group G by vector fields of the form (1.1.6) and to prolong them as in (1.1.9). The invariant equation is constructed using only the invariants I of the algebra $\mathfrak{g}$. These will be the solution of the linear PDE

$$\mathrm{pr}\hat{X}\, I(x_i, u_\alpha, u_{\alpha,x_i}, u_{\alpha,x_i x_k}, \dots) = 0.$$

To clarify the concepts involved, let us look at the example of the KdV equation

$$u_t + uu_x + u_{xxx} = 0 \tag{1.1.28}$$

Its Lie point symmetry algebra is well known [98, 153, 659, 771]. A standard basis is given by

$$\hat{P}_0 = \partial_t,\ \hat{P}_1 = \partial_x,\ \hat{B} = t\partial_x + \partial_u,\ \hat{D} = x\partial_x + 3t\partial_t - 2u\partial_u, \tag{1.1.29}$$

with non-zero commutation relations

$$[\hat{P}_0, \hat{B}] = \hat{P}_1,\ [\hat{P}_0, \hat{D}] = 3\hat{P}_0,\ [\hat{P}_1, \hat{D}] = \hat{P}_1,\ [\hat{B}, \hat{D}] = -2\hat{B}. \tag{1.1.30}$$

To identify (1.1.30) as an abstract Lie algebra [422] we calculate the dimensions of its characteristic series.
The derived series (DS), a sequence of subalgebras of $\mathfrak{g}$ of decreasing order, is defined by :$g^{(0)} = \mathfrak{g},\ g^{(k)} = [g^{(k-1)}, g^{(k-1)}]$.
In the case of (1.1.30) the ideals and their dimensions are

$$\{\hat{P}_1, \hat{P}_0, \hat{B}, \hat{D}\} \supset \{\hat{P}_1, \hat{P}_0, \hat{B}\} \supset \{\hat{P}_1\} \supset \{0\},\ [4, 3, 1, 0] \tag{1.1.31}$$

DS terminates ($g^{(k)} = 0, k \geq 3$), hence the algebra is solvable.
The lower central series (CS) is defined by: $g^1 = \mathfrak{g},\ g^k = [g^{k-1}, \mathfrak{g}]\ k > 1$. We have:

$$\{\hat{P}_1, \hat{P}_0, \hat{B}, \hat{D}\} \supset \{\hat{P}_1, \hat{P}_0, \hat{B}\} \supseteq \{\hat{P}_1, \hat{P}_0, \hat{B}\},\ [4, 3, 3, \dots] \tag{1.1.32}$$

CS does not terminate, hence the algebra is not nilpotent.
CS is formed from the center of the Lie algebra and the centers of a series of factor algebras (see[697]) and is not relevant in the present algebra (since its center is $\{0\}$).
The nilradical of (1.1.29), $\{\hat{P}_1, \hat{P}_0, \hat{B}\}$, is isomorphic to the Heisenberg algebra (the only 3-dimensional indecomposable nilpotent Lie algebra). The entire algebra is isomorphic to $S_{4,8}$ given in [697] with parameter $a = -\frac{2}{3}$.
In order to find equations invariant under the symmetry group of the KdV equation we must first choose the class of equations we want to consider. Let us choose this class as PDEs of order 3 with one dependent and two independent variables that includes the KdV. Our choice is

$$u_{xxx} - F(x, t, u, u_x, u_t) = 0, \tag{1.1.33}$$

where F is an arbitrary sufficiently smooth function. The prolonged vector field acting on (1.1.33) will have the form

$$\mathrm{pr}\,\hat{X} = \xi\partial_x + \tau\partial_t + \phi\partial_u + \phi^x\partial_{u_x} + \phi^t\partial_{u_t} + \phi^{xxx}\partial_{u_{xxx}}. \tag{1.1.34}$$

The second and third derivatives $u_{xx}, u_{xt}, u_{tt}, u_{xxt}, u_{xtt}$ and u_{ttt} do not figure in (1.1.33) and are hence the corresponding prolongation of $\hat{X}$ is not needed in (1.1.34).
We have

$$\begin{aligned} \phi^t &= \phi_t - \xi_t u_x + (\phi_u - \tau_t)u_t - \xi_u u_x u_t - \tau_u u_t^2 \\ \phi^x &= \phi_x + (\phi_u - \xi_x)u_x - \tau_x u_t - \xi_u u_x^2 - \tau_u u_x u_t \end{aligned} \tag{1.1.35}$$

$$
\begin{aligned}
\phi^{xxx} = {} & \phi_{xxx} + (3\phi_{xxu} - \xi_{xxx})u_x - \tau_{xxx}u_t \\
& + 3(\phi_{xuu} - \xi_{xxu})u_x^2 - 3\tau_{xxu}u_x u_t \\
& + (\phi_{uuu} - 3\xi_{xuu})u_x^3 - 3\tau_{xuu}u_x^2 u_t - \xi_{uuu}u_x^4 \\
& + 3(\phi_{xu} - \xi_{xx})u_{xx} - 3\tau_{xx}u_{tx} - \tau_{uuu}u_x^3 u_t \\
& + 3(\phi_{uu} - 3\xi_{xu})u_x u_{xx} - 6\tau_{xu}u_x u_{xt} \qquad (1.1.36)\\
& - 3\tau_{ux}u_t u_{xx} - 6\xi_{uu}u_x^2 u_{xx} \\
& - 3\tau_{uu}u_x u_t u_{xx} - 3\tau_{uu}u_x^2 u_{xt} - 3\xi_u u_{xx}^2 \\
& - 3\tau_u u_{tx}u_{xx} + (\phi_u - 3\xi_x)u_{xxx} - 3\tau_x u_{txx} \\
& - \tau_u u_t u_{xxx} - 4\xi_u u_x u_{xxx} - 3\tau_u u_x u_{xxt}
\end{aligned}
$$

Specifically for the algebra (1.1.29) the relevant prolongations are

$$
\begin{aligned}
& \text{pr}\,\hat{P}_1 = \partial_x,\ \text{pr}\,\hat{P}_0 = \partial_t,\ \text{pr}\,\hat{B} = t\partial_x + \partial_u - u_x\partial_{u_t} \\
& \text{pr}\,\hat{D} = x\partial_x + 3t\partial_t - 2u\partial_u - 3u_x\partial_{u_x} - 5u_t\partial_{u_t} - 5u_{xxx}\partial_{u_{xxx}} \qquad (1.1.37)
\end{aligned}
$$

The invariants of the Lie algebra will satisfy

$$\text{pr}\,\hat{X}I(x,t,u,u_x,u_t,u_{xxx}) = 0, \qquad (1.1.38)$$

for $\hat{X} = \hat{P}_1, \hat{P}_0, \hat{B}$ and $\hat{D}$.

The elements $\hat{P}_1$ and $\hat{P}_0$ restrict I to $I(u,u_x,u_t,u_{xxx})$. The element $\hat{D}$ restricts further to $I = I(J_1,J_2,J_3)$ with

$$J_1 = \frac{u_x^2}{u^3}, \quad J_2 = \frac{u_t^2}{u^5}, \quad J_3 = \frac{u_{xxx}^2}{u^5}. \qquad (1.1.39)$$

Finally the element $\hat{B}$ restricts the number of independent invariants to two which we choose to be

$$Z_1 = \frac{u_{xxx}}{u_x^{5/3}}, \quad Z_2 = \frac{u_t + uu_x}{u_x^{5/3}}. \qquad (1.1.40)$$

A general evolution equation invariant under the KdV group corresponding to the algebra (1.1.28) can be written as

$$\frac{Z_2}{Z_1} = F(Z_1) \quad \text{or} \quad u_t + uu_x = u_{xxx}F\left(\frac{u_{xxx}}{u_x^{5/3}}\right). \qquad (1.1.41)$$

The KdV is obtained for $F =$ constant, in the form (1.1.28) for $F = -1$. In general F is an arbitrary smooth function. For $F(Z_1) = A_k Z_1^k + B_1, k \in \mathbb{Z}$ the lowest nontrivial equation is obtained for $k = 1$:

$$u_t + uu_x = u_{xxx}\left[B_1 + A_1\frac{u_{xxx}}{u_x^{5/3}}\right]. \qquad (1.1.42)$$

Group invariant solutions. As an example of the construction of group invariant solutions we consider the KdV equation (1.1.28) and its Galilei invariant solutions, i.e. solutions invariant with respect to $\hat{B}$ (1.1.29). The independent invariants in this case are

$$y = t, \quad v = tu - x. \qquad (1.1.43)$$

From (1.1.43) we derive

$$u = \frac{1}{y}(x+v),\ u_x = \frac{1}{y}, \quad u_{xxx} = 0, \quad u_t = \frac{1}{y^2}\left[yv_y - v - x\right], \qquad (1.1.44)$$

where x is a parametric variable. Introducing (1.1.44) into (1.1.28) we get

(1.1.45) $$v_y = 0, \text{ i.e. } v = \delta$$

where δ is an arbitrary constant. So the general Galilei invariant solution is

$$u = \frac{1}{t}(x + \delta).$$

Classifications of invariant solutions. Given a K_0 dimensional Lie point algebra we have an infinity of group invariant solutions due to the presence of the K_0 constants. So in general it will be impossible to list all the possible group invariant solutions of the system at study. However not all these solutions will be independent as some will be related by group transformations. We will call the independent solutions, i.e. those which will not be related by group transformations, the *optimal system of solutions*.

Given a group G and a subgroup H, and element $g \in G$ such that $g \notin H$. g will transform the solutions invariants under H into another group invariant solution. So the optimal system will be given by those invariant solutions which are not related by any element $g \in G$. The problem of finding the optimal system of solutions is strictly related to the problem of finding the *optimal system of subgroups* H of G which are not related among themselves by any element $g \in G$. In correspondence with the optimal system of subgroups we can find an optimal system of subalgebras $\mathfrak{h} \in \mathfrak{g}$.

The general procedure to construct the s dimensional subalgebras $\mathfrak{h}$, with $s > 1$, of $\mathfrak{g}$ is outlined in [659] and given in [666]. A procedure to construct the one dimensional subalgebras $\mathfrak{h}$ of $\mathfrak{g}$ is given in [659]. The construction of the optimal system of solutions for the KdV equation (1.1.28) can be found in [403, 659] and for the motion of a two dimensional gas in [600].

Conservation laws and symmetries. An important notion for differential equations is that of conservation laws. Conservation laws like the conservation of energy or the conservation of momentum are very important for physical systems and play an important role in the description of their solutions. In Chapter 3 the notion of conservation laws is introduced to classify integrable non linear PDEs and to distinguish truly non linear integrable equations from those non linear PDEs linearizable by a transformation. In 1918 Emmy Noether [647] showed that for systems arising from a variational principle every conservation law of the system comes from a symmetry. Recently Ibragimov [414, 415] and Bluman and Anco [38, 98] extended this connection from systems arising from a variational principle to general systems for which the conservation laws can be found by a direct computational method similar to Lie's method for finding the symmetries. A complete description of these results can be found in [39]. Older results for obtaining conservations laws for integrable systems both in 1+1 and 2+1 dimensions can be found in [266].

Extensions. Many different extensions of Lie's original method of group invariant solutions exist. Among them we mention, first of all, *conditional symmetries* [49, 50, 99, 278, 280, 534]. For differential equations, they were introduced under several different names [99, 534, 662] in order to obtain dimensional reductions of PDEs, beyond those obtained by using ordinary Lie symmetries.

An interesting extension, mainly concerned with ODEs, is denoted λ-*symmetries* [179, 290, 293, 488, 624, 625, 650, 679] which help to provide reductions even when no Lie symmetries are available and have been shown to be related to *potential symmetries* [160, 161]. In the case of PDEs it has been called μ-symmetries.

Another valuable extension is the concept of *partial symmetries*. They correspond to the existence of a subset of solutions which, without necessarily being invariant, are mapped into each other by the transformation [178, 180]. Further extensions are given by *asymptotic symmetries* [289, 292, 509, 722], when extra symmetries are obtained in the asymptotic regime, or *approximate symmetries* [43, 70, 279] where one considers the symmetries of approximate solutions of a system depending on a small parameter.

2. What is a lattice

2.1. 1-dimensional lattices. A one dimensional lattice in a domain $\mathcal{D}$ of the line is a set of K points $P(x_k)$, $0 \leq k \leq (K-1)$ on a line. These points are characterized by their position. The origin on the line will be denoted by $x_0 = x$. The $(n+1)^{th}$ point will be denoted by its position x_n. So a natural lattice of K points in $\mathcal{D}$ will be given by the points $\{x_{k-1}\}$, $1 \leq k \leq K$. A dependent variable on this lattice is given by $u_n(x_n)$. A one dimensional scheme is given by

$$\{x_{k-1}, u_{k-1}(x_{k-1})\}, \quad 1 \leq k \leq K. \tag{1.2.1}$$

The value of the dependent variable in the point $x_0 = x$ will be just denoted by $u_0(x_0) = u$.

An alternative equivalent set of coordinates, as seen in [511], is given by:

$$\{x, u, p^{(1)}, p^{(2)}, p^{(3)}, p^{(4)}, \dots p^{(K-1)}, p^{(K)}, h_1, h_2, \dots h_K\} \tag{1.2.2}$$

with

$$\begin{aligned} h_k &= x_k - x_{k-1}, \qquad 1 \leq k \leq K, \quad p^{(\ell)} = [D_x]^{\ell} u, \; \ell \leq K-1, \\ D_x &= \frac{1}{h_1}[S_x - 1], \; S_x u_n = u_{n+1}, \quad n \in (0, K-1) \end{aligned} \tag{1.2.3}$$

In the continuous limit $h_k \to 0$, $u_0(x_0) = u(x)$ and $p^{(\ell)} \to \frac{d^{\ell} u(x)}{dx^{\ell}}$.

The passage from (1.2.1) to (1.2.2) corresponds to the passage from the set of the points on the lattice with a function defined on them to a reference point, the origin $x_0 = x$, and the function on it, $u_0(x_0) = u$, together with the distances among the points, h_k, and the ℓ discrete derivative of the function u, $p^{(\ell)}$. The two systems are related by a one to one correspondence. They are characterized by the same number of quantities, $2K$. For a generic $m \in (1, K)$ we have

$$\begin{aligned} x_m &= x + \sum_i^m h_i, \\ u_m &= u + \Big(\sum_{j=1}^m h_j\Big) p^{(1)} + \Big(\sum_{i,j=1,\, i<j}^m h_i h_j\Big) p^{(2)} \\ &\quad + \Big(\sum_{i,k,k=1,\, i<j<k}^m h_i h_j h_k\Big) p^{(3)} + \dots + \Big(\Pi_{j=1}^m h_j\Big) p^{(m)}. \end{aligned} \tag{1.2.4}$$

2.2. 2-dimensional lattices. A point $P(x, y)$ in a two dimensional plane will be characterized by two numbers. A 2-dimensional lattice will be a map $\mathbb{Z}^2 \to \mathbb{R}^2$ (see Fig. 1.1). The variables in $\mathbb{Z}^2$ n, m, will be often called the *computational coordinates*, and in $\mathbb{R}^2$, x, y, the *physical coordinates*. We will denote the physical coordinates of a point of indexes (n, m) as

$$x(n, m) = x_{n,m}, \quad y(n, m) = y_{n,m}. \tag{1.2.5}$$

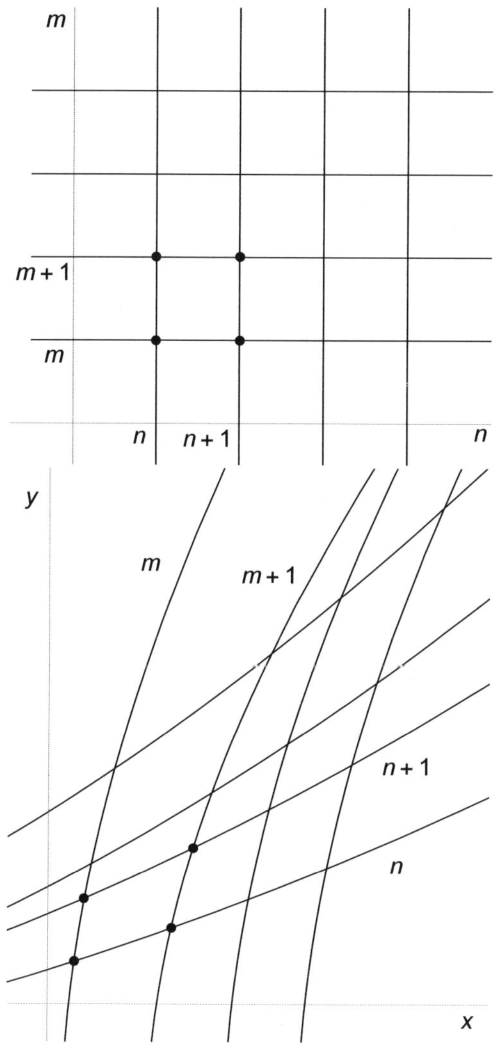

FIGURE 1.1. A 2-dimensional lattice, with lines n = constant and m = constant and an elementary cell in the index space (on the top) and in the physical space (on the bottom) [reprinted from [511]].

Four points $(x_{n,m}, y_{n,m})$, $(x_{n+1,m}, y_{n+1,m})$, $(x_{n,m+1}, y_{n,m+1})$ and $(x_{n+1,m+1}, y_{n+1,m+1})$, associated to (n, m), $(n + 1, m)$, $(n, m + 1)$, $(n + 1, m + 1)$ will define a quadrilateral in $\mathbb{R}^2$, once we impose some constraints on the way to construct the lattice. Let us write

$$\begin{aligned} x_{n+1,m} - x_{n,m} &= h^x_{n,m}, \\ x_{n,m+1} - x_{n,m} &= \sigma^x_{n,m}, \\ y_{n+1,m} - y_{n,m} &= \sigma^y_{n,m}, \\ y_{n,m+1} - y_{n,m} &= h^y_{n,m}, \end{aligned} \tag{1.2.6}$$

where $h^x_{n,m}$, $h^y_{n,m}$, $\sigma^x_{n,m}$ and $\sigma^y_{n,m}$, defined implicitly by (1.2.6), are given explicitly in Fig. 1.2. On them we must impose a consistency condition, since the point $(x_{n+1,m+1}, y_{n+1,m+1})$

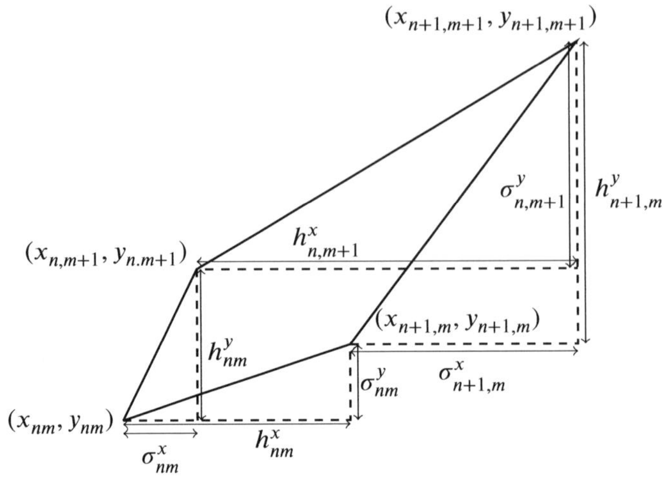

FIGURE 1.2. An elementary cell [reprinted from [511]].

can be reached following two different routes

$$\begin{aligned}(x_{n+1,m+1}, y_{n+1,m+1}) - (x_{n,m}, y_{n,m}) &= (x_{n+1,m+1}, y_{n+1,m+1}) - (x_{n+1,m}, y_{n+1,m}) \\ &\quad +(x_{n+1,m}, y_{n+1,m}) - (x_{n,m}, y_{n,m}) \\ &= (x_{n+1,m+1}, y_{n+1,m+1}) - (x_{n,m+1}, y_{n,m+1}) + (x_{n,m+1}, y_{n,m+1}) - (x_{n,m}, y_{n,m})\end{aligned}$$

that is we have

$$(\sigma^x_{n+1,m}, h^y_{n+1,m}) + (h^x_{n,m}, \sigma^y_{n,m}) = (h^x_{n,m+1}, \sigma^y_{n,m+1}) + (\sigma^x_{n,m}, h^y_{n,m}).$$

So, we have:

$$\begin{aligned} h^x_{n,m+1} - h^x_{n,m} &= \sigma^x_{n+1,m} - \sigma^x_{n,m}, \\ h^y_{n+1,m} - h^y_{n,m} &= \sigma^y_{n,m+1} - \sigma^y_{n,m}. \end{aligned} \tag{1.2.7}$$

However the vectors defining two adjoint sides of the quadrilateral starting at the point of indexes (n, m)

$$\begin{aligned}(x_{n+1,m}, y_{n+1,m}) - (x_{n,m}, y_{n,m}) &= (h^x_{n,m}, \sigma^y_{n,m}), \\ (x_{n,m+1}, y_{n,m+1}) - (x_{n,m}, y_{n,m}) &= (\sigma^x_{n,m}, h^y_{n,m}),\end{aligned}$$

must not be parallel, that is

$$h^x_{n,m} h^y_{n,m} - \sigma^x_{n,m} \sigma^y_{n,m} \neq 0. \tag{1.2.8}$$

Moreover the expression (1.2.8) must preserve the sign to avoid the folding of the lattice itself. If $\alpha_{n,m}$ is the angle of an $m =$ constant line with the x-axis and $\beta_{n,m}$ the angle of an $n =$ constant line with the same axis, we must have

$$\beta_{n,m} > \alpha_{n,m}.$$

As

$$\tan \alpha_{n,m} = \frac{\sigma^y_{n,m}}{h^x_{n,m}}, \qquad \tan \beta_{n,m} = \frac{h^y_{n,m}}{\sigma^x_{n,m}}, \tag{1.2.9}$$

we have

$$\frac{\sigma^y_{n,m}}{h^x_{n,m}} < \frac{h^y_{n,m}}{\sigma^x_{n,m}}, \quad \text{i.e.} \quad h^x_{n,m}h^y_{n,m} - \sigma^y_{n,m}\sigma^x_{n,m} > 0. \tag{1.2.10}$$

In the case when the quadrilateral is a parallelogram we have

$$\alpha_{n,m} = \alpha_{n,m+1}, \qquad \beta_{n,m} = \beta_{n+1,m},$$

$$\text{i.e.} \quad \frac{h^x_{n,m+1}}{\sigma^y_{n,m+1}} = \frac{h^x_{n,m}}{\sigma^y_{n,m}}, \quad \frac{h^y_{n+1,m}}{\sigma^x_{n+1,m}} = \frac{h^y_{n,m}}{\sigma^x_{n,m}}. \tag{1.2.11}$$

2.3. Differential and difference operators on the lattice. Let us consider the following Taylor series expansions around the point $u_{n,m}$. We have

$$\begin{aligned} u_{n+1,m} &= u_{n,m} + (x_{n+1,m} - x_{n,m})u_x + (y_{n+1,m} - y_{n,m})u_y + \cdots, \\ u_{n,m+1} &= u_{n,m} + (x_{n,m+1} - x_{n,m})u_x + (y_{n,m+1} - y_{n,m})u_y + \cdots, \end{aligned}$$

i.e.

$$\begin{aligned} u_{n+1,m} &= u_{n,m} + h^x_{n,m}u_x + \sigma^y_{n,m}u_y + O(h^2, h\sigma, \sigma^2), \\ u_{n,m+1} &= u_{n,m} + \sigma^x_{n,m}u_x + h^y_{n,m}u_y + O(h^2, h\sigma, \sigma^2). \end{aligned} \tag{1.2.12}$$

Inverting (1.2.12) at first order in h and σ and defining, at this order, $u_y = u^d_y$ and $u_x = u^d_x$ we get the approximate partial difference

$$\begin{aligned} u^d_x &= \frac{h^y_{n,m}}{h^x_{n,m}h^y_{n,m} - \sigma^x_{n,m}\sigma^y_{n,m}}\Delta_n u_{n,m} - \frac{\sigma^y_{n,m}}{h^x_{n,m}h^y_{n,m} - \sigma^x_{n,m}\sigma^y_{n,m}}\Delta_m u_{n,m}, \\ u^d_y &= -\frac{\sigma^x_{n,m}}{h^x_{n,m}h^y_{n,m} - \sigma^x_{n,m}\sigma^y_{n,m}}\Delta_n u_{n,m} + \frac{h^x_{n,m}}{h^x_{n,m}h^y_{n,m} - \sigma^x_{n,m}\sigma^y_{n,m}}\Delta_m u_{n,m}. \end{aligned}$$

where the corresponding shift and difference operator are given by

$$S_n \quad \text{such that} \quad S_n f_{n,m} = f_{n+1,m}, \quad \Delta_n = S_n - 1, \tag{1.2.13}$$

$$S_m \quad \text{such that} \quad S_m f_{n,m} = f_{n,m+1}, \quad \Delta_m = S_m - 1. \tag{1.2.14}$$

Definition 1. *We define the partial difference operators in the directions x and y for a general 2-dimensional lattice as*

$$D_x = \frac{1}{h^x_{n,m}h^y_{n,m} - \sigma^x_{n,m}\sigma^y_{n,m}}(h^y_{n,m}\Delta_n - \sigma^y_{n,m}\Delta_m), \tag{1.2.15}$$

$$D_y = \frac{1}{h^x_{n,m}h^y_{n,m} - \sigma^x_{n,m}\sigma^y_{n,m}}(-\sigma^x_{n,m}\Delta_n + h^x_{n,m}\Delta_m), \tag{1.2.16}$$

where $u^d_x = D_x u$ *and* $u^d_y = D_y u$. *The functions* $h^x_{n,m}$, $h^y_{n,m}$, $\sigma^x_{n,m}$ *and* $\sigma^y_{n,m}$ *are defined in (1.2.6). The lattice must satisfy compatibility conditions*

$$h^x_{n,m+1} - h^x_{n,m} = \sigma^x_{n+1,m} - \sigma^x_{n,m}, \quad h^y_{n+1,m} - h^y_{n,m} = \sigma^y_{n,m+1} - \sigma^y_{n,m},$$

and the non-parallel condition

$$h^x_{n,m}h^y_{n,m} - \sigma^x_{n,m}\sigma^y_{n,m} \neq 0,$$

with a definite sign, either positive or negative.

2.4. Grids and lattices in the description of difference equations. In the continuous case, the orthogonal systems of coordinates are well studied and well characterized. The most important are probably the cylindrical, parabolic and spherical ones [621]. However there are many skew non orthogonal systems which sometimes are used depending on the system at study.

Lattices (or grids) are at the base of the numerical methods used to solve PDEs. Often the numerical approximation of a differential equation is on an orthogonal lattice[35]. However, orthogonal lattices are in general too rigid and some deviations from orthogonality must be allowed. The numerical construction of the grid is often a problem as this is built in a dynamical way, and depends on the equation and the values of the gradient of the solutions. It may then happen that when trying to keep ortogonality the coordinate lines collapse, i.e. the Jacobian of the transformation becomes zero [238], rendering the method useless.

A practical reason to use orthogonal coordinates (or equivalently, orthogonal grids) is that when PDEs are solved in non-orthogonal lattices, one usually finds additional terms which should be implemented in the computer code [789].

It seems impossible to fulfill all good conditions for a lattice using only orthogonal coordinates.

In practical problems such as flows bounded by curved surfaces the calculations in skew coordinates, when the metric tensor has non zero off diagonal components, may be more accurate. Moreover skew coordinates may be more natural when doing the continuous limit of quad-graph equations, see Section 3.2.4.3.

Here in the following we present just a few more significant lattices leaving to [512] for more details.

2.4.1. *Cartesian lattices.* If the curves $n =$ constant and the curves $m =$ constant are straight lines we have a generalized Cartesian lattice. In this case the conditions are

$$\frac{\sigma^y_{n,m}}{h^x_{n,m}} = \frac{\sigma^y_{n+1,m}}{h^x_{n+1,m}}, \qquad \frac{\sigma^x_{n,m}}{h^y_{n,m}} = \frac{\sigma^x_{n,m+1}}{h^x_{n,m+1}}, \tag{1.2.17}$$

and thus

$$\sigma^x_{n+1,m} = \sigma^x_{n,m}, \quad \sigma^y_{n,m+1} = \sigma^y_{n,m}, \quad h^x_{n,m+1} = h^x_{n,m}, \quad h^y_{n+1,m} = h^y_{n,m}.$$

So the equations of the lattice are:

$$\begin{aligned} x_{n+1,m} - x_{n,m} &= h^x_n, & x_{n,m+1} - x_{n,m} &= \sigma^x_m, \\ y_{n+1,m} - y_{n,m} &= \sigma^y_n, & y_{n,m+1} - y_{n,m} &= h^y_m. \end{aligned} \tag{1.2.18}$$

Moreover the compatibility relations (1.2.7) are trivially satisfied. Note that the lines $m =$ constant form a fixed angle α_n with the x-axis and the lines $n =$ constant form a fixed angle β_m with the same axis and (1.2.9) is valid. The cells are rhomboids, but they can also be rectangles so that we have a *rotated Cartesian lattice*, and the orthogonality condition to be satisfied is

$$h^x_n \sigma^x_m + h^y_m \sigma^y_n = 0. \tag{1.2.19}$$

Eq. (1.2.19) can be written as

$$\frac{\sigma^y_n}{h^x_n} = -\frac{\sigma^x_m}{h^y_m} = \gamma$$

where γ, which is independent of n and m, is the tangent of $\alpha_n = \alpha$, the angle of rotation. The other angle is $\beta_m = \frac{\pi}{2} + \alpha$.

When α is equal to zero, that is, when the rotated Cartesian lattice is parallel to the coordinate axes, we have $\sigma_m^x = \sigma_n^y = 0$ and we get a *Cartesian lattice*. In this case, $x_{n,m}$ depends only on n and $y_{n,m}$ depends only on m, the quadrilateral cells are rectangles of generally different sizes with sides parallel to the axes. The equations of the lattice are

$$x_{n+1} - x_n = h_n^x, \quad y_{m+1} - y_m = h_m^y.$$

The lattice can be interpreted as a discretization of the Cartesian coordinate system $\{x, y\}$. We can thus deduce from the compatibility condition (1.2.7) the following theorem:

Theorem 2. *Necessary and sufficient condition to get a Cartesian orthogonal lattice is that* $\sigma_{n,m}^x = \sigma_{n,m}^y = 0$.

2.4.2. *Galilei invariant lattice.* This lattice is defined by

(1.2.20)
$$\begin{aligned} &t_{n,m+1} - t_{n,m} = \tau_1, \quad t_{n+1,m} - t_{n,m} = \tau_2, \\ &x_{n,m+1} - 2x_{n,m} + x_{n,m-1} = 0, \\ &(x_{n,m+1} - x_{n,m})\tau_2 - (x_{n+1,m} - x_{n,m})\tau_1 = \zeta. \end{aligned}$$

When the lattice (1.2.20) is considered together with the discretized heat equation

$$\frac{u_{n,m+1} - u_{n,m}}{\tau_2} = \tau_2^2 \frac{u_{n,m+1} - 2u_{n,m} + u_{n,m-1}}{\zeta^2}$$

it provides a Galilei invariant system. The lattice equation (1.2.20) can be solved. We obtain

(1.2.21)
$$t_{n,m} = \tau_1\, n + \tau_2\, m + t_0, \quad x_{n,m} = \sigma\tau_1\, n + \left(\frac{\sigma\tau_1\tau_2 - \zeta}{\tau_1}\right) m + x_0,$$

where σ, t_0 and x_0 are integration constants. The corresponding lattice is equally spaced and in general, nonorthogonal (see Figure 1.3). Indeed, the coordinate curves, corresponding to $m =$ constant and $n =$ constant, respectively, are

(1.2.22)
$$\begin{aligned} x - x_0 &= \sigma(t - t_0) - \frac{\zeta}{\tau_1} m, \\ x - x_0 &= \frac{\sigma\tau_1\tau_2 - \zeta}{\tau_1\tau_2}(t - t_0) + \frac{\zeta}{\tau_2} n. \end{aligned}$$

These are two families of straight lines, orthogonal only in the special case $(\sigma^2 + 1)\tau_1\tau_2 = \sigma\zeta$. If we choose

(1.2.23)
$$\sigma\tau_1\tau_2 - \zeta = 0$$

then the second family of coordinate lines in equation (1.2.22) is parallel to the x axis.

2.4.3. *Exponential lattice.* Let us now consider a lattice that is neither equally spaced, nor orthogonal, given by the equations

(1.2.24)
$$x_{n,m+1} - 2x_{n,m} + x_{n,m-1} = 0, \qquad x_{n+1,m} = (1 + c)\, x_{n,m},$$

(1.2.25)
$$t_{n,m+1} - t_{n,m} = 0, \quad t_{n+1,m} - t_{n,m} = h,$$

with $c \neq 0, -1$. These equations can be solved and we obtain the lattice

(1.2.26)
$$t_{n,m} = hn + t_0 \quad x_{n,m} = (1 + c)^n\, (\alpha m + \beta)$$

where t_0, α and β are integration constants. Thus while t grows by constant increments, x grows with increments which vary exponentially with time (see Figure 1.4). Numerically

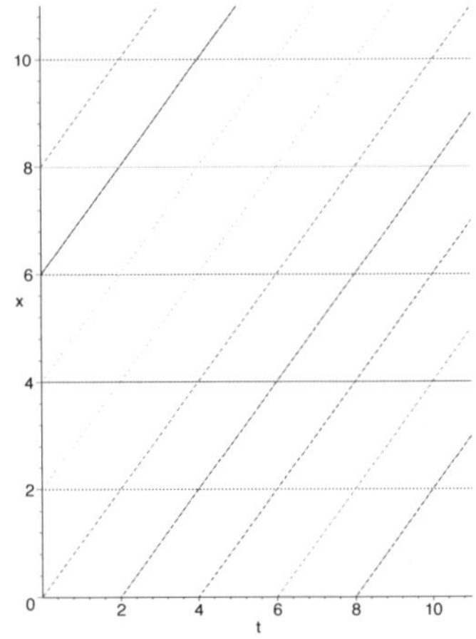

FIGURE 1.3. Variables (x, t) as functions of m and n for the lattice equations (1.2.20). The parameters and the integration constants are, respectively, $\tau_1 = 1, \tau_2 = 2, \zeta = 2$ and $\sigma = 1, x_0 = 0, t_0 = 0$ so that one family of coordinate lines is parallel to the t axis as (1.2.23) is satisfied [reprinted from [530]].

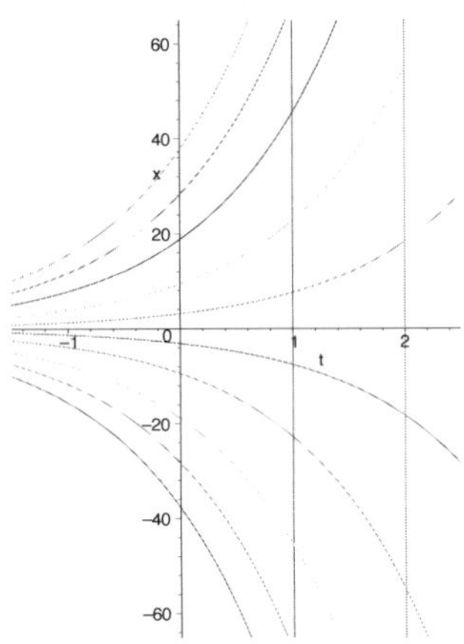

FIGURE 1.4. Variables (x, t) as functions of m and n for the lattice equations (1.2.24), (1.2.25). The parameters and the integration constants are, respectively, $c = \sqrt{2}, h = 1$ and $\alpha = \pi, \beta = 0, t_0 = 0$ [reprinted from [530]].

this type of lattice may be useful if we can solve the equation asymptotically for large values of t and are interested in the small t behavior.

2.4.4. *Polar coordinate systems.* The discretization of continuous orthogonal coordinate systems does not assure commutativity of difference operators. In fact, the elementary cells are not necessarily parallelograms. For instance, let us consider polar coordinates in the plane and restrict them to the first quadrant in the interior of the unit circle. A lattice adapted to these coordinates is given by (N, M are two arbitray positive integers)

$$x_{n,m} = \frac{n}{N}\cos\frac{\pi m}{2M}, \quad n = 1, \dots, N, \tag{1.2.27}$$
$$y_{n,m} = \frac{n}{N}\sin\frac{\pi m}{2M}, \quad m = 0, \dots, M.$$

The steps h and σ are:

$$\begin{aligned} h^x_{n,m} &= x_{n+1,m} - x_{n,m} = \frac{1}{N}\cos\frac{\pi m}{2M}, \\ h^y_{n,m} &= y_{n,m+1} - y_{n,m} = \frac{2n}{N}\sin\frac{\pi}{4M}\cos\frac{\pi(2m+1)}{4M}, \\ \sigma^x_{n,m} &= x_{n,m+1} - x_{n,m} = -\frac{2n}{N}\sin\frac{\pi}{4M}\sin\frac{\pi(2m+1)}{4M}, \\ \sigma^y_{n,m} &= y_{n+1,m} - y_{n,m} = \frac{1}{N}\sin\frac{\pi m}{2M}. \end{aligned} \tag{1.2.28}$$

Eq. (1.2.28) obviously satisfy the compatibility equations (1.2.7), but the elementary cells are not parallelograms (see Figure 1.5).

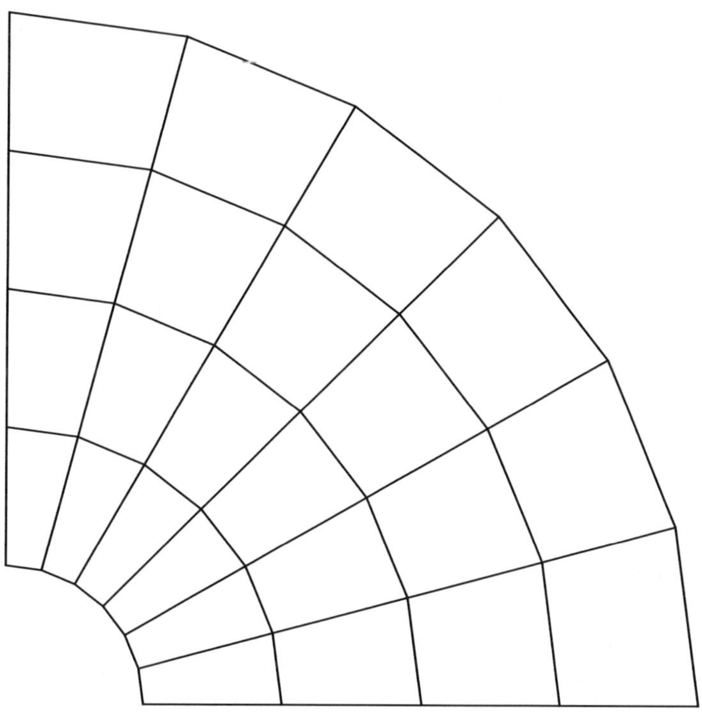

FIGURE 1.5. Noncommutative lattice: discretization of polar coordinates [reprinted from [511]].

This is an example of symmetry adapted lattice. The use of polar coordinates will certainly provide a convenient lattice for a difference equation invariant under the rotation group in the plane.

2.5. Clairaut–Schwarz–Young theorem on the lattices and its consequences. In the case of PDEs when we have derivatives of second order or higher we know that the following theorem holds for some well defined classes of functions [184, 741, 855]:

Theorem 3. *Let $f : \mathbb{R}^n \to \mathbb{R}$ be twice continuously differentiable on its domain of definition, $\mathbb{X} \in \mathbb{R}^n$. Then on the interior of its domain, the $n \times n$ matrix of second-order partial derivatives is symmetric,*

$$\frac{\partial^2 f}{\partial x_i \partial x_j} = \frac{\partial^2 f}{\partial x_j \partial x_i}, \quad \forall \quad i, j = 1, \cdots, n. \tag{1.2.29}$$

This is sometimes known as Schwarz's theorem, Clairaut's theorem, or Young's theorem from the mathematician who proved it. The very simple conditions required to have commuting partial derivatives for PDEs turns out to be nontrivial in the discrete case.

2.5.1. *Commutativity and non commutativity of difference operators.* When the lattice is arbitrary, the partial difference operators (1.2.15) will no longer commute among themselves. However they might commute for some particular lattices (apart from the obvious Cartesian one).

The commutator, when applied to $u_{n,m}$, is a linear combination of $u_{n,m}$ (which has a coefficient equal to 0), $u_{n+1,n}$, $u_{n,m+1}$, $u_{n+2,m}$, $u_{n,m+2}$ and $u_{n+1,m+1}$. All the coefficients must be zero if we want the operators to commute. The equations to be satisfied by the functions $h^x_{n,m}$, $h^y_{n,m}$, $\sigma^x_{n,m}$ and $\sigma^y_{n,m}$ are

$$h^y_{n+1,m}\sigma^x_{n,m} - h^y_{n,m}\sigma^x_{n+1,m} = 0, \tag{1.2.30}$$

$$h^x_{n,m}\sigma^y_{n,m+1} - h^x_{n,m+1}\sigma^y_{n,m} = 0, \tag{1.2.31}$$

$$-h^x_{n+1,m}h^y_{n,m} + h^x_{n+1,m}h^y_{n+1,m} + \sigma^x_{n,m}\sigma^y_{n+1,m} - \sigma^x_{n+1,m}\sigma^y_{n+1,m} = 0, \tag{1.2.32}$$

$$h^x_{n,m}h^y_{n,m+1} - h^x_{n,m+1}h^y_{n,m+1} - \sigma^x_{n,m+1}\sigma^y_{n,m} + \sigma^x_{n,m+1}\sigma^y_{n,m+1} = 0, \tag{1.2.33}$$

$$\begin{aligned} &h^x_{n,m+1}h^x_{n+1,m}h^y_{n,m}h^y_{n,m+1} - h^x_{n,m}h^x_{n+1,m}h^y_{n,m+1}h^y_{n+1,m} \\ &\quad + h^x_{n+1,m}h^y_{n+1,m}\sigma^x_{n,m+1}\sigma^y_{n,m} - h^x_{n+1,m}h^y_{n,m}\sigma^x_{n,m+1}\sigma^y_{n,m+1} \\ &\quad - h^x_{n,m+1}h^y_{n,m+1}\sigma^x_{n,m}\sigma^y_{n+1,m} + h^x_{n,m}h^y_{n,m+1}\sigma^x_{n+1,m}\sigma^y_{n+1,m} \\ &\quad - \sigma^x_{n,m+1}\sigma^x_{n+1,m}\sigma^y_{n,m}\sigma^y_{n+1,m} + \sigma^x_{n,m}\sigma^x_{n,m+1}\sigma^y_{n,m+1}\sigma^y_{n+1,m} = 0. \end{aligned} \tag{1.2.34}$$

Then we must satisfy the compatibility conditions (1.2.7) and the non vanishing of the denominator of the discrete derivative (1.2.15).

We assume that all the steps are different from zero. If $\sigma^x_{n,m} = \sigma^y_{n,m} = 0$, then the lattice is cartesian, as we proved before in Theorem 2. If $h^x_{n,m} = 0$ or $h^y_{n,m} = 0$, the limits are not correct. When solving (1.2.15, 1.2.30–1.2.34) with the condition that all steps are different from zero we get two cases:

(1) $h^y_{n+1,m} = h^y_{n,m}$, $h^x_{n,m+1} = h^x_{n,m}$.
This means that the step $h^x_{n,m}$ does not depend on m: $h^x_{n,m} = h^x_n$ and the step $h^y_{n,m}$ does not depend on n: $h^y_{n,m} = h^y_m$.
The step $\sigma^x_{n,m}$ does not depend on n:

$$\sigma^x_{n+1,m} = x_{n+1,m+1} - x_{n+1,m} = x_{n,m+1} - x_{n,m} = \sigma^x_{n,m} \equiv \sigma^x_m \tag{1.2.35}$$

and $\sigma^y_{n,m}$ does not depend on m: $\sigma^y_{n,m} = \sigma^y_n$. So we are in the case of the generalized Cartesian lattice (1.2.18). All of the steps are different from zero and depend on only one index. The compatibility conditions are identically satisfied. The geometry of the lattice is given in the following proposition:

Theorem 4. *The unit cells of the lattice defined by equation* (1.2.18) *are parallelograms.*

Proof:

$$\begin{aligned}
x_{n+1,m+1} - x_{n,m+1} &= h^x_n = x_{n+1,m} - x_{n,m},\\
y_{n+1,m+1} - y_{n,m+1} &= \sigma^y_n = y_{n+1,m} - y_{n,m},\\
x_{n+1,m+1} - x_{n+1,m} &= \sigma^x_m = x_{n,m+1} - x_{n,m},\\
y_{n+1,m+1} - y_{n+1,m} &= h^y_m = y_{n,m+1} - y_{n,m}.
\end{aligned}$$

□

This case obviously includes the generalized Cartesian lattices, but it is more general, since we do not impose the condition of straight coordinate lines.

(2) $h^y_{n+1,m} \neq h^y_{n,m}$, $h^x_{n,m+1} \neq h^x_{n,m}$. From (1.2.30, 1.2.31) we get

$$\frac{\sigma^x_{n+1,m}}{h^y_{n+1,m}} = \frac{\sigma^x_{n,m}}{h^y_{n,m}} = c_m, \quad \sigma^x_{n,m} = c_m h^y_{n,m},$$

$$\frac{\sigma^y_{n,m+1}}{h^x_{n,m+1}} = \frac{\sigma^y_{n,m}}{h^x_{n,m}} = d_n, \quad \sigma^y_{n,m} = d_n h^x_{n,m}.$$

Then (1.2.7, 1.2.15, 1.2.30–1.2.34) with the condition that all steps are different from zero reduces to

$$(h^x_{n,m} - h^x_{n,m+1})(1 - c_{m+1} d_n) = 0, \tag{1.2.36}$$

$$h^x_{n,m} - h^x_{n,m+1} = c_m (h^y_{n,m} - h^y_{n+1,m}), \tag{1.2.37}$$

$$h^x_{n+1,m}(1 - c_m d_{n+1}) = 0. \tag{1.2.38}$$

Eq. (1.2.38) implies that c_m, d_n are constant and one the inverse of the other. So, in this case, the condition (1.2.7), which now reads $h^x_{n,m} h^y_{n,m}(1 - c_m d_n) \neq 0$, is not satisfied.

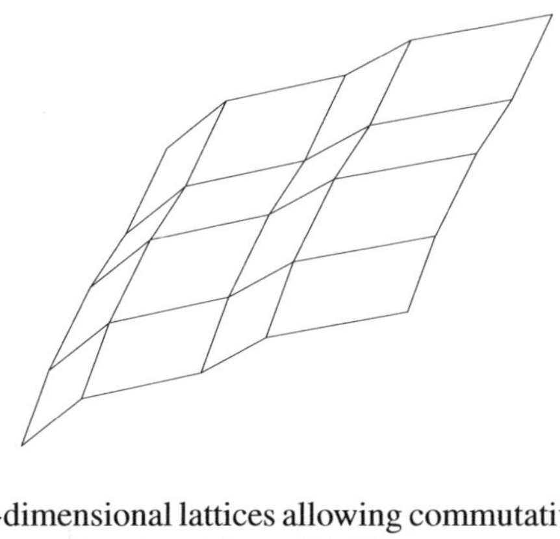

FIGURE 1.6. 2-dimensional lattices allowing commutativity of first order difference operators [reprinted from [511]].

We conclude summing up our results in a theorem:

Theorem 5. *The only 2-dimensional lattices satisfying the commutativity of the two partial difference operators are those formed by parallelograms (see Fig.* 1.6*).*

This class of lattices includes as particular cases the generalized Cartesian and rotated Cartesian. This result excludes some types of more general lattices, for instance those used in [232] in the study of invariant discretizations of the KdV equation or others arising from discretizations of orthogonal coordinate systems.

3. What is a difference equation

By a difference equation we mean a functional relation, linear or non linear, among functions calculated at different points of a lattice [12,34,239,426,441,796]. These systems appear in many applications. First of all they can be written down as discretizations of a differential equation when one is trying to solve it with a computer. In such a case one reduces the ODE to a recurrence relation, also denoted as Ordinary Difference Equation (OΔE):

$$\frac{du}{dx} = f(x,u) \qquad \Rightarrow \qquad v(n+1) = g(n, v(n)). \tag{1.3.1}$$

On the other hand we can consider dynamical systems defined on a lattice, i.e. systems where the real independent fields depend on a set of independent variables which vary partly on the integers and partly on the reals. For example we can consider

$$\frac{d^2u(n,t)}{dt^2} = F(t, u(n,t), u(n-1,t), .., u(n-a,t), u(n+1,t), .., u(n+b,t)). \tag{1.3.2}$$

These kind of equations, also denoted as Differential Difference Equations (DΔEs), can appear in many different settings. Among them they are associated to the evolution of many body problems, to the study of crystals, to biological and economical systems, etc..

As an example of possible applications we consider the problem of the transmission of energy in one dimensional molecular system, problem which is of particular relevance for understanding the functioning of physical systems of biological interest[150,319]. This is a particularly hot topic as some relevant biological processes require the transport of energy with low dispersion along essentially one dimensional chains, such as the spines in an α helix [204]. A mechanism for the nondispersive transport of vibrational energy along hydrogen bonded chains was proposed by Davydov [175,203,204,747] and its continuous limit for small lattice spacing gave rise to a Nonlinear Schrödinger equation (NLS)

$$iu_t + u_{xx} - 2\sigma|u|^2u = 0 \tag{1.3.3}$$

which has soliton solutions [9,12,83,215,234,443,461,467,584,585,623,630,649,707, 763,775,853,862]. If such soliton like solutions are valid also at biological temperatures is an open problem for researchers.

In the case of diatomic non linear lattices we can describe such systems by the equations

$$\begin{aligned} M_1\ddot{x}_n - k_1(y_n - x_n) + k_2(x_n - y_{n-1}) - \epsilon\beta_1(y_n - x_n)^2 & \\ +\epsilon\beta_2(x_n - y_{n-1})^2 = 0, & \\ M_2\ddot{y}_n + k_1(y_n - x_n) - k_2(x_{n+1} - y_n) + \epsilon\beta_1(y_n - x_n)^2 & \\ -\epsilon\beta_2(x_{n+1} - y_n)^2 = 0, & \end{aligned} \tag{1.3.4}$$

where M_1 and M_2 are the different values of the two atomic masses, ϵ is a small parameter while k_1, k_2, β_1 and β_2 are four constants of order 1. When $k_2 \ll k_1$ these equation represent a molecular chain with intramolecular interaction stronger than the intermolecular one. This is the case, for example, of an hydrogen-bonded polypeptide chain [672].

Further examples are connected to Quantum Gravity (QG), where one of the most fruitful recent approaches to the problem is based on the discretization of space–time [826]. In this way one hopes to obtain a non–trivial theory without the use of perturbation expansions. In the approach to QG introduced by Gambini and Pullin [296, 297] one considers a discrete action $S = \sum_{n=1}^{N} \mathcal{L}(q_n, q_{n+1})$. The Lagrange equations of motion are given by $\frac{\partial S}{\partial q_n} = \frac{\partial \mathcal{L}(q_{n-1}, q_n)}{\partial q_n} + \frac{\partial \mathcal{L}(q_n, q_{n+1})}{\partial q_n} = 0$ (here the definition of variational derivative on the lattice, given by (3.2.40) in Section 3.2.1 was used to get the equation of motion) and a simple example of Lagrangian is given by $\mathcal{L} = \frac{m}{2\tau}(q_{n+1} - q_n)^2 - \tau V(q_n)$. In this approach one then quantizes the systems by a discrete canonical transformation and then looks for the consequences of the theory by performing a quantum continuous limit.

Further examples of discrete equations are provided by recurrence relations, i.e. equations that recursively define a sequence of values once one or more initial values are given. Each term of the sequence is defined in term of the preceding ones. As an example one can consider the logistic map which describes the growth of a population: $x_{n+1} = rx_n(1 - x_n)$. These maps can appear as approximations of differential equations (see, for example, (1.3.1)) or as functional equations in various fields as statistics, engineering, etc. . A further example of equation of this kind appears when looking for invariant solutions of scale transformations $w(x,t) = a^k w(a^m x, a^n t)$. The problem of solving functional equations has been discussed by many mathematicians, see for example, D'Alambert, Cauchy, Abel, Hilbert, etc. [14, 15].

Summing up difference equations either appear in themselves and we would like to use Lie theory to get classes of exact solutions or we obtain them by discretizing the continuous equation in such a way to preserve the symmetries, i.e. create sets of discrete equations which provide numerical schemes approximating the continuous equation, see Section 1.4.2.

For simplicity in the following we will consider just the case of a scalar equation ($q = 1$) in at most two independent variables ($p = 2$) but, similar results can be obtained in the case of $p > 2$ independent and $q > 1$ dependent variables. We will denote these equations as Partial Difference Equations (PΔEs).

A PΔE in $\mathbb{R}^2$ is thus a functional relation for a field u at different points P_i in $\mathbb{R}^2$, i.e. $E = E(x, t, u(P_1), \dots, u(P_L)) = 0$. A DΔE is obtained by considering the points P_i uniformly spaced in one direction, say t, with spacing h_t, in such a way that we are allowed to consider the continuous limit when h_t goes to zero.

As we saw in Section 1.2.2 the points P_i in $\mathbb{R}^2$ can be labeled by two discrete indexes which characterize the points with respect to two independent directions, $P_{n,m}$ and can be displayed on lines characterized by the constancy of one index.
In Cartesian coordinates we have

$$P_{n,m} = (x_{n,m}, t_{n,m}) \tag{1.3.5}$$

and the function $u(P)$ reads

$$u_{P_{n,m}} = u(x_{n,m}, t_{n,m}) = u_{n,m}. \tag{1.3.6}$$

A difference scheme will be *a set of relations* among the values of $\{x, t, u(x,t)\}$ at a finite number, say L, of points in $\mathbb{R}^2$ $\{P_1, \dots, P_L\}$ around a reference point, say P_1. Some of these relations will define where the points are in $\mathbb{R}^2$ and others how $u(P)$ transforms in $\mathbb{R}^2$. In our case, as we have one only dependent variable and two independent variables we

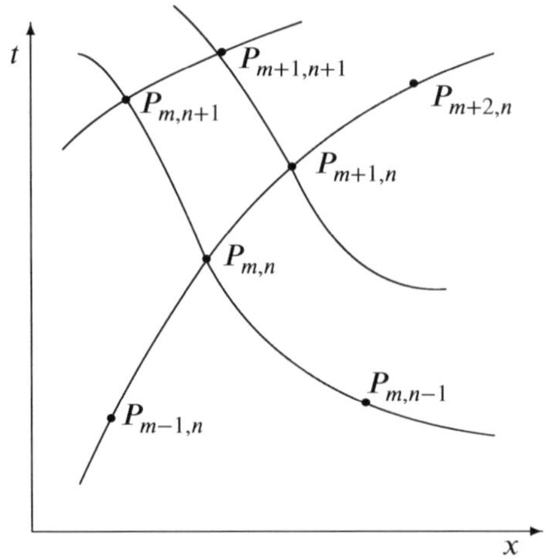

FIGURE 1.7. Points on a two dimensional lattice [reprinted from [520]].

expect to have at most five equations, four which define the two independent variables in the two independent directions in $\mathbb{R}^2$, and one the dependent variable in terms of the lattice points:

$$
\begin{aligned}
&E_a(\{x_{n+j,m+i}, t_{n+j,m+i}, u_{n+j,m+i}\}) = 0,\\
&1 \leq a \leq 5; \qquad -i_1 \leq i \leq i_2,\\
&-j_1 \leq j \leq j_2 \quad (i_1, i_2, j_1, j_2) \in Z^{\geq 0},\\
&i_1 + i_2 = \mathcal{N}, \qquad j_1 + j_2 = \mathcal{M}, \quad L = \mathcal{N} \cdot \mathcal{M}.
\end{aligned} \tag{1.3.7}
$$

System (1.3.7) must be such that, starting from L points we are able to calculate $\{x, t, u\}$ in all points of interest. So if we give four equations for the lattice, two for each independent direction, then these equations must be compatible among themselves. If the lattice is not defined a priory and is constructed dynamically then we can have less equations. Three equations may be sufficient if we solve a Cauchy problem.

3.1. Examples. Discrete wave equation $u_{xt} = 0$

$$\frac{1}{t_{n+1,m} - t_{n,m}} \left[\frac{u_{n+1,m+1} - u_{n+1,m}}{x_{n+1,m+1} - x_{n+1,m}} - \frac{u_{n,m+1} - u_{n,m}}{x_{n,m+1} - x_{n,m}} \right] = 0, \tag{1.3.8}$$

$$t_{n,m+1} - t_{n,m} = 0, \qquad x_{n+1,m} - x_{n,m} = 0. \tag{1.3.9}$$

Eq. (1.3.8) relates four lattice points at the vertex of a square, see Fig. 2.3 in Section 2.4.6, whose solution is given by $u_{n,m} = f(x_{n,m}) + g(t_{n,m})$ with, due to (1.3.9), $t_{n,m} = \alpha_n$ and $x_{n,m} = \beta_m$. If we define the functions $t_{n,m}$ and $x_{n,m}$ by adding two other equations, for example

$$t_{n+1,m} - t_{n,m} = h_m, \qquad x_{n,m+1} - x_{n,m} = k_n, \tag{1.3.10}$$

the compatibility of (1.3.9) and (1.3.10) implies $h_{m+1} = h_m$ and $k_{n+1} = k_n$, i.e. $h_m = h$ and $k_n = k$ constants.

If a continuous limit of (1.3.7) exists, then one of the equations will go over to a PDE and the others will be identically satisfied (generically $0 = 0$). We can also do partial continuous limits when only one of the independent variables become continuous while the other is still discrete. In this case only part of the lattice equations are identically satisfied and we obtain a DΔE for the dependent variable and an equation for the lattice variable. In the case of (1.3.8, 1.3.9), in the continuous limit (1.3.9) goes into ($0 = 0$, $0 = 0$) while (1.3.8) goes into $u_{xt} = 0$. Let us do a partial continuous limit then we require that $h_m \to 0$ in (1.3.10). In this limit, defining $t_{n,m} = t_m = t$, $x_{n,m} = x_m$ and $u_{n,m} = v(x_m, t) = v_m(t)$, (1.3.8) goes into $\dot{v}_{m+1} - \dot{v}_m = 0$.

Let us present now a further example of difference scheme. We consider a discretization of the heat equation $u_t = u_{xx}$ on a uniform orthogonal lattice:

$$\frac{u_{n+1,m} - u_{n,m}}{t_{n+1,m} - t_{n,m}} = \frac{u_{n,m+2} - 2u_{n,m+1} + u_{n,m}}{(x_{n,m+1} - x_{n,m})^2}, \tag{1.3.11}$$

$$x_{n,m+1} - x_{n,m} = h_x; \qquad t_{n,m+1} - t_{n,m} = 0, \tag{1.3.12}$$

$$x_{n+1,m} - x_{n,m} = 0; \qquad t_{n+1,m} - t_{n,m} = h_t, \tag{1.3.13}$$

where h_x, h_t are two a priory fixed constants which define the spacing between two neighboring points in the two directions of the orthogonal lattice. The two lattice equations (1.3.12, 1.3.13) are compatible as $x_{n+1,m+1}$ obtained by shifting n by one (1.3.12) is the same as what one obtain by shifting m by one in (1.3.13). Also (1.3.11) relates four points of the two dimensional lattice but in a different position with respect to the previous example. The example is simple; the lattice equations (1.3.12, 1.3.13) are compatible and can be solved explicitly to give

$$x_{m,n} = h_1 m + x_0 \quad t_{m,n} = h_2 n + t_0. \tag{1.3.14}$$

The choice $x_0 = t_0 = 0$, $h_1 = h_2 = 1$ identify x with m and t with n.

The two examples presented bring out several points:

(1) Four equations are needed to describe completely the lattice but in this case there is a compatibility condition. In the whole generality two equations are sufficient and provide the lattice starting from some initial conditions.
(2) Four points are needed for equations of second order in x, first in t. Only three figure in the lattice equation, namely $P_{m+1,n}$, $P_{m,n}$ and $P_{m,n+1}$. To get the fourth point, $P_{m-1,n}$, we shift m down by one unit the equations (1.3.12-1.3.11).
(3) An independence condition is needed to be able to solve for $x_{m+1,n}$, $t_{m+1,n}$, $x_{m,n+1}$, $t_{m,n+1}$ and $u_{m,n+1}$.

We need the more complicated two index notation to describe arbitrary lattices and to formulate the symmetry algorithm.

4. How do we find symmetries for difference equations

Lie point symmetries are characterized by transformations of the form:

$$\begin{aligned} \tilde{x} &= F_\epsilon(x,t,u) = x + \epsilon\, \xi(x,t,u) + \mathcal{O}(\epsilon^2), \\ \tilde{t} &= G_\epsilon(x,t,u) = t + \epsilon\, \tau(x,t,u) + \mathcal{O}(\epsilon^2), \\ \tilde{u} &= H_\epsilon(x,t,u) = u + \epsilon\, \phi(x,t,u) + \mathcal{O}(\epsilon^2), \end{aligned} \tag{1.4.1}$$

where ϵ, such that $|\epsilon| \leq 1$, is a group parameter defined in a domain D seated around the value $\epsilon = 0$, corresponding to the identity transformation. The transformation (1.4.1) is such that if $\{x, t, u\}$ satisfy the difference scheme (1.3.7), $\{\tilde{x}, \tilde{t}, \tilde{u}\}$ will be a solution of the same scheme. Such a transformation acts on the whole space of the independent and dependent variables $\{x, t, u\}$, at least in some neighborhood of P_1 including all points up to P_L. This means that the set of functions F_ϵ, G_ϵ and H_ϵ must be well behaved in the region where P_i, $i = 1, \ldots, L$ are defined and will determine the transformation in all points of the scheme. In the point P_1 of coordinates (x, t) where the dependent variable is u we define the infinitesimal generator as

$$\hat{X}_{P_1} = \xi(x,t,u)\partial_x + \tau(x,t,u)\partial_t + \phi(x,t,u)\partial_u \tag{1.4.2}$$

and then we prolong it to all other $L-1$ points of the scheme. Since the transformation is given by the same set of functions $\{F_\epsilon, G_\epsilon, H_\epsilon\}$ at all points P_i, the *prolongation* of $\hat{X}_{P_1}$ is obtained simply by evaluating $\hat{X}_{P_1}$ at all the points involved in the scheme. So

$$pr\hat{X} = \sum_{i=1}^{L} \hat{X}_{P_i}. \tag{1.4.3}$$

Consequently the invariance condition for the difference scheme (1.3.7) is:

$$pr\hat{X}\, E_a|_{E_a=0} = 0. \tag{1.4.4}$$

Eq.(1.4.4) is, in the case of PΔEs, a set of functional equations whose solution may be obtained, following Abel [2], by turning them into differential equations by successive derivation with respect to the independent variables $\{x, t, u\}$ at different points of the lattice [14, 15].

The solution of (1.4.4) provide the functions $\xi(x,t,u)$, $\tau(x,t,u)$ and $\phi(x,t,u)$, the infinitesimal coefficients of the local Lie point symmetry group. The transformation is obtained, as in the continuous case, by integrating the vector field, i.e. by solving the following system of differential equations:

$$\begin{aligned} \frac{d\tilde{x}}{d\epsilon} &= \xi(\tilde{x},\tilde{t},\tilde{u}), & \tilde{x}|_{\epsilon=0} &= x, \\ \frac{d\tilde{t}}{d\epsilon} &= \tau(\tilde{x},\tilde{t},\tilde{u}), & \tilde{t}|_{\epsilon=0} &= t, \\ \frac{d\tilde{u}}{d\epsilon} &= \phi(\tilde{x},\tilde{t},\tilde{u}), & \tilde{u}|_{\epsilon=0} &= u. \end{aligned} \tag{1.4.5}$$

In general we expect the infinitesimal coefficients ξ and τ to be determined by the lattice equations. So according to the form of the lattice, different symmetries can appear. In fact, in the case of the discrete heat equation (1.3.11), by applying the infinitesimal generator

$$\begin{aligned} \hat{X}_{n,m} &= \xi_{n,m}(x_{n,m}, t_{n,m}, u_{n,m})\partial_{x_{n,m}} + \\ &+ \tau_{n,m}(x_{n,m}, t_{n,m}, u_{n,m})\partial_{t_{n,m}} + \phi_{n,m}(x_{n,m}, t_{n,m}, u_{n,m})\partial_{u_{n,m}}, \end{aligned} \tag{1.4.6}$$

to the lattice equations (1.3.12,1.3.13) we get:

$$\begin{aligned} \xi(x_{n,m+1}, t_{n,m+1}, u_{n,m+1}) &= \xi(x_{n,m}, t_{n,m}, u_{n,m}), \\ \xi(x_{n+1,m}, t_{n+1,m}, u_{n+1,m}) &= \xi(x_{n,m}, t_{n,m}, u_{n,m}). \end{aligned}$$

From (1.3.11) $u_{n,m+1}$, $u_{n+1,m}$ and $u_{n,m}$ can be chosen as independent functions and thus we get $\xi = \xi(x,t)$. As $t_{n,m+1} = t_{n,m}$ and $x_{n,m+1} \neq x_{n,m}$ we get $\xi = \xi(t)$. As $x_{n+1,m} = x_{n,m}$ and $t_{n+1,m} \neq t_{n,m}$ we get that the only possible value for the function $\xi(x,t,u)$ is ξ=constant. In a similar fashion we derive that also $\tau(x,t,u)$ must be a constant. Then we have

$$\phi = u + s(x,t), \tag{1.4.7}$$

where $s(x,t)$ is a solution of the discrete heat equation (1.3.11), i.e. (1.4.7) is the linear superposition formula. Summarizing we get that the infinitesimal generators of the symmetries for the discrete heat equation (1.3.11) are given by

$$\hat{P}_0 = \partial_t; \quad \hat{P}_1 = \partial_x; \quad \hat{W} = u\partial_u; \quad \hat{S} = s(x,t)\partial_u. \tag{1.4.8}$$

Let us prove, in the case of OΔEs when we have just one discrete independent variable x_n and one dependent variable $u_n(x_n)$, that the prolongation formula given above (1.4.3) has the proper continuous limit (1.1.11). To do so we consider a prolonged vector field

$$\text{(1.4.9)} \qquad \begin{aligned} \mathrm{pr}^1 \hat{X} &= \xi(x_n, u_n)\partial_{x_n} + \phi(x_n, u_n)\partial_{u_n} \\ &+ \xi(x_{n+1}, u_{n+1})\partial_{x_{n+1}} + \phi(x_{n+1}, u_{n+1})\partial_{u_{n+1}}, \end{aligned}$$

depending on two neighboring points x_n and x_{n+1} and a function u_n on them as we want to approximate a first derivative. We can define the *new variables* $\tilde{x}_n$, $\tilde{u}_n$, h_{n+1} and $u_{x,n+1}$

$$\text{(1.4.10)} \qquad \tilde{x}_n = x_n, \quad \tilde{u}_n = u_n, \quad h_{n+1} = x_{n+1} - x_n, \quad u_{x,n+1} = \frac{u_{n+1} - u_n}{x_{n+1} - x_n}.$$

Rewriting the prolonged vector field (1.4.9) in the new variables, we get:

$$\text{(1.4.11)} \qquad \begin{aligned} \mathrm{pr}^1 \hat{X} = {} & \xi(\tilde{x}_n, \tilde{u}_n)\partial_{\tilde{x}_n} + \phi(\tilde{x}_n, \tilde{u}_n)\partial_{\tilde{u}_n} \\ & + [\xi(\tilde{x}_n + h_{n+1}, \tilde{u}_n + h_{n+1}u_{x,n+1}) - \xi(\tilde{x}_n, \tilde{u}_n)]\partial_{h_{n+1}} \\ & + \Big[\frac{\phi(\tilde{x}_n + h_{n+1}, \tilde{u}_n + h_{n+1}u_{x,n+1}) - \phi(\tilde{x}_n, \tilde{u}_n)}{h_{n+1}} \\ & - u_{x,n+1}\frac{\xi(\tilde{x}_n + h_{n+1}, \tilde{u}_n + h_{n+1}u_{x,n+1}) - \xi(\tilde{x}_n, \tilde{u}_n)}{h_{n+1}}\Big]\partial_{u_{x,n+1}}. \end{aligned}$$

When $h_{n+1} \to 0$, (1.4.11) and (1.1.11) become equal. Equation (1.4.11) gives a formula for the discrete prolongation

$$\text{(1.4.12)} \qquad \begin{aligned} \phi^{(1)} &= \Big[\frac{\phi(\tilde{x}_n + h_{n+1}, \tilde{u}_n + h_{n+1}u_{x,n+1}) - \phi(\tilde{x}_n, \tilde{u}_n)}{h_{n+1}} \\ &- u_{x,n+1}\frac{\xi(\tilde{x}_n + h_{n+1}, \tilde{u}_n + h_{n+1}u_{x,n+1}) - \xi(\tilde{x}_n, \tilde{u}_n)}{h_{n+1}}\Big]. \end{aligned}$$

We can also write down a recursive formula which provides the infinitesimal coefficient of the prolongation with respect to higher shifted variables, from which (1.4.12) can be obtained, see [521].

Let us assume a vector field $\hat{X} = \xi_n\partial_{x_n} + \phi_n\partial_{u_n}$ and its prolongation to N points

$$\text{(1.4.13)} \qquad \begin{aligned} \mathrm{pr}\hat{X} = {} & \xi_n\partial_{x_n} + \phi_n\partial_{u_n} + \xi_{n+1}\partial_{x_{n+1}} + \phi_{n+1}\partial_{u_{n+1}} \\ & + \xi_{n+2}\partial_{x_{n+2}} + \phi_{n+2}\partial_{u_{n+2}} + \ldots + \xi_{n+N}\partial_{x_{n+N}} + \phi_{n+N}\partial_{u_{n+N}}. \end{aligned}$$

Let us consider the following change of variable: from $\{x_n, u_n, x_{n+1}, u_{n+1}, x_{n+2}, u_{n+2}, \ldots x_{n+N}, u_{n+N}\}$ to $\{x_n, u_n, p_{n+1}, q_{n+2}, s_{n+3}, r_{n+4},$

$\dots p_{n+N}^{(N)}, h_{n+1}, h_{n+2}, \dots h_{n+N}\}$ where

$$
\begin{aligned}
p_{n+1} &= p_{n+1}^{(1)} = \frac{u_{n+1}-u_n}{x_{n+1}-x_n} \\
q_{n+2} &= p_{n+2}^{(2)} = \frac{p_{n+2}-p_{n+1}}{\frac{x_{n+2}-x_n}{2}} \\
s_{n+3} &= p_{n+3}^{(3)} = \frac{q_{n+3}-q_{n+2}}{\frac{x_{n+3}-x_n}{3}} \\
&\vdots \\
p_{n+N}^{(N)} &= \frac{p_{n+N}^{(n+N-1)} - p_{n+N-1}^{(n+N-1)}}{\frac{x_{n+N}-x_n}{N}} \\
h_{n+1} &= x_{n+1}-x_n \\
h_{n+2} &= x_{n+2}-x_{n+1} \\
&\vdots \\
h_{n+N} &= x_{n+N}-x_{n+N-1}.
\end{aligned}
$$

The prolongation of vector field (1.4.13) in these new variables then reads:

$$
\begin{aligned}
\mathrm{pr}\hat{X} &= \xi_n \partial_{x_n} + \phi_n \partial_{u_n} + \phi_n^{(1)} \partial_{p_{n+1}} + \phi_n^{(2)} \partial_{q_{n+2}} \\
&+ \phi_n^{(3)} \partial_{s_{n+3}} + \dots + \phi_n^{(N)} \partial_{p_{n+N}^{(N)}} + (\xi_{n+1}-\xi_n)\partial_{h_{n+1}} \\
&+ \dots + (\xi_{n+N}-\xi_{nN-1})\partial_{h_{n+N}}
\end{aligned}
$$

where, (see 1.4.12)

$$
\begin{aligned}
\phi_n^{(1)} &= \Delta(\phi_n) - p_{n+1}\Delta(\xi_n) = \frac{\phi_{n+1}-\phi_n}{h_{n+1}} - p_{n+1}\left(\frac{\xi_{n+1}-\xi_n}{h_{n+1}}\right), \\
\phi_n^{(2)} &= \Delta(\phi_n^{(1)}) - q_{n+2}\frac{1}{h_{n+2}+h_{n+1}}\left(h_{n+2}\frac{\xi_{n+2}-\xi_{n+1}}{h_{n+2}} + h_{n+1}\frac{\xi_{n+1}-\xi_n}{h_{n+1}}\right), \\
\phi^{(N)} &= \Delta(\phi_n^{(N-1)}) - p_{n+N}^{(N)}\frac{1}{\sum_{i=1}^N h_{n+i}}\left(\sum_{i=1}^N h_{n+i}\frac{\xi_{n+i}-\xi_{n+i-1}}{h_{n+i}}\right),
\end{aligned}
$$

where $\Delta(\phi_n)$ denotes the discrete derivative, i.e. $\Delta(\phi_n) = \frac{\phi_{n+1}-\phi_n}{h_{n+1}}$.

On the construction of symmetries of OΔEs see also [406].

4.1. Examples.

4.1.1. *Lie point symmetries of the discrete time Toda lattice.* The discrete time Toda equation [391] is one of the most well known completely integrable PΔEs considered in the literature [20, 384, 390, 392–394, 493, 637, 639, 645, 663, 728, 736, 737, 777] and is given by

$$
\begin{aligned}
\Delta_{Toda} &= e^{u_{n,m}-u_{n,m+1}} - e^{u_{n,m+1}-u_{n,m+2}} - \\
&- \alpha^2 (e^{u_{n-1,m+2}-u_{n,m+1}} - e^{u_{n,m+1}-u_{n+1,m}}) = 0.
\end{aligned}
\tag{1.4.14}
$$

It is a five points scheme. From the first two terms of (1.4.14) we can easily obtain the second difference of the function $u_{n,m}$ with respect to the discrete-time m. Thus, defining

$$
t = m\sigma_t; \qquad v_n(t) = u_{n,m}; \qquad \alpha = h_t^2,
\tag{1.4.15}
$$

we find that (1.4.14) when $h_t \to 0$ and $m \to \infty$ in such a way that t remains finite reduces to the well known Toda lattice:

$$\Delta^{(2)}_{Toda} = v_{n,tt} - e^{v_{n-1}-v_n} + e^{v_n - v_{n+1}} = 0. \tag{1.4.16}$$

The Toda lattice (1.4.16) is probably the best known and most studied DΔE. It plays, in the case of lattice equations, the same role as the Korteweg - de Vries equation for PDEs [796, 797]. It was obtained by Toda [798] when trying to explain the Fermi, Pasta and Ulam results [253] obtained when carrying out numerical experiments on the equipartition of energy in a non linear lattice of interacting oscillators. For more details see Section 2.1. As will be shown below in Chapter 2, (1.4.16) reduces, in the continuous limit, to the potential Korteweg-de Vries equation. It can be encountered in many applications from solid state physics to DNA biology, from molecular chain dynamics to chemistry [707].

Let us consider the Lie point symmetries of the discrete-time Toda lattice (1.4.14), on a fixed non transforming bidimensional lattice characterized by two lattice spacings in the two directions m and n, h_t and h_x.

If the lattice is uniform and homogeneous in both variables, we can represent the lattice by the following two equations:

$$x_{n,m} - nh_x = 0, \quad t_{n,m} - mh_t = 0. \tag{1.4.17}$$

Eqs. (1.4.17) from now on will be denoted as $\Delta_{Lattice} = 0$.

A Lie point symmetry is defined by giving its infinitesimal generators (1.4.6) which generates an infinitesimal transformation in the site (n, m) of its coordinates and of the function $u(x_{n,m}, t_{n,m}) = u_{n,m}$. The action of (1.4.6) on (1.4.14) is obtained, as explained before, by prolonging (1.4.6) to all points of the lattice. In this case the prolongation is obtained [530, 535, 538] by shifting (1.4.6) to the remaining 4 points

$$\mathrm{pr}\,\hat{X} = \hat{X}_{n,m} + \hat{X}_{n+1,m} + \hat{X}_{n,m+1} + \hat{X}_{n,m+2} + \hat{X}_{n-1,m+2}. \tag{1.4.18}$$

The invariance condition then reads:

$$\begin{aligned} &\mathrm{pr}\,\hat{X}\Delta_{Toda}|_{(\Delta_{Toda}=0,\Delta_{Lattice}=0)} = 0, \\ &\mathrm{pr}\,\hat{X}\Delta_{Lattice}|_{(\Delta_{Toda}=0,\Delta_{Lattice}=0)} = 0. \end{aligned} \tag{1.4.19}$$

The action of (1.4.18) on the lattice (1.4.17) gives $\xi_{n,m} = 0$, $\tau_{n,m} = 0$ and thus the variables x and t are invariant. When we act with (1.4.18) on the Toda equation (1.4.14), we get

$$\begin{aligned} &e^{u_{n,m}-u_{n,m+1}}[\phi_{n,m} - \phi_{n,m+1}] - e^{u_{n,m+1}-u_{n,m+2}}[\phi_{n,m+1} - \phi_{n,m+2}]- \\ &\quad - \alpha^2\{e^{u_{n-1,m+2}-u_{n,m+1}}[\phi_{n-1,m+2} - \phi_{n,m+1}]- \\ &\quad - e^{u_{n,m+1}-u_{n+1,m}}[\phi_{n,m+1} - \phi_{n+1,m}]\} = 0. \end{aligned} \tag{1.4.20}$$

If we differentiate (1.4.20) twice with respect to $u_{n,m+2}$ we get a differential equation for $\phi_{n,m+2}$ whose solution gives $\phi_{n,m} = c_1 e^{u_{n,m}} + c_2$, where c_1 and c_2 are two integration constants (that can depend on n and m). Introducing this result into (1.4.20) we get that c_1 must be equal to zero. Taking into account that, due to the form of the lattice, all points are independent we get that c_2 must be just a constant.

To sum up, the discrete time Toda lattice (1.4.14) considered on a fixed lattice has only a one-dimensional continuous symmetry group. It consists of the translation of the dependent variable u, i.e. $\tilde{u}_{n,m} = u_{n,m} + \kappa$ with κ constant. This symmetry is obvious from the beginning as (1.4.14) does not involve $u_{n,m}$ itself but only differences between values of u at different points of the lattice. Other transformations that leave the lattice and solutions invariant will be discrete [508]. In this case they are simply translations of x and t by integer multiples of the lattice spacing h_x and h_t.

The same conclusion holds in the general case of PΔEs on fixed lattices. Lie algebra techniques will provide transformations of the continuous dependent variables only, though the transformations can depend on the discrete independent variables.

In Chapter 2 we will see that the situation is completely different when generalized symmetries are considered.

If we will assume transforming lattices, i.e. discrete schemes, the number of point symmetries can increase, the scheme is more symmetrical.

4.1.2. *Lie point symmetries of DΔEs.* Let us now consider the more interesting case of DΔEs. For notational simplicity, let us restrict ourselves to scalar DΔEs for one real function $u_n(t)$ depending on one lattice variable n and one continuous real variable, t. Moreover, we will only be interested in DΔEs containing up to second order derivatives, as those, see for example (1.3.2, 1.3.4), are the ones of particular interest in applications to dynamical systems. We write such equations as

$$\Delta_n^{(2)} \equiv \Delta\big(t, n, u_{n+k}\Big|_{k=a_0}^{b_0}, u_{n+k,t}\Big|_{k=a_1}^{b_1}, u_{n+k,tt}\Big|_{k=a_2}^{b_2}\big) = 0, \qquad a_j \leq b_j \in \mathbb{Z}, \tag{1.4.21}$$

with $u_n \equiv u_n(t)$.

The lattice is uniform, time independent and fixed, the continuous variable t is the same at all points of the lattice. Thus to (1.4.21) we add the lattice equation

$$t_n - t_{n+1} = \Delta_t = 0, \quad \text{i.e. } t_n = t. \tag{1.4.22}$$

The Toda lattice equation (1.4.16) and the inhomogeneous Toda lattice [497]

$$\begin{aligned}\tilde{\Delta}_n^{(2)} = w_{\bar{t}\bar{t}}(n) - \frac{1}{2}w_{\bar{t}} + \frac{1}{4} - \frac{n}{2} + \left[\frac{1}{4}(n-1)^2 + 1\right] e^{w(n-1)-w(n)} \\ - \left[\frac{1}{4}n^2 + 1\right] e^{w(n)-w(n+1)} = 0,\end{aligned} \tag{1.4.23}$$

are examples of such equations. Other examples are given in Section 3.4.2 by (3.4.49) and (3.4.53) and in Section 2.3.2.7 by (2.3.161) and (2.3.165) (see also [497]). We are interested in Lie point transformations which leave the solution set of (1.4.21), (1.4.22) invariant. They have the form:

$$\tilde{t} = \Lambda_\epsilon\big(t, n, u_n(t)\big), \quad \tilde{u}_{\tilde{n}}(\tilde{t}) = \Omega_\epsilon\big(t, n, u_n(t)\big), \quad \tilde{n} = n \tag{1.4.24}$$

where ϵ represents a set of continuous group parameters.

Continuous transformations of the form (1.4.24) are generated by a Lie algebra of vector fields of the form:

$$\hat{X} = \tau_n\big(t, u_n(t)\big)\partial_t + \phi_n\big(t, u_n(t)\big)\partial_{u_n} \tag{1.4.25}$$

where n is treated as a discrete variable and we have $\tilde{n} = n$, when considering continuous transformations.

Invariance of the condition (1.4.22) implies that τ does not depend on n. This can be checked considering that symmetries define compatible flows [546].

As in the case of PDEs, for DΔEs the following invariance condition

$$\begin{aligned}\text{pr}^{(2)}\, \hat{X}\Delta_n^{(2)}\Big|_{\Delta_n^{(2)}=0,\Delta_t=0} = 0, \\ \text{pr}\, \hat{X}\Delta_t\Big|_{\Delta_n^{(2)}=0,\Delta_t=0} = 0,\end{aligned} \tag{1.4.26}$$

must be true if $\hat{X}$ is to belong to the Lie symmetry algebra of $\Delta_n^{(2)}$ and $\Delta_t = 0$. The symbol $\mathrm{pr}^{(2)} \hat{X}$ denotes the second prolongation of the vector field $\hat{X}$, i.e. in this case

$$
\begin{aligned}
\mathrm{pr}^{(2)} \hat{X} = \tau(t, u_n)\partial_t &+ \sum_{k=n-a}^{n+b} \phi_k(t, u_k)\partial_{u_k} \\
&+ \sum_{k=n-a_1}^{n+b_1} \phi_k^t(t, u_k, u_{k,t})\partial_{u_{k,t}} \\
&+ \sum_{k=n-a_2}^{n+b_2} \phi_k^{tt}(t, u_k, u_{k,t}, u_{k,tt})\partial_{u_{k,tt}}
\end{aligned}
\tag{1.4.27}
$$

with (see (1.1.35))

$$\phi_k^t(t, u_k, u_{k,t}) = D_t\phi_k(t, u_k) - [D_t\tau_k(t, u_k)]u_{k,t} \tag{1.4.28}$$

$$\phi_k^{tt}(t, u_k, u_{k,t}, u_{k,tt}) = D_t\phi_k^t(t, u_k, u_{k,t}) - [D_t\tau_k(t, u_k)]u_{k,tt} \tag{1.4.29}$$

where D_t is the total derivative (1.1.10) with respect to t. Here ϕ^t and ϕ^{tt} are the prolongation coefficients with respect to the continuous variable. The prolongation with respect to the discrete variable is reflected in the summation over k.

Eq. (1.4.26) is one equation with n as a discrete variable; thus we have a finite algorithm for obtaining the determining equations, a usually overdetermined system of linear partial differential equations for $\tau(t, u_n)$ and $\phi(n, t, u_n)$. We will call this approach the *intrinsic method* for obtaining symmetries of DΔEs.

A different approach consists of considering (1.4.21) as a system of coupled differential equations for the various functions $u_n(t)$. Thus, in general we have infinitely many equations for infinitely many functions. In this case the ansatz for the vector field $\hat{X}$ would be:

$$\hat{X} = \tau(t, \{u_j(t)\}_j)\partial_t + \sum_k \phi_k(t, \{u_j(t)\}_j)\partial_{u_k(t)} \tag{1.4.30}$$

where by $\{u_j(t)\}_j$ we mean the set of all $u_j(t)$ with j and k varying a priori over an infinite range. Calculating the second prolongation $\mathrm{pr}^{(2)} \hat{X}$ in a standard manner (see (1.4.27, 1.4.28,1.4.29)) and imposing

$$\mathrm{pr}^{(2)} \hat{X}\Delta_n^{(2)}\Big|_{(\Delta_j^{(2)}=0,\ \Delta_t=0)} = 0 \quad \forall n, j \tag{1.4.31}$$

we obtain, in general, an infinite system of determining equations for an infinite number of functions. Conceptually speaking, this second method, called the *differential equation method* in [538], may give rise to a larger symmetry group than the intrinsic method. In fact the intrinsic method yields purely point transformations, while the differential equation method can yield generalized symmetries with respect to the differences (but not the derivatives). In practice, in this example, it turns out that usually no higher order symmetries with respect to the discrete variable exist; then the two methods give the same result and the intrinsic method is simpler.

A third approach [681,683] consists of interpreting the variable n as a continuous variable and consequently the DΔE as a differential delay equation. We will call this method the *differential delay method*. In such an approach $u_{n+k}(t) \equiv \exp[\frac{k\,\partial}{\partial n}]\{u_n(t)\}$ and consequently the DΔE is interpreted as a PDE of infinite order. In such a case formula (1.4.26) is meaningless as we are not able to construct the infinite order prolongation of a vector field $\hat{X}$. The Lie symmetries are obtained by requiring that the solution set of the equation $\Delta_n^{(2)} = 0$ (1.4.16) be invariant under the infinitesimal transformation

$$\begin{aligned} \tilde{t} &= t + \epsilon\tau_n\big(t, u_n(t)\big), \\ \tilde{n} &= n + \epsilon\nu_n\big(t, u_n(t)\big), \\ \tilde{u}_{\tilde{n}}(\tilde{t}) &= u_n(t) + \epsilon\phi_n\big(t, u_n(t)\big). \end{aligned} \tag{1.4.32}$$

4.1.3. *Lie point symmetries of the Toda lattice.* Let us now apply the techniques introduced in Section 1.4.1.2 to the case of (1.4.16). In this case (1.4.26) reduces to an overdetermined system of determining equations obtained by equating to zero the coefficients of $[\upsilon_{n,t}]^k$, $k = 0, 1, 2, 3$ and of $\upsilon_{n\pm1}$. They imply

$$\tau = at + d, \quad \phi = b + 2an + ct, \qquad a, b, c, d \text{ real constants,} \tag{1.4.33}$$

corresponding to a four dimensional Lie algebra generated by the vector fields

$$\hat{D} = t\partial_t + 2n\partial_{\upsilon_n}, \quad \hat{T} = \partial_t, \quad \hat{W} = t\partial_{\upsilon_n}, \quad \hat{U} = \partial_{\upsilon_n}. \tag{1.4.34}$$

The group transformation which will leave (1.4.16) invariant is hence

$$\tilde{\upsilon}_n(\tilde{t}) = \upsilon_n(\tilde{t}e^{-\epsilon_4/2} - \epsilon_3) + \epsilon_2(\tilde{t}e^{-\epsilon_4/2} - \epsilon_3) + \epsilon_4 n + \epsilon_1 \tag{1.4.35}$$

where ϵ_j, $j = 1, 2, 3, 4$, are real group parameters.

To the transformation (1.4.35) we can add some discrete ones [506]:

$$\tilde{n} = n + N \quad N \in \mathbb{Z} \tag{1.4.36}$$

and

$$\big(t, \upsilon_n\big) \to \big(-t, \upsilon_n\big); \quad \big(t, \upsilon_n\big) \to \big(t, -\upsilon_{-n}\big). \tag{1.4.37}$$

We write the symmetry group of (1.4.16) as

$$G = G_D \rhd G_C \tag{1.4.38}$$

where G_D are the discrete transformations (1.4.36), (1.4.37) and the invariant subgroup G_C corresponds to the transformation (1.4.35). A complete classification of the one dimensional subgroups of G can be easily obtained [538].

In fact, if we complement the Lie algebra (1.4.34) by the vector field

$$\hat{Z} = \frac{\partial}{\partial n} \tag{1.4.39}$$

and require, at the end of the calculations, that the corresponding group parameter be integer, the commutation relations become

$$[\hat{Z}, \hat{D}] = 2\hat{U}; \quad [\hat{D}, \hat{T}] = -\hat{T}; \quad [\hat{D}, \hat{W}] = \hat{W}; \quad [\hat{T}, \hat{W}] = \hat{U}. \tag{1.4.40}$$

The one dimensional subalgebras are

$$\begin{gathered} \{\hat{Z} + a\hat{D} + b\hat{U}\}, \quad \{\hat{Z} + a\hat{T} + k\hat{W}\}, \quad \{\hat{Z} + \epsilon\hat{W}\}, \quad \{\hat{Z} + a\hat{U}\}, \\ \{\hat{T} + c\hat{W}\}, \quad \{\hat{D} + c\hat{U}\}, \quad \{\hat{U}\}, \quad \{\hat{Z}\}, \quad \{\hat{W}\}, \\ (a, b, c) \in \mathbb{R}; a \neq 0; k = 0, 1, -1; \epsilon = \pm 1. \end{gathered} \tag{1.4.41}$$

Nontrivial solutions, corresponding to reductions with respect to *continuous subgroups* $G_0 \subset G_C$, are obtained by considering invariance of the Toda lattice under $\{\hat{T} + c\hat{W}\}$, or

$\{\hat{D} + c\hat{U}\}$. They are

$$v_n(t) = p - \frac{1}{2}ct^2 - \sum_{j=1}^{n} \log(q - cj), \tag{1.4.42}$$

$$v_n(t) = p + 2(n + c)\log(t) - \sum_{j=0}^{n} \log[q + (2c - 1)j + j^2], \tag{1.4.43}$$

where p and q are arbitrary constants of integration.

Reduction by the purely *discrete subgroup*, $G_0 \subset G_D$, given in (1.4.36) implies the invariance of (1.4.16) under discrete translation of n and makes it possible to impose the periodicity condition

$$u(n + N, t) = u(n, t). \tag{1.4.44}$$

This reduces the DΔE (1.4.16) to an ODE (or a finite system of equations). For example for $N = 2$, we get a sine-Gordon type ODE

$$v_{tt} = -4 \sinh v, \tag{1.4.45}$$

while for $N = 3$, we get the Tzitzèika differential equation [805, 806]

$$v_{tt} = e^{-2v} - e^{v}. \tag{1.4.46}$$

Let us now consider a subgroup $G_0 \subset G$ that is not contained in G_C, nor in G_D, i.e. a *nonsplitting subgroup* of G. A reduction corresponding to $\hat{Z} + a\hat{D} + b\hat{U}$ implies the symmetry variables

$$y = te^{-an}, \quad v_n(t) = an^2 + bn + F(y) \tag{1.4.47}$$

and yields the differential delay equation

$$F''(y) = e^{-b}[\exp(F(ye^{a}) - F(y) + a) - \exp(F(y) - F(ye^{-a}) - a)]. \tag{1.4.48}$$

Using the subalgebra $\hat{Z} + a\hat{T} + k\hat{W}$ the symmetry variables are

$$y = t - an, \quad v_n(t) = \frac{k}{2a}t^2 + F(y), \tag{1.4.49}$$

and we get the differential delay equation

$$F''(y) = e^{F(y+a)-F(y)} - e^{F(y)-F(y-a)} - \frac{k}{a}. \tag{1.4.50}$$

Eq. (1.4.48) involves one independent variable y, but the function F and its derivatives are evaluated at the point y and at the dilated points ye^{a} and ye^{-a}. Eq. (1.4.50) is a differential delay equation which has interesting solutions, such as the soliton and periodic solutions of the Toda lattice (for $k = 0$).

The other two nonsplitting subgroups give rise to linear delay equations which can be solved explicitly.

This same calculation can also be carried out for the inhomogeneous Toda lattice (1.4.23). The symmetry algebra is

$$\begin{aligned} &\tilde{D} = 2\partial_{\tilde{t}} + \frac{1}{2}\partial_{w_n}, \quad \tilde{T} = e^{-\tilde{t}/2}\left[\partial_{\tilde{t}} - \left(w_n - \frac{1}{2}\right)\partial_{w_n}\right] \\ &\tilde{W} = 2e^{\tilde{t}/2}\partial_{w_n}, \quad \tilde{U} = \partial_{w_n}. \end{aligned} \tag{1.4.51}$$

These vector fields have the same commutation relations as those of the Toda lattice (1.4.16). This is a necessary condition for the existence of a point transformation between the two equations. In fact by comparing the two sets of vector fields, we get the following transformation which changes a solution $v_n(t)$ of equation (1.4.16) into a solution $w_n(\bar{t})$ of (1.4.23)

$$\bar{t} = 2\log\left(\frac{t}{2}\right),$$
$$w_n(\bar{t}) = v_n(t) - 2\left(n - \frac{1}{2}\right)\log(t) + \sum_{j=0}^{n}\left[\frac{1}{4}(j-1)^2 + 1\right]. \tag{1.4.52}$$

4.1.4. *Classification of DΔEs.* Group theoretical methods can also be used to classify equations according to their symmetry groups. This has been done in Section 1.1 in the case of PDEs [314] showing, for instance, that in the class of variable coefficient Korteweg-de Vries equations, the Korteweg-de Vries itself has the largest symmetry group. The same kind of results can also be obtained in the case of DΔEs.

Let us consider a class of equations involving nearest neighbor interactions [542]

$$\Delta_n = u_{n,tt}(t) - F_n(t, u_{n-1}(t), u_n(t), u_{n+1}(t)) = 0, \tag{1.4.53}$$

where F_n is non linear in $u_k(t)$ and coupled, i.e. such that $F_{n,u_k} \neq 0$ for some $k \neq n$.

We consider point symmetries only. The continuous transformations of the form (1.4.24) are again generated by a Lie algebra of vector fields of the form (1.4.25). To preserve the form of (1.4.53), we can reduce (1.4.25) to

$$\hat{X} = \tau(t)\partial_t + \left[\left(\frac{1}{2}\tau_{,t} + a_n(t)\right)u_n + b_n(t)\right]\partial_{u_n} \tag{1.4.54}$$

with $a_{n,t} = 0$ i.e $a_n(t) = a_n$. The determining equations reduce to

$$\frac{1}{2}\tau_{ttt}u_n + b_{n,tt} + (a_n - \frac{3}{2}\tau_t)F_n - \tau F_{n,t}$$
$$- \sum_{k=0,\pm 1}\left[\left(\frac{1}{2}\tau_t + a_{n+k}\right)u_{n+k} + b_{n+k}(t)\right]F_{n,u_{n+k}} = 0. \tag{1.4.55}$$

Our aim is to solve (1.4.55) with respect to both the form of the non linear equation, i.e. F_n, and the symmetry vector field $\hat{X}$, i.e. $\left(\tau(t), a_n, b_n(t)\right)$. In other words, for every non linear interaction F_n we wish to find the corresponding maximal symmetry group G. Associated with any symmetry group G there will be a whole class of non linear DΔEs related to each other by point transformations. To simplify the results, we will just look for the simplest element of a given class of non linear DΔEs, associated to a certain symmetry group. To do so we introduce so called *allowed transformations*, i.e. a set of transformations of the form

$$\tilde{t} = \tilde{t}(t), \quad \tilde{n} = n \quad u_n(t) = \Omega_n(\tilde{u}_n(\tilde{t}), t) \tag{1.4.56}$$

that transform (1.4.53) into a different one of the same type. By a straightforward calculation we find that the only allowed transformations (1.4.56) are given by

$$\tilde{t} = \tilde{t}(t), \quad \tilde{n} = n, \quad u_n(t) = \frac{A_n}{\sqrt{\tilde{t}_{,t}(t)}}\tilde{u}_n(\tilde{t}) + B_n(t) \tag{1.4.57}$$

with $B_n(t)$, A_n, $\tilde{t}(t)$ arbitrary functions of their arguments.

Under an allowed transformation, (1.4.53) is transformed into

$$\tilde{u}_{n,\tilde{t}\tilde{t}}(\tilde{t}) = \tilde{F}_n\left(n, \tilde{t}, \tilde{u}_{n+1}(\tilde{t}), \tilde{u}_n(\tilde{t}), \tilde{u}_{n-1}(\tilde{t})\right) \tag{1.4.58}$$

with

$$\tilde{F}_n = \frac{1}{\tilde{t}_t^{\frac{3}{2}} A_n}\Big\{ F_n(n,t,\{u_{n+1}(t),u_n(t),u_{n-1}(t)\}) - B_{n,tt} - \left[\frac{3}{4}\frac{\tilde{t}_{tt}^2}{\tilde{t}_t^{\frac{5}{2}}} - \frac{1}{2}\frac{\tilde{t}_{ttt}}{\tilde{t}_t^{\frac{3}{2}}}\right] A_n \tilde{u}_n(\tilde{t})\Big\} \tag{1.4.59}$$

and the symmetry generator (1.4.54) into

$$\hat{X} = [\tau(t)\tilde{t}_t]\partial_{\tilde{t}} + \Big\{\left[\frac{\tau(t)}{2}\frac{\tilde{t}_{tt}}{\tilde{t}_{,t}} + \frac{1}{2}\tau_t(t) + a_n\right]\tilde{u}_n + \frac{\tilde{t}_t^{\frac{1}{2}}}{A_n}\left[-\tau(t)B_{n,t}(t) + B_n(t)\left(\frac{1}{2}\tau_t(t) + a_n\right) + b_n(t)\right]\Big\}\partial_{\tilde{u}_n}. \tag{1.4.60}$$

We see that, up to an allowed transformation, every one-dimensional symmetry algebra associated to (1.4.53), can be represented by one of the following vector fields:

$$\hat{X}_1 = \partial_t + a_n^1 u_n \partial_{u_n} \quad \hat{X}_2 = a_n^2 u_n \partial_{u_n} \quad \hat{X}_3 = b_n(t)\partial_{u_n} \tag{1.4.61}$$

where a_n^j with $j = 1, 2$ are two arbitrary functions of n and $b_n(t)$ is an arbitrary function of n and t. The vector fields $\hat{X}_j$, $j = 1, 2, 3$ are the symmetry vectors of the Lie point symmetries of the following non linear DΔEs:

$$\begin{aligned} &\hat{X}_1 : \quad u_{n,tt} = e^{a_n^1 t} f_n(\xi_{n+1}, \xi_n, \xi_{n-1}), && \text{with } \xi_j = u_j e^{-a_j^1 t} \\ &\hat{X}_2 : \quad u_{n,tt} = u_n f_n(t, \eta_{n+1}, \eta_{n-1}), && \text{with } \eta_j = \frac{(u_j)^{a_n^2}}{(u_n)^{a_j^2}} \\ &\hat{X}_3 : \quad u_{n,tt} = \frac{b_{n,tt}}{b_n} u_n + f_n(t, \zeta_{n+1}, \zeta_{n-1}) && \text{with } \zeta_j = u_j b_n(t) - u_n b_j(t). \end{aligned} \tag{1.4.62}$$

These equations are still quite general, as they are written in terms of arbitrary functions depending on three continuous variables. More specific equations are obtained for larger symmetry groups [542].

The Toda equation (1.4.16) is included in a class of equations whose infinitesimal symmetry generators satisfy a four dimensional solvable symmetry algebra with a non abelian nilradical. The interactions in this class are given by

$$F_n(t, u_{n-1}(t), u_n(t), u_{n+1}(t)) = e^{-2\frac{u_{n+1}-u_n}{\gamma_{n+1}-\gamma_n}} f_n(\xi) \tag{1.4.63}$$

where $\xi = (\gamma_n(t) - \gamma_{n+1}(t))u_{n-1} + (\gamma_{n+1}(t) - \gamma_{n-1}(t))u_n + (\gamma_{n-1}(t) - \gamma_n(t))u_{n+1}$ and the function $\gamma_n(t)$ is such that $\gamma_{n+1}(t) \neq \gamma_n(t)$ and $\frac{\partial \gamma_n(t)}{\partial t} = 0$. The associated symmetry generators are:

$$\hat{X}_1 = \partial_{u_n}, \quad \hat{X}_2 = \partial_t, \quad \hat{X}_3 = t\partial_{u_n}, \quad \hat{Y} = t\partial_t + \gamma_n(t)\partial_{u_n}. \tag{1.4.64}$$

The Toda equation (1.4.16) is obtained by choosing $\gamma_n(t) = 2n$ and $f_n(\xi) = -1 + e^{\frac{1}{2}\xi}$. Among the equations of the class (1.4.53), the Toda equation does not have the largest group of point symmetries.

A complete list of all equations of the type (1.4.53) with nontrivial symmetry group is given in the original article [542] with the additional assumption that the interaction and the vector fields depend continuously on n. Here we just give two examples of interactions with symmetry groups with dimension seven. The first one is solvable, non nilpotent and its Lie algebra is given by

$$\begin{aligned}(1.4.65)\qquad &\hat{X}_1 = \partial_{u_n}, \quad \hat{X}_2 = (-1)^n \partial_{u_n}, \quad \hat{X}_3 = t\partial_{u_n}, \\ &\hat{X}_4 = (-1)^n t\partial_{u-n}, \quad \hat{X}_5 = (-1)^n u_n \partial_{u_n}, \quad \hat{X}_6 = \partial_t, \quad \hat{X}_7 = t\partial_t + 2u_n\partial_{u_n}.\end{aligned}$$

Eq. (1.4.65) is meaninfull only when n is an integer. The invariant equation is

$$(1.4.66)\qquad u_{n,tt} = \frac{\gamma_n}{u_{n-1} - u_{n+1}}.$$

This algebra was not included in [542] because of its non analytical dependence on n (in $\hat{X}_2$, $\hat{X}_4$ and $\hat{X}_5$).

The second symmetry algebra is nonsolvable. It contains the simple Lie algebra as a subalgebra. A basis of this algebra is

$$\begin{aligned}(1.4.67)\qquad &\hat{X}_1 = \partial_{u_n}, \quad \hat{X}_2 = t\partial_{u_n}, \quad \hat{X}_3 = b_n\partial_{u_n} \\ &\hat{X}_4 = b_n t\partial_{u_n}, \quad \hat{X}_5 = \partial_t, \quad \hat{X}_6 = t\partial_t + \frac{1}{2}u_n\partial_{u_n}, \quad \hat{X}_7 = t^2\partial_t + tu_n\partial_{u_n}\end{aligned}$$

with $b_{n,t} = 0$, $b_{n+1} \neq b_n$. The corresponding invariant non linear DΔE is:

$$(1.4.68)\qquad u_{n,tt} = \frac{\gamma_n}{[(b_{n+1} - b_n)u_{n-1} + (b_{n-1} - b_{n+1})u_n + (b_n - b_{n-1})u_{n+1}]^3}$$

where γ_n and b_n are arbitrary n-dependent constants.

In Section 3.4.2 we report the integrability conditions for equations to belong to the class (1.4.53) [852]. It will be shown that any equation of this class which has local generalized symmetries can be reduced by point transformations of the form

$$(1.4.69)\qquad \tilde{u}_n = \sigma_n(t, u_n), \quad \tilde{t} = \theta(t)$$

to either the Toda equation (1.4.16) or to the potential Toda equation

$$(1.4.70)\qquad u_{n,tt} = e^{u_{n+1} - 2u_n + u_{n-1}}.$$

4.1.5. *Lie point symmetries of the two dimensional Toda equation.* Let us now apply the techniques introduced in Section 1.4.1.2 to the Two Dimensional Toda System (TDTS)

$$(1.4.71)\qquad \Delta_{TDTS} = u_{n,xt} - e^{u_{n-1} - u_n} + e^{u_n - u_{n+1}} = 0$$

where $u_n = u_n(x,t)$. The TDTS was proposed and studied by Mikhailov [602] and Fordy and Gibbons [270]. See also [493]. It is an integrable DΔE, having a Lax pair, infinitely many conservation laws, Bäcklund transformations, soliton solutions, and all the usual attributes of integrability [3, 147, 265, 267, 355, 652, 732, 763] that we will encounter in Section 2.2 for KdV.

The continuous symmetries for (1.4.71) are obtained by considering the infinitesimal symmetry generator

$$(1.4.72)\qquad \hat{X} = \xi_n(x,t,u_n)\partial_x + \tau_n(x,t,u_n)\partial_t + \phi_n(x,t,u_n)\partial_{u_n}.$$

From the determining equation (1.4.26) we get

$$(1.4.73)\qquad \tau_n = f(t), \quad \xi_n = h(x), \quad \phi_n = (h_{,x} + f_{,t})\,n + g(t) + k(x),$$

where $f(t)$, $g(t)$, $h(x)$ and $k(x)$ are arbitrary C^∞ functions of one variable. A basis for the infinite dimensional symmetry algebra (1.4.72, 1.4.73) is given by

$$\begin{aligned} T(f) &= f(t)\partial_t + nf_{,t}\partial_{u_n}, \quad U(g) = g(t)\partial_{u_n}, \\ X(h) &= h(x)\partial_x + nh_{,x}\partial_{u_n}, \quad W(k) = k(x)\partial_{u_n}, \end{aligned} \tag{1.4.74}$$

where, to avoid redundancy, we must impose $k_{,x} \neq 0$.

The nonzero commutation relations are

$$\begin{aligned} &[T(f_1), T(f_2)] = T(f_1 f_{2,t} - f_{1,t} f_2), \quad [T(f), U(g)] = U(f g_{,t}), \\ &[X(h_1), X(h_2)] = X(h_1 h_{2,x} - h_{1,x} h_2), \\ &[X(h), W(k)] = \begin{cases} W(hk_{,x}), & (hk_{,x})_{,x} \neq 0, \\ cU(1), & (hk_{,x})_{,x} = 0, \quad hk_{,x} = c. \end{cases} \end{aligned} \tag{1.4.75}$$

Thus $\{T(f), U(g)\}$ form a Kac–Moody–Virasoro $\hat{u}(1)$ algebra, as do $\big(\{X(h), W(k), h_{,x} \neq 0, U(1)\}\big)$. However the two $\hat{u}(1)$ algebras are not disjoint.

This Kac–Moody–Virasoro character of the symmetry algebra is found also in the case of a $(2+1)$–dimensional Volterra equation [602, 603, 819]

$$\begin{aligned} a_{n,t} + \sigma^2 b_{n,x} &= a_n(a_{n-1}^2 - a_{n+1}^2) \\ (a_n a_{n-1})_x &= a_n b_{n-1} - a_{n-1} b_n, \end{aligned} \tag{1.4.76}$$

where $\sigma^2 = \pm 1$ and $a_n = a_n(x,t)$, $b_n = b_n(x,t)$. It is also characteristic of many other integrable equations involving three continuous variables, such as the Davey–Stewartson, Kadomtsev–Petviashvili or three–wave equations [83, 165, 201, 202, 461, 533, 595, 665, 829].

From the symmetry algebra we can construct the group of symmetry transformations which leave the TDTS (1.4.71) invariant and transform noninvariant solutions into new solutions. Moreover, we can use the subgroups to reduce the TDTS (1.4.71) to equations in a lower dimensional space.

4.2. Lie point symmetries preserving discretization of ODEs.

General Comments. In the previous Sections we assumed that a difference equation is given and we showed how to determine its symmetries.

Here we will discuss a different problem, namely the construction of difference equations and lattices corresponding to differential equations with a priory given symmetry groups. As an introductory material we limit ourselves to ODEs. More specifically, we start from a given ODE

$$E(x, y, \dot{y}, \ddot{y}, \ldots) = 0 \tag{1.4.77}$$

and its symmetry algebra $\mathfrak{g}$ of order ℓ, realized by vector fields of the form (1.1.6) with $p = q = 1$. We now wish to construct an Ordinary Difference System (OΔS),

$$\begin{aligned} E_a(\{x_k\}_{k=n}^{n+N}, \{u_k\}_{k=n}^{n+N}) = 0, \ a = 1, 2, \\ n, N \in \mathbb{Z}, \quad N \geq 0, \quad u_k \equiv u(x_k), \end{aligned} \tag{1.4.78}$$

approximating the ODE (1.4.77) and having the same Lie point symmetry algebra (and the same symmetry group). In general, the motivation for such a study is multifold. In physical applications the symmetry may actually be more important than the equation itself. A discrete scheme with the correct symmetries has a good chance of describing the physics correctly. This is specially true if the underlying phenomena really are discrete and the differential equations come from a continuous approximation. Furthermore, the existence of point symmetries for ODEs and OΔEs makes it possible to obtain explicit analytical solutions. Finally, one is expecting that a discretization respecting point symmetries should provide improved numerical methods [91, 92, 116, 117, 152, 352, 573, 711].

Let us at first outline the general method of discretization. If the ODE (1.4.77) is of order N we need a OΔS involving at least $N+1$ points

$$\{x_i, u_i; i = 1, \dots, N+1\}. \tag{1.4.79}$$

The procedure is as follows

(1) Take the Lie algebra $\mathfrak{g}$ of the symmetry group G of the ODE (1.4.77) and prolong the given vector fields $\{\hat{X}_1, \dots, \hat{X}_\ell\}$ to all $N+1$ points (1.4.79),

$$\mathrm{pr}\,\hat{X} = \sum_{k=n}^{n+N} \xi_k(x_k, u_k)\partial_{x_k} + \sum_{k=n}^{n+N} \phi_k(x_k, u_k)\partial_{u_k}. \tag{1.4.80}$$

(2) Find a basis for all invariants of the prolonged Lie algebra $\mathfrak{g}$ in the space (1.4.79) of independent and dependent variables. Such a basis will consist of $K < N$ functionally independent invariants

$$I_b = I_b(x_1, \dots, x_{N+1}, u_1, \dots, u_{N+1}), \quad 1 \le b \le K. \tag{1.4.81}$$

They are determined as the common solutions of the differential equations

$$pr\hat{X}_i I_b(x_1, \dots, x_{N+1}, u_1, \dots, u_{N+1}) = 0, \quad i = 1, \dots, \ell. \tag{1.4.82}$$

The actual number K satisfies

$$K = 2N + 2 - (\dim\mathfrak{g} - \dim\mathfrak{g}_0) \tag{1.4.83}$$

where $\mathfrak{g}_0$ is the Lie algebra of the subgroup $G_0 \subset G$, stabilizing the $N+1$ points (1.4.79), i.e. leaving them invariant.

We need at least two independent invariants of the form (1.4.81) to write an invariant difference scheme.

(3) If the number of invariants is not sufficient, we can make use of invariant manifolds. To find them, we first write out the matrix of coefficients of the prolonged vector fields $\{\hat{X}_1, \dots, \hat{X}_\ell\}$:

$$M = \begin{pmatrix} \xi_{1n} & \xi_{1n+1} & \cdots & \xi_{1N+n} & \phi_{1n} & \phi_{1n+1} & \cdots & \phi_{1N+n} \\ \vdots & \vdots & & \vdots & \vdots & \vdots & & \vdots \\ \xi_{\ell n} & \xi_{\ell n+1} & \cdots & \xi_{\ell N+n} & \phi_{\ell n} & \phi_{\ell n+1} & \cdots & \phi_{\ell N+n} \end{pmatrix} \tag{1.4.84}$$

and determine the manifolds on which the rank of M, Rank(M), satisfies

$$\mathrm{Rank}(M) < \min(\ell, 2N+2), \tag{1.4.85}$$

i.e. is less than maximal. The invariant manifolds are then obtained by requiring that (1.4.82) be satisfied on the manifold satisfying (1.4.85).

Example [116]. Let us consider the second order non linear ODE

$$x^2 u_{xx} + 4x u_x + 2u = (2xu + x^2 u_x)^{\frac{k-2}{k-1}}, \quad k \neq 0, \frac{1}{2}, 1, 2. \tag{1.4.86}$$

The choice of the parameter k is such that the equation is non singular, non linear and not linearizable. For these values of k the equation has a three dimensional symmetry algebra given by

$$\begin{aligned}\hat{X}^{(1)} &= \partial_x - \frac{2u}{x}\partial_u, \quad \hat{X}^{(2)} = \frac{1}{x^2}\partial_u, \\ \hat{X}^{(3)} &= x\partial_x + (k-2)u\partial_u, \\ [\hat{X}^{(1)}, \hat{X}^{(2)}] &= 0, \quad [\hat{X}^{(1)}, \hat{X}^{(3)}] = \hat{X}^{(1)}, \quad [\hat{X}^{(2)}, \hat{X}^{(3)}] = k\hat{X}^{(2)}.\end{aligned} \tag{1.4.87}$$

As the equation is an ODE of second order and has a three dimensional symmetry group which has an Abelian subalgebra, it will be solvable and its general solution is

$$u = \left(\frac{1}{k-1}\right)^{k-1}\frac{1}{k}\frac{x-x_0}{x^2} + \frac{u_0}{x}.$$

As the equation is of second order the minimum number of point necessary to describe it is three: (x, x_+, x_-), (u, u_+, u_-), where $x = x_n$, $x_\pm = x_{n\pm 1}$. The invariance condition reads:

$$\mathrm{pr}\hat{X}F(x_-, x, x_+, u_-, u, u_+) = 0, \tag{1.4.88}$$

where F, an apriori arbitrary function of its arguments, is an invariant and (1.4.88) must be satisfied by F for the prolongation (1.4.18) of all generators $\hat{X}$ given by (1.4.87)

$$\mathrm{pr}\hat{X}^{(1)} = \hat{X}^{(1)} + \partial_{x_-} - \frac{2u_-}{x_-}\partial_{u_-} + \partial_{x_+} - \frac{2u_+}{x_+}\partial_{u_+}, \text{ etc..} \tag{1.4.89}$$

As $\hat{X}^{(1)}$ contains the terms ∂_x with constant coefficients, the functions $I^{(1)} = x_+ - x$ and $I^{(2)} = x - x_-$ are two independent invariants together with any function of them. The only other infinitesimal generator which involves variation with respect to x is $\hat{X}^{(3)}$. The x–dependent part of the prolongation of $\hat{X}^{(3)}$ can be written in terms of $I^{(1)}$ and $I^{(2)}$, $\mathrm{pr}\hat{X}^{(3)} = I^{(1)}\partial_{I^{(1)}} + I^{(2)}\partial_{I^{(2)}}$. $I^{(1)}$ and $I^{(2)}$ are not invariants of $\mathrm{pr}\hat{X}^{(3)}$ but $\xi_1 = (x_+ - x)/(x - x_-)$ is.

A second invariant is obtained solving among the equations $\mathrm{pr}\hat{X}^{(1)}F = 0$ the characteristic differential equation $u_x = -u/x$ and its shifted ones. They provide two new invariants $I^{(3)} = (x_+)^2u_+$ and $I^{(4)} = x^2u$. We can express the variables x, x_+, u and u_+ in terms of $I^{(1)}$, $I^{(3)}$ and $I^{(4)}$ and then $\mathrm{pr}\hat{X}^{(2)} = (1/x^2)\partial_u + [1/(x_+)^2]\partial_{u_+}$ reads $\mathrm{pr}\hat{X}^{(2)} = \partial_{I^{(3)}} + \partial_{I^{(4)}}$ and $J^{(1)} = I^{(4)} - I^{(3)}$ is its invariant. In a similar way we get $\mathrm{pr}\hat{X}^{(3)} = kJ^{(1)}\partial_{J^{(1)}} + I^{(1)}\partial_{I^{(1)}}$ which has $\xi_2 = J^{(1)}/(I^{(1)})^k$ as an invariant. Taking into account the variables x, x_-, u and u_- we get in a similar way the partial invariants $I^{(5)} = (x_-)^2u_-$, $J^{(2)} = I^{(5)} - I^{(4)}$ and consequently the invariant $\xi_3 = J^{(2)}/(I^{(2)})^k$.

When we perform the continuous limit, $h_{n+1} = h_+ = I^{(1)}$ and $h_n = h = I^{(2)}$ go to zero while

$$\begin{aligned}u_+ &= u(x_+) = u(x) + h_+u_x + \frac{(h_+)^2}{2!}u_{xx} + \mathcal{O}((h_+)^3), \\ u_- &= u(x_-) = u(x) - hu_x + \frac{h^2}{2!}u_{xx} + \mathcal{O}(h^3).\end{aligned} \tag{1.4.90}$$

Combining ξ_1, ξ_2 and ξ_3 we get in the continuous limit

$$\begin{aligned}\frac{2\xi_1}{\xi_1 + 1}\left(\xi_2 - \frac{\xi_2}{\xi_1^{k-1}}\right) = (h_+)^{2-k}\Big[&(x^2u_{xx} + 4xu_x + 2u) + \\ &+ \frac{1}{3}(h_+ - h)(x^2u_{xxx} + 6xu_{xx} + 6u_x) + \mathcal{O}(h^2)\Big],\end{aligned} \tag{1.4.91}$$

$$\frac{1}{2}\big(\xi_2 + \frac{\xi_3}{\xi_1^{k-1}}\big)^{(k-2)/(k-1)} = (h_+)^{2-k}(x^2u_x + 2xu)^{(k-2)/(k-1)}$$

$$\Big[1 + (h_+ - h)\frac{k-2}{k-1}\frac{x^2u_{xx} + 4xu_x + 2u}{x^2u_x + 2xu} + \mathcal{O}(h^2)\Big].$$

We can thus write down in terms of the invariants ξ_1, ξ_2 and ξ_3 the difference equation

(1.4.92)
$$\frac{2\xi_1}{\xi_1 + 1}\big(\xi_2 - \frac{\xi_2}{\xi_1^{k-1}}\big) = \frac{1}{2}\big(\xi_2 + \frac{\xi_3}{\xi_1^{k-1}}\big)^{(k-2)/(k-1)},$$

which, taking into account (1.4.91), will approximate up to order h^2 the differential equation (1.4.86). The only invariant dependent just on the lattice variable is ξ_1 and thus an admissible lattice of the symmetry preserving discretization is given by

(1.4.93)
$$\xi_1 = K.$$

When $K \neq 1$ (1.4.93) will give a lattice up to order h. When $K = 1$ the lattice equation represent a uniform lattice and will approximate the continuous case up to order h^2.

Let us compare the discrete scheme provided by (1.4.92, 1.4.93), for $k = 3$ and $K = 1$

$$x_+^2u_+ - 2x^2u + x_-^2u_- = \frac{h^{\frac{3}{2}}}{\sqrt{2}}(x_+^2u_+ - x_-^2u_-)^{\frac{1}{2}}$$

$$x_+ - 2x + x_- = 0$$

with a Runge–Kutta defined on the same number of points [55]

$$(\tilde{u}_+ - 2\tilde{u} + \tilde{u}_-)\tilde{x}^2 + 2\tilde{x}\tilde{h}(\tilde{u}_+ - \tilde{u}_-) + 2\tilde{h}^2\tilde{u} = \tilde{h}^2(2\tilde{x}\tilde{u} + \tilde{x}^2\frac{\tilde{u}_+ - \tilde{u}_-}{2\tilde{h}})^{\frac{1}{2}}.$$

In both schemes the problem of obtaining u_+ from u and u_- is non linear and to solve it we need to apply a fixed point iteration up to convergence.

If we choose $x \in [1, 3]$ with $u(1) = \frac{13}{12}$ and $u_x(1) = -1$, the exact solution of (1.4.86) is $u(x) = \frac{x}{12} + \frac{1}{x^2}$. In the discrete scheme we consider the initial condition $u_0 = u(x = 1) = \frac{13}{12}$ and $u_1 = u(x = 1 + h) = \frac{1+h}{12} + \frac{1}{(1+h)^2}$. In Fig. 1.8 we present the differences of the discretization errors of the two methods with respect to the exact result. Both schemes have the same accuracy but the best result is obtained in the symmetry preserving scheme.

4.3. Group classification and solution of OΔEs.

4.3.1. *Symmetries of second order ODEs.* Let us now restrict to the case of a second order ODE

(1.4.94)
$$u_{xx} = F(x, u, u_x).$$

Lie gave a symmetry classification of (1.4.94) (over the field of complex numbers $\mathbb{C}$) [564, 566]. A similar classification over $\mathbb{R}$ is much more recent [580, 581].

The main classification results can be summed up as follows.

(1) The dimension $n = \dim \mathfrak{g}$ of the symmetry algebra of (1.4.94) can be $\dim \mathfrak{g} = 0, 1, 2, 3$ or 8.
(2) If we have $\dim \mathfrak{g} = 1$ we can decrease the order of (1.4.94) by one. If the dimension is $\dim \mathfrak{g} \geq 2$ we can integrate by quadratures.
(3) If we have $\dim \mathfrak{g} = 8$, then the symmetry algebra is $\boldsymbol{sl}(3, \mathbb{C})$, or $\boldsymbol{sl}(3, \mathbb{R})$, respectively. The equation can be transformed into $\ddot{y} = 0$ by a point transformation.

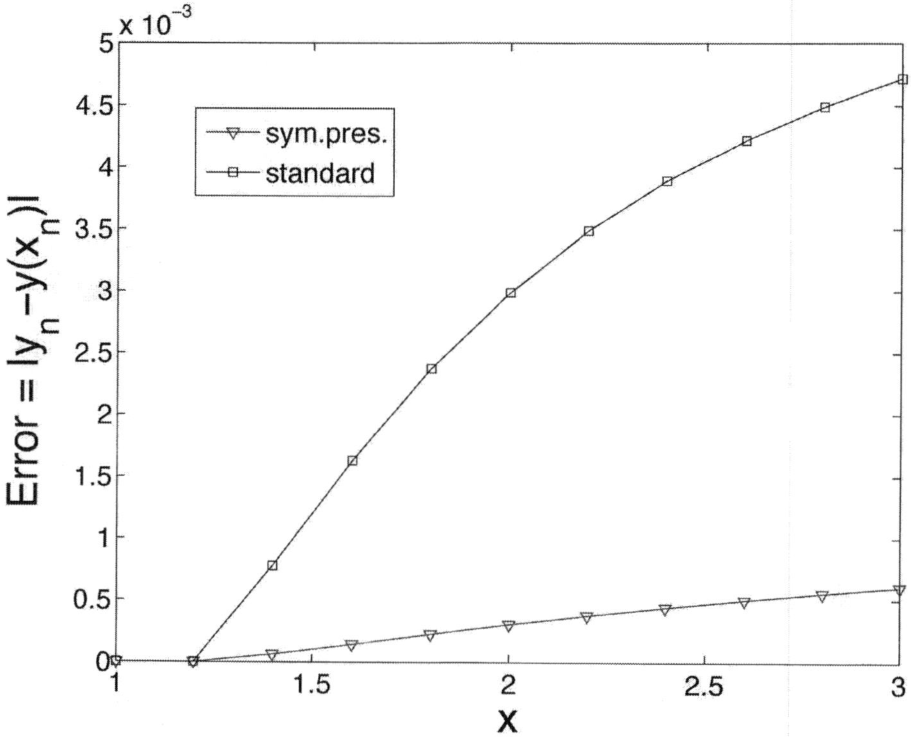

FIGURE 1.8. Discretization errors for the symmetry preserving scheme (1.4.92) and the standard scheme for (1.4.86), reprinted from [116].

Further symmetry results are due to E. Noether [647] and Bessel–Hagen [86]. Every ODE (1.4.94) can be interpreted as an Euler–Lagrange equation for some Lagrangian density

$$\mathcal{L} = \mathcal{L}(x, u, u_x). \tag{1.4.95}$$

The Euler Lagrange equation is

$$\frac{\partial \mathcal{L}}{\partial u} - D(\frac{\partial \mathcal{L}}{\partial u_x}) = 0, \tag{1.4.96}$$

where $D = D_x$ is the total derivative operator in the direction x (1.1.10). An *infinitesimal divergence symmetry*, or a Lagrangian symmetry is a vector field $\hat{X}$ (1.1.6) with $p = q = 1$ satisfying

$$\text{pr}\, \hat{X}(\mathcal{L}) + \mathcal{L}D(\xi) = D(V), \quad V = V(x, u), \tag{1.4.97}$$

where V is some function of x and u. A symmetry of the Lagrangian $\mathcal{L}$ is always a symmetry of the Euler–Lagrange equation (1.4.96), however equation (1.4.96) may have additional, non Lagrangian symmetries.

A relevant symmetry result is that if we have $\dim \mathfrak{g} = 1$, or $\dim \mathfrak{g} = 2$ for (1.4.94), then there always exists a Lagrangian having the same symmetry. For $\dim \mathfrak{g} = 3$, at least a two-dimensional subalgebra of the Lagrangian symmetries exists. For $\dim \mathfrak{g} = 8$ a four-dimensional solvable subalgebra of Lagrangian symmetries exists.

4.3.2. *Symmetries of the three-point difference schemes.* A symmetry classification of three-point difference schemes was performed quite recently [230, 231]. It is similar to Lie's classification of second order ODE's and goes over into this classification in the continuous limit. We shall now review the main results of the classification following the method outlined in Section 1.4.2.

Lie in [565] gave a classification of all finite dimensional Lie algebras that can be realized by vector fields of the form (1.1.21). This was done over the field $\mathbb{C}$ and thus amounts to a classification of finite dimensional subalgebras of diff$(2, \mathbb{C})$, the Lie algebra of the group of diffeomorphisms of the complex plane $\mathbb{C}^2$. A similar classification of finite dimensional subalgebras of diff$(2, \mathbb{R})$ exists [323], but we restrict ourselves to the simpler complex case.

We use the same notation for 3 neighboring points on the lattice as in the example presented in Section 1.4.2.

Let us now proceed by dimension of the symmetry algebras.

dim $\mathfrak{g} = 1$ **:** A single vector field can always be rectified into the form

$$\mathbf{A_{1,1}}: \qquad \hat{X}_1 = \frac{\partial}{\partial u} \tag{1.4.98}$$

The invariant ODE is

$$u_{xx} = F(x, u_x). \tag{1.4.99}$$

Putting $\dot{u} = y$ we obtain a first order ODE for y.

The difference invariants of $\hat{X}_1$ are

$$x,\ h_+ = x_+ - x,\ h_- = x - x_-,\ \eta_+ = u_+ - u,\ \eta_- = u - u_-. \tag{1.4.100}$$

Using these invariants we can introduce the discrete functions

$$\begin{aligned} u_x &= \frac{\eta_+}{h_+}, \quad u_{\underline{x}} = \frac{\eta_-}{h_-}, \\ u_{x\underline{x}} &= 2\frac{u_x - u_{\underline{x}}}{h_+ + h_-}, \ldots, \end{aligned} \tag{1.4.101}$$

and then we can write a difference scheme

$$u_{x\underline{x}} = F(x, \frac{u_x + u_{\underline{x}}}{2}, h_-), \quad h_+ = h_- G(x, \frac{u_x + u_{\underline{x}}}{2}, h_-). \tag{1.4.102}$$

This scheme goes into (1.4.99) if we require that the otherwise arbitrary functions F and G are such that

$$\lim_{h_- \to 0} F(x, \frac{u_x + u_{\underline{x}}}{2}, h_-) = F(x, u_x), \quad \lim_{h_- \to 0} G(x, \frac{u_x + u_{\underline{x}}}{2}, h_-) < \infty. \tag{1.4.103}$$

dim $\mathfrak{g} = 2$ **:** Precisely four equivalence classes of two-dimensional subalgebras of diff$(2, \mathbb{C})$ exist. Let us consider them separately.

$\mathbf{A_{2,1}}$ **:**

$$\hat{X}_1 = \partial_x, \quad \hat{X}_2 = \partial_u \tag{1.4.104}$$

The algebra $A_{2,1}$ is Abelian, the elements $\hat{X}_1$ and $\hat{X}_2$ are linearly not connected (linearly independent in any point (x, y)). The invariant ODE is

$$u_{xx} = F(u_x), \tag{1.4.105}$$

and can be immediately integrated.

An invariant difference scheme is given by any two relations between the invariants h_+, h_-, η_+, η_- of (1.4.100), for instance

(1.4.106) $$u_{x\underline{x}} = F(\frac{u_x + u_{\underline{x}}}{2}, h_-), \; h_+ = h_- G(\frac{u_x + u_{\underline{x}}}{2}, h_-),$$

with conditions (1.4.103) imposed on the functions F and G.

$\mathbf{A_{2,2}}$:

(1.4.107) $$\hat{X}_1 = \partial_u, \quad \hat{X}_2 = x\partial_x + u\partial_u$$

This Lie algebra in non-Abelian, the two elements are linearly not connected. The invariant ODE is

(1.4.108) $$u_{xx} = \frac{1}{x}F(u_x).$$

A basis for the difference invariance is

(1.4.109) $$\{xu_{x\underline{x}}, \; u_x + u_{\underline{x}}, \; \frac{h_+}{h_-}, \; \frac{h_-}{x}\}$$

so a possible invariant difference scheme is

(1.4.110) $$u_{x\underline{x}} = \frac{1}{x}F(\frac{u_x + u_{\underline{x}}}{2}, \frac{h_-}{x}), \; h_+ = h_- G(\frac{u_x + u_{\underline{x}}}{2}, \frac{h_-}{x}).$$

$\mathbf{A_{2,3}}$:

(1.4.111) $$\hat{X}_1 = \partial_u, \quad \hat{X}_2 = x\partial_u$$

The algebra is Abelian, the elements $\hat{X}_1$ and $\hat{X}_2$ are linearly connected. The invariant ODE is

(1.4.112) $$u_{xx} = F(x).$$

This equation is linear and hence has an eight dimensional symmetry algebra (of which $A_{2,3}$ is just a subalgebra).
The difference invariants are

(1.4.113) $$\{u_{x\bar{x}}, \; x, \; h_+, \; h_-\}$$

so the invariant difference scheme will also be linear (at least in the dependent variable u).

$\mathbf{A_{2,4}}$:

(1.4.114) $$\hat{X}_1 = \partial_u, \quad \hat{X}_2 = u\partial_u$$

The algebra is non-Abelian and isomorphic to $A_{2,2}$, but with linearly connected elements. The invariant ODE is again linear,

(1.4.115) $$u_{xx} = F(x)u_x,$$

as is the invariant difference scheme. Eq. (1.4.115) is invariant under the group $SL(3, \mathbb{C})$. Difference invariants are

(1.4.116) $$\{\xi = 2\frac{u_{x\underline{x}}}{u_x + u_{\underline{x}}}, \quad x, \; h_+, \; h_-\}$$

and a possible invariant OΔS is

(1.4.117) $$2\frac{u_{x\underline{x}}}{u_x + u_{\underline{x}}} = F(x, h_-), \quad G(x, h_+, h_-) = 0.$$

dim $\mathfrak{g} = 3$ **:** We will restrict ourselves to the case when the corresponding ODE is non linear. Hence we will omit all algebras that contain $A_{2,3}$ or $A_{2,4}$ subalgebras (they were considered in [230]).

$A_{3,1}$:

$$\hat{X}_1 = \partial_x, \quad \hat{X}_2 = \partial_u, \quad \hat{X}_3 = x\partial_x + ku\partial_u, \; k \neq 0, \frac{1}{2}, 1, 2 \tag{1.4.118}$$

The invariant ODE is

$$u_{xx} = u_x^{\frac{k-2}{k-1}}. \tag{1.4.119}$$

For $k = 1$ there is no invariant second order equation; for $k = 2$ the equation is linear, for $k = \frac{1}{2}$ it is transformable into a linear equation and has a larger symmetry group.

Difference invariants are

$$I_1 = \frac{h_+}{h_-}, \quad I_2 = u_x h_+^{1-k}, \quad I_3 = u_{\underline{x}} h_-^{1-k}. \tag{1.4.120}$$

A simple invariant difference scheme is

$$u_{x\underline{x}} = (\frac{u_x + u_{\underline{x}}}{2})^{[\frac{k-2}{k-1}} f(\frac{u_x + u_{\underline{x}}}{2} h_-^{1-k}), \tag{1.4.121}$$
$$h_+ = h_- g(\frac{u_x + u_{\underline{x}}}{2} h_-^{1-k}).$$

We shall see in Section 1.4.3.1 that in this case we can find other invariant schemes which may be more convenient for the integration.

$A_{3,2}$:

$$\hat{X}_1 = \partial_x, \quad \hat{X}_2 = \partial_u, \quad \hat{X}_3 = x\partial_x + (x+u)\partial_u, \tag{1.4.122}$$

The invariant ODE is

$$u_{xx} = e^{-u_x}. \tag{1.4.123}$$

Difference invariants in this case are

$$I_1 = \frac{h_+}{h_-}, \quad I_2 = h_+ e^{-u_x}, \quad I_3 = h_- e^{-u_{\underline{x}}}. \tag{1.4.124}$$

A possible invariant scheme is

$$u_{x\underline{x}} = e^{-\frac{u_x + u_{\underline{x}}}{2}} f(\sqrt{h_- h_+} e^{-\frac{u_x + u_{\underline{x}}}{2}}), \tag{1.4.125}$$
$$h_+ = h_- g(\sqrt{h_- h_+} e^{-\frac{u_x + u_{\underline{x}}}{2}}).$$

No further solvable three-dimensional subalgebras of diff$(2, \mathbb{C})$ exist (though there is another family for diff$(2, \mathbb{R})$ [230]).

Two inequivalent realizations of $sl(2, \mathbb{C})$ exist. Let us consider them separately.

$A_{3,3}$:

$$\hat{X}_1 = \partial_x, \quad \hat{X}_2 = 2x\partial_x + u\partial_u, \quad \hat{X}_3 = x^2\partial_x + xu\partial_u, \tag{1.4.126}$$

The corresponding invariant ODE is

$$u_{xx} = \frac{1}{u^3}, \tag{1.4.127}$$

and its general solution is

(1.4.128) $$u^2 = A(x-x_0)^2 + \frac{1}{A}, \qquad A \neq 0.$$

A convenient set of difference invariantsis

(1.4.129) $$\begin{aligned} I_1 &= \frac{h_+}{uu_+}, \quad I_2 = \frac{h_-}{h_+ + h_-}\frac{u_+}{u}, \\ I_3 &= \frac{h_+}{h_+ + h_-}\frac{u_-}{u}, \quad I_4 = \frac{h_-}{uu_-}. \end{aligned}$$

Any three of these are independent; the four satisfy the identity

(1.4.130) $$I_1 I_2 = I_3 I_4.$$

An invariant difference scheme can be written as

(1.4.131) $$2(I_2 + I_3 - 1) = I_1^2 I_2 \frac{I_2 + I_3}{I_3} f(I_1 I_2), \quad I_1 + I_4 = 4 I_1 I_2 g(I_1 I_2),$$

i.e.

(1.4.132) $$\begin{aligned} u_{x\bar{x}} &= \frac{1}{h_+ + h_-}\frac{1}{u^2}\left(\frac{h_+}{u_+} + \frac{h_-}{u_-}\right) f\left(\frac{1}{u^2}\frac{h_+ h_-}{h_+ + h_-}\right), \\ \frac{h_+}{u_+} + \frac{h_-}{u_-} &= \frac{4}{u}\frac{h_+ h_-}{h_+ + h_-} g\left(\frac{1}{u^2}\frac{h_+ h_-}{h_+ + h_-}\right). \end{aligned}$$

For $f = g = 1$ this scheme approximates the ODE (1.4.127).

$A_{3,4}$:

(1.4.133) $$\hat{X}_1 = \partial_u, \quad \hat{X}_2 = x\partial_x + u\partial_u, \quad \hat{X}_3 = x^2\partial_x + (-x^2 + u^2)\partial_u.$$

This algebra is again $sl(2, \mathbb{C})$ and can be transformed into

(1.4.134) $$\hat{Y}_1 = \partial_x + \partial_u, \quad \hat{Y}_2 = x\partial_x + u\partial_u, \quad \hat{Y}_3 = x^2\partial_x + u^2\partial_u.$$

The realization (1.4.134) (and hence also (1.4.133)) is imprimitive; (1.4.122) is primitive. Hence $A_{3.4}$ and $A_{3.3}$ are not equivalent. The invariant ODE for the algebra (1.4.133) is

(1.4.135) $$xu_{xx} = C(1 + {u_x}^2)^{\frac{3}{2}} + u_x(1 + {u_x}^2),$$

where C is a constant. The general integral of (1.4.135) can be written as

(1.4.136) $$(x-x_0)^2 + (u-u_0)^2 = \left(\frac{x_0}{C}\right)^2, \quad C \neq 0,$$

(1.4.137) $$x^2 + (u-u_0)^2 = x_0^2, C = 0,$$

where x_0 and u_0 are integration constants.

The difference invariants corresponding to the algebra (1.4.134) are

(1.4.138) $$\begin{aligned} I_1 &= \frac{x_+ - x}{x_+ x}(1 + u_x^2), \quad I_2 = \frac{x - x_-}{x_- x}(1 + u_{\underline{x}}^2), \\ I_3 &= -\frac{(x_+ - x)(x - x_-)}{2xx_+x_-}\{(h_+ u_x^2 + x_+ + x)u_{\underline{x}} + \\ &\qquad +(h_- u_{\underline{x}}^2 - x_- - x)u_x\}. \end{aligned}$$

An invariant scheme representing the ODE (1.4.135) can be written as

$$I_3 = C\left(\frac{I_1 + I_2}{2}\right)^{\frac{3}{2}}, \quad I_1 = I_2, \tag{1.4.139}$$

(this is not the most general such scheme).

4.3.3. *Lagrangian formalism and solutions of three-point OΔS.* In Section 1.4.3.1 we presented a Lagrangian formalism for the integration of second order ODE's. Let us now adapt it to OΔS [231].

The Lagrangian density (1.4.95) will now be a two-point function

$$\mathcal{L} = \mathcal{L}(x, u, x_+, u_+). \tag{1.4.140}$$

Instead of the Euler-Lagrange equation (1.4.96) we have two *quasi extremal equations* [219, 222, 223, 231] corresponding to *discrete variational derivatives* of $\mathcal{L}$ with respect to x and u independently

$$\begin{aligned} \frac{\delta \mathcal{L}}{\delta x} &= h_+ \frac{\partial \mathcal{L}}{\partial x} + h_- \frac{\partial \mathcal{L}^-}{\partial x} + \mathcal{L}^- - \mathcal{L} = 0, \\ \frac{\delta \mathcal{L}}{\delta u} &= h_+ \frac{\partial \mathcal{L}}{\partial u} + h_- \frac{\partial \mathcal{L}^-}{\partial u} = 0 \end{aligned} \tag{1.4.141}$$

where $\mathcal{L}^-$ is obtained by downshifting $\mathcal{L}$ (replacing n by $n-1$ everywhere, i.e. $\mathcal{L}^- = \mathcal{L}^-(x_-, u_-, x, u)$). The same definition of variational derivative will be considered later in Chapter 3 as is given by (3.2.40). In the continuous limit both quasi extremal equations (1.4.141) reduce to the same Euler-Lagrange equation. Thus, the two quasi extremal equations together can be viewed as an OΔS, where e.g. the difference between them defines the lattice.

The Lagrangian density (1.4.140) will be divergence invariant under the transformation generated by vector field $\hat{X}$, if it satisfies

$$\operatorname{pr} \hat{X}(\mathcal{L}) + \mathcal{L} D_+(\xi) = D_+(V), \tag{1.4.142}$$

for some function $V(x, u)$ where $D_+(f)$ is the discrete total derivative

$$D_+ f(x, u) = \frac{f(x+h, u(x+h)) - f(x, u)}{h}. \tag{1.4.143}$$

Each infinitesimal Lagrangian divergence symmetry operator $\hat{X}$ will provide a first integral of the quasi extremal equation

$$h_- \phi \frac{\partial \mathcal{L}^-}{\partial u} + h_- \xi \frac{\partial \mathcal{L}^-}{\partial x} + \xi \mathcal{L}^- - V = K \tag{1.4.144}$$

[231]. These first integrals will have the form

$$f_a(x, x_+, u, u_+) = K_a, \quad a = 1, \ldots. \tag{1.4.145}$$

Thus, if we have two first integrals, we are left with a two-point OΔS to solve. If we have three first integrals, then the quasi extremal equations reduce to a single two-point difference equation, e.g. involving just x_n and x_{n+1}. This can often be solved explicitly [614].

This procedure has been systematically applied to three-point OΔS in the original article [231]. For brevity we will just consider some examples here.

Let us first consider a two-dimensional Abelian Lie algebra and the corresponding invariant second order ODE:

$$\hat{X}_1 = \partial_x, \ \hat{X}_2 = \partial_u, \qquad u_{xx} = F(u_x). \tag{1.4.146}$$

This equation is the Euler-Lagrange equation for the Lagrangian

$$\mathcal{L} = u + \mathcal{G}(u_x), \quad \mathcal{G}_{xx} = \frac{1}{F}, \tag{1.4.147}$$

and both symmetries are Lagrangian ones

$$\mathrm{pr}\,\hat{X}_1\mathcal{L} + \mathcal{L}D_x(\xi_1) = 0, \quad \mathrm{pr}\,\hat{X}_2\mathcal{L} + \mathcal{L}D_x(\xi_2) = 1 = D_x(x). \tag{1.4.148}$$

The corresponding two first integrals are

$$J_1 = u + \mathcal{G}(u_x) - u_x\mathcal{G}_x(u_x), \tag{1.4.149}$$

$$J_2 = \mathcal{G}_x(u_x) - x \tag{1.4.150}$$

Introducing $\mathcal{H}$ as the inverse function of $\mathcal{G}_x$ from (1.4.149) we have

$$u_x = \mathcal{H}[J_2 + x], \qquad \mathcal{H}[J_2 + x] = [\mathcal{G}_x]^{-1}[J_2 + x]. \tag{1.4.151}$$

Substituting into (1.4.149), we obtain the general solution of (1.4.146) as

$$u(x) = J_1 - \mathcal{G}[\mathcal{H}[J_2 + x]] + (J_2 + x)\mathcal{H}[J_2 + x]. \tag{1.4.152}$$

Now let us consider the discrete case. We introduce the discrete Lagrangian analogue of (1.4.147) as

$$\mathcal{L} = \frac{u + u_+}{2} + \mathcal{G}(u_x) \tag{1.4.153}$$

for some smooth function $\mathcal{G}$ and where u_x is defined in (1.4.101). Eqs. (1.4.148) hold [with D_x interpreted as the discrete total derivative D_+ defined in (1.4.143)]. The two quasi extremal equations are

$$\frac{x_+ - x_-}{2} - D_+\mathcal{G}(u_x) + D_+\mathcal{G}(u_{\underline{x}}) = 0, \tag{1.4.154}$$

$$u_x D_+\mathcal{G}(u_x) - u_{\underline{x}} D_+\mathcal{G}(u_{\underline{x}}) - \mathcal{G}(u_x) + \mathcal{G}(u_{\underline{x}}) - \frac{u_+ - u_-}{2} = 0.$$

The two first integrals obtained using Noether's theorem in this case can be written as

$$D_+\mathcal{G}(u_x) - \frac{x + x_+}{2} = \mathcal{B} \tag{1.4.155}$$

$$-u_x D_+\mathcal{G}(u_x) + \mathcal{G}(u_x) + u + \frac{1}{2}(x_+ - x)u_x = \mathcal{A}. \tag{1.4.156}$$

In principle, these two first integrals can be solved to obtain

$$u_x = \mathcal{H}[\mathcal{B} + \frac{1}{2}(x_+ + x)], \quad u = \Phi(\mathcal{A}, \mathcal{B}, x, x_+), \tag{1.4.157}$$

where $\mathcal{H}[z] = [D_+\mathcal{G}]^{-1}(z)$ and Φ is obtained by solving (1.4.156), once $u_x = \mathcal{H}$ is substituted into this equation. A three-point difference equation for x_{n+2}, x_{n+1} and x_n, not involving u is obtained from the consistency condition

$$u_x = \frac{u_{n+1} - u_n}{x_{n+1} - x_n}.$$

In general this equation is difficult to solve. We shall follow a different procedure which is less general, but works well when the considered OΔS has a three dimensional solvable symmetry algebra with $\{\partial_x, \partial_u\}$ as a subalgebra. We add a third equation to the system (1.4.155, 1.4.156), namely

$$\frac{x_+ - x}{x - x_-} = 1 + \varepsilon. \tag{1.4.158}$$

The general solution of (1.4.158) is

$$x_n = (x_0 + B)(1+\varepsilon)^n - B \tag{1.4.159}$$

where x_0 and B are integration constants. We will identify B with the constant in (1.4.155), but leave ε as an arbitrary constant. Eq. (1.4.159) defines an exponential lattice (for $\varepsilon \neq 0$). Using (1.4.159) together with (1.4.155) and (1.4.156), we find

$$u_x = \mathcal{H}[(x_n + \mathcal{B})(1+\frac{\varepsilon}{2})], \quad \mathcal{H}[z] = [D_+\mathcal{G}]^{-1}(z) \tag{1.4.160}$$

$$\begin{aligned} u_n = {} & \mathcal{A} + (x_n + \mathcal{B})\mathcal{H}[(x_n+\mathcal{B})(1+\frac{\varepsilon}{2})] \\ & - \mathcal{G}[\mathcal{H}(x_n+\mathcal{B})(1+\frac{\varepsilon}{2})] \end{aligned} \tag{1.4.161}$$

There is no guarantee that equation (1.4.160) and (1.4.161) are compatible. However, let us consider the two special cases with three-dimensional solvable symmetry algebras, namely algebras $\mathbf{A}_{3,1}$ and $\mathbf{A}_{3,2}$ of Section 1.4.3.2.

$\mathbf{A}_{3,1}$**:** We choose $\mathcal{G}(u_x)$ to be

$$\mathcal{G}(u_x) = \frac{(k-1)^2}{k} u_x^{\frac{k}{k-1}}, \quad k \neq 0, 1 \tag{1.4.162}$$

From (1.4.160) and (1.4.161) we obtain

$$u_x = (\frac{1}{k-1})^{k-1} x_n^{k-1} (1+\frac{\varepsilon}{2})^{k-1} \tag{1.4.163}$$

$$u_n = \frac{1}{k}(\frac{1}{k-1})^{k-1}(x_n + B)^k (1+\frac{\varepsilon}{2})^{k-1}[1+(1-k)\frac{\varepsilon}{2}] \tag{1.4.164}$$

The consistency condition (for u_x to be the discrete derivative of u_n) provides us with a transcedental equation for ϵ:

$$[(1+\varepsilon)^k - 1][1+(1-k)\frac{\varepsilon}{2}] = k\epsilon. \tag{1.4.165}$$

In the continuous limit we take $\varepsilon \to 0$ and u_n given by (1.4.164) goes to the general solution of the ODE (1.4.117). In (1.4.165) terms of order ε^0, ε, and ε^2 cancel. The solution u_n coincides with the continuous limit up to terms of order ε^2.

We mention that in the special case $k = -1$ all three symmetries of the OΔS are Lagrangian ones and in this case (1.4.165) is identically satisfied for any ε.

$\mathbf{A}_{3,2}$**:** We choose $\mathcal{G}(u_x)$ to be

$$\mathcal{G}(u_x) = e^{u_x} \tag{1.4.166}$$

and obtain

$$u_x = \ln(x_n + B)(1+\frac{\varepsilon}{2}), \tag{1.4.167}$$

$$\begin{aligned} u_n = {} & (x_n + B)\ln(x_n + B) + A + \\ & +(x_n+B)[\ln(1+\frac{\varepsilon}{2}) - (1+\frac{\epsilon}{2})]. \end{aligned} \tag{1.4.168}$$

The expressions (1.4.167) and (1.4.168) are consistent if ϵ satisfies

$$\varepsilon(1+\frac{\varepsilon}{2}) - (1+\varepsilon)\ln(1+\varepsilon) = 0 \tag{1.4.169}$$

Again (1.4.168) coincides with its continuous limit up to terms of order ε^2 and in (1.4.169) terms of order ε^0, ε^1 and ε^2 cancel.

For the $sl(3,\mathbb{R})$ algebra $A_{3,3}$ all three symmetry operators $\hat{X}_1$, $\hat{X}_2$ and $\hat{X}_3$ correspond to Lagrangian symmetries. The corresponding OΔS is integrated in [231].

5. What we leave out on symmetries in this book

Due to the premature dead of two of the authors of this monograph, Ravil Yamilov and Pavel Winternitz we leave out of the book a substantial part of the outline of the book we wrote down when we started to write the book.

For completeness some of the most important concepts of Lie theory and its application to differential and difference equations which are left out of this monograph have been briefly presented up to now in this Introduction.

The book was to be consistent of 11 Chapters. We will indicate here the 9 Chapter left out with some of the relative references which were to form the first volume of the book and on which just sparse sections have been written.

(1) Lie groups and differential equations.
  - Conditional symmetries [186, 534].
  - λ and μ symmetries [178, 180, 290, 293, 488, 650]. Nonlocal symmetries [160, 161].

(2) Lie point symmetries of DΔEs.
  - Lie point symmetries of Fermi-Pasta-Ulam systems and reductions.
  - Lie point symmetries of the Krichever - Novikov equation and reductions [505, 547].
  - Lie point symmetries of generalized Toda lattices and reductions [464].

(3) Symmetries of differential delay equations [225–229].

(4) Contact transformations for difference systems [521, 527].

(5) λ and non local symmetries for difference systems [488, 510].

(6) Symmetry classification of molecular chains [320, 321, 465].
  - Double chains.
  - Diatomic chains [320].
  - Atomic chains in two discrete variables.

(7) Symmetry preserving discretization of differential equations as part of geometric integration [95, 131, 249, 560]
  - Ordinary difference schemes [91, 152, 352, 573, 711].
  - Invariant Lagrangians, solutions and their generalization [231].
  - Numerical tests [116, 117, 793].
  - Partial difference schemes: Liouville equation, elliptic Liouville equation, KdV equation, etc. [54, 93, 94, 96, 116, 130, 132, 221, 334, 482–484, 701, 702, 808, 809].
  - Preserving conditional symmetries. Example: the Boussinesq equation [514–516].
  - Invariant construction on moving frames [118, 251, 252, 661].

(8) Umbral calculus and generalized symmetries of linear difference equations [168, 191, 210, 212, 525, 526, 676, 717, 719].
  - Basic concepts of umbral calculus on uniform lattices.
  - Umbral calculus and symmetries of linear difference equations [526].
  - Examples: heat equation, first order equations, Airy equations.
  - Nonrelativistic quantum mechanics on a lattice [525, 570, 571].
  - Discretization of a nonrelativistic wave equation.
  - Umbral calculus on an exponential lattice: q-umbral calculus [487, 718].

- Examples: First order equations, the q-Airy function, symmetries of the q-heat equation.
- Symmetries of the q-Schrödinger equation[487].

(9) Nonlinear difference equations with superposition formulas.
- Lie's theorem on non linear ordinary differential equations with superposition formulas[40, 42, 44, 78, 79, 115, 294, 357, 466, 656, 657, 695, 759, 760, 766, 827, 828].
- Discretization preserving the non linear superposition formulas.
  (a) Riccati equation [673, 711].
  (b) Matrix Riccati equation [357, 657].
  (c) Equations related to orthogonal and symplectic groups.

6. Outline of the book

The main content of the present book is contained in Chapter 2 and Chapter 3. In the numerous Appendices we report results which are not essential for the presentation of the book but complement for the interested reader the material presented before. Both Chapters 2 and 3 start by discussing the PDE case before turning on to the DΔE and the PΔE case.

Chapter 2 is presenting integrability and symmetries of non linear PDEs, DΔEs and PΔEs in two independent variables. It starts by considering the case of PDEs as an introduction to the discrete case. It considers two cases, an integrable equation, the well known Korteweg de Vries equation and its integrable deformations and a linearizable equation, the also well known Burgers equation. In all cases we introduce Lax pairs and Bäcklund transformations and from those we obtain a hierarchy of non linear equations, non linear superposition formulae and Bianchi identities which provide a way to discretize. Then we study the Lie point and generalized symmetries as obtained from the commuting flows, the symmetry reduction of the non linear equation as a tool to get special solutions and the relation between symmetries and Bäcklund.

We will also present a brief review of the results by Bluman and Kumei on the symmetry approach to linearizability.

Following there is a Section on the relation between integrability of PDEs and superintegrability of ODEs.

Then we consider integrability of DΔEs. Among the integrable ones we consider the Toda equation, the Toda system and its inhomogeneous version, the Volterra equation and the discrete Nonlinear Schrödinger equation . We then consider the linearizable DΔE Burgers.

In the case of PΔEs we start from some well known integrable cases. We consider the discrete time Toda Lattice also known as Hirota Miwa equation, the discrete time Volterra equation, the discrete potential KdV equation given by the Bianchi identity of the KdV and the lattice version of the Schwarzian KdV. In this same Chapter we consider the Adler Bobenko Suris class of equations obtained by the compatibility around the cube, its extension given by Viallet, denoted Q_V, and their extension given by Boll. Most of the equations of the Boll classification are non autonomous, linearizable and Darboux integrable. As the prototype of linearizable PΔEs we consider the PΔE Burgers. Here we consider also the discrete extension of Bluman and Kumei linearization procedure where we look for the existence of an infinite dimensional Lie point symmetry, a leftover of the underlined linear equation.

Chapter 3 contains a detailed analysis of the symmetries as integrability criteria. At first we introduce the generalized symmetry method as developed for PDEs by A. B. Shabat and his school at the Russian Academy of Sciences firstly in Ufa and then at the Landau Institute in Chernogolovka. This theory has been extended to DΔEs by Yamilov and to PΔEs by Levi and Yamilov. We present all the tools necessary to carry out this program. Among the results one can find a curious theorem, the first of this kind, on the necessary shape for a given evolutionary DΔE to be integrable and not linearizable. In this theory one construct at first integrability conditions based on the the existence of an infinite number of formal generalized symmetries, formal conservation laws, formal Lax operators and formal recursion operators. This theory is applied to the classification of Volterra, Toda and relativistic Toda equations. Some results, comparable with those presented in the previous Chapter are obtained for non autonomous DΔEs. This Section on DΔEs ends with results on scalar evolutionary DΔEs of an arbitrary order and multi-component DΔEs.

The last part of this Chapter is devoted to the very recent results on the generalized symmetry method for PΔEs. The integrability conditions valid for these equations are presented and, due to its inherent difficulties, are limited to the existence of generalized symmetries involving just 5 points. Then we test classes of non linear PΔEs for the integrability and present some classification results.

Using the techniques introduced in the previous Section we formulate the integrability conditions for linearizable PΔEs involving three and four lattice points. In the first case the classification problem for multilinear linearizable PΔEs can be carried out up the end. In the case of quad-graph equations we can carry out the classification of complex autonomous multilinear PΔEs linearizable by a point transformation.

CHAPTER 2

Integrability and symmetries of nonlinear differential and difference equations in two independent variables

1. Introduction

There is no common opinion among the researchers in the field on the meaning of integrability. However there are various properties which are connected with integrability. Some of these properties are general and valid for any kind of system. For PDEs, for DΔEs, for PΔEs, for ultra discrete systems with many independent ($x \in \mathbb{R}^p, p \in Z^+$) or dependent variables ($u \in \mathbb{R}^q, q \in Z^+$), either real or complex or varying on the integers (see the corresponding entries of the Encyclopedia of Nonlinear Sciences [748] and of the Encyclopedia of Mathematical Physics [274]). We will indicate some of these properties in the following:

(1) Existence of a Lax pair, i.e. an overdetermined system of linear equations for a function ψ depending on two independent variables (x, t), on a function $u(x, t)$ and its derivatives and on a spectral parameter λ [60, 144–147, 149, 447, 460, 473]. The non linear PDE, DΔE or PΔE for $u(x, t)$ is obtained as their compatibility. At least two different formalisms have been introduced to deal with Lax pairs.

One by Peter Lax [470], originally in the case of two independent variables say x and t, where the Lax pair is given in terms of differential operators. One of the operators is a spectral operator for the function ψ and the other governs its t evolution. Their compatibility implies the given non linear system. The study of the spectral problem can be carried out by various techniques [12, 46, 147, 649, 674, 803].

A second formalism was introduced by Ablowitz, Kaup, Newell and Segur [5] and Zakharov and Shabat [863] and consists in the introduction of two matrix equations in such a way that their compatibility gives the non linear system for $u(x, t)$.

In the case of PDEs, DΔEs or PΔEs the structure of the Lax equations will be discussed later in the corresponding Sections. This notion has been extended to the case of more independent variables for which we refer to the corresponding literature [447, 749, 865].

(2) Existence of an infinity of independent conserved quantities [147, 384, 604, 608, 858].

(3) Existence of an infinity of generalized symmetries [262, 604, 858].

(4) Existence of Bäcklund transformations between solutions of the non linear system [140, 147, 472].

There exist tests for the integrability and classification techniques which depend in a crucial way on the kind of equation we are considering [383]. Among the integrability tests let us mention:

(1) The Painlevé test for ODEs or PDEs or OΔEs [10, 11, 192, 194, 196, 354, 456, 688, 690, 788, 825]. For Painlevé equations and their symmetries see also [522].
(2) The singularity confinement for discrete equations and mappings[329, 331, 333, 380, 386, 655, 689, 790].
(3) The algebraic entropy analysis of mappings, difference equation, differential delay equations, etc. [385, 816].
(4) The existence of generalized symmetries or formal symmetry test [604, 608, 858].
(5) The Laurent property of difference equations: all of the iterates are Laurent polynomials in the initial data. [437].
(6) Analysis of the degree of the iterates [81, 801]
(7) The diophantine integrability for discrete equations [353, 382].
(8) The application to discrete equations of the Nevanlinna theory of meromorphic functions [4, 692].

Among the classification techniques

(1) Generalized symmetry approach to the classification of PDE by Ibragimov, Mikhailov, Zhiber and Shabat [608]. On this point let us add the results obtained by Sanders and Wang on the classification of evolutionary homogeneous polynomials PDEs of all orders by symbolic manipulation techniques and number theory [729, 730].
(2) Classification of DΔEs by the generalized symmetry method [311, 312, 842].
(3) Classification of PΔEs by the compatibility around the cube hypothesis by Adler, Bobenko and Suris (ABS) [22, 23, 29, 112–114] with extensions by Hietarinta and Viallet [387].
(4) Classification of PΔEs by the generalized symmetry method [555–558]
(5) Classification of two-dimensional equations on the lattice via characteristic Lie rings by Habibullin et al. [347, 348, 868].

In the following we will mainly deal with the Lax technique [124–128] which can provide algebraically a lot of interesting structures for integrable systems in its two versions which, using a notion introduced by Calogero, we will call either C-integrable, integrable by a transformation of coordinates i.e. linearizable equations, or S-integrable, i.e. equations integrable by a spectral transform[141, 142]. We will show in the S-integrable case that the existence of a Lax pair allows us to construct a recursion operator for the associated non linear hierarchy of equations and Bäcklund and Darboux transformations [599, 716]. The spectral problem can be solved and allows us to solve a Cauchy problem. The derivation in the C-integrable case follows the same pattern but it is simpler and no spectral problem can be associated to it [502]. Moreover no infinite number of conserved quantities exists (see Section 3.2.4.1).

2. Integrability of PDEs

2.1. Introduction. The notion of integrability was firstly introduced in the case of PDEs in two independent variables and one dependent one. So this case will be considered here as an introduction to the discrete cases which will follow in Section 2.3.

The developments on the integrability of PDEs were partly motivated by the near-recurrence paradox that had been observed in a very early computer simulation of a non linear lattice by Fermi, Pasta, Ulam and Tsingou, at Los Alamos in 1955 [200, 253]. Those authors had observed long-time nearly recurrent behavior of a one-dimensional chain of anharmonic oscillators, in contrast to the rapid thermalization that had been expected. In a nice work in asymptotology Kruskal [455] derived the non linear Korteweg de Vries equation (KdV) [448]

$$u_t = u_{xxx} - 6u\,u_x, \quad u = u(x,t), \tag{2.2.1}$$

as an asymptotic continuous approximation of the Fermi, Pasta, Ulam and Tsingou model. The KdV [448] had been obtained by Korteweg and de Vries in 1895 as an approximation of shallow water waves. Gardner and Morizava derived it in 1960 for hydromagnetic waves [301], thus showing that the KdV is a universal model for non linear dispersive waves [120, 637].

In a pioneering computer simulation of the KdV, Zabusky and Kruskal [857] (with some assistance from Deem[205]) made the startling discovery of a "solitary wave" solution of the KdV equation that propagates nondispersively and regains its shape after a collision with other such waves. Such behavior is the opposite of thermalization. Because of the particle-like properties of such a solitary wave [772], they used the term *solitron* initially. However they found that the term had already been used by others. Since 1959, **Solitron** has been the name of an American industry leader in power semiconductors (Solitron Devices, inc.). So they named it *soliton*, a term that caught on almost immediately. That turned out to be at the heart of the phenomenon. Solitonic behavior suggested that the KdV equation must have conservation laws beyond the obvious conservation laws of mass, energy, and momentum. A fourth conservation law was discovered by Whitham [835] and a fifth one by Kruskal and Zabusky [856]. Several new conservation laws, up to 10, were presented by Miura[617, 618]. Miura also showed that many conservation laws also existed for a related equation known as the modified Korteweg de Vries equation (mKdV)

$$v_t = v_{xxx} - 6v^2 v_x. \tag{2.2.2}$$

With these conservation laws, Miura showed a connection (now called the Miura transformation) [316]

$$u = v_x - v^2 \tag{2.2.3}$$

between solutions of the KdV and mKdV equations. This was the clue that enabled Gardner, Greene, Kruskal and Miura (GGKM) [299, 300, 457, 774] to discover a general technique for constructing exact solutions of the KdV equation and understanding the origin of its conservation laws. This was the Inverse Spectral Transform (IST), a surprising and elegant method that demonstrates that the KdV admits an infinite number of commuting conserved quantities and thus is completely S-integrable. This discovery gave the modern basis for understanding the soliton phenomenon: the solitary wave is recreated after the collision in the outgoing state because this is the only way to satisfy all of the conservation laws. Soon after GGKM, Lax interpreted the IST in terms of isospectral deformations of a spectral problem and introduced the so-called "Lax pairs" [470].

Noether was the first to notice in 1918 [647] that one can extend point symmetries of differential equations by including in the transformation higher derivatives of the dependent variables, i.e. generalized symmetries. For a generic equation they are more rare than point symmetries. We can obtain an infinite number of local generalized symmetries [659] when the system is S- or C-integrable in the sense presented in Section 2.2.1 [3, 12, 147, 247, 630, 649, 858].

2.2. All you ever wanted to know about the integrability of the KdV equation and its hierarchy. Let us study in detail the integrability properties of the KdV equation (2.2.1) the prototype of the integrable PDEs with two independent continuous variables and one dependent one. Here we follow mainly [124–126, 147, 684].

When u goes to zero asymptotically faster than x^{-2}, so that

$$\int_{-\infty}^{\infty} dx(1+|x|)|u(x,t)| < \infty, \tag{2.2.4}$$

we can solve the Cauchy problem for (2.2.1) in terms of the solution of an associated spectral problem [60, 147].

Let us introduce a nontrivial Lax pair [470, 757], i.e an overdetermined system

$$L(u)\psi = \lambda\psi, \tag{2.2.5}$$

$$\psi_t = -M(u)\psi, \tag{2.2.6}$$

of linear equations for the complex function $\psi = \psi(x,t;\lambda)$. In (2.2.5, 2.2.6), $L(u)$ and $M(u)$ are linear operators in ∂_x whose coefficients are functions of u and its x-derivatives. For an explicit example see (2.2.11, 2.2.12). The compatibility of (2.2.5, 2.2.6), i.e. the request that the function ψ, solution of (2.2.5), evolves in t according to (2.2.6), is obtained by differentiating (2.2.5) with respect to t

$$L_t(u)\psi + L(u)\psi_t = \lambda_t\psi + \lambda\psi_t \tag{2.2.7}$$

and substituting for ψ_t (2.2.6). This implies an operator equation for $L(u)$ and $M(u)$, the so called Lax equation

$$L_t(u) = [L(u), M(u)] \tag{2.2.8}$$

if $\lambda_t = 0$ or

$$L_t(u) = [L(u), M(u)] + f(L(u), t) \tag{2.2.9}$$

if

$$\lambda_t = f(\lambda, t). \tag{2.2.10}$$

Eq. (2.2.9) is meaningful when $f(\lambda, t)$ is an entire function of its first argument. In (2.2.5) λ is an eigenvalue and in (2.2.5, 2.2.6) L and M are λ independent linear operators. The operator equations (2.2.8, 2.2.9) for given L and M will provide a differential equation for the function u. The function ψ, often called the wave or spectral function, depends on the independent variables (x,t), the dependent function $u(x,t)$ and on λ. If $\lambda_t = 0$ then the time evolution (2.2.6) is said to be *isospectral* as there is no evolution of λ in the time t, and λ is an integral of motion, together with all the functions which depend only on it. In all other cases, when $f(\lambda, t) \neq 0$, the evolution equations obtained from (2.2.9) are *non isospectral*.

In the particular case when $L(u)$ and $M(u)$ are given by

$$L(u) = -\partial_{xx} + u, \tag{2.2.11}$$

$$M(u) = -4\partial_{xxx} + 6u\partial_x + 3u_x, \tag{2.2.12}$$

with $\lambda_t = 0$, (2.2.8) turns out to be the KdV equation (2.2.1) as

$$\frac{dL(u)}{dt} = u_t(x,t), \quad \text{and } [L(u), M(u)] = u_{xxx} - 6u\,u_x.$$

As an alternative to the Lax pair formalism a matrix representation of the non linear integrable equation [379] has been given by Ablowitz et al. in 1974 [5] and independently by Zakharov and Shabat in 1979 [860, 861, 863]. In this approach to integrability, the overdetermined system of equations is given by the matrix equations [500]

$$\boldsymbol{\psi}_x = \boldsymbol{U}(\{u\}, \lambda)\,\boldsymbol{\psi}, \tag{2.2.13}$$

$$\boldsymbol{\psi}_t = \boldsymbol{V}(\{u\}, \lambda)\,\boldsymbol{\psi}, \tag{2.2.14}$$

where $\boldsymbol{\psi} = \boldsymbol{\psi}(x,t;\lambda)$ is a vector function and $\boldsymbol{U}(\{u\},\lambda)$ and $\boldsymbol{V}(\{u\},\lambda)$ are λ dependent matrices of order greater or equal to two and by $\{u\}$ we mean the function u and possibly some of its x derivatives. The compatibility of (2.2.13, 2.2.14) is given by the non linear equation

$$\boldsymbol{U}_t - \boldsymbol{V}_x + [\boldsymbol{U},\boldsymbol{V}] = 0. \tag{2.2.15}$$

The coefficients of the various powers of λ in (2.2.15) give the non linear equation in u. In the case of KdV the matrices $\boldsymbol{U}$ and $\boldsymbol{V}$ are 2×2 and are given by [630]

$$\boldsymbol{U} = \begin{pmatrix} -i\lambda & u \\ -1 & i\lambda \end{pmatrix}, \tag{2.2.16}$$

$$\boldsymbol{V} = \begin{pmatrix} -4i\lambda^3 + 2iu\lambda - u_x & 4\lambda^2 + 2i\lambda u_x - u_{xx} - 2u^2 \\ -4\lambda^2 + 2u & 4i\lambda^3 - 2iu\lambda + u_x \end{pmatrix}.$$

The matrix $\boldsymbol{U}$ given in (2.2.16) is a sub-case of the AKNS [5] $\boldsymbol{U}$ matrix

$$\boldsymbol{U} = \begin{pmatrix} -i\lambda & u \\ v & i\lambda \end{pmatrix} \tag{2.2.17}$$

when $v = -1$. The mKdV (2.2.2) is associated to (2.2.17) when $v = u^*$, where by u^* we mean the complex conjugate of u. The two equations, KdV and mKdV, are associated to two different reductions of the AKNS $\boldsymbol{U}$ matrix and for this reason we can find a Miura transformation between them.

We can interpret the linear equation (2.2.5) as a *spectral problem* [147, 445, 704]. In correspondence with every function $u(x,0)$ the asymptotic behavior of the corresponding solution $\psi(x,0;\lambda)$ of the spectral problem (2.2.5) can be constructed in a unique way [246] when u satisfies (2.2.4). The asymptotic behavior of the function $\psi(x,0;\lambda)$ provides what is called the spectrum $\mathcal{S}[0,\lambda]$ of the function $u(x,0)$. This construction is called the *direct problem*. We can look for a solution of the Cauchy problem of (2.2.8) with the initial condition given by $u(x,0)$. To the t–evolution of u given by (2.2.1), there will correspond the t–evolution of the wave function $\psi(x,t;\lambda)$ given by (2.2.6) and correspondingly a time evolution of the spectrum $\mathcal{S}[t,\lambda]$. The procedure of reconstructing the function $u(x,t)$ from the spectrum $\mathcal{S}[t,\lambda]$ is denoted *inverse problem* and is given by solving a Gel'fand, Levitan and Marchenko linear integral equation [60, 315, 445, 589].

In the particular case when (2.2.5) is the Schrödinger spectral problem (2.2.11) and when $u(x,0)$ vanishes at infinity faster than x^{-2}, so that (2.2.4) is satisfied, the solution ψ of (2.2.5, 2.2.11) for $\lambda > 0$, ($\lambda = k^2$, $k \in \mathbb{R}$), has the following asymptotic behaviour:

$$\psi(x,k) \to \quad e^{-ikx} + R(k)\, e^{ikx}, \qquad (x \to +\infty), \tag{2.2.18}$$

$$\psi(x,k) \to \quad T(k)\, e^{-ikx}, \qquad (x \to -\infty), \tag{2.2.19}$$

corresponding to an incoming wave from $+\infty$ which, due to the presence of the potential $u(x,0)$, is partly reflected to $+\infty$ and partly trasmitted to $-\infty$. $T(k)$ is the transmission coefficient and $R(k)$ is the reflection coefficient. The transmission and reflection coefficents are generically complex functions of k. If $u(x,0)$ is real, as we generically assume in applications, and also k is real, they have the following symmetry properties: $T(-k) = T^*(k)$, $R(-k) = R^*(k)$. Moreover they satisfy the unitarity condition, corresponding to the conservation of mass, $|T(k)|^2 + |R(k)|^2 = 1$.

The Schrödinger spectral problem (2.2.5, 2.2.11) may also have bound state solutions, i.e. solutions $\psi = \phi(x)$ which vanish at both $+\infty$ and $-\infty$. These solutions are obtained in correspondence with a discrete negative eigenvalue, $\lambda_j = -p_j^2$, with $p_j > 0$, corresponding to k_j imaginary, $k_j = ip_j$. If u vanishes asymptotically faster than x^{-2} then the number of the discrete eigenvalues $j = 1, 2, \ldots, N$ will be finite. An asymptotic behavior of u at $\pm\infty$ proportional to x^{-1} would instead yield an infinite number of discrete eigenvalues λ_j accumulating at $p_\infty = 0$ [147]. The solutions $\phi_j(x))$ of (2.2.11) corresponding to the jth bound state is square integrable and is normalized so that

$$\int_{-\infty}^{\infty} dx \phi_j^2 = 1.$$

Let us consider the bounded solutions of the Schrödinger spectral problem $f_j(x)$, characterized by the asymptotic behaviour $f_j(x) = e^{-p_j x}$ as $x \to +\infty$. If $u(x, 0)$ is real, then ϕ_j and f_j will be proportional to each other, $\phi_j(x) = c_j f_j(x)$, where

$$c_j = \lim_{x \to +\infty} [e^{p_j x} \phi_j(x)]. \tag{2.2.20}$$

The real quantity $\rho_j = c_j^2$ is denoted the normalization coefficient of the bound state of discrete negative eigenvalue λ_j and it plays an important role in the spectral problem as

$$\rho_j = \Big[\int_{-\infty}^{\infty} dx f_j^2(x) \Big]^{-1}. \tag{2.2.21}$$

Depending on the asymptotic behavior of the function $u(x, 0)$, the function $\psi(x, 0; ik_j)$ may not be well defined in the complex k-plane as $T(k)$ and $R(k)$ may have poles at $k = ip_j$. In such a case we have:

$$\lim_{k \to ip_j} [(k - ip_j)\psi(x, 0; k)] = c_j \phi_j(x, 0). \tag{2.2.22}$$

The spectrum $\mathcal{S}[t, \lambda]$ associated to the function $u(x, t)$, $\mathcal{S}[u, \lambda]$, is, by definition, the collection of data

$$\mathcal{S}[u, \lambda] = \big\{ R(k, t), \ -\infty < k < +\infty; \ p_j, \ \rho_j(t), \ j = 1, 2, \ldots N \big\}. \tag{2.2.23}$$

It is important to notice that if u vanishes asymptotically exponentially [147]

$$\lim_{x \to +\infty} \Big[u e^{2\mu x} \Big] = 0, \quad \mu > 0, \tag{2.2.24}$$

$R(k)$ is meromorphic in the Bargman strip $-\mu < \mathcal{I}(k) < +\mu$ [73, 147]. Inside the Bargman strip there is a one–to–one correspondence between the poles of $R(k)$ in the upper half k–plane and the discrete eigenvalues corresponding to the bound state ϕ_j, i.e.

$$\lim_{k \to ip_j} \Big[(k - ip_j) R(k) \Big] = i\rho_j. \tag{2.2.25}$$

Consequently, if we consider solutions of the KdV equation $u(x, t)$ which vanish asymptotically faster than exponentially, all the information on the spectrum $\mathcal{S}[u, \lambda]$ (2.2.23) is contained in the reflection coefficient $R(k)$. By analyzing its poles we obtain the bound states eigenvalues and the residue in the pole gives us the bound state normalization coefficients.

The t evolution of the reflection and transmission coefficients is obtained calculating (2.2.6) in the asymptotic regime in the variable x, when the function $\psi(x, t; k)$ is given by (2.2.18, 2.2.19) with $R = R(k, t)$ and $T = T(k, t)$. As the solution of (2.2.5) depends parametrically on t, the asymptotic behavior (2.2.18, 2.2.19) of the function $\psi(x, t; \lambda)$ is defined up to an arbitrary function of k and t, by $\Omega(k, t)$. So we have:

$$\psi(x, t; \lambda) \equiv \Omega(k, t)\psi(x, 0; k), \qquad |x| \to \infty. \tag{2.2.26}$$

Introducing (2.2.26) in (2.2.19) and taking into account that $M(u)$, given by (2.2.12), in this asymptotic limit is equal to

$$M(u) = -4\,\partial_{xxx}, \qquad |x| \to \infty, \tag{2.2.27}$$

we have:

$$\Omega_t\Big(e^{-ikx} + R(k,t)e^{ikx}\Big) + \Omega R_t(k,t)e^{ikx} = \tag{2.2.28}$$

$$= 4i\,\Omega\,k^3\Big(e^{-ikx} - R(k,t)e^{ikt}\Big), \qquad x \to \infty,$$

$$\Big(\Omega_t T(k,t) + \Omega T_t(k,t)\Big)e^{-ikx} = 4\Omega i k^3 T(k,t)e^{-ikx}, \;\; x \to -\infty. \tag{2.2.29}$$

As the functions (Ω, R, T) do not depend on x, from (2.2.28) we get:

$$\Omega_t = 4ik^3\Omega, \tag{2.2.30}$$

$$\Omega_t R + \Omega R_t = -4ik^3\Omega R, \tag{2.2.31}$$

and from (2.2.29):

$$\Omega_t T + \Omega T_t = 4ik^3\Omega T. \tag{2.2.32}$$

Eq. (2.2.30) defines the normalization function $\Omega(k,t)$ as $\Omega(k,t) = \Omega_0(k)e^{4ik^3t}$ and (2.2.31, 2.2.32) the evolution of the reflection and transmission coefficients:

$$R_t = -8ik^3 R, \qquad T_t - 0. \tag{2.2.33}$$

So, if we write down (2.2.6) in the x–asymptotic regime, when $u(x,t)$ vanishes, and substitute the eigenfunction ψ by its asymptotic value given by (2.2.18, 2.2.19), we find that, if $u(x,t)$ evolves according to the KdV equation (2.2.1) [147], the function $T(k,t)$ is conserved, i.e.

$$T(k,t) = T(k,0), \tag{2.2.34}$$

and the reflection coefficient $R(k,t)$ is given by

$$R(k,t) = e^{-8\,i\,k^3t}R(k,0). \tag{2.2.35}$$

From (2.2.25) it follows that the evolution of the normalization coefficients ρ_j and that of the discrete eigenvalues p_j follow from (2.2.35).

2.2.1. *The KdV hierarchy: recursion operator.* We show here that we can construct algorithmically a denumerable number of $M(u)$ operators associated to the $L(u)$ operator (2.2.11). This will be true *mutatis mutandis* for any *bona fide* $L(u)$ depending on derivatives or shift operators. Consequently we can construct a denumerable set of non linear equations associated to the same linear problem (2.2.5). We will call such equations a class or *hierarchy* of equations. To each equation of the hierarchy we can associate an $M(u)$ operator (2.2.6). The set of all $M(u)$ operators introduced in the Lax equation (2.2.8) together with the linear problem (2.2.5) defines the set of non linear evolution equations associated to $L(u)$. The set of non linear evolution equations are written down in term of a recursion operator. In the case of the Schrödinger spectral problem (2.2.11) the recursion operator $\mathcal{L}$ is given by [124, 147, 300, 446, 658, 675, 859]

$$\mathcal{L}\psi(x) = \psi_{xx}(x) - 4u(x,t)\psi(x) + 2u_x(x,t)\int_x^\infty dy\,\psi(y), \tag{2.2.36}$$

and the hierarchy of non linear equations reads:

$$u_t(x,t) = \alpha(\mathcal{L},t)u_x + \beta(\mathcal{L},t)[xu_x + 2u]. \tag{2.2.37}$$

The entire (with respect to the first argument) functions α and β characterize the equation of the hierarchy. If only the function α is present then, as we shall see below, $\lambda_t = 0$ and the hierarchy of equations is said to be isospectral. If the function β is different from zero then the hierarchy of non linear equations is said to be non isospectral as we have an evolution of the spectral parameter

$$\lambda_t = \beta(-4\,\lambda, t). \tag{2.2.38}$$

To (2.2.37) we can associate the following evolution of the reflection coefficient $R(k)$

$$\frac{d\,R(k,t)}{dt} = 2\,i\,k\,\alpha(-4k^2,\,t)R(k,t), \tag{2.2.39}$$

where by the symbol $\frac{d}{dt}$ we mean the total derivative with respect to t.

The class of equations when $\beta = 0$ is called the KdV hierarchy

$$u_t(x,t) = \alpha(\mathcal{L}, t)u_x(x,t), \tag{2.2.40}$$

$$\frac{\partial R(k,t)}{\partial t} = 2ik\alpha(-4k^2, t)R(k,t). \tag{2.2.41}$$

Eqs. (2.2.40) are all PDEs while when $\beta_\lambda \neq 0$ the equations (2.2.37) are integro differential. The first equations in the KdV hierarchy (2.2.40) are:

$$\begin{aligned} u_t &= u_x, \\ u_t &= u_{xxx} - 6uu_x, \\ u_t &= u_{xxxxx} - 10uu_{xxx} - 20u_xu_{xx} + 30u^2u_x. \end{aligned} \tag{2.2.42}$$

Many techniques have been developed in the past years to construct the recursion operator (2.2.36) which provides the hierarchy of equations containing the KdV. Here we use the *Lax technique* [124] which is a general approach based on the algebra of operators. The Lax technique allows to obtain not only the hierarchy of evolution equations but also the symmetries and the Bäcklund transformations. This technique is algorithmic and its basic assumptions are intuitive and simple. The basic ingredient of the Lax technique for the construction of the hierarchy (2.2.37) and the recursive operator (2.2.36) are the spectral problem, in the example we are considering, the Schrödinger equation (2.2.5, 2.2.11), and the Lax equation (2.2.8).

Eq. (2.2.37) contains two parts. One of them corresponds to isospectral deformations and is characterized by the Lax equation (2.2.8), the operator function $\alpha(\mathcal{L}, t)$ and the basic starting term u_x. The second one corresponds to non isospectral deformations and is characterized by the Lax equation (2.2.9), the operator function $\beta(\mathcal{L}, t)$ and the basic starting term $x\,u_x(x,t) + 2\,u(x,t)$.

Let us at first consider the isospectral case. The basic assumptions of the Lax technique are:

(1) The existence of a hierarchy of equations all sharing the same $L(u)$ operator.
(2) All equations satisfy the Lax equation (2.2.8).

From these two assumptions we imply the existence of a set of $M(u)$ operators each corresponding to one equation of the hierarchy. We postulate the existence of an operator, the recursion operator $\mathcal{L}$, which connects one equation to the following one. Consequently, given an $L(u)$ operator (2.2.11), to get the recursion operator $\mathcal{L}$ we assume that there are two operators $M(u)$ and $\tilde{M}(u)$, which satisfy the Lax equation (2.2.8) and give two subsequent equations of the same hierarchy. As on the left hand side of the Lax equation $L_t(u) = u_t$

is just a multiplicative operator we assume that the commutator $[L(u), M(u)]$ is equal to a multiplicative operator V which will depend on u and its x-derivatives. Then we have

$$L_t(u) = u_t = [L(u), M(u)] = V(u). \tag{2.2.43}$$

The commutator $[L(u), \tilde{M}(u)]$ will be equal to a different multiplicative operator $\tilde{V}$ also depending on u and its x-derivatives, corresponding to the next equation of the hierarchy, so that

$$L_t(u) = u_t = [L(u), \tilde{M}(u)] = \tilde{V}(u). \tag{2.2.44}$$

If a hierarchy of non linear equations associated to $L(u)$ exists the two functions V and $\tilde{V}$ (2.2.43, 2.2.44) must be related by a recursion operator $\mathcal{L}$. So we must have:

$$\tilde{V} = -\frac{1}{4}\mathcal{L}V + V^{(0)}, \tag{2.2.45}$$

where by $V^{(0)}$ we mean some V–independent term, that is just a function of the integration constants. Thus, starting from $M(u) = 0$, from (2.2.43), we get $V(u) = 0$ and from (2.2.45) $\tilde{V}(u) = V^{(0)}$. If $V(u)$ and $\tilde{V}(u)$ are related by (2.2.45), $\tilde{M}(u)$ and $M(u)$ must be also related. As $L(u)$ is a second order differential operator (2.2.11), we can assume in all generality

$$\tilde{M}(u) = L(u)M(u) + F(u)\partial_x + G(u), \tag{2.2.46}$$

where $F(u)$ and $G(u)$ are arbitrary scalar functions depending on u and possibly on its derivative with respect to x. Introducing (2.2.46) into (2.2.44) and taking into account (2.2.43), we get:

$$\begin{aligned}\tilde{V} &= L(u)V + [L(u), F(u)\partial_x + G(u)] = \\ &= -\Big[V + 2F_x(u)\Big]\partial_{xx} - \Big[2V_x + F_{xx}(u) + 2G_x(u)\Big]\partial_x + \\ &\quad - V_{xx} + uV - F(u)u_x - G_{xx}(u).\end{aligned} \tag{2.2.47}$$

Requiring that $\tilde{V}$ is, as $V(u)$, a multiplicative operator, we get, by setting equal to zero the coefficient of ∂_{xx} and ∂_x, two first order ODEs determining $F(u)$ and $G(u)$

$$V + 2F_x(u) = 0, \quad 2V_x + F_{xx}(u) + 2G_x(u) = 0. \tag{2.2.48}$$

From (2.2.48) we determine the functions $F(u)$ and $G(u)$ in terms of V and of two integration constants, $F^{(0)}$ and $G^{(0)}$:

$$\begin{aligned}F(u) &= F^{(0)} + \frac{1}{2}\int_x^{\infty} dyV(y), \\ G(u) &= G^{(0)} - \frac{3}{4}V.\end{aligned} \tag{2.2.49}$$

From (2.2.45, 2.2.47, 2.2.49) we get:

$$V^{(0)} = -F^{(0)}u_x, \tag{2.2.50}$$

and the recursion operator (2.2.36). Formulas (2.2.46) and (2.2.49) give a recursive relation which allows to construct $M(u)$ for any equation of the hierarchy. However the explicit form of $M(u)$ for $\alpha(\mathcal{L}, t) = \sum_{j=0}^{N} 4^j \alpha_j(t)\mathcal{L}^j$ of a sufficiently high order N will be very complicate. Luckily, in the construction of the symmetries of the KdV we will not need the explicit construction of $M(u)$ but only its asymptotic form, when $|x| \to \infty$ and $u \to 0$ with all its derivatives. In this way we will be able to construct the evolution of the spectrum $S[u, \lambda]$ for any equation of the hierarchy characterized by a function $\alpha(\mathcal{L}, t)$ for any given N.

From (2.2.50), in the asymptotic regime when $u \to 0$, we have $V^{(0)} = 0$. All functions $V(u)$ giving the higher equations of the hierarchy, will, in the asymptotic regime, also be

$V(u) = 0$. From (2.2.49) asymptotically $F(u) = F^{(0)}$ and, with no loss of generality we can take $G(u) = 0$. Thus

$$M(u) \to -\sum_{j=0}^{N} 4^j \alpha_j(t) \partial_{(2j+1)x}, \quad \text{as} \quad |x| \to \infty. \tag{2.2.51}$$

By comparing (2.2.50, 2.2.37) we have $F^{(0)} = -1$ and $\alpha(\mathcal{L}, t) = \sum_{j=0}^{N} 4^j \alpha_j(t) \mathcal{L}^j$. In a way similar to that used to get the KdV when $M(u)$ is given asymptotically by (2.2.27), we obtain (2.2.39).

Let us now consider the non isospectral case (2.2.9), when $\lambda_t = f(\lambda, t)$ with $f(\lambda, t)$ an entire function of its first argument. Then the Lax equation is given by (2.2.9) with the extra term $f(L(u), t)$. In this case, apart from the functions $M(u)$ and $\tilde{M}(u)$ we have to introduce in (2.2.43) and (2.2.44) the functions $N(u) = f(L(u), t)$ and $\tilde{N}(u) = \tilde{f}(L(u), t)$. So (2.2.43, 2.2.44) now read:

$$V(u) = [L(u), M(u)] + N(u), \quad \tilde{V}(u) = [L(u), \tilde{M}(u)] + \tilde{N}(u). \tag{2.2.52}$$

Let us require, as before, that $\tilde{M}(u)$ and $M(u)$ are related by (2.2.46) and that $\tilde{N}(u) = L(u)N(u) + h(t)L(u)$, so that $N(u)$ is just a function of $L(u)$ and t. Then from (2.2.52) we get instead of (2.2.47) the following equation:

$$\begin{aligned} \tilde{V} &= L(u)V + [L(u), F(u)\partial_x + G(u)] + hL(u) = \\ &= -\Big[V + 2F_x(u) + h(t)\Big]\partial_{xx} - \Big[2V_x + F_{xx}(u) + 2G_x(u)\Big]\partial_x + \\ &- \quad V_{xx} + uV - F(u)u_x - G_{xx}(u) + h(t)u(x,t). \end{aligned} \tag{2.2.53}$$

From (2.2.53) we get the same recursion operator (2.2.36). However $V^{(0)}$ is different as the definition of $F(u)$ is changed by the presence of the function $h(t)$. Solving the equations

$$V + 2F_x(u) + h(t) = 0, \quad 2V_x + F_{xx}(u) + 2G_x(u) = 0, \tag{2.2.54}$$

we have:

$$\begin{aligned} F(u) &= F^{(0)} - \frac{1}{2}h(t)x + \frac{1}{2}\int_x^\infty dyV(y), \\ G(u) &= G^{(0)} + \frac{1}{4}h(t) - \frac{3}{4}V. \end{aligned} \tag{2.2.55}$$

Introducing (2.2.55) in (2.2.53) we obtain as coefficient of $h(t)$, $V^{(0)} = xu_x + 2u$, i.e. the starting point of the non isospectral terms in the hierarchy (2.2.37).

As in the case of the isospectral evolution, we can consider now the evolution of the spectrum $S[u, \lambda]$. As before we require that $u \to 0$ as $|x| \to \infty$ and consequently all $V(u)$ will vanish asymptotically and only the $V(u)$–independent terms in $F(u)$ and $G(u)$ will be different from zero. The non zero terms which give rise to a non isospectral hierarchy are thus $F^{(0)} = -\frac{1}{2}h(t)x$ and $G^{(0)} = \frac{1}{4}h(t)$. Thus, choosing $\alpha(\mathcal{L}, t) = 0$, the Nth equation in the non isospectral hierarchy, $\beta(\mathcal{L}, t) = \sum_{j=0}^{N} \beta_j(t)\mathcal{L}^j$, is given by

$$M(u) \to \sum_{j=0}^{N} 4^j \beta_j(t) \partial_{(2j)x}\Big(-\frac{1}{2}x\partial_x + \frac{1}{4}\Big), \quad \text{as} \quad |x| \to \infty, \tag{2.2.56}$$

Due to the choice (2.2.45), in correspondence with a given a function $\beta(\mathcal{L}, t)$ the spectral parameter λ will evolve in t as

$$\lambda_t = \beta(-4\lambda, t). \tag{2.2.57}$$

Then with a calculation equivalent to the one done before for obtaining the evolution of the spectral data of the isospectral KdV hierarchy (2.2.40), we get

$$\frac{dR(k,t)}{dt}=0, \quad \text{i.e.} \quad \frac{\partial R(k,t)}{\partial t}+\beta(-4k^2,t)\frac{\partial R(k,t)}{\partial k}=0, \tag{2.2.58}$$

consistently with (2.2.39) as $\alpha(-4k^2,t)=0$.

2.2.2. *The Bäcklund transformations, Darboux operators and Bianchi identity for the KdV hierarchy.* Bäcklund transformations are discrete transformations (i.e. non linear mappings) depending on a free parameter that starting from a solution $u_1(x,t)$ of a non linear integrable PDE produce a new solution $u_2(x,t)$ of another non linear integrable PDE or of the same equation. When the Bäcklund transforms a solution into a new solution of the same equation we say that we have an auto-Bäcklund transformation. In the following, whenever clear, we will not differentiate between Bäcklund and auto-Bäcklund transformation. A Miura transformation is a non-auto-Bäcklund transformation. Bäcklund transformations commute amongst each other, allowing the definition of a superposition formula for solutions that endows the evolution equation with an integrability feature [504].

The Bäcklund transformations are obtained by *requiring the existence of a relation* between different solutions. We assume the existence of two essentially different solutions to the Lax equations (2.2.5, 2.2.6), $\psi_1(x,t;\lambda)$ and $\psi_2(x,t;\lambda)$. These two solutions will be associated to two different solutions of the non linear PDE described by the Lax equation (2.2.8), $u_1(x,t)$ and $u_2(x,t)$ and consequently two different Lax pairs $(L(u_1), M(u_1))$ and $(L(u_2), M(u_2))$. A relation between solutions implies the existence of an operator $\hat{D}(u_1,u_2)$, often called the *Darboux operator* which relate $\psi_1(x,t;\lambda)$ and $\psi_2(x,t;\lambda)$, i.e.

$$\psi_2(x,t;\lambda)=\hat{D}(u_1,u_2)\psi_1(x,t;\lambda). \tag{2.2.59}$$

Taking into account the Lax equations for $\psi_1(x,t;\lambda)$ and $\psi_2(x,t;\lambda)$, we get from (2.2.59) the following operator equations for $\hat{D}$:

$$L(u_2)\hat{D}(u_1,u_2)=\hat{D}(u_1,u_2)L(u_1), \tag{2.2.60}$$

$$\hat{D}_t(u_1,u_2)=\hat{D}(u_1,u_2)M(u_1)-M(u_2)\hat{D}(u_1,u_2). \tag{2.2.61}$$

In the matrix formalism the functions ψ_1 and ψ_2 in (2.2.59) are vectors while $\hat{D}=D$ is a λ dependent matrix. Moreover in correspondence with ψ_1 we have in (2.2.13) $(U(\{u_1\},\lambda), V(\{u_1\},\lambda))$ and with ψ_2 $(U(\{u_2\},\lambda), V(\{u_2\},\lambda))$. The differential equations (2.2.60) and (2.2.61) became the matrix equations

$$U(\{u_2\},\lambda)D(u_1,u_2;\lambda)=D_x(u_1,u_2;\lambda)+D(u_1,u_2;\lambda)U(\{u_1\},\lambda), \tag{2.2.62}$$

$$V(\{u_2\},\lambda)D(u_1,u_2;\lambda)=D_t(u_1,u_2;\lambda)+D(u_1,u_2;\lambda)V(\{u_1\},\lambda). \tag{2.2.63}$$

From (2.2.60, 2.2.61) we get, in analogy with the calculation carried out in the previous Section a class of Bäcklund transformations, which we will symbolically write as $B_j(\{u_1\},\{u_2\})=0$, characterized by a recursion operator Λ.

Let us start from two copies of the Schrödinger equation (2.2.5, 2.2.11) corresponding to two different functions $u_1(x,t)$ and $u_2(x,t)$,

$$\begin{aligned} L(u_1)\psi_1(x,t;\lambda) &= -\psi_{1,xx}(x,t;\lambda)+u_1(x,t)\psi_1(x,t;\lambda) \\ &= \lambda\psi_1(x,t;\lambda), \\ L(u_2)\psi_2(x,t;\lambda) &= -\psi_{2,xx}(x,t;\lambda)+u_2(x,t)\psi_2(x,t;\lambda)=\lambda\psi_2(x,t;\lambda). \end{aligned} \tag{2.2.64}$$

We require that (2.2.59) is satisfied. Then we get the operator equation (2.2.60) which is, for the Bäcklund transformations, the equivalent of the Lax equation (2.2.8) considered when constructing integrable equations. When u_1 and u_2 satisfy the same equation, i.e. they are characterized by the same Lax pair and $\hat{D}$ satisfies (2.2.60), then it can be shown that (2.2.61) is identically satisfied.

As for the derivation of the hierarchy of equations by the Lax technique, the class of Bäcklund transformations $B_j(\{u_1\},\{u_2\}) = 0$ is obtained by looking for an operator $\hat{D}(u_1,u_2)$ such that

$$L(u_2)\hat{D}(u_1,u_2) - \hat{D}(u_1,u_2)L(u_1) = V. \tag{2.2.65}$$

Moreover, we assume that there exists an operator $\tilde{D}(u_1,u_2)$ such that

$$L(u_2)\tilde{D}(u_1,u_2) - \tilde{D}(u_1,u_2)L(u_1) = \tilde{V}, \tag{2.2.66}$$

where (see (2.2.46)) we assume

$$\tilde{D}(u_1,u_2) = L(u_2)\hat{D}(u_1,u_2) + F(u_1,u_2)\partial_x + G(u_1,u_2). \tag{2.2.67}$$

Here V and $\tilde{V}$ are pure multiplicative operators depending on $u_1(x,t)$ and $u_2(x,t)$ and their derivatives. Bäcklund transformations are given by the equations $V = \tilde{V} = 0$ so that (2.2.60) is satisfied. As we did in the previous Section let us insert (2.2.67) into (2.2.66) and let us expand the obtained equation. In this way we get:

$$\begin{aligned}\tilde{V} &= L(u_2)V + L(u_2)[F\partial_x + G] - [F\partial_x + G]L(u_1) = \\ &= -[V + 2F_x]\partial_{xx} - [2V_x + F_{xx} + (u_1 - u_2)F + 2G_x]\partial_x \\ &\quad - V_{xx} + u_2V - Fu_{1,x} - G_{xx} - (u_1 - u_2)G, \\ &= -\frac{1}{4}\Lambda V + V^{(0)}.\end{aligned} \tag{2.2.68}$$

From (2.2.68), setting to zero the coefficients of ∂_x and ∂_{xx}, we get

$$\begin{aligned}F &= F^{(0)} + \frac{1}{2}\int_x^\infty dyV(y), \\ G &= G^{(0)} - \frac{3}{4}V - \frac{1}{2}F^{(0)}\int_x^\infty dy[u_2(y,t) - u_1(y,t)] \\ &\quad - \frac{1}{4}\int_x^\infty dz[u_2(z,t) - u_1(z,t)]\int_z^\infty dyV(y),\end{aligned} \tag{2.2.69}$$

where $F^{(0)}$ and $G^{(0)}$ are some x–independent integration constants. Then, from (2.2.68), taking into account (2.2.69) we get the initial condition and recursive operator for the Bäcklund transformations:

$$\begin{aligned}V^{(0)} &= -\frac{1}{2}F^{(0)}\,[u_{2,x} + u_{1,x}] + G^{(0)}\,[u_2 - u_1] \\ &\quad - \frac{1}{2}F^{(0)}\,[u_2 - u_1]\int_x^\infty dy\,[u_2(y,t) - u_1(y,t)],\end{aligned} \tag{2.2.70}$$

$$\begin{aligned}\Lambda V &= V_{xx} - 2[u_2 + u_1]V + [u_{2,x} + u_{1,x}]\int_x^\infty dy\,V(y) \\ &\quad + [u_2 - u_1]\int_x^\infty dz[u_2(z,t) - u_1(z,t)]\int_z^\infty dy\,V(y).\end{aligned} \tag{2.2.71}$$

The Bäcklund transformations $B_j(\{u_1\}, \{u_2\}) = 0$ are obtained by setting $\tilde{V} = 0$ in (2.2.68) when $V = 0$ and, in complete generality can be written as:

(2.2.72)
$$\gamma(\Lambda, t)[u_2 - u_1] = \\ = \delta(\Lambda, t)\Big\{ [u_{2,x} + u_{1,x}] + [u_2 - u_1] \int_x^{\infty} dy[u_2(y,t) - u_1(y,t)] \Big\},$$

where γ and δ are entire functions of their first argument.

Let $u_1(x,t)$ and $u_2(x,t)$ be related by the Bäcklund transformation (2.2.72) and let both $u_1(x,t)$ and $u_2(x,t)$ be exponentially bounded so that all information on the spectrum is contained in the reflection coefficient. Then a relation between the reflection coefficients exists, namely

(2.2.73)
$$R_2(k,t) = R_1(k,t)\Big[\frac{\gamma(\lambda,t) - 2ik\delta(\lambda,t)}{\gamma(\lambda,t) + 2ik\delta(\lambda,t)}\Big].$$

The elementary Bäcklund transformation is obtained by choosing the functions γ and δ as non zero constants. In such a case we can introduce a non zero constant p such that $\gamma = p\delta$ and rewrite (2.2.72) in term of the potential functions $u_j(x,t) = -w_{j,x}(x,t)$, $j = 1, 2$, where w_j satisfies the non linear PDE

(2.2.74)
$$w_{j,t} + w_{j,xxx} + 3\, w_{j,x}^2 = 0,$$

the potential KdV (pKdV). In this case, when $w_j(x,t)$, $j = 1, 2$ are asymptotically bounded, (2.2.72) becomes:

(2.2.75)
$$w_{1,x} + w_{2,x} = -\frac{1}{2}(w_2 - w_1)^2 - p(w_2 - w_1),$$

the well known elementary Bäcklund transformation for the KdV [480, 823].

Eq. (2.2.75) has been interpreted in [480, 812] as a DΔE by identifying $w_1(x) = v_n(x)$ and $w_2(x) = v_{n+1}(x)$ so that it reads

(2.2.76)
$$v_{n,x} + v_{n+1,x} = -\frac{1}{2}(v_{n+1} - v_n)^2 - p(v_{n+1} - v_n),$$

where now the constant p can be a function of n [812].

Starting from (2.2.75) we construct the *Bianchi identity* using the *Bianchi permutability theorem* [88] whose proof in the space of the reflection coefficient is trivial. In fact, given two Bäcklund transformations in the spectral space

(2.2.77)
$$R_1(k,t) = R(k,t)\Big[\frac{\gamma_1(\mu,t) - 2ik\delta_1(\mu,t)}{\gamma_1(\mu,t) + 2ik\delta_1(\mu,t)}\Big],$$

(2.2.78)
$$R_2(k,t) = R(k,t)\Big[\frac{\gamma_2(\mu,t) - 2ik\delta_2(\mu,t)}{\gamma_2(\mu,t) + 2ik\delta_2(\mu,t)}\Big],$$

we can construct two possible combinations

$$R_{12} = R_1(k,t)R_2(k,t) = R(k,t)\left[\frac{\gamma_1(\mu,t) - 2ik\delta_1(\mu,t)}{\gamma_1(\mu,t) + 2ik\delta_1(\mu,t)}\right]\left[\frac{\gamma_2(\mu,t) - 2ik\delta_2(\mu,t)}{\gamma_2(\mu,t) + 2ik\delta_2(\mu,t)}\right]$$

and

$$R_{21} = R_2(k,t)R_1(k,t) = R(k,t)\Big[\frac{\gamma_2(\mu,t) - 2ik\delta_2(\mu,t)}{\gamma_2(\mu,t) + 2ik\delta_2(\mu,t)}\Big]\Big[\frac{\gamma_1(\mu,t) - 2ik\delta_1(\mu,t)}{\gamma_1(\mu,t) + 2ik\delta_1(\mu,t)}\Big].$$

Now they turn out, trivially, to be equal

(2.2.79)
$$R_{12} = R_{21}.$$

In terms of the fields w the permutability theorem is not trivial. We have:

$$w_{1,x} + w_{,x} = -\frac{1}{2}(w - w_1)^2 - p_1(w - w_1), \tag{2.2.80}$$

$$w_{12,x} + w_{1,x} = -\frac{1}{2}(w_1 - w_{12})^2 - p_2(w_1 - w_{12}),$$

$$w_{2,x} + w_{,x} = -\frac{1}{2}(w - w_2)^2 - p_2(w - w_2), \tag{2.2.81}$$

$$w_{21,x} + w_{2,x} = -\frac{1}{2}(w_2 - w_{21})^2 - p_1(w_2 - w_{21}),$$

and thus, as from (2.2.79) $w_{12,x} = w_{21,x}$, the Bianchi identity for the KdV hierarchy reads

$$w_{12} = w - (p_1 - p_2)\frac{w_1 - w_2}{p_1 - p_2 - (w_1 - w_2)}. \tag{2.2.82}$$

Eq. (2.2.82) relates algebraically four solutions of the pKdV hierarchy w, w_1, w_2 and w_{12}. Thus we can call (2.2.82) also a *non linear superposition formula*. More on the theory of superposition formulas can be found in different settings, among others in [40, 42, 44, 77–79, 294, 356, 357, 565, 656, 657, 673, 759, 760, 804].

We can identify the four solutions of the potential KdV as $w = v_{n,m}$, $w_1 = v_{n+1,m}$, $w_2 = v_{n,m+1}$ and $w_{12} = v_{n+1,m+1}$ and thus (2.2.80) becomes the PΔE

$$v_{n+1,m+1} = v_{n,m} - (p_1 - p_2)\frac{v_{n+1,m} - v_{n,m+1}}{p_1 - p_2 - (v_{n+1,m} - v_{n,m+1})}, \tag{2.2.83}$$

known as the lattice potential KdV (lpKdV) (see Section 2.4.4).

2.2.3. *The conservations laws for the KdV equation.* Here we construct, following [147], the conservation laws associated to the KdV equation (2.2.1). A version valid for the whole KdV hierarchy in its matrix form can be found in [684].

Defining

$$\varepsilon \equiv \frac{1}{2p}, \quad y \equiv \frac{w_2 - w_1}{2\varepsilon}, \quad \eta \equiv -y_{xx} + 3y^2 + 2\varepsilon^2 y^3 \tag{2.2.84}$$

we get from (2.2.75)

$$y = u_1 - \varepsilon y_x - \varepsilon^2 y^2 \tag{2.2.85}$$

and the following conservation law

$$y_t = \eta_x. \tag{2.2.86}$$

The definitions of η and y in (2.2.84, 2.2.85) are polynomial expression in ε and we can introduce the following ε expansions for η and y

$$y = \sum_{m=0}^{M} \varepsilon^m y^{(m)} + o(\varepsilon^M), \tag{2.2.87}$$

$$\eta = \sum_{m=0}^{M} \varepsilon^m \eta^{(m)} + o(\varepsilon^M). \tag{2.2.88}$$

From (2.2.85) and (2.2.87) we get a unique definition of the coefficients appearing in (2.2.87)

$$y^{(0)} = u_1, \qquad y^{(1)} = -u_{1,x}, \tag{2.2.89}$$

$$y^{(m+1)} = -y_x^{(m)} - \sum_{j=0}^{m-1} y^{(j)}\, y^{(m-j-1)}, \; m = 1, 2, \cdots \tag{2.2.90}$$

From (2.2.90), taking into account (2.2.89), we get an expression for the higher coefficients of the expansion (2.2.87), i.e.

(2.2.91)
$$y^{(2)} = u_{1,xx} - u_1^2, \qquad y^{(3)} = -y_x^{(2)},$$
$$y^{(4)} = u_{1,xxxx} - 6\,u_1\,u_{1,xx} - 5u_{1,x}^2 + 2u_1^3, \cdots$$

As one can see from (2.2.85) if u_1 vanishes asymptotically

(2.2.92)
$$\lim_{x\to\pm\infty} u_1 = 0,$$

also all coefficients of the expansion of y will do so, i.e.

(2.2.93)
$$\lim_{x\to\pm\infty} y^{(m)} = 0.$$

From (2.2.84) and (2.2.87) we get a unique definition of the coefficients appearing in (2.2.88)

(2.2.94)
$$\eta^{(0)} = -u_{1,xx} + 3\,u_1^2, \qquad \eta^{(1)} = -\eta_x^{(0)},$$

(2.2.95)
$$\eta^{(m)} = -y_{xx}^{(m)} + 3\sum_{j=0}^{m} y^{(j)}\,y^{(m-j)}$$
$$+2\sum_{k=0}^{m-2}\sum_{j=0}^{m-k-2} y^{(k)}\,y^{(j)}\,y^{(m-k-j-2)}, \quad m = 2, 3, \cdots.$$

Then the coefficients of the expansion of the function η satisfy the boundary condition (2.2.93) when u_1 vanish asymptotically. As from (2.2.86) we have

(2.2.96)
$$y_t^{(m)} = \eta_x^{(m)},$$

we obtain

(2.2.97)
$$\int_{-\infty}^{+\infty} dx\, y^{(m)} = a_m$$

where a_m are constants which, for m odd, are zero as in this case $y^{(m)}$ is a total derivative of a function vanishing at infinity. For m even, however, $y^{(m)}$ is not a total derivative of a function vanishing at infinity and thus a_m is different from zero.

The conserved quantities for the KdV we constructed here are the same as those obtained by GGKM [617,618].

2.2.4. *The symmetries of the KdV hierarchy.* Let us now construct the symmetries of the KdV hierarchy. A partial result can be found in the work of Fuchsteiner [282,283].

The symmetries for any equation of the KdV hierarchy (2.2.40) are provided by flows commuting with the equations themselves [546,608,658] as we saw in Section 1.1. An infinite number of such symmetries is given by the equations

(2.2.98)
$$u_{\epsilon_\ell} = \mathcal{L}^\ell\, u_x, \quad u = u(x,t;\epsilon_\ell), \quad \ell = 0,1,2,\cdots$$

Here ϵ_ℓ is a group parameter and $\mathcal{L}^\ell u_x$ the characteristic of the symmetry. From the point of view of the spectral problem (2.2.5, 2.2.11) the equation (2.2.98) corresponds to an isospectral deformation as $\lambda_{\epsilon_\ell} = 0$. For any ϵ_ℓ, the solution of the Cauchy problem for (2.2.98), provides a solution $u(x,t;\epsilon_\ell)$ of one of the equations of the KdV hierarchy in terms of the initial condition $u(x,t;\epsilon_\ell = 0)$. However the construction of the group transformation from these symmetries can be done only for few values of ℓ. Which value of ℓ depends on the equation in the hierarchy we are considering. They are those values of ℓ which correspond to Lie point symmetries. In all other cases one can just use the symmetries (2.2.98) to carry out symmetry reduction i.e. reduce the equation under consideration to an ODE, or possibly a functional one. The proof of the validity of (2.2.98) as symmetries is easily given by taking into account the one-to-one correspondence between the equation and the spectrum (2.2.23), provided the asymptotic conditions (2.2.24) are satisfied. In this case we can biunivocally associate to both the KdV hierarchy (2.2.40) and the symmetries (2.2.98) an evolution of the reflection coefficient. In the case of the symmetries (2.2.98), we have:

$$\frac{\partial R}{\partial \epsilon_\ell} = 2ik\lambda^\ell R. \tag{2.2.99}$$

It is easy to prove that the flows of the corresponding reflection coefficients (2.2.41) and (2.2.99) commute and hence the same must be true for the corresponding non linear PDEs (2.2.40, 2.2.98).

We can extend the class of symmetries we constructed above by considering non isospectral deformations of the spectral problem (2.2.5, 2.2.11) [147]. Thus for the KdV hierarchy we have

$$u_{\epsilon_\ell} = \beta(\mathcal{L}, t)u_x, +\mathcal{L}^\ell[xu_x + 2u] \quad \ell = 0, 1, 2, \cdots, \tag{2.2.100}$$

where the function $\beta(\mathcal{L}, t)$ is obtained as an entire in $\mathcal{L}$ solution of the differential equation:

$$\beta(\mathcal{L}, t)_t = \mathcal{L}^\ell[\frac{2\mathcal{L}\partial\alpha(\mathcal{L}, t)}{\partial \mathcal{L}} + \alpha(\mathcal{L}, t)]. \tag{2.2.101}$$

In (2.2.101) $\beta(\mathcal{L}, t)$ is expressed in terms of the function $\alpha(\mathcal{L}, t)$ and $\alpha(\mathcal{L}, t)$ characterize the equation in the KdV hierarchy. In correspondence with (2.2.100) we have the evolution of the reflection coefficient, given by

$$\frac{dR}{d\epsilon_\ell} = 2ik\beta(\epsilon_\ell, t)R, \qquad k_{\epsilon_\ell} = k(\epsilon_\ell)^\ell. \tag{2.2.102}$$

As in the case of isospectral symmetries (2.2.98), the proof that the non isospectral flows (2.2.100) and (2.2.40) commute is reduced to the easier task of showing that the flows (2.2.102) and (2.2.41) in the space of the reflection coefficient commute.

The symmetries (2.2.98) are all given by differential equations but the non isospectral ones, (2.2.100), in general involve integrals as $xu_x + 2u$ is not a total derivative. Thus the application of the recursion operator (2.2.36) gives equations containing integral terms. The only case when a non isospectral symmetry of the KdV hierarchy is local is when ℓ is equal to the power s of the recursion operator which gives the equation in the KdV hierarchy. In this case we have the dilation symmetry for the sth equation in the KdV hierarchy.

In the case of the KdV (2.2.1) we have an extra symmetry, the well known Lie point Galilean boost, whose infinitesimal generator is

$$\hat{Z} = 6t\partial_x + \partial_u. \tag{2.2.103}$$

The symmetry of generator (2.2.103) is not related to any evolution of the reflection coefficient. Eq. (2.2.103) in evolutionary form reads:

$$\hat{Z}_e = (1 - 6tu_x)\partial_u. \tag{2.2.104}$$

2.2.5. *Lie algebra of the symmetries.* The structure of the symmetry algebra for the KdV hierarchy is obtained by computing the commutation relations between the symmetries.

The first result is that the isospectral symmetry generators for the KdV hierarchy, provided by (2.2.98), commute among themselves. The infinitesimal generators for the isospectral symmetries for the KdV hierarchy are

$$\hat{X}_\ell = \mathcal{L}^\ell u_x \partial_u. \tag{2.2.105}$$

In the previous Section we proved that (2.2.98) provides symmetries for a generic equation of the KdV hierarchy. Let us take now two infinitesimal generators of the symmetries with $\ell = n$ and $\ell = m$. We have

$$[\hat{X}_n, \hat{X}_m] = 0, \tag{2.2.106}$$

as, from (2.2.99), we have

$$\frac{d^2 R}{d\epsilon_n d\epsilon_m} = \frac{d^2 R}{d\epsilon_m d\epsilon_n}, \tag{2.2.107}$$

A simple way of obtaining the result given by (2.2.107) is to introduce symmetry generators in the space of the reflection coefficient. From (2.2.99) these generators are written as

$$\hat{\mathcal{X}}_\ell = 2ik\lambda^\ell R\,\partial_R. \tag{2.2.108}$$

From Lie theory the corresponding group transformations are obtained by solving the equations

$$\frac{d\tilde{R}}{d\epsilon_\ell} = 2i\tilde{k}\tilde{\lambda}^\ell \tilde{R}, \quad \frac{d\tilde{k}}{d\epsilon_\ell} = 0, \quad \tilde{R}(\tilde{k}, \epsilon_\ell = 0) = R, \quad \tilde{k}(\epsilon_\ell = 0) = k, \tag{2.2.109}$$

where (2.2.109) coincides with (2.2.99). In terms of the vector fields $\hat{\mathcal{X}}_\ell$ given by (2.2.108), (2.2.107) is

$$[\hat{\mathcal{X}}_\ell, \hat{\mathcal{X}}_m] = [2ik\lambda^\ell R\partial_R, 2ik\lambda^m R\partial_R] = 0. \tag{2.2.110}$$

So far, the use of the vector fields in the reflection coefficient space has just re-expressed a known result, namely (2.2.107) which is rewritten as (2.2.110). We now extend the use of vector fields in the reflection coefficient space to the case of the non isospectral symmetries (2.2.100). We restrict, for the sake of the simplicity of the exposition, to equations of the KdV hierarchy with no explicit dependence on time. That is, for the Nth equation of the KdV hierarchy, we have

$$\alpha(\lambda, t) = \lambda^N, \quad N \in \mathbf{Z}^+, \tag{2.2.111}$$

and then solving (2.2.101) we get

$$\beta(\lambda, t) = \lambda^{\ell+N}(1 + 2N)t. \tag{2.2.112}$$

The symmetry vector fields for the KdV hierarchy are now:

$$\hat{Y}_\ell = \{t(1+2N)\mathcal{L}^{\ell+N}u_x + \mathcal{L}^\ell[xu_x + 2u]\}\partial_u. \tag{2.2.113}$$

Taking into account (2.2.102) and (2.2.112) we can define the symmetry generators (2.2.113) in the reflection coefficient space as

$$\hat{\mathcal{Y}}_\ell = 2ik\lambda^{\ell+N}t(1+2N)R\partial_R - k\lambda^\ell\partial_k, \tag{2.2.114}$$

Commuting $\hat{\mathcal{Y}}_\ell$ with $\hat{\mathcal{Y}}_m$ we have:

$$[\hat{\mathcal{Y}}_\ell, \hat{\mathcal{Y}}_m] = (m - \ell)\hat{\mathcal{Y}}_{\ell+m}, \tag{2.2.115}$$

a Witt centerless Virasoro algebra [404, 429, 430, 438]. From the isomorphism between the spectral space and the space of the solutions, we conclude that the vector fields representing the symmetries of the studied evolution equations, satisfy the same commutation relations. Hence we have

$$[\hat{Y}_\ell, \hat{Y}_m] = (m - \ell)\hat{Y}_{\ell+m}. \tag{2.2.116}$$

In a similar manner we can work out the commutation relations between the symmetry generators $\hat{Y}_\ell$ and $\hat{X}_m$. In the reflection coefficient space we get:

$$[\hat{\mathcal{X}}_\ell, \hat{\mathcal{Y}}_m] = (1+2\ell)\hat{\mathcal{X}}_{\ell+m}, \tag{2.2.117}$$

and consequently for the symmetries

$$[\hat{X}_\ell, \hat{Y}_m] = (1+2\ell)\hat{X}_{\ell+m}. \tag{2.2.118}$$

Choosing $m = 1$ from (2.2.117) we get

$$[\hat{\mathcal{X}}_\ell, \hat{\mathcal{Y}}_1] = (1+2\ell)\hat{\mathcal{X}}_{\ell+1}, \tag{2.2.119}$$

which is a recursion relation for the isospectral symmetries generators. In this way, given the non local symmetry generator $\hat{Y}_1$ we are able to compute all generalized isospectral symmetries starting from the Lie point symmetry $\hat{X}_0$. Due to its special role $\hat{Y}_1$ has been called the *master symmetry* of the KdV hierarchy [264], see also [262, 284, 285, 652]. In (3.1.25) in Section 3.1 the master symmetry $\hat{Y}_1$ is given explicitly.

The generators $\hat{Y}_\ell$ and $\hat{\mathcal{Y}}_\ell$ (see (2.2.113), (2.2.114)) depend on the number N, which identifies the equation we are considering among the equations of the KdV hierarchy. Interestingly, the commutation relations involving the generators $\hat{X}$ and $\hat{Y}$ are the same for all N (see (2.2.106), (2.2.110), (2.2.115)–(2.2.118)).

In the case of the KdV (2.2.1) we have the extra symmetry $\hat{Z}$ given in evolutionary form by (2.2.104). The commutation relations between $\hat{Z}$ and the simpler generators of the isospectral and non isospectral ones $\hat{X}_\ell$ and $\hat{Y}_\ell$ are:

$$\begin{aligned} &[\hat{Z}, \hat{X}_0] = 0, \quad [\hat{Z}, \hat{X}_1] = -6\hat{X}_0, \qquad [\hat{Z}, \hat{X}_2] = -10\hat{X}_1, \\ &\qquad [\hat{Z}, \hat{X}_3] = -14\hat{X}_2, \qquad\qquad [\hat{Z}, \hat{Y}_0] = 2\hat{Z}. \end{aligned} \tag{2.2.120}$$

The commutation relations obtained above determine the structure of the infinite dimensional Lie symmetry algebra. For the KdV hierarchy can be written as:

$$L = L_0 \dotplus L_1, \quad L_0 = \{\hat{Y}_0, \hat{Y}_1, \cdots\}, \quad L_1 = \{\hat{X}_0, \hat{X}_1, \cdots\}. \tag{2.2.121}$$

The algebra L_0 is perfect, i.e. we have $[L_0, L_0] = L_0$. Let us point out that $\hat{X}_0$, $\hat{X}_1$ and $\hat{Y}_0$ are Lie point symmetries, all other are generalized symmetries.

In the case of the KdV equation we have

$$L = L_0 \dotplus L_2, \quad L_0 = \{\hat{Z}, \hat{Y}_0, \hat{Y}_1, \cdots\}, \quad L_2 = \{\hat{X}_0, \hat{X}_1, \cdots\}. \tag{2.2.122}$$

The algebra L_0 is still perfect. $\hat{X}_0$, $\hat{X}_1$, $\hat{Z}$ and $\hat{Y}_0$ are Lie point symmetries, all other are generalized symmetries. Explicitly, in the space time variables, $\hat{Z}$ is given in (2.2.103) while $\hat{X}_0$ and $\hat{X}_1$ are obtained from (2.2.105) and read

$$\hat{X}_0 = \partial_x, \quad \hat{X}_1 = \partial_t. \tag{2.2.123}$$

The generator $\hat{Y}_0$, obtained from (2.2.113) is the dilation infinitesimal generator of the Nth equation of the KdV hierarchy and reads

$$\hat{Y}_0 = (1+2N)t\partial_t + x\partial_x - 2u\partial_u. \tag{2.2.124}$$

2.2.6. *Relation between Bäcklund transformations and isospectral symmetries.* Here we present the connection between Bäcklund transformations and symmetries for the KdV hierarchy (2.2.40). As far as we know similar results have been obtained by Sato but we have not been able to find the precise reference. Those results are formulated in the following two theorems.

Theorem 6. *Let*

$$R(k,\epsilon) = \exp\left[2ik\epsilon\alpha(\lambda)\right] R(k,0), \;\; \lambda = -4k^2 \tag{2.2.125}$$

be an isospectral symmetry transformation in the spectral space of group parameter ϵ. *This transformation determines a Bäcklund transformation* (2.2.73)*, with*

$$\delta(\lambda) = \frac{1}{2k}\sin[k\epsilon\alpha(\lambda)], \tag{2.2.126}$$

and

$$\gamma(\lambda) = \cos[k\epsilon\alpha(\lambda)], \tag{2.2.127}$$

PROOF. The transformation (2.2.125) of the reflection coefficient $R(k,\epsilon)$ is obtained by integrating (2.2.99) where $\alpha(\lambda)$ is an entire function in λ which defines the equation in the KdV hierarchy. In order to identify the finite transformation (2.2.125) with a general Bäcklund transformation (2.2.73) of the reflection coefficient, we set

$$R(k,\epsilon) \equiv R_2(k), \qquad R(k,0) \equiv R_1(k). \tag{2.2.128}$$

We obtain

$$e^{2ik\epsilon\alpha(\lambda)} = \frac{1 - 2ik\beta(\lambda)}{1 + 2ik\beta(\lambda)}, \qquad \beta(\lambda) = \frac{\delta(\lambda)}{\gamma(\lambda)}. \tag{2.2.129}$$

We need to prove that $\beta(\lambda)$, defined in (2.2.129), depends only on λ and can be written as the ratio of two entire functions. We have

$$e^{2ik\epsilon\alpha(\lambda)} = \cos[2k\epsilon\alpha(\lambda)] + i\sin[2k\epsilon\alpha(\lambda)], \tag{2.2.130}$$

From (2.2.129, 2.2.130) we obtain two equations for $\beta(\lambda)$

$$\lambda\beta^2(\lambda) = \frac{\cos[2k\epsilon\alpha(\lambda)] - 1}{\cos[2k\epsilon\alpha(\lambda)] + 1}, \tag{2.2.131}$$

$$4k\beta(\lambda) = 2\frac{\sin[2k\epsilon\alpha(\lambda)]}{\cos[2k\epsilon\alpha(\lambda)] + 1} \tag{2.2.132}$$

Eqs.(2.2.131) and (2.2.132) are compatible and give the final result, namely (2.2.126) and (2.2.127). □

For $\epsilon \to 0$, we have that $\delta \to 0$ and $\gamma \to 1$, so that the identity transformation in the group corresponds to a trivial Bäcklund transformation (no transformation).

Theorem 7. *Let*

$$R_2(k) = \frac{1 - 2ikp}{1 + 2ikp} R_1(k), \tag{2.2.133}$$

where p *is a real constant parameter, be an elementary Bäcklund transformation in the spectral space. Then a symmetry transformation of the reflection coefficient is given by* (2.2.125) *with*

$$\alpha(\lambda) = \frac{1}{\epsilon k}\tan^{-1}[2pk] = \frac{2p}{\epsilon}\sum_{j=0}^{\infty}\frac{(p^2\lambda)^j}{2j+1}. \tag{2.2.134}$$

PROOF. We need to show that $\alpha(\lambda)$ in (2.2.134) does not depend on k and is an entire function of λ. Eq. (2.2.134) is obtained as the inverse of the ratio of (2.2.126) and (2.2.127) taking into account (2.2.129). By expanding the $\tan^{-1}$ in power series and using the definition of λ in terms of k we get the wanted result as pointed out in (2.2.134). □

From these two theorems it follows that a Bäcklund transformation corresponds to an infinite number of generalized isospectral symmetries and vice versa an isospectral symmetry corresponds to an infinite number of Bäcklund transformations. This result corroborates the idea presented in Chapter 3 that the existence of an infinite number of symmetries is associated to integrability.

2.2.7. *Symmetry reductions of the KdV equation.* We can perform the symmetry reduction of the KdV (2.2.1) with respect to its point symmetries of infinitesimal generator $\hat{X}_0$, $\hat{X}_1$, $\hat{Z}$ and $\hat{Y}_0$, its generalized isospectral and its non isospectral symmetries (2.2.98), (2.2.113). The reduction with respect to the Lie point symmetry group was carried out already in Section 1.1 where we find the solutions invariant with respect to a Lie point symmetry of (2.2.1). Here we consider just the case of the lowest generalized isospectral symmetry given in (2.2.98) with $\ell = 2$. This symmetry is

$$u_\epsilon = u_{xxxxx} - 10uu_{xxx} - 20u_xu_{xx} + 30u^2u_x. \tag{2.2.135}$$

The symmetry reduction is obtained [470] by requiring

$$u_\epsilon = 0. \tag{2.2.136}$$

The KdV itself (2.2.1) will provide the time evolution of the solution. Eq. (2.2.135) with the condition (2.2.136) is a fifth order non linear ODE which can be integrated twice. Under the hypothesis that the final solution be bounded asymptotically we get at first the fourth order ODE

$$u_{xxxx} - 10uu_{xx} - 5u_x^2 + 10u^3 = 0. \tag{2.2.137}$$

Eq. (2.2.137) belongs to the class F-V of the classification of higher order Painlevé equations of polinomial type presented by Cosgrove [195].

Considering the integrating factor $K_1 = u_{xxx} - 6uu_x$ we can integrate once (2.2.137) and we get the non linear third order ODE

$$\begin{aligned}\frac{1}{2}u_{xxx}^2 - 6uu_xu_{xxx} - 2uu_{xx}^2 + u_x^2u_{xx} + 10u^3u_{xx}\\ +15u^2u_x^2 - 12u^5 = 0.\end{aligned} \tag{2.2.138}$$

We can also consider the integrating factor $K_2 = u_x$ and we get

$$u_xu_{xxx} - \frac{1}{2}u_{xx}^2 - 5uu_x^2 + \frac{5}{2}u^4 = 0. \tag{2.2.139}$$

By solving (2.2.139) with respect to u_{xxx} and substituting the result in (2.2.138) we get a second order non linear ODE for u, polynomial in u_{xx}:

$$\begin{aligned}u_{xx}^4 - 10u\Big(u^3 + 2u_x^2\Big)u_{xx}^2 + 8u_x\Big(10u^3 + u_x^2\Big)u_{xx}\\ +u^2\Big(25u^6 - 76u^3u_x^2 - 20u_x^4\Big) = 0.\end{aligned} \tag{2.2.140}$$

Eq. (2.2.140) has two singular solutions given in term of elliptic functions

$$\begin{aligned}u &= \wp\big(\frac{1}{2^{1/3}}x + c_1, 0, 0\big)2^{1/3},\\ u &= \wp\big(\frac{10^{2/3}}{2\cdot 3^{3/2}}x + c_2, 0, 0\big)10^{1/3},\end{aligned} \tag{2.2.141}$$

where c_1 and c_2 are integration constants. Eq. (2.2.140) admits as solution the two soliton solution of KdV (2.2.1), as suggested by Lax in [470].

2.3. The cylindrical KdV, its hierarchy and Darboux and Bäcklund transformations. As we shall see in Section 3.4 one can find integrable DΔEs with n and t dependent coefficients. Here we show that also in the case of PDEs of the KdV-type we can find hierarchies of integrable equations with x dependent coefficients [473].

A few years ago Calogero and Degasperis [144] solved the Schrödinger spectral problem for asymptotically vanishing potentials $u(x)$ in the presence of a linear reference function

$$L\psi = -\psi_{xx} + [u(x) - g(x)]\psi = \lambda\psi, \quad \psi = \psi(x, u(x), \lambda), \; g(x) = -x. \tag{2.2.142}$$

The spectral problem considered allowed them to solve the Cauchy problem for the cylindrical KdV equation (cKdV) [145]

$$u_t + 6uu_x + u_{xxx} + \frac{u}{2t} = 0, \tag{2.2.143}$$

an important equation for its physical applications [51–53, 425, 475, 597].

Using the Lax technique, introduced for the construction of the recursion operator and Bäcklund transformations for the KdV equation, we can construct a hierarchy of equations, including the cKdV, with the function $g(x)$ allowing a potential $u(x)$ vanishing asymptotically. The function $g(x)$ in itself can be considered as part of a more general x and t dependent spectral parameter $\zeta(x,t) = \lambda(t) + g(x)$ (in [523] one can find an example of an x and t dependent spectral parameter).

We look for non linear PDEs written in Lax form

$$L_t = [L, M] + N, \tag{2.2.144}$$

where M is some operator which defines the t evolution of the eigenfunction of the L operator introduced in (2.2.142) and given in (2.2.6). The construction of the hierarchy of equations associated to (2.2.142) is obtained constructing the recursion operator obtained assuming the existence of two sets of operators $(M,\ N)$ and $(M',\ N')$ such that

$$[L, M] + N = V, \quad [L, M'] + N' = V', \tag{2.2.145}$$

where V and V' are multiplicative operators.

By assuming a relation between the two sets of operators M and N

$$M' = LM + F_1\partial_x + F_0, \quad N' = LN + G_1\partial_x + G_0, \tag{2.2.146}$$

where F_i, $i = 0, 1$, are a priori arbitrary functions of $u(x,t)$, $g(x)$ and V while G_i are x-independent constants as no equation will determine them. From (2.2.145, 2.2.146) we get:

$$F_1 = -\frac{1}{2}\int^x dx'V - \frac{1}{2}G_1 x + F_1^0, \tag{2.2.147}$$

$$F_0 = -\frac{1}{4}V + F_0^0, \tag{2.2.148}$$

$$\begin{aligned} V' = {} & -\frac{1}{4}V_{xx} + (u-g)V + \frac{1}{2}(u_x - g_x)\int^x dx'V + G_0 \\ & + G_1[u - g + \frac{1}{2}x(u_x - g_x)] - F_1^0(u_x - g_x), \end{aligned} \tag{2.2.149}$$

where F_0^0 and F_1^0 are x-independent constants.

From (2.2.149) we get the recursion operator $\mathcal{L}_c$ and the lowest equation of the hierarchy V_0,

$$u_t(x,t) = f(\mathcal{L}_c, t)V_0, \tag{2.2.150}$$

where f is an arbitrary entire function of the first argument. Now $\mathcal{L}_c$ is

$$\mathcal{L}_c V = -\frac{1}{4}V_{xx} + (u-g)V + \frac{1}{2}(u_x - g_x)\int^x dx' V, \tag{2.2.151}$$

and V_0

$$V_0 = G_0 + G_1(u + \frac{1}{2}xu_x) - F_1^0 u_x - G_1(g + \frac{1}{2}xg_x) + F_1^0 g_x. \tag{2.2.152}$$

In principle for any choice of G_0, G_1, F_1^0 and $g(x)$ we have a hierarchy of integrable PDEs. However if $u(x,t)$ vanish at infinity we need to require that V_0 vanishes asymptotically too. This is achieved if $V_0 = V_0(u)$. This condition defines V_0 and the admissible functions $g(x)$. We have:

$$V_0 = G_1(u + \frac{1}{2}xu_x) - F_1^0 u_x, \tag{2.2.153}$$

$$G_0 + F_1^0 g_x - G_1(g + \frac{1}{2}xg_x) = 0. \tag{2.2.154}$$

Eq. (2.2.154) has the following solution:

$$\text{if } G_1 \neq 0, \quad g(x) = -\frac{G_0}{G_1} + \Big(g_0 + \frac{G_0}{G_1}\Big)\Big(\frac{F_1^0 - \frac{1}{2}G_1 x_0}{F_1^{)} - \frac{1}{2}G_1 x}\Big)^2, \tag{2.2.155}$$

$$\text{if } G_1 = 0, \quad g(x) = -\frac{G_0}{F_1^0}(x - x_0) + g_0, \tag{2.2.156}$$

where x_0 is the arbitrary initial point of the integration and g_0 is an arbitrary integration constant. By proper choice of the constants involved, we have two possible solutions for the function $g(x)$,

$$g^1 = -x, \qquad g^2 = -\frac{2}{x^2}. \tag{2.2.157}$$

For g^1 we have the class of equations

$$u_t = f(\mathcal{L}_{c_1}, t)u_x, \quad \mathcal{L}_{c_1} V = -\frac{1}{4}V_{xx} + (u+x)V + \frac{1}{2}(u_x + 1)\int^x dx' V. \tag{2.2.158}$$

For g^2 we have the class of equations
(2.2.159)

$$u_t = f(\mathcal{L}_{c_2}, t)(u + \frac{1}{2}xu_x), \quad \mathcal{L}_{c_2} V = -\frac{1}{4}V_{xx} + \Big(u + \frac{2}{x^2}\Big)V + \frac{1}{2}\Big(u_x - \frac{4}{x^3}\Big)\int^x dx' V.$$

Choosing $f(z,t) = -4za + b$ in (2.2.158) we get a non linear PDE which is related to the cKdV

$$u_t = bu_x + a(u_{xxx} - 6uu_x - 4xu_x - 2u). \tag{2.2.160}$$

The higher equations of the hierarchy (2.2.158) will all be non local. Also all the equations of the hierarchy (2.2.159), apart from the lowest order one, will be non local.

It is easy to show that the Bäcklund transformation we derived for the KdV in (2.2.75) will not preserve the class of bounded solutions for the cKdV. In [474] a new Bäcklund transformation has been introduced to deal with this problem. In the literature has been called with different names, New Darboux Transformation [474], Darboux-Levi transformation [568, 713, 735] or Moutard transformation [59, 714] and in nuce it can be found in older articles by Kuznetsov [462, 463].

Let us consider a solution of (2.2.142) for $\lambda = \lambda_0$ a fixed value of the spectral parameter such that the wave function $\psi_0 = \psi(x, u_0(x), \lambda_0)$ corresponding to the potential $u_0(x)$ goes to zero asymptotically. So the "intermediate wave function"

$$F(x) = F(x, \lambda_0, \rho_0) = 1 + \rho_0 \int_x^\infty dy \psi_0^2(y, \lambda_0), \tag{2.2.161}$$

is well defined. Under these assumption we can define a new potential

$$u_1(x) = u_0(x) - 2\ln(F(x))_{xx} \tag{2.2.162}$$

and a new wave function

$$\psi^1(x, \lambda) = \frac{\left[\lambda - \lambda_0 - \frac{1}{2}\frac{F_{xx}}{\chi}\right]\psi(x, u_0(x), \lambda) + \frac{F_x}{\chi}\psi_x(x, u_0(x), \lambda)}{\lambda - \lambda_0} \tag{2.2.163}$$

such that the Schrödinger equation

$$-\psi^1_{xx} + (u_1(x) - g(x))\psi^1 = \lambda\psi^1 \tag{2.2.164}$$

is satisfied. The new Darboux transformation can be thought as the composition of a Darboux transformation of parameter λ_0 and one of λ_1 in the limit when λ_1 goes into λ_0 (see on this [59, 497]).

Introducing the integrated fields

$$v^0(x) = \int_x^\infty dy u_0(y), \quad v^1(x) = \int_x^\infty dy u_1(y), \tag{2.2.165}$$

we can rewrite the new Darboux transformation (2.2.164, 2.2.163, 2.2.162,2.2.142) as a new Bäcklund transformation

$$v^1_{xx} - v^0_{xx} = -\frac{1}{8}(v^1 - v^0)^3 - \left[v^1_x + v^0_x + 2g + 2\lambda_0\right](v^1 - v^0) + \frac{1}{2}\frac{(v^1_x - v^0_x)^2}{v^1 - v^0}. \tag{2.2.166}$$

This Bäcklund transformation, unlike (2.2.75) allows bounded solutions even for $g(x)$ diverging asymptotically provided $v^1 - v^0$ vanishes asymptotically. More details on the derivation of the new Darboux transformation can be found in [474].

We can use the new Darboux transformation to construct a solution for the Schrödinger equation in correspondence with $g(x) = -x$ for the cKdV. Starting from $u_0(x) = 0$ the solution of the Schrödinger equation is given by

$$\psi_0(x, 0, \lambda_0) = \mathrm{Ai}(y^0), \quad y^0 = x - \lambda_0. \tag{2.2.167}$$

As $\mathrm{Ai}(y^0)$ goes asymptotically to zero we can define the function $F(y^0)$

$$\begin{aligned} F(y^0) &= 1 + \rho_0 \int_x^\infty dy \, \mathrm{Ai}^2(y - \lambda_0) \\ &= 1 + \rho_0\Big(\mathrm{Ai}'^2(y^0) - y^0\, \mathrm{Ai}^2(y^0)\Big). \end{aligned} \tag{2.2.168}$$

Then the new potential is

$$u_1(y^0) = \rho_0\left[4\frac{\mathrm{Ai}'(y^0)\mathrm{Ai}(y^0)}{F(y^0)} + \frac{\mathrm{Ai}^4(y^0)}{F^2(y^0)}\right], \tag{2.2.169}$$

the same result as obtained by Calogero and Degasperis [144, 145] solving the spectral problem.

2.4. Integrable PDEs as infinite-dimensional superintegrable systems. The usual way of studying S-integrable PDEs (in the following denoted as soliton equations) is equivalent to considering them as infinite dimensional Hamiltonian systems [619, 664, 667].

Here we wish to point out that soliton equations have further general features relating them to finite dimensional systems that are not only integrable, but actually superintegrable.

Let us first sum up some results on finite dimensional *classical and quantum* superintegrable system [615, 791].

A classical system in an n–dimensional Riemanian space with Hamiltonian

$$H = \sum_{i,k=1}^{n} g_{ik}(\mathbf{x})p_i p_k + V(\mathbf{x}), \quad \mathbf{x} \in \mathbb{R}^n \tag{2.2.170}$$

is called completely integrable (or Liouville integrable) if it allows $n-1$ Poisson commuting integrals of motion (in addition to H)

$$\begin{aligned} X_n &= f_a(\mathbf{x}, \mathbf{p}), \quad a = 1, \cdots, n-1, \\ \frac{dX_a}{dt} &= \{H, X_a\}_p = 0, \ \{X_a, X_b\}_p = 0, \end{aligned} \tag{2.2.171}$$

where $\{,\}_p$ is the Poisson bracket and p_i are the momenta canonically conjugate to the coordinate x_i.

This system is *superintegrable* if it allows further integrals

$$\begin{aligned} Y_b &= f_b(\mathbf{x}, \mathbf{p}), \quad b = 1, \cdots, k \quad 1 \le k \le n-1, \\ \frac{dY_b}{dt} &= \{H, Y_b\}_p = 0. \end{aligned} \tag{2.2.172}$$

In addition, the integrals must satisfy the following requirements

(1) The integrals H, X_a, Y_b are well defined functions on phase space, i.e. polinomials or convergent power series on phase space (or an open submanifold of phase space).
(2) The integrals H, X_a are in involution, i.e. Poisson commute as indicated in (2.2.171) The integrals Y_b Poisson commute with H but not necessarily with each other, nor with X_a.
(3) The entire set of integrals is functionally independent, i.e., the Jacobian matrix satisfies

$$\text{rank}\frac{\partial(H, X_1, \cdots, X_{n-1}, Y_1, \cdots, Y_k)}{\partial(x_1, \cdots, x_n, p_1, \cdots, p_n)} = n + k \tag{2.2.173}$$

Superintegrable systems are interesting in classical physics for many reasons. Let us list some of them.

(1) Integrability makes it possible to introduce action-angle variables [47] and thus reduces motion to an N dimensional subspace of phase space (a torus if the trajectories are bounded). Superintegrability goes further and reduce the motion to an $n-k$ dimensional subspace. In the case of maximally superintegrability ($k = n-1$ in (2.2.172)) this implies that all finite trajectories are closed and the motion is periodic [629].
(2) The integrals of motion $\{H, X_a, Y_b\}$ form a Lie algebra under Poisson commutation. Usually it is infinite dimensional, exceptionally as in the case of the harmonic oscillator, it is finite dimensional (isomorphic to $su(n)$). If the integrals of

motions can be expressed in terms of polynomials in the momenta p_i (or convergent series in p_i) then it is more fruitful to view $\{H, X_a, Y_b\}$ as a finitely generated polynomial algebra [85, 260, 276, 277, 436, 583, 792, 864]. This algebra can be used to integrate the equations of motion.

(3) In the case of quadratic integrability (n independent Poisson commuting integrals that are at most quadratic in the moments) the Hamilton-Jacobi equation allows the separation of variables (in configuration space). Quadratic superintegrability then corresponds to multiseparability [276, 277, 431–436, 583, 615].

(4) It follows from Bertrand's theorem [864] that the only spherically symmetric potentials in E_n for which all finite trajectories are closed are $\omega^2 r^2$ and α/r [85]. Hence no other maximally superintegrable systems are spherically symmetric.

In quantum mechanics the situation is somewhat more complicate. No generally accepted definition of integrability exists, still less superintegrability.

We will restrict our considerations here to a system of the form (2.2.170), (2.2.171), (2.2.172) where the Hamiltonian and integrals $\{H, X_a, Y_b\}$ are operators obtained by the usual procedure of putting $p_k \to -i\hbar\frac{\partial}{\partial x_k}$. The Poisson brackets $\{,\}$ are replaced by Lie commutators. The conditions imposed on the integrals of motion in the classical case are replaced in the quantum case by:

(1) The integrals of motion $\{H, X_a, Y_b\}$ are well defined Hermitian operators in the enveloping algebra of the Heisenberg algebra H_n $\{\vec{x}, \vec{p}, \hbar\}$ (or in some generalization of the enveloping algebra).

(2) The integrals satisfy the Lie commutation relations

$$[H, X_a] = [H, Y_b] = 0, \quad [X_a, Y_b] = 0. \tag{2.2.174}$$

(3) Functional independence is replaced by polynomial independence. We require that no Jordan polynomial in H, X_a, Y_b should vanish identically.

We remark here that we are in the quantum case dealing with three algebras. The first is an associative algebra generated by X_a, Y_b and H with the product defined by the usual product of polynomials. The second is the Lie algebra with product $[X, Y] = XY - YX$ and the third is the special Jordan algebra with product $X \cdot Y = (XY + YX)/2$.

Among the properties of superintegrable systems in quantum mechanics we mention the following:

(1) Superintegrability leads to the degeneracy of energy levels of the Schrödinger equation, i.e more than one-dimensional eigenspaces of the Hamiltonian. Thus if an operator Z commutes with H and the function $\psi(\vec{x})$ is an eigenfunction of H, then $Z\psi(\vec{x})$ is also an eigenfunction for the same energy E. If Z is realized as a first order differential operator it will generate point transformations and correspond to geometrical symmetries of the Hamiltonian (like rotations for any spherically symmetric potential $V(r)$). If Z is realized by an operator of order $N \geq 2$ the corresponding symmetries will be associated with more specific interactions, e.g. the Kepler-Coulomb potential $V(r) = \alpha/r$ or the harmonic oscillator $V(r) = \alpha r^2$. The Kepler-Coulomb potential and the harmonic oscillator were the only two potentials known before 1940 that have degenerate energy levels not explained by geometrical symmetries. Fock coined the term "accidental degeneracy" for the case of the hydrogen atom[260], though he himself proved that this is no accident.

The two potentials are the only spherically symmetrical superintegrable ones that exist in the Euclidean space E_n, in particular in E_2 and E_3 (in classical mechanics this is a consequence of Bertrand's theorem).

(2) A systematic search for superintegrable systems of the type (2.2.170, 2.2.171, 2.2.172) in Euclidean spaces E_2 and E_3 was started in 1965 [276]. It was conducted in quantum mechanics and the integrals X_a and Y_b were restricted to being second order polynomials in the momenta, i.e. second order Hermitian differential operators. In more recent articles the integrals are polynomials of arbitrary order N in the momenta [678, 802].

Moreover, it was assumed that the Abelian subalgebra $\{X_1, \cdots, X_{n-1}\}$ consisted entirely of second order polynomials. This guarantees that the Schrödinger equation will allow the separation of variables in one or more of the systems of coordinates in which the Helmholtz equation allows separation of variables. In Euclidean space E_n the potential will then involve n arbitrary functions $f_i(\xi_i)$ of one variable each. The additional integrals $\{Y_1, \cdots, Y_k\}$ will impose further constraints on the functions $f_i(\xi_i)$ and on the coefficients in the integrals $\{Y_1, \cdots, Y_k\}$. These constraints have the form of a system of coupled ODEs. They are called "standard potentials" if all the functions $f_i(\xi_i)$ satisfy linear ODEs and "exotic potentials" if at least one of the $f_i(\xi_i)$ satisfy only non linear ODEs. So far it was shown that in E_2 all exotic potentials that allow separation of variables in Cartesian or polar coordinates and allow an additional integral of order $N \geq 3$ satisfy non linear ODEs that pass the Painlevé test (for $3 \leq N \leq 10$) [240, 242, 593, 678] and can be integrated in terms of the known (second order) Painlevé transcendent $P_I, \cdots, P_{VI}$ or elliptic functions. It has been conjectured that this Painlevé property holds for all values of N and also for higher dimentional Euclidean spaces.

(3) A conjecture, holding in all known examples, is that all maximally superintegrable systems ($2n-1$ independent integrals of motion in E_n) are exactly solvable [616, 710, 792]. That means that the Hamiltonian H can be block-diagonalized into finite dimensional blocks so that the energies can be calculated algebraically (without solving transcendental equations [792]). The wave functions are then polynomials in some chosen variables, multiplied by a common factor [792].

(4) Under Lie commutation the integrals of motion form a Lie algebra (since a commutator of two integrals is also an integral). This Lie algebra is usually infinite dimensional. It is more convenient to view the algebra as an associative one (under multiplication of the differential operators that realize the integrals). This algebra is finitely generated [615] and provides information on the energy spectrum and wave functions.

To sum up, the superintegrable systems are, by definition, also integrable. They have additional integrability properties which simplify the calculation of trajectories in classical mechanics. They simplify the calculation of energies and wave functions in quantum mechanics.

The main similarities between infinite dimensional superintegrables systems and soliton equations are the following:

(1) Both have more independent well defined integrals of motion than is necessary for integrability and these integrals form a non abelian algebra. Integrability in both cases is assured by the existence of a maximal subalgebra of n commuting integrals of motion. For finite dimensional systems n is equal to the number of degrees of freedom, for solitons equations $n \to \infty$.

(2) Soliton equations are similar to the maximally superintegrable finite dimensional ones also in the sense of their "solvability". Indeed the existence of a Lax pair makes it possible to obtain large classes of exact solutions of the soliton equations using linear techniques. For maximally superintegrable classical systems it is possible to calculate trajectories without using calculus. In the quantum case it is possible to reduce the calculation of energy levels to algebraic equations and to obtain solutions of the Schrödinger equation.

(3) The Painlevé property and Painlevé transcendent play an important role in both cases. In soliton theory the transcendent often figure in group invariant solutions. More important, the Painlevé test is an important tool in recognizing integrability.

Exotic potentials in quantum mechanics are, by definition, superintegrable potentials satisfying non linear equations. In all known examples these non linear equations have the Painlevé property [241, 591, 592]. Interestingly, in classical mechanics the equations for exotic potentials do not have the Painlevé property but can be integrated to provide implicit solutions that amount to algebraic equations for the potentials.

2.5. Integrability of the Burgers equation, the prototype of linearizable PDEs. The Burgers equation

$$u_t = u_{xx} + uu_x, \quad u = u(x,t), \tag{2.2.175}$$

is probably the best known C-integrable equation. It was introduced firstly by Bateman [74] and later considered by Burgers as a mathematical model in his study of the theory of turbulence [134, 135]. It is related to the Navier–Stokes momentum equation with the pressure term removed and can be found in various areas of mathematical physics, such as fluid mechanics, non linear acoustics, gas dynamics, traffic flow, etc. . It is the simplest partial differential equation that combines non linear effects with dissipation. From the mathematical point of view it is the prototype equation linearizable via a coordinate transformation.

Putting

$$u = 2v_x, \quad v = v(x,t), \tag{2.2.176}$$

we obtain from (2.2.175) the potential form of the Burgers equation, namely

$$v_t = v_{xx} + v_x^2. \tag{2.2.177}$$

Finally, setting

$$\psi = e^v, \quad \psi = \psi(x,t), \tag{2.2.178}$$

we obtain the linear heat equation for ψ, namely

$$\psi_t = \psi_{xx}. \tag{2.2.179}$$

In other words, the standard Burgers equation (2.2.175) is transformed into the heat equation by the Cole-Hopf transformation [190, 400]

$$u = 2\frac{\psi_x}{\psi}. \tag{2.2.180}$$

The potential Burgers equation (2.2.177) has an infinite dimensional Lie algebra that is "inherited" from the linear heat equation [659]. That of the Burgers equation (2.2.175) is five-dimensional. They both have infinitely many higher symmetries, and Bäcklund transformations, however only a finite number of conserved quantities.

One can prove [173, 502] that (2.2.175) is part of a hierarchy of equations defined in term of a recursion operator $\mathcal{L}_B$

$$u_t = \mathcal{L}_B^{\ell} u_x, \quad \mathcal{L}_B f = f_x + \frac{1}{2} u f - \frac{1}{2} u_x \int_x^{\infty} f(y,t) dy, \quad f = f(x,t). \tag{2.2.181}$$

Eq. (2.2.175) corresponds to $\ell = 1$ and the following two members of the hierarchy for $\ell = 2$ and $\ell = 3$ are

$$u_t = u_{xxx} + \frac{3}{2} u u_{xx} + \frac{3}{2} u_x^2 + \frac{3}{4} u^2 u_x, \tag{2.2.182}$$

$$u_t = u_{xxxx} + 2 u u_{xxx} + 5 u_x u_{xx} + \frac{3}{2} u^2 u_{xx} + 3 u u_x^2 + \frac{1}{2} u^3 u_x. \tag{2.2.183}$$

Eq. (2.2.175) can be associated to a Lax pair with a spectral problem without spectral parameter given by (2.2.179, 2.2.180), i.e.

$$L\psi = \psi_x - \frac{1}{2} u \psi = 0, \quad \psi = \psi(x,t;u), \tag{2.2.184}$$

$$\psi_t = -M\psi, \quad M = -\partial_{xx}, \tag{2.2.185}$$

whose compatibility is (2.2.175). The compatibility of (2.2.184, 2.2.185) can be written as a Lax equation in the isospectral (2.2.8) and non isospectral (2.2.9) form.

The same formalism which was used in the case of the KdV (2.2.1) can be used to construct here the recursion operator (2.2.181) and the corresponding initial condition, starting from the definition of the "spectral problem" given in (2.2.184). As in the case of KdV we start from (2.2.43) and (2.2.44) with $\tilde{M}$ defined by (2.2.46). As the "spectral problem" is different, (2.2.47) will be different. In this case we obtain the equations

$$\tilde{V} = V_x - \frac{1}{2} u V + G_x + \frac{1}{2} F u_x, \tag{2.2.186}$$

$$V + F_x = 0. \tag{2.2.187}$$

From (2.2.186) we see that in this case the function G is not defined and it can be chosen as a function of u and V in an arbitrary way. We choose G as

$$G_x \equiv uV - u_x \int_x^{\infty} dy V(y). \tag{2.2.188}$$

In this case we get the recursion operator $\mathcal{L}_B$. As before we solve (2.2.187) for the function F and, taking into account (2.2.188) we get $\mathcal{L}_B$ and

$$\begin{aligned} &\tilde{V} = \mathcal{L}_B V + V^{(0)}, \quad V^{(0)} = -\frac{1}{2} F_0 u_x, \quad \tilde{V} = -\frac{1}{2} u_t, \\ &\tilde{M} = LM + \Big[\int_x^{\infty} dy V(y) + F_0\Big]\partial_x - \int_x^{\infty} dy \Big[uV - u_y \int_y^{\infty} dz V(z)\Big], \\ &F = F_0 + \int_x^{\infty} dy V(y), \end{aligned} \tag{2.2.189}$$

where F_0 is an arbitrary integration constant. To obtain (2.2.181) we have to set $F_0 = -1$ in (2.2.189). The corresponding t evolution of the function ψ for (2.2.181) is obtained from (2.2.184) and (2.2.189) and reads:

$$\psi_t = \psi_{(\ell+1)x}. \tag{2.2.190}$$

In [173, 502] the recursion operator is different. We denote it $\tilde{\mathcal{L}}_B$. The derivation presented in [502] has the same starting points (2.2.186) and (2.2.187) but (2.2.187) is now not solved for V but for F. So we have:

$$V = -F_x, \quad \tilde{V} = -\tilde{F}_x, \tilde{F} = F_x - \frac{1}{2}uF - G + F_0. \tag{2.2.191}$$

Choosing

$$G = -uF \tag{2.2.192}$$

we get

$$\tilde{V} = \mathcal{S}\,\tilde{\mathcal{L}}_B\,F_0, \tag{2.2.193}$$

where F_0 as before is an integration constant and

$$\mathcal{S} = -\partial_x, \quad \tilde{\mathcal{L}}_B = \partial_x + \frac{1}{2}u. \tag{2.2.194}$$

The Burgers hierarchy now reads

$$u_t = \mathcal{S}\,\tilde{\mathcal{L}}_B^{\ell+1}\,F_0. \tag{2.2.195}$$

Eqs. (2.2.195) and (2.2.181) are the same set of equations and this proofs that, even if the recursion operator $\mathcal{L}_B$ is integro-differential, all the equations of the Burgers hierarchy are evolutionary PDEs. The recurrence operator $\tilde{\mathcal{L}}_B$, firstly presented without derivation in [173], can be found in Olver [659].

Let us notice that if we use the procedure introduced in [502] in the case of the Schrödinger spectral problem (2.2.11) we get the same recursion operator (2.2.36).

The Lax equation corresponding to *non isospectral deformation* of the hierarchy of equations (2.2.181) is given by (2.2.9). In this case, when no spectral problem is present, we have non autonomous equations with explicit dependence on x. From the equations corresponding to (2.2.54) and (2.2.55) we get as coefficient of $h(t)$, i.e. the starting point of the non isospectral hierarchy of the Burgers, $V^{(0)} = u + xu_x$. Then the non isospectral hierarchy of equations can be written as

$$u_t = h(t)\mathcal{L}_B^n\,[u + xu_x], \tag{2.2.196}$$

which, at difference with case of the KdV, is always a PDE. The corresponding evolution of the function ψ is

$$\psi_t = -M\psi, \qquad M\psi = -h(t)[x\psi_x]_{nx} = -h(t)[x\psi_{(n+1)x} + n\psi_{nx}]. \tag{2.2.197}$$

The simplest non isospectral PDEs are

$$\begin{aligned} u_t &= x\,u_x + u, \\ u_t &= x\,u_{xx} + 2\,u_x + \frac{1}{2}u^2 + x\,u\,u_x. \\ u_t &= x\,u_{xxx} + 3u_{xx} + \frac{3}{2}x\,u\,u_x + \frac{3}{2}x\,u_x^2 + \frac{7}{2}u\,u_x + \frac{3}{4}x\,u^2\,u_x + u^3 \end{aligned} \tag{2.2.198}$$

2.5.1. *Bäcklund transformation and Bianchi identity for the Burgers hierarchy of equations.* The Bäcklund transformations are obtained with the same procedure we used for the KdV hierarchy. Starting from (2.2.65) and (2.2.66), we get:

- A recursion operator for the Bäcklund

$$\Lambda_B\,f(x) = f_x + \frac{1}{2}\tilde{u}f - \frac{1}{2}u_x e^{H(x)} \int_x^\infty dy f(y)e^{-H(y)} \tag{2.2.199}$$

where

$$H_x = \frac{1}{2}(\tilde{u} - u). \tag{2.2.200}$$

- The class of Bäcklund transformations is given by

$$\alpha(\Lambda_B, t)(\tilde{u} - u) = \beta(\Lambda_B, t)u_x e^{H(x)}. \tag{2.2.201}$$

- Introducing the potential function $w(x, t)$ which goes asymptotically to zero, such that $u = w_x$, we get $H = \frac{1}{2}(\tilde{w} - w)$ and the simplest Bäcklund transformation reads:

$$\tilde{w}_x - w_x = p w_{xx} e^{\frac{1}{2}(\tilde{w}-w)}. \tag{2.2.202}$$

Eq. (2.2.202) is a transcendental relation between $\tilde{w}$ and w.

A simpler Bäcklund transformation is obtained using the approach introduced in [502] and presented for the PDEs in the previous Section. From (2.2.65), (2.2.66) and (2.2.46) with $u_1 = u$ and $u_2 = \tilde{u}$ we get the following two equations

$$V_x + \frac{1}{2}\tilde{u}V - \frac{1}{2}u_x F + G_x + \frac{1}{2}(\tilde{u} - u)G = \tilde{V}, \tag{2.2.203}$$

$$V + F_x + \frac{1}{2}(\tilde{u} - u)F = 0 \tag{2.2.204}$$

From (2.2.204) we extract V in term of F and define $\tilde{V} = -\left[\tilde{F}_x + \frac{1}{2}(\tilde{u} - u)\tilde{F}\right] = \hat{\Omega}\tilde{F}$. Then choosing $G = 0$ (2.2.203) gives a relation between $\tilde{F}$, F and a recursion operator $\hat{\Lambda}_B$ which reads

$$\tilde{F} = \hat{\Lambda}_B F + F_0, \quad \hat{\Lambda}_B = \partial_x + \frac{1}{2}\tilde{u}, \tag{2.2.205}$$

where F_0 is a constant. The class of Bäcklund transformation is obtained starting with $F = 0$. It is given by

$$\hat{\Omega} \sum_n \hat{\Lambda}_B^n F_n = 0, \tag{2.2.206}$$

where the F_n are arbitrary constants. The simplest Bäcklund transformation for the Burgers is obtained by taking only F_0 and F_1 in (2.2.206) and reads:

$$\tilde{u}_x = (\tilde{u} - u)(p - \tilde{u}), \quad p = -\frac{F_0}{F_1}. \tag{2.2.207}$$

Following [502] we can write the corresponding Bianchi identity:

$$p_2\, u_1 - p_1\, u_2 = u_{12}\left(u_1 - u_2 + p_2 - p_1\right), \tag{2.2.208}$$

which relate three solutions, u_1, u_2 and u_{12}, of the Burgers hierarchy of equations and is written in terms of two different parameters p_1 and p_2. From (2.2.208), given two solution u_1 and u_2 we are able to create a third u_{12}.

Eq. (2.2.207), by the identification $u = v_{n-1}$ and $\tilde{u} = v_n$, gives the DΔE

$$\dot{v}_n = (v_n - v_{n-1})(p - v_n), \tag{2.2.209}$$

were by a dot we mean the x derivative. Eq. (2.2.209), by defining $v_n = w_n + p$ can be rewritten as the DΔE Burgers (see (2.3.329) in Section 2.3.5)

$$\dot{w}_n = w_n\,(w_{n-1} - w_n). \tag{2.2.210}$$

Eq. (2.2.208), by the identification $u_1 = v_{nm}$, $u_2 = v_{n+1,m}$ and $u_{12} = v_{n,m+1}$, gives the three-point PΔE

$$p_2 v_{nm} - p_1 v_{n+1,m} = v_{n,m+1}\left(v_{nm} - v_{n+1,m} + p_2 - p_1\right). \tag{2.2.211}$$

2.5.2. *Symmetries of the Burgers equation.* In this case we cannot define a spectrum as no spectral parameter is present in the Lax pair. However the evolution of the wave function ψ of the Burgers spectral problem can play a similar role and can be used to define the symmetries by providing in a simple way the flows commuting with the equations of the Burgers hierarchy. In fact all the information on the equation of the Burgers hierarchy is included in the evolution of the wave function ψ. So by looking for the flows commuting with the time evolution of a given equation of the Burgers hierarchy we can construct its symmetries.

The Lie point symmetry algebra of the Burgers is given by (2.2.220) in Section 2.2.5.3.

Given an equation of the Burgers hierarchy (2.2.181), the evolution of the wave function ψ which is associated via the Cole-Hopf transformation to the potential $u(x,t)$ is given by (2.2.190). The isospectral symmetries will be given by those evolutions in the infinitesimal group parameter ϵ which commute with the time evolution of the equations of the Burgers hierarchy given by (2.2.190). They are the equations

$$u_{\epsilon_n} = \mathcal{L}_B^n u_x, \tag{2.2.212}$$

whose corresponding ϵ_n evolution of the wave function is

$$\psi_{\epsilon_n} = \psi_{(n+1)x}. \tag{2.2.213}$$

The commutativity of (2.2.181) and (2.2.212) is due to the commutativity of the differentials present on the right hand side of the expressions (2.2.190) and (2.2.213). So we have an infinite dimensional symmetry algebra of Abelian symmetries.

Can one construct also some non isospectral symmetries using the non autonomous equations (2.2.196)? As we did in the case of the KdV we could define the equation

$$u_{\epsilon_N} = h(t)\mathcal{L}_B^M u_x + \mathcal{L}_B^N [xu_x + u], \tag{2.2.214}$$

a combination of an isospectral term with a coefficient given by a t-dependent arbitrary function and an unknown power M of the recursive operator with a non isospectral one characterized by a power N of the recursive operator. The corresponding evolution of the wave function ψ is

$$\psi_{\epsilon_N} = h(t)\psi_{Mx} + (x\psi_x)_{Nx} = h(t)\psi_{Mx} + x\psi_{(N+1)x} + N\psi_{Nx}. \tag{2.2.215}$$

Due to the presence of the explicit x-dependent coefficient of the non isospectral ϵ_N evolution of the wave function, (2.2.215) and (2.2.190) will never commute. However one can show that the non isospectral hierarchy of equations (2.2.196) are *master symmetries*. Defining from (2.2.190) and (2.2.197) the infinitesimal generators

$$\hat{X}_\ell = \partial_x^\ell, \qquad \hat{Y}_N = x\partial_x^{N+1} + N\partial_x^N \tag{2.2.216}$$

corresponding respectively to symmetries and master symmetries, it is immediate to show the following commutation relation

$$[\hat{X}_\ell, \hat{Y}_N] = \ell \hat{X}_{N+\ell}, \tag{2.2.217}$$

which proves that (2.2.214) with $h(t) = 0$ are master symmetries.

2.5.3. *Symmetry reduction by Lie point symmetries.* As we saw in the case of (2.2.1) in Section 2.2.2.7, symmetry reduction is a tool for decreasing the number of independent variables in the equation and possibly integrate it.

If (1.1.22) is implemented, together with the PDE to be solved, the resulting system may have some further symmetries, inherited from the symmetry algebra $\mathfrak{g}$ of the original equation. More specifically, let $\hat{X}_e^\alpha = Q_\alpha \partial_u$ be the corresponding evolutionary vector fields. The symmetries that will survive, once the surface condition

$$u_\epsilon = Q(x, t, u, u_x, u_t, \dots) = 0. \tag{2.2.218}$$

is imposed, form a subalgebra $\mathfrak{g}_0 \subset \mathfrak{g}$, where $\mathfrak{g}_0$ is the normalizer algebra of $\hat{X}_e$:

$$\mathfrak{g}_0 = \left\{ \hat{Y}_e \subset \mathfrak{g} \,\middle|\, [\hat{Y}_e, \hat{X}_e] = \pi \hat{X}_e \right\}, \quad \pi \in \mathbb{R}. \tag{2.2.219}$$

Let us illustrate the situation using the continuous Burgers equation (2.2.175) as an example. The Lie point symmetry algebra $\mathfrak{g}$ of Burgers [275] in the usual vector field formalism has a basis given by

$$\begin{gathered} \hat{P}_0 = \partial_t, \quad \hat{P}_1 = \partial_x, \quad \hat{B} = t\partial_x - \partial_u, \quad \hat{D} = 2t\partial_t + x\partial_x - u\partial_u, \\ \hat{R} = t^2\partial_t + tx\partial_x - (tu + x)\,\partial_u. \end{gathered} \tag{2.2.220}$$

From the commutation relations of these vector fields, we see that $\mathfrak{g}$ is a semidirect sum of $sl(2, \mathbb{R})$ and an abelian Lie algebra:

$$\mathfrak{g} \sim \{\hat{P}_0, \hat{D}, \hat{R}\} \dot{+} \{\hat{P}_1, \hat{B}\}. \tag{2.2.221}$$

As an example let us look at the reductions of the continuous Burgers equation by time translations $\hat{P}_0$. Eq. (1.1.22) in this case is simply $u_\epsilon = u_t = 0$. The Burgers equation (2.2.175) reduces to the ODE

$$u_{xx} + uu_x = 0. \tag{2.2.222}$$

We obtain three types of solutions:

$$u = \frac{1}{x}, \qquad u = k \operatorname{arctanh}(kx), \quad u = k \arctan(kx) \tag{2.2.223}$$

depending on whether the first integral of the reduced equation is zero, positive or negative.

The normalizer of $\hat{P}_0$ in the invariance algebra is $\mathrm{nor}\{\hat{P}_0\} = \{\hat{D}, \hat{P}_1, \hat{P}_0\}$, so we can use either $\hat{D}$, or $\hat{P}_1$, to perform a further reduction of (2.2.222). A reduction by $\hat{P}_1$ leads to the trivial solution $u = u_0 =$ constant while the reduction under dilations provides the first of the three solutions in (2.2.223).

2.6. General ideas on linearization. As we saw Calogero [141] introduced a distinction between non linear PDEs which are "C-integrable" and "S-integrable", namely, equations that are linearizable by an appropriate change of variables (i.e. by an explicit redefinition of the dependent variable and maybe, in some cases also the independent variables), and those equations that are integrable via the Inverse Scattering Transform (IST) [147]. A complete analysis of the C–integrable non linear PDEs [141], i.e. those equations which are linearizable via a transformation, has been carried out by Calogero, Eckhaus and Ji [148]. In the cited reference Calogero used the asymptotic behavior, multiple scale expansions, to find equations belonging to these two classes of equations as this technique usually preserves integrability and linearizability [206, 207, 368–370, 374, 374, 517, 742–744]. However this approach, even if very fruitful is often very cumbersome and not always exhaustive (see for example the results of [374] for discrete equations).

A more intrinsic approach is based on the existence of symmetries.
C-integrable and S-integrable equations are associated to the well-known notion of higher symmetry together with the notion of IST. Moreover the distinction between C-integrable and S-integrable equations is mainly at the level of the conservations laws, see Section 3.2.4.1. Linearizable equations have no local conservations laws of arbitrary high order [850] and their Lax pair is fake. By a fake Lax pair we mean a Lax pair in which we have

no spectral parameter or it can be taken away simply[136, 137, 181–183, 340, 524, 596] as we saw in Section 2.2.5 and will see when we derive the DΔE and PΔE Burgers in Section 2.3.5.2 and 2.4.10.

2.6.1. *Linearization of PDEs through symmetries.* In [459] Bluman and Kumei introduce a series of theorems dealing with the conditions for a non linear PDE to be transformable into a linear one by *contact transformations*. Here, in the following, we will limit ourselves to the case of just *point transformations* as these will be the relevant ones in the discrete case where contact transformations effectively do not exist [521, 527]. In more recent works the same authors [101] extended the consideration to the case when we have *non invertible transformations* between a non linear and a linear PDE.

The basic observation is that a linear PDE,

$$\mathfrak{L}v(y) = \mathcal{F}(y), \tag{2.2.224}$$

where $\mathfrak{L}$ is a v–independent but possibly y–dependent linear operator and $\mathcal{F}(y)$ is the inhomogeneous term, has one point symmetry of infinitesimal symmetry generator

$$\hat{X} = w(y)\frac{\partial}{\partial v} \tag{2.2.225}$$

depending on a function w which satisfy the homogeneous equation

$$\mathfrak{L}w(y) = 0. \tag{2.2.226}$$

Any solution of (2.2.224) is the sum of a particular solution plus the general solution of the associated homogeneous equation. The existence of an infinitesimal generator of the form (2.2.225) is preserved when we transform a linear equation into a non linear one by an invertible point transformation. We present here the conditions for the existence of an invertible linearization mapping of a non linear PDE stated in [98] Section 2.4:

Theorem 8. *A non linear PDE*

$$\mathcal{E}_n(x, u, u_x, \cdots u_{nx}) = 0 \tag{2.2.227}$$

of order n for a scalar function u of an r–dimensional ($r \geq 2$) vector x, where by u_{kx} we mean the set of all derivative of $u(x)$ of order k, will be linearizable by a point transformation

$$w(y) = f(x, u), \qquad y_i = g_i(x, u), \quad i = 1, \cdots, r \tag{2.2.228}$$

to a linear equation (2.2.226) for w if it possesses a symmetry generator

$$\hat{X} = \sum_{i=1}^{r} \xi_i(x, u)\partial_{x_i} + \phi(x, u)\partial_u, \quad \xi_i(x, u) = \alpha_i(x, u)w(y), \tag{2.2.229}$$
$$\phi(x, u) = \sigma(x, u)w(y),$$

with σ and α_i given functions of their arguments and $w(y)$ an arbitrary solution of (2.2.226).

Following [459] we can state the sufficient conditions for the existence of an invertible linearization mapping of a non linear PDE. This theorem defines the transformation (2.2.228):

Theorem 9. *If a symmetry generator for the non linear PDE (2.2.227) exists, as specified in Theorem 8, the invertible transformation (2.2.228) which transforms (2.2.227) to the linear PDE (2.2.226) is given by*

$$y_i = \Phi_i(x, u), \quad i = 1, \cdots, r, \tag{2.2.230}$$

$$w = \Psi(x, u). \tag{2.2.231}$$

where $\Phi_i(x,u)$ are r functionally independent solutions, $i = 1, \cdots, r$, of the linear homogeneous first order PDE for a scalar function $\Phi(x,u)$

$$\sum_{i=1}^{r} \alpha_i(x,u)\Phi(x,u)_{x_i} + \sigma(x,u)\Phi(x,u)_u = 0, \tag{2.2.232}$$

and $\Psi(x,u)$ is a particular solution of the linear inhomogeneous first order PDE for a scalar function $\Psi(x,u)$

$$\sum_{i=1}^{r} \alpha_i(x,u)\Psi(x,u)_{x_i} + \sigma(x,u)\Psi(x,u)_u = 1. \tag{2.2.233}$$

If a given linearizable non linear PDE does not have local symmetries of the form (2.2.229), i.e. its local symmetries do not satisfy the criteria of Theorem 8, it could still happen, as shown in [98] Section 4.3, that a *nonlocally related system* has an infinite set of local symmetries that yields an invertible mapping of the nonlocally related system to some linear system of PDEs. Consequently, the invertible mapping of the nonlocally related system to a linear system will provide a non local (non invertible) mapping of the given non linear PDE to a linear PDE. This non invertible transformation will be a kind of Cole–Hopf transformation [190, 400]. In this case, however, we have to generalize Theorem 9 to take into account the fact that we are dealing with a system of equations.

Theorem 10. *Let us consider a system of non linear PDE*

$$\begin{aligned} &\mathcal{E}_n^{(1)}(x,u,v,u_x,v_x,\cdots u_{nx},v_{nx}) = 0,\\ &\mathcal{E}_n^{(2)}(x,u,v,u_x,v_x,\cdots u_{nx},v_{nx}) = 0 \end{aligned} \tag{2.2.234}$$

of order n for two scalar functions u and v of an r–dimensional ($r \geq 2$) vector x which possesses a symmetry generator

$$\begin{aligned} \hat{X} &= \sum_{i=1}^{r} \xi^i(x,u,v)\partial_{x_i} + \phi(x,u,v)\partial_u + \psi(x,u,v)\partial_v,\\ \xi^i(x,u,v) &= \sum_{j=1}^{2} \alpha_j^i(x,u,v)w^{(j)}(y),\\ \phi(x,u,v) &= \sum_{j=1}^{2} \beta_j(x,u,v)w^{(j)}(y),\\ \psi(x,u,v) &= \sum_{j=1}^{2} \gamma_j(x,u,v)w^{(j)}(y), \end{aligned} \tag{2.2.235}$$

with α_j^i, β_j and γ_j given functions of their arguments and the function $w = (w^{(1)}(y), w^{(2)}(y))$ satisfying the linear homogeneous equations

$$\mathfrak{L}(y)w(y) = 0, \tag{2.2.236}$$

with y an r–dimensional vector depending on u, v and the vector x and $\mathfrak{L}$ is a 2×2 matrix linear operator.

The invertible transformation

$$w^{(1)}(y) = F^{(1)}(x,u,v), \qquad w^{(2)}(y) = F^{(2)}(x,u,v), \qquad y = G(x,u,v), \tag{2.2.237}$$

which transforms (2.2.234) to the system of linear PDEs (2.2.236) is given by r functionally independent solutions $G_i(x,u,v)$ with $i = 1, \cdots, r$ of the linear homogeneous first order system of PDEs for a scalar function $\mathcal{G}(x,u,v)$

$$\sum_{i=1}^{r} \alpha_k^i(x,u,v)\mathcal{G}_{x_i} + \beta_k(x,u,v)\mathcal{G}_u + \gamma_k(x,u,v)\mathcal{G}_v = 0 \tag{2.2.238}$$

and by a particular solution of the linear inhomogeneous first order system of PDEs for the function $F = (F^{(1)}(x,u,v), F^{(2)}(x,u,v))$

$$\sum_{i=1}^{r} \alpha_k^i(x,u,v)F_{x_i}^{(j)} + \beta_k(x,u,v)F_u^{(j)} + \gamma_k(x,u,v)F_v^{(j)} = \delta_k^j, \tag{2.2.239}$$

with δ_k^j the standard Kronecker symbol.

For the sake of completeness and to clarify the application of the theorems presented above, in view of the discretization which will be presented in Section 2.4.11, we consider here one example of linearizable non linear PDEs belonging to each of the two cases presented above.

A non linear PDE linearizable by a point transformation. It is well know, see for example Olver book [659], that the potential Burgers equation

$$u_t = u_{xx} + (u_x)^2, \tag{2.2.240}$$

is linearizable by a point transformation. In fact the infinite dimensional part of the infinitesimal generator of its point symmetries is given by

$$\hat{X} = w(x,t)e^{-u}\partial_u, \tag{2.2.241}$$

where $w(x,t)$ satisfies the homogeneous linear heat equation $w_t - w_{xx} = 0$.

The conditions of Theorem 8 are satisfied with $\sigma = e^{-u}$ and $\alpha_i = 0$. We can apply Theorem 9 and we get $\Phi_1 = x$ and $\Phi_2 = t$ as from (2.2.232) $\Phi_u = 0$ while from (2.2.233) $\Psi(x,u)$ satisfies the equation $\Psi(x,u)_u = e^u$ i.e. from (2.2.231)

$$u = \log_e(w). \tag{2.2.242}$$

Eq. (2.2.242) is the linearizing transformation for the potential Burgers equation (2.2.240).

A non linear PDE linearizable by a non invertible transformation. The standard example in this class is the Burgers equation

$$u_t = u_{xx} - uu_x = [u_x - \frac{1}{2}u^2]_x, \tag{2.2.243}$$

linearizable by a Cole–Hopf transformation. As (2.2.243) has no infinite dimensional symmetry algebra but it is written as a conservation law we can introduce a potential function $v(x,t)$ and (2.2.243) can be written as the system

$$\begin{aligned} v_x &= 2u, \\ v_t &= 2u_x - u^2. \end{aligned} \tag{2.2.244}$$

Applying Theorem 10 we can find an infinite dimensional symmetry for equations of the form of (2.2.234). In fact, solving the determining equations, apart from terms corresponding to a finite dimensional algebra, we obtain an infinite dimensional dilation symmetry given by

$$\psi = 4\sigma(x,t)e^{\frac{v}{4}}, \qquad \phi = \frac{1}{2}\psi_x + u\psi_v = e^{\frac{v}{4}}[2\sigma_x + \sigma u] \tag{2.2.245}$$

where $\sigma(x,t)$ satisfies the linear heat equation $\sigma_t - \sigma_{xx} = 0$.

The linearizing transformation can be obtained from Theorem 10. Let us define $w^{(1)}(y) = \sigma(x,t)$, $w^{(2)}(y) = \sigma_x(x,t)$ and take as functionally independent solutions of (2.2.238) $G_1 = x$ and $G_2 = t$. As $\alpha_k^i = 0$ and $\gamma_1 = 4e^{\frac{v}{4}}$, $\gamma_2 = 0$, $\beta_1 = ue^{\frac{v}{4}}$ and $\beta_2 = 2e^{\frac{v}{4}}$, we get as a particular solution of (2.2.239)

(2.2.246) $$F^{(1)} = -e^{-\frac{v}{4}}, \quad F^{(2)} = \frac{1}{2}ue^{-\frac{v}{4}}.$$

Eq. (2.2.237) implies $\sigma = -e^{-\frac{v}{4}}$ and $\sigma_x = \frac{1}{2}ue^{-\frac{v}{4}}$ and from it we obtain as a linearizing transformation the Cole–Hopf transformation

$$u = -2\frac{\sigma_x}{\sigma}.$$

3. Integrability of DΔEs

3.1. Introduction. We have described the integrability procedure and the construction of the infinite dimensional symmetry algebra in the case of PDEs, where it was firstly introduced. This procedure has been extended to the case of DΔEs and PΔEs [6, 7, 27, 172, 214, 256, 257, 572, 586, 636, 640, 752, 755]. Not in all cases we will present the same level of details as we did for PDEs. More results can be found in the literature or can be left to the reader to implement.

In the case of an integrable DΔEs of order k, i.e. such that it depends on k shifted points in the one dimensional lattice

(2.3.1) $$u_{n,t}(t) = E_k(n, t, u_n(t), u_{n+1}(t), \dots, u_{n+k}(t)),$$

the linear operators L and M that describe its Lax pair are not finite dimensional differential operators but depend on the shift operator $S_n = S$ in the discrete variable n (1.2.13). For simplicity we just wrote down here evolutionary equations in t, but (2.3.1) could have higher order t derivatives as is the case of the Toda lattice (1.4.16). The Lax equations (2.2.8, 2.2.9) are still valid. The recursion operator $\mathscr{L}$ will depend on the shift operator S, rather than on x derivatives. This implies that higher equations of the hierarchy

(2.3.2) $$u_{n,t}(t) = E_{k_j}(n, t, u_n(t), u_{n+1}(t), \dots, u_{n+k_j}(t))$$

and higher symmetries will depend on points further away from the point n instead of depending on higher derivatives.

In the matrix formalism, in the linear equations the vector $\psi(x, t; \lambda)$ goes into the vector $\psi_n(t; \lambda)$. Eq. (2.2.13) becomes a matrix equation with an evolution in n and (2.2.13, 2.2.14) become:

(2.3.3) $$\psi_{n+1}(t; \lambda) = U(\{u_n(t)\}, \lambda)\, \psi_n(t; \lambda) = U_n^{(\lambda)}\, \psi_n(t; \lambda),$$

(2.3.4) $$\psi_{n,t}(t; \lambda) = V(\{u_n(t)\}, \lambda)\, \psi_n(t; \lambda) = V_n^{(\lambda)}\, \psi_n(t; \lambda).$$

In (2.3.3, 2.3.4) $U(\{u_n(t)\})$ and $V(\{u_n(t)\})$ are matrix functions and by the set $\{u_n(t)\}$ we mean u_n, its shifted values like, for example, $u_{n-1}(t)$ or $u_{n+1}(t)$ and possibly its t derivative if the DΔE depends on higher derivatives. The compatibility of (2.3.3) and (2.3.4) is given by

(2.3.5) $$D_t \psi_{n+1}(t, \lambda) = S\psi_{n,t}(t, \lambda)$$

and provides the DΔE

(2.3.6) $$U_{n,t}^{(\lambda)} + U_n^{(\lambda)} V_n^{(\lambda)} = V_{n+1}^{(\lambda)} U_n^{(\lambda)},$$

which is to be valid for any λ as was (2.2.15). By $V_{n+1}^{(\lambda)} = V(\{u_{n+1}(t)\}, \lambda)$ we mean $V_n^{(\lambda)}$ where $\{u_n(t)\}$ is substituted by the set $\{u_{n+1}(t)\}$ i.e $\{u_n(t)\}$ is shifted up by one, n is substituted everywhere by $n+1$. It is worthwhile to observe here, something that is not evident in the case of the Lax pair as discussed at the beginning of this Section. By the identification

$$u_n(t) = u_1(x,t), \qquad u_{n+1}(t) = u_2(x,t), \tag{2.3.7}$$
$$\psi_n(t;\lambda) = \psi_1(x,t;\lambda), \qquad \psi_{n+1}(t;\lambda) = \psi_2(x,t;\lambda)$$

(2.3.4, 2.3.6, 2.3.7) turn out to be identical to (2.2.59, 2.2.6, 2.2.61). So using Bäcklund transformations we can construct, as we have already shown in a previous Section, integrable DΔEs [471, 480, 750, 751, 812]. This concept is at the base of the Lax-Darboux scheme of integrable equations proposed by Mikhailov [451, 605] starting from [172, 471, 480, 812].

In the following we will see that the same construction we have carried out in the case of the KdV hierarchy can be carried out for the Toda, Volterra, discrete Nonlinear Schrödinger equations (dNLS) and DΔE Burgers equation, four of the best well known integrable DΔE associated to the discrete Schrödinger spectral problem [156–159, 257], to the discrete AKNS spectral problem [5, 9, 12] and linearizable [502].

3.2. The Toda lattice, the Toda system, the Toda hierarchy and their symmetries. The Toda lattice (1.4.16) was the first DΔE on the lattice shown to be integrable in a sequence of pioneering papers by Toda [798], derived immediately after the works of Gard ner, Green, Kruskal and Miura on the integration of the KdV (see also [796, 797] and [707] for applications). The *Toda lattice* (1.4.16) has been rewritten in the form of a system by Flaschka [256, 257]

$$\dot{a}_n = a_n(b_n - b_{n+1}), \qquad \dot{b}_n = a_{n-1} - a_n, \qquad a_n = a_n(t),\ b_n = b_n(t), \tag{2.3.8}$$

where

$$b_n = \dot{v}_n, \quad a_n = e^{v_n - v_{n+1}}. \tag{2.3.9}$$

Eq. (2.3.8) is called the *Toda system*. The Lax pair of the Toda system (2.3.8) is given by the discrete Schrödinger spectral problem [156, 158, 159, 586]

$$L\psi(n,t;\lambda) = \left[S^{-1} + b_n + a_n\,S\right]\psi(n,t;\lambda) = \lambda\psi(n,t;\lambda), \tag{2.3.10}$$

where λ is a t independent spectral parameter and the time evolution of the wave function $\psi(n,t;\lambda)$ is given by

$$\psi_t(n,t;\lambda) = -M\psi(n,t;\lambda) = -a_n S\psi(n,t;\lambda) = -a_n\,\psi(n+1,t;\lambda). \tag{2.3.11}$$

The Lie point symmetries of the Toda equation have been studied in Section 1.4.1.3. The discrete spectral problem (2.3.10) has been introduced and studied by Case and Kac [156–159] and subsequently for the solution of the Toda system (1.4.16) by Flaschka, Toda and Manakov [256, 257, 586, 796]. Let us impose the following boundary conditions on the fields a_n and b_n

$$\lim_{|n|\to\infty} a_n - 1 = \lim_{|n|\to\infty} b_n = 0. \tag{2.3.12}$$

Then by looking into solutions of (2.3.10) for the wave function with the asymptotic (in n):

$$\psi(n,t;z) \to \quad z^{-n} + R(z,t)z^n, \qquad \text{for} \quad n \to +\infty, \tag{2.3.13}$$
$$\psi(n,t;z) \to \quad T(z,t)z^{-n} \qquad \text{for} \quad n \to -\infty,$$

we can associate to (2.3.10) a spectrum $\mathcal{S}[a_n, b_n]$ [123, 256, 257, 586, 796] defined in the complex plane of the variable z ($\lambda = z + z^{-1}$):

$$\mathcal{S}[a_n, b_n] = \{R(z,t),\ z \in \mathbf{C}_1; z_j, \rho_j(t),\ |z_j| < 1, j = 1, 2, \dots, N\}. \tag{2.3.14}$$

Here $R(z,t)$ is the reflection coefficient, $T(z,t)$ is the transmission coefficient, $\mathbf{C}_1$ is the unit circle in the complex z plane, z_j are isolated points inside the unit disk and ρ_j are some complex functions of t related, as in the continuous case to the residues of $R(z,t)$ at the poles z_j.

To the spectral problem (2.3.10) we can associate a set of non linear DΔEs (the Toda system hierarchy), its symmetries and Bäcklund transformations. To do so we follow the procedure introduced in Section 2.2.2 in the case of the differential Schrödinger spectral problem with differential operators substituted by shift operators. The Lax equation (2.2.8) is still valid but (2.2.43, 2.2.44) are replaced by

$$\begin{aligned} L_t &= [L, M] = P_n S + Q_n, \\ L_t &= [L, \tilde{M}] = \tilde{P}_n S + \tilde{Q}_n, \end{aligned} \tag{2.3.15}$$

and the relation between M and $\tilde{M}$ (2.2.46) is replaced by

$$\tilde{M} = LM + F_n S + G_n. \tag{2.3.16}$$

Introducing (2.3.16) into (2.3.15) we get two difference equations for the functions F_n and G_n in terms of P_n, Q_n whose solutions give

$$\begin{aligned} F_n &= F^0 + a_n \sum_{j=n+1}^{\infty} \frac{P_j}{a_j}, \\ G_n &= G^0 - \sum_{j=n}^{\infty} Q_j, \end{aligned} \tag{2.3.17}$$

where F^0 and G^0 are constants. Eq. (2.3.17) provide us with the discrete equivalent of the recursion operator $\mathcal{L}_d$ [122, 214]

$$\mathcal{L}_d \begin{pmatrix} p_n \\ q_n \end{pmatrix} = \begin{pmatrix} p_n b_{n+1} + a_n(q_n + q_{n+1}) + (b_n - b_{n+1}) s_n \\ b_n q_n + p_n + s_{n-1} - s_n \end{pmatrix}. \tag{2.3.18}$$

The initial conditions P_n^0 and Q_n^0

$$P_n^0 = F^0 a_n (b_n - b_{n+1}), \quad Q_n^0 = F^0 (a_{n-1} - a_n). \tag{2.3.19}$$

In (2.3.18) s_n is a solution of the non homogeneous first order equation

$$s_{n+1} = \frac{a_{n+1}}{a_n}(s_n - p_n). \tag{2.3.20}$$

with boundary conditions

$$\lim_{|n| \to \infty} s_n = 0. \tag{2.3.21}$$

In conclusion the class of isospectral DΔEs associated to the discrete Schrödinger spectral problem (2.3.10) is given by

$$\begin{pmatrix} \dot{a}_n \\ \dot{b}_n \end{pmatrix} = f_1(\mathcal{L}_d, t) \begin{pmatrix} a_n(b_n - b_{n+1}) \\ a_{n-1} - a_n \end{pmatrix}, \tag{2.3.22}$$

where $f_1(\mathcal{L}_d, t)$ is an entire function of its first argument.

In a similar way, starting from (2.2.9), defining $N(a_n, b_n) = f(L(a_n, b_n), t)$ and $\tilde{N}(a_n, b_n) = L(a_n, b_n)N(a_n, b_n) + h(t)L(a_n, b_n)$ we get

$$P_n^0 = h^0 a_n[(2n+3)b_{n+1} - (2n-1)b_n], \quad Q_n^0 = h^0\{b_n^2 - 4 + 2[(n+1)a_n - (n-1)a_{n-1}]\}. \tag{2.3.23}$$

Then, the class of non isospectral deformations is

$$\begin{pmatrix} \dot{a}_n \\ \dot{b}_n \end{pmatrix} = f_1(\mathscr{L}_d, t)\begin{pmatrix} a_n(b_n - b_{n+1}) \\ a_{n-1} - a_n \end{pmatrix} + + g_1(\mathscr{L}_d, t)\begin{pmatrix} a_n[(2n+3)b_{n+1} - (2n-1)b_n] \\ b_n^2 - 4 + 2[(n+1)a_n - (n-1)a_{n-1}] \end{pmatrix}, \tag{2.3.24}$$

where $f_1(\mathscr{L}_d, t)$ and $g_1(\mathscr{L}_d, t) \neq 0$ are entire functions of their first argument.

For any equation of the hierarchy (2.3.24) we can write down an explicit evolution equation for the function $\psi(n, t; \lambda)$ [123, 127, 129] with

$$\dot{\lambda} = g_1(\lambda, t)\mu^2 \quad \mu = z^{-1} - z. \tag{2.3.25}$$

When a_n, b_n and s_n satisfy the boundary conditions (2.3.12, 2.3.21), the spectrum $S[a_n, b_n]$ defines the potentials in a unique way [257]. Thus, there is a one-to-one correspondence between the evolution of the potentials (a_n,b_n) of the discrete Schrödinger spectral problem (2.3.10), given by (2.3.24) and that of the reflection coefficient $R(z, t)$, given by

$$\frac{dR(z,t)}{dt} = \mu f_1(\lambda, t)R(z, t), \tag{2.3.26}$$

were, as before, $\frac{d}{dt}$ denotes the total derivative with respect to t. The boundedness of the solutions of (2.3.12) is necessary to get a hierarchy of non linear DΔEs with well defined evolution of the spectrum. As in the continuous case, $g_1 = 0$ corresponds to an *isospectral hierarchy* while the case when $g_1 \neq 0$ corresponds to a *non isospectral hierarchy*.

The Toda system is obtained from (2.3.24) by choosing $f_1(\lambda, t) = 1$, $g_1(\lambda, t) = 0$, i.e. it is an isospectral non linear DΔE and the evolution of the reflection coefficient is given by

$$\frac{\partial R(z,t)}{\partial t} = \mu R(z, t). \tag{2.3.27}$$

The Toda hierarchy is given by (2.3.22), i.e. (2.3.24) when $g_1(\lambda, t) = 0$. The evolution of its reflection coefficient is

$$\frac{\partial R(z,t)}{\partial t} = \mu f_1(\lambda, t)R(z, t). \tag{2.3.28}$$

The symmetries for the Toda system (2.3.8) are provided by all flows commuting with the equation itself. Let us at first consider a denumerable set of isospectral flows given by the following equations

$$\begin{pmatrix} a_{n,\epsilon_\ell} \\ b_{n,\epsilon_\ell} \end{pmatrix} = \mathscr{L}_d^{\ell}\begin{pmatrix} a_n(b_n - b_{n+1}) \\ a_{n-1} - a_n \end{pmatrix}. \tag{2.3.29}$$

Here ℓ is any positive integer. For any value of ℓ, if (2.3.29) commutes with (2.3.8), it is a symmetry of (2.3.8) and ϵ_ℓ is its continuous group parameter. We can associate to (2.3.29) an evolution of the reflection coefficient

$$\frac{\partial R}{\partial \epsilon_\ell} = \mu\lambda^{\ell} R. \tag{2.3.30}$$

By computing the compatibility condition between (2.3.27, 2.3.30) we get

$$\frac{\partial^2 R}{\partial \epsilon_\ell \partial t} = \frac{\partial^2 R}{\partial t \partial \epsilon_\ell}. \tag{2.3.31}$$

It follows that the flows (2.3.27) and (2.3.30) commute and hence, due to the one-to-one correspondence between the evolution of the equation and the spectrum $S[a_n, b_n]$, the same must be true for the Toda system (2.3.8) and the equations (2.3.29). This implies that for any value ℓ (2.3.29) are symmetries of the Toda system and ϵ_ℓ are their group parameters. From the point of view of the spectral problem (2.3.10), (2.3.29) corresponds to isospectral deformations, as (2.3.29) is obtained from (2.3.24) by choosing $t = \epsilon_\ell$, $f_1(\lambda, \epsilon_\ell) = \lambda^\ell$, $g_1(\lambda, \epsilon_\ell) = 0$, i.e. $\lambda_{\epsilon_\ell} = 0$. For any ϵ_ℓ, the solution of the Cauchy problem for (2.3.29), provides a solution of the Toda system (2.3.8) $[a_n(t, \epsilon_\ell), b_n(t, \epsilon_\ell)]$ in terms of the initial condition $[a_n(t, \epsilon_\ell = 0), b_n(t, \epsilon_\ell = 0)]$. The group transformation corresponding to the group parameter ϵ_ℓ can usually be written explicitly only for the lowest values of ℓ, when the symmetry is a Lie point symmetry and the DΔE (2.3.29) is solvable. In the case of the generalized symmetries, when ℓ does not correspond to a Lie point symmetry, the group action cannot be obtained. We can construct just a few classes of explicit group transformations corresponding to very specific solutions of the Toda lattice equation, namely the solitons, the rational solutions and the periodic solutions [318, 796]. In all cases one can use the symmetries (2.3.29) to perform a symmetry reduction, i.e. to reduce the equation under consideration to an OΔE, or maybe a functional one (see Section 2.3.2.6). This is done by looking for fixed points of the transformation, i.e. putting $a_{n,\epsilon_\ell} = 0$, $b_{n,\epsilon_\ell} = 0$.

We can extend the class of symmetries (2.3.29) by considering, as we did in the case of the KdV hierarchy, non isospectral deformations of the spectral problem (2.3.10) [261, 282, 284, 496, 563]. For the Toda system we have

$$\begin{aligned} \begin{pmatrix} a_{n,\epsilon_\ell} \\ b_{n,\epsilon_\ell} \end{pmatrix} &= 2\mathcal{L}_d^{\ell+1} t \begin{pmatrix} a_n(b_n - b_{n+1}) \\ a_{n-1} - a_n \end{pmatrix} \\ &+ \mathcal{L}_d^{\ell} \begin{pmatrix} a_n[(2n+3)b_{n+1} - (2n-1)b_n] \\ b_n^2 - 4 + 2[(n+1)a_n - (n-1)a_{n-1}] \end{pmatrix}. \end{aligned} \tag{2.3.32}$$

In correspondence with (2.3.32) we have the following evolution of the reflection coefficient (2.3.14)

$$\frac{dR}{d\epsilon_\ell} = 2\mu\lambda^{\ell+1} t R, \qquad \lambda_{\epsilon_\ell} = \mu^2 \lambda^\ell. \tag{2.3.33}$$

The proof that (2.3.32) are symmetries is done by showing that the flows (2.3.33) and (2.3.27) in the space of the reflection coefficients commute, i.e. (2.3.31) is satisfied also in this case.

In addition to the above two hierarchies of symmetries (2.3.29) and (2.3.32), we have constructed in Section 1.4.1.3 two further symmetries, which, however, do not satisfy the asymptotic boundary conditions (2.3.12). They are:

$$\begin{aligned} \begin{pmatrix} a_{n,\epsilon} \\ b_{n,\epsilon} \end{pmatrix} &= \begin{pmatrix} 0 \\ 1 \end{pmatrix}, \\ \begin{pmatrix} a_{n,\epsilon} \\ b_{n,\epsilon} \end{pmatrix} &= t \begin{pmatrix} a_n(b_n - b_{n+1}) \\ a_{n-1} - a_n \end{pmatrix} + \begin{pmatrix} 2a_n \\ b_n \end{pmatrix} = t \begin{pmatrix} a_{n,t} \\ b_{n,t} \end{pmatrix} + \begin{pmatrix} 2a_n \\ b_n \end{pmatrix}. \end{aligned} \tag{2.3.34}$$

As these Lie point symmetries do not satisfy the asymptotic boundary conditions (2.3.12), we cannot construct a reflection and transmission coefficients and write down the corresponding evolution equations (2.3.14).

In the following we will write down explicitly the lowest order symmetries for the Toda system (2.3.8), obtained from the hierarchies (2.3.29, 2.3.32). Then, the symmetries of the Toda lattice (1.4.16) are obtained from those of the Toda system (2.3.8) by using the transformation (2.3.9). The symmetries of the Toda lattice and the Toda system, corresponding to the isospectral and non isospectral flows, will have the same evolution of the reflection coefficient. The transformation (2.3.9) involves an integration and a summation (to obtain υ_n). The integration constant must be chosen so as to satisfy the following boundary conditions:

$$\lim_{|n|\to\infty} \upsilon_n = 0. \tag{2.3.35}$$

In the case of the exceptional symmetries such integration will provide an additional symmetry.

Taking $\ell = 0, 1$, and 2 in (2.3.29) we obtain the first three isospectral symmetries for the Toda system, namely:

$$\begin{aligned} a_{n,\epsilon_0} &= a_n(b_n - b_{n+1}), \\ b_{n,\epsilon_0} &= a_{n-1} - a_n, \end{aligned} \tag{2.3.36}$$

$$\begin{aligned} a_{n,\epsilon_1} &= a_n[b_n^2 - b_{n+1}^2 + a_{n-1} - a_{n+1}], \\ b_{n,\epsilon_1} &= a_{n-1}[b_n + b_{n-1}] - a_n[b_{n+1} + b_n], \end{aligned} \tag{2.3.37}$$

$$\begin{aligned} a_{n,\epsilon_2} &= a_n[b_n^3 - b_{n+1}^3 + a_n b_n - 2a_{n+1}b_{n+1} + a_{n-1}b_{n-1} + 2a_{n-1}b_n \\ &\quad -a_{n+1}b_{n+2} - a_n b_{n+1} - 2b_n + 2b_{n+1}], \\ b_{n,\epsilon_2} &= a_{n-1}[b_n^2 + b_{n-1}^2 + b_n b_{n-1} + a_{n-1} + a_{n-2} - 2] \\ &\quad -a_n[b_{n+1}^2 + b_n^2 + b_n b_{n+1} + a_{n+1} + a_n - 2]. \end{aligned} \tag{2.3.38}$$

The lowest non isospectral symmetry is obtained from (2.3.32), taking $\ell = 0$ and, at difference from the PDEs case, we get a local DΔE. It is:

$$\begin{aligned} a_{n,\nu} &= a_n\{2t[b_n^2 - b_{n+1}^2 + a_{n-1} - a_{n+1}] + (2n+3)b_{n+1} - (2n-1)b_n\}, \\ b_{n,\nu} &= 2t\{a_{n-1}(b_n + b_{n-1}) - a_n(b_{n+1} + b_n)\} + b_n^2 - 4 \\ &\quad +2[(n+1)a_n - (n-1)a_{n-1}]. \end{aligned} \tag{2.3.39}$$

The higher non isospectral symmetries, corresponding to $\ell > 0$, are all non local.

The exceptional symmetries (2.3.34) are:

$$a_{n,\mu_0} = 0, \quad b_{n,\mu_0} = 1, \tag{2.3.40}$$

$$a_{n,\mu_1} = 2a_n + t\dot{a}_n, \quad b_{n,\mu_1} = b_n + t\dot{b}_n. \tag{2.3.41}$$

The corresponding symmetries for the Toda lattice are

$$\upsilon_{n,\epsilon_0} = \dot{\upsilon}_n \tag{2.3.42}$$

$$\upsilon_{n,\epsilon_1} = \dot{\upsilon}_n^2 + e^{\upsilon_{n-1}-\upsilon_n} + e^{\upsilon_n - \upsilon_{n+1}} - 2 \tag{2.3.43}$$

$$\upsilon_{n,\epsilon_2} = \dot{\upsilon}_n^3 - 2\dot{\upsilon}_n + e^{\upsilon_{n-1}-\upsilon_n}(\dot{\upsilon}_{n-1} + 2\dot{\upsilon}_n) + e^{\upsilon_n-\upsilon_{n+1}}(\dot{\upsilon}_{n+1} + 2\dot{\upsilon}_n) \tag{2.3.44}$$

$$\upsilon_{n,\nu} = 2t\{\dot{\upsilon}_n^2 + e^{\upsilon_{n-1}-\upsilon_n} + e^{\upsilon_n-\upsilon_{n+1}} - 2\} - (2n-1)\dot{\upsilon}_n + w_n, \tag{2.3.45}$$

where $w_n = w_n(t)$ is defined by the following compatible system of equations:

$$w_{n+1} - w_n = -2\dot{\upsilon}_{n+1}, \qquad \dot{w}_n = 2(e^{\upsilon_n - \upsilon_{n+1}} - 1). \tag{2.3.46}$$

Under the assumption (2.3.35) we can integrate (2.3.46) and obtain a formal solution. That is, we can write w_n in the form of an infinite sum;

$$w_n = 2 \sum_{j=n+1}^{\infty} \dot{v}_j + \alpha, \tag{2.3.47}$$

where α is an arbitrary integration constant which can be interpreted as an additional symmetry. The exceptional symmetries read:

$$v_{n,\mu_1} = t\dot{v}_n - 2n \tag{2.3.48}$$

$$v_{n,\mu_0} = t \tag{2.3.49}$$

and the additional one, due to the integration, is

$$v_{n,\mu_{-1}} = 1. \tag{2.3.50}$$

3.2.1. *Symmetries for the Toda hierarchy.* The Nth equation of the Toda system hierarchy is obtained from (2.3.24) by choosing $f_1(\lambda, t) = \lambda^N$ and $g_1(\lambda, t) = 0$

$$\begin{pmatrix} a_{n,\epsilon_\ell} \\ b_{n,\epsilon_\ell} \end{pmatrix} = \mathcal{L}_d^{\;N} \begin{pmatrix} a_n(b_n - b_{n+1}) \\ a_{n-1} - a_n \end{pmatrix} \tag{2.3.51}$$

and from (2.3.26) the corresponding evolution of the reflection coefficient is given by

$$\frac{\partial R(z,t)}{\partial t} = \mu \lambda^N R(z,t). \tag{2.3.52}$$

As it is easy to prove the isospectral symmetries are given by (2.3.29) as (2.3.30) and (2.3.52) commute. The non isospectral symmetries are, however, not given by (2.3.32) as the result depends on the equation in the hierarchy we are considering. In this case (2.3.32) reads:

$$\begin{aligned} \begin{pmatrix} a_{n,\epsilon_\ell} \\ b_{n,\epsilon_\ell} \end{pmatrix} &= 2\mathcal{L}_d^{\;\ell}\, t\, [\mathcal{L}_d^{\;N+1}(1+N) - 4N\mathcal{L}_d^{\;N-1}] \begin{pmatrix} a_n(b_n - b_{n+1}) \\ a_{n-1} - a_n \end{pmatrix} \\ &+ \mathcal{L}_d^{\;\ell} \begin{pmatrix} a_n[(2n+3)b_{n+1} - (2n-1)b_n] \\ b_n^2 - 4 + 2[(n+1)a_n - (n-1)a_{n-1}] \end{pmatrix}. \end{aligned} \tag{2.3.53}$$

and the reflection coefficient evolves according to

$$\frac{dR}{d\epsilon_\ell} = 2\mu\lambda^{N+\ell}\, t\, [\lambda(1+N) - 4\frac{N}{\lambda}]R, \qquad \lambda_{\epsilon_\ell} = \mu^2 \lambda^\ell. \tag{2.3.54}$$

As in the case of the Toda system the non isospectral symmetry corresponding to $\ell = 0$ is local and it corresponds in the continuous limit to a dilation while the higher ones are non local. There might also in this case be some exceptional symmetries which do not satisfy the boundary conditions and thus do not have a spectral transform. When we consider the higher Toda given by (2.3.51) with $N = 1$ only an exceptional symmetry is obtained,

$$\begin{pmatrix} a_{n,\epsilon} \\ b_{n,\epsilon} \end{pmatrix} = t \begin{pmatrix} a_n(b_n - b_{n+1}) \\ a_{n-1} - a_n \end{pmatrix} + \begin{pmatrix} a_n \\ \frac{1}{2} b_n \end{pmatrix} = t \begin{pmatrix} a_{n,t} \\ b_{n,t} \end{pmatrix} + \begin{pmatrix} a_n \\ \frac{1}{2} b_n \end{pmatrix}. \tag{2.3.55}$$

When we take $N \geq 2$ no exceptional symmetries are obtained.

3.2.2. *The Lie algebra of the symmetries for the Toda system and Toda lattice.* To define the structure of the symmetry algebra for the Toda lattice we need to compute the commutation relations between the symmetries, as we did for KdV. Using the one-to-one correspondence between the integrable equations and the evolution equations for the reflection coefficients, we calculate the commutation relations between the symmetries and thus analyze the structure of the obtained infinite dimensional Lie algebra.

If we define

$$\mathscr{L}_d^{\ell} = \begin{pmatrix} \mathscr{L}_{d\,11}^{(\ell)} & \mathscr{L}_{d\,12}^{(\ell)} \\ \mathscr{L}_{d\,21}^{(\ell)} & \mathscr{L}_{d\,22}^{(\ell)} \end{pmatrix}, \tag{2.3.56}$$

we can write the generators for the isospectral symmetries as

$$\begin{aligned} \hat{X}_\ell^T = \; & \{\mathscr{L}_{d\,11}^{(\ell)}[a_n(b_n - b_{n+1})] + \mathscr{L}_{d\,12}^{(\ell)}(a_{n-1} - a_n)\}\partial_{a_n} \\ & + \{\mathscr{L}_{d\,21}^{(\ell)}[a_n(b_n - b_{n+1})] + \mathscr{L}_{d\,22}^{(\ell)}(a_{n-1} - a_n)\}\partial_{b_n}. \end{aligned} \tag{2.3.57}$$

The superscript T on the generator $\hat{X}$, is there to indicate that this is the symmetry generator for the Toda system (2.3.8). To these generators we can associate symmetry generators in the space of the reflection coefficients. These generators are written as

$$\hat{\mathscr{X}}_\ell^T = \mu\lambda^\ell R\partial_R. \tag{2.3.58}$$

In agreement with Lie theory, whenever R is an analytic function of ϵ_ℓ, the corresponding flows are given by solving the equations

$$\frac{d\tilde{R}}{d\epsilon_\ell} = \mu\lambda^\ell \tilde{R}, \qquad \frac{d\tilde{\lambda}}{d\epsilon_\ell} = 0, \qquad \tilde{R}(\epsilon_\ell = 0) = R, \qquad \tilde{\lambda}(\epsilon_\ell = 0) = \lambda. \tag{2.3.59}$$

By computing the corresponding commutation relation in the space of the reflection coefficient

$$[\hat{\mathscr{X}}_\ell^T, \hat{\mathscr{X}}_m^T] = [\mu\lambda^\ell R\partial_R, \mu\lambda^m R\partial_R] = 0, \tag{2.3.60}$$

one can prove that the isospectral symmetry generators (2.3.57) commute among themselves

$$[\hat{X}_\ell^T, \hat{X}_m^T] = 0. \tag{2.3.61}$$

So far, the use of the vector fields in the reflection coefficient space allows us to prove that the symmetries given by the isospectral flows commute. We now extend the use of vector fields in the space of the spectral data to the case of the non isospectral symmetries (2.3.32). Using the definition (2.3.56) we can introduce symmetry generators for the Toda system. They are:

$$\begin{aligned} \hat{Y}_\ell^T = \; & \{t[\mathscr{L}_{d\,11}^{(\ell+1)}[a_n(b_n - b_{n+1})] + \mathscr{L}_{d\,12}^{(\ell+1)}(a_{n-1} - a_n)] \\ & + \mathscr{L}_{d\,11}^{(\ell)}[a_n((2n+3)b_{n+1} - (2n-1)b_n)] \\ & + \mathscr{L}_{d\,12}^{(\ell)}[b_n^2 - 4 + 2(n+1)a_n - 2(n-1)a_{n-1}]\}\partial_{a_n} \\ & + \{t[\mathscr{L}_{d\,21}^{(\ell+1)}[a_n(b_n - b_{n+1})] + \mathscr{L}_{d\,22}^{(\ell+1)}(a_{n-1} - a_n)] \\ & + \mathscr{L}_{d\,21}^{(\ell)}[a_n((2n+3)b_{n+1} - (2n-1)b_n)] \\ & + \mathscr{L}_{d\,22}^{(\ell)}[b_n^2 - 4 + 2(n+1)a_n - 2(n-1)a_{n-1}]\}\partial_{b_n}. \end{aligned} \tag{2.3.62}$$

Taking into account (2.3.33), we can define the symmetry generators (2.3.62) in the space of the spectral data, as

$$\hat{\mathcal{Y}}_\ell^T = \mu\lambda^{\ell+1} t R\partial_R + \mu^2\lambda^\ell\partial_\lambda. \tag{2.3.63}$$

Commuting $\hat{\mathcal{Y}}_\ell^T$ with $\hat{\mathcal{Y}}_m^T$ we have:

$$[\hat{\mathcal{Y}}_\ell^T, \hat{\mathcal{Y}}_m^T] = (m-\ell)[\hat{\mathcal{Y}}_{\ell+m+1}^T - 4\hat{\mathcal{Y}}_{\ell+m-1}^T]. \tag{2.3.64}$$

From the relation between the reflection coefficients space and the space of the solutions, we conclude that the vector fields representing the non isospectral symmetries satisfy the commutation relations

$$[\hat{Y}_\ell^T, \hat{Y}_m^T] = (m-\ell)[\hat{Y}_{\ell+m+1}^T - 4\hat{Y}_{\ell+m-1}^T], \tag{2.3.65}$$

In a similar manner we can work out the commutation relations between the $\hat{Y}_\ell$ and $\hat{X}_m$ symmetry generators. We get:

$$[\hat{\mathcal{X}}_m^T, \hat{\mathcal{Y}}_\ell^T] = -(1+m)\hat{\mathcal{X}}_{\ell+m+1}^T + 4m\hat{\mathcal{X}}_{\ell+m-1}^T, \tag{2.3.66}$$

and consequently

$$[\hat{X}_m^T, \hat{Y}_\ell^T] = -(1+m)\hat{X}_{\ell+m+1}^T + 4m\hat{X}_{\ell+m-1}^T. \tag{2.3.67}$$

Relations like (2.3.65) and (2.3.67) can also be checked directly, but the use of the vector field in the reflection coefficient space is much more efficient.

As in the case of the KdV, if we take $\ell = 0$, the commutation relation (2.3.36) tell us that the commutator of $\hat{Y}_0^T$ with the generator of an isospectral symmetry provide a higher isospectral symmetry. In particular starting from the Lie symmetry $\hat{X}_0^T$ we get the generator of any higher isospectral generalized symmetry. Thus $\hat{Y}_0^T$ is a *master symmetry* for the Toda system and as in the case of Burgers, but opposed to the KdV case, it is local. The master symmetry will also be discussed in Section 3.2.7 together with discrete Miura transformations.

Let us now consider the commutation relations involving the exceptional symmetries (2.3.34). We write them as:

$$\hat{Z}_0^T = \partial_{b_n} \tag{2.3.68}$$

$$\hat{Z}_1^T = [2a_n + t\dot{a}_n]\partial_{a_n} + [b_n + t\dot{b}_n]\partial_{b_n}. \tag{2.3.69}$$

As mentioned above, the exceptional symmetries do not satisfy the asymptotic conditions (2.3.12). Hence we cannot write down the commutation relations in all generality for all symmetries simultaneously. We calculate explicitly the commutation relations involving, for example, $\hat{Z}_0^T$, $\hat{Z}_1^T$, $\hat{X}_0^T$, $\hat{X}_1^T$ and $\hat{Y}_0^T$. The non zero commutation relations are:

$$\begin{aligned}
&[\hat{X}_0^T, \hat{Z}_1^T] = -\hat{X}_0^T, \quad [\hat{Z}_0^T, \hat{Z}_1^T] = \hat{Z}_0^T \\
&[\hat{Y}_0^T, \hat{Z}_0^T] = -2\hat{Z}_1^T, \quad [\hat{Y}_0^T, \hat{Z}_1^T] = -\hat{Y}_0^T - 8\hat{Z}_0^T, \\
&[\hat{X}_1^T, \hat{Z}_0^T] = -2\hat{X}_0^T, \quad [\hat{X}_1^T, \hat{Z}_1^T] = -2\hat{X}_1^T, \\
&[\hat{X}_0^T, \hat{Y}_0^T] = -\hat{X}_1^T, \quad [\hat{X}_1^T, \hat{Y}_0^T] = -2\hat{X}_2^T + 4\hat{X}_0^T.
\end{aligned} \tag{2.3.70}$$

We will indicate by the superscript TL the symmetry generators for the Toda lattice(1.4.16). We have (see (2.3.49, 2.3.48))

$$\hat{Z}_0^{TL} = t\partial_{\upsilon_n} \tag{2.3.71}$$

$$\hat{Z}_1^{TL} = [t\dot{\upsilon}_n - 2n]\partial_{\upsilon_n} \tag{2.3.72}$$

and

$$\hat{Z}_{-1}^{TL} = \partial_{v_n} \tag{2.3.73}$$

in correspondence with (2.3.50). As (1.4.16) and (2.3.8) are just two different representations of the same system, the symmetry generators in the space of the spectral data are the same. Consequently the commutation relations between $\hat{X}_n^{TL}$ and $\hat{Y}_m^{TL}$ are given by (2.3.60, 2.3.65 and 2.3.67). The symmetries $\hat{X}_0^{TL}$, $\hat{X}_1^{TL}$ and $\hat{Y}_0^{TL}$, according to (2.3.42, 2.3.43, 2.3.45) are given by:

$$\begin{aligned} \hat{X}_0^{TL} &= \dot{v}_n\partial_{v_n}, \quad \hat{X}_1^{TL} = \left[\dot{v}_n^2 + e^{v_{n-1}-v_n} + e^{v_n-v_{n+1}} - 2\right]\partial_{v_n} \\ \hat{Y}_0^{TL} &= \left\{t[v_{n,t}^2 + e^{v_{n-1}-v_n} + e^{v_n-v_{n+1}} - 2] - (2n-1)v_{n,t} + w_n(t)\right\}\partial_{v_n} \\ & w_{n+1}(t) - w_n(t) = -2\dot{v}_{n+1}, \quad \dot{w}_n(t) = 2(e^{v_n-v_{n+1}} - 1). \end{aligned} \tag{2.3.74}$$

The nonzero commutation relations are:

$$\begin{aligned} &[\hat{X}_0^{TL}, \hat{Z}_0^{TL}] = -\hat{Z}_{-1}^{TL}, \quad [\hat{X}_0^{TL}, \hat{Z}_1^{TL}] = -\hat{X}_0^{TL}, \\ &[\hat{X}_0^{TL}, \hat{Y}_0^{TL}] = -\hat{X}_1^{TL} + \omega\hat{Z}_{-1}^{TL}, \\ &[\hat{X}_1^{TL}, \hat{Z}_0^{TL}] = -2\hat{X}_0^{TL}, \quad [\hat{X}_1^{TL}, \hat{Z}_1^{TL}] = -2\hat{X}_1^{TL} - 4\hat{Z}_{-1}^{TL}, \\ &[\hat{X}_1^{TL}, \hat{Y}_0^{TL}] = -2\hat{X}_2^{TL} + 4\hat{X}_0^{TL} + \sigma\hat{Z}_{-1}^{TL}, \\ &[\hat{Y}_0^{TL}, \hat{Z}_{-1}^{TL}] = \beta\hat{Z}_{-1}^{TL}, \quad [\hat{Y}_0^{TL}, \hat{Z}_0^{TL}] = -2\hat{Z}_1^{TL} + \gamma\hat{Z}_{-1}^{TL}, \\ &[\hat{Y}_0^{TL}, \hat{Z}_1^{TL}] = -\hat{Y}_0^{TL} - 8\hat{Z}_0^{TL} + \delta\hat{Z}_{-1}^{TL}, \quad [\hat{Z}_0^{TL}, \hat{Z}_1^{TL}] = \hat{Z}_0^{TL}, \end{aligned} \tag{2.3.75}$$

where $(\beta, \gamma, \delta, \omega, \sigma)$ are integration constants. The presence of these integration constants indicates that the symmetry algebra of the Toda lattice is not completely specified. The constants appear whenever the symmetry $\hat{Y}_0^{TL}$ is involved. The ambiguity is related to the ambiguity in the definition of $\hat{Y}_0^{TL}$ itself, i.e. in the solution of (2.3.74) for w_n. We fix these coefficients by requiring that one obtains the correct continuous limit, i.e. in the asymptotic limit, when h goes to zero and $n \to \infty$ in such a way that $hn = x$, a combination of the generators of the Toda lattice (1.4.16) and Toda system (2.3.8) goes over to the symmetry algebra of the pKdV (see Section 2.3.2.3).

The commutation relations obtained above determine the structure of the infinite dimensional Lie symmetry algebra. The first symmetry generators are given in (2.3.57), (2.3.62), (2.3.68), (2.3.69) and the corresponding commutation relations are given by (2.3.65), (2.3.67), (2.3.70). As one can see, the symmetry operators $\hat{Y}_k^T$ and $\hat{Z}_k^T$ are linear in t and the coefficient of t is an isospectral symmetry operator $\hat{X}_k^T$. Consequently, as the operators $\hat{X}_k^T$ commute amongst each other, the commutator of $\hat{X}_m^T$ with any of the $\hat{Y}_k^T$ or $\hat{Z}_k^T$ symmetries will not have any explicit time dependence and thus can be written in terms of $\hat{X}_n^T$ only. Thus the structure of the Lie algebra for the Toda system can be written as:

$$L = L_0 \ni L_1, \quad L_0 = \{\hat{h}, \hat{e}, \hat{f}, \hat{Y}_1^T, \hat{Y}_2^T, \cdots\}, L_1 = \{\hat{X}_0^T, \hat{X}_1^T, \cdots\} \tag{2.3.76}$$

where $\{\hat{h} = \hat{Z}_1^T, \hat{e} = \hat{Z}_0^T, \hat{f} = \hat{Y}_0^T + 4\hat{Z}_0^T\}$ denotes a $sl(2, \mathbb{R})$ subalgebra with $[\hat{h}, \hat{e}] = \hat{e}$, $[\hat{h}, \hat{f}] = -\hat{f}$, $[\hat{e}, \hat{f}] = 2\hat{h}$. The algebra L_0 is perfect, i.e. we have $[L_0, L_0] = L_0$. It is worthwhile to notice that $\hat{Z}_0^T$, $\hat{Z}_1^T$ and $\hat{X}_0^T$ are point symmetries while all the others are generalized symmetries. Indeed, all the other vector fields involve other values of the discrete variable than n or time derivatives of the fields.

For the Toda lattice the point symmetries are $\hat{X}_0^{TL}$, $\hat{Z}_0^{TL}$ and $\hat{Z}_1^{TL}$, as for the Toda system, plus the additional $\hat{Z}_{-1}^{TL}$. Taking into account (2.3.71–2.3.75), the structure of the

Lie algebra is the same as that of the Toda system with $L_0 = \{\hat{Z}_{-1}^{TL}, \hat{Z}_0^{TL}, \hat{Z}_1^{TL}, \hat{Y}_0^{TL}, \hat{Y}_1^{TL}, \hat{Y}_2^{TL}, \cdots\}$, $L_1 = \{\hat{X}_0^{TL}, \hat{X}_1^{TL}, \hat{X}_2^{TL}, \cdots\}$.

3.2.3. *Contraction of the symmetry algebras in the continuous limit.* It is well known [123, 129, 506, 507, 796], that the Toda lattice has the pKdV (2.2.74) as one of its possible continuous limits. In fact, by setting

$$v_n(t) = -\frac{1}{2}h\,u(x,\tau) \quad x = (n-t)h \quad \tau = -\frac{1}{24}h^3 t \tag{2.3.77}$$

we can write (1.4.16) as

$$(u_\tau - u_{xxx} - 3u_x^2)_x = \mathcal{O}(h^2) \tag{2.3.78}$$

i.e. the once differentiated pKdV. Let us now rewrite the symmetry generators in the new coordinate system defined by (2.3.77) and develop them for small h in Taylor series. We have:

$$\hat{X}_0^{TL} = \big\{-u_x(x,\tau)h - \frac{1}{24}u_\tau(x,\tau)h^3\big\}\partial_u \tag{2.3.79}$$

$$\hat{X}_1^{TL} = \big\{-2u_x(x,\tau)h - \frac{1}{3}u_\tau(x,\tau)h^3 + \mathcal{O}(h^5)\big\}\partial_u \tag{2.3.80}$$

$$\hat{X}_2^{TL} = \big\{-4u_x(x,\tau)h - \frac{7}{6}u_\tau(x,\tau)h^3 + \mathcal{O}(h^5)\big\}\partial_u \tag{2.3.81}$$

$$\hat{Y}_0^{TL} = \big\{2[u(x,\tau) + xu_x(x,\tau) + 3\tau u_\tau(x,\tau)] + \mathcal{O}(h)\big\}\partial_u \tag{2.3.82}$$

$$\hat{Z}_{-1}^{TL} = -\frac{2}{h}\partial_u, \qquad \hat{Z}_0^{TL} = \frac{48}{h^4}\tau\partial_u \tag{2.3.83}$$

$$\hat{Z}_1^{TL} = \big\{-\frac{96}{h^4}\tau + \frac{4}{h^2}[x + 6\tau u_x(x,\tau)] + \mathcal{O}(1)\big\}\partial_u. \tag{2.3.84}$$

Eqs. (2.3.80–2.3.82) are obtained using the evolution for u given by the pKdV (2.2.74).

The point symmetry generators written in the evolutionary form, for the pKdV (2.3.78) read:

$$\hat{P}_0 = u_\tau\partial_u, \quad \hat{P}_1 = u_x\partial_u, \quad \hat{B} = [x + 6\tau u_x]\partial_u, \tag{2.3.85}$$

$$\hat{D} = [u + xu_x + 3\tau u_\tau]\partial_u, \quad \hat{\Gamma} = \partial_u, \tag{2.3.86}$$

and their commutation table is:

(2.3.87)

	$\hat{P}_0$	$\hat{P}_1$	$\hat{B}$	$\hat{D}$	$\hat{\Gamma}$
$\hat{P}_0$	0	0	$-6\hat{P}_1$	$-3\hat{P}_0$	0
$\hat{P}_1$	0	0	$-\hat{\Gamma}$	$-\hat{P}_1$	0
$\hat{B}$	$6\hat{P}_1$	$\hat{\Gamma}$	0	$2\hat{B}$	0
$\hat{D}$	$3\hat{P}_0$	$\hat{P}_1$	$-2\hat{B}$	0	$-\hat{\Gamma}$
$\hat{\Gamma}$	0	0	0	$\hat{\Gamma}$	0

We can write down a linear combination of the generators of the Toda equation (2.3.79–2.3.84), so that in the continuous limit $h \to 0$ it goes over to the generators of the point symmetries of the pKdV (2.3.85, 2.3.86):

$$\tilde{P}_0 = \frac{4}{h^3}(2\hat{X}_0^{TL} - \hat{X}_1^{TL}), \quad \tilde{P}_1 = -\frac{1}{h}\hat{X}_0^{TL}, \quad \tilde{D} = \frac{1}{2}\hat{Y}_0^{TL}, \tag{2.3.88}$$

$$\tilde{B} = \frac{h^2}{4}(2\hat{Z}_0^{TL} + \hat{Z}_1^{TL}), \quad \tilde{\Gamma} = -\frac{h}{2}\hat{Z}_{-1}^{TL}. \tag{2.3.89}$$

Taking into account the commutation table among the generators $\hat{X}_0^{TL}$, $\hat{X}_1^{TL}$, $\hat{Z}_{-1}^{TL}$, $\hat{Z}_0^{TL}$, $\hat{Z}_1^{TL}$ and $\hat{Y}_0^{TL}$, given by (2.3.75) and the continuous limit of $\hat{X}_2^{TL}$ given by (2.3.81), we get:

(2.3.90)

	$\tilde{P}_0$	$\tilde{P}_1$	$\tilde{B}$	$\tilde{D}$	$\tilde{\Gamma}$
$\tilde{P}_0$	0	0	$-6\tilde{P}_1+\mathcal{O}(h^2)$	$-3\tilde{P}_0+\mathcal{O}(h^2)$	0
$\tilde{P}_1$	0	0	$-\tilde{\Gamma}+\mathcal{O}(h^2)$	$-\tilde{P}_1+\mathcal{O}(h^2)$	0
$\tilde{B}$	$6\tilde{P}_1-\mathcal{O}(h^2)$	$\tilde{\Gamma}-\mathcal{O}(h^2)$	0	$2\tilde{B}+\mathcal{O}(h^2)$	0
$\tilde{D}$	$3\tilde{P}_0-\mathcal{O}(h^2)$	$\tilde{P}_1-\mathcal{O}(h^2)$	$-2\tilde{B}-\mathcal{O}(h^2)$	0	$-\tilde{\Gamma}$
$\tilde{\Gamma}$	0	0	0	$\tilde{\Gamma}$	0

Table (2.3.90) is obtained by setting $\beta=-2$, $2\gamma+\delta=0$, and $\omega=\sigma=0$ in (2.3.75). Thus we have reobtained in the continuous limit $h\to 0$, all point symmetries of the pKdV. To do so we used not only the point symmetries $\hat{X}_0^{TL}$, $\hat{Z}_0^{TL}$, $\hat{Z}_{-1}^{TL}$ and $\hat{Z}_1^{TL}$ of the Toda lattice, but also the higher symmetries $\hat{X}_1^{TL}$, $\hat{Y}_0^{TL}$.

This procedure can be viewed as a new application of the concept of Lie algebra contraction. Lie algebra contraction were first introduced by Inönü and Wigner [411] in order to relate the group theoretical foundations of relativistic and nonrelativistic physics. The speed of light c was introduced as a parameter into the commutation relations of the Lorentz group. For $c\to\infty$ the Lorentz group "contracted" to the Galilei group. Lie algebra contraction thus relate different Lie algebras of the same dimension, but of different isomorphism classes. A systematic study of contractions, relating large families of nonisomorphic Lie algebras of the same dimension, based on Lie algebra grading, was initiated by Moody and Patera [620].

In general Lie algebra and Lie group contractions are extremely useful when describing the mathematical relation between different theories. The contraction parameter can be the Planck constant, when relating quantum systems to classical ones. It can be the curvature k of a space of constant curvature, which for $k\to 0$ goes to a flat space. The contraction will then relate special functions defined e.g. on spheres, to those defined in a Euclidean space [421].

In our case the contraction parameter is the lattice spacing h. Some novel features appear. First of all, we are contracting an infinite dimensional Lie algebra of generalized symmetries, that of the Toda lattice. The contraction leads to an infinite dimensional Lie algebra, not isomorphic to the first one. This "target algebra" is the Lie algebra of point and generalized symmetries of the pKdV. A particularly interesting feature is that the five dimensional Lie algebra of point symmetries of the pKdV is obtained from a subset of point and generalized symmetries of the Toda lattice. This 5 dimensional subset is not an algebra (it is not closed under commutations). It does contract into a Lie algebra in the continuous limit.

3.2.4. *Bäcklund transformations and Bianchi identities for the Toda system and Toda lattice.* In addition to the symmetry transformations presented in Section (2.3.2.2), the Toda system admits Bäcklund transformations [124, 147, 167, 214, 372, 396, 397, 471, 480, 497, 501].

In the discrete case, the technique is basically the same as for the PDEs, and the only difference is that all operators are written down in terms of the shift operators S. Taking into account (2.3.10) and (2.3.12) we can define a new solution of the Toda system hierarchy $\tilde{a}_n$ and $\tilde{b}_n$ having the same boundary conditions (2.3.12) and associated to the spectral problem

(2.3.91)
$$\tilde{L}\tilde{\psi}=(S^{-1}+\tilde{b}_n+\tilde{a}_nS)\tilde{\psi}=\lambda\tilde{\psi}.$$

We assume that ψ and $\tilde{\psi}$ are related by a Darboux operator

$$\tilde{\psi} = \hat{D}\psi \tag{2.3.92}$$

From (2.3.10), (2.3.91) and (2.3.92) we get that a Bäcklund transformation for the Toda system is given [see also (2.2.60)] by

$$\tilde{L}\hat{D} = \hat{D}L. \tag{2.3.93}$$

The calculation of the Bäcklund recursion operator and the hierarchy of Bäcklund transformations can be done in the same way, *mutatis mutandis* as it has been done in Section 2.2.2.2 for the KdV. To do so we define in this Chapter a new Darboux operator $\tilde{\hat{D}}$ such that

$$\tilde{L}\hat{D} - \hat{D}L = \sigma_n S + w_n, \quad \tilde{L}\tilde{\hat{D}} - \tilde{\hat{D}}L = \tilde{\sigma}_n S + \tilde{w}_n, \tag{2.3.94}$$

with

$$\tilde{\hat{D}} = \tilde{L}\hat{D} + f_n S + g_n. \tag{2.3.95}$$

The Bäcklund transformation is obtained by setting σ_n, w_n, $\tilde{\sigma}_n$ and $\tilde{w}_n$ equal to zero.

Introducing (2.3.95) into (2.3.94) and collecting terms containing the same power of the shift operator we get the system of equations

$$\begin{pmatrix} \tilde{\sigma}_n \\ \tilde{w}_n \end{pmatrix} = \Lambda_d \begin{pmatrix} \omega_n \\ w_n \end{pmatrix} + \begin{pmatrix} \sigma_n^0 \\ w_n^0 \end{pmatrix} \tag{2.3.96}$$

and

$$f_n = \tilde{\Pi}_n \Big(\sum_{j=n+1}^{\infty} \tilde{\Pi}_j^{-1} \sigma_j \Pi_{j+1} + f^0 \Big) \Pi_{n+1}^{-1}, \quad g_n = g^0 - \sum_{j=n}^{\infty} w_j, \tag{2.3.97}$$

$$\sigma_n^0 = f^0 \tilde{\Pi}_n \Pi_{n+1}^{-1} \left(\tilde{b}_n - b_{n+1} \right), \quad w_n^0 = g^0 \left(\tilde{a}_n - a_n \right). \tag{2.3.98}$$

The functions Π_n and $\tilde{\Pi}_n$ are given by

$$\Pi_n = \prod_{j=n}^{\infty} a_j, \quad \tilde{\Pi}_n = \prod_{j=n}^{\infty} \tilde{a}_j. \tag{2.3.99}$$

Λ_d is the recursion operator for the Bäcklund transformations associated to the discrete Schrödinger spectral problem

$$\Lambda_d \begin{bmatrix} p(n) \\ q(n) \end{bmatrix} = \begin{bmatrix} p(n)b_{n+1} + \tilde{a}_n[q(n) + q(n+1)] + \Sigma_n[\tilde{b}_n - b_{n+1}] \\ \qquad + [a_n - \tilde{a}_n] \sum_{j=n}^{\infty} p(j) \\ p(n) + \tilde{b}_n q(n) - \Sigma_n + \Sigma_{n-1} + [b_n - \tilde{b}_n] \sum_{j=n}^{\infty} q(j) \end{bmatrix} \tag{2.3.100}$$

and

$$\Sigma_n = \tilde{\Pi}_n \left[\sum_{j=n}^{\infty} \tilde{\Pi}_j^{-1} p(j) \Pi_{j+1} \right] \Pi_{n+1}^{-1}. \tag{2.3.101}$$

Then the class of Bäcklund transformations associated to the Toda system (2.3.8) is given by [128]

$$\gamma(\Lambda_d) \begin{pmatrix} \tilde{a}_n - a_n \\ \tilde{b}_n - b_n \end{pmatrix} = \delta(\Lambda_d) \begin{pmatrix} \tilde{\Pi}_n \Pi_{n+1}^{-1} (\tilde{b}_n - b_{n+1}) \\ \tilde{\Pi}_{n-1} \Pi_n^{-1} - \tilde{\Pi}_n \Pi_{n+1}^{-1} \end{pmatrix}, \tag{2.3.102}$$

where $\gamma(z)$ and $\delta(z)$ are entire functions of their argument.

In [129] it is proven that whenever (a_n, b_n) and $(\tilde{a}_n, \tilde{b}_n)$ satisfy the asymptotic conditions (2.3.12) and the Bäcklund transformations (2.3.102), the reflection coefficient satisfies the equation

$$\tilde{R}(\lambda) = \frac{\gamma(\lambda) - \delta(\lambda) z}{\gamma(\lambda) - \delta(\lambda)/z} R(\lambda). \tag{2.3.103}$$

When $\gamma(\lambda)$ and $\delta(\lambda)$ are constants and are such that $\delta = p\gamma$, the one soliton Bäcklund transformation obtained from (2.3.102) reads

$$\begin{pmatrix} \tilde{a}_n - a_n \\ \tilde{b}_n - b_n \end{pmatrix} = p \begin{pmatrix} \tilde{\Pi}_n \Pi_{n+1}^{-1} (\tilde{b}_n - b_{n+1}) \\ \tilde{\Pi}_{n-1} \Pi_n^{-1} - \tilde{\Pi}_n \Pi_{n+1}^{-1} \end{pmatrix}. \tag{2.3.104}$$

Eq. (2.3.104) is a trascendental functional relation among a_n, $\tilde{a}_n$, b_n and $\tilde{b}_n$ which can be simplified if instead of a_n and $\tilde{a}_n$ we use the dependent variables Π_n and $\tilde{\Pi}_n$ as

$$a_n = \Pi_n / \Pi_{n+1}, \qquad \tilde{a}_n = \tilde{\Pi}_n / \tilde{\Pi}_{n+1}. \tag{2.3.105}$$

For the Toda lattice (1.4.16) the one-soliton Bäcklund transformation obtained from (2.3.104) reads:

$$\dot{\tilde{\upsilon}}_n - \dot{\upsilon}_n = p \left\{ e^{\tilde{\upsilon}_{n-1} - \upsilon_n} - e^{\tilde{\upsilon}_n - \upsilon_{n+1}} \right\} \tag{2.3.106}$$

Eq. (2.3.106) is a (non linear) two point DΔE for $\tilde{\upsilon}_n$, when υ_n is a solution of (1.4.16).

Following the results presented for the Bäcklund of the KdV, the Bäcklund for the Toda lattice, and naturally also the one of the Toda system , (2.3.106) can be interpreted as a DΔE in two discrete and one continuous variables by defining

$$\upsilon_n(t) = w_{n,m}(t), \quad \tilde{\upsilon}_n(t) = w_{n,m+1}(t). \tag{2.3.107}$$

In the particular case of (2.3.106) it reads

$$\dot{w}_{n,m+1} - \dot{w}_{n,m} = p_m \left\{ e^{w_{n-1,m+1} - w_{n,m}} - e^{w_{n-1,m+1} - w_{n,m}} \right\}. \tag{2.3.108}$$

Formulas (2.3.102, 2.3.103) also provide more general transformations, i.e. higher order Bäcklund transformations. If the arbitrary functions $\gamma(\lambda)$ and $\delta(\lambda)$ are finite polynomials in λ, then we have a finite order Bäcklund transformation that can be interpreted as a composition of a finite number of one-soliton Bäcklund transformations.

From (2.3.103) one can show that the Bianchi permutability theorem is satisfied. Then the Bianchi identity for the Toda system reads

$$\begin{aligned} & p_1 \left[\frac{\Pi_n^{(1)}}{\Pi_{n+1}} (b^{(1)} - b_n) - \frac{\Pi_n^{(12)}}{\Pi_{n+1}^{(2)}} (b^{(12)} - b_n^{(2)}) \right] \\ & + p_2 \left[\frac{\Pi_n^{(12)}}{\Pi_{n+1}^{(1)}} (b^{(12)} - b_n^{(1)}) - \frac{\Pi_n^{(2)}}{\Pi_{n+1}} (b^{(2)} - b_n) \right] = 0, \end{aligned} \tag{2.3.109}$$

$$\begin{aligned} & p_1 \left[\frac{\Pi_{n-1}^{(1)}}{\Pi_n} - \frac{\Pi_n^{(1)}}{\Pi_{n+1}} - \frac{\Pi_{n-1}^{(12)}}{\Pi_n^{(2)}} + \frac{\Pi_n^{(12)}}{\Pi_{n+1}^{(2)}} \right] \\ & + p_2 \left[\frac{\Pi_{n-1}^{(12)}}{\Pi_n^{(1)}} - \frac{\Pi_n^{(12)}}{\Pi_{n+1}^{(1)}} - \frac{\Pi_{n-1}^{(2)}}{\Pi_n} + \frac{\Pi_n^{(2)}}{\Pi_{n+1}} \right] = 0. \end{aligned} \tag{2.3.110}$$

As in the case of the KdV (2.3.109, 2.3.110) relate 4 solutions of the Toda system (a_n, b_n), $(a_n^{(1)}, b_n^{(1)})$, $(a_n^{(2)}, b_n^{(2)})$, $(a_n^{(12)}, b_n^{(12)})$. In this case, Bianchi identities are a difference equation between the fields b_n and the products of the fields a_n as given in (2.3.99) in four points of the lattice. These Bianchi identities do not provide, as it was in the case of KdV, non linear superposition formulas.

In the case of the Toda lattice

$$p_1\left[e^{v^{(1)}_{n-1}-v_n}-e^{v^{(1)}_n-v_{n+1}}-e^{v^{(12)}_{n-1}-v^{(2)}_n}+e^{v^{(12)}_n-v^{(2)}_{n+1}}\right]$$
$$+p_2\left[e^{v^{(12)}_{n-1}-v^{(1)}_n}-e^{v^{(12)}_n-v^{(1)}_{n+1}}-e^{v^{(2)}_{n-1}-v_n}+e^{v^{(2)}_n-v_{n+1}}\right]=0 \tag{2.3.111}$$

The Bianchi identities (2.3.109, 2.3.110, 2.3.111) can be interpreted as three dimensional PΔEs. For example, in the case of the Toda lattice, defining

$$v_n=w_{n,m,\ell},\ v^{(1)}_n=w_{n,m+1,\ell},\ v^{(2)}_n=w_{n,m,\ell+1},\ v^{(12)}_n=w_{n,m+1,\ell+1}, \tag{2.3.112}$$

we get the 3D lattice equation

$$p_1\Big[e^{w_{n-1,m+1,\ell}-w_{n,m,\ell}}-e^{w_{n,m+1,\ell}-w_{n+1,m,\ell}}-e^{w_{n-1,m+1,\ell+1}-w_{n,m,\ell+1}}$$
$$+e^{w_{n,m+1,\ell+1}-w_{n+1,m,\ell+1}}\Big]+p_2\Big[e^{w_{n-1,m+1,\ell+1}-w_{n,m+1,\ell}}-e^{w_{n,m+1,\ell+1}-w_{n+1,m+1,\ell}}$$
$$-e^{w_{n-1,m,\ell+1}-w_{n,m,\ell}}+e^{w_{n,m,\ell+1}-w_{n+1,m,\ell}}\Big]=0, \tag{2.3.113}$$

a discrete Hirota type equation[254]. In the following Section we discuss how Bäcklund transformations are related to continuous symmetry transformations, allowing, albeit formally, an integration of an infinite number of the latter.

3.2.5. *Relation between Bäcklund transformations and isospectral symmetries.* A general isospectral higher symmetry of the Toda system is given by

$$\begin{pmatrix}a_{n,\epsilon}\\ b_{n,\epsilon}\end{pmatrix}=\phi(\mathscr{L}_d)\begin{pmatrix}a_n(b_n-b_{n+1})\\ a_{n-1}-a_n\end{pmatrix}, \tag{2.3.114}$$

where $\phi(z)$ is an entire function of its argument. The correlated spectum evolution is

$$\frac{dR(\lambda,\epsilon)}{d\epsilon}=\mu\phi(\lambda)R(\lambda,\epsilon). \tag{2.3.115}$$

These equations generalize (2.3.29), considered above. Eq. (2.3.115) can be formally integrated giving:

$$R(\lambda,\epsilon)=e^{\mu\phi(\lambda)\epsilon}R(\lambda,0). \tag{2.3.116}$$

Taking into account the already given definitions of λ and μ in terms of z

$$\lambda=\frac{1}{z}+z,\quad \mu=\frac{1}{z}-z,\quad \mu^2=\lambda^2-4, \tag{2.3.117}$$

$$z=\frac{\lambda-\mu}{2},\quad \frac{1}{z}=\frac{\lambda+\mu}{2}, \tag{2.3.118}$$

we can rewrite the general Bäcklund transformation (2.3.103) for the reflection coefficient as

$$\tilde{R}(\lambda)=\frac{2-(\lambda-\mu)\beta(\lambda)}{2-(\lambda+\mu)\beta(\lambda)}R(\lambda),\quad \beta(\lambda)=\frac{\delta(\lambda)}{\gamma(\lambda)}$$

In order to identify a general symmetry transformation with a Bäcklund transformation, and vice versa, we equate $R(\lambda,\epsilon)=\tilde{R}(\lambda)$ and $R(\lambda,0)=R(\lambda)$ and get

$$e^{\mu\phi(\lambda)\epsilon}=\frac{2-(\lambda-\mu)\beta(\lambda)}{2-(\lambda+\mu)\beta(\lambda)}. \tag{2.3.119}$$

Then $\phi(\lambda)$ in (2.3.116) is given by

$$\phi(\lambda)\epsilon = \frac{1}{\mu}\ln\left[\frac{2-(\lambda-\mu)\beta(\lambda)}{2-(\lambda+\mu)\beta(\lambda)}\right]. \tag{2.3.120}$$

The right hand side of (2.3.120) does not depend on μ. To prove this we will, in the following, analyze (2.3.119) in detail. Relations (2.3.117) allow us to separate the exponential in (2.3.119) into two entire components $E_0(\lambda)$ and $E_1(\lambda)$ as

$$e^{\mu\phi(\lambda)\epsilon} = \cosh[\mu\phi(\lambda)\epsilon] + \mu\Big(\frac{\sinh[\mu\phi(\lambda)\epsilon]}{\mu}\Big) = E_0(\lambda) + \mu E_1(\lambda). \tag{2.3.121}$$

Noticing that the rhs of (2.3.120) is an entire function of μ, developing μ^2 and identifying powers (0th and 1st) of μ, we get from (2.3.121) a system of two compatible equations

$$-(2-\lambda\beta)E_0 + (\lambda^2-4)\beta E_1 = -(2-\lambda\beta), \tag{2.3.122}$$

$$-\beta E_0 + (2-\lambda\beta)E_1 = \beta. \tag{2.3.123}$$

Eqs. (2.3.122, 2.3.123) provide us with explicit formulas relating a given general higher symmetry (characterized by ϕ, and thus E_0, E_1) with a general Bäcklund transformation (characterized by γ and δ, and thus by β):

$$\begin{aligned}\beta(\lambda) &= \frac{\delta(\lambda)}{\gamma(\lambda)} = \frac{2E_1}{E_0+\lambda E_1+1}\\ &= \frac{2\sinh[\mu\phi(\lambda)\epsilon]/\mu}{\cosh[\mu\phi(\lambda)\epsilon]+\lambda\sinh[\mu\phi(\lambda)\epsilon]/\mu+1}.\end{aligned} \tag{2.3.124}$$

From this equation we see that whatsoever be the symmetry, we find a Bäcklund transformation, i.e. for an arbitrary function ϕ we obtain the two entire functions γ and δ. Vice versa, given a general Bäcklund transformation, we can find the corresponding generalized symmetry

$$E_0 = -\frac{2(\beta^2-1)+\lambda\beta(2-\lambda\beta)}{2(\beta^2-\lambda\beta+1)}, \qquad E_1 = -\frac{(\lambda\beta-2)\beta}{2(\beta^2-\lambda\beta+1)}, \tag{2.3.125}$$

or more explicitly,

$$\phi(\lambda)\epsilon = \frac{1}{\mu}\sinh^{-1}\left[-\mu\frac{(\lambda\beta-2)\beta}{2(\beta^2-\lambda\beta+1)}\right]. \tag{2.3.126}$$

In the case of a one-soliton Bäcklund transformation with $\beta = 1$, we have:

$$E_0 = -\frac{\lambda}{2}, \quad E_1 = \frac{1}{2}.$$

and we can write $\phi(\lambda)$ as

$$\phi(\lambda)\epsilon = \frac{\sinh^{-1}\left[\sqrt{\lambda^2-4}/2\right]}{\sqrt{\lambda^2-4}}. \tag{2.3.127}$$

In this simple case we can write the symmetry in closed form as an infinite sequence of elementary symmetry transformations:

$$\phi(\lambda)\epsilon = \sum_{k=0}^{\infty}\left[\frac{(2k)!\pi}{k!(k-1)!2^{4k+2}}\lambda^{2k} + \frac{1}{2}\frac{k!(k+1)!}{(2k+2)!}\lambda^{2k+1}\right]. \tag{2.3.128}$$

In this way, the existence of a one-soliton transformation implies the existence of an infinite-order generalized symmetry.

Let us consider the symmetry given by $\phi(\lambda)\epsilon = 1$. Then (2.3.121) implies that $E_0 = \cosh\mu$ and $E_1 = \sinh\mu/\mu$. According to (2.3.124) the corresponding Bäcklund transformation is

$$\delta(\lambda) = 2\sinh\mu/\mu \tag{2.3.129}$$

$$\gamma(\lambda) = \cosh\mu + \lambda\sinh\mu/\mu + 1 \tag{2.3.130}$$

Thus in correspondence with a Lie point symmetry we have a Bäcklund transformation of infinite order.

3.2.6. *Symmetry reduction of a generalized symmetry of the Toda system.* From the first generalized isospectral symmetry of the Toda system (2.3.37) we derive a nontrivial reduction of the Toda system by setting $a_{n,\epsilon}$ and $b_{n,\epsilon}$ equal to zero. In this way we have to solve the system of equations

$$b_n^2 - b_{n+1}^2 + a_{n-1} - a_{n+1} = 0, \quad a_{n-1}\Big(b_n + b_{n-1}\Big) - a_n\Big(b_n + b_{n+1}\Big) = 0, \tag{2.3.131}$$

together with the Toda system itself. Eqs. (2.3.131) can be integrated to get

$$b_n^2 + a_{n-1} + a_n = \kappa_1, \tag{2.3.132}$$

$$a_n\Big(b_n + b_{n+1}\Big) = \kappa_0, \tag{2.3.133}$$

where κ_0 and κ_1 are some n-independent functions, possibly functions of t. In (2.3.133) we take with all generality $\kappa_0 \neq 0$ as otherwise the result is trivial. Eqs. (2.3.132, 2.3.133) are two coupled equations for the two fields a_n and b_n. They can be decoupled and give:

$$a_n\Big[\sqrt{\kappa_1 - a_n - a_{n-1}} + \sqrt{\kappa_1 - a_{n+1} - a_n}\Big] = \kappa_0, \tag{2.3.134}$$

$$b_n^2 + \frac{\kappa_0}{b_{n-1} + b_n} + \frac{\kappa_0}{b_n + b_{n+1}} = \kappa_1. \tag{2.3.135}$$

Using (2.3.8) we can write down two DΔEs for a_n and b_n

$$\dot{a}_n = a_n\Big[\sqrt{\kappa_1 - a_n - a_{n-1}} - \sqrt{\kappa_1 - a_{n+1} - a_n}\Big], \tag{2.3.136}$$

$$\dot{b}_n = \kappa_0\Big[\frac{1}{b_n + b_{n-1}} - \frac{1}{b_{n+1} + b_n}\Big]. \tag{2.3.137}$$

Eq. (2.3.137) is strictly related to the integrable DΔE pKdV [489] by defining $q_n(t) = b_n + b_{n+1}$ and $k_0 = 2p$. Eq. (2.3.136) seems to be new. As well as (2.3.137), (2.3.136) is integrable and, as the new 5 point DΔE obtained in [311, 312]

$$\dot{u}_n = \Big(u_{n+2}\sqrt{u_{n+1}^2 - 1} - u_{n-2}\sqrt{u_{n-1}^2 - 1}\Big), \tag{2.3.138}$$

it has an algebraic dependence on the field. Eq. (2.3.138), as shown in [307], goes in the continuous limit to the Kaup-Kupershmidt equation[271, 440]

$$q_t = q_{xxxxx} + 5q\,q_{xxx} + \frac{25}{2}q_x\,q_{xx} + 5q^2\,q_x. \tag{2.3.139}$$

From (2.3.132, 2.3.133) we get the difference equation

$$a_n^2[a_{n+1} + a_{n-1}] + \kappa_0^2 - 2\kappa_1 a_n = 0. \tag{2.3.140}$$

Taking into account the Toda system (2.3.8) and assuming that $\dot{\kappa}_0 = 0$ and $\dot{\kappa}_1 = 0$ we get the reduced difference equation

$$a_n a_{n-1} a_{n+1} + \kappa_0^2 + \frac{\kappa_2}{a_n} = 0, \tag{2.3.141}$$

where κ_2 is another integration constant. Eq. (2.3.141) can be reduced to a symmetric QRT-map introduced by Quispel, Roberts and Taylor [236]. The QRT - maps are the autonomous limit of a discrete Painlevé equation [687] whose integration can be found in [76, 686].

Introducing the new variables

$$a_{2m} = x_m, \quad a_{2m+1} = y_m, \tag{2.3.142}$$

corresponding to the split of the field a_n into two fields, one with only even and one with only odd values of the indexes, (2.3.141) reduces to

$$x_m^2 y_{m-1} y_m + \kappa_0^2 x_m - \kappa_2 = 0. \tag{2.3.143}$$

Multiplying (2.3.143) by $(y_m - y_{m-1})$ it is easy to see that we have an invariant

$$K(x_m, y_m) = x_m y_m + \kappa_0^2 \frac{x_m + y_m}{x_m y_m} - \kappa_2 \frac{1}{x_m y_m} \tag{2.3.144}$$

such that $K(x_m, y_m) = C$, where C is a constant, and

$$K(x_m, y_m) = K(x_m, y_{m-1}) = K(x_{m+1}, y_m). \tag{2.3.145}$$

Let us introduce an homographic transformation for x and y, which, as the system is symmetric, can be taken to be the same for x and y,

$$x = \frac{\alpha X + \beta}{\gamma X + 1}, \quad y = \frac{\alpha Y + \beta}{\gamma Y + 1}, \tag{2.3.146}$$

where α, β and γ are constant depending on the constant coefficients of the equation $K(x_m, y_m)$
$= C$. Then the biquadratic relation (2.3.144) can be reduced to

$$X^2 Y^2 + \tilde{\gamma}(X^2 + Y^2) + \tilde{\alpha} XY + 1 = 0, \tag{2.3.147}$$

where $\tilde{\alpha}$ and $\tilde{\gamma}$ are expressed in terms of the coefficients C, k_0 and k_2. Eq. (2.3.147) can be parametrized in terms of elliptic functions $X = \sqrt{k}\,\mathrm{sn}(z)$ and $Y = \sqrt{k}\,\mathrm{sn}(z + q)$ of modulus k and argument z. k satisfies a second order algebraic equation

$$k^2 + \left(\tilde{\gamma} + \frac{1}{\tilde{\gamma}} - \frac{\tilde{\alpha}^2}{4\tilde{\gamma}}\right)k + 1 = 0, \tag{2.3.148}$$

and q is obtained by solving the equation

$$k\,\mathrm{sn}^2(q) + 1 = 0. \tag{2.3.149}$$

In this way we have shown that the symmetry reduction of the Toda system is solved in terms of elliptic functions.

3.2.7. *The inhomogeneous Toda lattices.* In a way, parallel to the construction of the cKdV, we can construct inhomogeneous Toda lattices [476]. Results on this have been derived in [497].

An inhomogeneous Toda lattice can be constructed considering a simple non isospectral equation of the Toda lattice. Choosing in (2.3.24) $f_1 = 1$ and $g_1 = \alpha$ we get the inhomogeneous Toda System

$$\dot{a}_n = a_n\big[b_n\big(1 - \alpha(2n-1)\big) - b_{n+1}\big(1 - \alpha(2n+3)\big)\big], \tag{2.3.150}$$

$$\dot{b}_n = a_{n-1}\big(1 - \alpha(2n-1)\big) - a_n\big(1 - \alpha(n+1)\big) + \alpha\big(b_n^2 - 4\big). \tag{2.3.151}$$

Defining

$$b_n = \dot{v}_n, \quad a_n = e^{\left[\left(1-\alpha(2n-1)\right)v_n - \left(1-\alpha(2n+3)\right)v_{n+1}\right]}, \tag{2.3.152}$$

we get the following inhomogeneous Toda lattice

$$\ddot{v}_n = \big(1 - 2\alpha(n-1)\big)e^{\left[\left(1-\alpha(2n-3)\right)v_{n-1}-\left(1-\alpha(2n+1)\right)v_n\right]} - \big(1 - 2\alpha(n+1)\big)e^{\left[\left(1-\alpha(2n-1)\right)v_n-\left(1-\alpha(2n+3)\right)v_{n+1}\right]} + \alpha\big(\dot{v}_n^2 - 4\big). \tag{2.3.153}$$

Eq. (2.3.153) has explicit n-dependent coefficients and correspond to a velocity dependent force.

By allowing for more general boundary conditions than (2.3.12) we can obtain new classes of non linear Toda like equations with n-dependent coefficients. To do so let us introduce, as we did in the case of the cKdV, some reference potentials $g^1(n)$ and $g^2(n)$ such that

$$\tilde{a}_n(t) = a_n(t) + g_n^1, \quad \tilde{b}_n(t) = b_n(t) + g_n^2, \tag{2.3.154}$$

where $a(n,t)$ and $b(n,t)$ satisfy (2.3.12). Following the derivation of the cKdV hierarchy we get two equations, one for the fields and one for the admissible reference potentials. They are:

$$\begin{pmatrix} \dot{a}_n \\ \dot{b}_n \end{pmatrix} = \phi(\mathscr{L}_d, t)\Big[\begin{pmatrix} \alpha_0[(a_n + g_n^1)(b_n - b_{n+1})] \\ \alpha_0(a_{n-1} - a_n) + \gamma_0 b_n \end{pmatrix} + \begin{pmatrix} \delta_0[(a_n + g_n^1)(b_{n+1}(2n+3) - b_n(2n-1))] \\ \delta_0[b_n^2 + 2b_n g_n^2 - 4 + 2(n+1)a_n - 2(n-1)a_{n-1}] \end{pmatrix}\Big] \tag{2.3.155}$$

and

$$\alpha_0(g_n^2 - g_{n+1}^2) + 2\gamma_0 + \delta_0[g_{n+1}^2(2n+3) - g_n^2(2n-1)] = 0, \tag{2.3.156}$$

$$\alpha_0(g_{n-1}^1 - g_n^1) + \beta_0 + \gamma_0 g_n^2 + \delta_0[(g_n^2)^2 + 2(n+1)g_n^1 - 2(n-1)g_{n-1}^1] = 0, \tag{2.3.157}$$

where α_0, β_0, γ_0 and δ_0 are arbitrary constants, $\phi(\mathscr{L}, t)$ is an entire function of the first arguments and $\mathscr{L}$ is given in (2.3.18) with a_n substituted by $\tilde{a}_n$ and b_n by $\tilde{b}_n$.

Eqs. (2.3.156, 2.3.157) are a first order OΔE for g_n^1 and g_n^2 and can be easily solved. We get the following solutions:

(1) For $\alpha_0 = \beta_0$, $\gamma_0 = \delta_0 = 0$ we get

$$g_n^1 = n, \qquad g_n^2 = 0. \tag{2.3.158}$$

(2) For $\delta_0 = 0$, $\gamma_0 = \frac{1}{2}\alpha_0$ and $\beta = -\frac{1}{4}\alpha_0$ we get

$$g_n^1 = \frac{1}{4}n^2, \qquad g_n^2 = n. \tag{2.3.159}$$

(3) For $\alpha_0 = \beta_0 = \gamma_0 = 0$ we get

$$g_n^1 = -\frac{1}{4((2n+1)^2}, \qquad g_n^2 = \frac{1}{4n^2 - 1}. \tag{2.3.160}$$

In the case (2.3.158), defining $a_n = (n+1)\exp[v_n - v_{n+1}] - n$ and $b_n = \frac{\dot{v}_n}{\alpha_0}$, we get the inhomogeneous Toda like equation

$$\ddot{v}_n = \alpha_0^2\Big[ne^{v_{n-1}-v_n} - (n+1)e^{v_n - v_{n+1}} + 1\Big]. \tag{2.3.161}$$

The spectral problem associated to (2.3.161) reads

$$\psi_{n-1} + b_n\psi_n + (a_n + n)\psi_{n+1} = \lambda(t)\psi_n, \quad \lambda(t) = \alpha_0 t + \lambda_0, \tag{2.3.162}$$

where λ_0 is a complex parameter.

In the case (2.3.159), defining $a_n = (\frac{1}{4}n^2 + 1)\exp[\upsilon_n - \upsilon_{n+1}] - \frac{1}{4}n^2$ and $b_n = \frac{\dot{\upsilon}_n}{\alpha_0}$, we get the inhomogeneous Toda lattice like equation (1.4.23) we wrote in Section 1.4.1.2 and which here we repeat for the convenience of the reader

$$\ddot{\upsilon}_n = \alpha_0^2\Big\{[\frac{1}{4}(n-1)^2 + 1]e^{\upsilon_{n-1}-\upsilon_n} - [\frac{1}{4}n^2 + 1]e^{\upsilon_n - \upsilon_{n+1}} + \frac{1}{2}n - \frac{1}{2} + \frac{1}{2\alpha_0}\dot{\upsilon}_n\Big\}. \tag{2.3.163}$$

The spectral problem associated to (2.3.163) reads

$$\begin{aligned}\psi_{n-1} + (b_n + n)\psi_n + (a_n + \frac{1}{2}n^2)\psi_{n+1} &= \lambda(t)\psi_n,\\ \lambda(t) &= \frac{1}{2} + (\lambda_0 - \frac{1}{2})e^{\alpha_0 t/2},\end{aligned} \tag{2.3.164}$$

where λ_0 is a complex parameter.

In the last case (2.3.160), defining

$$a_n = \Big(1 - \frac{1}{4(2n+1)^2}\Big)e^{(2n+3)\upsilon_{n+1}-(2n-1)\upsilon_n} + \frac{1}{4(2n+1)^2}$$

and $b_n = \frac{\dot{\upsilon}_n}{\delta_0}$, we get the inhomogeneous Toda lattice like equation

$$\begin{aligned}\ddot{\upsilon}_n = \delta_0^2\Big\{&2(n+1)\Big[1 - \frac{1}{4(2n+1)^2}\Big]e^{[(2n+3)\upsilon_{n+1}-(2n-1)\upsilon_n]}\\ &- 2(n-1)\Big[1 - \frac{1}{4(2n-1)^2}\Big]e^{[(2n+1)\upsilon_n-(2n-3)\upsilon_{n-1}]}\\ &+ \frac{\dot{\upsilon}_n^2}{\delta_0^2} + 2\frac{\dot{\upsilon}_n}{\delta_0(4n^2-1)} - 4 + \frac{1}{(4n^2-1)^2}\Big\}.\end{aligned} \tag{2.3.165}$$

The spectral problem associated to (2.3.165) reads

$$\begin{aligned}&\psi_{n-1} + \Big(b_n + \frac{1}{4n^2-1}\Big)\psi_n + \Big(a_n - \frac{1}{4(2n+1)^2}\Big)\psi_{n+1} = \lambda(t)\psi_n,\\ &\lambda(t) = 2\coth[2\delta_0(t-t_0)],\end{aligned} \tag{2.3.166}$$

where t_0 is a real parameter.

As in the case of the cKdV in all cases we can construct Darboux and Bäcklund which preserve the class of solutions of the inhomogeneous Toda lattices we have constructed. We construct a two parameters Darboux transformation, for example using the dressing method introduced by Zakharov and Shabat [501, 863]. Then by going to the limit when the two parameters are equal we get a new one parameter Darboux transformation or Moutard transformation. The new solution of the inhomogeneous Toda lattices (2.3.163, 2.3.164, 2.3.165) and of the discrete Schrödinger spectral problem are given by

$$\tilde{\upsilon}_n = \upsilon_n - \ln\Big(1 + \frac{\psi_n(\mu)^2}{\psi_{n-1}(\mu)\psi_{n,\mu}(\mu) - \psi_{n-1,\mu}(\mu)\psi_n(\mu)}\Big) \tag{2.3.167}$$

where μ is a particular value of the parameter λ and

$$\begin{aligned}\tilde{\psi}_n(\lambda) &= \psi_n(\lambda) + \frac{1}{\lambda-\mu}\psi_n(\mu)\\ &\times\Big(\frac{\psi_n(\mu)\psi_{n+1}(\lambda) - \psi_{n+1}(\mu)\psi_n(\lambda)}{\psi_{n,\mu}(\mu)\psi_{n+1}(\mu) - \psi_{n+1,\mu}(\mu)\psi_n(\mu)}\Big).\end{aligned} \tag{2.3.168}$$

The Bäcklund transformation is obtained by eliminating the function $\psi_n(\mu)$ between (2.3.167) and the corresponding discrete spectral problem. We get

$$(2.3.169)\qquad \left(e^{\upsilon_n} - e^{\tilde{\upsilon}_n}\right)\left(e^{-\tilde{\upsilon}_{n+1}} - e^{-\upsilon_{n+1}}\right)\Big[\Big(\mu - \frac{\dot{\upsilon}_n}{x_0} - g_n^2\Big)\left(1 - e^{\upsilon_n - \tilde{\upsilon}_n}\right) + \frac{\dot{\upsilon}_n}{x_0} - \frac{\dot{\tilde{\upsilon}}_n}{x_0}\Big]^2 = 4(1 + g_n^1)\sinh^2(\upsilon_n - \tilde{\upsilon}_n)e^{2\upsilon_n}\left(e^{-\upsilon_{n+1}} - e^{-\tilde{\upsilon}_{n+1}}\right),$$

where x_0 is either α_0 or δ_0 according to the case we are considering. This Bäcklund transformation preserve the class of potentials of the inhomogeneous Toda lattice we are considering.

Further results on inhomogeneous non linear DΔEs will be found in Sections 2.3.3.5, 2.3.5 and 3.4.

3.3. Volterra hierarchy, its symmetries, Bäcklund transformations, Bianchi identity and continuous limit. Here we study the Volterra equation and its hierarchy of DΔEs [214, 402, 496, 586, 796] a subclass of the Toda system hierarchy obtained by setting $b_n(t) = 0$ in (2.3.24). The Volterra hierarchy is given by:

$$(2.3.170)\qquad \begin{aligned} \dot{a}_n &= g_1(\tilde{\mathscr{L}}, t)\{a_n(a_{n-1} - a_{n+1})\} \\ &\quad + g_2(\tilde{\mathscr{L}}, t)[a_n(a_n - (n-1)a_{n-1} + (n+2)a_{n+1} - 4)]. \end{aligned}$$

In (2.3.170) we have

$$(2.3.171)\qquad \tilde{\mathscr{L}} p_n = a_n(p_n + p_{n+1} + s_{n-1} - s_{n+1}),$$

where s_n is a bounded solution of the inhomogeneous first order difference equation (2.3.20). The recursion operator (2.3.171) is obtained considering the square of the recursion operator $\mathscr{L}_d$ of the Toda system (2.3.18) and then setting $b_n = 0$.

The simplest isospectral equation of the hierarchy (2.3.170), when $g_1 = 1$ and $g_2 = 0$, is the *Volterra equation* [821]

$$(2.3.172)\qquad \dot{a}_n = a_n(a_{n-1} - a_{n+1}).$$

The point symmetries of (2.3.172) can be found in [506, 507].

The Volterra isospectral hierarchy, from now on denoted as Volterra hierarchy, is obtained from (2.3.170) when $g_2 = 0$. It is associated to the discrete Schrödinger spectral problem

$$(2.3.173)\qquad \psi(n-1, t; \lambda) + a_n\psi(n+1, t; \lambda) = \lambda\psi(n, t; \lambda).$$

For any equation of the Volterra hierarchy we can write down an explicit evolution equation for the function $\psi(n, t; \lambda)$ [123, 127] such that λ does not evolve in time and the following boundary conditions

$$(2.3.174)\qquad \lim_{|n|\to\infty} a_n - 1 = \lim_{|n|\to\infty} s_n = 0,$$

on the fields a_n and s_n are satisfied. We can then associate to the discrete Schrödinger equation (2.3.173, 2.3.174) a spectrum defined in the complex plane of the variable z (2.3.117):

$$(2.3.175)\qquad S[a_n] = \{R(z, t),\ z \in \mathbf{C}_1; z_j, c_j(t),\ |z_j| < 1, j = 1, 2, \ldots, N\},$$

where $R(z, t)$ is the reflection coefficient, $\mathbf{C}_1$ is the unit circle in the complex z plane, z_j are isolated points inside the unit disk and c_j are some complex functions of t related to the residues of $R(z, t)$ at the poles z_j. When a_n and s_n satisfy the boundary conditions (2.3.174), the spectral data define the function $a_n(t)$ in a unique way. There is a one-to-one correspondence between the evolution of the potential a_n of the discrete Schrödinger spectral problem (2.3.173), given by (2.3.170) and that of the reflection coefficient $R(z,t)$, given by

$$\frac{dR(z,t)}{dt} = \mu\lambda g_1(\lambda^2,t)R(z,t), \quad \mu = z^{-1} - z, \quad \frac{d\lambda}{dt} = \mu^2\lambda g_2(\lambda^2,t). \tag{2.3.176}$$

The Volterra equation (2.3.172) is obtained for $g_1(\lambda^2,t) = 1$ and $g_2(\lambda^2,t) = 0$.

The symmetries for any equation of the Volterra (2.3.170) hierarchy are provided by all flows commuting with the equations themselves. An infinite number of such symmetries is provided by the equations

$$a_{n,\epsilon_\ell} = \tilde{\mathcal{L}}^\ell \left\{a_n(a_{n-1} - a_{n+1})\right\}. \tag{2.3.177}$$

Here ℓ is any positive integer and ϵ_ℓ is a group parameter. From the point of view of the spectral problem (2.3.173), (2.3.177) correspond to isospectral deformations, i.e. we have $\lambda_{\epsilon_\ell} = 0$. The proof that (2.3.177) are symmetries is easily given by taking into account the one-to-one correspondence between the equation and the spectrum (2.3.175) under the asymptotic conditions (2.3.174) as seen in detail in the case of the Toda lattice.

We can extend the class of symmetries by considering non isopectral deformations of the spectral problem (2.3.173) [496]. For any equation of the Volterra hierarchy, characterized by the evolution of the reflection coefficient (2.3.176), we have:

$$\begin{aligned} a_{n,\epsilon_\ell} &= h_1(\tilde{\mathcal{L}},t)\left[a_n(a_{n-1} - a_{n+1})\right] \\ &\quad + \tilde{\mathcal{L}}^\ell[a_n(a_n - (n-1)a_{n-1} + (n+2)a_{n+1} - 4)], \end{aligned} \tag{2.3.178}$$

where the function h_1 is obtained as a solution of the differential equation

$$h_1(\tilde{\mathcal{L}},t)_t = \tilde{\mathcal{L}}^\ell[\tilde{\mathcal{L}}(\tilde{\mathcal{L}} - 4)\frac{dg_1(\tilde{\mathcal{L}},t)}{d\tilde{\mathcal{L}}} + (\tilde{\mathcal{L}} - 2)g_1(\tilde{\mathcal{L}},t)]. \tag{2.3.179}$$

The function g_1 characterize the equation in the Volterra hierarchy (2.3.170) we are considering. In correspondence with (2.3.178) we have the following evolution of the reflection coefficient

$$\frac{dR}{d\epsilon_\ell} = \mu\lambda h_1(\lambda^2,t)R, \qquad \lambda_{\epsilon_\ell} = \frac{1}{2}\mu^2\lambda^{2\ell+1}, \tag{2.3.180}$$

As in the case of isospectral symmetries (2.3.177), we can easily prove that the non isospectral flows (2.3.178) commute with the corresponding hierarchy of evolution equations (2.3.170). This is done by showing that the flows (2.3.180) in the space of the reflection coefficients commute with that of the evolution equation (2.3.176).

For $g_1(\lambda^2,t) = \lambda^{2N}$, corresponding to the Nth-equation in the Volterra hierarchy , we are able to integrate (2.3.179) and we get

$$h_1(\lambda^2,t) = \lambda^{2\ell+2N}[(\lambda^2 - 2)(2N+1) - 4N]t. \tag{2.3.181}$$

In the case of the Volterra hierarchy we have only one exceptional symmetry, given by

$$a_{n,\epsilon} = g_3(\tilde{\mathcal{L}},t)[a_n(a_{n-1} - a_{n+1})] + a_n \tag{2.3.182}$$

where the function $g_3(\tilde{\mathcal{L}}_d,t)$ is to be determined directly for each equation of the hierarchy. As these exceptional symmetries do not satisfy the asymptotic boundary conditions (2.3.174), we cannot write a corresponding evolution equation for the reflection coefficient.

Let us now write down the lowest order symmetries for the Volterra equation (2.3.172)). They are:

$$a_{n,\epsilon_0} = a_n(a_{n-1} - a_{n+1}), \tag{2.3.183}$$

$$\begin{aligned} a_{n,\epsilon_1} = a_n\{&a_{n-1}(a_{n-2} + a_{n-1} + a_n - 2) \\ &- a_{n+1}(a_{n+2} + a_{n+1} + a_n - 2)\}, \end{aligned} \tag{2.3.184}$$

$$\begin{aligned} a_{n,\epsilon_2} = a_n\{&a_{n-1}[(a_n + a_{n-1})(a_{n-2} + a_{n-1} + a_n - 2) \\ &+ a_{n-2}(a_{n-3} + a_{n-2} + a_{n-1} - 2) - 2] \\ &\quad - a_{n+1}[(a_{n+1} + a_n)(a_{n+2} + a_{n+1} + a_n - 2) \\ &\qquad + a_{n+2}(a_{n+3} + a_{n+2} + a_{n+1} - 2) - 2]\}, \end{aligned} \tag{2.3.185}$$

$$\begin{aligned} a_{n,v} = a_n\{&t[a_{n-1}(a_{n-2} + a_{n-1} + a_n - 4) \\ &- a_{n+1}(a_{n+2} + a_{n+1} + a_n - 4)] \\ &\quad + a_n - (n-1)a_{n-1} + (n+2)a_{n+1} - 4\}, \end{aligned} \tag{2.3.186}$$

and the exceptional one

$$a_{n,\mu} = a_n + t\dot{a}_n. \tag{2.3.187}$$

3.3.1. *Bäcklund transformations.* The Bäcklund transformations for the Volterra hierarchy can be obtained from those of the Toda system (2.3.102) by imposing the reduction $b_n = 0$. Under this reduction (2.3.102) becomes

$$\gamma(\Lambda_d)\begin{pmatrix} \tilde{a}_n - a_n \\ 0 \end{pmatrix} = \delta(\Lambda_d)\begin{pmatrix} 0 \\ \tilde{\Pi}_{n-1}\Pi_n^{-1} - \tilde{\Pi}_n\Pi_{n+1}^{-1} \end{pmatrix}, \tag{2.3.188}$$

where Π_n and $\tilde{\Pi}_n$ are given by (2.3.99). Taking into account the form of the Bäcklund recursive operator for the Toda, Λ_d, we get:

$$\Lambda_d\begin{pmatrix} 0 \\ q_n \end{pmatrix} = \begin{pmatrix} \tilde{a}_n[q_n + q_{n+1}] \\ 0 \end{pmatrix}. \tag{2.3.189}$$

Then for $\gamma = \delta/p$, where δ and p are arbitrary constants,

$$\begin{aligned} \begin{pmatrix} \tilde{a}_n - a_n \\ 0 \end{pmatrix} &= p\Lambda_d\begin{pmatrix} 0 \\ \tilde{\Pi}_{n-1}\Pi_n^{-1} - \tilde{\Pi}_n\Pi_{n+1}^{-1} \end{pmatrix} = \\ &\begin{pmatrix} pa_n^2[\tilde{\Pi}_{n-1}\Pi_n^{-1} - \tilde{\Pi}_{n+1}\Pi_{n+2}^{-1}] \\ 0 \end{pmatrix}. \end{aligned} \tag{2.3.190}$$

Then the Bäcklund transformation adding one soliton to the solution of the Volterra hierarchy is given by

$$\tilde{a}_n - a_n = p\tilde{a}_n[\tilde{\Pi}_{n-1}\Pi_n^{-1} - \tilde{\Pi}_{n+1}\Pi_{n+2}^{-1}]. \tag{2.3.191}$$

The corresponding evolution of the reflection coefficient is

$$\tilde{R}(z) = z^4 R(z). \tag{2.3.192}$$

Defining $\Pi_n = \Pi_{n,m}$ and $\tilde{\Pi}_n = \Pi_{n,m+1}$ we can write (2.3.191) as a PΔE. We have

$$\frac{\Pi_{n,m+1}}{\Pi_{n+1,m+1}} - \frac{\Pi_{n,m}}{\Pi_{n+1,m}} = p\frac{\Pi_{n,m+1}}{\Pi_{n+1,m+1}}\left[\frac{\Pi_{n-1,m+1}}{\Pi_{n,m}} - \frac{\Pi_{n+1,m+1}}{\Pi_{n+2,m}}\right]. \tag{2.3.193}$$

As we saw the recursion operator for the Volterra hierarchy (2.3.171) is obtained by squaring the one of the Toda hierarchy (2.3.18) and setting $b_n = 0$. The approach of squaring the recursion operator of the Bäcklund transformation after setting to zero b_n and $\tilde{b}_n$ provide the expression

$$\Lambda_d^2 \begin{pmatrix} p_n \\ 0 \end{pmatrix} = \begin{pmatrix} a_n^2[p_n + p_{n+1} + \Sigma_{n-1} - \Sigma_n] + (a_n^1 - a_n^2)\sum_{j=n}^{\infty} \tilde{p}_j \\ \tilde{p}_n + \tilde{\Sigma}_{n-1} - \tilde{\Sigma}_n \end{pmatrix}, \tag{2.3.194}$$

where Σ_n is defined in (2.3.101). $\tilde{\Sigma}_n$ is given by (2.3.101) with p_n substituted by $\tilde{p}_n$, given by

$$\tilde{p}_n = (a_n - \tilde{a}_n)\sum_{j=n}^{\infty} p_j. \tag{2.3.195}$$

As $\tilde{p}_n + \tilde{\Sigma}_{n-1} - \tilde{\Sigma}_n \neq 0$ we are not able to obtain a recursion operator for the Bäcklund transformation for the Volterra hierarchy by just squaring the recursion operator of the Bäcklund of the Toda lattice. Thus this method does not provide any higher order Bäcklund transformation. To our knowledge no such recursion operator exists in the literature. However higher order soliton solutions could be obtained by applying iteratively the adding one soliton Bäcklund transformation.

From (2.3.191), as the Bianchi permutability theorem is true in the reflection coefficient space, Bianchi identity reads:

$$\begin{aligned} p_1\Big\{ a_n^{(1)}\Big[\frac{\Pi_{n-1}^{(1)}}{\Pi_n} - \frac{\Pi_{n+1}^{(1)}}{\Pi_{n+2}}\Big] - a_n^{(12)}\Big[\frac{\Pi_{n-1}^{(12)}}{\Pi_n^{(2)}} - \frac{\Pi_{n+1}^{(12)}}{\Pi_{n+2}^{(2)}}\Big]\Big\} \\ + p_2\Big\{ a_n^{(12)}\Big[\frac{\Pi_{n-1}^{(12)}}{\Pi_n^{(1)}} - \frac{\Pi_{n+1}^{(12)}}{\Pi_{n+2}^{(1)}}\Big] - a_n^{(2)}\Big[\frac{\Pi_{n-1}^{(2)}}{\Pi_n} - \frac{\Pi_{n+1}^{(2)}}{\Pi_{n+2}}\Big]\Big\} = 0. \end{aligned} \tag{2.3.196}$$

As for the Toda, this is a difference equation and so no superposition formula exists. From (2.3.196) we could obtain a 3 dimensional PΔE whose derivation we leave to diligent readers.

3.3.2. *Infinite dimensional symmetry algebra.* To define the structure of the symmetry algebra for the Volterra hierarchy we need to compute the commutation relations between the symmetries.

The first result is that the isospectral symmetry generators, provided by (2.3.177) for the Volterra hierarchy, commute amongst each other. We can write the generators for the isospectral symmetries of the Volterra hierarchy as

$$\hat{X}_\ell^V = \tilde{\mathcal{L}}^\ell [a_n(a_{n-1} - a_{n+1})]\partial_{a_n}. \tag{2.3.197}$$

The fact that (2.3.177) provides symmetries for a generic equation of the Volterra hierarchy, implies

$$[\hat{X}_\ell^V, \hat{X}_m^V] = 0. \tag{2.3.198}$$

A proof of (2.3.198) is given by

$$\frac{\partial^2 R}{\partial\epsilon_\ell \partial\epsilon_m} = \frac{\partial^2 R}{\partial\epsilon_m \partial\epsilon_\ell}, \tag{2.3.199}$$

which follows directly from (2.3.180). A natural way of representing the result given by (2.3.199) is to introduce symmetry generators in the space of the reflection coefficient.

These generators are written as

$$\hat{\mathcal{X}}_k^V = \mu\lambda^{2k+1} R\partial_R. \tag{2.3.200}$$

In terms of the vector fields $\hat{\mathcal{X}}_k^V$, (2.3.199) is written as

$$[\hat{\mathcal{X}}_\ell^V, \hat{\mathcal{X}}_m^V] = [\mu\lambda^{2\ell+1} R\partial_R, \mu\lambda^{2m+1} R\partial_R] = 0. \tag{2.3.201}$$

So far, the use of the vector fields in the reflection coefficient space has just reexpressed a known result, namely (2.3.199) is rewritten as (2.3.201). We now extend the use of vector fields in the reflection coefficient space to the case of the non isospectral symmetries (2.3.178). We restrict, for the sake of the simplicity of exposition, ourselves to the Nth equation of the Volterra hierarchy for which there is no explicit dependence on time. Thus we consider the case (2.3.181). The non isospectral symmetry vector fields for the Volterra hierarchy (2.3.181) are:

$$\begin{aligned}\hat{Y}_k^V = \; & \{t\tilde{\mathcal{L}}^{k+N}[(1+N)\tilde{\mathcal{L}}_d - 2(1+2N)][a_n(a_{n-1} - a_{n+1})] \\ & + \tilde{\mathcal{L}}_d{}^k [a_n(a_n - (n-1)a_{n-1} + (n+2)a_{n+1} - 4)]\}\partial_{a_n}.\end{aligned} \tag{2.3.202}$$

Taking into account (2.3.180) and (2.3.181) we can define the symmetry generators (2.3.202) in the reflection coefficient space

$$\hat{\mathcal{Y}}_k^V = \mu\lambda^{2k+2N+1} t[(1+N)\lambda^2 - 4N - 2]R\partial_R + \frac{1}{2}\mu^2\lambda^{2k+1}\partial_\lambda. \tag{2.3.203}$$

Commuting $\hat{\mathcal{Y}}_\ell^V$ with $\hat{\mathcal{Y}}_m^V$ we have:

$$[\hat{\mathcal{Y}}_\ell^V, \hat{\mathcal{Y}}_m^V] = (m-\ell)[\hat{\mathcal{Y}}_{\ell+m+1}^V - 4\hat{\mathcal{Y}}_{\ell+m}^V]. \tag{2.3.204}$$

From the isomorphism between the spectral space and the space of the solutions, we conclude that the vector fields representing the symmetries of the studied evolution equations, satisfy the same commutation relations. Hence we have

$$[\hat{Y}_\ell^V, \hat{Y}_m^V] = (m-\ell)[\hat{Y}_{\ell+m+1}^V - 4\hat{Y}_{\ell+m}^V]. \tag{2.3.205}$$

In a similar manner we can work out the commutation relations between the $\hat{Y}_\ell^V$ and $\hat{X}_m^V$ symmetry generators. We get:

$$[\hat{\mathcal{X}}_\ell^V, \hat{\mathcal{Y}}_m^V] = -(1+\ell)\hat{\mathcal{X}}_{\ell+m+1}^V + 2(2\ell+1)\hat{\mathcal{X}}_{\ell+m}^V, \tag{2.3.206}$$

and consequently

$$[\hat{X}_\ell^V, \hat{Y}_m^V] = -(1+\ell)\hat{X}_{\ell+m+1}^V + 2(2\ell+1)\hat{X}_{\ell+m}^V. \tag{2.3.207}$$

As in the case of the Toda equation $\hat{Y}_0^V$ is a master symmetry.

Let us now consider the commutation relations involving the exceptional symmetry (2.3.187). As mentioned before, this symmetry does not satisfy the asymptotic conditions (2.3.174). Hence we cannot write it in the space of the reflection coefficient and we cannot write down the commutation simultaneously for all equations in the hierarchy. Consequently we must consider each case separately. In the case of the equations of the Volterra hierarchy we have only one exceptional symmetry. For the Volterra equation it is (see (2.3.187))

$$\hat{Z}^V = [a_n + t\dot{a}_n]\partial_{a_n}. \tag{2.3.208}$$

From (2.3.183), (2.3.186) we obtain the lowest symmetries in the $\hat{X}^V$ and $\hat{Y}^V$ series. Commuting explicitly, we obtain

$$(2.3.209)\quad [\hat{Z}^V, \hat{X}_0^V] = \hat{X}_0^V, \quad [\hat{Z}^V, \hat{Y}_0^V] = \hat{Y}_0^V + 4\hat{Z}^V, \quad [\hat{Z}^V, \hat{X}_1^V] = 2(\hat{X}_0^V + \hat{X}_1^V),$$
$$[\hat{Y}_0^V, \hat{X}_0^V] = \hat{X}_1^V - 2\hat{X}_0^V, \quad [\hat{Y}_0^V, \hat{X}_1^V] = 2\hat{X}_2^V - 6\hat{X}_1^V.$$

The generators $\hat{Y}_k^V$ and $\hat{\mathcal{Y}}_k^V$ (see (2.2.132)–(2.3.203)) depend on the number N, which denotes the equation in the hierarchy. Interestingly, the commutation relations involving the generators $\hat{X}^V$ and $\hat{Y}^V$ are the same for all N (see (2.3.198), (2.3.201), (2.3.204)–(2.3.207)).

The commutation relations obtained above determine the structure of the infinite dimensional Lie symmetry algebras.

For the Volterra equation $\hat{X}_0^V$ and $\hat{Z}^V$ are point symmetries. All the other symmetries are higher ones. Taking into account (2.3.197), (2.3.202), (2.3.208) and (2.3.209), the structure of the Lie algebra is again $L = L_0 \ni\!\!\!+ L_1$ with $L_0 = \{\hat{Z}^V, \hat{Y}_0^V, \hat{Y}_1^V, \hat{Y}_2^V, \cdots\}$ and $L_1 = \{\hat{X}_0^V, \hat{X}_1^V, \cdots\}$

3.3.3. *Contraction of the symmetry algebras in the continuous limit.* Also in this case, as for the Toda, we can speak of a contraction of the Lie algebra when we consider the continuous limit.

The limit for the Volterra (2.3.172) is the KdV itself (2.2.1).

By setting

$$a_n(t) = 1 + h^2 q(x,\tau) \tag{2.3.210}$$
$$x = (n - 2t)h \tag{2.3.211}$$
$$\tau = -\frac{1}{3}h^3 t, \tag{2.3.212}$$

we can write (2.3.172) as

$$q_\tau = q_{xxx} + 6qq_x + \mathcal{O}(h^2), \tag{2.3.213}$$

i.e. the KdV up to higher order terms. Let us now rewrite the symmetry generators in the new coordinate system defined by (2.3.210)–(2.3.212) and develop them in Taylor series for small h. We have:

$$\hat{X}_0^V = \{-2hq_x(x,\tau) - \frac{1}{3}h^3 q_\tau(x,\tau)\}\partial_q \tag{2.3.214}$$
$$\hat{X}_1^V = \{-8hq_x(x,\tau) - \frac{10}{3}h^3 q_\tau(x,\tau) + \mathcal{O}(h^5)\}\partial_q \tag{2.3.215}$$
$$\hat{Y}_0^V = \{2[2q(x,\tau) + xq_x(x,\tau) + 3\tau q_\tau(x,\tau)] + \mathcal{O}(h)\}\partial_q \tag{2.3.216}$$
$$\hat{Z}^V = \{\frac{1}{h^2}[1 + 6\tau q_x(x,\tau)] + \mathcal{O}(1)\}\partial_q. \tag{2.3.217}$$

The symmetry generators, written in the evolutionary form, for the KdV (2.3.213) are:

$$\hat{P}_0 = q_\tau(x,\tau)\partial_q \tag{2.3.218}$$
$$\hat{P}_1 = q_x(x,\tau)\partial_q \tag{2.3.219}$$
$$\hat{B} = [1 + 6\tau q_x(x,\tau)]\partial_q \tag{2.3.220}$$
$$\hat{D} = [2q(x,\tau) + xq_x(x,\tau) + 3\tau q_\tau(x,\tau)]\partial_q, \tag{2.3.221}$$

and their commutation table is:

(2.3.222)

	$\hat{P}_0$	$\hat{P}_1$	$\hat{B}$	$\hat{D}$
$\hat{P}_0$	0	0	$-6\hat{P}_1$	$-3\hat{P}_0$
$\hat{P}_1$		0	0	$-\hat{P}_1$
$\hat{B}$			0	$2\hat{B}$
$\hat{D}$				0

We can write down linear combinations of the generators of the symmetries of the Volterra equation, (2.3.214)–(2.3.217), such that in the continuous limit they go over to the point symmetry generators of the KdV:

(2.3.223)
$$\tilde{P}_0 = \frac{1}{2h^3}(4\hat{X}_0^V - \hat{X}_1^V)$$

(2.3.224)
$$\tilde{P}_1 = \frac{1}{12h}(\hat{X}_1^V - 10\hat{X}_0^V)$$

(2.3.225)
$$\tilde{D} = \frac{1}{2}\hat{Y}_0^V$$

(2.3.226)
$$\tilde{B} = h^2\hat{Z}^V$$

Taking into account the commutation table between the generators $\hat{X}_0^V$, $\hat{X}_1^V$, $\hat{Z}^V$ and $\hat{Y}_0^V$,(2.3.207), (2.3.209) and the fact that the continuous limit of $\hat{X}_2^V$ is given by:

(2.3.227)
$$\hat{X}_2^V = [-32hq_x(x,\tau) - \frac{64}{3}h^3 q_\tau(x,\tau) + \mathcal{O}(h^5)]\partial_q$$

we get:

(2.3.228)

	$\tilde{P}_0$	$\tilde{P}_1$	$\tilde{B}$	$\tilde{D}$
$\tilde{P}_0$	0	0	$-6\tilde{P}_1 + \mathcal{O}(h^2)$	$-3\tilde{P}_0 + \mathcal{O}(h^2)$
$\tilde{P}_1$		0	$\mathcal{O}(h^4)$	$-\tilde{P}_1 + \mathcal{O}(h^2)$
$\tilde{B}$			0	$2\tilde{B} + \mathcal{O}(h^2)$
$\tilde{D}$				0

Comparing the commutation tables (2.3.222) and (2.3.228) we see that the infinite dimensional Lie algebra generated by $\hat{X}_0^V$, $\hat{X}_1^V$, $\hat{Z}^V$ and $\hat{Y}_0^V$, reduces, in the continuous limit, when h goes to 0, to the Lie algebra of the point symmetries of the KdV, a contraction.

3.3.4. *Symmetry reduction of a generalized symmetry of the Volterra equation.* Starting from the first higher symmetry of the Volterra equation (2.3.172), (2.3.184) the symmetry reduction is obtained by setting $a_{n_\epsilon} = 0$. In this way, after two integrations, we have a first order OΔE for a_n which reads

(2.3.229)
$$a_{n-1}[a_{n-1} + a_{n-1} + a_n - 2] = k_0 + k_1(-1)^n,$$

where k_0 and k_1 are two integration constants which may depend on t. A solution of the Volterra equation is obtained by solving (2.3.229) together with (2.3.172).

Eq. (2.3.229) can be reduced to an asymmetric QRT - map [236], the autonomous limit of a discrete Painlevé equation [687] whose integration can be found in [76, 686].

Let us introduce the new variables (2.3.142) corresponding to the split of the field a_n into two fields, one with only even and one with only odd values of the indexes. Then (2.3.229) reduces to the system

(2.3.230)
$$y_m[x_m + y_m + x_{m+1} - 2] - (k_0 + k_1) = 0,$$

(2.3.231)
$$x_{m+1}[y_m + y_{m+1} + x_{m+1} - 2] - (k_0 - k_1) = 0.$$

Multiplying (2.3.230) and (2.3.231) appropriately we obtain the invariant

$$K(x_m, y_m) = x_m y_m[x_m + y_m - 2] - (k_0 + k_1)x_m - (k_0 - k_1)y_m. \tag{2.3.232}$$

Eq.(2.3.232) is such that (2.3.145) is satisfied and thus we can say that $K(x_m, y_m) = C$. This is an asymmetric QRT equation which can be solved in a similar way as we did in the case of Toda system in Section 2.3.2.6.

Let us introduce the homographic transformation for x and y,

$$x = \frac{\alpha X + \beta}{\gamma X + 1}, \quad y = \frac{\delta Y + \omega}{\sigma Y + 1}, \tag{2.3.233}$$

where α, β, γ ω, δ and σ are constant depending on the constant coefficients of the equation $K(x_m, y_m) = C$. Then the biquadratic relation (2.3.232) can be reduced to (2.3.147) where $\tilde{\alpha}$ and $\tilde{\gamma}$ are expressed in terms of the coefficients of the biquadratic equation C, k_0 and k_1.

Eq. (2.3.147) can be parametrized in term of elliptic functions, as seen in Section 2.3.2.6. In this way we have shown that the symmetry reduction of the first higher symmetry of the Volterra equation has solutions given by elliptic functions.

3.3.5. *Inhomogeneous Volterra equations.* The simplest is obtained by just taking the lowest non isospectral equation (2.3.170) when $g_1 = p$ and $g_2 = q$; we get

$$\dot{a}_n = a_n\{q(a_n - 4) + a_{n+1}[p + q(n+2)] - a_{n-1}[p + q(n-1)]\}. \tag{2.3.234}$$

We can obtain new inhomogeneous equations of Volterra type following Section 2.3.2.7 taking into account that now $b_n = g_n^2 = 0$. Eq. (2.3.156) is satisfied when $\gamma_0 = 0$. Then (2.3.157) becomes

$$\alpha_0(g_{n-1}^1 + g_n^1) + \beta_0 + \delta_0[2(n+1)g_n^1 - 2(n-1)g_{n-1}^1] = 0, \tag{2.3.235}$$

while (2.3.155) reads

$$\dot{a}_n = \Phi(\tilde{\mathcal{L}}, t)[a_n + g_n^1]\{\alpha_0(a_{n+1} - a_{n-1}) + \delta_0[2(n+2)a_{n+1} - 2(n-1)a_{n-1} - 8 + 2a_n]\} \tag{2.3.236}$$

Solving (2.3.235) we get two possible values for g_n^1:

$$g_n^{1,1} = n \text{ for } \delta_0 = 0, \quad g_n^{1,2} = \frac{(-1)^n|(2n+1)}{n(n+1)} \text{ for } \alpha_0 = 0. \tag{2.3.237}$$

Eqs. (2.3.236, 2.3.237) give two hierarchies of inhomogeneous Volterra equations whose simplest members are,

$$\dot{a}_n = (a_n + n)[a_{n+1} - a_{n-1}], \tag{2.3.238}$$

$$\dot{a}_n = \left[a_n + \frac{(-1)^n|(2n+1)}{n(n+1)}\right] [2(n+2)a_{n+1} - 2(n-1)a_{n-1} - 8 + 2a_n]. \tag{2.3.239}$$

Eqs. (2.3.238, 2.3.239) have non trivial spectral problems and Bäcklund transformations which preserve the class of solutions.

3.4. Discrete Nonlinear Schrödinger equation, its symmetries, Bäcklund transformations and continuous limit. One of the most important integrable non linear equations is the NLS (1.3.3). Among the integrable non linear PDEs, the NLS determines, in the regime of weak nonlinearity, the slow amplitude modulation for a large class of equations [141]. The NLS has a discrete analogue [6, 8, 9], the dNLS

$$i\dot{q}_n + \frac{1}{2h^2}\Big(q_{n+1} - 2q_n + q_{n-1}\Big) = \frac{\sigma}{2}|q_n|^2(q_{n+1} + q_{n-1}), \tag{2.3.240}$$

where $q_n(t)$ is a complex variable of modulus $|q_n|$, h is an arbitrary constant taking the role of x lattice spacing and $\sigma = \pm 1$. The case with negative σ is called focusing and allows for bright soliton solutions (localized in space, and having spatial attenuation towards infinity) as well as breather solutions. The other case, with σ positive, is the defocusing dNLS which has dark soliton solutions (having constant amplitude at infinity, and a local spatial dip in amplitude).

The discretization which is usually preferred by the physicists is the not integrable "trivial one" [1, 9, 56, 203, 255, 491]

$$i\dot{q}_n + \frac{1}{2h^2}\Big(q_{n+1} - 2q_n + q_{n-1}\Big) = \sigma |q_n|^2 q_n. \tag{2.3.241}$$

The only essential difference between (2.3.240) and (2.3.241) lies in the non linear term. It is worthwhile to notice that both (2.3.240) and (2.3.241) satisfy the Theorem 34 presented in Section 3.2.4.1 but the first is integrable while the second not. This confirms the assertion that Theorem 34 is a necessary but not sufficient condition.

The dNLS equation belongs to the hierarchy of DΔEs [172] associated to the discrete Zakharov-Shabat spectral problem

$$\Phi_{n+1} = (Z + q_n(t)I_1 + r_n(t)I_2)\Phi_n \tag{2.3.242}$$

where $\Phi_n = \Phi_n(t, z)$ is a 2×2 matrix wavefunction,

$$Z = \begin{pmatrix} z & 0 \\ 0 & 1/z \end{pmatrix}, \quad I_1 = \begin{pmatrix} 0 & 1 \\ 0 & 0 \end{pmatrix}, \quad I_2 = \begin{pmatrix} 0 & 0 \\ 1 & 0 \end{pmatrix}, \tag{2.3.243}$$

z is the spectral parameter and $q_n(t)$ and $r_n(t)$ are two complex scalar functions. Eq. (2.3.240) is one of the members of the hierarchy of equations associated to the spectral problem (2.3.242) when $r_n(t) = -\sigma q_n^*(t)$. Eq. (2.3.240) is obtained when the time evolution of the matrix wave function Φ_n is given by

$$\Phi_{n,t} = M_n \Phi_n \tag{2.3.244}$$

with

$$M_n = \frac{i}{2h^2}\begin{pmatrix} 1 - z^2 - \sigma h^2 q_n q_{n-1}^* & q_{n-1}z^{-1} - zq_n \\ \sigma(zq_{n-1}^* - q_n^* z^{-1}) & -1 + z^{-2} + \sigma q_{n-1} q_n^* \end{pmatrix}. \tag{2.3.245}$$

3.4.1. *The dNLS hierarchy and its integrability.* Starting from the spectral problem (2.3.242) we can apply the Lax technique to get the hierarchy of non linear isospectral and non isospectral DΔEs (we leave to the diligent reader to construct it with the knowledge he acquired above in the case of KdV and Toda lattice). Here we present the final result (see the following references where this recurrence operator is constructed [121, 172, 494]):

$$\begin{pmatrix} \dot{r}_n \\ -\dot{q}_n \end{pmatrix} + \omega(\mathcal{L}_s, \mathcal{L}_s^{-1}, t)\begin{pmatrix} r_n \\ q_n \end{pmatrix} + \tilde{\omega}(\mathcal{L}_s, \mathcal{L}_s^{-1}, t)(2n+1)\begin{pmatrix} r_n \\ q_n \end{pmatrix} = 0. \tag{2.3.246}$$

In (2.3.246) $\omega(\mathcal{L}_s, \mathcal{L}_s^{-1}, t)$ and $\tilde{\omega}(\mathcal{L}_s, \mathcal{L}_s^{-1}, t)$ are entire functions of the recursion operator $\mathcal{L}_s$ and $\mathcal{L}_s^{-1}$. $\mathcal{L}_s$ is defined by

$$\mathcal{L}_s\begin{pmatrix} A_n \\ B_n \end{pmatrix} = \begin{pmatrix} A_{n-1} - r_{n-1}a_n S_n - r_n Q_n \\ B_{n+1} - q_{n+1}a_n S_{n+1} - q_n Q_{n+1} \end{pmatrix},$$

and its inverse by

$$\mathcal{L}_s^{-1}\begin{pmatrix} A_n \\ B_n \end{pmatrix} = \begin{pmatrix} A_{n+1} + r_{n+1}a_n S_{n+1} + r_n Z_{n+1} \\ B_{n-1} + q_{n-1}a_n S_n + q_n Z_n \end{pmatrix}.$$

S_n, Q_n and Z_n are solutions of the inhomogeneous first order equations:

$$S_{n+1} = S_n - \frac{q_n A_n - r_n B_n}{a_n}, \quad \text{where } a_n = 1 - r_n q_n \tag{2.3.247}$$

$$Q_{n+1} = Q_n - q_n A_{n-1} + r_n B_{n+1} \tag{2.3.248}$$

$$Z_{n+1} = Z_n - q_n A_{n+1} + r_n B_{n-1}. \tag{2.3.249}$$

Whenever $\tilde{\omega}$ is present, the hierarchy (2.3.246) corresponds to a non isospectral deformation of the discrete Zakharov and Shabat spectral problem (2.3.242), i.e. the spectral parameter z evolves in time according to the equation

$$z_t = \tilde{\omega}(z^2, \frac{1}{z^2}, t)z. \tag{2.3.250}$$

For any equation of the hierarchy (2.3.246) we can write down an explicit evolution equation for the matrix wave function $\Phi_n(t, z)$ [121, 494] figuring in (2.3.242). When the functions $q_n(t)$, $r_n(t)$, S_n, Q_n and Z_n are asymptotically bounded, i. e. when

$$\lim_{|n|\to\infty} q_n(t) = \lim_{|n|\to\infty} r_n(t) = \lim_{|n|\to\infty} S_n = \lim_{|n|\to\infty} Q_n = \lim_{|n|\to\infty} Z_n = 0, \tag{2.3.251}$$

we can associate to (2.3.242) a spectrum $\mathcal{S}[q_n, r_n]$, defined in the complex plane of the variable z by

$$\begin{aligned} \mathcal{S}[q_n, r_n] : \Big\{ &(R^+(t, z), T^+(z)) \; (|z| > 1), \; (R^-(t, z), T^-(z)) \; (|z| < 1); \\ &z_j^+ \; (|z_j^+| > 1), \; C_j^+(t), \; z_j^- \; (|z_j^-| < 1), \; C_j^-(t), \; j = 1, 2, \ldots, N \Big\}, \end{aligned} \tag{2.3.252}$$

through the asymptotic behaviour of the solution to the linear problem (2.3.242)

$$\begin{aligned} n \to +\infty : \quad \Phi_n(t, z) &\to \begin{pmatrix} z^n & R^-(t, z)z^n \\ R^+(t, z)z^{-n} & z^{-n} \end{pmatrix}, \\ n \to -\infty : \quad \Phi_n(t, z) &\to \begin{pmatrix} T^+(z)z^n & 0 \\ 0 & T^-(z)z^{-n} \end{pmatrix}. \end{aligned} \tag{2.3.253}$$

z_j^- (resp. z_j^+) are isolated points inside (resp. outside) the unit disk while C_j^- (resp. C_j^+) are some complex functions of t related to the residues of $R^-(z, t)$ (resp. $R^+(z, t)$) at the poles z_j^- (resp. z_j^+). When $q_n(t)$, $r_n(t)$, S_n, Q_n and Z_n satisfy the boundary conditions (2.3.251), the spectral data $\mathcal{S}[q_n, r_n]$ define the potentials (q_n, r_n) in a unique way. There is a one-to-one correspondence between the evolution of the potentials (q_n, r_n) of the discrete Zakharov and Shabat spectral problem (2.3.242), given by (2.3.246) and the evolution of the reflection coefficients $R^\pm(t, z)$, given by

$$\frac{dR^\pm(t, z)}{dt} \pm \omega(z^2, z^{-2}, t)R^\pm(t, z) = 0, \tag{2.3.254}$$

where the t-evolution of z is given in (2.3.250). It turns out that the transmission coefficients $T^\pm(z)$ are constants of the motion. In (2.3.254) and below, d/dt denotes the total derivative with respect to t. For a function depending on t and z we have $d/dt = \partial/\partial t + (dz/dt)\partial/\partial z$.

The hierarchy (2.3.246) can be reduced to the dNLS hierarchy by setting

$$r_n = \sigma q_n^*. \tag{2.3.255}$$

In such a case we get

$$\begin{pmatrix} \sigma \dot{q}_n^* \\ -\dot{q}_n \end{pmatrix} + \omega(\mathcal{L}_s, \mathcal{L}_s^{-1}, t) \begin{pmatrix} \sigma q_n^* \\ q_n \end{pmatrix} + \tilde{\omega}(\mathcal{L}_s, \mathcal{L}_s^{-1}, t)(2n+1) \begin{pmatrix} \sigma q_n^* \\ q_n \end{pmatrix} = 0 \tag{2.3.256}$$

with

$$\omega(z^2, z^{-2}, t) = \omega_1(z^2, t) - \omega_1^*(z^{*-2}, t),$$
$$\tilde{\omega}(z^2, z^{-2}, t) = \tilde{\omega}_1(z^2, t) - \tilde{\omega}_1^*(z^{*-2}, t), \tag{2.3.257}$$

where ω_1, and $\tilde{\omega}_1$ are entire functions of their first two arguments, and the star * denotes complex conjugation. Under this reduction the spectrum $S[q_n, q_n^*]$ can be defined in terms of a single function as the following relations holds [172, 494]

$$R^+(t, z) = -\sigma[R^-(t, \frac{1}{z^*})]^*, \tag{2.3.258}$$

$$z_j^+ = (\frac{1}{z_j^-})^*, \tag{2.3.259}$$

$$C_j^+(t) = \sigma(z_j^+)^2(C_j^-(t))^*. \tag{2.3.260}$$

From now on we will denote R^- as R, and its evolution is given by

$$\frac{dR}{dt} = \omega(z^2, z^{-2}, t)R, \qquad \frac{dz}{dt} = z\tilde{\omega}(z^2, z^{-2}, t). \tag{2.3.261}$$

As examples of non linear equations, let us consider at first the case when $\omega_1(z^2, t) = \alpha_0 + \alpha_1 z^2 + \alpha_2 z^4$ (α_j, $j = 0, 1, 2$ constants) and $\tilde{\omega}_1(z^2, t) = 0$, i.e. an isospectral deformation of the discrete spectral problem (2.3.242) with the reduction (2.3.255). In this case the non linear evolution equation reads

$$\begin{aligned} \dot{q}_n = (\alpha_0 - \alpha_0^*)q_n &+ (1 - \sigma|q_n|^2)\left(\alpha_1 q_{n+1} - \alpha_1^* q_{n-1}\right) + (1 - \sigma|q_n|^2)\cdot \\ &\cdot \left\{\alpha_2[q_{n+2}(1 - \sigma|q_{n+1}|^2) - \sigma q_{n+1}(q_n^* q_{n+1} + q_n q_{n-1}^*)]\right. \\ &\left. - \alpha_2^*[q_{n-2}(1 - \sigma|q_{n-1}|^2) - \sigma q_{n-1}(q_n^* q_{n-1} + q_n q_{n+1}^*)]\right\}. \end{aligned} \tag{2.3.262}$$

In correspondence with (2.3.262) R evolves according to the equation:

$$\frac{dR}{dt} + \left[\alpha_0 - \alpha_0^* + \alpha_1 z^2 - \alpha_1^* \frac{1}{z^2} + \alpha_2 z^4 - \alpha_2^* \frac{1}{z^4}\right] R = 0, \qquad \frac{dz}{dt} = 0. \tag{2.3.263}$$

The dNLS (2.3.240) is obtained from (2.3.262) by choosing $\alpha_0 = -\alpha_1 = i/h^2$ and $\alpha_2 = 0$ and thus the time evolution of its spectrum is given by

$$\frac{dR}{dt} = \frac{i}{h^2}\left[z^2 + \frac{1}{z^2} - 2\right] R, \qquad \frac{dz}{dt} = 0. \tag{2.3.264}$$

As a complementary example, let us consider a non isospectral (n-dependent) equation obtained by considering $\omega_1(z^2, t) = 0$ and $\tilde{\omega}_1(z^2, t) = \tilde{\alpha}_0 + \tilde{\alpha}_1 z^2$. In this case we get

$$\begin{aligned} \dot{q}_n = (\tilde{\alpha}_0 - \tilde{\alpha}_0^*)(2n+1)q_n &+ \tilde{\alpha}_1\left[(2n+3)(1-\sigma|q_n|^2)q_{n+1} + 2\sigma q_n S_n^*\right] \\ &+ \tilde{\alpha}_1^*\left[(2n-1)(1-\sigma|q_n|^2)q_{n-1} + 2\sigma q_n S_n\right], \end{aligned} \tag{2.3.265}$$

where

$$S_{n+1} - S_n = -q_n q_{n+1}^*. \tag{2.3.266}$$

Corresponding to (2.3.265) R satisfies the following equation

$$\frac{dR}{dt} = 0, \qquad \frac{dz}{dt} = z[\tilde{\alpha}_0 - \tilde{\alpha}_0^* + \tilde{\alpha}_1 z^2 - \tilde{\alpha}_1^* \frac{1}{z^2}]. \tag{2.3.267}$$

So (2.3.265) is a non local DΔE.

Bäcklund transformations. The hierarchy of Bäcklund transformations for the dNLS has been derived using the Wronskian technique in [172]. However in their article there are misprints which makes it difficult to construct the Bäcklund explicitly. A simpler result can be obtained constructing the simplest Darboux matrix associated to (2.3.242). Following the incomplete results presented in [720] we get

$$d_0\tilde{q}_n - d_0^* q_{n+1} = F_n\Big[a_0 q_n - a_0^* \tilde{q}_{n+1}\Big], \tag{2.3.268}$$

where a_0 and d_0 are two complex constants and $F_n = F_n(q_n, \tilde{q}_n)$ is a solution of the OΔE

$$F_{n-1} = F_n\left[\frac{1+\sigma h^2|q_n|^2}{1+\sigma h^2|\tilde{q}_n|^2}\right]. \tag{2.3.269}$$

The Bäcklund (2.3.268) is obtained as compatibility of (2.3.242) with the transformation of the matrix wave function Φ_n

$$\tilde{\Phi}_n = D_n(q_n, \tilde{q}_n)\Phi_n. \tag{2.3.270}$$

The matrix $D_n(q_n, \tilde{q}_n)$ introduced in (2.3.270) is often denoted Darboux matrix. It is given by

$$D_n = \begin{pmatrix} \frac{a_0}{z}F_{n-1} + a_0 F_{n-1} + d_0^* z + d_0^* z^2 & h\{z[d_0^* q_n - a_0^* \tilde{q}_n F_{n-1}] - a_0^* \tilde{q}_n F_{n-1} + d_0^* q_n\} \\ \sigma h\{z[a_0 \tilde{q}_n^* F_{n-1} - d_0 q_n^*] + a_0 \tilde{q}_n^* F_{n-1} - d_0 q_n^*\} & z^2 a_0^* F_{n-1} + z a_0^* F_{n-1} + d_0 + \frac{d_0}{z} \end{pmatrix}. \tag{2.3.271}$$

For (2.3.268) the Bianchi permutability theorem is valid and thus we can construct a Bianchi identity, a PΔE involving four solutions of the dNLS. In the same way as for the Bianchi identity for the Toda lattice it can be rewritten as a PΔE in three indexes. We leave to the diligent reader to write it explicitly.

3.4.2. *Lie point symmetries of the dNLS.* Let us calculate the Lie point symmetries of dNLS (2.3.240). Decomposing q_n in real and imaginary parts $q_n = u_n + i v_n$, the dNLS (2.3.240) becomes

$$\begin{aligned} \dot{u}_n &= \frac{1}{h^2}\left\{2v_n - \left[1-\sigma(u_n^2+v_n^2)\right](v_{n+1}+v_{n-1})\right\} \\ \dot{v}_n &= \frac{1}{h^2}\left\{\left[1-\sigma(u_n^2+v_n^2)\right](u_{n+1}+u_{n-1}) - 2u_n\right\}. \end{aligned} \tag{2.3.272}$$

In (2.3.272) the dependent variables $(u_n(t), v_n(t))$ are defined in the space of the independent variables (x_n, t) where x_n defines the points of a lattice while t is a continuous "time". It is convenient to characterize each point on the space of the independent variables by two indexes, say m, n, where m parametrizes the "time" while n characterizes the position in the lattice $(x_{n,m}, t_{n,m})$. In such a way the system (2.3.272) reads:

$$\begin{aligned} \frac{du_{n,m}}{dt_{n,m}} &= \frac{1}{h^2}\left\{2v_{n,m} - \left[1-\sigma(u_{n,m}^2+v_{n,m}^2)\right](v_{n+1,m}+v_{n-1,m})\right\} \\ \frac{dv_{n,m}}{dt_{n,m}} &= \frac{1}{h^2}\left\{\left[1-\sigma(u_{n,m}^2+v_{n,m}^2)\right](u_{n+1,m}+u_{n-1,m}) - 2u_{n,m}\right\}. \end{aligned} \tag{2.3.273}$$

The generator of the point symmetry at the point of indexes (n, m) is given by

$$\hat{X}_{n,m} = \xi_{n,m}\partial_{x_{n,m}} + \tau_{n,m}\partial_{t_{n,m}} + \phi^1_{n,m}\partial_{u_{n,m}} + \phi^2_{n,m}\partial_{v_{n,m}} \tag{2.3.274}$$

where $\xi_{n,m}$, $\tau_{n,m}$, $\phi^1_{n,m}$ and $\phi^2_{n,m}$, are functions of $(x_{n,m}, t_{n,m}, u_{n,m}, v_{n,m})$. The prolongation of $\hat{X}_{n,m}$ is given by

$$\mathrm{pr}\hat{X}_{n,m} = \sum_{n,m}\left[\hat{X}_{n,m} + \phi^{1,t}_{n,m}\partial_{\dot{u}_{n,m}} + \phi^{2,t}_{n,m}\partial_{\dot{v}_{n,m}}\right] \tag{2.3.275}$$

where $\phi^{1,t}_{n,m}$ and $\phi^{2,t}_{n,m}$ are given by

$$\phi^{1,t}_{n,m} = D_{t_{n,m}} \phi^1_{n,m} - \dot{u}_{n,m} D_{t_{n,m}} \tau_{n,m} - u_{n,m,x_{n,m}} D_{t_{n,m}} \xi_{n,m},$$
$$\phi^{2,t}_{n,m} = D_{t_{n,m}} \phi^2_{n,m} - \dot{v}_{n,m} D_{t_{n,m}} \tau_{n,m} - v_{n,m,x_{n,m}} D_{t_{n,m}} \xi_{n,m},$$

where by a dot we indicate the derivative with respect to $t_{n,m}$. In (2.3.275) the sum is extended to all points of the lattice present in the equation.

We can associate to the integrable dNLS (2.3.273) a lattice given by the following equations:

$$x_{n+1,m} - x_{n,m} = h$$
$$x_{n,m+1} = x_{n,m} \qquad (2.3.276)$$
$$t_{n+1,m} = t_{n,m}.$$

The application of the prolonged generator (2.3.274) to (2.3.273, 2.3.276) gives the defining equations for the symmetry, that must hold on solutions of (2.3.273, 2.3.276). The system of equations obtained by applying (2.3.274) to the lattice (2.3.276) implies that $\xi_{n,m}$ is constant and $\tau_{n,m} = \tau_{n,m}(t_{n,m})$. Applying (2.3.274) to the dNLS we obtain that there are only three independent intrinsic point symmetries:

$$\hat{X}_{n,m} = a\partial_{x_{n,m}} + b\partial_{t_{n,m}} + c(v_{n,m}\partial_{u_{n,m}} - u_{n,m}\partial_{v_{n,m}}) \qquad (2.3.277)$$

which, going back to the notation of (2.3.240) read:

$$\hat{X}_0 = q_n\partial_{q_n} - q_n^*\partial_{q_n^*}, \qquad \hat{Z} = \partial_n, \qquad \hat{T} = \partial_t. \qquad (2.3.278)$$

Note that the form of the two last symmetries, involving just the independent variables, depend strictly on the form of the lattice (2.3.276). Selecting other type of lattice would change $\hat{Z}$ and $\hat{T}$.

3.4.3. *Generalized symmetries of the dNLS.* Generalized symmetries of the dNLS can be constructed by considering flows

$$\frac{d}{d\epsilon} q_n = f(t, n, q_n, q_{n+1}, q_{n-1}, \ldots) \qquad (2.3.279)$$

in the group parameter ϵ commuting with (2.3.240). Due to the one-to-one correspondence between the evolution of q_n and that of the reflection coefficient R, commuting flows acting in the solution space of the evolution equation have counterparts in the form of symmetries acting in the space of the reflection coefficient. Thus we can define the symmetries by looking for commuting flows of the reflection coefficients $R(t, z, \epsilon)$. In the case of the non isospectral symmetries we have

$$\frac{d}{d\epsilon} R(t, z, \epsilon) = O(z^2, z^{-2}, t)R(t, z, \epsilon), \qquad \frac{d}{d\epsilon} z(t, \epsilon) = z\tilde{O}(z^2, z^{-2}, t) \qquad (2.3.280)$$

to commute with those of the dNLS (2.3.264). From this commutation we get:

$$O(z^2, z^{-2}, t) = -\frac{2it}{h^2}\left(z^2 - z^{-2}\right)\tilde{O}(z^2, z^{-2}, t) + \bar{O}(z^2, z^{-2}), \qquad \frac{d\tilde{O}}{dt} = 0 \qquad (2.3.281)$$

where $\bar{O}$ is an arbitrary entire function of z^2 and z^{-2} of the form (2.3.257) and $\tilde{O}$ is constrained so that O is of the form (2.3.257).

We can distinguish various types of symmetries, corresponding to isospectral or non isospectral flows. In the isospectral case we have that $\tilde{O} = 0$ and then from (2.3.281), (2.3.257) we get that O is given in terms of an arbitrary entire function of z^2 and z^{-2}. In this case we can define a base of symmetry generators in the spectral space as

$$\hat{X}_j^s = z^{2j} R\partial_R, \tag{2.3.282}$$

with j an arbitrary integer.

In the non isospectral case, $\tilde{O} \neq 0$ and $O = 0$, and we can choose the following base of symmetry generators:

$$\hat{Y}_j^s = z^{2j} \left[\frac{2it}{h^2}(z^2 - z^{-2})R - zR_z \right] \partial_R. \tag{2.3.283}$$

We present explicitly the simplest symmetries of the dNLS, which we will consider later when dealing with its continuous limit and the contraction of its infinite-dimensional algebra of generalized symmetries to the Lie point symmetries:

$$\begin{aligned}
\hat{X}_0^s &= q_n\partial_{q_n} - q_n^*\partial_{q_n^*},\\
\hat{X}_1^s &= (1-\sigma|q_n|^2)q_{n+1}\partial_{q_n} - (1-\sigma|q_n|^2)q_{n-1}^*\partial_{q_n^*}\\
\hat{X}_{-1}^s &= (1-\sigma|q_n|^2)q_{n-1}\partial_{q_n} - (1-\sigma|q_n|^2)q_{n+1}^*\partial_{q_n^*}\\
\hat{X}_2^s &= (1-\sigma|q_n|^2)\left[q_{n+2}(1-\sigma|q_{n+1}|^2) - \sigma q_{n+1}(q_n^*q_{n+1}+q_{n-1}^*q_n)\right]\partial_{q_n}\\
&\quad - (1-\sigma|q_n|^2)\left[q_{n-2}^*(1-\sigma|q_{n-1}|^2) - \sigma q_{n-1}^*(q_n^*q_{n+1}+q_{n-1}^*q_n)\right]\partial_{q_n^*}\\
\hat{X}_{-2}^s &= (1-\sigma|q_n|^2)\left[q_{n-2}(1-\sigma|q_{n-1}|^2) - \sigma q_{n-1}(q_n^*q_{n-1}+q_{n+1}^*q_n)\right]\partial_{q_n}\\
&\quad - (1-\sigma|q_n|^2)\left[q_{n+2}^*(1-\sigma|q_{n+1}|^2) - \sigma q_{n+1}^*(q_nq_{n+1}^*+q_{n-1}q_n^*)\right]\partial_{q_n^*}\\
\hat{Y}_0^s &= \left[-\frac{2it}{h^2}(1-\sigma|q_n|^2)(q_{n+1} - q_{n-1}) - (2n+1)q_n\right]\partial_{q_n}\\
&\quad + \left[-\frac{2it}{h^2}(1-\sigma|q_n|^2)(q_{n+1}^* - q_{n-1}^*) + (2n+1)q_n^*\right]\partial_{q_n^*}\\
\hat{Y}_1^s &= \left\{\frac{2it}{h^2}(1-\sigma|q_n|^2)\left[q_{n+2} - \sigma(q_{n+1}^*q_{n+2} + q_n^*q_{n+1} + q_{n-1}^*q_n)q_{n+1}\right]\right.\\
&\quad \left. - \frac{2it}{h^2}q_n + (2n+3)(1-\sigma|q_n|^2)q_{n+1} + 2\sigma q_nS_n^*\right\}\partial_{q_n}\\
&\quad + \left\{-\frac{2it}{h^2}(1-\sigma|q_n|^2)\left[q_{n-2}^* - \sigma(q_{n-1}q_{n-2}^* + q_nq_{n-1}^* + q_{n+1}q_n^*)q_{n-1}^*\right]\right.\\
&\quad \left. + \frac{2it}{h^2}q_n^* - (2n-1)(1-\sigma|q_n|^2)q_{n-1}^* - 2\sigma q_n^*S_n^*\right\}\partial_{q_n^*}\\
\hat{Y}_{-1}^s &= \left\{-\frac{2it}{h^2}(1-\sigma|q_n|^2)\left[q_{n-2} - \sigma(q_{n-1}^*q_{n-2}+q_n^*q_{n-1}+q_{n+1}^*q_n)q_{n-1}\right]\right.\\
&\quad \left. + \frac{2it}{h^2}q_n + (2n-1)(1-\sigma|q_n|^2)q_{n-1} + 2\sigma q_nS_n\right\}\partial_{q_n}\\
&\quad + \left\{\frac{2it}{h^2}(1-\sigma|q_n|^2)\left[q_{n+2}^* - \sigma(q_{n+1}q_{n-+}^* + q_nq_{n+1}^* + q_{n-1}q_n^*)q_{n+1}^*\right]\right.\\
&\quad \left. - \frac{2it}{h^2}q_n^* - (2n+3)(1-\sigma|q_n|^2)q_{n+1}^* - 2\sigma q_n^*S_n\right\}\partial_{q_n^*}
\end{aligned} \tag{2.3.284}$$

where the function S_n appearing in symmetries $\hat{Y}_1^s$ and $\hat{Y}_{-1}^s$ was defined in (2.3.266). Taking into account the temporal evolution of dNLS (2.3.240) we have

$$\dot{S}_n = \frac{i}{h^2}\left[|q_n|^2 - (1-\sigma|q_n|^2)q_{n-1}^*q_{n+1}\right].$$

Note that $\hat{X}_0^s$ is a point symmetry, and only one additional independent point symmetry can be obtained, $\hat{T}^s = 2\hat{X}_0^s - \hat{X}_1^s - \hat{X}_{-1}^s = \dot{q}_n\partial_{q_n} - \dot{q}_n^*\partial_{q_n^*}$. The remaining independent symmetries are generalized symmetries, and $\hat{Y}_i^s$ with $|i| \geq 1$ are non-local symmetries. The lattice point symmetry $\hat{Z}^s = \partial_n$ in (2.3.278) cannot be expressed as a symmetry (2.3.279) depending only on a finite number of fields q_i.

The structure of the algebra L of infinitesimal symmetries of dNLS can be inferred from the commutation relations

$$[\hat{X}^s_\ell, \hat{X}^s_j] = 0, \qquad [\hat{X}^s_\ell, \hat{Y}^s_j] = -2\ell \hat{X}^s_{\ell+j} \tag{2.3.285}$$

$$[\hat{Y}^s_\ell, \hat{Y}_j] = -2(\ell-j)\hat{Y}^s_{\ell+j}. \tag{2.3.286}$$

The subalgebra L_1 generated by $\hat{X}^s_\ell$, $\ell \in \mathbb{Z}$, is abelian. The symmetries $\hat{Y}^s_\ell$, $\ell \in \mathbb{Z}$, given by (2.3.286), also generate a subalgebra L_0, which is perfect, i. e. $[L_0, L_0] = L_0$. The structure of the whole algebra is that of of semidirect sum

$$L = L_0 \ni\!\!\!\!+ L_1,$$

where $L_1 = \{\hat{X}^s_0, \hat{X}^s_1, \hat{X}^s_2, \cdots\}$ is an Abelian ideal. The non local infinitesimal generator $\hat{Y}^s_1$ is the master symmetry for the Lie algebra L_1 as one can see from (2.3.285).

3.4.4. *Continuous limit of the symmetries of the dNLS.* The continuous limit

$$q_n(t) = hu(x,t), \qquad x = nh \tag{2.3.287}$$

transforms the dNLS equation (2.3.240) into the NLS equation (1.3.3)

$$iu_t + u_{xx} - 2\sigma|u|^2 u = O(h^2).$$

Now we study the relation between the symmetries of (2.3.240) and those of its continuous limit, (1.3.3). In particular we are interested in finding precursors of the point symmetries of NLS in the symmetry algebra of (2.3.240). Let us consider the point symmetries of NLS written in the evolutionary formalism:

$$\begin{aligned} y_1 &= u\partial_u - u^*\partial_{u^*}, \\ y_2 &= u_x\partial_u + u^*_x\partial_{u^*}, \\ y_3 &= u_t\partial_u + u^*_t\partial_{u^*}, \\ y_4 &= (ixu - 2tu_x)\partial_u - (ixu^* + 2tu^*_x)\partial_{u^*}, \\ y_5 &= (u + 2tu_t + xu_x)\partial_u + (u^* + 2tu^*_t + xu^*_x)\partial_{u^*}. \end{aligned} \tag{2.3.288}$$

This is a five-dimensional algebra, while we have seen that the point symmetry subalgebra of dNLS (2.3.240) is *three*-dimensional, generated by the vector fields $\hat{X}^s_0$ and T^s and the lattice symmetry $\hat{Z}^s = \partial_n$ (cf. (2.3.278)). The remaining two independent point symmetries of NLS must be recovered from the continuous limit of some generalized symmetry of (2.3.240).

Taking into account that under the transformation (2.3.287), the function S_n appearing in (2.3.284) reduces to

$$S_n \to -h\int |u|^2 dx + h^2 \int (u_x u^* - uu^*_x)dx + \mathcal{O}(h^3),$$

the appropriate combinations of the symmetries, such that in the continuous limit we get (2.3.288), are

$$
\begin{aligned}
&\hat{Z}_1 \equiv \hat{X}_0^s = y_1, \qquad \hat{Z}_2 \equiv \frac{\hat{X}_1 -^s \hat{X}_0^s}{h} = y_2 + \mathcal{O}(h)\\
&\hat{Z}_3 \equiv -\frac{i}{h^2}\hat{T}^s = y_3 + \mathcal{O}(h^2)\\
&\hat{Z}_4 \equiv -ih\frac{\hat{Y}_0^s + \hat{X}_0^s}{2} = y_4 + \mathcal{O}(h^2), \qquad \hat{Z}_5 \equiv \frac{\hat{Y}_1^s - \hat{Y}_{-1}^s}{4} = y_5 + \mathcal{O}(h)
\end{aligned}
$$

The point symmetries $\hat{X}_0^s$ and $\hat{T}^s$ of the dNLS equation produce, in the continuous limit, the point symmetries y_1 and y_3 of NLS. NLS point symmetries y_2, y_4 and y_5 have to be recovered from non point symmetries $\hat{X}_1^s$, $\hat{Y}_0^s$, $\hat{Y}_1^s$ and $\hat{Y}_{-1}^s$ of dNLS besides $\hat{X}_0^s$. Thus, a contraction of the Lie algebra of symmetries of dNLS is occurring [371, 373].

The non-zero elements of the commutator tables for the symmetries of (2.3.240) and for the NLS symmetries are, respectively

	$\hat{Z}_4$	$\hat{Z}_5$
$\hat{Z}_2$	$i\hat{Z}_1 + ih\hat{Z}_2$	$-\hat{Z}_2+\frac{ih}{2}\hat{Z}_3-\frac{h^2}{2}\hat{Z}_6$
$\hat{Z}_3$	$2\hat{Z}_2+ih\hat{Z}_3$	$-2\hat{Z}_3-\frac{h^2}{2}i\hat{Z}_7$
$\hat{Z}_4$	0	$\hat{Z}_4+\frac{h}{2}i\hat{Z}_1+\frac{h^2}{2}\hat{Z}_8-\frac{h^3}{2}\hat{Z}_3$
$\hat{Z}_5$	$-[\hat{Z}_4, \hat{Z}_5]$	0

,

	y_4	y_5
y_2	iy_1	$-y_2$
y_3	$2y_2$	$-2y_3$
y_4	0	y_4
y_5	$-y_4$	0

where $\hat{Z}_6 \equiv \hat{(}X_2^s + 3\hat{X}_0^s - 3\hat{X}_1^s - \hat{X}_{-1}^s)/h^3$, $\hat{Z}_7 \equiv (\hat{X}_{-2}^s + 6\hat{X}_0^s - 4\hat{X}_1^s - 4\hat{X}_{-1}^s + \hat{X}_2^s)/h^4$ and $\hat{Z}_8 \equiv i(\hat{Y}_1^s + \hat{Y}_{-1}^s + 2\hat{Y}_0^s - 2\hat{X}_{-1}^s - 2\hat{X}_1^s + 4\hat{X}_0^s)/(2h)$ are combinations with well defined continuous limit. We can see from the previous tables how the point-symmetry subalgebra of (1.3.3), generated by y_1, y_2, y_3, y_4 and y_5, is the image under a contraction of the set generated by $\hat{Z}_1$, $\hat{Z}_2$, $\hat{Z}_3$, $\hat{Z}_4$ and $\hat{Z}_5$. This set contains the subalgebra of point algebras of (2.3.240), but it is not an algebra in itself.

3.4.5. *Symmetry reductions.* In this Section we present results obtained by carrying out the symmetry reduction of the dNLS with respect to few Lie point symmetries. For completeness we firstly present the corresponding results for the continuous NLS equation (1.3.3).

Continuous case It is convenient now to rewrite the Lie point symmetries (2.3.288) of the NLS equation (1.3.3) using the polar representation of the complex function $u(x,t)$, i.e. $u(x,t) = \rho(x,t)\exp i\phi(x,t)$.

$$
\begin{aligned}
y_1 &= -i\partial_\phi, \qquad y_2 = \rho_x\partial_\rho + \phi_x\partial_\phi, \qquad y_3 = \rho_t\partial_\rho + \phi_t\partial_\phi,\\
y_4 &= -2t\partial_x + x\partial_\phi, \qquad y_5 = x\partial_x + 2t\partial_t - \rho\partial_\rho.
\end{aligned} \tag{2.3.289}
$$

The symmetry reductions are as follows.

- In the case of y_2 the NLS equation (1.3.3) reduces to

$$
\rho_t = 0; \qquad \phi_t = -2\sigma\rho^2, \tag{2.3.290}
$$

  which can be easily solved and give $u(x,t) = \rho_0 e^{i(\phi_0 + 2\varepsilon\rho_0^2 t)}$, ρ_0 and ϕ_0 being arbitrary integration constants.
- In the case of y_3 the NLS equation (1.3.3) reduces to

$$
2\rho_x\phi_x + \rho\phi_{xx} = 0; \qquad \rho_{xx} = \rho\phi_x^2 + 2\sigma\rho^3 \tag{2.3.291}
$$

which can be reduced to an elliptic equation for the variable $v = \rho^2$

$$v_x = 2\sqrt{\sigma v^3 + K_1 v - K_2^2} \tag{2.3.292}$$

where K_1 and K_2 are integration constants and ϕ is such that $\phi_x = K_2/v$. This case with an appropriate choice of the constants K_1 and K_2 reduces to the soliton solution of the NLS.

- In the case of y_4 the invariant variables are:

$$\rho(x,t) = \rho_0(t); \qquad \phi(x,t) = \phi_0(t) - \frac{x^2}{4t}, \tag{2.3.293}$$

and the NLS equation (1.3.3) is solved by $u(x,t) = \frac{K_1}{\sqrt{t}} \cdot \exp(i(K_2 - 2\sigma K_1^2 \log(t) - \frac{x^2}{4t}))$, where K_1 and K_2 are integration constants.

- Finally in the case of y_5 the invariant variables are:

$$\rho(x,t) = \frac{\rho_0(\eta)}{\sqrt{t}}; \qquad \phi(x,t) = \phi_0(\eta); \qquad \eta = \frac{x}{\sqrt{t}}. \tag{2.3.294}$$

In terms of these variables the NLS equation (1.3.3) reduces to the system

$$\rho_0 + \eta\rho_{0,\eta} + 4\rho_{0,\eta}\phi_{0,\eta} + 2\rho_0\phi_{0,\eta\eta} = 0 \tag{2.3.295}$$

$$\rho_{0,\eta\eta} - \frac{1}{2}\eta\rho_0\phi_{0,\eta} - \rho_0\phi_{0,\eta}^2 - 2\sigma\rho_0^3 = 0. \tag{2.3.296}$$

Defining the new variable $Y(\eta)$ so that $\phi_{0,\eta} = -\frac{1}{4}(\eta + \frac{Y}{Y_\eta})$ and $\rho_0^2 = Y_\eta$, (2.3.295) is identically satisfied and (2.3.296) reduces to

$$Y_{\eta\eta}^2 + 4(\eta Y_\eta - Y)^2 - 16\epsilon Y_\eta^3 - 4\sigma\mu^2 Y_\eta = 0, \tag{2.3.297}$$

where μ is an arbitrary integration constant. For $\sigma = -1$ it can be shown [108] that it has the only solution $Y = 0$ while for $\sigma = 1$ (2.3.297) can be reduced to the Painlevé IV equation [418]

$$WW_{\eta\eta} = \frac{1}{2}W_\eta^2 - 6W^4 + 8\eta W^3 - 2\eta^2 W^2 - \frac{1}{2}(\mu - 1)^2, \tag{2.3.298}$$

by defining:

$$Y = \frac{1}{2}W(W - \eta)^2 + \frac{1}{8W}(W_\eta^2 - 2W_\eta - \mu^2 + 1).$$

Discrete case

The symmetries that we will use to find reductions of dNLS are $\hat{Z}_2$ and $\hat{Z}_3$ while the remaining cases are left to the diligent reader.

Reduction by $\hat{Z}_2$ The reduction is obtained by solving the following equation:

$$(1 - \sigma|q_n|^2)(q_{n+1} - q_{n-1}) = 0. \tag{2.3.299}$$

One solution is given by $q_{n+1} = q_{n-1}$, i. e.

$$q_n = \alpha(t) + (-1)^n \beta(t).$$

As for this solution $q_{2n} = \alpha(t) + \beta(t) = \tilde{a}(t)$, $q_{2n+1} = \alpha(t) - \beta(t) = \tilde{b}(t)$, from (2.3.240) we get that $\tilde{a}$ and $\tilde{b}$ must satisfy the following equations

$$i\dot{\tilde{a}} = \frac{2}{h^2}\left[\tilde{a} - (1 - \sigma|\tilde{a}|^2)\tilde{b}\right]$$
$$i\dot{\tilde{b}} = \frac{2}{h^2}\left[\tilde{b} - (1 - \sigma|\tilde{b}|^2)\tilde{a}\right]$$

Defining $\tilde{a} = \exp(-\frac{2i}{h^2}t)a(t)$ and $\tilde{b} = \exp(-\frac{2i}{h^2}t)b(t)$ we get

$$i\dot{a} + \frac{2}{h^2}(1 - \sigma|a|^2)b = 0, \qquad i\dot{b} + \frac{2}{h^2}(1 - \sigma|b|^2)a = 0, \tag{2.3.300}$$

where $a = \rho_a \exp(i\phi_a)$, $b = \rho_b \exp(i\phi_b)$. Defining $B = \rho_b\rho_a \sin(\phi_a - \phi_b)$, (2.3.300) can be solved in terms of an elliptic integral

$$\dot{B} = -\frac{1}{h^2}\sqrt{B^4 + 8(\sigma K_0 - 2)B^2 + K_1}, \tag{2.3.301}$$

where K_0 and K_1 are two integration constants. In terms of the elliptic function B, ρ_a and ρ_b are given by

$$\rho_a^2 = \frac{1}{8}\sigma B^2 - \frac{h^2}{8}\dot{B} + \frac{K_0}{2},$$
$$\rho_b^2 = \frac{1}{8}\sigma B^2 + \frac{h^2}{8}\dot{B} + \frac{K_0}{2}$$

and the phases ϕ_a and ϕ_b are obtained by quadrature from the following equations:

$$\dot{\phi}_a = \frac{\dot{\rho}_a}{\rho_a}\frac{\sqrt{\rho_a^2\rho_b^2 - B^2}}{B},$$
$$\dot{\phi}_b = -\frac{\dot{\rho}_b}{\rho_b}\frac{\sqrt{\rho_a^2\rho_b^2 - B^2}}{B}.$$

Realizing that (2.3.300) admits two conserved quantities

$$(1 - \sigma\rho_a^2)(1 - \sigma\rho_b^2) = C_1, \tag{2.3.302}$$
$$\rho_a\rho_b \cos(\phi_b - \phi_a) = C_2 \tag{2.3.303}$$

we can give an alternative description of the solutions, readily interpretable in terms of a non linear two-body chain [449]. One can parametrize the algebraic curve given by (2.3.302) in terms of the evolution of elliptic functions as a system $\dot{\theta} = f(\theta, \phi)$, $\dot{\phi} = g(\theta, \phi)$ (where $\phi = \phi_b - \phi_a$). For example, for $C_1 = 2$ we have

$$\rho_a = \frac{1}{\sqrt{2}}\frac{\mathrm{sn}(\sqrt{2}\theta, 1/2)}{\mathrm{dn}(\sqrt{2}\theta, 1/2)}, \qquad \rho_b = \mathrm{cn}(\sqrt{2}\theta, 1/2)$$

(where sn, dn and cn are Jacobi elliptic functions [13]) and the evolution is given by

$$\dot{\theta} = \frac{2}{h^2}\sin\phi,$$
$$\dot{\phi} = \frac{2}{h^2}\frac{\rho_b^2 - \rho_a^2}{\rho_a\rho_b}\cos\phi = \frac{4\sqrt{2}}{h^2}\frac{\mathrm{cn}(2\sqrt{2}\theta, 1/2)}{\mathrm{sn}(2\sqrt{2}\theta, 1/2)}\cos\phi.$$

The second conserved quantity (2.3.303) provides us with the equation of the orbits

$$\frac{\mathrm{dn}(2\sqrt{2}\theta, 1/2) - 1}{\mathrm{sn}(2\sqrt{2}\theta, 1/2)} \cos\phi = \frac{\mathrm{sn}(\sqrt{2}\theta, 1/2)\,\mathrm{cn}(\sqrt{2}\theta, 1/2)}{\mathrm{dn}(\sqrt{2}\theta, 1/2)} \cos\phi = \sqrt{2}\, C_2, \tag{2.3.304}$$

which is represented in Figure 2.1.

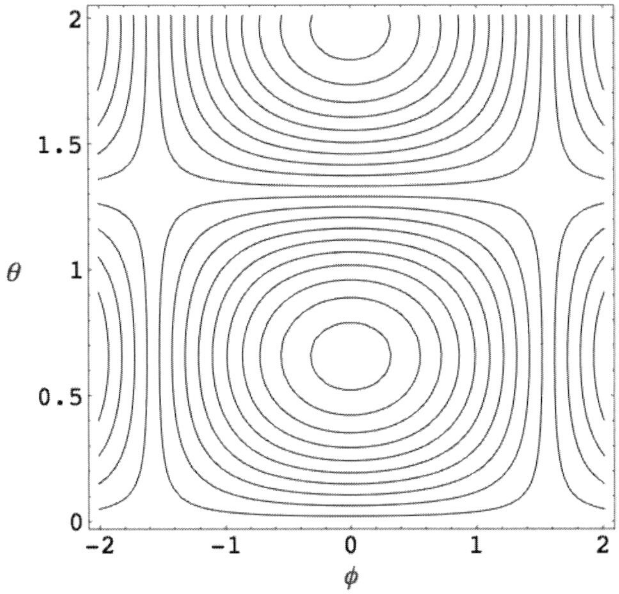

FIGURE 2.1. Level curves of (2.3.304), θ vs. ϕ for $\hat{Y}_2$-reduced dNLS, for $C_1 = 2$ and different values of C_2 [reprinted from [367]].

If $\sigma = +1$ there are additional solutions. The reduced equation (2.3.299) is equivalent the the following prescription

- A point n in the lattice with $|q_n| = 1$, has neighbours with arbitrary values q_{n+1}, q_{n-1}.
- A point n in the lattice with $|q_n| \neq 1$ has neighbours with equal values $q_{n+1} = q_{n-1}$.

A typical solution is of the form:

$$q_{2n} = e^{i\theta(t)}, \quad q_{2n+1} = \rho_n(t) e^{i\phi_n(t)}. \tag{2.3.305}$$

Substituting (2.3.305) into (2.3.240) we get for $\theta(t)$, $\rho_n(t)$ and $\phi_n(t)$ the equations

$$\begin{aligned} \dot{\theta} &= -\frac{2}{h^2}, \quad \dot{\rho}_n = -\frac{2}{h^2}(1 - \rho_n^2)\sin(\theta - \phi_n), \\ \dot{\phi}_n &= -\frac{2}{h^2} + \frac{2}{h^2}\frac{1 - \rho_n^2}{\rho_n}\cos(\theta - \phi_n). \end{aligned} \tag{2.3.306}$$

Let us notice that in these equations the dependence on n is parametric. It is possible to find the general solution of (2.3.306), given by

$$\theta(t) = -\frac{2}{h^2}t + \theta_0, \tag{2.3.307}$$

$$\rho_n(t) = 1 + A_n + r_n^2(t), \tag{2.3.308}$$

$$\sin\phi_n(t) = \frac{-r_n \cos\theta \pm \sqrt{1+A_n}\sin\theta}{1 + A_n + r_n^2} \tag{2.3.309}$$

where A_n is an arbitrary function of n, and there are three different expressions for $q_n(t)$ depending on the value of A_n

$$q_n^{(1)}(t) = \sqrt{|A_n|}\frac{e^{\frac{4\sqrt{|A_n|}}{h^2}(t-t_n)}+1}{1-e^{\frac{4\sqrt{|A_n|}}{h^2}(t-t_n)}}, \qquad \text{if } -1 < A_n < 0,$$

$$q_n^{(2)}(t) = -\frac{h^2}{2(t-t_n)}, \qquad \text{if } A_n = 0,$$

$$q_n^{(3)}(t) = \sqrt{|A_n|}\tan\left[\frac{2A_n}{h^2}(t-t_n)\right] m \qquad \text{if } 0 < A_n,$$

with t_n an arbitrary function of n.

So the general reduced solution when $\sigma = +1$ can be described as a piecewise function q_n on the lattice, as it is discussed in [177]. Every piece or "domain" is characterized by a sequence of points with an equal value $q_n = e^{i\theta(t)}$, interspaced with a sequence of points with values $q_n = \rho_n(t)e^{i\phi_n(t)}$ of modulus and phase given by (2.3.308, 2.3.309). At the extremes of each domain we find points where q_n has modulus 1; different domains are characterized by a priori different values of the phases. See Fig. 2.2 for an example.

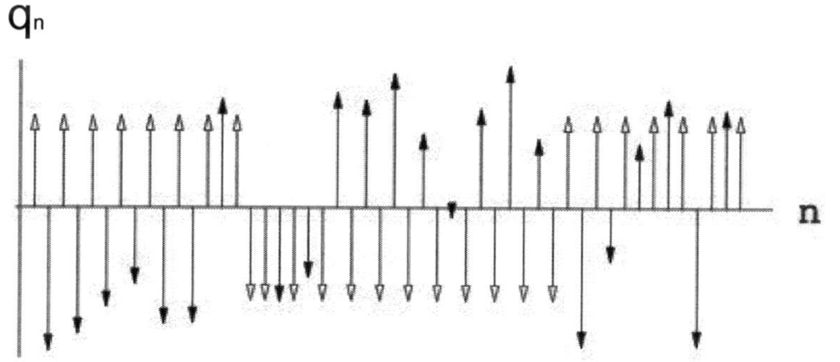

FIGURE 2.2. Schematic plot of q_n at a given time t, showing three "domains", solution of the $\hat{Z}_2$-reduced dNLS. The white arrowheads correspond to the real values of the points of modulus 1 that define the domains. The black arrowheads correspond to the real values of the points inside a domain of modulus and phase given by (2.3.308, 2.3.309) [reprinted from [367]].

Reduction by $\hat{\mathbf{Z}}_3$ The symmetry reduction is obtained by solving the following equation

$$(1-\sigma|q_n|^2)(q_{n-1}+q_{n+1})-2q_n=0. \tag{2.3.310}$$

Taking into account the dNLS equation (2.3.240), (2.3.310) implies $\dot{q}_n=0$. Writing q_n in polar coordinates as

$$q_n=\rho_n\exp(i\theta_n) \tag{2.3.311}$$

we have that

$$(1-\sigma\rho_n^2)\left[\rho_{n+1}\sin(\theta_{n+1}-\theta_n)-\rho_{n-1}\sin(\theta_n-\theta_{n-1})\right]=0 \tag{2.3.312}$$

$$(1-\sigma\rho_n^2)\left[\rho_{n+1}\cos(\theta_{n+1}-\theta_n)-\rho_{n-1}\cos(\theta_n-\theta_{n-1})\right]=2\rho_n \tag{2.3.313}$$

We can see that if $\sigma\rho_n^2\neq 1$ (2.3.312) reads

$$\rho_{n+1}\sin(\theta_{n+1}-\theta_n)=\rho_{n-1}\sin(\theta_n-\theta_{n-1}) \tag{2.3.314}$$

which can be once integrated to get

$$\sin(\theta_{n+1}-\theta_n)=\frac{C}{\rho_{n+1}\rho_n} \tag{2.3.315}$$

where C is an arbitrary integration constant. Substituting (2.3.315) into (2.3.313) and taking into account that

$$\cos(\theta_{n+1}-\theta_n)=\frac{1}{\rho_{n+1}\rho_n}\sqrt{\rho_{n+1}^2\rho_n^2-C^2}$$

we get the following OΔE for ρ_n^2

$$\sqrt{\rho_n^2\rho_{n+1}^2-C^2}+\sqrt{\rho_n^2\rho_{n-1}^2-C^2}=\frac{2\rho_n^2}{1-\sigma\rho_n^2}. \tag{2.3.316}$$

Alternatively, substituting ρ_{n+1} and ρ_{n-1} from (2.3.315) in (2.3.313) we obtain:

$$C(1-\sigma\rho_n^2)\left[\text{ctan}(\theta_{n+1}-\theta_n)+\text{ctan}(\theta_n-\theta_{n-1})\right]=2\rho_n^2. \tag{2.3.317}$$

Then

$$\rho_n^2=\frac{C\left[\text{ctan}(\theta_{n+1}-\theta_n)+\text{ctan}(\theta_n-\theta_{n-1})\right]}{2+\sigma C\left[\text{ctan}(\theta_{n+1}-\theta_n)+\text{ctan}(\theta_n-\theta_{n-1})\right]} \tag{2.3.318}$$

and substituting (2.3.318) in (2.3.314) we find an equation for the phases

$$\frac{\text{ctan}(\theta_{n+2}-\theta_{n+1})+\text{ctan}(\theta_{n+1}-\theta_n)}{2+\sigma C[\text{ctan}(\theta_{n+2}-\theta_{n+1})+\text{ctan}(\theta_{n+1}-\theta_n)]}\sin^2(\theta_{n+1}-\theta_n)=$$
$$=\frac{\text{ctan}(\theta_n-\theta_{n-1})+\text{ctan}(\theta_{n-1}-\theta_{n-2})}{2+\sigma C[\text{ctan}(\theta_n-\theta_{n-1})+\text{ctan}(\theta_{n-1}-\theta_{n-2})]}\sin^2(\theta_n-\theta_{n-1}).$$

3.5. The DΔE Burgers. A DΔE will be linearizable by a discrete Cole-Hopf transformation

$$u_n(t) = \frac{\psi_{n+1}(t)}{\psi_n(t)} \tag{2.3.319}$$

if the Lax operator (2.2.184) is given by

$$L_n^d = S - u_n, \quad L_n^d \psi_n(t) = 0. \tag{2.3.320}$$

We define the t evolution of the function $\psi_n(t)$ as

$$\dot{\psi}_n(t) = -M_n \psi_n(t), \tag{2.3.321}$$

where M_n is an operator in S whose coefficients depend on u_n and its shifted values. The Lax equation (2.2.8) is still valid and, *mutatis mutandis*, we can assume the existence of a hierarchy of M_n operators so that we have

$$\dot{L}_n^d(u_n) = \left[L_n^d(u_n),\, M_n\right] = V_n, \tag{2.3.322}$$

$$\dot{L}_n^d(\tilde{u}_n) = \left[L_n^d(\tilde{u}_n),\, \tilde{M}_n\right] = \tilde{V}_n. \tag{2.3.323}$$

Defining

$$\tilde{M}_n = L_n^d(\tilde{u}_n)M_n + F_n S + G_n \tag{2.3.324}$$

and taking into account that (2.3.322) and (2.3.323) are operator equations valid on $\psi_n(t)$ we obtain the following relation between V_n and $\tilde{V}_n$:

$$\tilde{V}_n = -u_n V_n + \left[V_{n+1} - u_n F_n + u_{n+1} F_n + G_{n+1} - G_n\right] S + \left(F_{n+1} - F_n\right) S^2. \tag{2.3.325}$$

We can define two operators $\mathcal{N}_b$ and $\mathcal{L}_b$ so that

$$\tilde{V}_n = \mathcal{N}_b \tilde{F}_n = u_n(S-1)\tilde{F}_n, \quad \tilde{F}_n = \mathcal{L}_b F_n + F^{(0)} = u_n F_{n+1} + F^{(0)}, \tag{2.3.326}$$

where $F^{(0)}$ is a constant with respect to n. The hierarchy of autonomous DΔEs Burgers is given by

$$\dot{u}_n = \mathcal{N}_b \sum_{k=0}^{N} \mathcal{L}_b^k F^{(k)}, \tag{2.3.327}$$

where $F^{(k)}$ are constants with respect to n, but may depend on t. The corresponding M_n operators are

$$M_n = -\sum_{k=0}^{N} F^{(k)} S^k. \tag{2.3.328}$$

The simplest isospectral equations of the hierarchy are

$$\dot{u}_n = u_n(u_n - u_{n+1}), \tag{2.3.329}$$

$$\dot{u}_n = u_n u_{n+1}(u_{n+2} - u_n). \tag{2.3.330}$$

As one can see from (2.3.329) and (2.3.330) the shifts are non symmetric with respect to the index n, an indication of the linearizability of the DΔEs we derived (see Section 3.2.4.1). Similar results could be obtained using negative shifts when the Lax operator is written in terms of S^{-1}. In this case (2.3.329) will read:

$$\dot{u}_n = u_n(u_n - u_{n-1}), \tag{2.3.331}$$

$$\dot{u}_n = u_n u_{n-1}(u_{n-2} - u_n). \tag{2.3.332}$$

Eq. (2.3.329) turns out to be the Bäcklund transformation of the PDE Burgers as given in Section 2.2.5 by (2.2.210).

Non isospectral equations can be obtained with the same procedure as for the continuous Burgers, *mutatis mutandis*. We obtain

$$\dot{u}_n = \mathcal{N}_b \sum_{k=0}^{K} \mathcal{L}_b^k n G^{(k)}, \quad M = \sum_{k=0}^{K} G^{(k)} S^k n, \tag{2.3.333}$$

and the simplest equations are

$$\begin{aligned} \dot{u}_n &= u_n \Big[u_{n+1}(n+2) - u_n(n+1) \Big], \\ \dot{u}_n &= u_n u_{n+1} \Big[u_{n+2}(n+3) - u_n(n+2) \Big]. \end{aligned} \tag{2.3.334}$$

3.5.1. *Bäcklund transformations for the DΔE Burgers and its non linear superposition formula.* We will use here the same procedure as we used to obtain the Bäcklund transformation for the KdV and the Toda lattice. We assume the existence of two different solutions u_n and $\tilde{u}_n$ of the spectral problem (2.3.320) where the functions ψ_n and $\tilde{\psi}_n$ are related by a Darboux operator $D_n(u_n, \tilde{u}_n)$. The existence of a hierarchy of Bäcklund transformations implies the existence of a hierarchy of Darboux operators. So defining a new Darboux operator as $\tilde{D}_n(u_n, \tilde{u}_n)$ we impose the following equations

$$L_n^d(\tilde{u}_n) D - D L_n^d(u_n) = V_n, \quad L_n^d(\tilde{u}_n)\tilde{D} - \tilde{D} L_n^d(u_n) = \tilde{V}_n, \tag{2.3.335}$$

and

$$\tilde{D} = L_n^d(\tilde{u}_n) D + F_n S + G_n. \tag{2.3.336}$$

Inserting (2.3.320) into the second of the equations in (2.3.333), taking into account the first one of the equations in (2.3.335) and the fact that

$$S\psi_n(t) = u_n \psi_n(t), \quad S^2 \psi_n(t) = u_n u_{n+1} \psi_n(t) \tag{2.3.337}$$

we get

$$\tilde{V}_n = \Lambda_n^b V_n + V_n^{(0)}, \quad \Lambda_n^b = u_n S - \tilde{u}_n, \quad V_n^{(0)} = G^{(0)}(u_n - \tilde{u}_n) + F^{(0)} u_n (u_{n+1} - \tilde{u}_n), \tag{2.3.338}$$

where $F^{(0)}$ and $G^{(0)}$ are two independent constants. The hierarchy of Bäcklund transformations is given by

$$\sum_{k=0} (\Lambda_n^b)^k F^{(k)} u_n (u_{n+1} - \tilde{u}_n) + \sum_{\ell=0} (\Lambda_n^b)^\ell G^{(\ell)} (u_n - \tilde{u}_n) = 0. \tag{2.3.339}$$

The simplest elementary Bäcklund transformation is

$$u_n - \tilde{u}_n = p u_n (u_{n+1} - \tilde{u}_n), \quad p = -\frac{F^{(0)}}{G^{(0)}}. \tag{2.3.340}$$

The superposition formula of elementary Bäcklund transformations provides an algebraic relation between three solutions

$$u_n^{12} = \frac{u_n^1 (1 - p_2 u_{n+1}^1) - u_n^2 (1 - p_1 u_{n+1}^2)}{p_1 u_n^2 - p_2 u_n^1}, \tag{2.3.341}$$

where, given two solutions u_n^1 and u_n^2, we obtain algebraically a third one, u_n^{12}.

By the identification $u_n = u_{nm}$ and $\tilde{u}_n = u_{n,m+1}$ (2.3.340) became a PΔE in a lattice plane

$$u_{nm} - u_{n,m+1} = p u_{nm} (u_{n+1,m} - u_{n,m+1}), \tag{2.3.342}$$

which we will encounter in Section 2.4.9 when discussing PΔEs. Identifying $u_n^1 = u_{nm\ell}$, $u_n^2 = u_{n,m+1,\ell}$ and $u_n^{12} = u_{n,m,\ell+1}$ (2.3.341) becomes a linearizable PΔE in three dimensional lattice space,

$$u_{n,m,\ell+1}\Big[p_1 u_{n,m+1,\ell} - p_2 u_{nm\ell}\Big] = \Big[u_{nm\ell}(1 - p_2 u_{n+1,m,\ell}) - u_{n,m+1,\ell}(1 - p_1 u_{n+1,m+1,\ell})\Big]. \tag{2.3.343}$$

3.5.2. *Symmetries for the DΔE Burgers.* As in the case of Burgers equation, due to the linearizability, also in the case of the DΔE Burgers we do not have a working spectral problem and a spectrum which evolves linearly with time. However, as in the continuous case, the evolution of the fake wave function ψ [136, 137, 181–183, 524, 596] is linear and takes, in the case of C-integrable equations, the role of the reflection coefficients when we were looking for the symmetries in the case of S-integrable equations. So by looking for commuting flows in the Burgers hierarchy we can construct its symmetries.

Given an equation of the Burgers hierarchy (2.3.327), the evolution of the wave function ψ is given by (2.3.321) and (2.3.328). The isospectral symmetries will be given by those autonomous evolution equations in the infinitesimal group parameter ϵ which commute with the time evolution of the equations of the DΔE Burgers hierarchy given by (2.3.327). They are the equations

$$u_{\epsilon_\ell} = \mathcal{N}_b \mathcal{L}_b^\ell H_\ell, \tag{2.3.344}$$

whose corresponding ϵ evolution of the wave function is

$$\psi_{\epsilon_\ell} = H_\ell \psi_{n+\ell}. \tag{2.3.345}$$

The commutativity of (2.3.323) and (2.3.345) is due to the commutativity of the shifts present on the right hand side of the expressions (2.3.328) and (2.3.340). So we have an infinite dimensional symmetry algebra of Abelian symmetries.

Can one construct also non isospectral symmetries using the non isospectral equations (2.3.333)? As we did in the case of the KdV we could define the equation

$$u_{n\,\epsilon_{\tilde\ell}} = h(t)\mathcal{N}_b \mathcal{L}_b^K A_K + \mathcal{N}_b \mathcal{L}_b^{\tilde\ell}\, nH_\ell, \tag{2.3.346}$$

a combination of an isospectral term with a coefficient given by a t-dependent arbitrary function and a non isospectral term characterized by a power $\tilde\ell$ of the recursive operator. The corresponding evolution of the wave function ψ is

$$\psi_{n,\epsilon_{\tilde\ell}} = h(t)A_K\psi_{n+K} + (\tilde\ell + n)H_\ell\psi_{n+\tilde\ell}. \tag{2.3.347}$$

The compatibility between the symmetry (2.3.347) and the Nth equation of the DΔE Burgers hierarchy is satisfied when

$$h(t) = A_N\, H_\ell\, N\, t. \tag{2.3.348}$$

This implies that we can associate to any equation of Burgers hierarchy a hierarchy of local non isospectral symmetries.

4. Integrability of PΔEs

4.1. Introduction. Good reviews on integrable PΔEs and on their physical and numerical applications can be found in the literature [151, 176, 269, 777, 796, 798]. We will just consider here the minimal amount of ideas and results necessary to make this section selfcontained.

The situation is slightly different from the results presented above for DΔEs.

Let us consider a PΔE for one dependent variable depending on two independent discrete variables n and m:

$$E(n, m, u_{n,m}, u_{n+1,m}, u_{n,m+1}, u_{n+1,m+1}, \ldots) = 0. \tag{2.4.1}$$

The equation in this case has shifts in both directions in the plane characterized by the indexes n and m. The shift operator in the two directions are indicated as S_n and S_m, defined in (1.2.13, 1.2.14) in Section 1.2.3. For a comparison with the previous results we will assume that m takes the role of a discretized time and n takes the role of a discretized space. Generalities on lattice on the plane and equations on them have been presented in Section 1.2.1 and 1.3.

The Lax pair in this case [360, 361] involves two linear discrete operators, $L_{n,m}$ which satisfies (2.2.5) and $M_{n,m}$ depending on the shift operators in n, with coefficients depending on $\{u_{n,m}\}$, i.e. $u_{n,m}$ and possibly its shifted values both in n and m. The linear equation (2.2.6) governing the time evolution of the spectral function is given in this case by

$$\psi_{n,m+1} = -M_{n,m}(u)\psi_{n,m}. \tag{2.4.2}$$

Due to (2.4.2) the Lax equation for PΔEs in the isospectral regime, when $\lambda_{m+1} = \lambda_m$, now reads:

$$L_{n,m+1}M_{n,m} = M_{n,m}L_{n,m}. \tag{2.4.3}$$

In the non isospectral case, when $\lambda_{m+1} = f_m(\lambda_m)$, with $f_m(z)$ an entire function of its argument, we have

$$L_{n,m+1}M_{n,m} = M_{n,m}f_m(L_{n,m}). \tag{2.4.4}$$

In the AKNS, Zakharov and Shabat matrix formalism, see [501], the overdetermined system of equations which defines a PΔE is given by

$$\psi_{n+1,m} = U_{n,m}(\{u_{n,m}\}, \lambda)\, \psi_{n,m}, \tag{2.4.5}$$

$$\psi_{n,m+1} = V_{n,m}(\{u_{n,m}\}, \lambda)\, \psi_{n,m}. \tag{2.4.6}$$

The compatibility of (2.4.5, 2.4.6) implies

$$\begin{aligned} &U_{n,m+1}(\{u_{n,m+1}\}, \lambda)\, V_{n,m}(\{u_{n,m}\}, \lambda) \\ &= V_{n+1,m}(\{u_{n+1,m}\}, \lambda)\, U_{n,m}(\{u_{n,m}\}, \lambda) \end{aligned} \tag{2.4.7}$$

It is worthwhile to notice that by the definition $u_{n,m} = u_n$ and $u_{n,m+1} = \tilde{u}_n$ (2.4.7) is equivalent to a Bäcklund transformation for a DΔE as considered before in Section 2.3. The Lax pair (2.4.5, 2.4.6) and consequently its compatibility (2.4.7) are symmetric under the exchange of n with m. So we can get the same connection of (2.4.7) with Bäcklund transformation also by defining $u_{n,m} = u_m$ and $u_{n+1,m} = \tilde{u}_m$.

Few results are known on generalized symmetries of PΔEs [481, 582, 633, 699]. We will present results on the discrete time Toda lattice [391, 481], the discrete time Volterra equation, the lattice pKdV [12, 22, 138, 489, 637], the lattice Schwartzian KdV [22, 490, 632, 637, 641], the discrete Burgers equation [502] and equations on quad-graph including the ABS equations [22, 29, 114].

Nonlinear integrable PDEs or DΔEs appeared in the previous Sections in the form of hierarchies of equations [124, 129, 147], all characterized by a common spectral problem and by the existence of a recursion operator. Equations belonging to the same hierarchy share many properties due to the common spectral problem. Among them are the existence of Bäcklund transformations [61] and generalized symmetries.

Many integrable non linear PΔEs have been considered in the literature [7, 84, 390–393, 451, 613, 639, 663], but up to now, few examples of hierarchies of non linear PΔEs are known. Recently Mikhailov following [555,556] developed a theory of PΔEs with generalized symmetries, conservation laws and formal recursion operators. However the recursion operator provide an infinity of symmetries and not hierarchies of PΔEs.

Here we show how, by applying the technique previously used for obtaining hierarchies of PDEs or DΔEs, we can get hierarchies of PΔEs together with Bäcklund transformations and generalized symmetries. More specifically we will show how, by constructing the recursion operator, we will produce hierarchies of PΔEs. We will determine the symmetries of the PΔE, making use of their integrability properties.

The ABS equations are the result of a classification procedure called the Compatibily around the Cube (CaC) [102,634,642,644,818,824]. We present the equations, their integrability and/or their linearizability, symmetries and Bäcklund transformations. However, except in the special cases presented before, as far as we know no recursion operator or hierarchy of equations is known. Recently hierarchies of symmetries of the ABS equations have been presented in [611].

4.2. Discrete time Toda lattice, its hierarchy, symmetries, Bäcklund transformations and continuous limit.

4.2.1. *Construction of the discrete time Toda lattice hierarchy.* Let us consider the discrete Schrödinger spectral problem we introduced in (2.3.10) with the potentials a and b depending on two discrete indexes n and m. So we have

$$L_{n,m}\psi_{n,m} \equiv \psi_{n-1,m} + a_{n,m}\psi_{n+1,m} + b_{n,m}\psi_{n,m} = \lambda\psi_{n,m}, \tag{2.4.8}$$

where $a_{n,m}$ and $b_{n,m}$ for any m tend respectively to 1 and to 0 as $|n|$ goes to ∞ and $\psi_{n,m} = \psi_{n,m}(\lambda)$. In (2.4.8) λ is an m-independent spectral parameter for isospectral evolutions. Here, as in the case of the Toda lattice, λ is expressed in terms of a variable z by (2.3.117). As shown in (2.4.2), the time–evolution is discrete and, for convenience we write it as

$$\psi_{n,m+1} = \psi_{n,m} - M_{n,m}\psi_{n,m}, \tag{2.4.9}$$

where $M_{n,m}$ is an operator function of the n shift operator S_n defined in (1.2.13) and possibly on $\{u_{n,m}\}$. An integrable non linear PΔE can be written in operator form as

$$L_{n,m+1} - L_{n,m} = L_{n,m+1}M_{n,m} - M_{n,m}L_{n,m} \tag{2.4.10}$$

with $L_{n,m}$ given by (2.4.8).

For $L_{n,m}$ given by (2.4.8) we have

$$L_{n,m+1} - L_{n,m} = (a_{n,m+1} - a_{n,m})\, S_n + b_{n,m+1} - b_{n,m}. \tag{2.4.11}$$

We use the by now standard Lax technique [124], in a way similar to the construction of the Toda lattice hierarchy. We construct a hierarchy of non linear PΔEs by requiring that

$$L_{n,m+1}M_{n,m} - M_{n,m}L_{n,m} = U_{n,m}S_n + V_{n,m}. \tag{2.4.12}$$

and

$$L_{n,m+1}\tilde{M}_{n,m} - \tilde{M}_{n,m}L_{n,m} = \tilde{U}_{n,m}S_n + \tilde{V}_{n,m} \tag{2.4.13}$$

$$\tilde{M}_{n,m} = L_{n,m+1}M_{n,m} + F_{n,m}S_n + G_{n,m}, \tag{2.4.14}$$

where $\tilde{M}_{n,m}$ is a new operator and $U_{n,m}$, $V_{n,m}$, $\tilde{U}_{n,m}$, $\tilde{V}_{n,m}$, $F_{n,m}$ and $G_{n,m}$ are scalar functions. Imposing the compatibility condition of (2.4.8, 2.4.12-2.4.14) we get the following hierarchy of equations

$$\begin{pmatrix} a_{n,m+1} - a_{n,m} \\ b_{n,m+1} - b_{n,m} \end{pmatrix} = f_m^1(\mathscr{L}_{n,m}^{\Delta}) \begin{pmatrix} (b_{n,m+1} - b_{n+1,m})\frac{\pi_{n,m+1}}{\pi_{n+1,m}} \\ \frac{\pi_{n-1,m+1}}{\pi_{n,m}} - \frac{\pi_{n,m+1}}{\pi_{n+1,m}} \end{pmatrix} + f_m^2(\mathscr{L}_{n,m}^{\Delta}) \begin{pmatrix} a_{n,m+1} - a_{n,m} \\ b_{n,m+1} - b_{n,m} \end{pmatrix}. \tag{2.4.15}$$

Here f_m^1 and f_m^2 are entire functions of their argument and $\mathscr{L}_{n,m}^{\Delta}$ is the recursion operator of the hierarchy, obtained from (2.4.8,2.4.12-2.4.14) and given by:

$$\mathscr{L}_{n,m}^{\Delta} \begin{pmatrix} p_{n,m} \\ q_{n,m} \end{pmatrix} = \begin{pmatrix} a_{n,m+1}\sigma_{n+2,m} - a_{n,m}\sigma_{n,m} \\ p_{n-1,m} + \Sigma_{n-1,m}\frac{\pi_{n-1,m+1}}{\pi_{n,m}} - \Sigma_{n,m}\frac{\pi_{n,m+1}}{\pi_{n+1,m}} \end{pmatrix} + \begin{pmatrix} b_{n,m+1}p_{n,m} + (+b_{n,m+1} - b_{n+1,m})\Sigma_{n,m}\frac{\pi_{n,m+1}}{\pi_{n+1,m}} \\ +b_{n,m+1}q_{n,m} + (b_{n,m+1} - b_{n,m})\sigma_{n,m} \end{pmatrix}. \tag{2.4.16}$$

The starting points $\begin{pmatrix} (b_{n,m+1} - b_{n+1,m})\frac{\pi_{n,m+1}}{\pi_{n+1,m}} \\ \frac{\pi_{n-1,m+1}}{\pi_{n,m}} - \frac{\pi_{n,m+1}}{\pi_{n+1,m}} \end{pmatrix}$ and $\begin{pmatrix} a_{n,m+1} - a_{n,m} \\ b_{n,m+1} - b_{n,m} \end{pmatrix}$ are obtained as coefficients of the integration constants for the functions $F_{n,m}$ and $G_{n,m}$. The function $\pi_{n,m}$ is given by

$$\pi_{n,m} = \Pi_{j=n}^{\infty} a_{j,m}, \tag{2.4.17}$$

while $\sigma_{n,m}$ and $\Sigma_{n,m}$ are defined as the bounded solutions of the equations

$$\begin{aligned} \sigma_{n+1,m} - \sigma_{n,m} &= q_{n,m} \\ \Sigma_{n+1,m} - \Sigma_{n,m} &= -p_{n+1,m}\frac{\pi_{n+2,m}}{\pi_{n+1,m+1}}. \end{aligned} \tag{2.4.18}$$

The boundedness of the solutions of (2.4.18) is necessary to get a hierarchy of non linear PΔEs with well defined evolution of the spectra.

The class of non linear PΔEs (2.4.15) is strictly related to the Bäcklund transformations for the Toda system. In fact it is fundamentally obtained [471, 480] by setting $(a_n, b_n) = (a_{n,m}, b_{n,m})$ and $(\tilde{a}_n, \tilde{b}_n) = (a_{n,m+1}, b_{n,m+1})$.

Let us define, as in the case of the Toda lattice, the reflection and transmission coefficients $R_m(z)$ and $T_m(z)$ in terms of the asymptotic behavior in n of the function $\psi_{n,m}$

$$\begin{aligned} \lim_{n\to\infty} \psi_{n,m}(z) &= \phi_m(z)\left[z^{-n} + R_m z^m\right], \\ \lim_{n\to-\infty} \psi_{n,m}(z) &= \phi_m(z) T_m z^{-n}, \end{aligned} \tag{2.4.19}$$

where ϕ_m is an appropriate normalization function depending just on m and z. In the case of a generic equation of the discrete time Toda lattice hierarchy (2.4.15) the discrete evolution of the reflection coefficient is

$$R_{m+1} = \frac{1 - f_m^2(\lambda) - z f_m^1(\lambda)}{1 - f_m^2(\lambda) - \frac{f_m^1(\lambda)}{z}} R_m, \tag{2.4.20}$$

while the transmission coefficient T_m does not evolve in m.

At difference from the case of hierarchies of PDEs or DΔEs, the recursion operator (2.4.16) depends on both $(a_{n,m}, b_{n,m})$ and $(a_{n,m+1}, b_{n,m+1})$. Thus, in order to write the non linear PΔE as an evolution equation in which we explicit the fields at the time $m + 1$ in terms of those at the time m, we must write down the complete system of equations and then solve, if possible, for the fields at the time $m+1$. It is not guaranteed that this can always be done since often the equation provide an implicit evolution in the discrete time.

Let us write down, as an example, the simplest member of the hierarchy (2.4.15). Choosing $f_m^2 = 0$ and $f_m^1 = \alpha$ in (2.4.15) we get

$$a_{n,m+1} - a_{n,m} = \alpha(b_{n,m+1} - b_{n+1,m})\frac{\pi_{n,m+1}}{\pi_{n+1,m}}, \tag{2.4.21}$$

$$b_{n,m+1} - b_{n,m} = \alpha(\frac{\pi_{n-1,m+1}}{\pi_{n,m}} - \frac{\pi_{n,m+1}}{\pi_{n+1,m}}). \tag{2.4.22}$$

Solving (2.4.21, 2.4.22) for $b_{n+1,m} - b_{n,m}$ and taking into account the boundary conditions for the fields $a_{n,m}$ and $b_{n,m}$, we get

$$b_{n,m} = \alpha + \frac{1}{\alpha} - \alpha\frac{\pi_{n-1,m+1}}{\pi_{n,m}} - \frac{\pi_{n,m}}{\alpha\pi_{n,m+1}}. \tag{2.4.23}$$

Substituting (2.4.23) into (2.4.21) we obtain a single equation of higher order for the field $\pi_{n,m}$:

$$\Delta_{Toda} = \pi_{n-1,m+2} - \frac{1}{\alpha^2}\pi_{n,m} - \pi^2_{n,m+1}(\frac{1}{\pi_{n+1,m}} - \frac{1}{\alpha^2\pi_{n,m+2}}) = 0, \tag{2.4.24}$$

which, for $\pi_{n,m} = e^{u_{n,m}}$ reads

$$e^{u_{n,m}-u_{n,m+1}} - e^{u_{n,m+1}-u_{n,m+2}} = \alpha^2(e^{u_{n-1,m+2}-u_{n,m+1}} - e^{u_{n,m+1}-u_{n+1,m}}), \tag{2.4.25}$$

i.e. a form similar to the well known discrete time Toda lattice equation [330, 391, 778]. On the left hand side of (2.4.25) we have, by expanding the exponential terms, the second difference of the function $u_{n,m}$ with respect to the discrete time m. Thus, defining

$$t = m\omega, \qquad v_n(t) = u_{n,m}, \qquad \alpha = \omega^2 \tag{2.4.26}$$

we find that (2.4.25) reduces to the continuous-time Toda lattice equation (1.4.16)

$$\ddot{v}_n = e^{v_{n-1}-v_n} - e^{v_n-v_{n+1}} + \mathcal{O}(\omega). \tag{2.4.27}$$

Eq. (2.4.25) has the Lax pair [481]

$$\psi_{n-1,m} + \left(\alpha + \frac{1}{\alpha} - \alpha e^{u_{n-1,m+1}-u_{n,m}} - \frac{e^{u_{n,m}-u_{n,n+1}}}{\alpha}\right)\psi_{n,m} \tag{2.4.28}$$
$$+e^{u_{n,m}-u_{n+1,m}}\psi_{n+1,m} = \lambda\psi_{n,m},$$

$$\psi_{n,m+1} = \psi_{n,m} - \alpha e^{u_{n,m+1}-u_{n+1,m}}\psi_{n+1,m}. \tag{2.4.29}$$

From (2.4.20) we get the evolution of the reflection coefficient R_m and T_m

$$R_{m+1} = \frac{1-\alpha z}{1-\frac{\alpha}{z}}R_m, \quad T_{m+1} = T_m. \tag{2.4.30}$$

4.2.2. *Isospectral and non isospectral generalized symmetries for the discrete time Toda lattice.* The Lie point symmetries of (2.4.25) have been considered in the Introduction, Section 1.4.2. See also [423] where the symmetries are obtain by Eastabrook-Wahlquist technique [823].

Infinitesimal symmetries for the discrete time Toda lattice can be obtained as commuting flows, i.e. an infinitesimal symmetry is obtained when its flow in the group parameter ϵ and the discrete evolution equations commute. The symmetries must thus be obtained by looking into the hierarchy of non linear DΔEs associated to the Schrödinger spectral problem (2.4.8). The non linear PΔEs commuting with the discrete time Toda lattice turn out not to form a group of symmetry transformations associated to (2.4.25) but they provide us with the associated Bäcklund transformations.

To discuss these issues, as we saw before in the case of the KdV and Toda lattice, it is easier to work in the space of the spectral parameter where the non linear evolution of the fields is substituted by the linear evolution of the reflection coefficient. The two spaces are in one to one correspondence for fields which are asymptotically bounded [257]. In such a situation the discrete time Toda lattice equation (2.4.25) is represented by the following m evolution of the reflection coefficient $R_m(z, \epsilon)$ and transmission coefficient $T_m(z, \epsilon)$:

$$R_{m+1}(z, \epsilon) = \frac{1 - z\alpha}{1 - \frac{\alpha}{z}} R_m(z, \epsilon), \quad T_{m+1}(z, \epsilon) = T_m(z, \epsilon). \tag{2.4.31}$$

where ϵ is the infinitesimal group parameter (see (2.4.30)).

Any isospectral deformation ($\frac{dz}{d\epsilon_\ell} = 0$) of the discrete Schrödinger spectral problem (2.4.8) provide the isospectral symmetries

$$\begin{pmatrix} a_{n,m} \\ b_{n,m} \end{pmatrix}_{,\epsilon_\ell} = (\mathscr{L}_d)^\ell \begin{pmatrix} a_{n,m}(b_{n,m} - b_{n+1,m}) \\ a_{n-1,m} - a_{n,m} \end{pmatrix}. \tag{2.4.32}$$

The recursion operator $\mathscr{L}_d$ is given by (2.3.18) with $(a_n, b_n, q_n, p_n, s_n)$ depending on the discrete time m and on the group parameter ϵ_ℓ. The index ℓ of ϵ_ℓ denotes the fact that this symmetry is given by the ℓth equation of the Toda lattice hierarchy (2.4.32). In correspondence with (2.4.32) we have an evolution (in ϵ_ℓ) of the reflection coefficient associated to the discrete Schrödinger spectral problem (2.4.8), i.e.

$$\frac{\partial R_m(z, \epsilon_\ell)}{\partial \epsilon_\ell} = \mu \lambda^\ell R_m(z, \epsilon_\ell) \tag{2.4.33}$$

with λ and μ defined as in (2.3.117). It is easy to prove that the flows (2.4.21) and (2.4.32) commute by checking that the corresponding flows of the reflection coefficients, given by (2.4.31) and (2.4.33), commute. Moreover the symmetries (2.4.32) corresponding to $\ell = \ell_1$ and $\ell = \ell_2$ commute among themselves as one can see from (2.4.33).

A less obvious calculation has to be done to get the non isospectral symmetries of the discrete time Toda lattice equation. In this case we have:

$$\begin{aligned} \begin{pmatrix} a_{n,m} \\ b_{n,m} \end{pmatrix}_{,\epsilon_\ell} &= f_m^\ell(\mathscr{L}) \begin{pmatrix} a_{n,m}(b_{n,m} - b_{n+1,m}) \\ a_{n-1,m} - a_{n,m} \end{pmatrix} \\ &+ \mathscr{L}^\ell \begin{pmatrix} a_{n,m}[(2n+3)b_{n+1,m} - (2n-1)b_{n,m}] \\ b_{n,m}^2 - 4 + 2[(n+1)a_{n,m} - (n-1)a_{n-1,m}] \end{pmatrix}. \end{aligned} \tag{2.4.34}$$

The function $f_m^\ell(\lambda)$ depends on the equation under consideration and, for the discrete time Toda lattice, is obtained as a solution of the difference equation:

$$f_{m+1}^\ell(\lambda) - f_m^\ell(\lambda) = -2\lambda^\ell \frac{2\alpha^2 - \alpha\lambda}{1 + \alpha^2 - \alpha\lambda}. \tag{2.4.35}$$

Up to an arbitrary inessential constant the function $f_m^\ell(\lambda)$ is given by:

$$f_m^\ell(\lambda) = -2m\lambda^\ell \frac{2\alpha^2 - \alpha\lambda}{1 + \alpha^2 - \alpha\lambda}. \tag{2.4.36}$$

The proof that the flow (2.4.34) with f_m^ℓ given by (2.4.36) commutes with that of (2.4.21) is easily obtained in the space of the spectrum. The reflection coefficient associated to

(2.4.34) satisfies the equation

$$\frac{dR_m(z,\epsilon_\ell)}{d\epsilon_\ell} = \mu f_m^\ell(\lambda) R_m(z,\epsilon_\ell), \qquad \lambda_{\epsilon_\ell} = \mu^2 \lambda^\ell. \tag{2.4.37}$$

were, on the l.h.s. of (2.4.37) we have the total derivative of $R_m(z,\epsilon_\ell)$ with respect to ϵ_ℓ.

Both the isospectral (2.4.32) and non isospectral (2.4.34) symmetries involve the dependent variable in different points of the lattice and, even if the continuous limit will correspond to Lie point symmetries [371, 373], they are effectively generalized symmetries. As such they are not integrable, i.e. we are not able to get from them the group transformations. However they can be used to provide solutions of the discrete Toda via symmetry reduction. On the problem of symmetry reduction for PΔEs see also [799].

4.2.3. *Symmetry reductions for the discrete time Toda lattice.* As an example of these symmetry reductions let us write down the simplest non isospectral symmetry obtained for $\ell = 0$ and $\alpha = 1$

$$\begin{aligned} \begin{pmatrix} a_{n,m} \\ b_{n,m} \end{pmatrix}_{,\epsilon_0} &= -2m \begin{pmatrix} a_{n,m}(b_{n,m} - b_{n+1,m}) \\ a_{n-1,m} - a_{n,m} \end{pmatrix} \\ &+ \begin{pmatrix} a_{n,m}[(2n+3)b_{n+1,m} - (2n-1)b_{n,m}] \\ b_{n,m}^2 - 4 + 2[(n+1)a_{n,m} - (n-1)a_{n-1,m}] \end{pmatrix}. \end{aligned} \tag{2.4.38}$$

Taking into account (2.4.17), we can rewrite (2.4.38) as the system

$$\begin{aligned} (\pi_{n,m})_{,\epsilon_0} &= \pi_{n,m}\{-(2m+2n+1)b_{n,m} + 2\sum_{j=n}^{\infty} b_{j,m}\} \\ (b_{n,m})_{,\epsilon_0} &= b_{n,m}^2 - 4 + 2[(n+m+1)a_{n,m} - (n+m-1)a_{n-1,m}]. \end{aligned} \tag{2.4.39}$$

In view of (2.4.23), $b_{n,m}$ can be rewritten in terms of $\pi_{n,m}$ and its shifted values.

A symmetry reduction with respect to the symmetry given by (2.4.39) is obtained by solving the discrete time Toda lattice (2.4.24) together with the equation we get by equating to zero the r.h.s. of (2.4.39), i.e.

$$\begin{aligned} (2m+2n-1)b_{n,m} - (2m+2n+3)b_{n+1,m} &= 0, \\ a_{n,m}[2(n+1)+2m] - a_{n-1,m}[2(n-1)+2m] &= 4 - b_{n,m}^2. \end{aligned} \tag{2.4.40}$$

The general solution is given by

$$\begin{aligned} b_{n,m} &= \frac{b_m^0}{(2m+2n-1)(2m+2n+1)}, \\ a_{n,m} &= \frac{1}{(2n+2m+2)(2n+2m)}\Bigg[a_m^0 + 4n(2m+1+n) \\ &\qquad + \frac{(b_m^0)^2}{4(2m+2n+1)^2}\Bigg]. \end{aligned} \tag{2.4.41}$$

Using (2.4.21, 2.4.22) with $\alpha = 1$, we get two equations for b_m^0 and a_m^0, the reduced equations. The interested reader can carry out easily other possible reductions of the discrete time Toda with respect to both the Lie point and generalized symmetries.

4.2.4. *Bäcklund transformations and symmetries for the discrete time Toda lattice.* Bäcklund transformations are obtained by the same kind of formulas as those used to get the PΔEs when the new functions $(\tilde{a}_n, \tilde{b}_n)$ correspond to $(a_{n,m+1}, b_{n,m+1})$. With this identification the class of Bäcklund transformations associated to the discrete time Toda lattice hierarchy reads:

$$\delta(\Lambda_d^{\Delta})\begin{pmatrix}(\tilde{b}_{n,m}-b_{n+1,m})\frac{\tilde{\pi}_{n,m}}{\pi_{n+1,m}}\\ \frac{\tilde{\pi}_{n-1,m}}{\pi_{n,m}}-\frac{\tilde{\pi}_{n,m}}{\pi_{n+1,m}}\end{pmatrix}=\gamma(\Lambda_d^{\Delta})\begin{pmatrix}\tilde{a}_{n,m}-a_{n,m}\\ \tilde{b}_{n,m}-b_{n,m}\end{pmatrix}, \tag{2.4.42}$$

where Λ_d^{Δ} is the Bäcklund recursion operator, obtained in the same way as $\mathcal{L}^{\Delta}$, and given by:

$$\Lambda_d^{\Delta}\begin{pmatrix}p_{n,m}\\ q_{n,m}\end{pmatrix}=\begin{pmatrix}\tilde{a}_{n,m}(q_{n,m}+q_{n+1,m})+(a_{n,m}-\tilde{a}_{n,m})\tilde{P}_{n,m}\\ p_{n,m}+\tilde{\Sigma}_{n-1,m}-\tilde{\Sigma}_{n,m}+\tilde{b}_{n,m}q_{n,m}\end{pmatrix} + \begin{pmatrix}b_{n+1,m}p_{n,m}+(\tilde{b}_{n,m}-b_{n+1,m})\tilde{\Sigma}_{n,m}\\ (b_{n,m}-\tilde{b}_{n,m})\tilde{P}_{n,m}\end{pmatrix}. \tag{2.4.43}$$

Above, $\tilde{\Sigma}_{n,m}$ and $\tilde{P}_{n,m}$ are now defined as the bounded solutions to the following difference equations:

$$\begin{aligned}&\tilde{P}_{n,m}-\tilde{P}_{n+1,m}=q_{n,m}\\ &\tilde{\Sigma}_{n,m}\frac{\pi_{n+1,m}}{\tilde{\pi}_{n,m}}-\tilde{\Sigma}_{n+1,m}\frac{\pi_{n+2,m}}{\tilde{\pi}_{n+1,m}}=p_{n,m}\frac{\pi_{n+1,m}}{\tilde{\pi}_{n,m}}.\end{aligned} \tag{2.4.44}$$

In (2.4.42) γ and δ are entire functions of their arguments. Eq. (2.4.43) corresponds asymptotically to

$$\tilde{R}_m=\frac{\gamma(\lambda)-z\delta(\lambda)}{\gamma(\lambda)-\frac{\delta(\lambda)}{z}}R_m. \tag{2.4.45}$$

The simplest Bäcklund transformation is obtained by choosing $\gamma=1$ and δ constant and reads:

$$\begin{aligned}\tilde{a}_{n,m}-a_{n,m}&=\delta(\tilde{b}_{n,m}-b_{n+1,m})\frac{\tilde{\pi}_{n,m}}{\pi_{n+1,m}},\\ \tilde{b}_{n,m}-b_{n,m}&=\delta[\frac{\tilde{\pi}_{n-1,m}}{\pi_{n,m}}-\frac{\tilde{\pi}_{n,m}}{\pi_{n+1,m}}].\end{aligned} \tag{2.4.46}$$

It is worthwhile to recall that while the composition of two Bäcklund transformations is still a Bäcklund transformation, usually of higher order, the Bäcklund transformations do not form a Lie group as the product of two Bäcklund transformations does not give a Bäcklund transformation of the same form as the original ones. However the Bäcklund transformations form a kind of group [504]. Moreover, the theorems presented in Section 2.3.2.5 [372] for the Toda lattice equation are valid also in this case, i.e. any Bäcklund transformation can be written as a superposition of an infinite number of symmetries and viceversa. As the equations are already discrete in all variables, the Bäcklund transformations do not provide any new information. They are discrete flows commuting with the equations but not having the properties of a Lie group.

4.3. Discrete time Volterra equation. From the discrete time Toda lattice we can construct a discrete time Volterra equation by applying one time the recursion operator $\mathscr{L}_{n,m}^{\Delta}$ of (2.4.16) on the starting point of the discrete time Toda hierarchy $\begin{pmatrix}(b_{n,m+1}-b_{n+1,m})\frac{\pi_{n,m+1}}{\pi_{n+1,m}}\\ \frac{\pi_{n-1,m+1}}{\pi_{n,m}}-\frac{\pi_{n,m+1}}{\pi_{n+1,m}}\end{pmatrix}$ and setting $b_{n,m}$ equal to zero together with its consequences for $p_{n,m}$ and $\Sigma_{n,m}$. In conclusion choosing $f_m^1(z)=\alpha_m z$ and $f_m^2(z)=0$ in (2.4.15) we get

$$a_{n,m+1}-a_{n,m}=\alpha_m\left[\frac{\pi_{n-1,m+1}}{\pi_{n+1,m}}-\frac{\pi_{n,m+1}}{\pi_{n+2,m}}\right]. \tag{2.4.47}$$

Taking into account the relation between $\pi_{n,m}$ and $a_{n,m}$ given by (2.4.17) we can write (2.4.47) as an equation for the function $\pi_{n,m}$ alone

$$\frac{\pi_{n,m+1}}{\pi_{n+1,m+1}} - \frac{\pi_{n,m}}{\pi_{n+1,m}} = \alpha_m \left[\frac{\pi_{n-1,m+1}}{\pi_{n+1,m}} - \frac{\pi_{n,m+1}}{\pi_{n+2,m}} \right]. \tag{2.4.48}$$

In correspondence with (2.4.48) we have the following evolution of the reflection coefficient, that we can derive from (2.4.20):

$$R_{m+1} = \frac{1 - \alpha_m z \lambda}{1 - \frac{\alpha_m}{z} \lambda} R_m. \tag{2.4.49}$$

In Section 2.3.3.1 we showed that we are not able to construct a hierarchy of Bäcklund transformations for the Volterra equation. This is because the reduction from the Toda system to the Volterra is not possible for the square of the recursion operator Λ_d (2.3.100). Thus we will not be able to construct a hierarchy of discrete time Volterra equations.

4.3.1. *Continuous limit of the discrete time Volterra equation.* Let us rewrite the positive definite function $\pi_{n,m}$ (as $a_{n,m}$ tends asymptotically to one and $\pi_{n,m}$ is defined by (2.4.17)) as

$$\pi_{n,m} = e^{v_{n,m}}. \tag{2.4.50}$$

Then (2.4.48) becomes

$$e^{v_{n,m+1}-v_{n+1,m+1}} - e^{v_{n,m}-v_{n+1,m}} = \alpha_m \left[e^{v_{n-1,m+1}-v_{n+1,m}} - e^{v_{n,m+1}-v_{n+2,m}} \right]. \tag{2.4.51}$$

Let us introduce a continuous variable t as $t = m\varepsilon$ where $\varepsilon \to 0$ as $m \to \infty$ in such a way their product is finite. In this limit α_m will transform into $\alpha(t)$. Then we can introduce a new function u depending on a discrete index n and the continuous variable t in such a way that $v_{n,m} = u_n(t)$. Then (2.4.51) becomes

$$\begin{aligned} &e^{u_n(t+k)-u_{n+1}(t+k)} - e^{u_n(t)-u_{n+1}(t)} \\ &\qquad = \alpha(t) \left[e^{u_{n-1}(t+k)-u_{n+1}(t)} - e^{u_n(t+k)-u_{n+2}(t)} \right]. \end{aligned} \tag{2.4.52}$$

Defining $\alpha(t) = \varepsilon\beta(t)$ then at the first order in ε when $\varepsilon \to 0$, we get

$$\dot{u}_n - \dot{u}_{n+1} = \beta(t) \left[e^{u_{n-1}-u_n} - e^{u_{n+1}-u_{n+2}} \right] + \mathcal{O}(\varepsilon). \tag{2.4.53}$$

We can now introduce the function

$$a_n(t) = e^{u_n - u_{n+1}} \tag{2.4.54}$$

and (2.4.53) becomes (2.3.172) in Section 2.3.3, the Volterra equation for $\beta(t) = 1$.

4.3.2. *Symmetries for the discrete Volterra equation.* The isospectral symmetries of the discrete time Volterra equation (2.4.47)are given by the Volterra hierarchy of DΔEs with a_n going over to $a_{n,m}$ i.e. equations (2.3.177) whose reflection coefficient evolve in the group parameter ϵ_ℓ according to (2.3.176) with $t \to \epsilon_\ell$ and $g_1(\lambda^2, t) \to \lambda^{2\ell}$. The non isospectral ones are obtained commuting (2.3.178), with $g_2(\tilde{\mathcal{L}}, t)$ substituted by the entire function $g_m^2(\tilde{\mathcal{L}})$, with (2.4.47). This same commutation can naturally be carried out at the level of the reflection coefficients (2.3.180) and (2.4.49). So we get the equation for $g_m^2(\lambda^2)$

$$\begin{aligned} g_{m+1}^2(\lambda^2) - g_m^2(\lambda^2) &= -2\alpha_m \lambda^{2\ell}(\lambda^2 - 4) \\ &\frac{\left(\alpha_m \lambda^2 - \lambda^2 + 2\right)\left(\lambda^4 + \lambda^3 \mu - 4\,\lambda^2 - 2\,\lambda\,\mu + 2\right)}{\left(\alpha_m{}^2 \lambda^2 - \alpha_m \lambda^2 + 1\right)\left(\lambda^2 + \lambda\,\mu - 2\right)}, \end{aligned} \tag{2.4.55}$$

where the function μ and its relation to λ are given in (2.3.117). As the right hand side of (2.4.55) is not entire in λ^2 we are not able to find a function g_m^2 and thus a non isospectral symmetry. Eq. (2.3.178) with $h_1 = 0$ and $k = 0$ will be a local master symmetry for the discrete time Volterra equation (2.4.47) but not a symmetry.

4.4. Lattice version of the potential KdV, its symmetries and continuous limit.

4.4.1. *Introduction.* The lattice version of the pKdV (lpKdV) was obtained (2.2.83) as the superposition formula for the KdV equation [637]. We write it again here for the convenience of the reader:

$$(p - q + u_{n,m+1} - u_{n+1,m})(p + q - u_{n+1,m+1} + u_{n,m}) = p^2 - q^2. \tag{2.4.56}$$

This equation involves just four u points which lay on a two dimensional orthogonal infinite lattice and are the vertices of an elementary square, see Fig. 2.3. In (2.4.56) $u_{n,m}$ is the dynamical field variable, which we assume to be real, at site $(m, n) \in \mathbb{Z} \times \mathbb{Z}$ while $(p, q) \in \mathbb{R} \times \mathbb{R}$ are two non zero parameters. As we will see in the following they are related to the lattice steps α and β between the points (see Fig. 2.3 in Section 2.4.6) and will go to zero when we carry out the continuous limit to the pKdV (2.2.74). $u_{n,m}$, p and q can also be complex quantities and p and q can depend on m.

Since we have two discrete independent variables we can perform, following Nijhoff and Capel [637], the continuous limit in two steps by shrinking the corresponding lattice step to zero.

The first step transform the discrete index m into a continuous variable $t = m\,\varepsilon$ in such a way that equation (2.4.56) becomes an evolutionary DΔE for the unknown function $u_n(t)$ depending on a continuous variable t and a discrete index n.

The basic of the method is a Taylor expansion in a parameter ε of $u_{n,m}$ around a particular solution u_0 of the equation. Both the index m and the corresponding parameter q will depend on ε. As a particular solution it is convenient to use a simple function which, in this case, is given by

$$u_0 = pn + qm.$$

By the change of variables:

$$u_{n,m} = v_{n,m} - u_0, \tag{2.4.57}$$

(2.4.56) becomes:

$$(p - q + v_{n,m+1} - v_{n+1,m})(p + q + v_{n,m} - v_{n+1,m+1}) = p^2 - q^2, \tag{2.4.58}$$

which has as a solution $v_0 = 0$.

We start the continuous limit by requiring that m goes to infinity and ε goes to zero in such a way that $t = m\varepsilon$ is finite. We define a new function $V_n(t) = v_{n,m}$ such that

$$v_{n,m+j} = V_n(t) + j\varepsilon \dot{V}_n(t) + \mathcal{O}(\varepsilon^2). \tag{2.4.59}$$

Moreover we define the parameter q as

$$q = \varepsilon^{-1}. \tag{2.4.60}$$

Substituting (2.4.59), (2.4.60) in (2.4.58), the ε^{-1} term vanishes while the zero order term yields the equation

$$\dot{V}_n + \dot{V}_{n+1} = (V_{n+1} - V_n)[2p - (V_{n+1} - V_n)]. \tag{2.4.61}$$

Eq. (2.4.61) is a non local DΔE which has derivatives with respect to t in two different points of the lattice n and $n+1$. To obtain a local evolutionary DΔE, i.e with just a derivative at one lattice point, we have to mix the indexes in the lattice (a skew limit as is called in [384]). Take $\ell = n + m$ and introduce a new variable, $w_{\ell,m}$ such that

(2.4.62) $$v_{n+i,m+j} = w_{\ell+i+j,m+j}$$

Then (2.4.58) is transformed into

$$(p - q + w_{\ell+1,m+1} - w_{\ell+1,m})(p + q + w_{\ell,m} - w_{\ell+2,m+1}) = p^2 - q^2,$$

In this case, defining as above $U_\ell(t) = w_{\ell,m}$ with $q = p + \varepsilon$, the ε^0 term vanishes and at first order we get:

(2.4.63) $$\dot{U}_\ell = \frac{U_{\ell-1} - U_{\ell+1}}{U_{\ell-1} - U_{\ell+1} + 2p}.$$

Eq. (2.4.63) is an evolutionary DΔE, with terms at points $\ell - 1$, ℓ and $\ell + 1$ which thus satisfies Yamilov's condition for S-integrability given in Theorem 34 in Section 3.2.4.1. Defining

(2.4.64) $$r_\ell \equiv 2p - U_{\ell+2} + U_\ell$$

we can rewrite (2.4.63) as the DΔE in q_k:

(2.4.65) $$\dot{r}_\ell = 2p\left(\frac{1}{r_{\ell-1}} - \frac{1}{r_{\ell+1}}\right).$$

Eq. (2.4.65) has already been presented in [107] and it is associated to the discrete Schrödinger spectral problem

(2.4.66) $$\psi_{\ell+2} = r_\ell \psi_{\ell+1} + \lambda \psi_\ell,$$

where $\lambda \in \mathbb{C}$ is the spectral parameter.

By defining $s_\ell \equiv (2p)/r_\ell$, (2.4.65) can be also written as

(2.4.67) $$\dot{s}_\ell = s_\ell^2 (s_{\ell+1} - s_{\ell-1}),$$

the modified Volterra equation (V_1), also called discrete KdV equation [626]. By setting $a_\ell \equiv s_\ell s_{\ell-1}$, (2.4.67) can be transformed into the Volterra equation (2.3.172) (see Section 3.3.1.2).

The second continuous limit of (2.4.56) is performed by taking in (2.4.63)

(2.4.68) $$U_\ell(\tau) \equiv w(x,t), \qquad x \equiv \frac{2}{p}\left(\ell + \frac{\tau}{p}\right), \qquad t \equiv \frac{2}{p^3}\left(\frac{\ell}{3} + \frac{\tau}{p}\right).$$

Eqs. (2.4.68) can be inverted to give

(2.4.69) $$\ell = \frac{p}{4}(3x - p^2 t), \qquad \tau = \frac{p^2}{4}(p^2 t - x).$$

In the limit $p \to \infty$, $\ell \to \infty$, $\tau \to \infty$, when x and t are finite, (2.4.63) is transformed in the pKdV equation (2.2.74).

In [637] the integrability of the lpKdV (2.4.56) is established by writing down its Lax pair in the Zakharov-Shabat form

(2.4.70a) $$\Psi_{n+1,m}(\sigma) = U_{n,m}(\sigma)\Psi_{n,m}(\sigma),$$

(2.4.70b) $$\Psi_{n,m+1}(\sigma) = V_{n,m}(\sigma)\Psi_{n,m}(\sigma),$$

where we define $\Psi_{n,m}(\sigma) \equiv (\psi^1_{n,m}(\sigma), \psi^2_{n,m}(\sigma))^T$ and σ plays the role of a spectral parameter. The matrices $U_{n,m}(\sigma)$ and $V_{n,m}(\sigma)$ are

$$U_{n,m}(\sigma) = \begin{pmatrix} p - u_{n+1,m} & 1 \\ \sigma^2 - p^2 + (p + u_{n,m})(p - u_{n+1,m}) & p + u_{n,m} \end{pmatrix},$$

and

$$V_{n,m}(\sigma) = \begin{pmatrix} q - u_{n,m+1} & 1 \\ \sigma^2 - q^2 + (q + u_{n,m})(q - u_{n,m+1}) & q + u_{n,m} \end{pmatrix}.$$

We can rewrite the Lax equations (2.4.70a, 2.4.70b) in scalar form. We get as a spectral problem (2.4.66) with $\lambda = \sigma^2 - p^2$ and $\ell = n$

$$r_n = r_{n,m} = 2p - u_{n+2,m} + u_{n,m}, \tag{2.4.71}$$

where m enters parametrically. The m–evolution of the wave function is given by

$$\psi_{n,m+1}(\lambda) = \psi_{n+1,m}(\lambda) + (q - p + u_{n+1,m} - u_{n,m+1})\,\psi_{n,m}(\lambda), \tag{2.4.72}$$

where $\psi_{n,m}(\lambda) \equiv \psi^1_{n,m}(\sigma)$. Eq. (2.4.72) is not easily expressed in term of $r_{n,m}$.

4.4.2. *Solution of the discrete spectral problem associated with the lpKdV equation.* Let us consider the direct and inverse problems associated with the spectral problem (2.4.66). Our results are a generalization of those contained in [109] with $\ell = n$. We refer to [109] for proofs and technical details. Here we just present the formulas we will need to obtain the time evolution of spectral data necessary for the calculation of the symmetries of (2.4.56). In (2.4.66) the function $\psi_n \equiv \psi_{n,m}(\lambda, \epsilon)$ depends parametrically on the variables m, the time of the lpKdV, and ϵ, the continuous symmetry parameter.

To solve the direct problem we assume that the solutions $u_{n,0}$ of (2.4.56) go asymptotically to an arbitrary constant in agreement with the difference equation. Then r_n, given by (2.4.71) will go asymptotically to $2p$.

Following [109] we rewrite the spectral problem (2.4.66) in terms of a field η_n vanishing asymptotically as $|n| \to \infty$ as

$$(p + \mathrm{i}k)\chi_{n+2} - 2p\chi_{n+1} + (p - \mathrm{i}k)\chi_n = \eta_n \chi_{n+1}, \tag{2.4.73}$$

where

$$\sigma \equiv \mathrm{i}k \quad \lambda \equiv -k^2 - p^2, \quad r_n \equiv \eta_n + 2p, \quad \psi_n \equiv (p + \mathrm{i}k)^n \chi_n. \tag{2.4.74}$$

The Jost functions $\mu_n^{\pm}$ of the spectral problem (2.4.73) can be defined in terms of the potential η_n and of the discrete complex *exponential* function

$$E_n = [(p + \mathrm{i}k/(p - \mathrm{i}k)]^n$$

through the following discrete integral equations:

$$\mu_n^+ = 1 - \frac{1}{2\mathrm{i}k} \sum_{j=n+1}^{+\infty} \left[1 + E_{j-n}\right] \eta_{j-1} \mu_j^+, \tag{2.4.75a}$$

$$\mu_n^- = 1 + \frac{1}{2\mathrm{i}k} \sum_{j=-\infty}^{n} \left[1 + E_{j-n}\right] \eta_{j-1} \mu_j^-. \tag{2.4.75b}$$

For a potential η_n decaying sufficiently rapidly to zero at large $|n|$ the Jost solution μ_n^+ is an analytic function of k for $\mathrm{Im}(k) > 0$ and μ_n^- for $\mathrm{Im}(k) < 0$ such that

$$\lim_{n\to\pm\infty} \mu_n^{\pm} = 1, \qquad \mathrm{Im}(k) \gtrless 0, \tag{2.4.76}$$

For $\mathrm{Im}(k) = 0$, assuming that (2.4.75) can be solved and their solutions are unique, we obtain

$$\mu_n^{\pm}(k) = a^{\pm}(k)\mu_n^{\mp}(k) + E_n b^{\pm}(k)\mu_n^{\mp}(-k),$$

which define the spectral data $a^{\pm}(k)$ and $b^{\pm}(k)$. Due to the analyticity property of the Jost solutions it is possible to prove that $a^{+}(k)$ can be analytically extended to $\mathrm{Im}(k) > 0$ and $a^{-}(k)$ to $\mathrm{Im}(k) < 0$. The functions

$$[a^{\pm}(k)]^{-1} \equiv [T^{\pm}()],$$

play the role of the transmission coefficient (its poles are related to the soliton solutions of the evolution equations associated to the spectral problem (2.4.73)) and the functions

$$R^{\pm}(k) \equiv \frac{b^{\pm}(k)}{a^{\pm}(k)} \tag{2.4.77}$$

are the reflection coefficients. Taking into account the limits (2.4.76), we get for $\mathrm{Im}(k)= 0$:

$$\mu_n^{\pm} \sim [T^{\pm}(k)]^{-1}\left[1 + E_n R^{\pm}(k)\right], \qquad n \to \mp\infty. \tag{2.4.78}$$

So to a given potential η_n we can associate in a unique way the spectral data, obtained as a solution of the spectral problem (2.4.73), $T_m^{\pm}(k,\epsilon)$ and $R_m^{\pm}(k,\epsilon)$. As the spectral problem (2.4.66) is a linear OΔE in n, the solution $\chi_n = \chi_{n,m}$ is defined only in its n–dependence. From the linearity of (2.4.66) it follows that $\chi_{n,m}$ is defined up to an arbitrary constant Ω, which in our case can be a function of all the other variables of the problem, i.e. m and k.

Assuming that the function $a^{+}(k)$ has N simple zeros at $k = k_j^{+}$, $1 \le j \le N$, we denote by $\{c_j^{+}\}_{j=1}^{N}$ the set of the N residues of the transmission function $T^{+}(k_j^{+})$. Taking into account (2.4.76, 2.4.78) and using the Cauchy–Green formula we are able to reconstruct in a unique way the Jost solutions. We refer to [109] for further details and for a study of the convergence of the series involved. Taking into account the definitions (2.4.74), the m–evolution of the spectral function $\chi_{n,m}(k)$ reads:

$$\chi_{n,m+1} = (p + \mathrm{i}k)\chi_{n+1,m} + (q - p + u_{n+1,m} - u_{n,m+1})\chi_{n,m}. \tag{2.4.79}$$

Let us now introduce the m–normalization function Ω_m in such a way that

$$\chi_{n,m} = \Omega_m \mu_{n,m}^{+}. \tag{2.4.80}$$

Introducing (2.4.80) into (2.4.79) and taking into account the asymptotic behaviour of the Jost functions (2.4.78, 2.4.76), we find from (2.4.72) the following discrete time evolution for the spectral data for the equation (2.4.56):

$$T_{m+1}^{\pm}(k) = T_m^{\pm}(k), \qquad R_{m+1}^{\pm}(k) = \left(\frac{q - ik}{q + ik}\right) R_m^{\pm}(k). \tag{2.4.81}$$

The time evolutions (2.4.81) of the spectral data are linear and can be easily integrated. Defining the exponential function

$$\mathcal{E}_m = \left(\frac{q - \mathrm{i}k}{q + \mathrm{i}k}\right)^m,$$

we get:

$$T_m^{\pm}(k) = T_0^{\pm}(k), \qquad R_m^{\pm}(k) = \mathcal{E}_m R_0^{\pm}(k), \tag{2.4.82}$$

i.e. the transmission coefficient is invariant under the evolution of the lpKdV (2.4.56) while the reflection coefficient adquires a m–dependent *exponential* factor.

4.4.3. *Symmetries of the lpKdV equation.* The symmetries of the lpKdV (2.4.56) are obtained as flows (in the group parameter space)

$$u_{n,m,\epsilon} = F(n, m, u_{n,m}, u_{n\pm1,m}, u_{n,m\pm1}, \ldots), \tag{2.4.83}$$

commuting with the equation itself. If the function F depends just on $(n, m, u_{n,m})$ then we have a Lie point symmetry, otherwise we have generalized symmetries. In the case of non linear discrete equations the point symmetries are not very common (see Section 1.4.1.1) but, if the equation is integrable and there exists a Lax pair, we can construct an infinity of generalized symmetries. The Lie point symmetries for the lpKdV will be constructed here explicitly. They will also be presented later in Section 2.4.6.3 as this equation is part of the ABS classification.

Let us start by constructing, using the standard technique introduced before, the Lie point symmetries. We will be interested in Lie point symmetries which leave the lattice, characterized by the lattice spacing α and β, invariant. As no independent continuous variable is present the infinitesimal generator is just given by

$$\hat{X}_{n,m} = \phi_{n,m}(u_{n,m})\partial_{u_{n,m}} \tag{2.4.84}$$

Applying (2.4.84) to (2.4.56) we get the following determining equation

$$\begin{aligned}(\phi_{n,m} - \phi_{n+1,m+1})(p - q + u_{n,m+1} - u_{n+1,m})\\ + (p + q - u_{n+1,m+1} + u_{n,m})(\phi_{n,m+1} - \phi_{n+1,m}) = 0,\end{aligned} \tag{2.4.85}$$

to be valid when the equation (2.4.56) is satisfied. Taking (2.4.56) into account it follows that only three of the four different fields present in (2.4.85) are independent and, in all generality we can take them to be $u_{n,m}$, $u_{n,m+1}$ and $u_{n+1,m}$.

Differentiating (2.4.85) with respect to $u_{n,m}$ we get

$$\frac{d\phi_{n,m}}{du_{n,m}} = \frac{d\phi_{n+1,m+1}}{du_{n+1,m+1}}. \tag{2.4.86}$$

Differentiating (2.4.86) with respect to $u_{n,m+1}$ we get $\frac{d^2\phi_{n,m}(u_{n,m})}{du^2_{n,m}} = 0$, i.e.

$$\phi_{n,m} = \mathcal{K}^0_{n,m} + \mathcal{K}^1_{n,m}u_{n,m}. \tag{2.4.87}$$

Introducing (2.4.87) into (2.4.86) we obtain

$$\mathcal{K}^1_{n,m} = \mathcal{K}^1_{n+1,m+1}, \text{ i.e. } \mathcal{K}^1_{n,m} = \mathcal{K}^1_{n-m}. \tag{2.4.88}$$

Introducing (2.4.87) into (2.4.85) we get explicit equations for $u_{n,m}$, $u_{n,m+1}$ and $u_{n+1,m}$. From the various powers of $u_{n,m}$, $u_{n,m+1}$ and $u_{n+1,m}$ we get a set of coupled linear PΔEs for $\mathcal{K}^1_{n,m}$ and $\mathcal{K}^0_{n,m}$ like (2.4.88), which give

$$\mathcal{K}^1_{n,m} = \beta, \quad \mathcal{K}^0_{n,m} = \gamma + \delta(-1)^{n+m} + \alpha(-1)^{n-m} + (qn + pm)\beta. \tag{2.4.89}$$

So the lpKdV (2.4.56) admits a four dimensional group of point symmetries whose infinitesimal generators are

$$\begin{aligned}\hat{X}^1 = \partial_{u_{n,m}}, \quad \hat{X}^2 = (-1)^{n+m}\partial_{u_{n,m}}, \quad \hat{X}^3 = (-1)^{n-m}\partial_{u_{n,m}},\\ \hat{X}^4 = [u_{n,m} + (pm + qn)]\partial_{u_{n,m}}.\end{aligned} \tag{2.4.90}$$

Apart from these symmetries we have a particularly interesting discrete symmetry which involves the exchange of n and m together with the exchange of p and q.

An infinity of generalized symmetries of the lpKdV are obtained as DΔEs associated with the spectral problem (2.4.66). They are obtained constructing the infinity of equations associated to the spectral problem (2.4.66). As there is a one to one correspondence between the equations and the spectral data, we will look for the commutativity of the flows in the space of the spectral data, where the equations are linear. In the following to simplify the formulas we will replace the derivative with respect to the group parameter ϵ by a *dot*, i.e. $u_{n,m,\epsilon} \equiv \dot{u}_{n,m}$

For convenience we rewrite equation (2.4.66) as:

$$L_{n,m}\psi_{n,m} = \lambda\psi_{n,m}, \qquad L_{n,m} \equiv S^2 - (\eta_{n,m} + 2p)S. \tag{2.4.91}$$

The definition of the shift operator $S = S_n$ is given in (1.2.13), the eigenvalue $\lambda \in \mathbb{C}$ is defined in (2.4.74) and from (2.4.71, 2.4.74) $\eta_{n,m} \equiv u_{n,m} - u_{n+2,m}$ is a bounded potential. Starting from (2.4.91) we can apply the *Lax technique*, and obtain the recursion operator $\mathcal{L}_\ell$ and the hierarchy of non linear evolution equations associated to it as it has been done in [107].

The isospectral hierarchy of symmetries is obtained requiring that $\dot{\lambda}(\epsilon) = 0$. It corresponds to looking at DΔEs whose Lax equation reads

$$\dot{L}_{n,m} = [L_{n,m}, M_{n,m}]. \tag{2.4.92}$$

The *Lax technique* provides the relation

$$\tilde{V}_{n,m} = \mathcal{L}_\ell V_{n,m} + V^{(0)}_{n,m}, \tag{2.4.93}$$

where $\tilde{V}_{n,m} = \dot{\eta}_{n,m}$. $V^{(0)}_{n,m}$ is some given functions of $\eta_{n,m}$ and of a certain number of arbitrary integration constants a, b and c

$$\begin{aligned} V^{(0)}_{n,m} = (\eta_{n,m} + 2p)\Big[a(-1)^n + b\Big(\eta_{n,m} + 4p + 2\sum_{k=1}^{\infty}(-1)^k\eta_{k,m}\Big) \\ + c(-1)^n\Big(\eta_{n,m} + 2p + 4pn + 2\sum_{k=1}^{\infty}\eta_{k,m}\Big)\Big]. \end{aligned} \tag{2.4.94}$$

The recursion operator $\mathcal{L}_\ell$, is defined by

$$\mathcal{L}_\ell \equiv -(\eta_{n,m} + 2p)\Delta\tilde{\Delta}^{-1}(\eta_{n,m} + 2p)S\Delta^{-1}\tilde{\Delta}^{-1}, \tag{2.4.95}$$

where

$$\Delta \equiv S - 1, \qquad \tilde{\Delta} \equiv S + 1,$$
$$\Delta^{-1} = -\sum_{k=0}^{\infty} S^k, \qquad \tilde{\Delta}^{-1} = \sum_{k=0}^{\infty}(-1)^k S^k.$$

Choosing $V_{n,m} = 0$ a first equation is given by $\tilde{V}_{n,m} = V^{(0)}_{n,m}$. Thus we get the following isospectral hierarchy of equations:

$$\dot{\eta}_{n,m} = g_1(\mathcal{L}_\ell)V^{(0)}_{n,m}, \tag{2.4.96}$$

where g_1 is an entire function of its argument. Eq. (2.4.96) involves at least a summation if b and c are different from zero and thus is not local. As was shown in [107] we can always obtain a class of local equations when we consider $g_1 = g_1(\mathcal{L}_\ell^{-1})$. From (2.4.95) we have

$$\mathcal{L}_\ell^{-1} = -S^{-1}\Delta\tilde{\Delta}\frac{1}{\eta_{n,m} + 2p}(\Delta)^{-1}\tilde{\Delta}\frac{1}{\eta_{n,m} + 2p}.$$

As $\Delta\alpha = 0$ when α is an arbitrary complex constant, $\Delta^{-1}0 = \alpha$ where Δ^{-1} is the formal inverse operator of Δ. Consequently

$$\dot{\eta}_{n,m} = g_1(\mathcal{L}_\ell^{-1})0, \tag{2.4.97}$$

will provide local equations, the so called *inverse hierarchy*. The first equation we obtain in this inverse hierarchy by choosing $g_1(z) = z$ is (2.4.65) when $\alpha = 2p$ and we take into account the definition (2.4.74). The next equation is obtained by choosing $g_1(z) = z^2$. We have

$$\dot{u}_{n,m} = \tilde{\mathscr{L}}_n^{-1} \frac{1}{2p - u_{n+1,m} - u_{n-1,m}} + \beta_k. \tag{2.4.98}$$

Here the β_k's are some integration constants to be defined in such a way that $u_{n,m}$, asymptotically bounded, is a compatible solution of (2.4.98). Taking into account the definition (2.4.74), we have:

$$\begin{aligned}\dot{\eta}_{n,m} = {} & \frac{2p}{\eta_{n-1,m} + 2p}\left[\frac{1}{\eta_{n-2,m} + 2p} + \frac{2}{\eta_{n-1,m} + 2p} + \frac{1}{\eta_{n,m} + 2p}\right] \\ & - \frac{2p}{\eta_{n+1,m} + 2p}\left[\frac{1}{\eta_{n,m} + 2p} + \frac{2}{\eta_{n+1,m} + 2p} + \frac{1}{\eta_{n+2,m} + 2p}\right].\end{aligned} \tag{2.4.99}$$

Eq. (2.4.99) can be written in terms of $r_{n,m}$

$$\begin{aligned}\dot{r}_{n,m} = {} & \frac{2p}{r_{n-1,m}}\left[\frac{1}{r_{n-2,m}} + \frac{2}{r_{n-1,m}} + \frac{1}{r_{n,m}}\right] \\ & - \frac{2p}{r_{n+1,m}}\left[\frac{1}{r_{n,m}} + \frac{2}{r_{n+1,m}} + \frac{1}{r_{n+2,m}}\right],\end{aligned} \tag{2.4.100}$$

and by an integration we get, defining by γ an integration constant,

$$\begin{aligned}\dot{u}_{n,m} = {} & \frac{2p}{2p + u_{n-1,m} - u_{n+1,m}}\left[\frac{1}{2p + u_{n-2,m} + u_{n,m}} + \frac{2}{2p + u_{n-1,m} - u_{n+1,m}}\right. \\ & \left. \gamma + \frac{1}{2p + u_{n,m} - u_{n+2,m}}\right] - \frac{2}{p}\end{aligned} \tag{2.4.101}$$

the higher equation of (2.4.63).

In correspondence with (2.4.101) we obtain the evolution of the reflection coefficient

$$\dot{R}_m = \left[1 - E_k\right] R_m. \tag{2.4.102}$$

Taking into account (2.4.81) and (2.4.102) it follows the compatibility condition

$$S_m\, \dot{R}_m = \dot{R}_{m+1} \qquad \forall\, k \in \mathbb{N}, \tag{2.4.103}$$

where the m-shift operator is defined as in (1.2.13), i.e. (2.4.98) is a symmetry of the lpKdV (2.4.56). Moreover, we also have the compatibility

$$\frac{d}{d\epsilon_h}\left(\frac{dR_m}{d\epsilon_k}\right) = \frac{d}{d\epsilon_k}\left(\frac{dR_m}{d\epsilon_h}\right) \qquad \forall\, k, h \in \mathbb{N}, \tag{2.4.104}$$

i.e. the symmetries commute among themselves.

The non isospectral hierarchy of symmetries is obtained requiring that $\lambda_\epsilon = f(\lambda)$ and corresponds to considering equations whose Lax equation reads

$$\dot{L}_{n,m} = [L_{n,m}, M_{n,m}] + f(L_{n,m}). \tag{2.4.105}$$

In this case $V_{n,m}^{(0)}$ is given by

$$V_{n,m}^{(0)} = h(\eta_{n,m} + 2p). \tag{2.4.106}$$

The function $V_{n,m}^{(0)}$ always diverges asymptotically. However, also in this case, we can consider the inverse hierarchy, when $g_2 = g_2(\mathscr{L}_\ell^{-1})$. As the solution of the OΔE $\Delta f_{n,m}(\epsilon) =$

$\beta_m(\epsilon)$ is given by $f_{n,m}(\epsilon) = \beta_m(\epsilon)n + \gamma_m(\epsilon)$, starting from $V^{(0)}_{n,m}$, a well defined non isospectral hierarchy of equations is given by

$$\dot{u}_{n,m} = \mathscr{L}_\ell^{-k}(\mathscr{L}_\ell^{-1} + \frac{1}{p^2})(\eta_{n,m} + 2p), \quad k \in \mathbb{N}. \tag{2.4.107}$$

In this case the first equation we obtain in this inverse hierarchy by choosing $k = 1$ is a non isospectral deformation of (2.4.63)

$$\dot{\eta}_{n,m} = \frac{2n-1}{\eta_{n-1,m} + 2p} - \frac{2n+3}{\eta_{n+1,m} + 2p} + \frac{\eta_{n,m} + 2p}{p^2}. \tag{2.4.108}$$

The higher order equations in the non isospectral hierarchy are all non local. In correspondence with (2.4.108) we have the following evolution of the spectral data:

$$\dot{T}^{\pm}_m(k) = 0, \quad \dot{R}^{\pm}_m(k) = -\frac{ik}{p(p^2 + k^2)} R^{\pm}_m(k). \tag{2.4.109}$$

From (2.4.108) we get the master symmetry for the lpKdV, which reads

$$\hat{Y} = \Big(\frac{n\,p}{u_{n+1,m} - u_{n-1,m} - 2p} + \frac{1}{2}p\,n - \frac{1}{4}u_{n,m}\Big)\partial_{u_{n,m}}. \tag{2.4.110}$$

Taking into account the discrete symmetry of the problem we can extend (2.4.110) to

$$\hat{Y} = \Big(\frac{n\,p}{u_{n+1,m} - u_{n-1,m} - 2p} + \frac{m\,q}{u_{n,m+1} - u_{n,m-1} - 2q} + \frac{1}{2}[p\,n + q\,m - u_{n,m}]\Big)\partial_{u_{n,m}}, \tag{2.4.111}$$

which is a symmetry of the lpKdV.

4.5. Lattice version of the Schwarzian KdV. The lattice version of the Schwarzian KdV

$$w_t = w_{xxx} - \frac{3\,w_{xx}^2}{2\,w_x}, \tag{2.4.112}$$

(lSKdV), is given by the non linear PΔE [637,641]:

$$\begin{aligned} Q \equiv\ & \alpha_1\,(u_{n,m} - u_{n,m+1})\,(u_{n+1,m} - u_{n+1,m+1}) \\ & - \alpha_2\,(u_{n,m} - u_{n+1,m})\,(u_{n,m+1} - u_{n+1,m+1}) = 0. \end{aligned} \tag{2.4.113}$$

Eq. (2.4.113) involves just four lattice points on a two dimensional orthogonal lattice situated at the vertices of an elementary square (see Fig. 2.3 in Section 2.4.6). It is a lattice equation on quad-graphs belonging to the classification presented in [22] (see Section 2.4.6), where the CaC is used as a tool to establish its integrability. It is a subcase of the first element of the Q–list (see Section 2.4.6) [22], namely Q_1, with $\delta = 0$.

As far as we know the lSKdV equation has been introduced for the first time by Nijhoff, Quispel and Capel in 1983 [641]. A review of results about the lSKdV equation can be found in [632,637].

Since we have two discrete independent variables, i.e. n and m, we can perform the continuous limit in two steps. Each step is achieved by shrinking the corresponding lattice step to zero and sending to infinity the number of points of the lattice.

In the first step, setting $\alpha_1 \equiv q^2$, $\alpha_2 \equiv p^2$, we define $u_{n,m} \equiv y_\ell(\tau)$, where $\ell \equiv n + m$ and $\tau \equiv \delta\, m$, $\delta \equiv p - q$, i.e. a skew limit as in the case of the lpKdV. Considering the limits

$m \to \infty$, $\delta \to 0$ in such a way that τ is finite, we get the DΔE

$$y_{\ell,\tau} = \frac{2\,(y_{\ell+1} - y_\ell)\,(y_{\ell-1} - y_\ell)}{p\,(y_{\ell-1} - y_{\ell+1})}. \tag{2.4.114}$$

Eq. (2.4.114) is a subcase of the Yamilov discretization of the Krichever-Novikov equation (2.4.129) with $A_0 = \frac{2}{p}$, $B_0 = -\frac{2}{p} y_\ell$, $C_0 = \frac{2}{p} y_\ell^2$.

The second step is performed by taking $y_\ell(\tau) \equiv w(x,t)$ in (2.4.114), with $x \equiv 2\,(\ell + \tau/p)/p$ and $t \equiv 2\,(\ell/3 + \tau/p)/p^3$. If we carry out the limit $p \to \infty$, $\ell \to \infty$, $\tau \to \infty$, in such a way that x and t remain finite, then (2.4.114) is transformed into the continuous Schwarzian KdV equation (2.4.112).

4.5.1. *The integrability of the lSKdV equation.* Eq. (2.4.113) has been obtained firstly by the direct linearization method [641]. In [637] one can find its associated spectral problem, which, as this equation is part of the ABS classification, can be obtained using a well-defined procedure [102, 103, 634] (see Section 2.4.6).

Its Lax pair is given by the overdetermined system of matrices (2.4.5, 2.4.6) of the Zakharov & Shabat matrix formalism with $U_{n,m}$ and $V_{n,n}$ given by

$$U_{n,m} = \begin{pmatrix} 1 & u_{n,m} - u_{n+1,m} \\ \lambda\,\alpha_1\,(u_{n,m} - u_{n+1,m})^{-1} & 1 \end{pmatrix},$$

and

$$V_{n,m} = \begin{pmatrix} 1 & u_{n,m} - u_{n,m+1} \\ \lambda\,\alpha_2\,(u_{n,m} - u_{n,m+1})^{-1} & 1 \end{pmatrix}.$$

For these functions $U_{n,m}$ and $V_{n,m}$ we can rewrite (2.4.5, 2.4.6) in scalar form in terms of just one field $\psi_{n,m}(\lambda)$:

$$\begin{aligned}(u_{n,m} - u_{n+1,m})\,\psi_{n+2,m} &+ (u_{n+2,m} - u_{n,m})\,\psi_{n+1,m} \\ &+ \left(1 - \lambda\,\alpha_1\right)(u_{n+1,m} - u_{n+2,m})\,\psi_{n,m} = 0,\end{aligned} \tag{2.4.115a}$$

$$\begin{aligned}(u_{n,m} - u_{n,m+1})\,\psi_{n,m+2} &+ (u_{n,m+2} - u_{n,m})\,\psi_{n,m+1} \\ &+ \left(1 - \lambda\,\alpha_2\right)(u_{n,m+1} - u_{n,m+2})\,\psi_{n,m} = 0.\end{aligned} \tag{2.4.115b}$$

It is worthwhile to observe here that (2.4.113) and the Lax equations (2.4.115) are invariant under the discrete symmetry obtained by interchanging at the same time n with m and α_1 with α_2.

To get meaningful Lax equations, the field $u_{n,m}$ cannot go asymptotically to a constant c but must be written as $u_{n,m} \equiv v_{n,m} + \beta_0\,m + \alpha_0\,n$, where α_0 and β_0 are constants related to α_1 and α_2 by the condition $\alpha_1\,\beta_0^2 = \alpha_2\,\alpha_0^2$, and $v_{n,m}$ goes asymptotically to a constant. Under this transformation of the dependent variable, the lSKdV equation and its Lax equations read

$$\begin{aligned}&\alpha_1\,(v_{n,m} - v_{n,m+1} - \beta_0)\,(v_{n+1,m} - v_{n+1,m+1} - \beta_0) \\ &= \alpha_2\,(v_{n,m} - v_{n+1,m} - \alpha_0)\,(v_{n,m+1} - v_{n+1,m+1} - \alpha_0),\end{aligned} \tag{2.4.116}$$

$$(1 + v^{(n)}_{n,m})\,\psi_{n+2,m} - (2 + v^{(n)}_{n,m})\,\psi_{n+1,m} + (1 - \lambda\,\alpha_1)\,\psi_{n,m} = 0, \tag{2.4.117a}$$

$$(1 + v^{(m)}_{n,m})\,\psi_{n,m+2} - (2 + v^{(m)}_{n,m})\,\psi_{n,m+1} + (1 - \lambda\,\alpha_2)\,\psi_{n,m} = 0. \tag{2.4.117b}$$

The functions $v^{(n)}_{n,m}$ and $v^{(m)}_{n,m}$ are defined by

$$v^{(n)}_{n,m} \equiv \frac{v_{n+2,m} - 2\,v_{n+1,m} + v_{n,m}}{v_{n+1,m} - v_{n,m} + \alpha_0}, \qquad v^{(m)}_{n,m} \equiv \frac{v_{n,m+2} - 2\,v_{n,m+1} + v_{n,m}}{v_{n,m+1} - v_{n,m} + \beta_0},$$

where, as $v_{n,m} \to c$, with $c \in \mathbb{R}$, as n and m go to infinity, $v_{n,m}^{(n)}$ and $v_{n,m}^{(m)}$ go to zero.

The symmetries of (2.4.113) are given by compatible evolutions in the group parameter ϵ. They can be constructed starting from (2.4.117a) by the Lax technique requiring the existence of a set of operators M_n such that [124]

$$L_n \psi_n = \lambda \psi_n, \qquad \psi_{n,\epsilon} = -M_n \psi_n, \qquad L_{n,\epsilon} = [\, L_n, M_n \,].$$

with

$$L_n = (1 + v_{n,m}^{(n)}(t))\, S_n^2 - (2 + v_{n,m}^{(n)}(t))\, S_n.$$

Here λ is a spectral parameter. If $\lambda_\epsilon = 0$ the class of DΔEs one so obtains will be called isospectral, while if $\lambda_\epsilon \neq 0$ it will be called non isospectral.

As the field $v_{n,m}^{(n)}(t)$ appears multiplying both S_n and S_n^2 the expression of the recursive operator turns out to be extremely complicated, containing triple sums and products of the dependent fields. So we look for transformations of the spectral problem (2.4.117a) which reduce it to a simpler form in which the potential will appear just once. There are two different discrete spectral problems involving three lattice points. The discrete Schrödinger spectral problem introduced by Case [157],

$$\phi_{n-1} + a_n \phi_{n+1} + b_n \phi_n = \lambda \phi_n, \tag{2.4.118}$$

which is associated with the Toda and Volterra DΔEs [371, 373] and the asymmetric discrete Schrödinger spectral problem introduced by Shabat et al. [109, 750],

$$\phi_{n+2} = \frac{2p}{s_n} \phi_{n+1} + \lambda \phi_n. \tag{2.4.119}$$

The latter one, considered in (2.4.66), has been used to solve the lpKdV equation [626]. In (2.4.118, 2.4.119) the functions a_n, b_n, s_n may depend parametrically on a *continuous variable* ϵ but also on a *discrete variable* m. As all three spectral problems (2.4.117a, 2.4.118, 2.4.119) involve just three points on the lattice, we can relate them by a gauge transformation $\psi_n \equiv f_n(\sigma)\, g_n(\{v_{n,m}\})\, \phi_n$, where $\{v_{n,m}\} \equiv (v_{n,m}, v_{n\pm1,m}, v_{n,m\pm1}, \ldots)$. These transformations give rise to a Miura transformation between the involved fields. For instance, when we transform (2.4.117a) into the discrete Schrödinger spectral problem (2.4.118) we get

$$b_n \equiv b_{n,m} = 0, \tag{2.4.120a}$$

$$a_n \equiv a_{n,m} = \frac{4\,(v_{n+1,m} - v_{n,m} + \alpha_0)^2}{(v_{n+2,m} - v_{n,m} + 2\,\alpha_0)\,(v_{n+1,m} - v_{n-1,m} + 2\,\alpha_0)}. \tag{2.4.120b}$$

If we transform the spectral problem given in (2.4.117a) into the asymmetric discrete Schrödinger spectral problem (2.4.119) the relation between the fields $s_n \equiv s_{n,m}$ and $v_{n,m}$ is more involved as it is expressed in terms of infinite products.

We will use in the following (2.4.120b) and the equivalent one obtained by transforming (2.4.117b) into (2.4.118) which will define a field $\tilde{a}_m$ given by (2.4.120b) with n and m and α_0 and β_0 interchanged. These transformations will be used to build the generalized symmetries of the lSKdV (2.4.116) from the non linear DΔEs associated with the spectral problem (2.4.118) [371, 373] with b_n given by (2.4.120a).

4.5.2. *Point symmetries of the lSKdV equation.* Let us construct the Lie point symmetries of (2.4.113), with $\alpha_1 \neq \alpha_2$, using the technique introduced in [544]. The Lie point symmetries we obtain in this way turn out to be the same as those for (2.4.116).

The Lie symmetries of the lSKdV equation (2.4.113) are given by those continuous transformations which leave the equation invariant. From the infinitesimal point of view they are obtained by requiring the infinitesimal invariant condition

$$\text{pr}\,\widehat{X}_{n,m}\,Q\Big|_{Q=0} = 0, \tag{2.4.121}$$

where, as we keep the lattice invariant,

$$\widehat{X}_{n,m} = \Phi_{n,m}(u_{n,m})\partial_{u_{n,m}}. \tag{2.4.122}$$

By $pr\,\widehat{X}_{n,m}$ we mean the prolongation of the infinitesimal generator $\widehat{X}_{n,m}$ to the other three points appearing in $Q = 0$, i.e. $u_{n+1,m}$, $u_{n,m+1}$ and $u_{n+1,m+1}$.

Solving the equation $Q = 0$ w.r.t. $u_{n+1,m+1}$ and substituting it in (2.4.121) we get a functional equation for $\Phi_{n,m}(u_{n,m})$. Looking at its solutions in the form $\Phi_{n,m}(u_{n,m}) = \sum_{k=0}^{\gamma} \Phi_{n,m}^{(k)}\, u_{n,m}^k$, $\gamma \in \mathbb{N}$, we see that in order to balance the leading order in $u_{n,m}$, if $\alpha_1 \neq \alpha_2$, γ cannot be greater than 2, and thus must belong to the interval $[\,0, 2\,]$. Equating now to zero the coefficients of the powers of $u_{n,m}$, $u_{n+1,m}$ and $u_{n,m+1}$ we get an overdetermined system of determining equations. Solving the resulting difference equations we find that the functions $\Phi_{n,m}^{(i)}$'s, $i = 0, 1, 2$, must be constants. Hence the infinitesimal generators of the algebra of Lie point symmetries are given by

$$\widehat{X}_{n,m}^{(0)} = \partial_{u_{n,m}}, \qquad \widehat{X}_{n,m}^{(1)} = u_{n,m}\,\partial_{u_{n,m}}, \qquad \widehat{X}_{n,m}^{(2)} = u_{n,m}^2\,\partial_{u_{n,m}}.$$

The generators $\widehat{X}_{n,m}^{(i)}$, $i = 0, 1, 2$, span the Lie algebra $sl(2)$:

$$\left[\widehat{X}_{n,m}^{(0)}, \widehat{X}_{n,m}^{(1)}\right] = \widehat{X}_{n,m}^{(0)}, \qquad \left[\widehat{X}_{n,m}^{(1)}, \widehat{X}_{n,m}^{(2)}\right] = \widehat{X}_{n,m}^{(2)},$$

$$\left[\widehat{X}_{n,m}^{(0)}, \widehat{X}_{n,m}^{(2)}\right] = 2\,\widehat{X}_{n,m}^{(1)}.$$

We can write down the group transformation by integrating the DΔE

$$\tilde{u}_{n,m,\epsilon} = \Phi_{n,m}(\tilde{u}_{n,m}(\epsilon)), \tag{2.4.123}$$

with the initial condition $\tilde{u}_{n,m}(\epsilon = 0) = u_{n,m}$. We get the Möbius transformation [102, 103, 637]

$$\tilde{u}_{n,m}(\epsilon_0, \epsilon_1, \epsilon_2) = \frac{(\epsilon_0 + u_{n,m})\, e^{\epsilon_1}}{1 - \epsilon_2\,(\epsilon_0 + u_{n,m})\, e^{\epsilon_1}},$$

where the ϵ_i's are the group parameters associated with the infinitesimal generators $\widehat{X}^{(i)}$, $i = 0, 1, 2$.

We finally notice that, in the case when $\alpha_1 = \alpha_2$ (2.4.113) reduces to the product of two linear discrete wave equations:

$$(u_{n,m} - u_{n+1,m+1})\,(u_{n+1,m} - u_{n,m+1}) = 0,$$

which is trivially solved by taking $u_{n,m} = f_{n\pm m}$ and the Lie point symmetries belong to an infinite dimensional Lie algebra.

4.5.3. *Generalized symmetries of the lSKdV equation.* A generalized symmetry is obtained when the function $\Phi_{n,m}$ appearing in (2.4.122) depends on $\{u_{n,m}\}$ and not only on $u_{n,m}$. A way to obtain it, is to look at those DΔEs (2.4.123) associated with (2.4.118) which are compatible with (2.4.113). From (2.4.120) we see that the lSKdV equation can be associated with the discrete Schrödinger spectral problem when $b_{n,m} = 0$, i.e. when the associated hierarchy of differential difference equations is given by the Volterra hierarchy [371]. So, applying the Miura transformation (2.4.120) to the DΔEs of the Volterra hierarchy we can obtain the symmetries of the lSKdV equation. The Miura transformation (2.4.120) preserves the integrability of the Volterra hierarchy if $v_{n,m} \to c$, with $c \in \mathbb{R}$.

The procedure to get the generalized symmetries for the lSKdV is better shown on a specific example, the case of the Volterra equation itself, an isospectral deformation of (2.4.118) given by (2.3.172) with $a_n(t) \equiv a_{n,m}(\epsilon_0)$. Let us substitute the Miura transformation, given by (2.4.120b), into (2.3.172) and let us assume that

$$v_{n,m,\epsilon_0} = F_{n,m}(v_{n-1,m}, v_{n,m}, v_{n+1,m}).$$

Eq. (2.3.172) is thus a functional equation for $F_{n,m}$ which can be solved as we did in the previous Section by comparing powers at infinity or by transforming it into an overdetermined system of linear PDEs [14, 15]. In this way we get, up to a point transformation,

$$F_{n,m} = \frac{4\,(v_{n,m} - v_{n-1,m} + \alpha_0)\,(v_{n,m} - v_{n+1,m} - \alpha_0)}{v_{n+1,m} - v_{n-1,m} + 2\,\alpha_0}. \tag{2.4.124}$$

Eq. (2.4.124) is nothing else but (2.4.114) *mutatis mutandis*. One can verify that (2.4.124) is a generalized symmetry of the lSKdV equation (2.4.116) by proving that $\Phi_{n,m} = F_{n,m}(v_{n-1,m}, v_{n,m}, v_{n+1,m})$ satisfies (2.4.121).

If we start from a higher equation of the isospectral Volterra hierarchy (2.3.170) with $a_n(t) \equiv a_{n,m}(\epsilon_1)$ we get a second generalized symmetry of the lSKdV equation requiring $F_{n,m} = F_{n,m}(v_{n-2,m}, v_{n-1,m}, v_{n,m}, v_{n+1,m}, v_{n+2,m})$. It reads

$$\begin{aligned} F_{n,m} &= \frac{(v_{n,m} - v_{n-1,m} + \alpha_0)\,(v_{n,m} - v_{n+1,m} - \alpha_0)}{(v_{n+1,m} - v_{n-1,m} + 2\,\alpha_0)^2} \\ &\quad \times \left[\frac{(v_{n+2,m} - v_{n+1,m} + \alpha_0)\,(v_{n-1,m} - v_{n,m} - \alpha_0)}{v_{n+2,m} - v_{n,m} + 2\,\alpha_0} \right. \\ &\qquad \left. - \frac{(v_{n-1,m} - v_{n-2,m} + \alpha_0)\,(v_{n,m} - v_{n+1,m} - \alpha_0)}{v_{n,m} - v_{n-2,m} + 2\,\alpha_0} \right]. \end{aligned} \tag{2.4.125}$$

This procedure could be clearly carried out for any of the equations of the Volterra hierarchy [371] presented in Section 2.3.3 and we would obtain by the Miura transformation a hierarchy of isospectral symmetries for the lSKdV equation.

If we consider the non isospectral hierarchy the only local equation is (see Section 2.3.3.5)

$$a_{n,m,\epsilon} = a_{n,m}\,[\,a_{n,m} - (n-1)\,a_{n-1,m} + (n+2)\,a_{n+1,m} - 4\,]$$

obtained from (2.3.234) by setting $p = 0$ and $q = 1$. It provides up to a Lie point symmetry, two local equations:

$$v_{n,m,\epsilon_0} = v_{n,m} + \alpha_0\, n, \tag{2.4.126}$$

$$v_{n,m,\epsilon_1} = \frac{4\,n\,(v_{n,m} - v_{n-1,m} + \alpha_0)\,(v_{n,m} - v_{n+1,m} - \alpha_0)}{v_{n+1,m} - v_{n-1,m} + 2\,\alpha_0}. \tag{2.4.127}$$

One can easily show that (2.4.126) is not a symmetry of the lSKdV equation but it commutes with all its known symmetries. Eq. (2.4.127) is a master symmetry [287]: it does not commute with the lSKdV equation but commuting it with (2.4.124) one gets (2.4.125) and commuting it with (2.4.125) one gets a higher order symmetry. So through it one can reconstruct the hierarchy of isospectral generalized symmetries of the lSKdV equation.

In the construction of generalized symmetries for the DΔE Volterra (see Section 2.3.3) one was able to construct a symmetry from the master symmetry (2.4.127) by combining it with a second isospectral symmetry (2.4.125) multiplied by t. This seems not to be the case for PΔEs. As was shown in Section 2.4.4.3 for the case of the lpKdV equation, there is no combination of (2.4.127) with isospectral symmetries which gives us a symmetry of (2.4.116).

As the lSKdV equation admits a discrete symmetry corresponding to an exchange of n with m and α_1 with α_2, one can construct another class of generalized and master symmetries by considering the equations obtained from the spectral problem (2.4.118) in the m lattice variable depending on the potential $\tilde{a}_m$. In this way we get:

$$\begin{aligned} v_{n,m,\tilde{e}_0} &= \frac{4\,(v_{n,m}-v_{n,m-1}+\beta_0)\,(v_{n,m}-v_{n,m+1}-\beta_0)}{v_{n,m+1}-v_{n,m-1}+2\,\beta_0}, \\ v_{n,m,\tilde{e}_1} &= \frac{(v_{n,m}-v_{n,m-1}+\beta_0)\,(v_{n,m}-v_{n,m+1}-\beta_0)}{(v_{n,m+1}-v_{n,m-1}+2\,\beta_0)^2} \\ &\quad\times\left[\frac{(v_{n,m+2}-v_{n,m+1}+\beta_0)\,(v_{n,m-1}-v_{n,m}-\beta_0)}{v_{n,m+2}-v_{n,m}+2\,\beta_0}\right. \\ &\quad\left.-\frac{(v_{n,m-1}-v_{n,m-2}+\beta_0)\,(v_{n,m}-v_{n,m+1}-\alpha_0)}{v_{n,m}-v_{n,m-2}+2\,\beta_0}\right]. \end{aligned}$$

and

$$\begin{aligned} v_{n,m,\bar{e}_0} &= v_{n,m}+\beta_0\,m, \\ v_{n,m,\bar{e}_1} &= \frac{4\,m\,(v_{n,m}-v_{n,m-1}+\beta_0)\,(v_{n,m}-v_{n,m+1}-\beta_0)}{v_{n,m+1}-v_{n,m-1}+2\,\beta_0}. \end{aligned}$$

A different class of symmetries can be obtained applying the following theorem which provides a constructive tool to obtain generalized symmetries for the lSKdV equation (2.4.113).

Theorem 11. *Let $Q(v_{n,m}, v_{n\pm1,m}, v_{n,m\pm1}, \ldots; \alpha_1, \alpha_2) = 0$ be an integrable PΔE invariant under the discrete symmetry $n \leftrightarrow m$, $\alpha_1 \leftrightarrow \alpha_2$. Let $\widehat{Z}_n$ be a differential operator*

$$\widehat{Z}_n \equiv Z_n(v_{n,m}, v_{n\pm1,m}, v_{n,m\pm1}, \ldots; \alpha_1, \alpha_2)\,\partial_{v_{n,m}},$$

such that

$$pr\,\widehat{Z}_n\,Q\Big|_{Q=0} = a\,g_{n,m}(v_{n,m}, v_{n\pm1,m}, v_{n,m\pm1}, \ldots; \alpha_1, \alpha_2),$$

where $g_{n,m}(v_{n,m}, v_{n\pm1,m}, v_{n,m\pm1}, \ldots; \alpha_1, \alpha_2)$ is invariant under the discrete symmetry $n \leftrightarrow m$, $\alpha_1 \leftrightarrow \alpha_2$ and a is an arbitrary constant. Then we have

$$\left(\frac{1}{a}\,pr\,\widehat{Z}_n - \frac{1}{b}\,pr\,\widehat{Z}_m\right)\,Q\Big|_{Q=0} = 0,$$

where the operator $\widehat{Z}_m \equiv Z_m(v_{n,m}, v_{n,m\pm1}, v_{n\pm1,m}, \ldots; \alpha_2, \alpha_1)\,\partial_{u_{n,m}}$ is obtained from $\widehat{Z}_n$ under $n \leftrightarrow m$, $\alpha_1 \leftrightarrow \alpha_2$, so that

$$pr\,\widehat{Z}_m\,Q\Big|_{Q=0} = b\,g_{n,m}(v_{n,m}, v_{n\pm1,m}, v_{n,m\pm1}, \ldots; \alpha_2, \alpha_1),$$

with b a constant. So

$$\widehat{Z}_{n,m} \equiv \frac{1}{a}\,\widehat{Z}_n - \frac{1}{b}\,\widehat{Z}_m$$

is a symmetry of $Q = 0$.

Using Theorem 11 it is easy to show that from the master symmetry (2.4.127) we can construct a generalized symmetry, given by

$$v_{n,m,\epsilon} = \frac{4\,n\,(v_{n,m}-v_{n-1,m}+\alpha_0)\,(v_{n,m}-v_{n+1,m}-\alpha_0)}{v_{n+1,m}-v_{n-1,m}+2\,\alpha_0} + \frac{4\,m\,(v_{n,m}-v_{n,m-1}+\beta_0)\,(v_{n,m}-v_{n,m+1}-\beta_0)}{v_{n,m+1}-v_{n,m-1}+2\,\beta_0}.$$

The above symmetry has been implicitly used, together with point symmetries, by Nijhoff and Papageorgiou [638] to perform the similarity reduction of the lSKdV equation and get a discrete analogue of the Painlevé II equation. Something like Theorem 11 has been used in Section 2.4.4.3 to get from master symmetries generalized symmetries for lpKdV.

4.6. Volterra type DΔEs and the ABS classification. Here we present the equations of the well-known ABS list [22, 29] and show that they possess all the properties of integrability. This is the first time in which we are presented with a list of integrable PΔEs not expressed as a hierarchy. More we will see in Chapter 3. The equations of the ABS list are all S-integrable while the extensions presented by Boll are, in most of the cases, C-integrable. Most Boll equations are Darboux integrable and can be solved completely (see Section 2.4.7).

Many results can be found in the literature on solutions of the ABS equations, see for example [63–67, 389, 635].

In the following we write down the simplest three-point symmetries and the Lax pairs of the ABS list of equations. Then in Section 2.4.8, comparing symmetries, we will be able to show that the ABS list does not cover all integrable quad-graph equations. We show there that one can find other integrable discrete equations differing essentially from the equations of the ABS classification and Boll extension.

Whenever possible, for example when the PΔEs have no explicit dependence on n and m, we will use a simplified notation. In this case we will not write down n and m so that $u_{n,m} = u_{0,0}$, $u_{n+1,m} = u_{1,0}$, $u_{n,m-1} = u_{0,-1}$, etc. We will use a similar simplified notation also in the case of DΔEs.

We also show that there is a close connection between the symmetries of the discrete equations of the ABS list and the DΔEs of the Volterra type considered in Sections 2.3.3 and 3.3.1.

Let us introduce the Krichever-Novikov equation

$$u_t = \frac{1}{4}u_{xxx} - \frac{3}{2u_x}\,[u_{xx}^2 - 4\,P(u)], \tag{2.4.128}$$

where $P(u)$ is an arbitrary fourth order degree polynomial of its argument with constant coefficients. Then one can show that all three point symmetries of the ABS equations correspond to particular cases of a discrete analogue of the Krichever-Novikov equation (2.4.128) [454], the Yamilov discretization of the Krichever-Novikov equation (YdKN), contained in (V_4) as part of the classification of Volterra type equations presented in Section 3.3.1.2 [492, 842]:

$$u_{0,\epsilon} = \frac{R(u_1,u_0,u_{-1})}{u_1-u_{-1}}, \qquad R(u_1,u_0,u_{-1}) = A_0u_1u_{-1} + B_0(u_1+u_{-1}) + C_0, \tag{2.4.129}$$

where

$$A_0 = c_1 u_0^2 + 2c_2 u_0 + c_3,$$
$$B_0 = c_2 u_0^2 + c_4 u_0 + c_5,$$
$$C_0 = c_3 u_0^2 + 2c_5 u_0 + c_6.$$

We consider here the completely autonomous case when both the non linear PΔEs and their generalized symmetries are autonomous. The form of any relation does not depend on the point (n, m) in this case. In Section 2.4.7.5 we will also consider its non autonomous extension (2.2.51, 2.4.199).

The ABS equations take the form

$$F(u_{0,0}, u_{1,0}, u_{0,1}, u_{1,1}; \alpha, \beta) = 0, \tag{2.4.130}$$

where α and β are two constants related to the lattice spacing in the two independent directions of the plane (see Fig.2.3).

The list of the ABS equations has been obtained in [22] by using the CaC property.

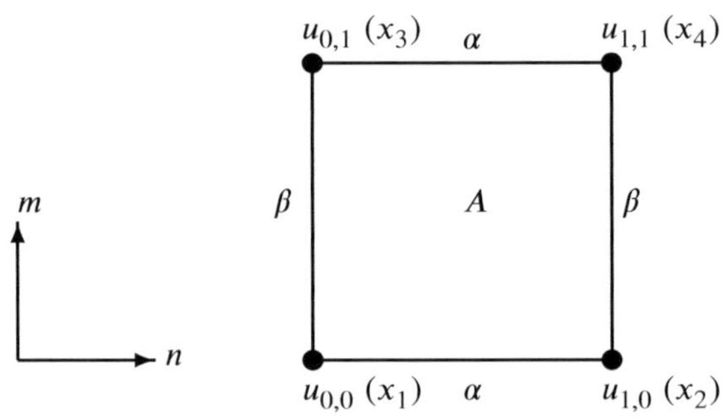

FIGURE 2.3. A square lattice (quad-graph)

The main idea of this consistency method is the following:

(1) One starts from a square lattice and defines the three variables $u_{i,j}$ on the vertices (see Fig. 2.3). By solving $F = 0$ one obtains an expression for the fourth one which, if $F = 0$ is multilinear, is rational.
(2) One adjoins a third direction, say k, and imagines the map giving $u_{1,1,1}$ as being the composition of maps on the various planes (see Fig. 2.4). There exist **three different ways** to obtain $u_{1,1,1}$ and the consistency constraint is that they all lead to the **same result**.
(3) Two further constraints have been introduced by Adler, Bobenko and Suris to carry out the classification:
    - *D_4-symmetry*:
$$\begin{aligned} F(u_{0,0}, u_{1,0}, u_{0,1}, u_{1,1}; \alpha, \beta) &= \pm F(u_{0,0}, u_{0,1}, u_{1,0}, u_{1,1}; \beta, \alpha) \\ &= \pm F(u_{1,0}, u_{0,0}, u_{1,1}, u_{0,1}; \alpha, \beta). \end{aligned}$$
    - *Tetrahedron property*: $u_{1,1,1}$ is independent of $u_{0,0,0}$.
(4) The equations are classified according to the following equivalence group:
    - A Möbius transformation.
    - Simultaneous **point** change of all variables.

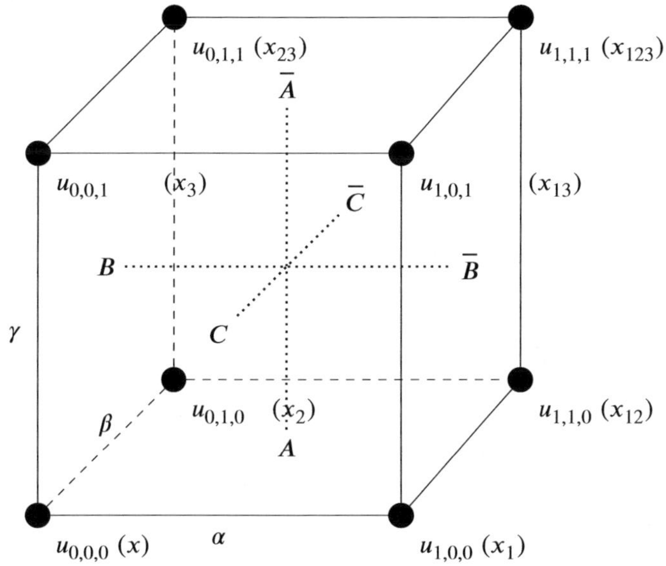

FIGURE 2.4. Three-dimensional consistency (equations on a cube)

As a result of this procedure all equations possess a discrete symmetry, the exchange of the first with the second index as well as a proper exchange of the constants α and β. The compatible equations are integrable by construction, as the CaC provides them with Lax pairs and Bäcklund transformations (as we will see in Section 2.4.6.2) [22, 29, 102, 634].

The ABS equations have the form (2.4.130) and are affine linear, i.e. the function F is a polynomial of degree one in each argument. The ABS list traditionally consists of two lists, the H and Q PΔEs.

The ABS list

$$H_1: \quad (u_{0,0} - u_{1,1})(u_{1,0} - u_{0,1}) - \alpha + \beta = 0$$

$$H_2: \quad (u_{0,0} - u_{1,1})(u_{1,0} - u_{0,1}) + (\beta - \alpha)(u_{0,0} + u_{1,0} + u_{0,1} + u_{1,1}) \\ -\alpha^2 + \beta^2 = 0$$

$$H_3: \quad \alpha(u_{0,0}u_{1,0} + u_{0,1}u_{1,1}) - \beta(u_{0,0}u_{0,1} + u_{1,0}u_{1,1}) + \delta(\alpha^2 - \beta^2) = 0$$

$$Q_1: \quad \alpha(u_{0,0} - u_{0,1})(u_{1,0} - u_{1,1}) - \beta(u_{0,0} - u_{1,0})(u_{0,1} - u_{1,1}) \\ +\delta^2\alpha\beta(\alpha - \beta) = 0$$

$$Q_2: \quad \alpha(u_{0,0} - u_{0,1})(u_{1,0} - u_{1,1}) - \beta(u_{0,0} - u_{1,0})(u_{0,1} - u_{1,1}) + \\ \alpha\beta(\alpha - \beta)(u_{0,0} + u_{1,0} + u_{0,1} + u_{1,1}) - \alpha\beta(\alpha - \beta)(\alpha^2 - \alpha\beta + \beta^2) = 0$$

$$Q_3: \quad (\beta^2 - \alpha^2)(u_{0,0}u_{1,1} + u_{1,0}u_{0,1}) + \beta(\alpha^2 - 1)(u_{0,0}u_{1,0} + u_{0,1}u_{1,1}) -$$

$$\alpha(\beta^2-1)(u_{0,0}u_{0,1}+u_{1,0}u_{1,1})-\frac{\delta^2(\alpha^2-\beta^2)(\alpha^2-1)(\beta^2-1)}{4\alpha\beta}=0$$

$$\begin{aligned} Q_4: \quad & a_0u_{0,0}u_{1,0}u_{0,1}u_{1,1}+\\ & a_1(u_{0,0}u_{1,0}u_{0,1}+u_{1,0}u_{0,1}u_{1,1}+u_{0,1}u_{1,1}u_{0,0}+u_{1,1}u_{0,0}u_{1,0})+\\ & a_2(u_{0,0}u_{1,1}+u_{1,0}u_{0,1})+\bar{a}_2(u_{0,0}u_{1,0}+u_{0,1}u_{1,1})+\\ & \tilde{a}_2(u_{0,0}u_{0,1}+u_{1,0}u_{1,1})+a_3(u_{0,0}+u_{1,0}+u_{0,1}+u_{1,1})+a_4=0 \end{aligned}$$

The coefficients of the Q_4 equation are connected to α and β by the relations

$$\begin{aligned} a_0&=a+b, \quad a_1=-a\beta-b\alpha, \quad a_2=a\beta^2+b\alpha^2,\\ \bar{a}_2&=\frac{ab(a+b)}{2(\alpha-\beta)}+a\beta^2-\left(2\alpha^2-\frac{g_2}{4}\right)b,\\ \tilde{a}_2&=\frac{ab(a+b)}{2(\beta-\alpha)}+b\alpha^2-\left(2\beta^2-\frac{g_2}{4}\right)a,\\ a_3&=\frac{g_3}{2}a_0-\frac{g_2}{4}a_1, \quad a_4=\frac{g_2^2}{16}a_0-g_3a_1, \end{aligned}$$

where $a^2=r(\alpha)$, $b^2=r(\beta)$, where $r(x)=4x^3-g_2x-g_3$. The coefficients g_2, g_3, δ are arbitrary constants. The parameter δ in the H_3, Q_1 and Q_3 equations can be rescaled, so that one can assume without loss of generality that either $\delta=0$ or $\delta=1$. The arbitrary constants α,β are lattice parameters. These parameters may, in general, depend on the discrete variables n,m. For all equations of the above list, α and β are some concrete numbers.

The equation Q_4 was obtained before by Adler studying the Bäcklund transformations of the Krichever Novikov (2.4.128) in [17].

The original ABS list contains two further equations: A_1 and A_2. We exclude them from consideration, as those equations are related to the Q equations of the ABS list by non autonomous point transformations. Namely, any solution $u_{n,m}$ of the A_1 equation is transformed into a solution $\tilde{u}_{n,m}$ of Q_1 by the transformation $u_{n,m}=(-1)^{n+m}\tilde{u}_{n,m}$. Solutions $u_{n,m}$ of the A_2 equation are transformed into solutions $\tilde{u}_{n,m}$ of the equation Q_3 with $\delta=0$ by $u_{n,m}=\left(\tilde{u}_{n,m}\right)^{(-1)^{n+m}}$.

By a proper limiting procedure all equations of the ABS list are contained in Q_4 [635].

Denoting by A (see Fig. 2.3) any equation of the ABS list, we can define the six accompanying biquadratics, given by

$$A^{i,j}\equiv A^{i,j}(x_i,x_j)=A_{,x_m}A_{,x_n}-AA_{,x_mx_n}, \tag{2.4.132}$$

where $\{m,n\}$ is the complement of $\{i,j\}$ in $\{1,2,3,4\}$. In the Q-type equations the biquadratics are non degenerate, i.e. all different. In the H−type equations some of the biquadratics are degenerate.

In Fig. 2.4 $\bar{A}$ represents the equation A, one of the seven ABS equations, with x substituted by x_3, x_1 by x_{13}, x_2 by x_{23} and x_{12} by x_{123}. On the faces B and $\bar{B}$ we have a relation between a solution of A and one of $\bar{A}$, i.e. an auto–Bäcklund transformation for A. By going over to projective space the auto–Bäcklund transformation will provide the Lax pair.

It should be remarked that many of the above discrete equations were known before Adler, Bobenko and Suris presented their classification, as, for instance, the lpKdV and lSKdV [17, 637, 682] we considered in Section 2.4.4 and 2.4.5.

4.6.1. *The derivation of the Q_V equation.* Consider a field u defined on a two-dimensional square lattice (see Fig. 2.3). Let us assume that at each vertex of the lattice, the value of u is related to the value at neighbouring vertices by a multilinear relation

$$Q = p_1 u_{0,0} u_{1,0} u_{0,1} u_{1,1} + u_{0,0}\, u_{1,0}\Big(p_2 u_{0,1} + p_3 u_{1,1}\Big) + u_{0,1}\, u_{1,1}\Big(p_4 u_{0,1} + p_5 u_{0,0}\Big) + p_6 u_{0,0} u_{1,0} + p_7 u_{1,0} x_{0,1} + + p_8 u_{1,0} u_{1,1} + p_9 u_{0,0} u_{0,1} + p_{10} u_{0,0} u_{1,1} + p_{11} u_{0,1} u_{1,1} + p_{12} u_{1,0} + p_{13} u_{0,0} + p_{14} u_{0,1} + p_{15} u_{1,1} + p_{16} = 0 \tag{2.4.133}$$

so that any of the four corner values can be rationally expressed in terms of the three others. Eq. (2.4.133), being multilinear, is the simplest relation linking the values of u at the four corners of an elementary square plaquette as presented in Fig. 2.3.

In the following we will analyze (2.4.133) with the algebraic entropy integrability test to find for which values of the constants p_i, $i = 1, \cdots 16$ it will turn out to be integrable.

Algebraic entropy

The notion of algebraic entropy was introduced by Bellom and Viallet [82], see also the review by Gubbiotti [337], to define a global index of the complexity of a discrete time dynamical system with rational evolution (the state at time $t+1$ is expressible rationally in terms of the state at time t). It is not attached to any particular domain of initial conditions and reflects its asymptotic behavior.
The space of initial data of the evolutions defined by relation (2.4.133) is infinite dimensional as an infinity of initial data given on a line is necessary to calculate the values at all points of the lattice. The simplest possible choice is to take as initial line the two axis (see Fig. 2.3). Then by an iteration of the evolution, given by (2.4.133), we can calculate the values off the axis. From the discrete time dynamical system we construct a sequence of degrees d_k of growth defined as the maximum degree of the homogeneous polynomials describing the system in projective space for the various discrete variables after k iterations in terms of the initial data. Then the algebraic entropy is defined by the formula

$$\eta = \lim_{k\to\infty} \frac{1}{k} \log(d_k). \tag{2.4.134}$$

The outcome of numerous experiments, as well as of what is know for maps [248, 385], leads to the claim [387] that according to the growth of d_k we have

- Linear growth: The equation is linearizable.
- Polynomial growth: The equation is integrable.
- Exponential growth: The equation is chaotic.

Then *integrability of the lattice map is equivalent to the vanishing of its algebraic entropy.*

From Q_4 to Q_V

Let us apply the algebraic entropy calculation to Q_4. Since we use computer algebra to evaluate the sequence of degrees, it is more efficient to work with integer coefficients. It is easy to find integer coefficients verifying the conditions fulfilled by $\{a_0, \ldots, a_4\}$. For example, choosing $r(z) = 4\ z^3 - 32\ z + 4$ and the points $(a, A) = (0, 2)$, $(c, C) = (3, 4)$, $(b, B) = (a, A) \oplus (c, C) = (-26/9, -2/27)$, we get for Q_4 the sequence $\{d_k^{Q_4}\} = \{1, 3, 7, 13, 21, 31, 43, 57, 73, 91, 111, \ldots\}$, which correspond to a quadratic growth

$$d_k = 1 + k\,(k-1) \tag{2.4.135}$$

The most general form of (2.4.133) having the same symmetries as Q_4 is:

$$\begin{aligned} Q_V &= a_1 u_{0,0}u_{1,0}u_{0,1}u_{1,1} + a_2\big(u_{0,0}u_{0,1}u_{1,1} + u_{1,0}u_{0,1}u_{1,1} + u_{0,0}u_{1,0}u_{1,1} \\ &+ u_{0,0}u_{0,1}u_{1,0}\big) + a_3\left(u_{0,0}u_{1,1} + u_{1,0}u_{0,1}\right) + a_4\left(u_{0,0}u_{0,1} + u_{1,0}u_{1,1}\right) \\ &+ a_5\left(u_{0,0}u_{1,0} + u_{0,1}u_{1,1}\right) + a_6\left(u_{0,0} + u_{1,0} + u_{0,1} + u_{1,1}\right) + a_7 = 0, \end{aligned} \tag{2.4.136}$$

with no constraint on the coefficients $\{a_1, \dots, a_7\}$. Q_{V} [815] is the most general multilinear equation on a quad-graph possessing Klein discrete symmetries, i.e. such that:

$$\begin{aligned} Q\left(u_{n+1,m}, u_{n,m}, u_{n+1,m+1}, u_{n,m+1}\right) &= \tau Q\left(u_{n,m}, u_{n+1,m}, u_{n,m+1}, u_{n+1,m+1}\right), \\ Q\left(u_{n,m+1}, u_{n+1,m+1}, u_{n,m}, u_{n,m+1}\right) &= \tau' Q\left(u_{n,m}, u_{n+1,m}, u_{n,m+1}, u_{n+1,m+1}\right), \end{aligned} \tag{2.4.137}$$

where $(\tau, \tau') = (\pm 1, \pm 1)$.

For arbitrary values of the coefficients $\{a_1, \dots, a_7\}$ we get for Q_V the same quadratic growth as for Q_4, when the parameters are constrained:

$$\{d_k^{Q_V}\} = \{1, 3, 7, 13, 21, 31, 43, 57, 73, 91, 111, \dots\}. \tag{2.4.138}$$

Eq. (2.4.138) fits (2.4.135) and gives the generating function

$$g(s) = \sum_{k=0}^{\infty} d_k\, s^k = \frac{1+s^2}{(1-s)^3},$$

as was checked for a number of randomly chosen coefficients. This result indicates the integrability of Q_V, with 7 free homogeneous coefficients. The integrability of Q_V has later been confirmed by showing the existence of a point fractional-linear transformation which reduce it to Q_4. It was shown that it has a recursion operator for its generalized symmetries [611, 612]. It is worthwhile to notice that the sequence of degrees verifies also a finite recursion relation

$$d_k - 3\, d_{k-1} + 3\, d_{k-2} - d_{k-3} = 0$$

This means that the global behaviour of the sequence of degrees is dictated by a local condition.

4.6.2. *Lax pair and Bäcklund transformations for the ABS equations.* The algorithmic procedure described in [22, 102, 119, 359, 634] and briefly sketched above produces a 2×2 matrix Lax pair for the ABS equations, thus ensuring their integrability. It may be written as

$$\Psi_{1,0} = L(u_{0,0}, u_{1,0}; \alpha, \lambda)\Psi_{0,0}, \qquad \Psi_{0,1} = M(u_{0,0}, u_{0,1}; \beta, \lambda)\Psi_{0,0}, \tag{2.4.139}$$

with $\Psi = (\psi(\lambda), \phi(\lambda))^T$, where the lattice parameter λ plays the role of the spectral parameter. We shall use the following notation

$$L(u_{0,0}, u_{1,0}; \alpha, \lambda) = \frac{1}{\ell}\begin{pmatrix} L_{11} & L_{12} \\ L_{21} & L_{22} \end{pmatrix}, \quad M(u_{0,0}, u_{0,1}; \beta, \lambda) = \frac{1}{t}\begin{pmatrix} M_{11} & M_{12} \\ M_{21} & M_{22} \end{pmatrix},$$

where $\ell = \ell_{0,0} = \ell(u_{0,0}, u_{1,0}; \alpha, \lambda)$, $t = t_{0,0} = t(u_{0,0}, u_{0,1}; \beta, \lambda)$,
$L_{ij} = L_{ij}(u_{0,0}, u_{1,0}; \alpha, \lambda)$ and $M_{ij} = M_{ij}(u_{0,0}, u_{0,1}; \beta, \lambda)$, $i, j = 1, 2$. The matrix M can be obtained from L by replacing α with β and shifting along direction 2 instead of 1. In Table 2.1 we give the entries of the matrix L for the ABS equations.

Note that ℓ and t are computed by requiring that the compatibility condition between L and M produces the ABS equations $H_1 - H_3$ and $Q_1 - Q_4$. The term ℓ can be factorized as

$$\ell_{0,0} = f(\alpha, \lambda)[\rho(u_{0,0}, u_{1,0}; \alpha)]^{1/2}, \tag{2.4.140}$$

TABLE 2.1. Matrix L for the ABS equations (in equation Q_4 $a^2 = r(\alpha)$, $b^2 = r(\lambda)$, $r(x) = 4x^3 - g_2x - g_3$) [reprinted from [492] licensed under Creative Commons Attribution-ShareAlike 4.0 International (https://creativecommons.org/licenses/by-sa/4.0/)].

	L_{11}	L_{12}	L_{21}	L_{22}
H_1	$u_{0,0} - u_{1,0}$	$(u_{0,0} - u_{1,0})^2 + \alpha - \lambda$	1	$u_{0,0} - u_{1,0}$
H_2	$u_{0,0} - u_{1,0} + \alpha - \lambda$	$2(\alpha - \lambda)(u_{0,0} + u_{1,0}) +$ $(u_{0,0} - u_{1,0})^2 + \alpha^2 - \lambda^2$	1	$u_{0,0} - u_{1,0} - \alpha + \lambda$
H_3	$\lambda u_{0,0} - \alpha u_{1,0}$	$\lambda(u_{0,0}^2 + u_{1,0}^2) - 2\alpha u_{0,0}u_{1,0} +$ $+\delta(\lambda^2 - \alpha^2)$	α	$\alpha u_{0,0} - \lambda u_{1,0}$
Q_1	$\lambda(u_{1,0} - u_{0,0})$	$-\lambda(u_{1,0} - u_{0,0})^2 + \delta\alpha\lambda(\alpha - \lambda)$	$-\alpha$	$\lambda(u_{1,0} - u_{0,0})$
Q_2	$\lambda(u_{1,0} - u_{0,0}) +$ $+\alpha\lambda(\alpha - \lambda)$	$-\lambda(u_{1,0} - u_{0,0})^2 +$ $+2\alpha\lambda(\alpha - \lambda)(u_{1,0} + u_{0,0}) -$ $-\alpha\lambda(\alpha - \lambda)(\alpha^2 - \alpha\lambda + \lambda^2)$	$-\alpha$	$\lambda(u_{1,0} - u_{0,0}) -$ $-\alpha\lambda(\alpha - \lambda)$
Q_3	$\alpha(\lambda^2 - 1)u_{0,0} -$ $-(\lambda^2 - \alpha^2)u_{1,0}$	$-\lambda(\alpha^2 - 1)u_{0,0}u_{1,0} +$ $+\delta(\alpha^2 - \lambda^2)(\alpha^2 - 1) \cdot$ $\cdot (\lambda^2 - 1)/(4\alpha\lambda)$	$\lambda(\alpha^2 - 1)$	$(\lambda^2 - \alpha^2)u_{0,0} -$ $-\alpha(\lambda^2 - 1)u_{1,0}$
Q_4	$-a_1u_{0,0}u_{1,0} -$ $-a_2u_{1,0} - \tilde{a}_2u_{0,0} -$ $- a_3$	$-\bar{a}_2u_{0,0}u_{1,0} -$ $-a_3(u_{0,0} + u_{1,0}) -$ $-a_4$	$a_0u_{0,0}u_{1,0} +$ $+a_1(u_{0,0} + u_{1,0})$ $+ \bar{a}_2$	$a_1u_{0,0}u_{1,0} + a_2u_{0,0} +$ $+\tilde{a}_2u_{1,0} + a_3$

where the function $f = f(\alpha, \lambda)$ is an arbitrary normalization factor. The functions $f = f(\alpha, \lambda)$ and $\rho = \rho_{0,0} = \rho(u_{0,0}, u_{1,0}; \alpha)$ for equations $H_1 - H_3$ and $Q_1 - Q_4$ are given in Table 2.2. A formula similar to (2.4.140) holds also for the factor t.

The scalar Lax pairs for the ABS equations may be immediately computed from (2.4.139). Let us write the scalar equation for the second component ϕ of the vector Ψ (the use of the first component would give similar results). For $H_1 - H_3$ and $Q_1 - Q_3$ it reads

$$(\rho_{1,0})^{1/2}\phi_{2,0} - (u_{2,0} - u_{0,0})\phi_{1,0} + (\rho_{0,0})^{1/2}\mu\phi_{0,0} = 0, \tag{2.4.141}$$

where the explicit expressions of $\mu = \mu(\alpha, \lambda)$ are given in Table 2.2. The corresponding scalar equation for equation Q_4 is always a second order difference equation as (2.4.141) but its coefficients are not so simple and will not be presented here. For those interested we refer to the original reference [634].

We can write down the second order scalar difference equation in terms of the coefficients of the matrix spectral problem (2.4.139). We have:

$$\mathcal{A}\phi_{2,0} + \mathcal{B}\phi_{1,0} + \mathcal{C}\phi_{0,0} = 0,$$

$$\mathcal{A} = L_{12}\ell_{1,0}, \quad \mathcal{B} = -\Big(L_{12}S_1L_{11} + L_{22}S_1L_{12}\Big),$$

$$\mathcal{C} = \frac{S_1L_{12}}{\ell_{0,0}}\Big[L_{11}L_{22} - L_{12}L_{21}\Big],$$

where S_1 is the shift in the first index.

The Bäcklund transformations for the ABS class of equations can be found in [837]. We present its result without proof and the diligent reader can check it there.

Proposition 1. *The system*

$$B_d(u, \tilde{u}, \lambda) := \begin{cases} F(u_{(0,0)}, u_{(1,0)}, \tilde{u}_{(0,0)}, \tilde{u}_{(1,0)}; \alpha, \lambda) = 0 \\ F(u_{(0,0)}, u_{(0,1)}, \tilde{u}_{(0,0)}, \tilde{u}_{(0,1)}; \beta, \lambda) = 0 \end{cases} \tag{2.4.142}$$

defines an auto-Bäcklund transformation for the ABS equation (2.4.130)

TABLE 2.2. Functions f, ρ and μ for the ABS equations (Here $c^2 = r(\lambda)$) [reprinted from [492]licensed under Creative Commons Attribution-ShareAlike 4.0 International (https://creativecommons.org/licenses/by-sa/4.0/)].

	$f(\alpha,\lambda)$	$\rho(u_{0,0},u_{1,0};\alpha)$	$\mu(\alpha,\lambda)$
H_1	-1	1	$\lambda-\alpha$
H_2	-1	$u_{0,0}+u_{1,0}+\alpha$	$2(\lambda-\alpha)$
H_3	$-\lambda$	$u_{0,0}u_{1,0}+\delta\alpha$	$\frac{\alpha^2-\lambda^2}{\alpha\lambda^2}$
Q_1	λ	$(u_{1,0}-u_{0,0})^2-\delta^2\alpha^2$	$\frac{\lambda-\alpha}{\lambda}$
Q_2	λ	$(u_{1,0}-u_{0,0})^2-2\alpha^2(u_{1,0}+u_{0,0})+$ $+\alpha^4$	$\frac{\lambda-\alpha}{\lambda}$
Q_3	$\alpha(1-\lambda^2)$	$\alpha(u_{0,0}^2+u_{1,0}^2)-(\alpha^2+1)u_{0,0}u_{1,0}+$ $+\frac{\delta(\alpha^2-1)^2}{4\alpha}$	$\frac{\alpha^2-\lambda^2}{\alpha^2(1-\lambda^2)}$
Q_4	$(\alpha-\lambda)c^{1/2}\times$ $\times\left[2a+c+\frac{1}{4}\left(\frac{a+c}{\alpha-\lambda}\right)^3-\right.$ $\left.-\frac{3a(a+c)}{\alpha-\lambda}\right]^{1/2}$	$(u_{0,0}u_{1,0}+\alpha u_{0,0}+\alpha u_{1,0}+g_2/4)^2-$ $-(u_{0,0}+u_{1,0}+\alpha)(4\alpha u_{0,0}u_{1,0}-g_3)$	$-$

4.6.3. *Symmetries of the ABS equations.* We present here the Lie point symmetries and the three-point generalized symmetries of the ABS equations [699, 800, 837]. We give the corresponding generators, using the symbol of each equation employed in the list (2.4.131).

- **H_1**
  Point symmetries : $\hat{X}_1 = \partial_{u_{0,0}}$, $\hat{X}_2 = (-1)^{n-m}\partial_{u_{0,0}}$, $\hat{X}_3 = (-1)^{n-m}u_{0,0}\partial_{u_{0,0}}$.
  Three-point generalized symmetries :

(2.4.143a) $$\hat{V}_1 = \frac{1}{u_{1,0}-u_{-1,0}}\partial_{u_{0,0}},\quad \hat{V}_2 = n\hat{V}_1 + \frac{u_{0,0}}{2(\alpha-\beta)}\partial_{u_{0,0}},$$

(2.4.143b) $$\hat{V}_3 = \frac{1}{u_{0,1}-u_{0,-1}}\partial_{u_{0,0}},\quad \hat{V}_4 = m\hat{V}_3 - \frac{u_{0,0}}{2(\alpha-\beta)}\partial_{u_{0,0}}.$$

- **H_2**
  Point symmetries : $\hat{X}_1 = (-1)^{n+m}\partial_{u_{0,0}}$.

Three-point generalized symmetries :

(2.4.143c) $$\hat{V}_1 = \frac{u_{1,0} + 2u_{0,0} + u_{-1,0} + 2\alpha}{u_{1,0} - u_{-1,0}} \partial_{u_{0,0}},$$

(2.4.143d) $$\hat{V}_2 = n\,\hat{V}_1 + \frac{2u_{0,0} + \beta}{2(\alpha - \beta)}\,\partial_{u_{0,0}},$$

(2.4.143e) $$\hat{V}_3 = \frac{u_{1,0} + 2u_{0,0} + u_{0,-1} + 2\beta}{u_{0,1} - u_{0,-1}} \partial_{u_{0,0}},$$

(2.4.143f) $$\hat{V}_4 = m\,\hat{V}_3 - \frac{2u_{0,0} + \alpha}{2(\alpha - \beta)}\,\partial_{u_{0,0}}.$$

- **H_3**
  (1) $\delta = 0$.
  Point symmetries : $\hat{X}_1 = u_{0,0}\partial_{u_{0,0}}$, $\hat{X}_2 = (-1)^{n+m} u_{0,0}\partial_{u_{0,0}}$.
  Three-point generalized symmetries :

(2.4.143g) $$\hat{V}_1 = \frac{u_{0,0}(u_{1,0} + u_{-1,0})}{u_{1,0} - u_{-1,0}} \partial_{u_{0,0}},$$

(2.4.143h) $$\hat{V}_2 = \frac{u_{0,0}(u_{0,1} + u_{0,-1})}{u_{0,1} - u_{0,\ 1}} \partial_{u_{0,0}}.$$

  (2) $\delta \neq 0$.
  Point symmetries : $\hat{X}_1 = (-1)^{n+m} u_{0,0}\partial_{u_{0,0}}$.
  Three-point generalized symmetries :

(2.4.143i) $$\hat{V}_1 = \frac{u_{0,0}(u_{1,0} + u_{-1,0}) + 2\,\alpha\,\delta}{u_{1,0} - u_{-1,0}} \partial_{u_{0,0}},$$

(2.4.143j) $$\hat{V}_2 = \frac{u_{0,0}(u_{0,1} + u_{0,-1}) + 2\,\beta\,\delta}{u_{0,1} - u_{0,-1}} \partial_{u_{0,0}}.$$

- **Q_1**
  (1) $\delta = 0$
  Point symmetries : $\hat{X}_1 = u_{0,0}^2\partial_{u_{0,0}}$, $\hat{X}_2 = u_{0,0}\partial_{u_{0,0}}$, $\hat{X}_3 = \partial_{u_{0,0}}$.
  Three-point generalized symmetries :

(2.4.143k) $$\hat{V}_1 = \frac{(u_{1,0} - u_{0,0})(u_{0,0} - u_{-1,0})}{u_{1,0} - u_{-1,0}} \partial_{u_{0,0}},$$

(2.4.143l) $$\hat{V}_2 = \frac{(u_{0,1} - u_{0,0})(u_{0,0} - u_{0,-1})}{u_{0,1} - u_{0,-1}} \partial_{u_{0,0}}.$$

  (2) $\delta \neq 0$
  Point symmetries : $\hat{X}_1 = \partial_{u_{0,0}}$.
  Three-point generalized symmetries :

(2.4.143m) $$\hat{V}_1 = \frac{(u_{1,0} - u_{0,0})(u_{0,0} - u_{-1,0}) + \alpha^2\delta^2}{u_{1,0} - u_{-1,0}} \partial_{u_{0,0}},$$

(2.4.143n) $$\hat{V}_2 = \frac{(u_{0,1} - u_{0,0})(u_{0,0} - u_{0,-1}) + \beta^2\delta^2}{u_{0,1} - u_{0,-1}} \partial_{u_{0,0}}.$$

- **$\mathbf{Q_2}$**
Three-point generalized symmetries :

(2.4.143o) $$\hat{V}_1 = \frac{(u_{1,0} - u_{0,0})(u_{0,0} - u_{-1,0}) + (u_{1,0} + 2u_{0,0} + u_{-1,0})\alpha^2 - \alpha^4}{u_{1,0} - u_{-1,0}} \partial_{u_{0,0}},$$

(2.4.143p) $$\hat{V}_2 = \frac{(u_{0,1} - u_{0,0})(u_{0,0} - u_{0,-1}) + (u_{0,1} + 2u_{0,0} + u_{0,-1})\beta^2 - \beta^4}{u_{0,1} - u_{0,-1}} \partial_{u_{0,0}}.$$

- **$\mathbf{Q_3}$**
Point symmetries : If $\delta = 0$, then it admits one point symmetry with generator $\hat{X}_1 = u_{0,0}\partial_{u_{0,0}}$. Otherwise, there are no point symmetries.
Three-point generalized symmetries :

(2.4.143q) $$\hat{V}_1 = \frac{2\alpha(\alpha^2 + 1)u_{0,0}(u_{1,0} + u_{-1,0}) - 4\alpha^2(u_{0,0}^2 + u_{1,0}u_{-1,0}) - (\alpha^2 - 1)^2\delta^2}{u_{1,0} - u_{-1,0}} \partial_{u_{0,0}},$$

(2.4.143r) $$\hat{V}_2 = \frac{2\beta(\beta^2 + 1)u_{0,0}(u_{0,1} + u_{0,-1}) - 4\beta^2(u_{0,0}^2 + u_{0,1}u_{0,-1}) - (\beta^2 - 1)^2\delta^2}{u_{0,1} - u_{0,-1}} \partial_{u_{0,0}}.$$

- **$\mathbf{Q_4}$**
Three-point generalized symmetries :

(2.4.143s) $$\hat{V}_1 = \frac{(u_{1,0} - u_{-1,0})f_{,u_{1,0}}(u_{0,0}, u_{1,0}, \alpha) - 2f(u_{0,0}, u_{1,0}, \alpha)}{u_{1,0} - u_{-1,0}} \partial_{u_{0,0}},$$

(2.4.143t) $$\hat{V}_2 = \frac{(u_{0,1} - u_{0,-1})f_{,u_{0,1}}(u_{0,0}, u_{0,1}, \beta) - 2f(u_{0,0}, u_{0,1}, \beta)}{u_{0,1} - u_{0,-1}} \partial_{u_{0,0}},$$

where

$$f(u_{0,0}, u_{1,0}, \alpha) = \left(u_{0,0}u_{1,0} + \alpha(u_{0,0} + u_{1,0}) + \frac{g_2}{4}\right)^2 - (u_{0,0} + u_{1,0} + \alpha)(4\alpha u_{0,0}u_{1,0} - g_3).$$

- **$\mathbf{Q_V}$**
Three-point generalized symmetries :

(2.4.143u) $$\hat{V}_n = \left(\frac{h_1(u_{0,0}, u_{1,0})}{u_{1,0} - u_{-1,0}} - \frac{1}{2}\frac{\partial h_1(u_{0,0}, u_{1,0})}{\partial u_{1,0}}\right)\partial_{u_{0,0}},$$

(2.4.143v) $$\hat{V}_{-n} = \left(\frac{h_1(u_{-1,0}, u_{0,0})}{u_{1,0} - u_{-1,0}} + \frac{1}{2}\frac{\partial h_1(u_{-1,0}, u_{0,0})}{\partial u_{-1,0}}\right)\partial_{u_{0,0}},$$

(2.4.143w) $$\hat{V}_m = \left(\frac{h_2(u_{0,0}, u_{0,1})}{u_{0,1} - u_{0,-1}} - \frac{1}{2}\frac{\partial h_2(u_{0,0}, u_{0,1})}{\partial u_{0,1}}\right)\partial_{u_{0,0}},$$

(2.4.143x) $$\hat{V}_{-m} = \left(\frac{h_2(u_{0,-1}, u_{0,0})}{u_{0,1} - u_{0,-1}} - \frac{1}{2}\frac{\partial h_2(u_{0,-1}, u_{0,0})}{\partial u_{0,-1}}\right)\partial_{u_{0,0}},$$

(2.4.143y) $$V_n = \frac{h_1(u_{0,0}, u_{1,0}) + h_1(u_{-1,0}, u_{0,0})}{u_{1,0} - u_{-1,0}} \partial_{u_{0,0}},$$

(2.4.143z) $$V_m = \frac{h_2(u_{0,0}, u_{0,1}) + h_2(u_{0-1}, u_{0,0})}{u_{0,1} - u_{0,-1}} \partial_{u_{0,0}}$$

where

$$h_1(x,y) = (a_3 + a_1 xy + a_2(x+y))(a_7 + a_3 xy + a_6(x+y)) \\ - (a_6 + a_5 x + (a_4 + a_2 x)y)(a_6 + a_4 x + (a_5 + a_2 x)y),$$

$$h_2(x,y) = (a_5 + a_1 xy + a_2(x+y))(a_7 + a_5 xy + a_6(x+y)) \\ - (a_6 + a_4 x + (a_3 + a_2 x)y)(a_6 + a_3 x + (a_4 + a_2 x)y).$$

The three point symmetries we presented above for the discrete equations of the ABS list have been constructed in [699, 800] while the three-point symmetries of Q_V in [339, 611, 837].

In [492, 611, 612, 837] we can find the recursion operator and the master symmetries for the symmetries of the ABS equations.

The three-point generalized symmetries of the ABS equations are given by $D\Delta E$, subcases of YdKN (2.4.129). By defining $v_i = u_{i,j}$ and $\tilde{v}_i = u_{i,j+1}$, the equations of the ABS list are nothing else but Bäcklund transformations for particular subcases of the YdKN [471, 492]. Here we present for the various equations of the ABS classification the coefficients c_i, $1 \le i \le 6$, of the YdKN (2.4.129) corresponding to three-point generalized symmetries in the n direction. They read

$$\begin{array}{llllllll}
\mathbf{H_1}: & c_1 = 0, & c_2 = 0, & c_3 = 0, & c_4 = 0, & c_5 = 0, & c_6 = 1, \\
\mathbf{H_2}: & c_1 = 0, & c_2 = 0, & c_3 = 0, & c_4 = 0, & c_5 = 1, & c_6 = 2\alpha, \\
\mathbf{H_3}: & c_1 = 0, & c_2 = 0, & c_3 = 0, & c_4 = 1, & c_5 = 0, & c_6 = 2\alpha\delta, \\
\mathbf{Q_1}: & c_1 = 0, & c_2 = 0, & c_3 = -1, & c_4 = 1, & c_5 = 0, & c_6 = \alpha^2\delta^2, \\
\mathbf{Q_2}: & c_1 = 0, & c_2 = 0, & c_3 = 1, & c_4 = -1, & c_5 = -\alpha^2, & c_6 = \alpha^4, \\
\mathbf{Q_3}: & c_1 = 0, & c_2 = 0, & c_3 = -4\alpha^2, & c_4 = 2\alpha(\alpha^2+1), & c_5 = 0, & c_6 = -(\alpha^2-1)^2\delta^2, \\
\mathbf{Q_4}: & c_1 = 1, & c_2 = -\alpha, & c_3 = \alpha^2, & c_4 = \frac{g_2}{4} - \alpha^2, & c_5 = \frac{\alpha g_2}{4} + \frac{g_3}{2}, & c_6 = \frac{g_2^2}{16} + \alpha g_3.
\end{array}$$

In the case of Q_V we have:

$$\begin{aligned} c_1 &= a_1 a_5 - a_2^2, \quad c_2 = \frac{1}{2}\Big[a_1 a_6 - a_2(a_3 + a_4 - a_5)\Big], \\ c_3 &= a_2 a_6 - a_3 a_4, \quad c_4 = \frac{1}{2}\Big[a_1 a_7 + a_5^2 - a_3^2 - a_4^2\Big], \\ c_5 &= \frac{1}{2}\Big[a_2 a_7 - a_6(a_3 + a_4 - a_5)\Big], \quad c_6 = a_5 a_7 - a_6^2. \end{aligned} \tag{2.4.144}$$

We can then state the following proposition:

Proposition 2. *The ABS equations $H_1 - H_3$ and $Q_1 - Q_V$ correspond to Bäcklund transformations of the particular cases of the YdKN equation* (2.4.129) *listed above. Equation* (2.4.129) *with the substitution $u_i \to u_{i,0}$ provides the three-point generalized symmetries in the n-direction of the ABS equations with a constant α and $\beta = \beta_m$. Eq,* (2.4.129) *with the substitution $u_i \to u_{0,i}$, $\alpha \to \beta$ provides symmetries in the m-direction for the case $\alpha = \alpha_n$ and a constant β.*

The ABS equations do not exhaust all the possible Bäcklund transformations for the YdKN equation as the whole parameter space is not covered [492, 837]. Moreover, in the list of integrable $D\Delta E$ of Volterra type [842], there are equations different from the YdKN which may also have Bäcklund transformations of the form (3.5.1). So we have space for new integrable $P\Delta E$ which can be searched by using the formal symmetry approach (see Section 3.6.1).

An extension of the 3D consistency approach has been proposed by Adler, Bobenko and Suris [23] allowing different equations in the different faces of the cube. We will dealt with it in next Section.

4.7. Extension of the ABS classification: Boll results. In [23] Adler, Bobenko and Suris considered a more general perspective in the classification problem. They assumed that the faces of the consistency cube A, B, C and $\bar{A}$, $\bar{B}$ and $\bar{C}$ could carry a priori different quad equations without assuming either the D_4 symmetry or the tetrahedron property. They considered six-tuples of (a priori different) quad equations assigned to the faces of a 3D cube:

$$
\begin{aligned}
&A\left(x, x_1, x_2, x_{12}; \alpha_1, \alpha_2\right) = 0, \qquad && \bar{A}\left(x_3, x_{13}, x_{23}, x_{123}; \alpha_1, \alpha_2\right) = 0,\\
&B\left(x, x_2, x_3, x_{23}; \alpha_3, \alpha_2\right) = 0, && \bar{B}\left(x_1, x_{12}, x_{13}, x_{123}; \alpha_3, \alpha_2\right) = 0,\\
&C\left(x, x_1, x_3, x_{13}; \alpha_1, \alpha_3\right) = 0, && \bar{C}\left(x_2, x_{12}, x_{23}, x_{123}; \alpha_1, \alpha_3\right) = 0,
\end{aligned}
\tag{2.4.145}
$$

see Fig. 2.4. Such a six-tuple is defined to be *3D consistent* if, for arbitrary initial data x, x_1, x_2 and x_3, the three values for x_{123} (calculated by using $\bar{A} = 0$, $\bar{B} = 0$ and $\bar{C} = 0$) coincide. As a result in [23] they reobtained the Q-type equations of [22] and some new quad equations of type H which turn out to be deformations of those presented in (2.4.131).

In [112–114], Boll, starting from [23], classified all the consistent equations on the quad-graph possessing the tetrahedron property without any other additional assumption. The results were summarized by Boll in a set of theorems presented in [114], from Theorem **3.9** to Theorem **3.14**, listing all the consistent six-tuples configurations (2.4.145) up to $(\text{Möb})^8$, the group of independent Möbius transformations of the eight fields on the vertexes of the consistency three dimensional cube, Fig 2.4. In [336] Boll equations are reobtained and analyzed in detail for its integrability . All these equations fall into three disjoint families:

- Q-type (no degenerate biquadratic),
- H^4-type (four biquadratics are degenerate),
- H^6-type (all of the six biquadratics are degenerate).

It's worth noticing that the classification results hold locally, i.e. the equations are valid on a single quadrilateral cell or on a single cube. The non secondary problem which has been solved is the embedding of the single cell/single cube equations in a $2D$/$3D$ lattice, so as to preserve the $3D$ consistency. This was discussed in [23] introducing the concept of Black–White (BW) lattice. To get the lattice equations one needs to embed (2.4.145) into a $\mathbb{Z}^2$ lattice with an elementary cell of dimension greater than one. In such a case the generic equation Q on a quad-graph is extended to a lattice and the lattice equation will have Lax pair and Bäcklund transformation. To do so, following [112], one reflects the square with respect to the normal to its right and top sides and then complete a 2×2 lattice by reflecting again one of the obtained equation with respect to the other direction[1]. Such a procedure is graphically described in Fig. 2.5, and at the level of the quad equation this corresponds to constructing the three equations obtained from $Q = Q(x, x_1, x_2, x_{12}; \alpha_1, \alpha_2) = 0$ by all

[1]Let us note that, whatsoever side we reflect, the result of the last reflection is the same.

possible flipping of its fields:

$$Q = Q(x, x_1, x_2, x_{12}, \alpha_1, \alpha_2) = 0, \tag{2.4.146a}$$

$$|Q = Q(x_1, x, x_{12}, x_2, \alpha_1, \alpha_2) = 0, \tag{2.4.146b}$$

$$\underline{Q} = Q(x_2, x_{12}, x, x_1, \alpha_1, \alpha_2) = 0, \tag{2.4.146c}$$

$$|\underline{Q} = Q(x_{12}, x_2, x_1, x, \alpha_1, \alpha_2) = 0. \tag{2.4.146d}$$

By paving the whole $\mathbb{Z}^2$ with such equations we get a PΔE, which we can in principle study with the known methods. Since *a priori* $Q \neq |Q \neq \underline{Q} \neq |\underline{Q}$ the obtained lattice will be a four stripe lattice, i.e. an extension of the BW lattice considered in [388, 839]. This gives rise to lattice equations with two-periodic coefficients for an unknown function $u_{n,m}$, with $(n, m) \in \mathbb{Z}^2$:

$$\begin{aligned} &\chi_n\chi_m Q(u_{n,m}, u_{n+1,m}, u_{n,m+1}, u_{n+1,m+1}; \alpha_1, \alpha_2) \\ &\quad + \chi_{n+1}\chi_m |Q(u_{n,m}, u_{n+1,m}, u_{n,m+1}, u_{n+1,m+1}; \alpha_1, \alpha_2) \\ &\quad\quad + \chi_n\chi_{m+1}\underline{Q}(u_{n,m}, u_{n+1,m}, u_{n,m+1}, u_{n+1,m+1}; \alpha_1, \alpha_2) \\ &\quad\quad\quad + \chi_{n+1}\chi_{m+1}|\underline{Q}(u_{n,m}, u_{n+1,m}, u_{n,m+1}, u_{n+1,m+1}; \alpha_1, \alpha_2) = 0, \end{aligned} \tag{2.4.147}$$

where

$$\chi_k = \frac{1 + (-1)^k}{2}, \quad k = n, \text{or } m \tag{2.4.148}$$

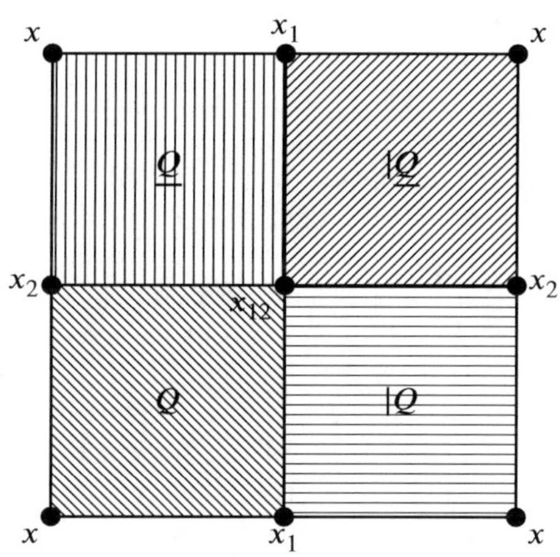

FIGURE 2.5. The "four colors" lattice

Let us notice that if Q possess the symmetries of the square, i.e. it is invariant under the action of D_4 one has:

$$Q = |Q = \underline{Q} = |\underline{Q}, \tag{2.4.149}$$

Eq. (2.4.149) implies that the elementary cell is actually of dimension one, and one falls into the case of the ABS classification. Beside the symmetry group of the square, D_4 [22, 29] there are two others relevant discrete symmetries for quad equations (2.4.130). One is the rhombic symmetry, which holds when:

$$
\begin{aligned}
Q(x, x_1, x_2, x_{12}, \alpha_1, \alpha_2) &= \sigma Q(x, x_2, x_1, x_{12}, \alpha_2, \alpha_1) \\
&= \epsilon Q(x_{12}, x_1, x_2, x, \alpha_2, \alpha_1),
\end{aligned}
\qquad (\sigma, \epsilon) \in \pm 1 \tag{2.4.150}
$$

Equations with rhombic symmetries have been introduced and classified in [23]. From their explicit form it is possible to show that they have the property:

$$Q = |\underline{Q}, \qquad \underline{Q} = |Q. \tag{2.4.151}$$

The other kind of relevant discrete symmetry for quad equations is the trapezoidal symmetry [23] given by:

$$Q(x, x_1, x_2, x_{12}, \alpha_1, \alpha_2) = Q(x_1, x, x_{12}, x_2, \alpha_1, \alpha_2). \tag{2.4.152}$$

Eq. (2.4.152) implies:

$$Q = |Q, \qquad \underline{Q} = |\underline{Q}. \tag{2.4.153}$$

Geometrically the trapezoidal symmetry is an invariance with respect to the axis parallel to (x_1, x_{12}). There might be a trapezoidal symmetry also with respect to the reflection around an axis parallel to (x_2, x_{12}), but this can be reduced to the previous one by a rotation. So there is no need to treat such symmetry but it is sufficient to consider (2.4.153).

A detailed study of all the lattices derived from the *rhombic* H^4 family, including the construction of their three-leg forms, Lax pairs, Bäcklund transformations and infinite hierarchies of generalized symmetries, was presented in [839].

The procedure presented above for the embedding of the equations defined on a cell into a 3D consistent lattice is due to Boll [112, 114]. Different embeddings in $3D$ consistent lattices resulting either in integrable or non integrable equations are discussed by Hietarinta and Viallet in [388] using the algebraic entropy analysis.

After the first results of ABS [22, 29] there have been various attempts to reduce the requirements imposed on consistent quad equations. Four non tetrahedral models, three of them with D_4 symmetry, were presented in [381, 382]. All of these models turn out to be more or less trivially linearizable [693]. Other consistent systems of quadrilateral lattice equations non possessing the tetrahedral property were studied in [23, 62].

In the following we study the independent lattice equations CaC not already considered in the literature [22, 23, 29, 839], i.e. those possessing the trapezoidal symmetry or with no symmetry at all. In Section 2.4.7.1 we list all independent equations defined on a cell and in Appendix B show in detail how one can extend the Möbius symmetry which classify Boll lattice equations defined on a four color lattice. In Section 2.4.7.2 we present all the independent lattice equations obtained in this way and in Appendix C.1 we analyze them from the point of view of the algebraic entropy showing that most of the new equations are linearizable. In Section 2.4.7.3 we explicitly linearize a few equations as an example of the procedure we have to use.

4.7.1. *Independent equations on a single cell.* In **Theorems 3.9 – 3.14** [114], Boll classified up to a $(\text{Möb})^8$ symmetry every 3D consistent six-tuples of equations with the tetrahedron property. Here we consider all the independent quad equations defined on a single cell not of type Q (Q_1^π, Q_2^π, Q_3^π and Q_4) or rhombic H^4 (${}_rH_1^\pi$, ${}_rH_2^\pi$ and ${}_rH_3^\pi$) as these two families have been already studied extensively [22, 23, 29, 839]. By independent we mean that the equations are defined up to a $(\text{Möb})^4$ symmetry on the fields, rotations, translations and inversions of the reference system. By reference system we mean those two vectors applied on the point x which define the two oriented directions n and m upon which the elementary square is constructed. The vertex of the square lying on direction n

(m) is then indicated by x_n (x_m). The remaining vertex is then called x_{nm}. In Fig. 2.3 one can see an elementary square where $n = 1$ and $m = 2$ or viceversa.

The list of independent quad equations defined on a single cell presented in the following expands the analogous one given by **Theorems 2.8-2.9** in [114], where the author does not distinguish between different arrangements of the fields (see Fig. 2.3) x_k, $k = 1, \dots, 4$ over the four corners of the elementary square. Different choices reflect in different bi-quadratic's patterns and, for any system presented in **Theorems 2.8-2.9** in [114], it is easy to see that a maximum of three different choices may arise up to rotations, translations and inversions.

Theorem 12. *All the independent consistent quad equations not of type Q or* rhombic H^4 *are given, up to* $(\text{Möb})^4$ *transformations of the fields* x, x_i, x_j *and* x_{ij} *(see Fig. 2.4) and rotations, translations and inversions of the reference system, by nine different representatives, three of* H^4*-type and six of* H^6*-type. We list them with their quadruples of discriminants and we identify the six-tuple where the equation appears by the theorem number indicated in* [114] *in the form* ***3.a.b***, *where* ***b*** *is the order of the six-tuple into the theorem* ***3.a***.

The trapezoidal equations of type H^4 *are:*

${}_tH_1^{\pi}$, $\left(\pi^2, \pi^2, 0, 0\right)$: *Eq. B of* ***3.10.1***.

$$\left(x - x_2\right)\left(x_3 - x_{23}\right) - \alpha_2(1 + \pi^2 x_3 x_{23}) = 0. \tag{2.4.154a}$$

${}_tH_2^{\pi}$, $\left(1 + 4\pi x, 1 + 4\pi x_2, 1, 1\right)$: *Eq. B of* ***3.10.2***.

$$\begin{aligned}&\left(x - x_2\right)\left(x_3 - x_{23}\right) + \alpha_2(x + x_2 + x_3 + x_{23})\\ &+ \left(\alpha_2 + \alpha_3\right)^2 - \alpha_3^2 + \frac{\pi\alpha_2^3}{2} +\\ &+ \frac{\pi\alpha_2}{2}\left(2x_3 + 2\alpha_3 + \alpha_2\right)\left(2x_{23} + 2\alpha_3 + \alpha_2\right) = 0.\end{aligned} \tag{2.4.154b}$$

${}_tH_3^{\pi}$, $\left(x^2 - 4\delta^2\pi^2, x_2^2 - 4\delta^2\pi^2, x_3^2, x_{23}^2\right)$: *Eq. B of* ***3.10.3***.

$$\begin{aligned}&e^{2\alpha_2}\left(x x_{23} + x_2 x_3\right) - \left(x x_3 + x_2 x_{23}\right) -\\ &- e^{2\alpha_3}\left(e^{4\alpha_2} - 1\right)\left(\delta^2 + \frac{\pi^2 x_3 x_{23}}{e^{4\alpha_3 + 2\alpha_2}}\right) = 0.\end{aligned} \tag{2.4.154c}$$

The trapezoidal equations of type H^6 *are:*

D_1, $(0, 0, 0, 0)$: *Eq. A of* ***3.12.1*** *and* ***3.13.1***.

$$x + x_1 + x_2 + x_{12} = 0. \tag{2.4.155a}$$

This equation is not invariant under any exchange of the fields.

${}_1D_2$, $\left(\delta_1^2, \left(\delta_1\delta_2 + \delta_1 - 1\right)^2, 1, 0\right)$: *Eq. A of* ***3.12.2***.

$$\delta_2 x + x_1 + \left(1 - \delta_1\right)x_2 + x_{12}\left(x + \delta_1 x_2\right) = 0. \tag{2.4.155b}$$

D_3, $(4x, 1, 1, 1)$: *Eq. A of* ***3.12.3***.

$$x + x_1 x_2 + x_1 x_{12} + x_2 x_{12} = 0. \tag{2.4.155c}$$

This equation is invariant under the exchange $x_1 \leftrightarrow x_2$.

${}_1D_4$, $\left(x^2 + 4\delta_1\delta_2\delta_3, x_1^2, x_{12}^2, x_2^2\right)$: *Eq. A of* ***3.12.4***.

$$x x_{12} + x_1 x_2 + \delta_1 x_1 x_{12} + \delta_2 x_2 x_{12} + \delta_3 = 0. \tag{2.4.155d}$$

This equation is invariant under the simultaneous exchanges $x_1 \leftrightarrow x_2$ *and* $\delta_1 \leftrightarrow \delta_2$.

${}_2D_2$, $\left(\delta_1^2, 0, 1\left(\delta_1\delta_2+\delta_1-1\right)^2\right)$: *Eq. C of **3.13.2**.*

$$\delta_2 x+\left(1-\delta_1\right)x_3+x_{13}+x_1\left(x+\delta_1 x_3-\delta_1\lambda\right)-\delta_1\delta_2\lambda=0. \tag{2.4.155e}$$

${}_3D_2$, $\left(\delta_1^2, 0, \left(\delta_1\delta_2+\delta_1-1\right)^2, 1\right)$: *Eq. C of **3.13.3**.*

$$\delta_2 x+x_3+\left(1-\delta_1\right)x_{13}+x_1\left(x+\delta_1 x_{13}-\delta_1\lambda\right)-\delta_1\delta_2\lambda=0. \tag{2.4.155f}$$

${}_2D_4$, $\left(x^2+4\delta_1\delta_2\delta_3, x_1^2, x_2^2, x_{12}^2\right)$: *Eq. A of **3.13.5**.*

$$xx_1+\delta_2 x_1x_2+\delta_1 x_1 x_{12}+x_2x_{12}+\delta_3=0. \tag{2.4.155g}$$

This equation is invariant under the simultaneous exchanges $x_2\leftrightarrow x_{12}$ and $\delta_1\leftrightarrow\delta_2$

Differently from the rhombic H^4 equations, which are π-deformations of the H equations in the ABS classification [22] and hence, in the limit $\pi\to 0$, have D_4 symmetries, the trapezoidal H^4 equations in the limit $\pi\to 0$ keep their discrete symmetry. Such class is then completely new with respect to the ABS classification and the "deformed" and the "undeformed" equations share the same properties.

Everything is just written on a single cell and no dynamical system over the entire lattice exists. The problem of the embedding in a $2D/3D$-lattice is discussed in Appendix A following [336].

4.7.2. *Independent equations on the $2D$-lattice.* Using the results presented in Appendix A we will now extend to the $2D$-lattice all the systems listed in Theorem 12 of Section 2.4.7.1. Independence is now understood to be up to $\widehat{(\text{Möb})}^4$ symmetry and rotations, translations and inversions of the reference system. The transformations of the reference system are taken to be acting on the discrete indexes rather than on the reference frame. For sake of compactness we shall omit the hats on the Möbius transformations when clear.

Theorem 13. *All the independent non linear, $2D$-dynamical systems not of type Q or* rhombic H^4 *which are consistent on the $3D$-lattice are given, up to $\widehat{(\text{Möb})}^4$ transformations of the fields $u_{n,m}$, $u_{n,m+1}$, $u_{n+1,m}$ and $u_{n+1,m+1}$, rotations, translations and inversions of the discrete indexes n and m, by nine non autonomous representatives, three of trapezoidal H^4 type and six of H^6 type.*

The H^4 type equations are:

$$\begin{aligned}{}_tH_1^\pi:\ &\left(u_{n,m}-u_{n+1,m}\right)\cdot\left(u_{n,m+1}-u_{n+1,m+1}\right)\\ &-\alpha_2\pi^2\left(\chi_m u_{n,m+1}u_{n+1,m+1}+\chi_{m+1}u_{n,m}u_{n+1,m}\right)-\alpha_2=0,\end{aligned} \tag{2.4.156a}$$

$$\begin{aligned}{}_tH_2^\pi:\ &\left(u_{n,m}-u_{n+1,m}\right)\left(u_{n,m+1}-u_{n+1,m+1}\right)\\ &+\alpha_2\left(u_{n,m}+u_{n+1,m}+u_{n,m+1}+u_{n+1,m+1}\right)\\ &+\frac{\pi\alpha_2}{2}\left(2\chi_m u_{n,m+1}+2\alpha_3+\alpha_2\right)\left(2\chi_m u_{n+1,m+1}+2\alpha_3+\alpha_2\right)\\ &+\frac{\pi\alpha_2}{2}\left(2\chi_{m+1}u_{n,m}+2\alpha_3+\alpha_2\right)\left(2\chi_{m+1}u_{n+1,m}+2\alpha_3+\alpha_2\right)\\ &+\left(\alpha_3+\alpha_2\right)^2-\alpha_3^2-2\pi\alpha_2\alpha_3\left(\alpha_3+\alpha_2\right)=0\end{aligned} \tag{2.4.156b}$$

$$\begin{aligned}{}_tH_3^\pi:\ &\alpha_2\left(u_{n,m}u_{n+1,m+1}+u_{n+1,m}u_{n,m+1}\right)\\ &-\left(u_{n,m}u_{n,m+1}+u_{n+1,m}u_{n+1,m+1}\right)-\alpha_3\left(\alpha_2^2-1\right)\delta^2\\ &-\frac{\pi^2(\alpha_2^2-1)}{\alpha_3\alpha_2}\left(\chi_m u_{n,m+1}u_{n+1,m+1}+\chi_{m+1}u_{n,m}u_{n+1,m}\right)=0,\end{aligned} \tag{2.4.156c}$$

*These equations arise from the B equation of the cases **3.10.1**, **3.10.2** and **3.10.3** in* [114] *respectively.*

The H^6 type equations are:

$$D_1: u_{n,m} + u_{n+1,m} + u_{n,m+1} + u_{n+1,m+1} = 0. \tag{2.4.157a}$$

$$\begin{aligned} {}_1D_2: \quad & \left(\chi_{n+m+1} - \delta_1 \chi_n \chi_{m+1} + \delta_2 \chi_n \chi_m\right) u_{n,m} \\ & + \left(\chi_{n+m} - \delta_1 \chi_{n+1} \chi_{m+1} + \delta_2 \chi_{n+1} \chi_m\right) u_{n+1,m} \\ & + \left(\chi_{n+m} - \delta_1 \chi_n \chi_m + \delta_2 \chi_n \chi_{m+1}\right) u_{n,m+1} \\ & + \left(\chi_{n+m+1} - \delta_1 \chi_{n+1} \chi_m + \delta_2 \chi_{n+1} \chi_{m+1}\right) u_{n+1,m+1} \\ & + \delta_1 \left(\chi_{m+1} u_{n,m} u_{n+1,m} + \chi_m u_{n,m+1} u_{n+1,m+1}\right) \\ & + \chi_{n+m} u_{n,m} u_{n+1,m+1} + \chi_{n+m+1} u_{n+1,m} u_{n,m+1} = 0, \end{aligned} \tag{2.4.157b}$$

$$\begin{aligned} {}_2D_2: \quad & \left(\chi_{m+1} - \delta_1 \chi_n \chi_{m+1} + \delta_2 \chi_n \chi_m - \delta_1 \lambda \chi_{n+1} \chi_m\right) u_{n,m} \\ & + \left(\chi_{m+1} - \delta_1 \chi_{n+1} \chi_{m+1} + \delta_2 \chi_{n+1} \chi_m - \delta_1 \lambda \chi_n \chi_m\right) u_{n+1,m} \\ & + \left(\chi_m - \delta_1 \chi_n \chi_m + \delta_2 \chi_n \chi_{m+1} - \delta_1 \lambda \chi_{n+1} \chi_{m+1}\right) u_{n,m+1} \\ & + \left(\chi_m - \delta_1 \chi_{n+1} \chi_m + \delta_2 \chi_{n+1} \chi_{m+1} - \delta_1 \lambda \chi_n \chi_{m+1}\right) u_{n+1,m+1} \\ & + \delta_1 \left(\chi_{n+m+1} u_{n,m} u_{n+1,m+1} + \chi_{n+m} u_{n+1,m} u_{n,m+1}\right) \\ & + \chi_m u_{n,m} u_{n+1,m} + \chi_{m+1} u_{n,m+1} u_{n+1,m+1} - \delta_1 \delta_2 \lambda = 0, \end{aligned} \tag{2.4.157c}$$

$$\begin{aligned} {}_3D_2: \quad & \left(\chi_{m+1} - \delta_1 \chi_{n+1} \chi_{m+1} + \delta_2 \chi_n \chi_m - \delta_1 \lambda \chi_{n+1} \chi_m\right) u_{n,m} \\ & + \left(\chi_{m+1} - \delta_1 \chi_n \chi_{m+1} + \delta_2 \chi_{n+1} \chi_m - \delta_1 \lambda \chi_n \chi_m\right) u_{n+1,m} \\ & + \left(\chi_m - \delta_1 \chi_{n+1} \chi_m + \delta_2 \chi_n \chi_{m+1} - \delta_1 \lambda \chi_{n+1} \chi_{m+1}\right) u_{n,m+1} \\ & + \left(\chi_m - \delta_1 \chi_n \chi_m + \delta_2 \chi_{n+1} \chi_{m+1} - \delta_1 \lambda \chi_n \chi_{m+1}\right) u_{n+1,m+1} \\ & + \delta_1 \left(\chi_{n+1} u_{n,m} u_{n,m+1} + \chi_n u_{n+1,m} u_{n+1,m+1}\right) \\ & + \chi_{m+1} u_{n,m+1} u_{n+1,m+1} + \chi_m u_{n,m} u_{n+1,m} - \delta_1 \delta_2 \lambda = 0, \end{aligned} \tag{2.4.157d}$$

$$\begin{aligned} D_3: \quad & \chi_n \chi_m u_{n,m} + \chi_{n+1} \chi_m u_{n+1,m} + \chi_n \chi_{m+1} u_{n,m+1} \\ & + \chi_{n+1} \chi_{m+1} u_{n+1,m+1} + \chi_{m+1} u_{n,m} u_{n+1,m} \\ & + \chi_{n+1} u_{n,m} u_{n,m+1} + \chi_{n+m+1} u_{n,m} u_{n+1,m+1} \\ & + \chi_{n+m} u_{n+1,m} u_{n,m+1} + \chi_n u_{n+1,m} u_{n+1,m+1} \\ & + \chi_m u_{n,m+1} u_{n+1,m+1} = 0, \end{aligned} \tag{2.4.157e}$$

$$\begin{aligned} {}_1D_4: \quad & \delta_1 \left(\chi_{n+1} u_{n,m} u_{n,m+1} + \chi_n u_{n+1,m} u_{n+1,m+1}\right) \\ & + \delta_2 \left(\chi_{m+1} u_{n,m} u_{n+1,m} + \chi_m u_{n,m+1} u_{n+1,m+1}\right) \\ & + u_{n,m} u_{n+1,m+1} + u_{n+1,m} u_{n,m+1} + \delta_3 = 0, \end{aligned} \tag{2.4.157f}$$

$$\begin{aligned} {}_2D_4: \quad & \delta_1 \left(\chi_{n+1} u_{n,m} u_{n,m+1} + \chi_n u_{n+1,m} u_{n+1,m+1}\right) \\ & + \delta_2 \left(\chi_{n+m+1} u_{n,m} u_{n+1,m+1} + \chi_{n+m} u_{n+1,m} u_{n,m+1}\right) \\ & + u_{n,m} u_{n+1,m} + u_{n,m+1} u_{n+1,m+1} + \delta_3 = 0. \end{aligned} \tag{2.4.157g}$$

*The equations ${}_1D_2$, ${}_1D_4$, ${}_2D_2$ and D_3 arise from the A equation in the cases **3.12.2, 3.12.3, 3.12.4** and **3.13.5** respectively. Instead ${}_2D_2$ and ${}_3D_2$ arise from the C equation in the cases **3.13.2** and **3.13.3** respectively.*

To write down the explicit form of (2.4.157) we used (A.19, A.21) and the fact that the following identities holds:

$$f_{n,m} = \chi_n \chi_m, \quad \lfloor f_{n,m} = \chi_{n+1}\chi_m, \\ \underline{f}_{n,m} = \chi_n \chi_{m+1}, \quad \lfloor \underline{f}_{n,m} = \chi_{n+1}\chi_{m+1}. \tag{2.4.158}$$

As mentioned in Appendix A if we apply this procedure to an equation of rhombic type we get a result consistent with [839].

For completeness we present also the other equations obtained by Adler, Bobenko and Suris in [23] and presented by Boll [113, 114]. Once written on the $\mathbb{Z}^2_{(n,m)}$ lattice, according to [839], the three equations belonging to the rhombic H^4 class have the form:

$$\begin{aligned} {}_rH_1^{\pi}: \quad & (u_{n,m} - u_{n+1,m+1})(u_{n+1,m} - u_{n,m+1}) - (\alpha - \beta) \\ & + \pi(\alpha - \beta)\left(\chi_{n+m}\, u_{n+1,m}u_{n,m+1} + \chi_{n+m+1}\, u_{n,m}u_{n+1,m+1}\right) = 0, \end{aligned} \tag{2.4.159a}$$

$$\begin{aligned} {}_rH_2^{\pi}: \quad & (u_{n,m} - u_{n+1,m+1})(u_{n+1,m} - u_{n,m+1}) \\ & + (\beta - \alpha)(u_{n,m} + u_{n+1,m} + u_{n,m+1} + u_{n+1,m+1}) - \alpha^2 + \beta^2 \\ & - \pi\,(\beta - \alpha)^3 - \pi\,(\beta - \alpha)\,\left(2\chi_{n+m+1}u_{n,m} + 2\chi_{n+m}u_{n+1,m} + \alpha + \beta\right) \cdot \\ & \cdot \left(2\chi_{n+m+1}u_{n+1,m+1} + 2\chi_{n+m}u_{n,m+1} + \alpha + \beta\right) = 0, \end{aligned} \tag{2.4.159b}$$

$$\begin{aligned} {}_rH_3^{\pi}: \quad & \alpha(u_{n,m}u_{n+1,m} + u_{n,m+1}u_{n+1,m+1}) \\ & - \beta(u_{n,m}u_{n,m+1} + u_{n+1,m}u_{n+1,m+1}) + (\alpha^2 - \beta^2)\delta \\ & - \frac{\pi(\alpha^2 - \beta^2)}{\alpha\beta}\left(\chi_{n+m}\, u_{n+1,m}u_{n,m+1} + \chi_{n+m+1}\, u_{n,m}u_{n+1,m+1}\right) = 0, \end{aligned} \tag{2.4.159c}$$

4.7.3. *Examples.* Here we consider in detail the ${}_tH_1^{\pi}$ (2.4.156a) and ${}_1D_2$ (2.4.157b) equations and shows the explicit form of the quadruple of matrices coming from the CaC, the non autonomous equations which give the consistency on $\mathbb{Z}^3$ and the effective Lax pair. Finally we confirm the predictions of the algebraic entropy analysis presented in Appendix C, showing how they can be explicitly linearized. Results on other cases can be found in [344, 345].

Example 1: ${}_tH_1^{\pi}$

To construct the Lax pair for (2.4.156a) we have to deal with **Case 3.10.1** in [114]. The sextuple we consider is:

$$\begin{aligned} A = {} & \alpha_2\left(x - x_1\right)\left(x_2 - x_{12}\right) - \alpha_1\left(x - x_2\right)\left(x_1 - x_{12}\right) \\ & + \pi^2\alpha_1\alpha_2\left(\alpha_1 - \alpha_2\right), \end{aligned} \tag{2.4.160a}$$

$$B = \left(x - x_2\right)\left(x_3 - x_{23}\right) - \alpha_2\left(1 + \pi^2 x_3 x_{23}\right) = 0, \tag{2.4.160b}$$

$$C = \left(x - x_1\right)\left(x_3 - x_{13}\right) - \alpha_1\left(1 + \pi^2 x_3 x_{13}\right) = 0, \tag{2.4.160c}$$

$$\bar{A} = \alpha_2\left(x_{13} - x_3\right)\left(x_{123} - x_{23}\right) - \alpha_1\left(x_{13} - x_{123}\right)\left(x_3 - x_{123}\right), \tag{2.4.160d}$$

$$\bar{B} = \left(x_1 - x_{12}\right)\left(x_{13} - x_{123}\right) - \alpha_2\left(1 + \pi^2 x_{13}x_{123}\right) = 0, \tag{2.4.160e}$$

$$\bar{C} = \left(x_2 - x_{12}\right)\left(x_{23} - x_{123}\right) - \alpha_1\left(1 + \pi^2 x_{23}x_{123}\right) = 0, \tag{2.4.160f}$$

In this sextuple (2.4.156a) originates from the B equation.

We now make the following identifications

$$\begin{array}{lllll} A: & x \to u_{p,n} & x_1 \to u_{p+1,n} & x_2 \to u_{p,n+1} & x_{12} \to u_{p+1,n+1} \\ B: & x \to u_{n,m} & x_2 \to u_{n+1,m} & x_3 \to u_{n,m+1} & x_{23} \to u_{n+1,m+1} \\ C: & x \to u_{p,m} & x_1 \to u_{p+1,m} & x_3 \to u_{p,m+1} & x_{13} \to u_{p+1,m+1} \end{array} \tag{2.4.161}$$

so that in any equation we can suppress the dependence on the appropriate parametric variables. On $\mathbb{Z}^3$ we get the following triplet of equations:

$$\begin{aligned} \tilde{A} = {} & \alpha_2 \left(u_{p,n} - u_{p+1,n}\right)\left(u_{p,n+1} - u_{p+1,n+1}\right) \\ & - \alpha_1 \left(u_{p,n} - u_{p,n+1}\right)\left(u_{p+1,n} - u_{p+1,n+1}\right) \\ & + \pi^2 \alpha_1 \alpha_2 \left(\alpha_1 - \alpha_2\right) \chi_m, \end{aligned} \tag{2.4.162a}$$

$$\begin{aligned} \tilde{B} = {} & \left(u_{n,m} - u_{n+1,m}\right)\left(u_{n,m+1} - u_{n+1,m+1}\right) - \alpha_2 \\ & - \alpha_2 \pi^2 \left(\chi_m u_{n,m+1} u_{n+1,m+1} + \chi_{m+1} u_{n,m} u_{n+1,m}\right), \end{aligned} \tag{2.4.162b}$$

$$\begin{aligned} \tilde{C} = {} & \left(u_{p,m} - u_{p+1,m}\right)\left(u_{p,m+1} - u_{p+1,m+1}\right) - \alpha_1 \\ & - \alpha_1 \pi^2 \left(\chi_m u_{p,m+1} u_{p+1,m+1} + \chi_{m+1} u_{p,m} u_{p+1,m}\right). \end{aligned} \tag{2.4.162c}$$

Then with the usual method we find the following Lax pair:

$$\begin{aligned} \tilde{L}_{n,m} = {} & \begin{pmatrix} u_{n,m+1} & -u_{n,m} u_{n,m+1} + \alpha_1 \\ 1 & -u_{n,m} \end{pmatrix} \\ & - \pi^2 \alpha_1 \begin{pmatrix} -\chi_{m+1} u_{n,m} & 0 \\ 0 & \chi_m u_{n,m+1} \end{pmatrix}, \end{aligned} \tag{2.4.163a}$$

$$\begin{aligned} \tilde{M}_{n,m} = {} & \begin{pmatrix} \alpha_1 \left(u_{n,m} - u_{n+1,m}\right) + \alpha_2 u_{n+1,m} & -\alpha_2 u_{n,m} u_{n+1,m} \\ \alpha_2 & \alpha_1 \left(u_{n,m} - u_{n+1,m}\right) - \alpha_2 u_{n,m} \end{pmatrix} \\ & - \pi^2 \alpha_1 \alpha_2 \left(\alpha_1 - \alpha_2\right) \chi_m \begin{pmatrix} 0 & 1 \\ 0 & 0 \end{pmatrix}. \end{aligned} \tag{2.4.163b}$$

Let us now turn to the linearization procedure. In (2.4.156a) we must set $\alpha_2 \neq 0$, otherwise the equation degenerates and trivialize to

$$\left(u_{n,m} - u_{n+1,m}\right)\left(u_{n,m+1} - u_{n+1,m+1}\right) = 0.$$

Let us define $u_{n,2k} = w_{n,k}$, $u_{n,2k+1} = z_{n,k}$; then we have the following system of two coupled autonomous difference equations

$$\left(w_{n,k} - w_{n+1,k}\right)\left(z_{n,k} - z_{n+1,k}\right) - \pi^2 \alpha_2 z_{n,k} z_{n+1,k} - \alpha_2 = 0, \tag{2.4.164a}$$

$$\left(w_{n,k+1} - w_{n+1,k+1}\right)\left(z_{n,k} - z_{n+1,k}\right) - \pi^2 \alpha_2 z_{n,k} z_{n+1,k} - \alpha_2 = 0. \tag{2.4.164b}$$

Subtracting (2.4.164b) from (2.4.164a), we obtain

$$\left(w_{n,k} - w_{n+1,k} - w_{n,k+1} + w_{n+1,k+1}\right)\left(z_{n,k} - z_{n+1,k}\right) = 0. \tag{2.4.165}$$

At this point the solution of the system bifurcates:

- **Case 1**: if $z_{n,k} = f_k$, where f_k is a generic function of its argument, equation (2.4.165) is satisfied and from (2.4.164a) or (2.4.164b) we have that $\pi \neq 0$ and, solving for f_k, one gets

$$f_k = \pm \frac{\mathrm{i}}{\pi}. \tag{2.4.166}$$

- **Case 2**: if $z_{n,k} \neq f_k$, with f_k given in (2.4.166), one has $w_{n,k} = g_n + h_k$, where g_n and h_k are arbitrary functions of their argument. Hence (2.4.164b) and (2.4.164a) reduce to

$$\pi^2 z_{n,k} z_{n+1,k} + \kappa_n \left(z_{n,k} - z_{n+1,k} \right) + 1 = 0, \quad \kappa_n = \frac{g_{n+1} - g_n}{\alpha_2}. \tag{2.4.167}$$

Two sub-cases emerge:
  - **Sub-case 2.1**: if $\pi = 0$, (2.4.167) then $\kappa_n \neq 0$, so that, solving,

$$z_{n+1,k} - z_{n,k} = \frac{1}{\kappa_n},$$

  we get

$$z_{n,k} = \begin{cases} j_k + \sum_{l=n_0}^{n-1} \frac{1}{\kappa_l}, & n \geq n_0 + 1, \\ j_k - \sum_{l=n}^{n_0-1} \frac{1}{\kappa_l}, & n \leq n_0 - 1, \end{cases}$$

  where $j_k = z_{n_0,k}$ is a generic integration function of its argument.
  - **Sub-case 2.2**: if $\pi \neq 0$, (2.4.167) is a discrete Riccati equation which can be linearized by the Möbius transformation $z_{n,k} = \frac{\mathrm{i}}{\pi} \frac{y_{n,k}-1}{y_{n,k}+1}$ to

$$\left(\mathrm{i}\kappa_n - \pi \right) y_{n+1,k} = \left(\mathrm{i}\kappa_n + \pi \right) y_{n,k}. \tag{2.4.168}$$

  In (2.4.168) $\kappa_n \neq \pm \mathrm{i}\pi$, because otherwise $y_{n,k} = 0$, and $z_{n,k} = -\mathrm{i}/\pi$. Eq. (2.4.168) implies

$$y_{n,k} = \begin{cases} j_k \prod_{l=n_0}^{n-1} \frac{\mathrm{i}\kappa_l + \pi}{\mathrm{i}\kappa_l - \pi}, & n \geq n_0 + 1, \\ j_k \prod_{l=n}^{n_0-1} \frac{\mathrm{i}\kappa_l - \pi}{\mathrm{i}\kappa_l + \pi}, & n \leq n_0 - 1, \end{cases}$$

  where $j_k = y_{n_0,k}$ is another arbitrary integration function of its argument.

In conclusion we have always integrated the original system.

Let us note that in the case $\pi = 0$ (2.4.156a) becomes

$$\left(u_{n,m} - u_{n+1,m} \right) \left(u_{n,m+1} - u_{n+1,m+1} \right) - \alpha_2 = 0. \tag{2.4.169}$$

So the contact Möbius-type transformation

$$u_{n+1,m} - u_{n,m} = \sqrt{\alpha_2} \frac{1 - w_{n,m}}{1 + w_{n,m}}, \tag{2.4.170}$$

brings (2.4.156a) into the following first order linear equation:

$$w_{n,m+1} + w_{n,m} = 0. \tag{2.4.171}$$

Then

$$u_{n,m} = \begin{cases} k_m + \sqrt{\alpha_2} \sum_{l=n_0}^{l=n-1} \frac{1-(-1)^m w_l}{1+(-1)^m w_l}, & n \geq n_0 + 1, \\ k_m - \sqrt{\alpha_2} \sum_{l=n}^{l=n_0-1} \frac{1-(-1)^m w_l}{1+(-1)^m w_l}, & n \leq n_0 - 1. \end{cases} \tag{2.4.172}$$

Here $k_m = u_{n_0,m}$ and w_n, are two arbitrary integration functions.

Example 2: $_1D_2$

Now we consider the equation $_1D_2$ (2.4.157b). The sextuple of equations given by **Case 3.12.2** in [114] is:

$$A = \delta_2 x + x_1 + (1-\delta_1)\, x_2 + x_{12}\,(x+\delta_1 x_2)\,, \tag{2.4.173a}$$

$$\begin{aligned} B = {} & (x - x_3)\,(x_2 - x_{23}) \\ & + \sigma\left[x + x_3 - \delta_1\,(x_2 + x_{23})\right] + \delta_1\sigma, \end{aligned} \tag{2.4.173b}$$

$$\begin{aligned} C = {} & (x - x_3)\,(x_1 - x_{13}) + \delta_1\,(\delta_1\delta_2 + \delta_1 - 1)\,\sigma^2 \\ & - \sigma\left[(\delta_1\delta_2 + \delta_1 - 1)\,(x + x_3) + \delta_1\,(x_1 + x_{13})\right], \end{aligned} \tag{2.4.173c}$$

$$\bar{A} = \delta_2 x_3 + x_{13} + (1-\delta_1)\, x_{23} + x_{123}\,(x + \delta_1 x_{23})\,, \tag{2.4.173d}$$

$$\begin{aligned} \bar{B} = {} & (x_1 - x_{13})\,(x_{12} - x_{123}) \\ & + \sigma\left[2\delta_2\,(\delta_1 - 1) + (\delta_1 - 1 - \delta_1\delta_2) - 2\delta_1 x_{12} x_{123}\right], \end{aligned} \tag{2.4.173e}$$

$$\bar{C} = (x_1 - x_{23})\,(x_{12} - x_{123}) - \sigma\,(2\delta_2 + x_{12} + x_{123})\,. \tag{2.4.173f}$$

The triplet of consistent dynamical systems on the $3D$-lattice is:

$$\begin{aligned} \tilde{A} = {} & (\chi_{p+n+1} - \delta_1 \chi_p \chi_{n+1} + \delta_2 \chi_p \chi_n)\, u_{p,n} \\ & + (\chi_{p+n} - \delta_1 \chi_{p+1} \chi_{n+1} + \delta_2 \chi_{p+1} \chi_n)\, u_{p+1,n} \\ & + (\chi_{p+n} - \delta_1 \chi_p \chi_n + \delta_2 \chi_p \chi_{n+1})\, u_{p,n+1} \\ & + (\chi_{p+n+1} - \delta_1 \chi_{p+1} \chi_n + \delta_2 \chi_{p+1} \chi_{n+1})\, u_{p+1,n+1} \\ & + \delta_1 (\chi_{n+1} u_{p,n} u_{p+1,n} + \chi_n u_{p,n+1} u_{p+1,n+1}) \\ & + \chi_{p+n} u_{p,n} u_{p+1,n+1} + \chi_{p+n+1} u_{p+1,n} u_{p,n+1}, \end{aligned}$$

$$\begin{aligned} \tilde{B} = {} & \sigma\left\{\left[(\delta_1 - 1)\,\chi_n - \delta_1\right]\chi_p + (\delta_1 - 1 - \delta_1\delta_2)\,\chi_{p+1}\chi_{n+1}\right\}(u_{n,m} + u_{n,m+1}) \\ & + \sigma\left\{\left[(\delta_1 - 1)\,\chi_{n+1} - \delta_1\right]\chi_p + (\delta_1 - 1 - \delta_1\delta_2)\,\chi_{p+1}\chi_n\right\}(u_{n+1,m} + u_{n+1,m+1}) \\ & - 2\delta_1\sigma\chi_{p+1}\,(\chi_{n+1} u_{n,m} u_{n,m+1} + \chi_n u_{n+1,m} u_{n+1,m+1}) + \\ & + (u_{n,m} - u_{n,m+1})\,(u_{n+1,m} - u_{n+1,m+1}) + \delta_1\sigma^2\chi_p + 2\,(\delta_1 - 1)\,\delta_2\sigma\chi_{p+1}, \end{aligned}$$

$$\begin{aligned} \tilde{C} = {} & \sigma\left\{\left[(1 - \delta_1\delta_2)\,\chi_p - \delta_1\right]\chi_n + \chi_{p+1}\chi_{n+1}\right\}(u_{p,m} + u_{p,m+1}) + 2\delta_2\sigma\chi_{n+1} \\ & + \sigma\left\{\left[(1 - \delta_1\delta_2)\,\chi_{p+1} - \delta_1\right]\chi_n + \chi_p\chi_{n+1}\right\}(u_{p+1,m} + u_{p+1,m+1}) \\ & + (u_{p,m} - u_{p,m+1})\,(u_{p+1,m} - u_{p+1,m+1}) + \delta_1\,(\delta_1 - 1 + \delta_1\delta_2)\,\sigma^2\chi_n. \end{aligned}$$

We leave out the Lax pair for $_1D_2$ as they are too complicate to write down and not worth while the effort for the reader. If necessary one can always write them down using the standard procedure outlined before in Section 2.4.6.2.

Let us now turn to the linearization procedure. Notice that there is no combination of the parameters δ_1 and δ_2 such that (2.4.157b) becomes autonomous. So we are naturally induced to introduce the following four fields:

$$\begin{aligned} w_{n,m} &= u_{2n,2m}, & y_{n,m} &= u_{2n+1,2m}, \\ v_{n,m} &= u_{2n,2m+1} & z_{n,m} &= u_{2n+1,2m+1}. \end{aligned} \tag{2.4.174}$$

which transform (2.4.157b) into the following system of four coupled autonomous difference equations:

(2.4.175a) $$\left(1-\delta_1\right) v_{n,m}+\delta_2 w_{n,m}+y_{n,m}+\left(\delta_1 v_{n,m}+w_{n,m}\right) z_{n,m}=0,$$

(2.4.175b) $$\begin{aligned}\left(1-\delta_1\right) v_{n+1,m}+\delta_2 w_{n+1,m}+y_{n,m}&\\ +\left(\delta_1 v_{n+1,m}+w_{n+1,m}\right) z_{n,m}&=0,\end{aligned}$$

(2.4.175c) $$\left(1-\delta_1\right) v_{n,m}+\delta_2 w_{n,m+1}+y_{n,m+1}+\left(\delta_1 v_{n,m}+w_{n,m+1}\right) z_{n,m}=0,$$

(2.4.175d) $$\begin{aligned}\left(1-\delta_1\right) v_{n+1,m}+\delta_2 w_{n+1,m+1}+y_{n,m+1}&\\ +\left(\delta_1 v_{n+1,m}+w_{n+1,m+1}\right) z_{n,m}&=0.\end{aligned}$$

Let us solve (2.4.175a) with respect to $y_{n,m}$:

(2.4.176) $$y_{n,m}=-\left(1-\delta_1\right) v_{n,m}-\delta_2 w_{n,m}-\left(\delta_1 v_{n,m}+w_{n,m}\right) z_{n,m}$$

and let us insert $y_{n,m}$ into (2.4.175b) in order to get an equation solvable for $z_{n,m}$. This is possible *iff* $\delta_1 v_{n,m}+w_{n,m}\neq f_t$, with f_t a generic function of t, since in this case the coefficient of $z_{n,m}$ is zero. Then the solution of the system (2.4.175) bifurcates.

Case 1 Assume that $\delta_1 v_{n,m}+w_{n,m}\neq f_t$, then we can solve with respect to $z_{n,m}$ the expression obtained inserting (2.4.176) into (2.4.175b). We get:

(2.4.177) $$z_{n,m}=-\frac{\left(1-\delta_1\right)\left(v_{n+1,m}-v_{n,m}\right)+\delta_2\left(w_{n+1,m}-w_{n,m}\right)}{\delta_1\left(v_{n+1,m}-v_{n,m}\right)+w_{n+1,m}-w_{n,m}}.$$

Now we can substitute (2.4.176) and (2.4.177) together with their difference consequences into (2.4.175c) and (2.4.175d) and we get two equations for $w_{n,m}$ and $v_{n,m}$:

(2.4.178a) $$\begin{aligned}\left(\delta_1\delta_2+\delta_1-1\right)\big[&w_{n+1,m+1}v_{n,m+1}w_{n,m}+v_{n+1,m+1}w_{n,m+1}w_{n+1,m}\\ &-v_{n+1,m+1}w_{n,m+1}w_{n,m}+w_{n,m+1}v_{n,m}w_{n+1,m+1}\\ &+v_{n,m}w_{n+1,m+1}w_{n+1,m}-w_{n,m+1}v_{n+1,m}w_{n+1,m+1}\\ &-w_{n,m+1}^2v_{n,m}+w_{n,m+1}^2v_{n+1,m}\\ &-w_{n+1,m+1}v_{n,m+1}w_{n+1,m}-v_{n,m}w_{n+1,m+1}w_{n,m}\\ &-v_{n,m}w_{n,m+1}w_{n+1,m}+v_{n,m}w_{n,m+1}w_{n,m}\\ &-\delta_1(v_{n,m}v_{n,m+1}w_{n+1,m}+w_{n,m+1}v_{n,m}v_{n,m+1}\\ &-w_{n+1,m+1}v_{n,m+1}v_{n,m}+w_{n+1,m+1}v_{n,m+1}v_{n+1,m}\\ &+v_{n,m}v_{n+1,m+1}w_{n,m}-v_{n,m}v_{n+1,m+1}w_{n+1,m}\\ &-v_{n,m}v_{n,m+1}w_{n,m}-w_{n,m+1}v_{n+1,m}v_{n,m+1})\big],\end{aligned}$$

$$(2.4.178b)\quad \begin{aligned}(\delta_1\delta_2+\delta_1-1)\big[&w^2_{n+1,m+1}v_{n+1,m}+w_{n+1,m+1}v_{n,m+1}w_{n+1,m}\\ &-w_{n+1,m+1}v_{n,m+1}w_{n,m}-w^2_{n+1,m+1}v_{n,m}\\ &-w_{n,m+1}v_{n+1,m}w_{n+1,m+1}+w_{n,m+1}v_{n,m}w_{n+1,m+1}\\ &-v_{n+1,m+1}w_{n,m+1}w_{n+1,m}+v_{n+1,m}w_{n,m+1}w_{n+1,m}\\ &+v_{n+1,m+1}w_{n,m+1}w_{n,m}-v_{n+1,m}w_{n+1,m+1}w_{n+1,m}\\ &+v_{n+1,m}w_{n+1,m+1}w_{n,m}-v_{n+1,m}w_{n,m+1}w_{n,m}\\ &+\delta_1(-v_{n+1,m}v_{n+1,m+1}w_{n+1,m}+v_{n+1,m}v_{n+1,m+1}w_{n,m}\\ &-w_{n+1,m+1}v_{n,m}v_{n+1,m+1}-v_{n+1,m}v_{n,m+1}w_{n,m}\\ &-v_{n+1,m+1}w_{n,m+1}v_{n+1,m}+w_{n+1,m+1}v_{n+1,m}v_{n+1,m+1}\\ &+v_{n+1,m}v_{n,m+1}w_{n+1,m}+v_{n+1,m+1}w_{n,m+1}v_{n,m})\big].\end{aligned}$$

If

$$(2.4.179)\qquad \delta_1(1+\delta_2)=1,$$

then (2.4.178) are identically satisfied. If $\delta_1\left(1+\delta_2\right)\neq 1$, adding (2.4.178a) and (2.4.178b), we obtain:

$$(2.4.180)\qquad \begin{aligned}&\left(w_{n,m}-w_{n+1,m}-w_{n,m+1}+w_{n+1,m+1}\right)\cdot\\ &\qquad\left(v_{n+1,m}-v_{n,m}\right)\left(\delta_1+\delta_1\delta_2-1\right)=0.\end{aligned}$$

Supposing $\delta_1+\delta_1\delta_2\neq 1$ we can annihilate the first or the second factor. If we set equal to zero the second factor, we get $v_{n,m}=g_m$ with g_m an arbitrary function of m alone. Substituting this result into (2.4.178a) or (2.4.178b), we find that they are identically satisfied provided $g_m=g_0$, with g_0 constant. Then the only non-trivial case is when $\delta_1+\delta_1\delta_2\neq 1$, and $v_{n,m}\neq g_m$. In this case $w_{n,m}$ solves the discrete wave equation, i.e. $w_{n,m}=h_n+l_m$. Substituting $w_{n,m}$ into (2.4.178) we get a single equation for $v_{n,m}$:

$$(2.4.181)\qquad \begin{aligned}\left(h_n-h_{n+1}\right)\big[&\left(v_{n+1,m+1}-v_{n+1,m}\right)\left(h_n+l_{m+1}\right)\\ &-\left(v_{n,m+1}-v_{n,m}\right)\left(h_{n+1}+l_{m+1}\right)\\ &+\delta_1\left(v_{n,m}v_{n+1,m+1}-v_{n+1,m}v_{n,m+1}\right)\big]=0.\end{aligned}$$

Eq. (2.4.181) is satisfied if $h_n=h_0$, with h_0 a constant. Therefore we have non-trivial cases only if $h_n\neq h_0$.

Case 1.1 We have a great simplification if in addition to $h_n\neq h_0$ we have $\delta_1=0$. In this case (2.4.181) is linear:

$$(2.4.182)\qquad \begin{aligned}&\left(v_{n+1,m+1}-v_{n+1,m}\right)\left(h_n+l_{m+1}\right)\\ &\qquad-\left(v_{n,m+1}-v_{n,m}\right)\left(h_{n+1}+l_{m+1}\right)=0.\end{aligned}$$

Eq. (2.4.182) can be easily integrated twice to give:

$$v_{n,m}=\begin{cases} j_n+\sum_{k=m_0}^{m-1}\left(h_n+l_{k+1}\right)i_k, & m\geq m_0+1,\\ j_n-\sum_{k=m}^{m_0-1}\left(h_n+l_{k+1}\right)i_k, & m\leq m_0-1,\end{cases}$$

with i_m and $j_n=v_{n,m_0}$ are arbitrary integration functions.

Case 1.2 Now let us suppose again $h_n\neq h_0$, $\delta_1\neq 0$ but let us choose $l_m=l_0$, with l_0 a constant. Performing the translation $\theta_{n,m}=v_{n,m}+\left(h_n+l\right)/\delta_1$, from (2.4.181) we get:

$$(2.4.183)\qquad \theta_{n,m}\theta_{n+1,m+1}-\theta_{n+1,m}\theta_{n,m+1}=0.$$

Eq. (2.4.183) is linearizable via a Cole-Hopf transformation $\Theta_{n,m} = \theta_{n+1,m}/\theta_{n,m}$ as $v_{n,m}$ cannot be identically zero. This linearization yields the general solution $\theta_{n,m} = \mathscr{S}_n \mathscr{T}_m$ with $\mathscr{S}_n$ and $\mathscr{T}_m$ arbitrary functions of their argument.

Case 1.3 Finally if $h_n \neq h$, $\delta_1 \neq 0$ and $l_m \neq l_0$, we perform the transformation

(2.4.184) $$\theta_{n,m} = \frac{1}{\delta_1}\left[\left(l_m - l_{m+1}\right) v_{n,m} - h_n - l_{m+1}\right].$$

Then from (2.4.181) we get:

(2.4.185) $$\theta_{n,m}\left(1+\theta_{n+1,m+1}\right) - \theta_{n+1,m}\left(1+\theta_{n,m+1}\right) = 0,$$

which, as $v_{n,m}$ cannot be identically zero, is easily linearized via the Cole-Hopf transformation $\Theta_{n,m} = (1+\theta_{n,m+1})/\theta_{n,m}$ to $\Theta_{n+1,n} - \Theta_{n,m} = 0$. Then we get for $\theta_{n,m}$ the linear equation:

(2.4.186) $$\theta_{n,m+1} - p_m \theta_{n,m} + 1 = 0,$$

where p_m is an arbitrary integration function. The general solution of (2.4.186) is given by:

(2.4.187) $$\theta_{n,m} = \begin{cases} \left(u_n - \sum_{l=m_0}^{m-1} \prod_{j=m_0}^{l} p_j^{-1}\right) \prod_{k=m_0}^{m-1} p_k, & m \geq m_0 + 1, \\ u_n \prod_{k=m}^{m_0-1} p_k^{-1} + \sum_{l=m}^{m_0-1} \prod_{j=m}^{l} p_j^{-1}, & m \leq m_0 - 1, \end{cases}$$

where $u_n = \theta_{n,m_0}$ is an arbitrary integration function.

Case 2 We now suppose:

(2.4.188) $$w_{n,m} = f_m - \delta_1 v_{n,m},$$

where f_m is a generic function of its argument. Inserting (2.4.176) and (2.4.188) and their difference consequences into (2.4.175), we get:

(2.4.189) $$\left(v_{n+1,m} - v_{n,m}\right)\left(\delta_1 + \delta_1\delta_2 - 1\right) = 0,$$

and the two relations:

(2.4.190a) $$\begin{aligned} f_{m+1} z_{n,m+1} &+ \left[\delta_1\left(v_{n,m} - v_{n,m+1}\right) - f_{m+1}\right] z_{n,m} \\ &+ \left(\delta_1 - 1\right)\left(v_{n,m} - v_{n,m+1}\right), \end{aligned}$$

(2.4.190b) $$\begin{aligned} f_{m+1} z_{n,m+1} &+ \left[\delta_1\left(v_{n+1,m} - v_{n+1,m+1}\right) - f_{t+1}\right] z_{n,m} \\ &+ \left(\delta_1 - 1\right)\left(v_{n+1,m} - v_{n,m+1}\right) + \delta_1\delta_2\left(v_{n+1,m+1} - v_{n,m+1}\right). \end{aligned}$$

Hence in (2.4.189) we have a bifurcation.

Case 2.1 If we annihilate the second factor in (2.4.189) we get $\delta_1\left(1+\delta_2\right) = 1$, i.e. $\delta_1 \neq 0$. Then adding (2.4.190a) and (2.4.190b) we obtain:

(2.4.191) $$\left(v_{n,m} - v_{n+1,m} - v_{n,m+1} + v_{n+1,m+1}\right)\left(1 - \delta_1 + \delta_1 z_{n,m}\right) = 0.$$

It seems that we are facing a new bifurcation. However annihilating the second factor, i.e. assuming that $z_{n,m} = 1 - 1/\delta_1$ gives a trivial case, since (2.4.190) are identically satisfied. Therefore we may assume that $z_{n,m} \neq 1 - 1/\delta_1$. This implies that $v_{n,m} = h_n + k_m$, where h_n and k_t are generic integration functions of their argument. Inserting it in (2.4.190) we can obtain the following linear equation for $z_{n,m}$:

(2.4.192) $$f_{m+1} z_{n,m+1} + \left(\delta_1 j_m - f_{m+1}\right) z_{n,m} + \left(\delta_1 - 1\right) j_m = 0,$$

with $j_m = k_{m+1} - k_m$. This equation can be solved. It gives:

$$z_{n,m} = (-1)^m \left(\delta_1 - 1\right) \prod_{m'=0}^{m-1} \frac{\delta_1 j_{m'} - f_{m'+1}}{f_{m'+1}} \sum_{m''=0}^{t-1} \frac{j_{m'} (-1)^{m''}}{f_{m''+1} \prod_{m'=0}^{m''} \frac{\delta_1 j_{m'} - f_{m'+1}}{f_{m'+1}}}$$

$$+ (-1)^m z_{n,0} \prod_{m'=0}^{m-1} \frac{\delta_1 j_{m'} - f_{m'+1}}{f_{m'+1}}.$$

Case 2.2 Now we annihilate the first factor in (2.4.189) i.e. $\delta_1 \left(1 + \delta_2\right) \neq 1$ and $v_{n,m} = l_m$, where l_m is an arbitrary function of its argument. From (2.4.190) we obtain (2.4.192) with $j_m = l_{m+1} - l_m$.

In conclusion we have always integrated the original system using an explicit linearization through a series of transformations and bifurcations.

As a final remark we observe that every transformation used in the linearization procedure both for ${}_tH_1^\pi$ (2.4.156a) and for ${}_1D_2$ (2.4.157b) is bi-rational in the fields and their shifts (like Cole-Hopf-type transformations). This, in fact, has to be expected, since the algebraic entropy test is valid only if we allow transformations which preserve the algebrogeometric structure underlying the evolution procedure [817]. Indeed there are examples on one-dimensional lattice of equations chaotic according to the algebraic entropy, but linearizable using some transcendental transformations [332]. So exhibiting the explicit linearization and showing that it can be attained by bi-rational transformations is indeed a very strong evidence of the algebraic entropy conjecture [387]. However this does not prevent that the equations can be linearized through transcendental transformations. In fact if $\pi = 0$ the ${}_tH_1^\pi$ equation (2.4.156a) can be linearized through the transcendental contact transformation:

$$u_{n,m} - u_{n+1,m} = \sqrt{\alpha_2} e^{z_{n,m}}, \tag{2.4.193}$$

i.e.

$$z_{n,m} = \log \frac{u_{n,m} - u_{n+1,m}}{\sqrt{\alpha_2}},$$

were the function log stands for the principal value of the complex logarithm (the principal value is intended for the square root too). The transformation (2.4.193) brings (2.4.156a) into the following family of first order linear equations:

$$z_{n,m+1} + z_{n,m} = 2\mathrm{i}\pi\kappa, \quad \kappa = 0, 1. \tag{2.4.194}$$

However this kind of transformation does not prove the result of the algebraic entropy and the method explained in this Section should be considered the correct one.

4.7.4. *The non autonomous Q_V equation.* To construct a non autonomous Q_V equation [343], say $Q_\mathrm{V}^{(n,m)}$, we need to extend the Klein discrete symmetry (2.4.137) to an equation with two-periodic coefficients. A way to do so is to consider a multilinear function Q with two-periodic coefficients in n and m such that the following equations holds:

$$\begin{aligned} &Q\left(u_{n+1,m}, u_{n,m}, u_{n+1,m+1}, u_{n,m+1}; (-1)^n, (-1)^m\right) = \\ &\quad \tau Q\left(u_{n,m}, u_{n+1,m}, u_{n,m+1}, u_{n+1,m+1}; -(-1)^n, (-1)^m\right), \\ &Q\left(u_{n,m+1}, u_{n+1,m+1}, u_{n,m}, u_{n+1,m}; (-1)^n, (-1)^m\right) = \\ &\quad \tau' Q\left(u_{n,m}, u_{n+1,m}, u_{n,m+1}, u_{n+1,m+1}; (-1)^n, -(-1)^m\right), \end{aligned} \tag{2.4.195}$$

with $(\tau,\tau')=(\pm 1,\pm 1)$ If a non autonomous equation Q satisfies the discrete symmetries (2.4.195) we will say that Q admits a *non autonomous Klein symmetry*. The name follows from the fact that if Q is autonomous then the discrete symmetry (2.4.195) reduces to the Klein one (2.4.137). Furthermore all the equations belonging to the Boll's classification satisfy these symmetry conditions (2.4.195) when $\tau=\tau'=1$.

Let us consider now the most general multilinear equation in the lattice variables with two-periodic coefficients:

$$\begin{aligned} Q = {} & p_1 u_{n,m}u_{n+1,m}u_{n,m+1}u_{n+1,m+1} \\ & + p_2 u_{n,m}u_{n,m+1}u_{n+1,m+1} + p_3 u_{n+1,m}u_{n,m+1}u_{n+1,m+1} \\ & + p_4 u_{n,m}u_{n+1,m}u_{n+1,m+1} + p_5 u_{n,m}u_{n+1,m}u_{n,m+1} \\ & + p_6 u_{n,m}u_{n+1,m} + p_7 u_{n,m+1}u_{n+1,m+1} + p_8 u_{n,m}u_{n+1,m+1} \\ & + p_9 u_{n+1,m}u_{n,m+1} + p_{10} u_{n+1,m}u_{n+1,m+1} + p_{11} u_{n,m}u_{n,m+1} \\ & + p_{12} u_{n,m} + p_{13} u_{n+1,m} + p_{14} u_{n,m+1} + p_{15} u_{n+1,m+1} + p_{16} = 0. \end{aligned} \tag{2.4.196}$$

In (2.4.196) the $p_i, i=1,\cdots,16$ coefficients have the following expression:

$$p_i = p_{i,0} + p_{i,1}(-1)^n + p_{i,2}(-1)^m + p_{i,3}(-1)^{n+m}, \quad i=1,\dots,16.$$

If we impose the non autonomous Klein symmetry condition (2.4.195) to Q with $\tau=\tau'=1$ the 64 coefficients of (2.4.196) turn out to be related among themselves and we can choose among them 16 independent coefficients. In term of the 16 independent coefficients (2.4.196) reads:

$$\begin{aligned} Q = {} & a_1 u_{n,m}u_{n+1,m}u_{n,m+1}u_{n+1,m+1} \\ & + \left[a_{2,0} - (-1)^n a_{2,1} - (-1)^m a_{2,2} + (-1)^{n+m} a_{2,3}\right] u_{n,m}u_{n,m+1}u_{n+1,m+1} \\ & + \left[a_{2,0} + (-1)^n a_{2,1} - (-1)^m a_{2,2} - (-1)^{n+m} a_{2,3}\right] u_{n+1,m}u_{n,m+1}u_{n+1,m+1} \\ & + \left[a_{2,0} + (-1)^n a_{2,1} + (-1)^m a_{2,2} + (-1)^{n+m} a_{2,3}\right] u_{n,m}u_{n+1,m}u_{n+1,m+1} \\ & + \left[a_{2,0} - (-1)^n a_{2,1} + (-1)^m a_{2,2} - (-1)^{n+m} a_{2,3}\right] u_{n,m}u_{n+1,m}u_{n,m+1} \\ & + \left[a_{3,0} - (-1)^m a_{3,2}\right] u_{n,m}u_{n+1,m} \\ & + \left[a_{3,0} + (-1)^m a_{3,2}\right] u_{n,m+1}u_{n+1,m+1} \\ & + \left[a_{4,0} - (-1)^{n+m} a_{4,3}\right] u_{n,m}u_{n+1,m+1} \\ & + \left[a_{4,0} + (-1)^{n+m} a_{4,3}\right] u_{n+1,m}u_{n,m+1} \\ & + \left[a_{5,0} - (-1)^n a_{5,1}\right] u_{n+1,m}u_{n+1,m+1} \\ & + \left[a_{5,0} + (-1)^n a_{5,1}\right] u_{n,m}u_{n,m+1} \\ & + \left[a_{6,0} + (-1)^n a_{6,1} - (-1)^m a_{6,2} - (-1)^{n+m} a_{6,3}\right] u_{n,m} \\ & + \left[a_{6,0} - (-1)^n a_{6,1} - (-1)^m a_{6,2} + (-1)^{n+m} a_{6,3}\right] u_{n+1,m} \\ & + \left[a_{6,0} + (-1)^n a_{6,1} + (-1)^m a_{6,2} + (-1)^{n+m} a_{6,3}\right] u_{n,m+1} \\ & + \left[a_{6,0} - (-1)^n a_{6,1} + (-1)^m a_{6,2} - (-1)^{n+m} a_{6,3}\right] u_{n+1,m+1} \\ & + a_7 = 0. \end{aligned} \tag{2.4.197}$$

Upon the substitution $a_{2,1} = a_{2,2} = a_{2,3} = a_{3,2} = a_{4,3} = a_{5,1} = a_{6,1} = a_{6,2} = a_{6,3} = 0$, (2.4.197) reduces to the $Q_{\rm V}$ equation (2.4.136). Therefore we will call Q given by (2.4.197) the non autonomous $Q_{\rm V}^{(n,m)}$ equation.

Eq.	$a_{3,0}$	$a_{3,2}$	$a_{4,0}$	$a_{4,3}$	$a_{5,0}$	$a_{5,1}$	$a_{6,0}$
${}_rH_1^\pi$	1	0	$\frac{1}{2}\pi(\alpha-\beta)$	$\frac{1}{2}\pi(\alpha-\beta)$	-1	0	0
${}_rH_2^\pi$	1	0	$2\pi(\beta-\alpha)$	$2\pi(\beta-\alpha)$	-1	0	$-(\alpha-\beta)(\pi\alpha+1+\pi\beta)$
${}_rH_3^\pi$	α	0	$\frac{1}{2}\frac{\pi\left(\beta^2-\alpha^2\right)}{\alpha\beta}$	$\frac{1}{2}\frac{\pi\left(\beta^2-\alpha^2\right)}{\alpha\beta}$	$-\beta$	0	0
${}_tH_1^\pi$	$-\frac{1}{2}\alpha_2\pi^2$	$-\frac{1}{2}\alpha_2\pi^2$	-1	0	1	0	0
${}_tH_2^\pi$	$\pi\alpha_2$	$\pi\alpha_2$	-1	0	1	0	$\frac{1}{2}\alpha_2\left[2+\pi(2\alpha_2+\alpha_3)\right\|$
${}_tH_3^\pi$	$\frac{1}{2}\frac{\pi^2\left(1-\alpha_2{}^2\right)}{\alpha_3\alpha_2}$	$\frac{1}{2}\frac{\pi^2\left(1-\alpha_2{}^2\right)}{\alpha_3\alpha_2}$	α_2	0	-1	0	0
${}_1D_2$	$\frac{1}{2}\delta_1$	$\frac{1}{2}\delta_1$	$\frac{1}{2}$	$-\frac{1}{2}$	0	0	$\frac{1}{2}-\frac{1}{4}(\delta_1-\delta_2)$
${}_2D_2$	$\frac{1}{2}$	$-\frac{1}{2}$	$\frac{1}{2}\delta_1$	$\frac{1}{2}\delta_1$	0	0	$\frac{1}{2}-\frac{1}{4}(\delta_1-\delta_2+\delta_1\lambda)$
${}_3D_2$	$\frac{1}{2}$	$-\frac{1}{2}$	0	0	$\frac{1}{2}\delta_1$	$-\frac{1}{2}\delta_1$	$\frac{1}{2}-\frac{1}{4}(\delta_1-\delta_2+\delta_1\lambda)$
D_3	$\frac{1}{2}$	$\frac{1}{2}$	$\frac{1}{2}$	$\frac{1}{2}$	$\frac{1}{2}$	$-\frac{1}{2}$	$\frac{1}{4}$
${}_1D_4$	$\frac{1}{2}\delta_2$	$\frac{1}{2}\delta_2$	1	0	$\frac{1}{2}\delta_1$	$-\frac{1}{2}\delta_1$	0
${}_2D_4$	1	0	$\frac{1}{2}\delta_2$	$\frac{1}{2}\delta_2$	$\frac{1}{2}\delta_1$	$-\frac{1}{2}\delta_1$	0

Eq.	$a_{6,1}$	$a_{6,2}$	$a_{6,3}$	a_7
rH_1^π	0	0	0	$\beta-\alpha$
${}_rH_2^\pi$	0	0	$\pi\left(\beta^2-\alpha^2\right)$	$-(\alpha-\beta)\left(2\pi\alpha^2+\alpha+2\pi\beta^2+\beta\right)$
${}_rH_3^\pi$	0	0	0	$\delta\left(\alpha^2-\beta^2\right)$
${}_tH_1^\pi$	0	0	0	$-\alpha_2$
${}_tH_2^\pi$	0	$\pi\alpha_2\alpha_3+\frac{1}{2}\pi\alpha_2{}^2$	0	$\alpha_2\left[\alpha_2+2\alpha_3+\pi\left(\alpha_2+\alpha_3\right)^2\right]$
${}_tH_3^\pi$	0	0	0	$\delta^2\alpha_3\left(1-\alpha_2^2\right)$
${}_1D_2$	$\frac{1}{4}(\delta_2-\delta_1)$	$-\frac{1}{4}(\delta_1+\delta_2)$	$\frac{1}{2}-\frac{1}{4}(\delta_1+\delta_2)$	0
${}_2D_2$	$\frac{1}{4}(\delta_1\lambda-\delta_1+\delta_2)$	$\frac{1}{2}-\frac{1}{4}(\delta_1-\delta_1\lambda+\delta_2)$	$-\frac{1}{4}(\delta_1+\delta_1\lambda+\delta_2)$	$-\delta_1\delta_2\lambda$
${}_3D_2$	$\frac{1}{4}(\delta_1+\delta_1\lambda+\delta_2)$	$\frac{1}{2}-\frac{1}{4}(\delta_1-\delta_1\lambda+\delta_2)$	$\frac{1}{4}(\delta_1-\delta_1\lambda-\delta_2)$	$-\delta_1\delta_2\lambda$
D_3	$\frac{1}{4}$	$-\frac{1}{4}$	$-\frac{1}{4}$	0
${}_1D_4$	0	0	0	δ_3
${}_2D_4$	0	0	0	δ_3

TABLE 2.3. Identification of the coefficients of the non autonomous Q_{V} equation with those of the Boll's equations (2.4.159,2.4.156, 2.4.157). Since $a_1 = a_{2,i} = 0$ for every equation these coefficients are absent in the Table [reprinted from [343] licensed under Creative Commons NonCommercial 4.0 International License (https://creativecommons.org/licenses/by-nc/4.0/)].

If we impose the non autonomous Klein symmetry condition (2.4.195) with the choice $\tau = 1$ and $\tau' = -1$ we will get an expression which can be reduced to (2.4.197) by multiplying by $(-1)^n$ and redefining the coefficients. In an analogous manner the two remaining cases $\tau = -1$, $\tau' = 1$ and $\tau = \tau' = -1$ can be identified with the case $\tau = \tau' = 1$ multiplying by $(-1)^n$ and $(-1)^{n+m}$ respectively and redefining the coefficients. Therefore the only equation belonging to the class of the lattice equation possessing the non autonomous Klein symmetries is just the non autonomous $Q_{\mathrm{V}}^{(n,m)}$ equation (2.4.197).

We note that the non autonomous $Q_{\mathrm{V}}^{(n,m)}$ equation contains as particular cases the rhombic H^4 equations, the trapezoidal H^4 equations (2.4.156) and the H^6 equations (2.4.157). The explicit identification of the coefficients of such equations is given in Table 2.3.

4.7.5. *Symmetries of Boll equations.* It was proved in Section 2.6.1.3 [492] that the three point symmetries of the equations belonging to the ABS classification [22], found systematically in [699], are all particular cases of the YdKN equation (2.4.129). Here we will show that the three point generalized symmetries of all the equations coming from the classification of Boll [23, 112–114], which extends the ABS one [22], are all particular cases of the YdKN or the non autonomous YdKN. In [550] we presented a non autonomous

YdKN

$$\dot{u}_n = f_n = \frac{A_n u_{n+1} u_{n-1} + B_n(u_{n+1} + u_{n-1}) + C_n}{u_{n+1} - u_{n-1}}, \tag{2.4.198}$$

where

$$\begin{aligned} A_n &= a_n u_n^2 + 2b_n u_n + c_n , \\ B_n &= \tilde{b}_n u_n^2 + d_n u_n + \tilde{e}_n , \\ C_n &= \tilde{c}_n u_n^2 + 2\delta_n u_n + f_n , \end{aligned} \tag{2.4.199}$$

where a_n, d_n, f_n must be constant, while b_n, c_n, e_n must be two-periodic constants so that $\tilde{b}_n = b_{n+1}$, $\tilde{c}_n = c_{n+1}$ and $\tilde{e}_n = e_{n+1}$. In particular we will present the symmetries of all the classes of equations H^4 and H^6, noting that the symmetries of the rhombic H^4 were found firstly in [839]. In Appendix C we present further indications of the integrability of the non autonomous YdKN (2.4.198,2.4.199) based on the algebraic entropy test and use the same criterion to prove the integrability of the other non autonomous equations of the H^4 and H^6 classes.

Rhombic H^4 equations. The generalized symmetries of the rhombic H^4 equations (2.4.159) [839] are given by the following generators:

$$\hat{X}_n^{r,H_1^\pi} = \frac{1 - \pi\left(\chi_{n+m} u_{n+1,m} u_{n-1,m} + \chi_{n+m+1} u_{n,m}^2\right)}{u_{n+1,m} - u_{n-1,m}} \partial_{u_{n,m}}, \tag{2.4.200a}$$

$$\hat{X}_m^{r,H_1^\pi} = \frac{1 - \pi\left(\chi_{n+m} u_{n,m+1} u_{n,m-1} + \chi_{n+m+1} u_{n,m}^2\right)}{u_{n,m+1} - u_{n,m-1}} \partial_{u_{n,m}}, \tag{2.4.200b}$$

$$\begin{aligned} \hat{X}_n^{r,H_2^\pi} = \Bigg[& \frac{\left(1 - 4\pi\alpha\chi_{n+m+1}\right)\left(u_{n+1,m} + u_{n-1,m}\right) - 4\pi\chi_{n+m} u_{n+1,m} u_{n-1,m}}{u_{n+1,m} - u_{n-1,m}} \\ & + \frac{2\alpha - 4\pi\alpha^2 - 4\pi\chi_{n+m+1} u_{n,m}^2 + \left(1 - 4\pi\alpha\chi_{n+m+1}\right) u_{n,m}}{u_{n+1,m} - u_{n-1,m}} \Bigg] \partial_{u_{n,m}} \end{aligned} \tag{2.4.200c}$$

$$\begin{aligned} \hat{X}_m^{r,H_2^\pi} = \Bigg[& \frac{\left(1 - 4\pi\beta\chi_{n+m+1}\right)\left(u_{n,m+1} + u_{n,m-1}\right) - 4\pi\chi_{n+m} u_{n,m+1} u_{n,m-1}}{u_{n,m+1} - u_{n,m-1}} \\ & + \frac{2\beta - 4\pi\beta^2 - 4\pi\chi_{n+m+1} u_{n,m}^2 + \left(1 - 4\pi\beta\chi_{n+m+1}\right) u_{n,m}}{u_{n,m+1} + u_{n,m-1}} \Bigg] \partial_{u_{n,m}} \end{aligned} \tag{2.4.200d}$$

$$\hat{X}_n^{r,H_3^\pi} = \left[\frac{1}{2} \frac{u_{n,m}\left(u_{n+1,m} + u_{n-1,m}\right) + 2\delta\alpha}{u_{n+1,m} - u_{n-1,m}} - \frac{\pi}{\alpha} \frac{\left(\chi_{n+m} u_{n+1,m} u_{n-1,m} + \chi_{n+m+1} u_{n,m}^2\right)}{u_{n+1,m} - u_{n-1,m}} \right] \partial_{u_{n,m}} \tag{2.4.200e}$$

$$\hat{X}_m^{r,H_3^\pi} = \left[\frac{1}{2} \frac{u_{n,m}\left(u_{n,m+1} + u_{n,m-1}\right) + 2\delta\beta}{u_{n,m+1} + u_{n,m-1}} - \frac{\pi}{\beta} \frac{\left(\chi_{n+m} u_{n,m+1} u_{n,m-1} + \chi_{n+m+1} u_{n,m}^2\right)}{u_{n,m+1} + u_{n,m-1}} \right] \partial_{u_{n,m}}. \tag{2.4.200f}$$

As stated in [839] the fluxes of the symmetries (2.4.200) are readily identified with the corresponding cases of the non autonomous YdKN equation (2.4.198, 2.4.199), see Table 2.4.

Eq.	k	a	b_k	c_k	d	e_k	f
${}_rH_1^\pi$	n	0	0	$-\pi\chi_{n+m}$	0	0	1
	m	0	0	$-\pi\chi_{n+m}$	0	0	1
${}_rH_2^\pi$	n	0	0	$-4\pi\chi_{n+m}$	0	$1-4\pi\alpha\chi_{n+m+1}$	$2\alpha-4\pi\alpha^2$
	m	0	0	$-4\pi\chi_{n+m}$	0	$1-4\pi\beta\chi_{n+m+1}$	$2\beta-4\pi\beta^2$
${}_rH_3^\pi$	n	0	0	$-\frac{\pi\chi_{n+m}}{\alpha}$	$\frac{1}{2}$	0	$\delta\alpha$
	m	0	0	$-\frac{\pi\chi_{n+m}}{\beta}$	$\frac{1}{2}$	0	$\delta\beta$

TABLE 2.4. Identification of the coefficients in the symmetries of the rhombic H^4 equations with those of the non autonomous YdKN equation [reprinted from [336]].

Trapezoidal H^4 equations. We can easily calculate the three-point generalized symmetries of ${}_tH_2^\pi$ (2.4.156b) and of ${}_tH_3^\pi$ (2.4.156c):

(2.4.201a)
$$\hat{X}_n^{{}_tH_2^\pi} = \left[\frac{(u_{n,m}+\pi\alpha_2^2\chi_m)(u_{n+1,m}+u_{n-1,m})-u_{n+1,m}u_{n-1,m}}{u_{n+1,m}-u_{n-1,m}} - \frac{u_{n,m}^2-2\pi\chi_m\alpha_2^2u_{n,m}-\alpha_2^2+4\pi\chi_m\alpha_2^3+8\pi\chi_m\alpha_2^2\alpha_3+\pi^2\chi_m\alpha_2^4}{u_{n+1,m}-u_{n-1,m}} \right] \partial_{u_{n,m}},$$

(2.4.201b)
$$\hat{X}_m^{{}_tH_2^\pi} = \left[\frac{\left[\frac{1}{2}-\pi(\alpha_2+\alpha_3)\chi_m\right](u_{n,m+1}+u_{n,m-1})-\pi\chi_m u_{n,m+1}u_{n,m-1}}{u_{n,m+1}-u_{n,m-1}} - \frac{\pi\chi_{m+1}u_{n,m}^2-\left[1-2\pi(\alpha_2+\alpha_3)\chi_{m+1}\right]u_{n,m}+\alpha_3+\pi\left(\alpha_2+\alpha_3\right)^2}{u_{n,m+1}-u_{n,m-1}} \right] \partial_{u_{n,m}},$$

(2.4.201c)
$$\hat{X}_n^{{}_tH_3^\pi} = \left[\frac{\frac{1}{2}\alpha_2(1+\alpha_2^2)u_{n,m}(u_{n+1,m}+u_{n-1,m})-\alpha_2^2u_{n+1,m}u_{n-1,m}}{u_{n+1,m}-u_{n-1,m}} - \frac{\alpha_2^2u_{n,m}^2+\pi^2\delta^2(1-\alpha_2^2)^2\chi_m}{u_{n+1,m}-u_{n-1,m}} \right] \partial_{u_{n,m}},$$

(2.4.201d)
$$\hat{X}_m^{{}_tH_3^\pi} = \left[\frac{\frac{1}{2}\alpha_3u_{n,m}(u_{n,m+1}+u_{n,m-1})-\pi^2\chi_m u_{n,m+1}u_{n,m-1}}{u_{n,m+1}-u_{n,m-1}} - \frac{\pi^2\chi_{m+1}u_{n,m}^2+\alpha_3^2\delta^2}{u_{n,m+1}-u_{n,m-1}} \right] \partial_{u_{n,m}},$$

The symmetries in the n and m directions and the linearizations of the ${}_tH_1^\pi$ equation (2.4.156a) have been presented in [340]. Their peculiarity is that they are determined by two arbitrary functions of one continuous variable and one lattice index and by arbitrary functions of the lattice indexes. This is the first time that we find a lattice equation whose generalized symmetries depend on arbitrary functions. Almost surely this peculiarity is related to the very specific way in which ${}_tH_1^\pi$ is linearizable. Here we present only the sub-cases which are related to the YdKN equation in its autonomous or non autonomous form.

The general symmetry in the n direction is:

$$(2.4.202)\qquad \hat{X}_n^{{}_iH_1^\pi} = \chi_m \left\{ \frac{\alpha_2\left(\upsilon^2+\pi^2\alpha_2^2\right)}{(r-\upsilon)(r+\upsilon)} B_n\left(\frac{\alpha_2}{r}\right) - \frac{\alpha_2\left(r^2+\pi^2\alpha_2^2\right)}{(r-\upsilon)(r+\upsilon)} B_{n-1}\left(\frac{\alpha_2}{\upsilon}\right) \right.$$
$$\left. + \left[u_{n,m} - \frac{\left(r^2+\pi^2\alpha_2^2\right)\upsilon}{(r-\upsilon)(r+\upsilon)}\right]\alpha + \gamma_m \right\} \partial_{u_{n,m}} + \chi_{m+1}\left[\frac{s^2t^2}{(s-t)(s+t)}\left(B_n(s)\right.\right.$$
$$\left.\left. -B_{n-1}(t)\right) - \frac{s^2 t}{(s-t)(s+t)}\alpha + \delta_m\right]\left(1+\pi^2u_{n,m}^2\right)\partial_{u_{n,m}}, \quad r \doteq u_{n+1,m}-u_{n,m},$$
$$s \doteq \frac{u_{n+1,m}-u_{n,m}}{1+\pi^2 u_{n+1,m}u_{n,m}}, \quad t \doteq \frac{u_{n,m}-u_{n-1,m}}{1+\pi^2 u_{n-1,m}u_{n,m}}, \quad \upsilon \doteq u_{n,m}-x_{u-1,m},$$

where $B_n(x)$, γ_m and δ_m are generic functions of their arguments and α is an arbitrary parameter. When $B_n(x) = -1/x$, $\alpha = \gamma_m = \delta_m = 0$, we get

$$(2.4.203)\qquad \hat{X}_n^{{}_iH_1^\pi} = \left[\frac{\left(u_{n+1,m}-u_{n,m}\right)\left(u_{n,m}-u_{n-1,m}\right)}{u_{n+1,m}-u_{n-1,m}} - \chi_m \frac{\pi^2\alpha_2^2}{u_{n+1,m}-u_{n-1,m}}\right]\partial_{u_{n,m}}.$$

The general symmetry in the m direction is:

$$(2.4.204)\qquad \hat{X}_m^{{}_iH_1^\pi} = [\chi_m \left(B_m\left(\frac{u_{n,m+1}-u_{n,m-1}}{1+\pi^2 u_{n,m+1}u_{n,m-1}}\right) + \kappa_m\right)$$
$$+ \chi_{m+1}\left(1+\pi^2 u_{n,m}^2\right)\left(C_m\left(u_{n,m+1}-u_{n,m-1}\right)+\lambda_m\right)]\partial_{u_{n,m}}.$$

When $B_m(t) = 1/t, C_m(t) = 1/t$ and $\kappa_m = \lambda_m = 0$ (2.4.204) becomes

$$(2.4.205)\qquad \hat{X}_m^{{}_iH_1^\pi} = [\chi_m \frac{1+\pi^2 u_{n,m+1}u_{n,m-1}}{u_{n,m+1}-u_{n,m-1}} + \chi_{m+1}\frac{1+\pi^2u_{n,m}^2}{u_{n,m+1}-u_{n,m-1}}]\partial_{u_{n,m}}.$$

Let us notice that the symmetries (2.4.201, 2.4.203) in the n direction are sub-cases of the original YdKN equation, see Table 2.5. As χ_m and χ_{m+1} depend on the other lattice index, they can be treated like a parameter which is either 0 or 1.

H^6 **equations.** Now we consider the equations of the family H^6 introduced in [113, 114]. The three forms of the equation D_2 (2.4.157b,2.4.157c,2.4.157d), ${}_iD_2$ $i = 1,2,3$, possess the following three-point generalized symmetries in the n direction and three-point

Eq.	k	a	b_k	c_k	d
${}_tH_1^\pi$	n	0	0	-1	1
	m	0	0	$\pi^2\chi_m$	0
${}_tH_2^\pi$	n	0	0	-1	1
	m	0	0	$-\pi\chi_m$	0
${}_tH_3^\pi$	n	0	0	$-\alpha_2^2$	$\frac{1}{2}\alpha_2(1+\alpha_2^2)$
	m	0	0	$-\pi^2\chi_m$	$\frac{1}{2}\alpha_3$

Eq.	e_k	f
${}_tH_1^\pi$	0	$-\pi^2\alpha_2^2\chi_m$
	0	2
${}_tH_2^\pi$	$\pi\alpha_2^2\chi_m$	$\alpha_2^2-\pi\alpha_2^2\left(4\alpha_2+8\alpha_3+\pi\alpha_2^2\right)\chi_m$
	$\frac{1}{2}-\pi(\alpha_2+\alpha_3)\chi_{m+1}$	$-\alpha_3-\pi\left(\alpha_2+\alpha_3\right)^2$
${}_tH_3^\pi$	0	$-\pi^2\delta^2\chi_m(1-\alpha_2^2)^2$
	0	$-\alpha_3^2\delta^2$

TABLE 2.5. Identification of the coefficients in the symmetries of the trapezoidal H^4 equations with those of the YdKN equation. In the direction n the YdKN is autonomous while in the m direction is non autonomous. Here the symmetries of ${}_tH_1^\pi$ in the m direction are the subcase (2.4.205) of (2.4.204) while those in the n direction are the subcase (2.4.203) of (2.4.202) [reprinted from [336]].

generalized symmetries in the m direction:

$$\hat{X}_n^{1D_2} = \left[\frac{\left(\chi_n\chi_{m+1}-\delta_1\chi_n\chi_m\right)\left(u_{n,m+1}+u_{n-1,m}\right)}{u_{n+1,m}-u_{n-1,m}} + \frac{\left(\chi_n\chi_{m+1}-\delta_1\chi_n\chi_m-\delta_1\delta_2\right)\chi_{n+1}\chi_m u_{n-1,m}}{u_{n+1,m}-u_{n-1,m}} + \frac{(\chi_{n+m}-\delta_1\chi_m-\delta_1\delta_2\chi_n\chi_m)u_{n,m}+\delta_2\chi_{m+1}}{u_{n+1,m}-u_{n-1,m}}\right]\partial_{u_{n,m}}, \tag{2.4.206a}$$

$$\hat{X}_m^{1D_2} = \left[\frac{\delta_1\chi_{n+1}\chi_m u_{n,m+1}u_{n,m-1}+\chi_{n+m+1}(u_{n,m+1}+u_{n,m-1})}{u_{n,m+1}-u_{n,m-1}} + \frac{\delta_1\chi_m u_{n,m+1}+\delta_1\delta_2\chi_{n+1}\chi_m u_{n,m-1}+\delta_1\chi_{n+1}\chi_{m+1}u_{n,m}^2}{u_{n,m+1}-u_{n,m-1}} + \frac{\left[\chi_{n+m}+\delta_1\left(\chi_n\chi_{m+1}-\chi_{n+1}\chi_{m+1}\right)+\delta_1\delta_2\chi_{n+1}\chi_{m+1}\right]u_{n,m}}{u_{n,m+1}-u_{n,m-1}} - \frac{\delta_2(\delta_1-1)\chi_{n+1}}{u_{n,m+1}-u_{n,m-1}}\right]\partial_{unm}, \tag{2.4.206b}$$

(2.4.206c) $$\hat{X}_n^{2\,D_2} = \left[\frac{\left(\chi_{n+1}\chi_{m+1}\delta_1 + \chi_{n+1}\chi_{m+1}\delta_1\delta_2 - \chi_{n+1}\chi_{m+1}\right)u_{n+1,m}}{u_{n+1,m} - u_{n-1,m}} + \frac{\left(\chi_n\chi_m\delta_1 - \chi_n\chi_{m+1}\right)u_{n-1,m}}{u_{n+1,m} - u_{n-1,m}} + \frac{\left(\delta_1\chi_{n+m+1} - \chi_{m+1} + \delta_1\delta_2\chi_n\chi_{m+1}\right)u_{n,m} - (\delta_1 - 1)\chi_m}{u_{n+1,m} - u_{n-1,m}}\right]\partial_{u_{n,m}},$$

(2.4.206d) $$\hat{X}_m^{2\,D_2}\left[\frac{\chi_{n+1}\chi_{m+1}\delta_1 u_{n,m+1}u_{n,m-1} + \left(\delta_1\delta_2\chi_{n+1}\chi_{m+1} + \chi_n\chi_{m+1}\right)u_{n,m+1}}{u_{n,m+1} - u_{n,m-1}} + \frac{+\left(\delta_1\chi_n\chi_m + \chi_{n+1}\chi_{m+1} - \delta_1\chi_{n+1}\chi_{m+1}\right)u_{n,m-1}}{u_{n,m+1} - u_{n,m-1}} + \frac{\delta_1\chi_{n+1}\chi_m u_{n,m}^2 + \left[\chi_n\chi_{m+1} + (\delta_2 - 1)\chi_{n+1}\chi_m + \chi_m\right]u_{n,m}}{u_{n,m+1} - u_{n,m-1}} + \frac{\delta_2(1-\delta_2)\chi_{n+1} - \delta_1\lambda\chi_n}{u_{n,m+1} - u_{n,m-1}}\right]\partial_{u_{n,m}},$$

(2.4.206e) $$\hat{X}_n^{3\,D_2} = \left[\frac{\left(\delta_1\chi_n\chi_{m+1} + \delta_1\delta_2\chi_n\chi_{m+1} - \chi_n\chi_{m+1}\right)u_{n+1,m}}{u_{n+1,m} - u_{n-1,m}} + \frac{\left(\chi_n\chi_m\delta_1 - \chi_{n+1}\chi_{m+1}\right)u_{n-1,m}}{u_{n+1,m} - u_{n-1,m}} + \frac{\left(\delta_1\chi_{n+1}\chi_{m+1}\delta_2 + \chi_{n+1}\delta_1 - \chi_{m+1}\right)u_{n,m} + (1-\delta_1)\chi_m}{u_{n+1,m} - u_{n-1,m}}\right]\partial u_{n,m},$$

(2.4.206f) $$\hat{X}_m^{3\,D_2} = \left[\frac{\left(1 - \delta_1 - \delta_1\delta_2\right)\chi_n\chi_{m+1}u_{n,m+1} + \delta_2\chi_{n+1}}{u_{n,m+1} - u_{n,m-1}} + \frac{\left(\chi_{n+1}\chi_{m+1} - \chi_n\chi_m\delta_1\right)u_{n,m-1}}{u_{n,m+1} - u_{n,m-1}} + \frac{\left(\chi_m - \delta_1\chi_n - \delta_1\delta_2\chi_n\chi_m\right)u_{n,m}}{u_{n,m+1} - u_{n,m-1}} - \frac{\lambda\delta_1(1 - \delta_1 - \delta_1\delta_2)\chi_n}{u_{n,m+1} - u_{n,m-1}}\right]\partial_{unm}.$$

It can be readily proved that these symmetries are not non autonomous YdKN equations (2.4.198, 2.4.199), however the equations ${}_iD_2$ possess also the following point symmetries:

(2.4.207a) $$\hat{Y}_1^{1\,D_2} = \left(\chi_n\chi_m + \chi_n\chi_{m+1} + \chi_{n+1}\chi_m\right)u_{n,m}\partial_{u_{n,m}},$$

(2.4.207b) $$\hat{Y}_2^{1\,D_2} = \left[\delta_1\chi_n\chi_m + [1 - \delta_1(1+\delta_2)]\chi_{n+1}\chi_m + \chi_n\chi_{m+1}\right]\partial_{u_{n,m}},$$

(2.4.207c) $$\hat{Y}_1^{2D_2} = \left[\left(\chi_n\chi_m + \chi_n\chi_{m+1} + \chi_{n+1}\chi_m\right)u_{n,m} - \lambda\chi_n\chi_{m+1} + \lambda[1-\delta_1(1+\delta_2)]\chi_{n+1}\chi_{m+1}\right]\partial_{u_{n,m}},$$

(2.4.207d) $$\hat{Y}_2^{2D_2} = \left[\delta_1\chi_n\chi_m + \chi_n\chi_{m+1}[1-\delta_1(1+\delta_2)]\chi_{n+1}\chi_{m+1}\right]\partial_{u_{n,m}},$$

(2.4.207e) $$\hat{Y}_1^{3D_2} = \left[\left(\chi_n\chi_m + \chi_n\chi_{m+1} + \chi_{n+1}\chi_m\right)u_{n,m} - \lambda\chi_{n+1}\chi_{m+1} + \lambda[1-\delta_1(1+\delta_2)]\chi_{n+1}\chi_{m+1}\right]\partial_{u_{n,m}},$$

(2.4.207f) $$\hat{Y}_2^{3D_2} = \left[\delta_1\chi_n\chi_m + [1-\delta_1(1+\delta_2)]\chi_n\chi_{m+1} - \chi_{n+1}\chi_{m+1}\right]\partial_{u_{n,m}}.$$

As the symmetries (2.4.206) are not in the form of the YdKN equation (2.4.198, 2.4.199), we look for a linear combination:

(2.4.208) $$\hat{Z}_j^{iD_2} = \hat{X}_j^{iD_2} + K_1\hat{Y}_1^{iD_2} + K_2\hat{Y}_2^{iD_2}, \quad j = n, m;\ i = 1, 2, 3,$$

such that the resulting symmetries of equations ${}_iD_2$ will be in the form (2.4.198, 2.4.199).

Indeed it turns out that this is the case and the resulting identification with the proper constants K_1 and K_2 is displayed in Table 2.6. The fact that the ${}_iD_2$ equations admit point symmetries and generalized symmetries makes them a unique case among the equations of Boll classification.

The D_3 equation (2.4.157e) admits only the following three-point generalized symmetries[2]:

(2.4.209a) $$\hat{X}_n^{D_3} = \left[\frac{\chi_n\chi_m u_{n+1,m}u_{n-1,m} + \frac{1}{2}\left(\chi_{m+1} - \chi_{n+1}\chi_m\right)u_{n,m}(u_{n+1,m}+u_{n-1,m})}{u_{n+1,m}-u_{n-1,m}} + \frac{\chi_{n+1}\chi_m u_{n,m}^2 + \left(\chi_{m+1} - \chi_n\chi_m\right)u_{n,m}}{u_{n+1,m}-u_{n-1,m}}\right]\partial_{u_{n,m}},$$

(2.4.209b) $$\hat{X}_m^{D_3} = \left[\frac{\chi_n\chi_m u_{n,m+1}u_{n,m-1} + \frac{1}{2}\left(\chi_{n+1} - \chi_n\chi_{m+1}\right)u_{n,m}(u_{n,m+1}+u_{n,m-1})}{u_{n,m+1}-u_{n,m-1}} + \frac{\chi_n\chi_{m+1} u_{n,m}^2 + \left(\chi_{n+1} - \chi_n\chi_m\right)u_{n,m}}{u_{n,m+1}-u_{n,m-1}}\right]\partial_{u_{n,m}}$$

and no point symmetries. Also the two forms of D_4 possess only the following three point generalized symmetries:

(2.4.209c) $$\hat{X}_n^{1D_4} = \left[\frac{-\delta_1\chi_n u_{n+1,m}u_{n-1,m} - \frac{1}{2}u_{n,m}(u_{n+1,m}+u_{n-1,m})}{u_{n+1,m}-u_{n-1,m}} + \frac{-\delta_1\chi_{n+1}u_{n,m}^2 + \delta_2\delta_3\chi_m}{u_{n+1,m}-u_{n-1,m}}\right]\partial_{u_{n,m}},$$

---

[2]Note that the equation D_3 (2.4.157e) is invariant under the exchange $n \leftrightarrow m$ so the symmetry $X_m^{D_3}$ (2.4.209b) can be obtained from the symmetry $X_n^{D_3}$ (2.4.209b) performing such exchange.

Eq.	k	a	b_k	c_k	d
${}_1D_2$	n	0	0	0	0
	m	0	0	$-\chi_{n+1}\chi_m\delta_1$	0
${}_2D_2$	n	0	0	0	0
	m	0	0	$-\delta_1\chi_{n+1}\chi_{m+1}$	0
${}_3D_2$	n	0	0	0	0
	m	0	0	0	0

Eq.	e_k
${}_1D_2$	$\frac{1}{2}[\delta_1(1+\delta_2)-1]\chi_n\chi_m+\frac{1}{2}\chi_{n+1}\chi_m-\frac{1}{2}\chi_{n+1}\chi_{m+1}$
	$\frac{1}{2}(\delta_1(1-\delta_2-1)\chi_{n+1}\chi_{m+1}-\frac{1}{2}\chi_n\chi_m-\frac{1}{2}\delta_1\chi_{n+1}\chi_m$
${}_2D_2$	$\frac{1}{2}[1-\delta_1(1+\delta_2)]\chi_n\chi_{m+1}+\frac{1}{2}\chi_{n+1}\chi_{m+1}-\frac{1}{2}\delta_1\chi_{n+1}\chi_m$
	$\frac{1}{2}[\delta_1(1-\delta_2)-1]\chi_{n+1}\chi_m-\frac{1}{2}\chi_n\chi_m-\frac{1}{2}\delta_1\chi_{n+1}\chi_m$
${}_3D_2$	$\frac{1}{2}[\delta_1(1+\delta_2)-1]\chi_{n+1}\chi_{m+1}+\frac{1}{2}\chi_n\chi_{m+1}+\frac{1}{2}\delta_1\chi_{n+1}\chi_m$
	$\frac{1}{2}[\delta_1(1-\delta_2)-1]\chi_n\chi_m-\frac{1}{2}\chi_n\chi_{m+1}+\frac{1}{2}\delta_1\chi_{n+1}\chi_m$

Eq.	f	K_1	K_2
${}_1D_2$	$-\delta_2\chi_{m+1}$	0	$-1/2$
	$\delta_2(\delta_1-1)\chi_{n+1}$	0	$-1/2$
${}_2D_2$	$(\delta_1-1)\chi_m$	0	$-1/2$
	$\delta_2\left[\delta_1-1\right]\chi_{n+1}+\lambda\delta_1\chi_n$	0	$-1/2$
${}_3D_2$	$(1-\delta_1)\chi_m$	0	$1/2$
	$\delta_1\lambda[-\delta_1(1+\delta_2)]\chi_n-\delta_2\chi_{n+1}$	0	$1/2$

TABLE 2.6. Identification of the coefficients of the symmetries of the ${}_iD_2$ equations and value of the constants K_1 and K_2 in (2.4.208) in order to obtain non autonomous YdKN equations.

$$\hat{X}_m^{{}_1D_4}=\left[\frac{\chi_{m+1}u_{n,m+1}u_{n,m-1}+\frac{1}{2}u_{n,m}(u_{n,m+1}+u_{n,m-1})}{u_{n,m+1}-u_{n,m-1}}\right.$$
$$\left.+\frac{\delta_2\chi_m u_{n,m}^2-\delta_1\delta_3\chi_n}{u_{n,m+1}-u_{n,m-1}}\right]\partial_{u_{n,m}}, \tag{2.4.209d}$$

Eq.	k	a	b_k	c_k	d
D_3	n	0	0	$\chi_n\chi_m$	0
	m	0	0	$\chi_n\chi_m$	0
${}_1D_4$	n	0	0	$-\delta_1\left(\chi_n\chi_m+\chi_n\chi_{m+1}\right)$	$-\frac{1}{2}$
	m	0	0	$\delta_2\left(\chi_n\chi_m+\chi_{n+1}\chi_m\right)$	$\frac{1}{2}$
${}_2D_4$	n	0	0	$-\chi_n\chi_m\delta_1\delta_2$	$\frac{1}{2}$
	m	0	0	$\delta_2\left(\chi_n\chi_m+\chi_{n+1}\chi_{m+1}\right)$	$\frac{1}{2}$

Eq.	e_k	f
D_3	$\frac{1}{2}\left(\chi_n\chi_{m+1}+\chi_{n+1}\chi_{m+1}-\chi_n\chi_m\right)$	0
	$\frac{1}{2}\left(\chi_{n+1}\chi_m+\chi_{n+1}\chi_{m+1}-\chi_n\chi_m\right)$	0
${}_1D_4$	0	$\delta_2\delta_3\chi_m$
	0	$-\delta_1\delta_3\chi_n$
${}_2D_4$	0	δ_3
	0	$-\delta_1\delta_3\chi_n$

TABLE 2.7. Identification of the coefficients of the symmetries (2.4.209) for D_3, ${}_1D_4$ and ${}_2D_4$ with those of a non autonomous YdKN [reprinted from [336]].

$$\hat{X}_n^{{}_2D_4} = \left[\frac{-\delta_1\delta_2\chi_n\chi_m u_{n+1,m}u_{n-1,m}+\frac{1}{2}u_{n,m}(u_{n+1,m}+u_{n-1,m})}{u_{n+1,m}-u_{n-1,m}} + \frac{-\delta_1\delta_2\chi_{n+1}\chi_m u_{n,m}^2+\delta_3}{u_{n+1,m}-u_{n-1,m}}\right]\partial_{u_{n,m}}, \tag{2.4.209e}$$

$$\hat{X}_m^{{}_2D_4} = \left[\frac{\delta_2\chi_{n+m}u_{n,m+1}u_{n,m-1}+\frac{1}{2}u_{n,m}(u_{n,m+1}+u_{n,m-1})}{u_{n,m+1}-u_{n,m-1}} + \frac{\delta_2\chi_{n+m+1}u_{n,m}^2-\delta_1\delta_3\chi_n}{u_{n,m+1}-u_{n,m-1}}\right]\partial_{u_{n,m}}, \tag{2.4.209f}$$

and no point symmetries. Again the fluxes of the symmetries (2.4.209) can be readily identified with some specific form of the non autonomous YdKN equations (2.4.198, 2.4.199), see Table 2.7.

Symmetries of the $Q_{\mathrm{V}}^{(n,m)}$ equation. The generalized three-point symmetries of Q_{V} are given by (2.4.143u, ⋯, 2.4.143z) which, in Section 2.4.6.3, are shown to be subcases of the YdKN (2.4.129).

So, when dealing with the symmetries of the $Q_{\mathrm{V}}^{(n,m)}$ we should look for them as subcases of the non autonomous YdKN equation (2.4.198, 2.4.199). As an example let us consider the function $Q_{\mathrm{V}}(x, u, y, z; (-1)^n, (-1)^m)$ given by a non autonomization of (2.4.136) with respect to a strict Klein symmetry just as in (2.4.136). Choosing in (2.4.136) the coefficients as

$$a_1 = 1 + (-1)^n,\ a_2 = (-1)^n,\ a_5 = -1 + (-1)^n,\ a_4 = (-1)^n,$$

$$a_3 = 1 + 2(-1)^n,\ a_6 = 1 + (-1)^n,\ a_7 = 4 + 2(-1)^n,$$

the symmetries of this non autonomized Q_{V} (2.4.144) provide a non autonomous YdKN. In this case, performing the algebraic entropy test the equation turns out to be integrable. Its generalized symmetries, however, are not necessarily in the form of a non autonomous YdKN equation. A different non autonomous choice of the coefficients of (2.4.136), gives, by the algebraic entropy test, a non integrable equation. We thus conjecture the existence of a non autonomous extension of Q_V related to the non autonomous YdKN (2.4.198,2.4.199) in the same way as Q_V is related to YdKN. For more details see Example 2 in Section 3.4.1.2. Eq. (2.4.198, 2.4.199) give us just three equations which have many possible solutions.

Using

(2.4.210)
$$\frac{du_{n,m}}{dt} = \frac{h_n}{u_{n+1,m} - u_{n-1,m}} - \frac{1}{2}\partial_{u_{n+1,m}} h_n.$$

where:

(2.4.211)
$$\begin{aligned} h_n(u_{n,m}, u_{n+1,m}) = {} & Q_V^{(n,0)} \partial_{u_{n,m+1}} \partial_{u_{n+1,m+1}} Q_V^{(n,0)} \\ & - \left(\partial_{u_{n,m+1}} Q_V^{(n,0)}\right)\left(\partial_{u_{n+1,m+1}} Q_V^{(n,0)}\right) \end{aligned}$$

or

(2.4.212)
$$\frac{du_{n,m}}{dt} = \frac{h_m}{u_{n,m+1} - u_{n,m-1}} - \frac{1}{2}\partial_{u_{n,m+1}} h_m.$$

where:

(2.4.213)
$$\begin{aligned} h_m(u_{n,m}, u_{n,m+1}) = {} & Q_V^{(0,m)} \partial_{u_{n+1,m}} \partial_{u_{n+1,m+1}} Q_V^{(0,m)} \\ & - \left(\partial_{u_{n+1,m}} Q_V^{(0,m)}\right)\left(\partial_{u_{n+1,m+1}} Q_V^{(0,m)}\right) \end{aligned}$$

we get a version of the non autonomous YdKN (2.4.198, 2.4.199) or with n substituted by m. The proof that the non autonomous YdKN (2.4.198, 2.4.199) is effectively a symmetry of the non autonomous $Q_{\mathrm{V}}^{(n,m)}$ (2.4.197) encounters serious computational difficulties.

We can prove by a direct computation its validity for the following sub-cases:

- When $Q_{\mathrm{V}}^{(n,m)}$ equation is non autonomous with respect to one direction only, either n or m. All the trapezoidal ${}_tH^4$ equations belong to these two sub-classes;
- For all the H^6 equations, which are non autonomous in both directions.

Its validity for the autonomous Q_V and for all the rhombic ${}_rH^4$ equations was already showed before in Sections 2.4.7.4 and 2.4.7.5 [837, 839]. However we cannot prove its validity for the general case (2.4.197).

Here in the following we compute the connection formulas with the non autonomous YdKN (2.4.198, 2.4.199) for $Q_{\rm V}^{(n,m)}$ (2.4.197). For the n directional symmetry we have:

$$
\begin{aligned}
(2.4.214)\quad \alpha &= a_1 a_{3,0} - a_{2,0}^2 \\
&\quad + a_{2,1}^2 - a_{2,2}^2 + a_{2,3}^2 - (-1)^m(2a_{2,0}a_{2,2} - 2a_{2,1}a_{2,3} + a_1 a_{3,2}),
\end{aligned}
$$

$$
\begin{aligned}
\beta_0 &= \frac{1}{2}\{a_{2,0}(a_{3,0} - a_{5,0} - a_{4,0}) + a_1 a_{6,0} + a_{2,2}a_{3,2} - a_{2,3}a_{4,3} - a_{2,1}a_{5,1} \\
&\quad - (-1)^m[a_{2,2}(a_{5,0} + a_{3,0} + a_{4,0}) + a_{2,3}a_{5,1} + a_1 a_{6,2} + a_{2,0}a_{3,2} \\
&\quad + a_{2,1}a_{4,3}]\},
\end{aligned}
$$

$$
\begin{aligned}
\beta_1 &= \frac{1}{2}\{a_{2,1}(a_{3,0} - a_{4,0} + a_{5,0}) + a_{2,3}a_{3,2} - a_{2,2}a_{4,3} + a_{2,0}a_{5,1} - a_1 a_{6,1} \\
&\quad + (-1)^m[a_1 a_{6,3} - a_{2,3}(a_{3,0} + a_{4,0} - a_{5,0}) - a_{2,1}a_{3,2} - a_{2,0}a_{4,3} \\
&\quad + a_{2,2}a_{5,1}]\},
\end{aligned}
$$

$$
\begin{aligned}
\gamma_0 &= a_{2,0}a_{6,0} - a_{4,0}a_{5,0} - a_{2,1}a_{6,1} - a_{2,3}a_{6,3} + a_{2,2}a_{6,2} \\
&\quad - (-1)^m[a_{2,2}a_{6,0} - a_{4,3}a_{5,1} - a_{2,3}a_{6,1} + a_{2,0}a_{6,2} - a_{2,1}a_{6,3}],
\end{aligned}
$$

$$
\begin{aligned}
\gamma_1 &= a_{4,0}a_{5,1} + a_{2,1}a_{6,0} - a_{2,0}a_{6,1} + a_{2,3}a_{6,2} - a_{2,2}a_{6,3} \\
&\quad + (-1)^m[a_{2,2}a_{6,1} - a_{4,3}a_{5,0} - a_{2,3}a_{6,0} - a_{2,1}a_{6,2} + a_{2,0}a_{6,3}],
\end{aligned}
$$

$$
\begin{aligned}
\lambda &= \frac{1}{2}[a_{3,0}^2 - a_{4,0}^2 - a_{5,0}^2 + a_1 a_7 - a_{3,2}^2 + a_{4,3}^2 + a_{5,1}^2 \\
&\quad - 4(-1)^m(a_{2,2}a_{6,0} + a_{2,3}a_{6,1} + a_{2,0}a_{6,2} + a_{2,1}a_{6,3}],
\end{aligned}
$$

$$
\begin{aligned}
\delta_0 &= \frac{1}{2}\{a_{6,0}[a_{3,0} - a_{4,0} - a_{5,0}] + a_{2,0}a_7 + a_{5,1}a_{6,1} - a_{3,2}a_{6,2} + a_{4,3}a_{6,3} \\
&\quad + (-1)^m[a_{3,2}a_{6,0} + a_{4,3}a_{6,1} + a_{5,1}a_{6,3} - a_{6,2}(a_{3,0} + a_{4,0} + a_{5,0}) - a_{2,2}a_7]\},
\end{aligned}
$$

$$
\begin{aligned}
\delta_1 &= \frac{1}{2}\{a_{6,1}[a_{3,0} - a_{4,0} + a_{5,0}] - a_{5,1}a_{6,0} + a_{4,3}a_{6,2} - a_{3,2}a_{6,3} - a_{2,1}a_7 \\
&\quad + (-1)^m[a_{4,3}a_{6,0} + a_{3,2}a_{6,1} - a_{5,1}a_{6,2} + a_{6,3}(a_{5,0} - a_{3,0} - a_{4,0}) + a_{2,3}a_7\},
\end{aligned}
$$

$$
\epsilon = a_{3,0}a_7 - a_{6,0}^2 - a_{6,2}^2 + a_{6,3}^2 + a_{6,1}^2 - (-1)^m(2a_{6,0}a_{6,2} - 2a_{6,1}a_{6,3} - a_{3,2}a_7).
$$

The non autonomous $Q_{\rm V}^{(n,m)}$ is not symmetric in the exchange of n and m so its symmetries in the m direction are different for their dependence on the coefficients and so the connection formulas in the m direction are:

$$
\begin{aligned}
(2.4.215)\quad \alpha &= a_1 a_{5,0} - a_{2,0}^2 \\
&\quad - a_{2,1}^2 + a_{2,2}^2 + a_{2,3}^2 + (-1)^n(2a_{2,0}a_{2,1} - 2a_{2,2}a_{2,3} + a_1 a_{5,1}),
\end{aligned}
$$

$$
\begin{aligned}
\beta_0 &= \frac{1}{2}\{a_{2,0}(a_{5,0} - a_{3,0} - a_{4,0}) + a_1 a_{6,0} - a_{2,2}a_{3,2} - a_{2,3}a_{4,3} + a_{2,1}a_{5,1} \\
&\quad + (-1)^n[a_{2,1}(a_{5,0} + a_{3,0} + a_{4,0}) + a_{2,3}a_{3,2} + a_1 a_{6,1} + a_{2,0}a_{5,1} \\
&\quad + a_{2,2}a_{4,3}]\},
\end{aligned}
$$

$$\beta_1 = \frac{1}{2}\{a_{2,2}(a_{4,0} - a_{3,0} - a_{5,0}) - a_{2,3}a_{5,1} + a_{2,1}a_{4,3} - a_{2,0}a_{3,2} + a_1a_{6,2}$$
$$+ (-1)^n[a_1a_{6,3} + a_{2,3}(a_{3,0} - a_{4,0} - a_{5,0}) + a_{2,1}a_{3,2} - a_{2,0}a_{4,3}$$
$$- a_{2,2}a_{5,1}]\},$$

$$\gamma_0 = a_{2,0}a_{6,0} - a_{4,0}a_{3,0} + a_{2,1}a_{6,1} - a_{2,3}a_{6,3} - a_{2,2}a_{6,2}$$
$$- (-1)^n[a_{2,2}a_{6,3} + a_{4,3}a_{3,2} + a_{2,3}a_{6,2} - a_{2,0}a_{6,1} - a_{2,1}a_{6,0}],$$

$$\gamma_1 = a_{2,1}a_{6,3} - a_{4,0}a_{3,2} + a_{2,0}a_{6,2} - a_{2,3}a_{6,1} - a_{2,2}a_{6,0}$$
$$- (-1)^n[a_{2,2}a_{6,1} + a_{4,3}a_{3,0} + a_{2,3}a_{6,0} - a_{2,1}a_{6,2} - a_{2,0}a_{6,3}],$$

$$\lambda = \frac{1}{2}[a_{5,0}^2 - a_{4,0}^2 - a_{3,0}^2 + a_1a_7 + a_{3,2}^2 + a_{4,3}^2 - a_{5,1}^2$$
$$+ 4(-1)^n(a_{2,1}a_{6,0} + a_{2,0}a_{6,1} + a_{2,3}a_{6,2} + a_{2,2}a_{6,3})],$$

$$\delta_0 = \frac{1}{2}\{a_{6,0}[a_{5,0} - a_{4,0} - a_{3,0}] + a_{2,0}a_7 - a_{5,1}a_{6,1} + a_{3,2}a_{6,2} + a_{4,3}a_{6,3}$$
$$- (-1)^n[a_{3,2}a_{6,3} + a_{4,3}a_{6,2} + a_{5,1}a_{6,0} - a_{6,1}(a_{3,0} + a_{4,0} + a_{5,0}) - a_{2,1}a_7]\},$$

$$\delta_1 = \frac{1}{2}\{-a_{6,2}[a_{3,0} - a_{4,0} + a_{5,0}] + a_{3,2}a_{6,0} - a_{4,3}a_{6,1} + a_{5,1}a_{6,3} + a_{2,2}a_7$$
$$+ (-1)^n[a_{4,3}a_{6,0} - a_{3,2}a_{6,1} + a_{5,1}a_{6,2} + a_{6,3}(a_{3,0} - a_{5,0} - a_{4,0}) + a_{2,3}a_7\},$$

$$\epsilon = a_{5,0}a_7 - a_{6,0}^2 + a_{6,2}^2 + a_{6,3}^2 - a_{6,1}^2 - (-1)^n(2a_{6,2}a_{6,3} - 2a_{6,0}a_{6,1} + a_{5,1}a_7).$$

4.7.6. *Darboux integrability of trapezoidal H^4 and H^6 families of lattice equations: first integrals* [336, 345]. In the continuous case, a hyperbolic partial differential equation (PDE) in two variables

$$u_{xt} = f\left(x, t, u, u_t, u_x\right) \tag{2.4.216}$$

is said to be *Darboux integrable* if it possesses two independent first integrals T, X depending only on derivatives with respect to one variable:

$$\begin{aligned} T &= T\left(x, t, u, u_t, \dots, u_{nt}\right), & \left.\frac{dT}{dx}\right|_{u_{xt}=f} &\equiv 0, \\ X &= X\left(x, t, u, u_x, \dots, u_{mx}\right), & \left.\frac{dX}{dt}\right|_{u_{xt}=f} &\equiv 0, \end{aligned} \tag{2.4.217}$$

where $u_{kt} = \partial^k u/\partial t^k$ and $u_{kx} = \partial^k u/\partial x^k$ for every $k \in \mathbb{N}$. A Darboux integrable equation is C-integrable [141, 198, 765]. The method is based on the linear theory developed by Euler and Laplace [243, 469] and extended to the non linear case in the 19th and early 20th centuries [197, 199, 313, 324, 813]. The method was then used at the end of the 20th century mainly by Russian mathematicians as a source of new exactly solvable PDEs in two variables [765, 867, 869–873]. We note that in many papers Darboux integrability is defined as the stabilization to zero of the so-called Laplace chain of the linearized equation. It can be proved that the two definitions are equivalent [41, 428, 873].

The most famous Darboux integrable equation is the Liouville equation [567]:

$$u_{xt} = e^u \tag{2.4.218}$$

which possesses the two following first integrals:

(2.4.219) $$X = u_{xx} - \frac{1}{2}u_x^2, \quad T = u_{tt} - \frac{1}{2}u_t^2.$$

In the discrete setting Darboux integrability was introduced in [28], where it was used to obtain a discrete analogue of the Liouville equation (2.4.218). In [309] we find a list of Darboux integrable discrete equations on square lattice.

As in the continuous case, we say that a quad-graph equation, possibly non autonomous:

(2.4.220) $$Q_{n,m}\left(u_{n,m}, u_{n+1,m}, u_{n,m+1}, u_{n+1,m+1}\right) = 0,$$

is *Darboux integrable* if there exist two independent first integrals, one containing only shifts in the n direction and the other containing only shifts in the m direction. This means that there exist two functions:

(2.4.221a) $$W_1 = W_{1,n,m}(u_{n+l_1,m}, u_{n+l_1+1,m}, \ldots, u_{n+k_1,m}), \quad l_1 < k_1$$

(2.4.221b) $$W_2 = W_{2,n,m}(u_{n,m+l_2}, u_{n,m+l_2+1}, \ldots, u_{n,m+k_2}), \quad l_2 < k_2$$

such that the relations

(2.4.222a) $$(S_n - I)W_2 = 0,$$

(2.4.222b) $$(S_m - I)W_1 = 0$$

hold true identically on the solutions of (2.4.220). By I we denote the identity operator $If_{n,m} = f_{n,m}$ and S_n (and consequently S_m where the shift is in m) is given by (1.2.13). The numbers $k_i - l_i$, where $i = 1, 2$, are the *order* of the first integrals W_i.

We notice that the existence of first integrals implies that the transformations:

(2.4.223a) $$u_{n,m} \to \tilde{u}_{n,m} = W_{1,n,m},$$

(2.4.223b) $$u_{n,m} \to \hat{u}_{n,m} = W_{2,n,m}$$

bring the quad-graph equation (2.4.220) into trivial linear equations (2.4.222) [28]

(2.4.224a) $$\tilde{u}_{n,m+1} - \tilde{u}_{n,m} = 0,$$

(2.4.224b) $$\hat{u}_{n+1,m} - \hat{u}_{n,m} = 0.$$

Therefore any Darboux integrable equation is linearizable in two different ways. The transformations (2.4.223) along with the relations (2.4.224) imply

(2.4.225a) $$W_{1,n,m} = \lambda_n,$$

(2.4.225b) $$W_{2,n,m} = \rho_m,$$

where λ_n and ρ_m are arbitrary functions of the lattice variables n and m, respectively. The relations (2.4.225) can be seen as OΔE which must be satisfied by any solution $u_{n,m}$ of (2.4.220). However the transformations (2.4.223) and the OΔEs (2.4.225) may be quite complicated.

In the case of the trapezoidal H^4 and the H^6 equations [344] the equations (2.4.225) are valid and thus are linearizable. Therefore we can use Darboux integrability in order to obtain the general solutions of these equations.

To get the first integrals let us consider the operator

(2.4.226) $$Y_{-1} = S_m \frac{\partial}{\partial u_{n,m-1}} S_m^{-1}$$

and apply it to (2.4.222b), we obtain:

(2.4.227) $$Y_{-1} W_1 = 0.$$

The application of the operator Y_{-1} is to be understood in the following sense: first we must apply S_m^{-1} and then using the equation (2.4.220) express $u_{n+i,m-1}$ in terms of $u_{n+j,m}$ and $u_{n,m-1}$, considered as independent variables. Then we can differentiate with respect to $u_{n,m-1}$ and safely apply S_m [305]. Taking in (2.4.227) the coefficients of the various powers of $u_{n,m+1}$, we obtain a system of PDEs for W_1. If this is sufficient to determine W_1 up to arbitrary functions of a single variable, then we are done. Otherwise we can add other equations by considering the "higher-order" operators

$$Y_{-k} = S_m^k \frac{\partial}{\partial u_{n,m-1}} S_m^{-k}, \quad k \in \mathbb{N}, \tag{2.4.228}$$

which annihilate the difference consequence of (2.4.222b) given by $S_m^k W_1 = W_1$ and gives

$$Y_{-k} W_1 \equiv 0, \quad k \in \mathbb{N}, \tag{2.4.229}$$

with the same computational prescriptions as given above. We can add equations until we find a non-constant function[3] W_1 which depends on a *single* combination of the variables $u_{n,m+j_1}, \ldots, u_{n,m+k_1}$. If we find a non-constant solution W_1 of the equations generated by (2.4.226) and possibly (2.4.228), then we must insert it back into (2.4.222b) to specify it.

In the same way first integrals in the m-direction W_2 can be found by considering the operators

$$Z_{-k} = S_n^k \frac{\partial}{\partial u_{n-1,m}} S_n^{-k}, \quad k \in \mathbb{N}, \tag{2.4.230}$$

which provide the equations

$$Z_{-k} W_2 \equiv 0, \quad k \in \mathbb{N}. \tag{2.4.231}$$

In the case of non autonomous equations with two-periodic coefficients, we can assume that a decomposition analogue of the quad-graph equation (2.4.147) holds for the first integrals:

$$\begin{aligned} W_i &= \chi_n \chi_m W_i^{(+,+)} + \chi_{n+1} \chi_m W_i^{(-,+)} \\ &\quad + \chi_n \chi_{m+1} W_i^{(+,-)} + \chi_{n+1} \chi_{m+1} W_i^{(-,-)}, \end{aligned} \tag{2.4.232}$$

with χ_k given by (2.4.148). We can then derive from (2.4.229, 2.4.231) a set of equations for the functions $W_i^{(\pm,\pm)}$ by considering the even/odd points on the lattice. The final form of the functions W_i will be then fixed by substituting in (2.4.222) and separating again.

As an example of this procedure let us consider in detail the problem of finding the first integrals of the ${}_tH_1^{\pi}$ equation (2.4.156a) whose solution was obtained in Section 2.4.7.3 by direct inspection [341]. Since all the H^4 equations and the ${}_tH_1^{\pi}$, in particular, are non autonomous only in the direction m, we can consider a simplified version of (2.4.232):

$$W_i = \chi_m W_i^{(+)} + \chi_{m+1} W_i^{(-)}. \tag{2.4.233}$$

If we assume that $W_1 = W_{1,n,m}\left(u_{n,m}, u_{n+1,m}\right)$, then, separating the even and odd terms with respect to m in (2.4.227), we find the following equations:

$$\frac{\partial W_1^{(+)}}{\partial u_{n+1,m}} + \frac{\partial W_1^{(+)}}{\partial u_{n,m}} = 0, \tag{2.4.234a}$$

$$\left(1 + \pi^2 u_{n+1,m}^2\right) \frac{\partial W_1^{(-)}}{\partial u_{n+1,m}} + \left(1 + \pi^2 u_{n,m}^2\right) \frac{\partial W_1^{(-)}}{\partial u_{n,m}} = 0. \tag{2.4.234b}$$

[3] Obviously constant functions are trivial first integrals.

Their solution is:

$$W_1 = \chi_m F\left(u_{n+1,m} - u_{n,m}\right) + \chi_{m+1} G\left(\frac{u_{n+1,m} - u_{n,m}}{1 + \pi^2 u_{n,m} u_{n+1,m}}\right), \tag{2.4.235}$$

where F and G are arbitrary functions of their argument. Inserting (2.4.235) into the difference equation (2.4.222b), we obtain that F and G must satisfy the following identity:

$$G(\xi) = F\left(\frac{\alpha_2}{\xi}\right). \tag{2.4.236}$$

This yields the first integral

$$W_1 = \chi_m F\left(\frac{\alpha_2}{u_{n+1,m} - u_{n,m}}\right) + \chi_{m+1} F\left(\frac{u_{n+1,m} - u_{n,m}}{1 + \pi^2 u_{n,m} u_{n+1,m}}\right). \tag{2.4.237}$$

For the m-direction we may also suppose that our first integral $W_2 = W_{2,m}\left(u_{n,m}, u_{n,m+1}\right)$ is of the first order or a two-point first integral. It easy to see from (2.4.226) that we get only the trivial solution $W_2 = \text{constant}$. Therefore we consider the case of $W_2 = W_{2,m}\left(u_{n,m-1}, u_{n,m}, u_{n,m+1}\right)$, a three-point first integral. From (2.4.231) with $k = 1$, separating the even and odd terms with respect to m, we obtain:

$$\alpha_2\left(1 + \pi^2 u_{n,m+1}^2\right) \frac{\partial W_2^{(+)}}{\partial u_{n,m+1}} - \left[\left(u_{n,m} - u_{n+1,m}\right)^2 + \pi^2 \alpha_2^2\right] \frac{\partial W_2^{(+)}}{\partial u_{n,m}} + \alpha_2\left(1 + \pi^2 u_{n,m-1}^2\right) \frac{\partial W_2^{(+)}}{\partial u_{n,m-1}} = 0, \tag{2.4.238a}$$

$$\alpha_2\left(1 + \pi^2 u_{n+1,m}^2\right) \frac{\partial W_2^{(-)}}{\partial u_{n,m+1}} - \left(u_{n,m} - u_{n+1,m}\right)^2 \frac{\partial W_2^{(-)}}{\partial u_{n,m}} + \alpha_2\left(1 + \pi^2 u_{n+1,m}^2\right) \frac{\partial W_2^{(-)}}{\partial u_{n,m-1}} = 0. \tag{2.4.238b}$$

Taking the coefficients with respect to $u_{n+1,m}$ we have:

$$W_2 = \chi_m \tilde{F}\left(\frac{1 + \pi^2 u_{n,m+1} u_{n,m-1}}{u_{n,m+1} - u_{n,m-1}}\right) + \chi_{m+1} \tilde{G}\left(u_{n,m+1} - u_{n,m-1}\right). \tag{2.4.239}$$

Inserting (2.4.239) into (2.4.222a) we do not have any further restriction on the form of the first integral. So we conclude that we have two independent first integrals in the m-direction, as it was observed in [341].

The fact that, when successful, the above procedure gives arbitrary functions has to be understood as a restatement of the trivial property that any autonomous function of a first integral is again a first integral. So, in general, one does not need first integrals depending on arbitrary functions. Therefore we can take these arbitrary functions in the first integrals to be linear function in their arguments.

With these simplifying assumptions we can consider as first integrals for the ${}_tH_1^\pi$ equation the functions

$$W_1 = \chi_m \frac{\alpha_2}{u_{n+1,m} - u_{n,m}} + \chi_{m+1} \frac{u_{n+1,m} - u_{n,m}}{1 + \pi^2 u_{n,m} u_{n+1,m}}, \tag{2.4.240a}$$

$$W_2 = \chi_m \alpha \frac{1 + \pi^2 u_{n,m+1} u_{n,m-1}}{u_{n,m+1} - u_{n,m-1}} + \chi_{m+1} \beta\left(u_{n,m+1} - u_{n,m-1}\right), \tag{2.4.240b}$$

where α_2 and β are two arbitrary constants. This was the form in which the first integrals for the ${}_tH_1^\pi$ equation (2.4.156a) were presented in [341]. In what follows we will write down the first integrals of the others H^4 and H^6 equations according to the above prescription.

First integrals for the H^4 and H^6 equations. Here we consider the ${}_tH_2^\pi$, ${}_tH_3^\pi$ equations and the whole family of the H^6 equations. We will not present the details of the calculations, since they are algorithmic and they can be implemented in any Computer Algebra System available (we have implemented them in `Maple`).

Trapezoidal H^4 equations. We now present the first integrals of the trapezoidal ${}_tH_2^\pi$, ${}_tH_3^\pi$ equations in both directions.

${}_tH_2^\pi$ equation (2.4.156b)**.** For ${}_tH_2^\pi$ we have a four-point, third order first integral in the n-direction:

(2.4.241a)

$$W_1 = \chi_m \frac{\left(-u_{n+1,m}+u_{n-1,m}\right)\left(u_{n,m}-u_{n+2,m}\right)}{\pi^2\alpha_2^4+4\pi\alpha_2^3+\left[\left(8\alpha_3-2u_{n,m}-2u_{n+1,m}\right)\pi-1\right]\alpha_2^2+\left(u_{n,m}-u_{n+1,m}\right)^2} - \chi_{m+1}\frac{\left(-u_{n+1,m}+u_{n-1,m}\right)\left(u_{n,m}-u_{n+2,m}\right)}{\left(-u_{n-1,m}+u_{n,m}+\alpha_2\right)\left(u_{n+1,m}+\alpha_2-u_{n+2,m}\right)}$$

and a five-point, fourth order first integral in the m-direction:

$$W_2 = \chi_m\alpha\frac{\left(u_{n,m-1}-u_{n,m+1}\right)^2\left(u_{n,m+2}-u_{n,m}\right)\left(u_{n,m}-u_{n,m-2}\right)}{\left[\left(\left(\alpha_2+\alpha_3+u_{n,m-1}\right)^2\pi-u_{n,m-1}+\alpha_3-u_{n,m}\right)\cdot\left(\left(\alpha_3+\alpha_2+u_{n,m+1}\right)^2\pi-u_{n,m+1}+\alpha_3-u_{n,m}\right)\right]} + \chi_{m+1}\beta\frac{\begin{bmatrix}-\pi\left(u_{n,m-2}-u_{n,m+2}\right)u_{n,m}^2-\left(\alpha_3+\alpha_2\right)^2\left(u_{n,m-2}-u_{n,m+2}\right)\pi\\+\left(-2\left(u_{n,m-2}-u_{n,m+2}\right)\left(\alpha_3+\alpha_2\right)\pi+u_{n,m-1}-u_{n,m+2}-u_{n,m+1}+u_{n,m-2}\right)u_{n,m}\\+\left(-\alpha_3+u_{n,m+1}\right)u_{n,m-2}+u_{n,m+2}\left(\alpha_3-u_{n,m-1}\right)\end{bmatrix}}{\left(-u_{n,m+2}+u_{n,m}\right)\left(-u_{n,m-2}+u_{n,m}\right)\left(u_{n,m-1}-u_{n,m+1}\right)} \tag{2.4.241b}$$

${}_tH_3^\pi$ equation (2.4.156c)**.** ${}_tH_3^\pi$ has a four-point, third order first integral in the n-direction:

$$W_1 = \chi_m\frac{\left(u_{n-1,m}-u_{n+1,m}\right)\left(-u_{n+2,m}+u_{n,m}\right)}{{\alpha_2}^4\pi^2\delta^2-\alpha_2(1+\alpha_2^2)u_{n+1,m}u_{n,m}+\left({u_{n,m}}^2+u_{n+1,m}^2-2\pi^2\delta^2\right){\alpha_2}^2+\pi^2\delta^2} - \chi_{m+1}\frac{\left(u_{n+1,m}-u_{n-1,m}\right)\left(u_{n+2,m}-u_{n,m}\right)}{\alpha_2\left(-u_{n-1,m}+\alpha_2u_{n,m}\right)\left(-u_{n+2,m}+u_{n+1,m}\alpha_2\right)} \tag{2.4.242a}$$

and a five-point, fourth order first integral in the m-direction:

$$W_2 = \chi_m\alpha\frac{\left(u_{n,m-1}-u_{n,m+1}\right)^2\left(u_{n,m+2}-u_{n,m}\right)\left(u_{n,m}-u_{n,m-2}\right)}{\left[\left(\delta^2\alpha_3^2+u_{n,m-1}^2\pi^2-\alpha_3u_{n,m-1}u_{n,m}\right)\left(\delta^2\alpha_3^2+u_{n,m+1}^2\pi^2-\alpha_3u_{n,m}u_{n,m+1}\right)\right]} - \chi_{m+1}\beta\frac{\begin{bmatrix}-u_{n,m}^2\alpha_3u_{n,m-1}+u_{n,m}^2\alpha_3u_{n,m+1}-u_{n,m}^2\pi^2u_{n,m+2}\\+u_{n,m}^2\pi^2u_{n,m-2}+\alpha_3u_{n,m}u_{n,m-1}u_{n,m+2}-\alpha_3u_{n,m}u_{n,m-2}u_{n,m+1}\\-\delta^2{\alpha_3}^2u_{n,m+2}+\delta^2{\alpha_3}^2u_{n,m-2}\end{bmatrix}}{\left(u_{n,m}-u_{n,m+2}\right)\left(-u_{n,m-2}+u_{n,m}\right)\left(u_{n,m-1}-u_{n,m+1}\right)}. \tag{2.4.242b}$$

Remark 1. *The first integrals of the* ${}_tH_2^{\pi}$ *and* ${}_tH_3^{\pi}$ *equations have the same order in each direction and they share the important property that in the direction m, which is the direction of the non autonomous factors* χ_m, χ_{m+1} *the* W_2 *integrals are built up from two different "sub"-integrals as in the known case of the* ${}_tH_1^{\pi}$ *equation.*

H^6 **equations** (2.4.157)**.** The formulas for the first integrals of the family H^6 introduced in [112–114] are:

${}_1D_2$ **equation** (2.4.157b)**.** For ${}_1D_2$ we have the following three-point, second order first integrals:

$$
\begin{aligned}
W_1 = {} & \chi_n\chi_m\alpha\frac{\left[\left(1+\delta_2\right)u_{n,m}+u_{n+1,m}\right]\delta_1-u_{n,m}}{\left[\left(1+\delta_2\right)u_{n,m}+u_{n-1,m}\right]\delta_1-u_{n,m}} \\
& +\chi_n\chi_{m+1}\alpha\frac{1+\left(u_{n+1,m}-1\right)\delta_1}{1+\left(u_{n-1,m}-1\right)\delta_1} \\
& +\chi_{n+1}\chi_m\beta\left(u_{n+1,m}-u_{n-1,m}\right) \\
& -\chi_{n+1}\chi_{m+1}\beta\frac{\left(u_{n+1,m}-u_{n-1,m}\right)\left[1-\left(1-u_{n,m}\right)\delta_1\right]}{\delta_2+u_{n,m}},
\end{aligned} \tag{2.4.243a}
$$

$$
\begin{aligned}
W_2 = {} & \chi_n\chi_m\alpha\frac{u_{n,m+1}-u_{n,m-1}}{u_{n,m}+\delta_1 u_{n,m-1}} \\
& +\chi_n\chi_{m+1}\beta\left(u_{n,m+1}-u_{n,m-1}\right) \\
& -\chi_{n+1}\chi_m\alpha\frac{u_{n,m+1}-u_{n,m-1}}{1+\delta_1\left(u_{n,m+1}-1\right)} \\
& -\chi_{n+1}\chi_{m+1}\beta\frac{u_{n,m+1}-u_{n,m-1}}{\delta_2+u_{n,m}}.
\end{aligned} \tag{2.4.243b}
$$

${}_2D_2$ **equation** (2.4.157c)**.** For ${}_2D_2$ we have the following three-point, second order first integrals:

$$
\begin{aligned}
W_1 = {} & \chi_n\chi_m\alpha\frac{\delta_2+u_{n+1,m}}{\delta_2+u_{n-1,m}} \\
& +\chi_n\chi_{m+1}\alpha\frac{\left(1-\left(1+\delta_2\right)\delta_1\right)u_{n,m}+u_{n+1,m}}{\left(1-\left(1+\delta_2\right)\delta_1\right)u_{n,m}+u_{n-1,m}} \\
& +\chi_{n+1}\chi_m\beta\frac{\left(u_{n+1,m}-u_{n-1,m}\right)\left(u_{n,m}+\delta_2\right)}{1+\left(-1+u_{n,m}\right)\delta_1} \\
& -\chi_{n+1}\chi_{m+1}\beta\left(u_{n+1,m}-u_{n-1,m}\right),
\end{aligned} \tag{2.4.244a}
$$

$$
\begin{aligned}
W_2 = {} & \chi_n\chi_m\alpha\left(u_{n,m+1}-u_{n,m-1}\right) \\
& -\chi_n\chi_{m+1}\beta\frac{u_{n,m+1}-u_{n,m-1}}{\left(\lambda-u_{n,m}\right)\delta_1-u_{n,m-1}}
\end{aligned} \tag{2.4.244b}
$$

$$- \chi_{n+1}\chi_m\alpha\frac{u_{n,m+1}-u_{n,m-1}}{1+\left(-1+u_{n,m}\right)\delta_1}$$
$$- \chi_{n+1}\chi_{m+1}\beta\frac{u_{n,m+1}-u_{n,m-1}}{u_{n,m+1}+\delta_2}.$$

${}_3D_2$ **equation** (2.4.157d). For ${}_3D_2$ we have the following three-point, second order first integrals:

(2.4.245a)
$$W_1 = \chi_n\chi_m\alpha\frac{\left(u_{n-1,m}+\delta_2\right)\left[1+\left(u_{n+1,m}-1\right)\delta_1\right]}{\left(u_{n+1,m}+\delta_2\right)\left[1+\left(u_{n-1,m}-1\right)\delta_1\right]}$$
$$+ \chi_n\chi_{m+1}\alpha\frac{u_{n,m}+\left(1-\delta_1-\delta_1\delta_2\right)u_{n-1,m}}{u_{n,m}+\left(1-\delta_1-\delta_1\delta_2\right)u_{n+1,m}}$$
$$+ \chi_{n+1}\chi_m\beta\left(u_{n+1,m}-u_{n-1,m}\right)\left(\delta_2+u_{n,m}\right)$$
$$- \chi_{n+1}\chi_{m+1}\beta\left(u_{n+1,m}-u_{n-1,m}\right),$$

(2.4.245b)
$$W_2 = \chi_n\chi_m\alpha\left(u_{n,m+1}-u_{n,m-1}\right)$$
$$- \chi_n\chi_{m+1}\beta\frac{u_{n,m+1}-u_{n,m-1}}{\lambda\left(1+\delta_2\right)\delta_1{}^2-\left[\left(1+\delta_2\right)u_{n,m-1}+u_{n,m}+\lambda\right]\delta_1+u_{n,m-1}}$$
$$+ \chi_{n+1}\chi_m\alpha\left(u_{n,m-1}-u_{n,m+1}\right)\left[1+\left(u_{n,m}-1\right)\delta_1\right]$$
$$+ \chi_{n+1}\chi_{m+1}\beta\frac{u_{n,m+1}-u_{n,m-1}}{\left(\delta_2+u_{n,m+1}\right)\left[1+\left(1-\delta_1\right)u_{n,m-1}\right]}.$$

D_3 **equation** (2.4.157e). For D_3 we have the following four-point, third order first integrals:

(2.4.246a)
$$W_1 = \chi_n\chi_m\alpha\frac{\left(u_{n+1,m}-u_{n-1,m}\right)\left(u_{n+2,m}-u_{n,m}\right)}{u_{n+1,m}^2-u_{n,m}}$$
$$+ \chi_n\chi_{m+1}\alpha\frac{\left(u_{n+1,m}-u_{n-1,m}\right)\left(u_{n+2,m-u_{n,m}}\right)}{u_{n,m}+u_{n-1,m}}$$
$$- \chi_{n+1}\chi_m\beta\frac{\left(u_{n+1,m}-u_{n-1,m}\right)\left(u_{n+2,m}-u_{n,m}\right)}{u_{n+1,m}-u_{n,m}^2}$$
$$+ \chi_{n+1}\chi_{m+1}\beta\frac{\left(u_{n+1,m}-u_{n-1,m}\right)\left(u_{n+2,m}-u_{n,m}\right)}{u_{n+1,m}+u_{n+2,m}},$$

(2.4.246b)
$$W_2 = \chi_n\chi_m\alpha\frac{\left(u_{n,m+1}-u_{n,m-1}\right)\left(u_{n,m+2}-u_{n,m}\right)}{u_{n,m+1}^2-u_{n,m}}$$
$$- \chi_n\chi_{m+1}\beta\frac{\left(u_{n,m+1}-u_{n,m-1}\right)\left(u_{n,m+2}-u_{n,m}\right)}{u_{n,m+1}-u_{n,m}^2}$$

$$+ \chi_{n+1}\chi_m\alpha\frac{\left(u_{n,m+1}-u_{n,m-1}\right)\left(u_{n,m+2}-u_{n,m}\right)}{u_{n,m}+u_{n,m-1}}$$
$$+ \chi_{n+1}\chi_{m+1}\beta\frac{\left(u_{n,m+1}-u_{n,m-1}\right)\left(u_{n,m+2}-u_{n,m}\right)}{u_{n,m+1}+u_{n,m+2}}.$$

Remark 2. *The equation D_3 is invariant under the exchange of lattice variables $n \leftrightarrow m$. Therefore its W_2 first integral* (2.4.246b) *can be obtained from the W_1 one* (2.4.246a) *simply by exchanging the indexes n and m.*

${}_1D_4$ **equation** (2.4.157f)**.** For ${}_1D_4$ we have the following four-point, third order first integrals:

(2.4.247a)
$$W_1 = \chi_n\chi_m\alpha\frac{u_{n+1,m}^2\delta_1+u_{n+1,m}u_{n+2,m}+u_{n-1,m}\left(u_{n,m}-u_{n+2,m}\right)-\delta_2\delta_3}{u_{n+1,m}\left(\delta_1+u_{n,m}\right)-\delta_2\delta_3}$$
$$+ \chi_n\chi_{m+1}\alpha\frac{\left(u_{n,m}-u_{n+2,m}+\delta_1u_{n+1,m}\right)u_{n-1,m}+u_{n+1,m}u_{n+2,m}}{\left(u_{n,m}+\delta_1u_{n-1,m}\right)u_{n+1,m}}$$
$$+ \chi_{n+1}\chi_m\beta\frac{\left(u_{n+1,m}-u_{n-1,m}\right)\left(u_{n+2,m}-u_{n,m}\right)}{u_{n,m}^2\delta_1+u_{n+1,m}u_{n,m}-\delta_2\delta_3}$$
$$+ \chi_{n+1}\chi_{m+1}\beta\frac{\left(u_{n+1,m}-u_{n-1,m}\right)\left(u_{n+2,m}-u_{n,m}\right)}{u_{n,m}\left(u_{n+2,m}\delta_1+u_{n+1,m}\right)},$$

(2.4.247b)
$$W_2 = \chi_n\chi_m\alpha\frac{\left(u_{n,m+2}-u_{n,m}\right)u_{n,m-1}+\delta_1\delta_3-\delta_2u_{n,m+1}^2-u_{n,m+1}u_{n,m+2}}{\delta_1\delta_3-u_{n,m}u_{n,m+1}-\delta_2u_{n,m+1}^2}$$
$$- \chi_n\chi_{m+1}\beta\frac{\left(u_{n,m+1}-u_{n,m-1}\right)\left(u_{n,m+2}-u_{n,m}\right)}{\delta_1\delta_3-\delta_2{u_{n,m}}^2-u_{n,m}u_{n,m+1}}$$
$$+ \chi_{n+1}\chi_m\alpha\frac{\left(u_{n,m}-u_{n,m+2}+\delta_2u_{n,m+1}\right)u_{n,m-1}+u_{n,m+1}u_{n,m+2}}{\left(u_{n,m}+\delta_2u_{n,m-1}\right)u_{n,m+1}}$$
$$+ \chi_{n+1}\chi_{m+1}\beta\frac{\left(u_{n,m+1}-u_{n,m-1}\right)\left(u_{n,m+2}-u_{n,m}\right)}{u_{n,m}\left(u_{n,m+2}\delta_2+u_{n,m+1}\right)}.$$

${}_2D_4$ **equation** (2.4.157g)**.** For ${}_2D_4$ we have the following four-point, third order first integrals:

(2.4.248a)
$$W_1 = \chi_n\chi_m\alpha\frac{\begin{bmatrix}\left(u_{n,m}-u_{n+2,m}-\delta_1\delta_2u_{n-1,m}\right)u_{n+1,m}^2\\ +u_{n+1,m}u_{n+2,m}u_{n-1,m}+\delta_3u_{n-1,m}\end{bmatrix}}{\left(\delta_2u_{n+1,m}^2\delta_1-\delta_3-u_{n,m}u_{n+1,m}\right)u_{n-1,m}}$$
$$- \chi_n\chi_{m+1}\alpha\frac{u_{n+2,m}u_{n-1,m}+\left(-u_{n+2,m}+u_{n,m}\right)u_{n+1,m}+\delta_3}{u_{n-1,m}u_{n,m}+\delta_3}$$

$$- \chi_{n+1}\chi_m\beta \frac{\left(u_{n+1,m} - u_{n-1,m}\right)\left(u_{n+2,m} - u_{n,m}\right) u_{n,m}}{u_{n+2,m}\left(\delta_2\delta_1 u_{n,m}{}^2 - u_{n,m}u_{n+1,m} - \delta_3\right)}$$
$$+ \chi_{n+1}\chi_{m+1}\beta \frac{\left(u_{n+1,m} - u_{n-1,m}\right)\left(u_{n+2,m} - u_{n,m}\right)}{u_{n+1,m}u_{n+2,m} + \delta_3},$$

$$W_2 = \chi_n\chi_m\alpha \frac{\left(u_{n,m+2} - u_{n,m}\right) u_{n,m-1} + \delta_1\delta_3 - \delta_2 u_{n,m+1}^2 - u_{n,m+1}u_{n,m+2}}{\delta_1\delta_3 - \delta_2 u_{n,m+1}^2 - u_{n,m}u_{n,m+1}} \tag{2.4.248b}$$
$$- \chi_n\chi_{m+1}\beta \frac{\left(u_{n,m+1} - u_{n,m-1}\right)\left(u_{n,m+2} - u_{n,m}\right)}{\delta_1\delta_3 - \delta_2 u_{n,m}^2 - u_{n,m}u_{n,m+1}}$$
$$+ \chi_{n+1}\chi_m\alpha \frac{u_{n,m+2}\delta_2 u_{n,m} + u_{n,m-1}u_{n,m} + u_{n,m+1}u_{n,m+2} - u_{n,m}u_{n,m+1}}{u_{n,m+2}\left(\delta_2 u_{n,m} + u_{n,m-1}\right)}$$
$$+ \chi_{n+1}\chi_{m+1}\beta \frac{\left(u_{n,m+1} - u_{n,m-1}\right)\left(u_{n,m+2} - u_{n,m}\right)}{\left(\delta_2 u_{n,m+1} + u_{n,m+2}\right) u_{n,m-1}}.$$

Remark 3. *The first integrals of the H^6 equations are rather peculiar. We have that all the H^6 equations possess two different integrals in every direction. This is due to the presence of two arbitrary constants α and β in the expressions of the first integrals. We believe that this reflects the fact that the H^6 equations on the lattice have two-periodic coefficients in both directions.*

Remark 4. *These results confirms the outcome of the algebraic entropy test presented in Appendix C* [339].

4.7.7. *Darboux integrability of trapezoidal H^4 and H^6 families of lattice equations: general solutions* [336, 344]. Here we show that from the knowledge of the first integrals and from the properties of the equations it is possible to construct, maybe after some complicate algebra, the general solutions of all the trapezoidal H^4 (2.4.156) and H^6 (2.4.157) equations. By general solution we mean a representation of the solution of any of the equations in (2.4.156) and (2.4.157) in terms of the right number of arbitrary functions of one lattice variable n or m. Since the trapezoidal H^4 (2.4.156) and H^6 equations (2.4.157) are quad-graph equations, i.e., the discrete analogue of second-order hyperbolic PDEs, the general solution must contain an arbitrary function in the n direction and another one in the m direction, i.e., a general solution is an expression of the form

$$u_{n,m} = F_{n,m}(a_n, b_m), \tag{2.4.249}$$

where a_n and b_m are arbitrary functions of their discrete variable. Initial conditions are then imposed through substitution in the equation (2.4.249). Nonlinear equations usually possesses also other kinds of solutions, as singular solutions which satisfy only a specific set of initial values. Among the general solutions, in the range of validity of their parameters, we may have also periodic solutions. Periodic initial values will reflect into periodic solution which will arise by fixing properly the arbitrary functions. As an example let us consider the $(N, -M)$ reduction of a quad-equation (2.4.220), with $N, M \in \mathbb{N}^+$ coprime [669, 680]. This implies the following condition

$$u_{n+N,m-M} = u_{n,m}. \tag{2.4.250}$$

If we possess the general solution of the quad-graph equation in the form (2.4.249) then the periodicity condition (2.4.250) is equivalent to

$$F_{n+N,m-M}(a_{n+N}, b_{m-M}) = F_{n,m}(a_n, b_m). \tag{2.4.251}$$

The existence of the associated periodic solution is subject to the ability to solve formula (2.4.251). When the integers N and M are not coprime it can be done: taking $K = \gcd(N, M)$ we have just to decompose the reduction condition into K superimposed staircases and convert the scalar condition (2.4.250) to a vector condition for K fields. The reduction will be possible if the associated system possesses a solution.

To obtain the desired solution we will need only the W_1 integrals derived in Section 2.4.7.6 [345] and the fact that the relation (2.4.222b) implies $W_1 = \xi_n$ with ξ_n an arbitrary function of n. The equation $W_1 = \xi_n$ can be interpreted as an OΔE in the n direction depending parametrically on m. Then from every W_1 integral we can derive two different OΔEs, one corresponding to m even and one corresponding to m odd. In both the resulting equations we can get rid of the two-periodic terms by considering the cases n even and n odd and defining

$$u_{2k,2l} = v_{k,l}, \qquad u_{2k+1,2l} = w_{k,l}, \tag{2.4.252a}$$

$$u_{2k,2l+1} = y_{k,l}, \qquad u_{2k+1,2l+1} = z_{k,l}. \tag{2.4.252b}$$

This transformation brings both equations to a system of coupled PΔEs. This reduction to a system is the key ingredient in the construction of the general solutions for the trapezoidal H^4 (2.4.156) and H^6 equations (2.4.157).

We note that the transformation (2.4.252) can be applied to the trapezoidal $H^4$and H^6 equations.This reduce these non autonomous equations with two-periodic coefficients into autonomous systems of four equations. We recall that in this way some examples of direct linearization (i.e., without the knowledge of the first integrals) were produced in [339]. Finally we note that if we apply the even/odd splitting of the lattice variables given by (2.4.252) to describe a general solution we will need two arbitrary functions in both directions, i.e., we will need a total of four arbitrary functions which will imply constraints on them.

In practice to construct these general solutions, we need to solve Riccati equations and non autonomous linear equations which, in general, cannot be solved in closed form. Using the fact that these equations contain arbitrary functions we introduce new arbitrary functions so as to solve these equations. This is usually done reducing to total difference, i.e., to OΔEs which can be trivially solved. Let us assume we are given the difference equation

$$u_{n+1,m} - u_{n,m} = f_n, \tag{2.4.253}$$

depending parametrically on another discrete index m. Then if we can express the function f_n as a discrete derivative

$$f_n = g_{n+1} - g_n,$$

then the solution of equation (2.4.253) is simply

$$u_{n,m} = g_n + \gamma_m,$$

where γ_m is an arbitrary function of the discrete variable m. This is the simplest possible example of reduction to total difference. The general solutions will then be expressed in terms of these new arbitrary functions obtained reducing to total differences and in terms of a finite number of discrete integrations. The solutions of the simple OΔE

$$u_{n+1} - u_n = f_n, \tag{2.4.254}$$

is reduced to consider u_n as the unknown and f_n as an assigned function. We note that the discrete integration (2.4.254) is the discrete analogue of the differential equation $u'(x) = f(x)$.

To give a very simple example of the method of solution we consider its application to the prototypical Darboux integrable equation: the discrete wave equation

$$u_{n+1,m+1} + u_{n,m} = u_{n+1,m} + u_{n,m+1}. \tag{2.4.255}$$

It is easy to check that the discrete wave equation (2.4.255) is Darboux integrable with two two point first-order first integrals

$$W_1 = u_{n+1,m} - u_{n,m}, \tag{2.4.256a}$$

$$W_2 = u_{n,m+1} - u_{n,m}. \tag{2.4.256b}$$

From the first integrals (2.4.256) it is possible to construct the well known discrete d'Alembert solution. We can write $W_1 = \xi_n$ with ξ_n arbitrary function of its argument. Then we have

$$u_{n+1,m} - u_{n,m} = \xi_n. \tag{2.4.257}$$

This means that choosing the arbitrary function as $\xi_n = a_{n+1} - a_n$, with a_n arbitrary function of its argument, we transform (2.4.257) into the total difference

$$u_{n+1,m} + a_{n+1} = u_{n,m} + a_n,$$

which readily implies

$$u_{n,m} = a_n + \alpha_m,$$

where α_m is an arbitrary function of its argument. This is the discrete analog of the d'Alembert solution of the wave equation.

To summarize, we present the following theorem:

Theorem 14. *The trapezoidal* H^4 (2.4.156) *and* H^6 *equations* (2.4.157) *are exactly solvable and we can represent the solution in terms of a finite number of discrete integrations* (2.4.254).

The proof of Theorem 14 is carried out in [344] except for ${}_tH_1^\pi$ equation (2.4.156a) which will be treated in this Section in the following as an example [345].

Remark 5. *The H equations of the ABS classification* [22] *and their rhombic deformations* [23, 112, 839] *should not be Darboux integrable. This can be confirmed directly excluding the existence of integrals up to a certain order as it was done in* [304, 305] *for other equations. Moreover it was proved rigorously in* [708] *using the* gcd*-factorization method that all the equations of the ABS list* [22, 29] *possess quadratic growth of the degrees. At heuristic level a similar result was presented in* [339] *for the rhombic* H^4 *equations. According to the algebraic entropy conjecture these result means that the ABS equations and the rhombic* H^4 *equations are S-integrable, but not linearizable. Since Darboux integrability for lattice equations implies linearizability we expect that these equations will not possess first integrals of any order. So the results obtained in* [345] *and in this book about the trapezoidal* H^4 *and* H^6 *equations do not imply anything for H equations and their rhombic deformations.*

The first integrals, even those of higher order, can be used to find the general solutions. In the case of ${}_tH_1^\pi$ (2.4.156a), the first integrals are given by (2.4.240) and have been first presented in [341]. We wish to solve the ${}_tH_1^\pi$ equation using both first integrals. We are going to construct those general solutions, slightly modifying the construction scheme from [306].

Let us start from the integral W_1 (2.4.240a). This is a two-point, first order integral. This implies that the ${}_tH_1^\pi$ equation (2.4.156a) can be rewritten as

$$(S_m - I)\left(\chi_m \frac{\alpha_2}{u_{n+1,m} - u_{n,m}} + \chi_{m+1} \frac{u_{n+1,m} - u_{n,m}}{1 + \pi^2 u_{n,m} u_{n+1,m}}\right) = 0. \tag{2.4.258}$$

From (2.4.258) we can derive the general solution of (2.4.156a) itself. In fact (2.4.258) implies:

$$\chi_m \frac{\alpha_2}{u_{n+1,m} - u_{n,m}} + \chi_{m+1} \frac{u_{n+1,m} - u_{n,m}}{1 + \pi^2 u_{n,m} u_{n+1,m}} = \xi_n, \tag{2.4.259}$$

where ξ_n is an arbitrary function of n. This is a first order difference equation in the n-direction in which m plays the role of a parameter. For this reason we can safely separate the two cases: m even and m odd.

Case $m = 2k$ In this case (2.4.259) is reduced to the linear equation

$$u_{n+1,2k} - u_{n,2k} = \frac{\alpha_2}{\xi_n} \tag{2.4.260}$$

which has the solution

$$u_{n,2k} = \theta_{2k} + \omega_n, \tag{2.4.261}$$

where θ_{2k} is an arbitrary function and ω_n is the solution of the simple OΔE

$$\omega_{n+1} - \omega_n = \frac{\alpha_2}{\xi_n}, \qquad \omega_0 = 0. \tag{2.4.262}$$

Case $m = 2k + 1$ In this case (2.4.259) is reduced to the discrete Riccati equation:

$$\xi_n \pi^2 u_{n,2k+1} u_{n+1,2k+1} - u_{n+1,2k+1} + u_{n,2k+1} + \xi_n = 0. \tag{2.4.263}$$

By using the Möbius transformation

$$u_{n,2k+1} = \frac{\imath}{\pi} \frac{1 - v_{n,2k+1}}{1 + v_{n,2k+1}}, \tag{2.4.264}$$

this equation can be recast into the linear equation

$$(\imath + \pi \pi_n)\, v_{n+1,2k+1} - (\imath - \pi \xi_n)\, v_{n,2k+1} = 0. \tag{2.4.265}$$

If we introduce a new function κ_n, such that

$$\frac{\kappa_{n+1}}{\kappa_n} = \frac{\imath - \pi \xi_n}{\imath + \pi \xi_n}, \tag{2.4.266}$$

then we have that the general solution of (2.4.265) is written as:

$$v_{n,2k+1} = \kappa_n \theta_{2k+1}, \tag{2.4.267}$$

where θ_{2k+1} is an arbitrary function. Using (2.4.264) and (2.4.266) we then obtain:

$$u_{n,2k+1} = \frac{\imath}{\pi} \frac{1 - \kappa_n \theta_{2k+1}}{1 + \kappa_n \theta_{2k+1}}, \qquad \xi_n = \frac{\imath}{\pi} \frac{\kappa_n - \kappa_{n+1}}{\kappa_n + \kappa_{n+1}}. \tag{2.4.268}$$

So we have the general solution of (2.4.156a) in the form:

(2.4.269)
$$u_{n,m} = \chi_m \left(\theta_m + \omega_n\right) + \chi_{m+1} \frac{\iota}{\pi} \frac{1 - \kappa_n \theta_m}{1 + \kappa_n \theta_m},$$

where θ_m,κ_n are arbitrary functions, ω_n is expressed in term of ξ_n by (2.4.262), and ξ_n is defined in term of κ_n by (2.4.268).

Now we pass to consider the integral W_2 in the direction m (2.4.240b). This case is more interesting, as now we are dealing with a three-point, second order integral. For this problem we can choose $\alpha = \beta = 1$. Our starting point is the relation (2.4.225b), i.e. $W_2 = \rho_m$, from which we can derive two different equations, one for the even and one for the odd m. Choosing $m = 2k$ and $m = 2k + 1$, we obtain the following two equations:

(2.4.270a)
$$1 + \pi^2 u_{n,2k+1} u_{n,2k-1} = \rho_{2k} \left(u_{n,2k+1} - u_{n,2k-1}\right),$$

(2.4.270b)
$$u_{n,2k+2} - u_{n,2k} = \rho_{2k+1}.$$

So the system consists of two *uncoupled* equations.

The first one (2.4.270a) is a discrete Riccati equation which can be linearized through the non autonomous Möbius transformation:

(2.4.271)
$$u_{n,2k-1} = \frac{1}{v_{n,k}} + \alpha_k, \quad \rho_{2k} = \frac{1 + \pi^2 \alpha_{k+1} \alpha_k}{\alpha_{k+1} - \alpha_k},$$

from which we obtain:

(2.4.272)
$$\left(1 + \pi^2 \alpha_{k+1}^2\right) v_{n,k+1} + \pi^2 \alpha_{k+1} = \left(1 + \pi^2 \alpha_k^2\right) v_{n,k} + \pi^2 \alpha_k.$$

Eq. (2.4.272) is equivalent to a total difference and therefore its solution is given by:

(2.4.273)
$$v_{n,k} = \frac{\theta_n - \pi^2 \alpha_k}{1 + \pi^2 \alpha_k^2},$$

with an arbitrary function θ_n. Putting $\alpha_k = \kappa_{2k-1}$, we obtain the solution for $u_{n,2k-1}$:

(2.4.274)
$$u_{n,2k-1} = \frac{1 + \kappa_{2k-1} \theta_n}{\theta_n - \pi^2 \kappa_{2k-1}}.$$

The second equation is just a linear OΔE which can be written as a total difference, by the substitution $\rho_{2k+1} = \kappa_{2k+2} - \kappa_{2k}$. We get:

(2.4.275)
$$u_{n,2k} = \omega_n + \kappa_{2k}.$$

The resulting solution reads:

(2.4.276)
$$u_{n,m} = \chi_m \left(\omega_n + \kappa_m\right) + \chi_{m+1} \frac{1 + \kappa_m \theta_n}{\theta_n - \pi^2 \kappa_m}.$$

This solution depends on three arbitrary functions. This is because we started from a second order first integral, which is just a consequence of the discrete equation. This means that there must be a relation between θ_n and ω_n. This relation can be retrieved by inserting (2.4.276) into (2.4.156a). As a result we obtain the following definition for ω_n:

(2.4.277)
$$\omega_n - \omega_{n+1} = \alpha_2 \frac{\pi^2 + \theta_n \theta_{n+1}}{\theta_{n+1} - \theta_n},$$

which gives us the final expression for the solution of (2.4.156a) up to the discrete integration given by (2.4.277). If we consider different first integrals the corresponding general solution will be the same as they are transformable one into the other.

As a final remark we note that it has been proved in [28] that Darboux integrable systems possess generalized symmetries depending on arbitrary functions of the first integrals. However, in case of the trapezoidal H^4 and H^6, the explicit form of symmetries depending on arbitrary functions is known only for the ${}_tH_1^\pi$ equation (2.4.156a) [340–342]. This poses the challenging problem of finding the explicit form of such generalized symmetries. These symmetries will be highly nontrivial, especially in the case of the ${}_tH_2^\pi$ and ${}_tH_3^\pi$ equations (2.4.156), where the order of the first integrals is particularly high.

4.8. Integrable example of quad-graph equations not in the ABS or Boll class. Here we show, using a simple example, that effectively there are integrable PΔEs which possess hierarchies of generalized symmetries of the form (2.4.130) and which are not included in the ABS lists.

As it is well-known [822], the modified Volterra equation (3.2.185), which here we write as

$$u_{n,t} = (u_n^2 - 1)(u_{n+1} - u_{n-1}) \tag{2.4.278}$$

is transformed into the Volterra equation (2.3.172) $v_{n,t} = v_n(v_{n+1} - v_{n-1})$ by two discrete Miura transformations:

$$v_n^{\pm} = (u_{n+1} \pm 1)(u_n \mp 1). \tag{2.4.279}$$

For any solution u_n of (2.4.278), one obtains by the transformations (2.4.279) two solutions v_n^+, v_n^- of the Volterra equation. From a solution of the Volterra equation v_n one obtains two solutions u_n and $\tilde{u}_n$ of the modified Volterra equation. The composition of the Miura transformations (2.4.279)

$$v_n = (u_{n+1} + 1)(u_n - 1) = (\tilde{u}_{n+1} - 1)(\tilde{u}_n + 1) \tag{2.4.280}$$

provides a Bäcklund transformation for (2.4.278). Eq. (2.4.280) allows one to construct, starting with a solution u_n of the modified Volterra equation (2.4.278), a new solution $\tilde{u}_n$.

Introducing for any index n $u_n = u_{n,m}$ and $\tilde{u}_n = u_{n,m+1}$, where m is a new index, we can rewrite the Bäcklund transformation (2.4.280) as a quad-graph equation of the form (2.4.130). At the point $(0,0)$ it reads:

$$(u_{1,0} + 1)(u_{0,0} - 1) = (u_{1,1} - 1)(u_{0,1} + 1). \tag{2.4.281}$$

Eq. (2.4.281) does not belong to the ABS classification, as it is not invariant under the exchange of n and m and does not satisfy the 3D–consistency property [555]. The modified Volterra equation (2.4.278) can then be interpreted as a three-point generalized symmetry of (2.4.281) involving only shifts in the n direction:

$$u_{0,0,\epsilon} = (u_{0,0}^2 - 1)(u_{1,0} - u_{-1,0}). \tag{2.4.282}$$

There exists also a generalized symmetry involving only shifts in the m direction, given in Section 3.3.1.2 by (V_2) with $P(u^2) = u^2 - 1$

$$u_{0,0,\mu} = (u_{0,0}^2 - 1)\left(\frac{1}{u_{0,1} + u_{0,0}} - \frac{1}{u_{0,0} + u_{0,-1}}\right). \tag{2.4.283}$$

Eq.(2.4.283), together with (2.4.282), to the complete list of integrable Volterra type equations presented in Section 3.3.1.2[842, 850]. Both equations have a hierarchy of generalized symmetries which, by construction, must be compatible with (2.4.281). Symmetries of (2.4.282) can be obtained in many ways, see e.g. [850]. Symmetries of (2.4.283) can be constructed, using the master symmetry presented in [169]. The simplest generalized symmetries of (2.4.282) and (2.4.283) are given by the following equations:

$$u_{0,0,\epsilon'} = (u_{0,0}^2-1)((u_{1,0}^2-1)(u_{2,0}+u_{0,0})-(u_{-1,0}^2-1)(u_{0,0}+u_{-2,0})),$$

$$u_{0,0,\mu'} = \frac{u_{0,0}^2-1}{(u_{0,1}+u_{0,0})^2}\left(\frac{u_{0,1}^2-1}{u_{0,2}+u_{0,1}}+\frac{u_{0,0}^2-1}{u_{0,0}+u_{0,-1}}\right) - \frac{u_{0,0}^2-1}{(u_{0,0}+u_{0,-1})^2}\left(\frac{u_{0,0}^2-1}{u_{0,1}+u_{0,0}}+\frac{u_{0,-1}^2-1}{u_{0,-1}+u_{0,-2}}\right).$$

As it can be checked by direct calculation, these equations are five-point symmetries of (2.4.281).

Moreover, (2.4.281) possesses two conservation laws (3.6.10) characterized by the following functions $p_{0,0}, q_{0,0}$:

(2.4.284) $$p_{0,0}^+ = \log\frac{u_{0,0}+u_{0,1}}{u_{0,0}+1}, \qquad q_{0,0}^+ = -\log(u_{0,0}+1),$$

(2.4.285) $$p_{0,0}^- = \log\frac{u_{0,0}+u_{0,1}}{u_{0,1}-1}, \qquad q_{0,0}^- = \log(u_{0,0}-1).$$

It is easy to check that (3.6.10) is identically satisfied by (2.4.284) and (2.4.285) on the solutions of (2.4.281). Eq. (2.4.281) possess also non autonomous conservation laws, however, conservation laws of this kind will not be discussed here.

A more general form of both (2.4.280, 2.4.281) is given by

(2.4.286) $$v_{n,m} = (u_{n+1,m}+\alpha_m)(u_{n,m}-\alpha_m) = (u_{n+1,m+1}-\alpha_{m+1})(u_{n,m+1}+\alpha_{m+1}),$$

where α_m is an m-dependent function. For any m the function $u_{n,m}$ satisfies the modified Volterra equation given by (V_1) in Section 3.3.1.2 with $P(u) = u^2 - \alpha_m$ where α_m are arbitrary functions. $v_{n,m}$, for any m, is a solution of the Volterra equation. Using (2.4.286) and starting from an initial solution $v_{n,0}$, we can construct new solutions of the Volterra equation:

$$v_{n,0} \to u_{n,1} \to v_{n,1} \to u_{n,2} \to v_{n,2} \to \cdots .$$

The Lax pair for (2.4.286) is given by

$$L_{n,m} = \begin{pmatrix} \lambda-\lambda^{-1} & -v_{n,m} \\ 1 & 0 \end{pmatrix},$$

which corresponds to the standard scalar spectral problem of the Volterra equation (2.3.173) written in matrix form, and by

$$M_{n,m} = \frac{1}{u_{n,m+1}-\alpha_{m+1}} \cdot \begin{pmatrix} (\lambda-\lambda^{-1})(u_{n,m+1}-\alpha_{m+1}) & 2\alpha_{m+1}(u_{n,m+1}^2-\alpha_{m+1}^2) \\ -2\alpha_{m+1} & (\lambda-\lambda^{-1})(u_{n,m+1}+\alpha_{m+1}) \end{pmatrix}.$$

This Lax pair satisfies the Lax equation (2.4.10). By setting $\alpha_m = 1$ we get a Lax pair for (2.4.281). A different Lax pair for this equation has been constructed in [555].

Eq. (2.4.286) is a direct analog of well-known dressing chain [812]

(2.4.287) $$u_{m+1,x} + u_{m,x} = u_{m+1}^2 - u_m^2 + \alpha_{m+1} - \alpha_m$$

which provides a way of constructing potentials $v_m = u_{m,x} - u_m^2 - \alpha_m$ for the discrete Schrödinger spectral problem [750, 755]. See also [480]. The Lax pair given above is analogous to the one of (2.4.287) presented in [755].

4.9. The completely discrete Burgers equation. We can construct the hierarchy of completely discrete Burgers equation in the same way, *mutatis mutandis*, as we did for the discrete time Toda hierarchy. We start from (2.3.320) with

$$L_{nm}(u_{nm}) = S_n - u_{nm}, \quad L_{nm}(u_{nm})\psi_{nm} = 0, \tag{2.4.288}$$

and consider a discrete time evolution given by (2.4.9) which we repeat here for the convenience of the reader

$$\psi_{n,m+1} = \psi_{n,m} - M_{n,m}\psi_{n,m}.$$

The Lax equation is given by (2.4.10)

$$L_{n,m+1} - L_{n,m} = L_{n,m+1}M_{n,m} - M_{n,m}L_{n,m}$$

with

$$L_{n,m+1} - L_{nm} = u_{nm} - u_{n,m+1}. \tag{2.4.289}$$

The starting point of the construction of the Burgers hierarchy is given by (2.4.12, 2.4.13, 2.4.5, 2.4.6). Taking into account (2.3.337) we get

$$\begin{aligned} \tilde{V}_{n,m} &= \Lambda^{dB}_{n,m}V_{nm} + V^{(0)}_{n,m}, \quad \Lambda^{dB}_{nm} = u_{n,m}S_n - u_{n,m+1}, \\ \tilde{M}_{n,m} &= L_{n,m+1}M_{n,m} + F_{n,m}S_n + G_{n,m}, \\ V^{(0)}_{n,m} &= F^{(0)}u_{nm}\Big(u_{n+1,m} - u_{nm}\Big) + G^{(0)}\Big(u_{n,m} - u_{n,m+1}\Big), \end{aligned} \tag{2.4.290}$$

where $G_{nm} = G^{(0)}$ and $F_{nm} = F^{(0)}$ are arbitrary summation constants. Then the hierarchy of Burgers equations is

$$u_{nm} - u_{n,m+1} = f_{nm}(\Lambda^{dB}_{nm})u_{nm}(u_{n+1,m} - u_{n,m+1}) + g_{nm}(\Lambda^{dB}_{nm})(u_{nm} - u_{n,m+1}). \tag{2.4.291}$$

The simplest equation of the hierarchy (2.4.291) is a non linear discrete wave equation

$$u_{n,m+1} - u_{n,m} = u_{nm}(u_{n+1,m} - u_{n,m+1}), \quad f_{nm}(z) = 1, \; g_{nm}(z) = 0, \tag{2.4.292}$$

which we encountered in (2.3.342) when dealing with the DΔE of Burgers type in Section 2.3.5. The second equation, obtained by choosing $f_{nm}(z) = z$, $g_{nm} = 0$ in (2.4.291), is a completely discrete Burgers equation

$$u_{n,m+1} - u_{n,m} = u_{nm}\Big[u_{n,m+1}(u_{n+1,m} - u_{n,m+1}) - u_{n+1,m}(u_{n+2,m} - u_{n+1,m+1})\Big]. \tag{2.4.293}$$

The corresponding discrete evolution of the wave function is respectively

$$\psi_{n,m+1} = (1 - S_n)\psi_{n,m} \equiv (1 - u_{n,m})\psi_{n,m}, \tag{2.4.294}$$

and

$$\begin{aligned} \psi_{n,m+1} &= (1 - S_n^2 + u_{n,m+1}S_n)\psi_{n,m} \\ &\equiv [1 - u_{n,m}(u_{n+1,m} - u_{n,m+1})]\psi_{n,m}. \end{aligned} \tag{2.4.295}$$

Up to a parametric m dependence, the spectral problem for the partial difference Burgers (2.4.288) is the same as that for the differential difference Burgers. So the space part of the Bäcklund transformation will be the same.

In principle the symmetries could be given by the equations of the hierarchy of the differential difference Burgers we presented above. It is easy to prove that the evolution of the wave function in the group parameter and that in m commute for (2.4.294) but not for (2.4.295) as the evolution presented in (2.4.295) has an explicit dependence on the field $u_{n,m+1}$.

Can one write other equations which have the differential difference Burgers equations as symmetries? To do so we need to construct equations whose evolution of the spectral problem in m do not depend explicitly on $u_{n,m}$. These new partial difference Burgers are given by

$$u_{n,m+1} - u_{n,m} = \left(u_{n,m+1} - u_{n+j,m}\right) \prod_{k=0}^{j-1} u_{n+k,m}, \tag{2.4.296}$$

where j is an integer number which characterizes the equation in the hierarchy. The corresponding m evolution of the wave function $\psi_{n,m}$ is given by

$$\psi_{n,m+1} = \psi_{n,m} - \psi_{n+j,m}. \tag{2.4.297}$$

The equations (2.4.296) and any combination of them for any integer value of j have the differential difference Burgers hierarchy of equations (2.3.327) with t substituted by the group parameter ϵ as symmetries.

4.10. The discrete Burgers equation from the discrete heat equation. In [258, 259, 532] we can find a discrete versions of the heat equation on a two-dimensional uniform lattice

$$\Delta_m \phi = \Delta_n^2 \phi, \qquad \phi = \phi_{n,m}, \tag{2.4.298}$$

where the difference operators in the discrete variables n and m are defined here by

$$\Delta_n = \frac{1}{h_x}(S_n - 1), \qquad \Delta_m = \frac{1}{h_t}(S_m - 1). \tag{2.4.299}$$

In (2.4.299) h_t and h_x are the lattice spacing in the two independent variables t and x of indexes respectively m and n. Eq. (2.4.298) is shown in [258, 259] to possess a symmetry algebra of generalized symmetries given by

$$\phi_{\epsilon_1} = \Delta_m \phi, \tag{2.4.300}$$

$$\phi_{\epsilon_2} = \Delta_n \phi, \tag{2.4.301}$$

$$\phi_{\epsilon_3} = 2t S_m^{-1} \Delta_n \phi + x S_n^{-1} \phi + \tfrac{1}{2} h_x S_n^{-1} \phi, \tag{2.4.302}$$

$$\phi_{\epsilon_4} = 2t S_m^{-1} \Delta_m \phi + x S_n^{-1} \Delta_x \phi + \left(1 - \tfrac{1}{2} S_n^{-1}\right) \phi, \tag{2.4.303}$$

$$\begin{aligned} \phi_{\epsilon_5} = {} & t^2 S_m^{-2} \Delta_m \phi + t x S_m^{-1} S_n^{-1} \Delta_n \phi + \tfrac{1}{4} x^2 S_n^{-2} \phi \\ & + t \left(S_m^{-2} - \tfrac{1}{2} S_m^{-1} S_n^{-1}\right) \phi - \tfrac{1}{16} h_x^2 S_n^{-2} \phi, \end{aligned} \tag{2.4.304}$$

$$\phi_{\epsilon_6} = \phi, \tag{2.4.305}$$

where

$$t = h_t m, \qquad x = h_x n. \tag{2.4.306}$$

In the continuous limit (2.4.300,…,2.4.305) go into the usual generators of the symmetries of the heat equation [659], i.e time translations, space translations, Galilei transformations, dilations, projective transformations and the multiplication by a constant.

From (2.4.298), using the Cole–Hopf transformation

$$\Delta_n \phi = u \phi, \qquad u = u_{n,m}, \tag{2.4.307}$$

we can derive a new discrete Burgers which possess a finite symmetry algebra of generalized symmetries.

The completely discrete Burgers as a compatibility condition. To derive a new Burgers we first use (2.4.307) to rewrite (2.4.298) as

$$\Delta_m \phi = [\Delta_n u + u S_n u]\phi. \tag{2.4.308}$$

We then require the compatibility of (2.4.307) and (2.4.308), i.e. $\Delta_m \Delta_n \phi = \Delta_n \Delta_m \phi$ and obtain an equation on $u(x,t)$, which we shall call the "new discrete Burgers equation" [394]:

$$\Delta_m u = \frac{1 + h_x u}{1 + h_t[\Delta_n u + u S_n u]} \Delta_n(\Delta_n u + u S_n u). \tag{2.4.309}$$

Eq. (2.4.309) can be rewritten in many different forms, for instance we can use (2.4.299) to eliminate all discrete derivatives in terms of shift operators S_n, S_m and spacings h_x, h_t.

The continuous limit of (2.4.309) is obtained by taking $h_x \to 0$, $h_t \to 0$ when n and m diverge but t and x given by (2.4.306) remain finite. We have

$$\Delta_m \to \frac{\partial}{\partial t}, \qquad S_m \to 1, \tag{2.4.310}$$

and similarly for Δ_n and S_n. In the continuous limit (2.4.309) goes into the Burgers equation in the form (2.2.175).

We mention that a related discrete Burgers appeared in a different context in [394], and that an ultradiscrete version of it found an application to traffic flow modeling in [646]. Using the Lax equation (2.4.298, 2.4.307) we can derive a Bäcklund transformation for the new discrete Burgers equation, in the same way as was done in the semidiscrete case [502] in Section 2.3.5.1. The Bäcklund transformation relates a solution u of (2.4.309) to a new solution $\tilde{u}$ of the same equation:

$$\tilde{u} = \frac{pu + (S_n u)(1 + h_x u)}{p + 1 + h_x u}, \tag{2.4.311}$$

where p is an arbitrary constant.

4.10.1. *Symmetries of the new discrete Burgers.* To obtain the symmetries of the new discrete Burgers (2.4.309) we proceed with the same strategy as used for deriving the equation itself. We start from the symmetries of the discrete heat equation (2.4.300,...,2.4.305) [258, 259, 532], apply the Cole-Hopf transformation (2.4.307) and express the result in terms of the function u.

Each of the symmetries (2.4.300,...,2.4.305) can be acted upon by arbitrary functions of the shift operators $f(S_n, S_m)$, obtaining further symmetries. However, in the continuous limit $f(S_n, S_m) \to f(1,1) =$ constant, so all these higher symmetries reduce to the six original ones when $h_x \to 0$ and $h_t \to 0$. Among these symmetries of the heat equation two are particularly relevant, namely

$$\phi_\epsilon = \phi_t, \tag{2.4.312}$$

$$\phi_\epsilon = \phi_x. \tag{2.4.313}$$

These are the usual t and x translations, that can be obtained from (2.4.300, 2.4.301) using the well known formula obtained by expanding the logarithm of $e^{\partial_x}\phi = S_x \phi$,

$$\phi_z = \sum_{k=0}^{\infty} \frac{(-1)^k}{k+1}(S_z - 1)^k \Delta_z \phi \tag{2.4.314}$$

(with $z = x$, or $z = t$).

All symmetries of the heat equation can be written symbolically as

$$\phi_\epsilon = \mathfrak{S}\phi, \tag{2.4.315}$$

where $\mathfrak{S} = \mathfrak{S}(x, t, \phi, S_n, S_m, \Delta_n, \Delta_m, \partial_x, \partial_t)$ is a linear operator that can in each case be read off from (2.4.300,…,2.4.313).

We use the Cole-Hopf transformation (2.4.307) to transform symmetries of the heat equation (2.4.298) into those of the new discrete Burgers equation (2.4.309). Let us first prove a general result.

Theorem 15. *Let* (2.4.315) *represent a symmetry of the discrete heat equation* (2.4.298). *Then the same operator* $\mathfrak{S}$ *provides a symmetry of the new discrete Burgers equation*(2.4.309) *via the formula*

$$u_\epsilon = (1 + h_x u)\Delta_n \left(\frac{\mathfrak{S}\phi}{\phi}\right), \tag{2.4.316}$$

where $(\mathfrak{S}\phi/\phi)$ *can be (and must be) expressed entirely in terms of* $u(x, t)$*, its variations and their shifted values using* (2.4.307).

Proof: Let us assume that the symmetry (2.4.315) and the Cole-Hopf transformation (2.4.307) be compatible. From the obvious result

$$\frac{\partial}{\partial \epsilon}(\Delta_n \phi) = \Delta_n \phi_\epsilon \tag{2.4.317}$$

we get

$$u_\epsilon = \frac{\Delta_n(\mathfrak{S}\phi) - u\mathfrak{S}\phi}{\phi}. \tag{2.4.318}$$

On the other hand, a direct calculation yields

$$\begin{aligned} \Delta_n \left(\frac{\mathfrak{S}\phi}{\phi}\right) &= \frac{1}{h_x}\left[\frac{S_n(\mathfrak{S}\phi)}{S_n\phi} - \frac{\mathfrak{S}\phi}{\phi}\right] \\ &= \frac{1}{h_x(S_n\phi)\phi}\left\{\phi\left[S_n(\mathfrak{S}\phi) - \mathfrak{S}\phi\right] - \left[(S_n\phi)(\mathfrak{S}\phi) - \phi(\mathfrak{S}\phi)\right]\right\}, \end{aligned} \tag{2.4.319}$$

$$= \frac{1}{S_n\phi}\left[\Delta_n(\mathfrak{S}\phi) - u(\mathfrak{S}\phi)\right]. \tag{2.4.320}$$

Multiplying by $h_x u$ and using the Cole-Hopf transformation again we obtain

$$h_x u \Delta_n \left(\frac{\mathfrak{S}\phi}{\phi}\right) = \frac{\Delta_n(\mathfrak{S}\phi) - u(\mathfrak{S}\phi)}{\phi} - \Delta_n \left(\frac{\mathfrak{S}\phi}{\phi}\right). \tag{2.4.321}$$

Using (2.4.318), we replace the first term on the right hand side of (2.4.321) by u_ϵ, and obtain (2.4.316). □

In order to show that the fraction $(\mathfrak{S}\phi/\phi)$ can be expressed in terms of the function u, it is sufficient to write $\Delta_m\phi$, $\Delta_n\phi$, $S_n\phi$, $S_m\phi$, $S_n^{-1}\phi$, etc, as expressions depending on u, times ϕ (see (2.4.300,…,2.4.305)). The necessary formulas are obtained from (2.4.307) and (2.4.308), namely

$$\begin{aligned} \Delta_n\phi &= \phi u, & \Delta_m\phi &= \phi v, \\ S_n\phi &= \phi(1 + h_x u), & S_m\phi &= \phi(1 + h_t v), \\ S_n^{-1}\phi &= \phi S_n^{-1}\frac{1}{1 + h_x u}, & S_m^{-1}\phi &= \phi S_m^{-1}\frac{1}{1 + h_t v}, \end{aligned} \tag{2.4.322}$$

where we have introduced the simplifying notation

$$v = \Delta_n u + u S_n u. \tag{2.4.323}$$

Applying Theorem 20 to the one-dimensional subalgebras $\phi_{\epsilon_1},\ldots,\phi_{\epsilon_6}$ given by (2.4.300,
…,2.4.305) we obtain thc corresponding symmetries of the new discrete Burgers equation. A basis for this Lie algebra is given by the following flows:

$$u_{\epsilon_1} = (1+h_t v)\Delta_m u, \tag{2.4.324}$$

$$u_{\epsilon_2} = (1+h_x u)\Delta_n u, \tag{2.4.325}$$

$$u_{\epsilon_3} = (1+h_x u)\Delta_n \left[2tS_m^{-1}\frac{u}{1+h_t v} + \left(x+\tfrac{1}{2}h_x\right) S_n^{-1}\frac{1}{1+h_x u}\right], \tag{2.4.326}$$

$$u_{\epsilon_4} = (1+h_x u)\Delta_n \left[2tS_m^{-1}\frac{v}{1+h_t v} + xS_n^{-1}\frac{u}{1+h_x u} - \tfrac{1}{2}S_n^{-1}\frac{1}{1+h_x u}\right], \tag{2.4.327}$$

$$\begin{aligned} u_{\epsilon_5} = (1+h_x u)\Delta_n &\left[t^2 S_m^{-1}\left(\frac{1}{1+h_t v}S_m^{-1}\frac{v}{1+h_t v}\right)\right. \\ &+ txS_n^{-1}\left(\frac{1}{1+h_x u}S_m^{-1}\frac{u}{1+h_t v}\right) + \frac{1}{4}\left(x^2 - \frac{h_x^2}{4}\right) S_n^{-1}\left(\frac{1}{1+h_x u}\cdot\right. \\ &\left. S_n^{-1}\frac{1}{1+h_x u}\right) + tS_m^{-1}\left(\frac{1}{1+h_t v}S_m^{-1}\frac{1}{1+h_t v}\right) \\ &\left. -\frac{1}{2}tS_n^{-1}\left(\frac{1}{1+h_x u}S_m^{-1}\frac{1}{1+h_t v}\right)\right], \end{aligned}$$

$$u_{\epsilon_6} = 0, \tag{2.4.328}$$

were the quantity v is defined in (2.4.323).

Thus, the six-dimensional symmetry algebra of the discrete heat equation gives rise to a five-dimensional symmetry algebra of the discrete Burgers equation. The same is true in the continuous case.

The fact that the flows (2.4.324,…,2.4.10.1) commute with the flow of the discrete Burgers equation (2.4.309) was also checked directly on a computer (using Mathematica).

In the continuous limit, (2.4.324,…,2.4.10.1) go over correctly into the well known symmetries of the usual Burgers equation (2.2.175) given in (2.2.220), see Section 2.2.5.2, namely time translations, space translations, Galilei boosts, dilations and projective transformations. The commutation relations in the discrete case are the same as in the continuous case.

We can show directly that the usual space and time translations are also symmetries of the new discrete Burgers equation:

$$u_{\epsilon_t} = u_t, \tag{2.4.329}$$

$$u_{\epsilon_x} = u_x. \tag{2.4.330}$$

Indeed, it is easy to check that the corresponding ϵ-flows commute with the t-flow given by (2.4.309).

The "higher" symmetries of the heat equation given by $\phi_\mu = S_n^a\phi$ will give new symmetries. For $a=-1$ we have

$$u_\mu = (1+h_x u)\Delta_n\left(\frac{S_n\phi}{\phi}\right),$$

i.e.

$$u_\mu = (1 + h_x u) S_n \Delta_n \left(\frac{1}{1 + h_x u} \right). \tag{2.4.331}$$

In the continuous limit this symmetry goes into $u_\mu = 0$, i.e. it becomes trivial.

4.10.2. *Symmetry reduction for the new discrete Burgers equation.* We have shown that all the symmetries of the discrete Burgers equation (2.4.309) can be written in the form (2.4.316) with $(\mathfrak{S}\phi/\phi)$ expressed in terms of u. This allows us to write all the reduction formulas in the form $\Delta_n (\mathfrak{S}\phi/\phi) = 0$. Hence, we can in all cases integrate once and write the reduction equations, (i.e. the surface condition) as

$$\frac{\mathfrak{S}\phi}{\phi} = K_m \tag{2.4.332}$$

and then rewrite (2.4.332) in terms of u. In general S is a linear combination of all the symmetry operators for the heat equation, i.e. the operators on the right hand sides of (2.4.300),...,(2.4.305). Instead of performing a general subalgebra analysis [830] as was done for the Burgers equation in Section 2.2.5.3, we shall just look at the individual basis elements of the Lie algebra.

Time translations. We rewrite (2.4.324) as

$$u_{\epsilon_1} = (1 + h_x u) \Delta_n \left(\frac{\Delta_m \phi}{\phi} \right). \tag{2.4.333}$$

Eq. (2.4.332) then is

$$\Delta_m \phi = K_m \phi, \tag{2.4.334}$$

or in terms of u:

$$v = \Delta_n u + u S_n u = K_m. \tag{2.4.335}$$

The Burgers equation (2.4.309) can be written as

$$\Delta_m u = \frac{1 + h_x u}{1 + h_t v} \Delta_n v. \tag{2.4.336}$$

Hence, in view of (2.4.335) we have

$$\Delta_m u = 0, \qquad K_m = K_0 = \text{constant}. \tag{2.4.337}$$

Since ϕ satisfies the heat equation we rewrite (2.4.334) as

$$\Delta_n^2 \phi = K\phi. \tag{2.4.338}$$

This is a linear difference equation with constant coefficients, and we can easily solve it, putting $\phi = a^n$ and finding a.

For $K \neq 0$ the general solution of (2.4.338) is

$$\phi = c_1 (1 + \sqrt{K} h_x)^{x/h_x} + c_2 (1 - \sqrt{K} h_x)^{x/h_x}, \tag{2.4.339}$$

where c_1 and c_2 are arbitrary real constants for $K > 0$ and are complex, satisfying $c_2 = \bar{c_1}$ for $K < 0$.

For $K = 0$ the solution of (2.4.338) is

$$\phi = c_1 + c_2 x, \qquad c_1, c_2 \in \mathbb{R}. \tag{2.4.340}$$

In all cases the corresponding invariant solution of the new discrete Burgers equation is obtained via the Cole-Hopf transformation as $u = \Delta_n\phi/\phi$. In particular (2.4.340) yields a solution invariant under time traslation that can be written as

$$u = \frac{1}{x + \mu h_x}, \qquad \mu = \text{constant}. \tag{2.4.341}$$

Space translations. In this case the result is trivial. From (2.4.325) we have directly $\Delta_n u = 0$, from the new discrete Burgers equation $\Delta_m u = 0$ and hence $u = \text{constant}$.

Galilei invariance. Substituting $\mathfrak{G}$ from (2.4.302) into (2.4.332) we obtain the linearized reduced equation

$$\begin{aligned} 2(t+h_t)\Delta_n\phi + \left(x + \tfrac{1}{2}h_x\right) S_n^{-1}\phi + h_t\left(x + \tfrac{1}{2}h_x\right) S_n^{-1}\Delta_n^2\phi \\ = K_m(\phi + h_t\Delta_n^2\phi). \end{aligned} \tag{2.4.342}$$

In terms of u the reduced equation is obtained from (2.4.322) and is

$$\begin{aligned} 2t\, S_n u + x - K_m + 2th_x u\, S_n u + h_t\left(\tfrac{7}{2}S_n u + \tfrac{7}{2}h_x u\, S_n u + xu\, S_n u \right. \\ \left. + x\Delta_n u - \tfrac{3}{2}u\right) + \tfrac{3}{2}h_x - K_m\left[h_x u + h_t\left(S_n\Delta_n u + uS_n^2 u - uS_n u\right)\right. \\ \left. + S_n u S_n^2 u + h_x u\, S_n u S_n^2 u\right] = 0. \end{aligned} \tag{2.4.343}$$

Eq. (2.4.343) is a difference equations in one variable (namely n) with parametric dependence in m. It is a linear second order equation with variable coefficients. So it is not so easily solvable.

The situation is similar for the dilations (2.4.327) and the projective transformation (2.4.10.1). So, in those cases, we shall just present the reduced equations for u.

Dilation invariance. The reduced equation for u is:

$$\begin{aligned} \tfrac{1}{2} + K_m - u\left[x + h_x(2 - K_m) - v_1\left(2t - \tfrac{3}{2}h_t\right)\right] \\ - u(S_m v_1)\left[2t + 4h_x h_t - xh_t - K_m h_x h_t\right] \\ - (\Delta_n v_1)\left(2t - xh_t\right) - h_t(u - K_m)S_n v_1 = 0, \end{aligned} \tag{2.4.344}$$

where

$$v_0 = u, \qquad v_1 = \Delta_n u + uS_n u. \tag{2.4.345}$$

Projective invariance. The reduced equation for u is

$$\begin{aligned} \left(t + 2h_t\right)^2\left(v_1 + 2h_x v_2 + h_x^2 v_3\right) \\ + \left(t + 2h_t\right)(x + h_x)\left(u + h_x v_1 + h_t v_2 + h_x h_t v_3\right) \\ + \tfrac{1}{4}\left[(x + 2h_x)^2 - \tfrac{1}{4}h_x^2\right]\left(1 + 2h_t v_1 + h_t^2 v_3\right) \\ + \left(t + 2h_t\right)\left(\tfrac{1}{2} + \tfrac{3}{2}h_x u + h_x^2 v_1 - \tfrac{1}{2}h_t v_1 - \tfrac{1}{2}h_x h_t v_2\right) \\ = K_m\left(1 + 2h_x u + h_x^2 v_1 + 2h_t v_1 + 4h_x h_t v_2 + 2h_t h_x^2 v_3 \right. \\ \left. + h_t^2 v_3 + 2h_t^2 h_x v_4 + h_t^2 h_x^2 v_5\right), \end{aligned} \tag{2.4.346}$$

were $v_0 = u$ and v_i, $i = 1, \cdots, 5$ are obtained from the recurrence relation $v_{j+1} = \Delta_n v_j + uS_n v_j$.

4.11. Linearization of PΔEs through symmetries. As we saw in Section 2.2.6.1 Kumei and Bluman [459], based on the analysis of the symmetry properties of linear PDEs, proved a Lie theory on the linearizability of non linear PDEs. For a recent extended review see [98]. Following the analogy of the continuous case we formulate here a theorem for the linearization of PΔEs by symmetry [519]. Partial results in this direction for the case of PΔEs defined on a fixed lattice have been obtained by Quispel and collaborators [139, 724, 726].

In this approach to the problem of linearization of PΔEs we will consider the case when the grid is preassigned and assumed to be fixed, with constant lattice spacing. Later we will consider the case of a discrete scheme. Moreover for simplicity we will consider autonomous equations defined on a two dimensional grid so that there is no privileged position and we can write the dependent variables just in terms of the shifts with respect to the reference point $u_{n,m} = u_{0,0}$ on the lattice.

A PΔE of order $N \cdot N'$ for a function $u_{n,m}$ will be a relation between $N \cdot N'$ points in the two dimensional grid, i.e.

$$\mathcal{E}_{N \cdot N'}\Big(u_{0,0}, u_{1,0}, \cdots, u_{N,0}, u_{0,1}, \cdots, u_{N,1}, \cdots, u_{N,N'}\Big) = 0. \tag{2.4.347}$$

A continuous symmetry for equations of the form (2.4.347), where the lattice is fixed, i.e. the two independent variables $x_{n,m}$ and $t_{n,m}$ are completely specified as $x_{n,m} = h_x n + x_0$ and $t_{n,m} = h_t m + t_0$ with h_x, h_t, x_0 and t_0 given constants, is given just by dilations

$$\hat{X}_{n,m} = \chi_{n,m}(u_{n,m})\partial_{u_{n,m}}. \tag{2.4.348}$$

It is easy to show that a linear PΔE of order $N \cdot N'$ for a function $v_{n,m}$

$$\mathcal{F}_{N \cdot N'} = b(n,m) + \sum_{(i,j)=(0,0)}^{(N,N')} a^{i,j}(n,m) v_{n+i,m+j} = 0, \tag{2.4.349}$$

has always the symmetry

$$\hat{X}_{n,m} = \phi_{n,m}\partial_{v_{n,m}}, \tag{2.4.350}$$

where $\phi_{n,m}$ is a solution of the homogeneous part of (2.4.349)

$$\sum_{(i,j)=(0,0)}^{(N,N')} a^{i,j}(n,m)\phi_{n+i,m+j} = 0. \tag{2.4.351}$$

It is not at all obvious, however, that an equation (2.4.347) having a symmetry (2.4.350) is linear when the function $\phi_{n,m}$ satisfies a homogeneous linear equation. We leave to Section 2.4.11.2 the proof of this proposition. The symmetry (2.4.350) corresponds to the superposition principle for linear equations. If the non linear equation (2.4.347) is linearizable by a point transformation then the symmetry (2.4.350) must be preserved. This is the content of the Theorem 8 we presented in Section 2.2.6.1 in the case of PDEs and this will still be valid here. So we can state the following theorem:

Theorem 16. *An autonomous linear PΔE (2.4.347) is linearizable if it has a point symmetry of the form*

$$\hat{X}_{n,m} = \alpha_{n,m}(u_{n,m})\phi_{n,m}\partial_{v_{n,m}}, \tag{2.4.352}$$

where the function $\phi_{n,m}$ satisfies the linear PΔE (2.4.351).

The proof of Theorem 16 and of the following Theorem 17 are the same as those presented by Bluman and Kumei [101] in the continuous case and so we will not repeat them here. It does not depend on the fact that the equation is a PDE.

As in the case of PDEs we can present the following theorem which provides the transformation which reduces the equation to a linear one. In this case, as the independent variables are not changed, the transformation is given just by a dilation. So we have:

Theorem 17. *The point transformation which linearizes the non linear PΔE (2.4.347)*

$$v_{n,m} = \Psi_{n,m}(u_{n,m}) \tag{2.4.353}$$

is obtained by solving the differential equation

$$\alpha_{n,m}(u_{n,m}) \frac{d\Psi_{n,m}(u_{n,m})}{du_{n,m}} = 1, \tag{2.4.354}$$

were $\alpha_{n,m}(u_{n,m})$ *appears in Theorem* 16.

As in the continuous case, if (2.4.347) has no symmetries of the form considered in Theorem 16 we can introduce some potential variables. On the lattice there are infinitely many ways to try to extend the symmetries by introducing a potential variable as there are infinitely many ways to define a first derivative. Thus it seems to be advisable to check the equation with a linearizability criterion like the algebraic entropy [327, 814] before looking for potential variables.

The simplest way to introduce a potential symmetry is by writing the difference equation (2.4.347) as a system

$$v_{n+1,m} = \mathcal{E}^{(1)}_{n,m}(u_{n,m}, \cdots), \qquad v_{n,m+1} = \mathcal{E}^{(2)}_{n,m}(u_{n,m}, \cdots). \tag{2.4.355}$$

In such a way by looking for the compatibility of the two equations (2.4.355), we get

$$\mathcal{E}_{N\cdot N'}\Big(u_{0,0}, u_{1,0}, \cdots, u_{N,0}, u_{0,1}, \cdots, u_{N,1}, \cdots, u_{N,N'}\Big) = \mathcal{E}^{(1)}_{n,m+1} - \mathcal{E}^{(2)}_{n+1,m}. \tag{2.4.356}$$

It is easy to show in full generality that the symmetries for (2.4.355) and for (2.4.347) are the same.

A different way to introduce potential symmetries is by considering the following system,

$$v_{n+1,m} - v_{n,m} = \mathcal{E}^{(1)}_{n,m}(u_{n,m}, \cdots), \qquad v_{n,m+1} - v_{n,m} = \mathcal{E}^{(2)}_{n,m}(u_{n,m}, \cdots). \tag{2.4.357}$$

In such a way

$$\begin{aligned} &\mathcal{E}_{N\cdot N'}\Big(u_{0,0}, u_{1,0}, \cdots, u_{N,0}, u_{0,1}, \cdots, u_{N,1}, \cdots, u_{N,N'}) \\ &\quad = [\mathcal{E}^{(1)}_{n,m+1} - \mathcal{E}^{(1)}_{n,m}] - [\mathcal{E}^{(2)}_{n+1,m} - \mathcal{E}^{(2)}_{n,m}]. \end{aligned} \tag{2.4.358}$$

i.e. the non linear difference equation is written as a discrete conservation law. As (2.4.357) are a system, to construct the symmetries we have to generalize the linearization theorem as we did in the continuous case.

Theorem 18. *Let us consider a system of non linear PΔEs*

$$\mathcal{F}^{(1)}_{n,m}(u_{n,m}, \cdots, v_{n,m}, \cdots) = 0, \quad \mathcal{F}^{(2)}_{n,m}(u_{n,m}, \cdots, v_{n,m}, \cdots) = 0 \tag{2.4.359}$$

of order $N \cdot N'$ for two scalar functions $u_{n,m}$ and $v_{n,m}$ of two indexes n and m which possesses a symmetry generator

$$\hat{X} = \phi_{n,m}(u_{n,m}, v_{n,m})\partial_{u_{n,m}} + \psi_{n,m}(u_{n,m}, v_{n,m})\partial_{v_{n,m}}, \tag{2.4.360}$$
$$\phi_{n,m}(u_{n,m}, v_{n,m}) = \sum_{j=1}^{2} \beta^{(j)}_{n,m}(u_{n,m}, v_{n,m}) w^{(j)}_{n,m},$$
$$\psi_{n,m}(u_{n,m}, v_{n,m}) = \sum_{j=1}^{2} \gamma^{(j)}_{n,m}(u_{n,m}, v_{n,m}) w^{(j)}_{n,m},$$

with $\beta^{(j)}_{n,m}$ and $\gamma^{(j)}_{n,m}$ given functions of their arguments and the function $w_{n,m} = (w^{(1)}_{n,m}, w^{(2)}_{n,m})$ satisfying the linear equations

$$\mathfrak{L}_{n,m} w_{n,m} = 0, \tag{2.4.361}$$

with $\mathfrak{L}_{n,m}$ a linear operator with coefficients depending only on n and m.

The invertible transformation

$$w^{(1)}_{n,m} = K^{(1)}_{n,m}(u_{n,m}, v_{n,m}), \qquad w^{(2)}_{n,m} = K^{(2)}_{n,m}(u_{n,m}, v_{n,m}), \tag{2.4.362}$$

which transforms (2.4.359) to the system of linear PΔEs (2.4.361) is given by a particular solution of the linear inhomogeneous first order system of PΔEs for the function $K = (K^{(1)}_{n,m}(u_{n,m}, v_{n,m}), K^{(2)}_{n,m}(u_{n,m}, v_{n,m}))$

$$\beta^{(k)}_{n,m}(u_{n,m}, v_{n,m}) K^{(j)}_{n,m} + \gamma^{(k)}_{n,m}(u_{n,m}, v_{n,m}) K^{(j)}_{n,m} = \delta^{j}_{k}, \tag{2.4.363}$$

where δ^j_k is the standard Kronecker symbol.

4.11.1. *Examples.* Here we present a few examples of linearizable PΔEs. For concreteness and for comparing with the previous sections we limit ourselves to the case when the non linear PΔE involves at most four lattice points.

Classification of quad-graph equations linearizable by point transformations. We consider a general autonomous PΔE defined on a square lattice:

$$\mathcal{F}(u_{0,0}, u_{0,1}, u_{1,0}, u_{1,1}) = 0. \tag{2.4.364}$$

If $u_{1,1}$ is present in (2.4.364) we can assume that we can rewrite the equation as

$$u_{1,1} = F(u_{0,0}, u_{0,1}, u_{1,0}). \tag{2.4.365}$$

Following Theorem 16 we look for an infinitesimal symmetry generator of the form

$$\hat{X}_{0,0} = \alpha_{0,0}(u_{0,0})\phi_{0,0}\partial_{u_{0,0}}, \tag{2.4.366}$$

where the function ϕ solves a linear homogeneous equation, i.e.

$$\hat{\mathfrak{L}}\phi_{0,0} = 0, \qquad \hat{\mathfrak{L}} = a + bS_n + cS_m + dS_nS_m, \tag{2.4.367}$$

with S_n and S_m given in (1.2.13, 1.2.14) and a, b, c and d constants. If $d \neq 0$ then we can write (2.4.367) as

$$\phi_{1,1} = -\frac{1}{d}[a\phi_{0,0} + b\phi_{1,0} + c\phi_{0,1}]. \tag{2.4.368}$$

In this setting $u_{0,i}$, $u_{j,0}$, $\phi_{0,i}$ and $\phi_{j,0}$, with $i, j = 0, 1$ are independent variables. If (2.4.366) is a generator of the symmetries of (2.4.365) then we must have

(2.4.369)
$$\mathrm{pr}\hat{X}F|_{F=0} = 0 \leftrightarrow F_{,u_{0,0}}\phi_{0,0}\alpha_{0,0}(u_{0,0}) + F_{,u_{1,0}}\phi_{1,0}\alpha_{1,0}(u_{1,0}) \\ +F_{,u_{0,1}}\phi_{0,1}\alpha_{0,1}(u_{0,1}) = \phi_{1,1}\alpha_{1,1}(u_{1,1})|_{(u_{1,1}=F,\hat{\mathfrak{L}}\phi_{0,0}=0)} \\ = -\frac{1}{d}[a\phi_{0,0} + b\phi_{1,0} + c\phi_{0,1}]\alpha_{1,1}(F(u_{0,0}, u_{0,1}, u_{1,0})).$$

As $\phi_{0,0}$, $\phi_{1,0}$ and $\phi_{0,1}$ are independent variables, we obtain from (2.4.369) three equations relating the function α, intrinsic of the symmetry, with the function F, intrinsic of the non linear equation:

(2.4.370)
$$a\alpha_{1,1}(F) + dF_{,u_{0,0}}\alpha_{0,0}(u_{0,0}) = 0, \\ b\alpha_{1,1}(F) + dF_{,u_{1,0}}\alpha_{1,0}(u_{1,0}) = 0, \quad c\alpha_{1,1}(F) + dF_{,u_{0,1}}\alpha_{0,1}(u_{0,1}) = 0.$$

As in (2.4.370), up to a constant, the first term is the same for all three equations, we can rewrite them as a system of PDE's for the function F depending on α

(2.4.371)
$$\frac{1}{a}F_{,u_{0,0}}\alpha_{0,0}(u_{0,0}) = \frac{1}{b}F_{,u_{1,0}}\alpha_{1,0}(u_{1,0}) = \frac{1}{c}F_{,u_{0,1}}\alpha_{0,1}(u_{0,1}),$$

which can be solved on the characteristic, giving F as a function of the symmetry variable

(2.4.372)
$$\xi = ag(u_{0,0}) + bg(u_{1,0}) + cg(u_{0,1}), \qquad \alpha(x) = \frac{1}{g_{,x}(x)}.$$

Introducing this result in Theorem 17 we get $\psi(u_{0,0}) = \int^{u_{0,0}} g_x(x)dx = g(u_{0,0}) + \kappa$, with κ an arbitrary integration constant. Then (2.4.370) gives that any linearizable non linear PΔE on a four-point lattice must be written as

(2.4.373)
$$dF_{,\xi} + \alpha(F(\xi)) = 0 \to F = g^{-1}(\frac{\xi - \xi_0}{d}),$$

where by $g^{-1}(x)$ we mean the inverse of the function $g(x)$ given in (2.4.372).

Let us notice that from (2.4.371) we can get the six linearizability necessary conditions we introduced in [746] to classify linearizable, multilinear equations on a four-point lattice, that is

(2.4.374a)
$$A\left(x, u_{0,1}\right) \doteq \frac{F_{,u_{0,0}}}{F_{,u_{1,0}}}|_{u_{0,0}=u_{1,0}=x} = \frac{a}{b}, \quad \forall x,\ u_{0,1},$$

(2.4.374b)
$$B\left(x, u_{1,0}\right) \doteq \frac{F_{,u_{0,0}}}{F_{,u_{0,1}}}|_{u_{0,0}=u_{0,1}=x} = \frac{a}{c}, \quad \forall x,\ u_{1,0},$$

(2.4.374c)
$$C\left(x, u_{0,0}\right) \doteq \frac{F_{,u_{0,1}}}{F_{,u_{1,0}}}|_{u_{1,0}=u_{0,1}=x} = \frac{c}{b}, \quad \forall x,\ u_{0,0},$$

(2.4.374d)
$$\frac{\partial}{\partial u_{0,1}}\frac{F_{,u_{0,0}}}{F_{,u_{1,0}}} = 0, \quad \forall u_{0,0},\ u_{1,0},\ u_{0,1},$$

(2.4.374e)
$$\frac{\partial}{\partial u_{1,0}}\frac{F_{,u_{0,0}}}{F_{,u_{0,1}}} = 0, \quad \forall u_{0,0},\ u_{1,0},\ u_{0,1},$$

(2.4.374f)
$$\frac{\partial}{\partial u_{0,0}}\frac{F_{,u_{0,1}}}{F_{,u_{1,0}}} = 0, \quad \forall u_{0,0},\ u_{1,0},\ u_{0,1}.$$

So linearizable equations on four-point lattice are characterized by a function $g(x)$ and its inverse. As a trivial example we can choose $g(x) = e^x$ and we get that the non linear equation $u_{1,1} = \log(\alpha e^{u_{0,0}} + \beta e^{u_{1,0}} + \gamma e^{u_{0,1}} + k)$ linearizes to $\psi_{1,1} = \alpha\psi_{0,0} + \beta\psi_{1,0} + \gamma\psi_{0,1}$. The corresponding function α is $\alpha(x) = e^{-x}$ and the linearizing transformation is $\psi_{0,0} = e^{u_{0,0}} + \kappa$.

In [374] we have shown that there is a multilinear equation (3.7.124) on the square lattice belonging to the Q^+ class which is linearizable. It is interesting to find the corresponding function $g(x)$ in terms of which we can linearize it. The function F in this case is a fraction of a second order polynomial over a third order polynomial. The only function g which provides this structure is $g(x) = \frac{1}{\ell_1 x+\ell_0}$ which gives $F = -\frac{1}{\ell_1}\left[\frac{d}{\xi-\xi_0} + \ell_0\right]$ where $\xi = \frac{a}{\ell_1 u_{0,0}+\ell_0} + \frac{b}{\ell_1 u_{1,0}+\ell_0} + \frac{c}{\ell_1 u_{0,1}+\ell_0}$. In this case the linearizing transformation is the linear fractional function $\psi_{0,0}(u_{0,0}) = \frac{1}{\ell_1 u_{0,0}+\ell_0} + \kappa$, where κ is an arbitrary constant.

Linearizable potential equations. For the sake of simplicity we set in (2.4.357) $\mathcal{E}^{(2)}_{n,m} = u_{n,m}$. If we want the equation (2.4.347) to be on the square we have to choose $\mathcal{E}^{(1)}_{n,m} = g_{n,m}(u_{n,m}, u_{n+1,m})$. The application of the prolongation of the infinitesimal generator (2.4.360) to the second equation in (2.4.357) gives [544]

(2.4.375)
$$\phi_{0,0}(u_{0,0}, v_{0,0}) = \psi_{0,1}(u_{0,1}, v_{0,1}) - \psi_{0,0}(u_{0,0}, v_{0,0}), \quad \rightarrow \psi_{0,0}(u_{0,0}, v_{0,0}) = \psi_{0,0}(v_{0,0}).$$

Then the prolongation of the infinitesimal generator (2.4.360) applied to the first equation in (2.4.357) gives

(2.4.376)
$$\psi_{1,0}(v_{1,0}) - \psi_{0,0}(v_{0,0}) = [\psi_{0,1}(v_{0,1}) - \psi_{0,0}(v_{0,0})]\frac{\partial g_{0,0}}{\partial u_{0,0}} + [\psi_{1,1}(v_{1,1}) - \psi_{1,0}(v_{1,0})]\frac{\partial g_{0,0}}{\partial u_{1,0}},$$

where

(2.4.377)
$$v_{1,0} = v_{0,0} + g_{0,0}(u_{0,0}, u_{1,0}), \quad v_{0,1} = v_{0,0} + u_{0,0}, \quad v_{1,1} = v_{0,0} + u_{1,0} + g_{0,0}(u_{0,0}, u_{1,0}).$$

To comply with Theorem 18 we look for an infinitesimal coefficient of the infinitesimal generator (2.4.360) of the form

$$\psi_{0,0} = w^{(1)}_{0,0}\gamma^{(1)}_{0,0}(v_{0,0}) + w^{(2)}_{0,0}\gamma^{(2)}_{0,0}(v_{0,0}), \tag{2.4.378}$$

where the functions $w^{(1)}_{n,m}$ and $w^{(2)}_{n,m}$ satisfy a linear PΔE on the square

$$\begin{aligned} w^{(1)}_{0,0} &= a^{(1)}_{0,0}w^{(1)}_{0,1} + a^{(2)}_{0,0}w^{(1)}_{1,0} + a^{(3)}_{0,0}w^{(1)}_{1,1}, \\ w^{(2)}_{0,0} &= b^{(1)}_{0,0}w^{(2)}_{0,1} + b^{(2)}_{0,0}w^{(2)}_{1,0} + b^{(3)}_{0,0}w^{(2)}_{1,1}. \end{aligned} \tag{2.4.379}$$

Introducing (2.4.378, 2.4.379) into (2.4.376) and taking into account that we can always choose $w^{(1)}_{0,1}$, $w^{(1)}_{1,0}$, $w^{(1)}_{1,1}$, $w^{(2)}_{0,1}$, $w^{(2)}_{1,0}$ and $w^{(2)}_{1,1}$ as independent variables we get the following system of coupled equations for $\gamma^{(1)}$

$$\gamma^{(1)}_{1,0}(v_{0,0} + g_{0,0})\left(1 + \frac{\partial g_{0,0}}{\partial u_{1,0}}\right) - a^{(2)}_{0,0}\gamma^{(1)}_{0,0}(v_{0,0})\left(1 + \frac{\partial g_{0,0}}{\partial u_{0,0}}\right) = 0, \tag{2.4.380}$$

$$\gamma^{(1)}_{0,1}(v_{0,0} + u_{0,0})\frac{\partial g_{0,0}}{\partial u_{0,0}} + a^{(1)}_{0,0}\gamma^{(1)}_{0,0}(v_{0,0})\left(1 + \frac{\partial g_{0,0}}{\partial u_{0,0}}\right) = 0, \tag{2.4.381}$$

$$\gamma^{(1)}_{1,1}(v_{0,0} + u_{1,0} + g_{0,0})\frac{\partial g_{0,0}}{\partial u_{1,0}} + a^{(3)}_{0,0}\gamma^{(1)}_{0,0}(v_{0,0})\left(1 + \frac{\partial g_{0,0}}{\partial u_{0,0}}\right) = 0, \tag{2.4.382}$$

and similar ones for the function $\gamma^{(2)}_{n,m}(v_{0,0})$. Adding (2.4.380) multiplied by $a^{(3)}_{0,0}$ to (2.4.382) multiplied by $a^{(1)}_{0,0}$ we get

$$a^{(3)}_{0,0}\gamma^{(1)}_{1,0}(v_{0,0}+g_{0,0})\big(1+\frac{\partial g_{0,0}}{\partial u_{1,0}}\big)+a^{(2)}_{0,0}\gamma^{(1)}_{1,1}(v_{0,0}+u_{1,0}+g_{0,0})\frac{\partial g_{0,0}}{\partial u_{1,0}}=0. \quad (2.4.383)$$

Eq. (2.4.383) is similar to (2.4.381). Upshifting by one the first index in (2.4.381) and comparing the result with (2.4.383) we get a discrete equation for $g_{n,m}$

$$\frac{a^{(3)}_{0,0}\left(1+\frac{\partial g_{0,0}}{\partial u_{1,0}}\right)}{a^{(1)}_{1,0}\left(1+\frac{\partial g_{1,0}}{\partial u_{1,0}}\right)}=a^{(2)}_{0,0}\frac{\frac{\partial g_{0,0}}{\partial u_{1,0}}}{\frac{\partial g_{1,0}}{\partial u_{1,0}}}. \quad (2.4.384)$$

In (2.4.384) we have the function $g_{1,0}=g_{1,0}(u_{1,0},u_{2,0})$ and if $\frac{\partial^2 g_{1,0}}{\partial u_{1,0}\partial u_{2,0}}\neq 0$ we get a linear differential equation for $g_{0,0}$ whose solution is $g_{0,0}=g^{(2)}_{0,0}(u_{0,0})+g^{(1)}_{0,0}(u_{0,0})u_{1,0}$. Introducing this solution in (2.4.384) we get $g_{0,0}=g^{(0)}_{0,0}+g^{(1)}_{0,0}(u_{0,0})u_{1,0}+g^{(2)}_{0,0}u_{0,0}$, i.e. a linear equation. By choosing, in place of (2.4.379) the most general linear coupled system of difference equations on the square lattice for $w^{(1)}$ and $w^{(2)}$, we would get the same result. So the introduced potential equation (2.4.357) does not provide linearizable discrete equations.

A discrete linearizable Burgers equation has been presented before in Section 2.4.9 [502] and in Section 2.4.10 [376, 377]. It will be discussed in Section 3.7.

In Section 2.4.9 we can find the discrete equation (2.4.292) and its Lax pair which we rewrite explicitly here as

$$\psi_{m,n+1}=u_{m,n}\psi_{m,n}, \qquad \psi_{m+1,n}=\frac{1}{1+u_{m+1,n}}\psi_{m,n}. \quad (2.4.385)$$

Eqs. (2.4.385) suggest to rewrite (2.4.357) as

$$v_{n+1,m}/v_{n,m}=\mathcal{E}^{(1)}_{m,n}(u_{n,m},\cdots), \qquad v_{n,m+1}/v_{n,m}=\mathcal{E}^{(2)}_{n,m}(u_{n,m},\cdots). \quad (2.4.386)$$

Eqs. (2.4.357, 2.4.386) are transformable one into the other by defining $v_{n,m}=\log(w_{n,m})$ and redefining appropriately the functions $\mathcal{E}^{(1)}_{n,m}$ and $\mathcal{E}^{(2)}_{n,m}$. However in doing so, if $w_{n,m}$ satisfies a linear equation, this will not be the case for $v_{n,m}$. So the fact that the ansatz (2.4.357) does not give rise to linearizable equations is not in contradiction with the fact that (2.4.292) is linearizable.

The compatibility of (2.4.386) implies

$$\mathcal{E}^{(1)}_{n,m+1}(u_{n,m+1},\cdots)\mathcal{E}^{(2)}_{n,m}(u_{n,m},\cdots)=\mathcal{E}^{(2)}_{n+1,m}(u_{n+1,m},\cdots)\mathcal{E}^{(1)}_{n,m}(u_{n,m},\cdots). \quad (2.4.387)$$

If (2.4.387) is constrained to be an equation on the square lattice, then we must have $\mathcal{E}^{(1)}_{n,m}(u_{n,m},\cdots)=\mathcal{E}^{(1)}_{n,m}(u_{n,m},u_{n+1,m})$ and $\mathcal{E}^{(2)}_{n,m}(u_{n,m},\cdots)=\mathcal{E}^{(2)}_{n,m}(u_{n,m},u_{n,m+1})$. Moreover with no loss of generality we can set $\mathcal{E}^{(2)}_{n,m}(u_{n,m},u_{n,m+1})=u_{n,m}$.

Let us look for the symmetries of (2.4.386). Applying the infinitesimal generator (2.4.360) to the right hand equation in (2.4.386) we get

$$\psi_{0,0}=\psi_{0,0}(v_{0,0}), \quad \phi_{0,0}=\frac{\psi_{0,1}(v_{0,1})-u_{0,0}\psi_{0,0}(v_{0,0})}{v_{0,0}}. \quad (2.4.388)$$

Then the determining equation associated to the left hand equation in (2.4.386) is given by

$$\psi_{1,0}(v_{1,0})=\left[\frac{\partial\mathcal{E}^{(1)}_{0,0}}{\partial u_{0,0}}\phi_{0,0}+\frac{\partial\mathcal{E}^{(1)}_{0,0}}{\partial u_{1,0}}\phi_{1,0}\right]v_{0,0}+\mathcal{E}^{(1)}_{0,0}\psi_{0,0}(v_{0,0}), \quad (2.4.389)$$

where the functions $\phi_{i,j}$ are expressed in term of the functions $\psi_{i,j}$ through (2.4.388).

As we look for linearizable equations, from Theorem 18 it follows that we must have:

$$\psi_{0,0}(v_{0,0}) = \sum_{j=1}^{2} w_{0,0}^{(j)} \gamma_{0,0}^{(j)}(v_{0,0}), \tag{2.4.390}$$

where the discrete functions $w_{0,0}^{(j)}$ satisfies a linear PΔE on the square. We can assume that the coefficient of $w_{1,1}^{(j)}$ is always different from zero so that we have

$$\begin{aligned} w_{1,1}^{(1)} &= a^{(1)} w_{0,0}^{(1)} + b^{(1)} w_{0,1}^{(1)} + c^{(1)} w_{1,0}^{(1)} + d^{(1)} w_{0,0}^{(2)} + e^{(1)} w_{0,1}^{(2)} + f^{(1)} w_{1,0}^{(2)}, \\ w_{1,1}^{(2)} &= a^{(2)} w_{0,0}^{(1)} + b^{(2)} w_{0,1}^{(1)} + c^{(2)} w_{1,0}^{(1)} + d^{(2)} w_{0,0}^{(2)} + e^{(2)} w_{0,1}^{(2)} + f^{(2)} w_{1,0}^{(2)}. \end{aligned} \tag{2.4.391}$$

In such a case the variables $w_{0,0}^{(j)}$, $w_{1,0}^{(j)}$ and $w_{0,1}^{(j)}$, $j = 1, 2$, are independent and (2.4.389) splits in three couples of equations relating the functions $\gamma_{0,0}^{(j)}(v_{0,0})$, $j = 1, 2$, with the function $\mathcal{E}_{0,0}^{(1)}$

$$\gamma_{1,0}^{(1)} \mathcal{E}_{0,0}^{(1)} = \frac{\partial \mathcal{E}_{0,0}^{(1)}}{\partial u_{1,0}} \left[b^{(1)} \gamma_{1,1}^{(1)} + b^{(2)} \gamma_{1,1}^{(2)} - u_{1,0} \gamma_{1,0}^{(1)} \right], \tag{2.4.392}$$

$$\gamma_{1,0}^{(2)} \mathcal{E}_{0,0}^{(1)} = \frac{\partial \mathcal{E}_{0,0}^{(1)}}{\partial u_{1,0}} \left[e^{(1)} \gamma_{1,1}^{(1)} + e^{(2)} \gamma_{1,1}^{(2)} - u_{1,0} \gamma_{1,0}^{(2)} \right], \tag{2.4.393}$$

$$\mathcal{E}_{0,0}^{(1)} \frac{\partial \mathcal{E}_{0,0}^{(1)}}{\partial u_{0,0}} \gamma_{0,0}^{(1)} u_{0,0} = \frac{\partial \mathcal{E}_{0,0}^{(1)}}{\partial u_{1,0}} \left[a^{(1)} \gamma_{1,1}^{(1)} + a^{(2)} \gamma_{1,1}^{(2)} \right] + \left[\mathcal{E}_{0,0}^{(1)} \right]^2 \gamma_{0,0}^{(1)}, \tag{2.4.394}$$

$$\mathcal{E}_{0,0}^{(1)} \frac{\partial \mathcal{E}_{0,0}^{(1)}}{\partial u_{0,0}} \gamma_{0,0}^{(2)} u_{0,0} = \frac{\partial \mathcal{E}_{0,0}^{(1)}}{\partial u_{1,0}} \left[d^{(1)} \gamma_{1,1}^{(1)} + d^{(2)} \gamma_{1,1}^{(2)} \right] + \left[\mathcal{E}_{0,0}^{(1)} \right]^2 \gamma_{0,0}^{(2)}, \tag{2.4.395}$$

$$\mathcal{E}_{0,0}^{(1)} \frac{\partial \mathcal{E}_{0,0}^{(1)}}{\partial u_{0,0}} \gamma_{0,1}^{(1)} + \frac{\partial \mathcal{E}_{0,0}^{(1)}}{\partial u_{1,0}} \left[c^{(1)} \gamma_{1,1}^{(1)} + c^{(2)} \gamma_{1,1}^{(2)} \right] = 0, \tag{2.4.396}$$

$$\mathcal{E}_{0,0}^{(1)} \frac{\partial \mathcal{E}_{0,0}^{(1)}}{\partial u_{0,0}} \gamma_{0,1}^{(2)} + \frac{\partial \mathcal{E}_{0,0}^{(1)}}{\partial u_{1,0}} \left[f^{(1)} \gamma_{1,1}^{(1)} + f^{(2)} \gamma_{1,1}^{(2)} \right] = 0, \tag{2.4.397}$$

where $v_{0,1} = u_{0,0} v_{0,0}$, $v_{1,0} = \mathcal{E}_{0,0}^{(1)} v_{0,0}$ and $v_{1,1} = u_{1,0} \mathcal{E}_{0,0}^{(1)} v_{0,0}$ due to (2.4.386).

As $\mathcal{E}_{0,0}^{(1)}$ is a function of $u_{0,0}$ and $\gamma_{0,0}^{(j)}$ is a function of $v_{0,0}$ we get from (2.4.392, 2.4.393)

$$\frac{\mathcal{E}_{0,0}^{(1)} + u_{1,0} \frac{\partial \mathcal{E}_{0,0}^{(1)}}{\partial u_{1,0}}}{\frac{\partial \mathcal{E}_{0,0}^{(1)}}{\partial u_{1,0}}} = \kappa_0 = \frac{b^{(1)} \gamma_{1,1}^{(1)} + b^{(2)} \gamma_{1,1}^{(2)} - u_{1,0} \gamma_{1,0}^{(1)}}{\gamma_{1,0}^{(1)}} = \frac{e^{(1)} \gamma_{1,1}^{(1)} + e^{(2)} \gamma_{1,1}^{(2)} - u_{1,0} \gamma_{1,0}^{(2)}}{\gamma_{1,0}^{(2)}}. \tag{2.4.398}$$

From (2.4.394, 2.4.395) we get

$$\frac{\mathcal{E}_{0,0}^{(1)} - u_{0,0} \mathcal{E}_{0,0}^{(1)} \frac{\partial \mathcal{E}_{0,0}^{(1)}}{\partial u_{0,0}}}{\frac{\partial \mathcal{E}_{0,0}^{(1)}}{\partial u_{1,0}}} = \kappa_1 = -\frac{a^{(1)} \gamma_{1,1}^{(1)} + a^{(2)} \gamma_{1,1}^{(2)} - u_{1,0} \gamma_{1,0}^{(1)}}{\gamma_{1,0}^{(1)}} = \frac{d^{(1)} \gamma_{1,1}^{(1)} + d^{(2)} \gamma_{1,1}^{(2)} - u_{1,0} \gamma_{1,0}^{(2)}}{\gamma_{1,0}^{(2)}}, \tag{2.4.399}$$

while from (2.4.396, 2.4.397) we get
(2.4.400)

$$\frac{\mathcal{E}^{(1)}_{0,0}\frac{\partial \mathcal{E}^{(1)}_{0,0}}{\partial u_{0,0}}}{\frac{\partial \mathcal{E}^{(1)}_{0,0}}{\partial u_{1,0}}} = \kappa_2 = -\frac{c^{(1)}\gamma^{(1)}_{1,1} + c^{(2)}\gamma^{(2)}_{1,1} - u_{1,0}\gamma^{(1)}_{1,0}}{\gamma^{(1)}_{1,0}} = \frac{f^{(1)}\gamma^{(1)}_{1,1} + f^{(2)}\gamma^{(2)}_{1,1} - u_{1,0}\gamma^{(2)}_{1,0}}{\gamma^{(2)}_{1,0}}.$$

When $\frac{\partial \mathcal{E}^{(1)}_{0,0}}{\partial u_{1,0}} \neq 0$, solving the equations for $\mathcal{E}^{(1)}_{0,0}$ from (2.4.398, 2.4.399, 2.4.400) we get

(2.4.401)
$$\mathcal{E}^{(1)}_{0,0} = \frac{\kappa_2 u_{0,0} + \kappa_1}{\kappa_0 - u_{10}}.$$

With no loss of generality we can set $\gamma^{(2)} = 0$ and then $e^{(j)} = d^{(j)} = f^{(j)} = 0$, $j = 1, 2$, $w^{(2)} = 0$ and the equations (2.4.398, 2.4.399, 2.4.400) are compatible if $\kappa_2\kappa_0 a^{(1)} = \kappa_1 b^{(1)} c^{(1)}$. The resulting class of linearizable PΔEs (2.4.401) is an extension of the Burgers equation (2.4.292)

(2.4.402)
$$(\kappa_0 - u_{1,0})(\kappa_2 u_{0,1} + \kappa_1)u_{0,0} - (\kappa_0 - u_{1,1})(\kappa_2 u_{0,0} + \kappa_1)u_{1,0} = 0,$$

which reduces to it when $\kappa_0 \neq 0$, $\kappa_1 = 1$ and $\kappa_2 = 0$. In (2.4.402) in all generality κ_1 can be taken to be either 0 or 1. Other two Burgers equations are obtained taking $\kappa_0 \neq 0$, $\kappa_1 = 0$ and $\kappa_2 \neq 0$ or $\kappa_0 = 0$, $\kappa_1 = 1$ and $\kappa_2 \neq 0$. All the these three Burgers equations can be transformed to $\left(1 + u_{0,0}\right) u_{1,0} = \left(1 + u_{0,1}\right) u_{0,0}$ and wc rccover the results obtained in Section 3.7.1 [745]. Moreover, if $\kappa_2 \neq 0$, $\kappa_1 = 1$ and $\kappa_0 \neq 0$, by the transformation $u_{0,0} = \kappa_0 \frac{e_1 - o_2}{e_1 - e_2} \frac{\tilde{u}_{0,0} + e_2}{\tilde{u}_{0,0} + o_2}$, where (e_j, o_j), $j = 1, 2$ are arbitrary parameters, $\tilde{u}_{0,0}$ will satisfy the Hietarinta equation [381, 769]

(2.4.403)
$$\frac{u_{0,0} + e_2}{u_{0,0} + e_1}\frac{u_{1,1} + o_2}{u_{1,1} + o_1} = \frac{u_{1,0} + e_2}{u_{1,0} + o_1}\frac{u_{0,1} + o_2}{u_{0,1} + e_1},$$

with $\frac{(o_1 - e_2)(e_1 - o_2)}{(e_1 - e_2)(o_1 - o_2)} = -\kappa_2\kappa_0$.

When $\frac{\partial \mathcal{E}^{(1)}_{0,0}}{\partial u_{1,0}} = 0$, we must have $\mathcal{E}^{(1)}_{0,0}\gamma^{(j)}_{1,0} = 0$ which has no nontrivial solution.

4.11.2. *Necessary and sufficient conditions for a PΔE to be linear.* Let us consider the case when (2.4.347) is defined on four points i.e.

(2.4.404)
$$F_{n,m}(u_{n,m}, u_{n+1,m}, u_{n,m+1}, u_{n+1,m+1}) = 0,$$

(see Fig.2.3).

Theorem 19. *Necessary and sufficient conditions for a discrete equation* (2.4.404) *to have a symmetry of infinitesimal generator* (2.4.350)*, is that it is linear. The infinitesimal symmetry coefficient* $\phi_{n,m}$ *must satisfy the following linear homogeneous discrete equation*

(2.4.405)
$$\begin{aligned}\mathfrak{L}\Big(\phi_{n,m}, \phi_{n+1,m}, \phi_{n,m+1}, \phi_{n+1,m+1}\Big) &= \phi_{n+1,m+1} - a_{n,m}\phi_{n,m}\\ &\quad - b_{n,m}\phi_{n,m+1} - c_{n,m}\phi_{n+1,m}\\ &= 0.\end{aligned}$$

PROOF. It is almost immediate to prove that a linear PΔE defined on four lattice points (2.4.404) has a symmetry (2.4.350) where $\phi_{n,m}$ satisfies a homogeneous linear equation. To obtain this result it is just sufficient to solve the invariance condition

(2.4.406)
$$\mathrm{pr}\hat{X} F_{n,m}\Big|_{F_{n,m}=0} = 0.$$

Not so easy is the proof that an equation which has such a symmetry (2.4.350) must be linear. In full generality (2.4.405) can be rewritten as

$$\phi_{n+1,m+1} = \mathscr{G}\Big(n, m, \phi_{n,m}, \phi_{n+1,m}, \phi_{n,m+1}\Big), \tag{2.4.407}$$

and, by assumption, (2.4.404) does not depend on $\phi_{n,m}$ and (2.4.407) on $u_{n,m}$.

Let us prolong the symmetry generator (2.4.350) to all points contained in (2.4.404)

$$\mathrm{pr}\hat{X} = \phi_{n,m}\partial_{u_{n,m}} + \phi_{n+1,m}\partial_{u_{n+1,m}} + \phi_{n,m+1}\partial_{u_{n,m+1}} + \phi_{n+1,m+1}\partial_{u_{n+1,m+1}}. \tag{2.4.408}$$

The most generic equation (2.4.404) having the symmetry (2.4.408) will be written in terms of its invariants

$$F_{n,m}\Big(K_1, K_2, K_3\Big) = 0, \tag{2.4.409}$$

with

$$K_1 = \frac{u_{n,m+1}}{\phi_{n,m+1}} - \frac{u_{n,m}}{\phi_{n,m}}, \quad K_2 = \frac{u_{n+1,m}}{\phi_{n+1,m}} - \frac{u_{n,m}}{\phi_{n,m}}, \quad K_3 = \frac{u_{n+1,m+1}}{\phi_{n+1,m+1}} - \frac{u_{n,m}}{\phi_{n,m}}. \tag{2.4.410}$$

As (2.4.409) depends on n, m we can with no loss of generality replace the invariants (2.4.410) in (2.4.409) by the functions

$$\tilde{K}_1 = u_{n,m+1} - u_{n,m}\frac{\phi_{n,m+1}}{\phi_{n,m}}, \quad \tilde{K}_2 = u_{n+1,m} - u_{n,m}\frac{\phi_{n+1,m}}{\phi_{n,m}}, \tag{2.4.411}$$

$$\tilde{K}_3 = u_{n+1,m+1} - u_{n,m}\frac{\mathscr{G}\Big(n, m, \phi_{n,m}, \phi_{n+1,m}, \phi_{n,m+1}\Big)}{\phi_{n,m}}.$$

Invariance of (2.4.409) then requires

$$\frac{\partial F_{n,m}}{\partial \phi_{n,m}} = \frac{\partial F_{n,m}}{\partial \phi_{n+1,m}} = \frac{\partial F_{n,m}}{\partial \phi_{n,m+1}} = 0,$$

i.e.

$$\frac{\partial F_{n,m}}{\partial \tilde{K}_1}\Big(u_{n,m}\frac{\phi_{n,m+1}}{\phi^2_{n,m}}\Big) + \frac{\partial F_{n,m}}{\partial \tilde{K}_2}\Big(u_{n,m}\frac{\phi_{n+1,m}}{\phi^2_{n,m}}\Big) + \frac{\partial F_{n,m}}{\partial \tilde{K}_3}\Big(\frac{\mathscr{G}}{\phi^2_{n,m}} - \frac{\mathscr{G}_{,\phi_{n,m}}}{\phi_{n,m}}\Big)u_{n,m} = 0, \tag{2.4.412}$$

$$\frac{\partial F_{n,m}}{\partial \tilde{K}_1}\Big(-\frac{u_{n,m}}{\phi_{n,m}}\Big) + \frac{\partial F_{n,m}}{\partial \tilde{K}_3}\Big(-\frac{\mathscr{G}_{,\phi_{n,m+1}}}{\phi_{n,m}}\Big)u_{n,m} = 0, \tag{2.4.413}$$

$$\frac{\partial F_{n,m}}{\partial \tilde{K}_2}\Big(-\frac{u_{n,m}}{\phi_{n,m}}\Big) + \frac{\partial F_{n,m}}{\partial \tilde{K}_3}\Big(-\frac{\mathscr{G}_{,\phi_{n+1,m}}}{\phi_{n,m}}\Big)u_{n,m} = 0. \tag{2.4.414}$$

As $\frac{\partial F_{n,m}}{\partial \tilde{K}_j} \neq 0$ and (2.4.412, 2.4.413, 2.4.414) are a homogeneous system of algebraic equations, the determinant of the coefficients must be zero. Consequently the function $\mathscr{G}$ must satisfy the first order linear PDE

$$\mathscr{G} - \phi_{n,m}\mathscr{G}_{,\phi_{n,m}} - \phi_{n,m+1}\mathscr{G}_{,\phi_{n,m+1}} - \phi_{n+1,m}\mathscr{G}_{,\phi_{n+1,m}} = 0 \tag{2.4.415}$$

i.e. $\mathscr{G}$ is given by

$$\mathscr{G} = \phi_{n,m}f(\xi, \tau), \qquad \xi = \frac{\phi_{n+1,m}}{\phi_{n,m}}, \quad \tau = \frac{\phi_{n,m+1}}{\phi_{n,m}}, \tag{2.4.416}$$

where $f(\xi, \tau)$ is an arbitrary function of its arguments.

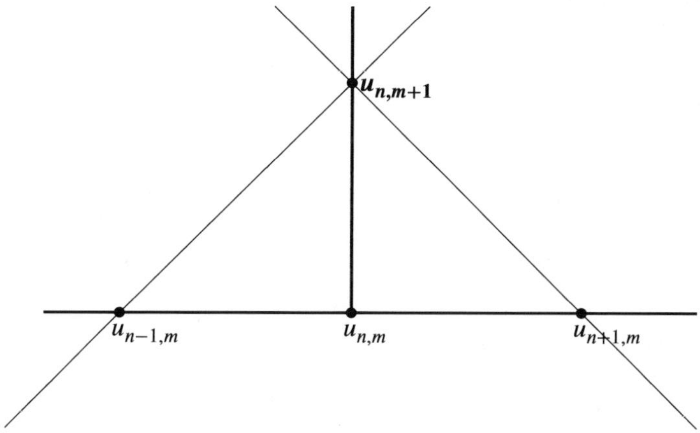

FIGURE 2.6. Four points on a triangle.

For $\mathscr{G}$ given by (2.4.416) the system (2.4.412, 2.4.413, 2.4.414) reduces to the following two equations for $\frac{\partial F_{n,m}}{\partial \tilde{K}_j}$

(2.4.417)
$$\frac{\partial F_{n,m}}{\partial \tilde{K}_1} + f_\xi \frac{\partial F_{n,m}}{\partial \tilde{K}_3} = 0, \quad \frac{\partial F_{n,m}}{\partial \tilde{K}_2} + f_\tau \frac{\partial F_{n,m}}{\partial \tilde{K}_3} = 0,$$

whose solution is obtained by solving (2.4.417) on the characteristics

(2.4.418)
$$F_{n,m} = F_{n,m}(L),$$
$$L = u_{n+1,m+1} - f u_{n,m} - f_{,\xi}\Big(u_{n,m+1} - u_{n,m}\frac{\phi_{n,m+1}}{\phi_{n,m}}\Big) - f_{,\tau}\Big(u_{n+1,m} - u_{n,m}\frac{\phi_{n+1,m}}{\phi_{n,m}}\Big).$$

Requiring that $F_{n,m}$ be independent of $\phi_{n,m}$, $\phi_{n+1,m}$ and $\phi_{n,m+1}$ we get $f_{,\xi\xi} = f_{,\tau\tau} = f_{,\xi\tau} = 0$ i.e.

(2.4.419)
$$f = a_{n,m} + b_{n,m}\xi + c_{n,m}\tau, \quad \mathscr{G} = a_{n,m}\phi_{n,m} + b_{n,m}\phi_{n,m+1} + c_{n,m}\phi_{n+1,m},$$

(2.4.420)
$$L = u_{n+1,m+1} - a_{n,m}u_{n,m} - b_{n,m}u_{n,m+1} - c_{n,m}u_{n+1,m}.$$

$F_{n,m} = 0$ is an (non autonomous, maybe transcendental) equation for L which, when solved, gives $L = d_{n,m}$, where $d_{n,m}$ stands for the set of the zeros of the equation (in addition to n and m possibly dependent on a set of parameters). In conclusion $u_{n,m}$ must satisfy the linear equation

(2.4.421)
$$u_{n+1,m+1} - a_{n,m}u_{n,m} - b_{n,m}u_{n,m+1} - c_{n,m}u_{n+1,m} - d_{n,m} = 0$$

□

Remark 6. *The proof of Theorem* 19 *does not depends on the position of the four lattice points considered in Fig.* 2.3. *The same result is also valid if the four points are put on the triangle shown in Fig.* 2.6, *i.e.*

(2.4.422)
$$F_{n,m}\Big(u_{n-1,m}, u_{n,m}, u_{n+1,m}, u_{n,m+1}\Big) = 0.$$

4.11.3. *Four-point linearizable lattice schemes.* In this Section we provide the symmetry conditions under which a scheme, as introduced in Section 1.3 is linearizable. We limit ourselves to the case when the equation and the lattice are defined on a quad-graph (see Fig. 2.3), i.e. we consider one scalar equation for a continuous function of two (continuous) variables: $u_{m,n} = u(x_{m,n}, t_{m,n})$ defined on a four-point lattice. How to find symmetries for such systems was considered and discussed in Section 1.4.

Symmetries of a linear partial difference scheme. To be able to linearize a difference scheme (1.3.7) [520] using the knowledge of its symmetries we must be able to characterize the symmetries of a linear scheme. To do so here we prove a theorem on the structure of the symmetries of a linear partial difference scheme:

Theorem 20. *Necessary and sufficient conditions for three difference equations* $\mathfrak{E}_{m,n} = 0$, $\mathfrak{F}_{m,n} = 0$ *and* $\mathfrak{G}_{m,n} = 0$ *defined on four points* $\{(m,n),(m+1,n),(m,n+1),(m+1,n+1)\}$ *for a scalar function* $u_{m,n}(x_{m,n}, t_{m,n})$ *and the lattice variables* $x_{m,n}$ *and* $t_{m,n}$ *to be linear is that they are invariant with respect to the following infinitesimal generator*

$$\hat{X}_{m,n} = \upsilon_{m,n}\partial_{u_{m,n}} + \sigma_{m,n}\partial_{x_{m,n}} + \eta_{m,n}\partial_{t_{m,n}}. \tag{2.4.423}$$

The infinitesimal symmetry coefficients $\upsilon_{m,n}$, $\sigma_{m,n}$, $\eta_{m,n}$ *satisfy three linear autonomous equations which we can write for* $n = m = 0$ *as* $\upsilon_{1,1} = \mathfrak{e}_{0,0}$, $\chi_{1,1} = \mathfrak{f}_{0,0}$ *and* $\tau_{1,1} = \mathfrak{g}_{0,0}$. *The functions* $\mathfrak{e}$, $\mathfrak{f}$ *and* $\mathfrak{g}$ *are given by:*

$$\begin{aligned}
\mathfrak{e}_{0,0} &= a_1\upsilon_{0,0} + a_2\upsilon_{0,1} + a_3\upsilon_{1,0} + a_4\sigma_{0,0} + a_5\sigma_{0,1} + a_6\sigma_{1,0} + a_7\eta_{0,0}\\
&\quad + a_8\eta_{0,1} + a_9\eta_{1,0},\\
\mathfrak{f}_{0,0} &= b_1\upsilon_{0,0} + b_2\upsilon_{0,1} + b_3\upsilon_{1,0} + b_4\sigma_{0,0} + b_5\sigma_{0,1} + b_6\sigma_{1,0} + b_7\eta_{0,0}\\
&\quad + b_8\eta_{0,1} + b_9\eta_{1,0},\\
\mathfrak{g}_{0,0} &= c_1\upsilon_{0,0} + c_2\upsilon_{0,1} + c_3\upsilon_{1,0} + c_4\sigma_{0,0} + c_5\sigma_{0,1} + c_6\sigma_{1,0} + c_7\eta_{0,0}\\
&\quad + c_8\eta_{0,1} + c_9\eta_{1,0},
\end{aligned} \tag{2.4.424}$$

where $a_1, \cdots, c_9$ *depend only on the lattice indexes and where, here and in the following, for the sake of simplicity we set in any discrete variable on the square* $z_{m+i,n+j} = z_{i,j}$. *The linear equations* $\mathfrak{E}_{m,n} = 0$, $\mathfrak{F}_{m,n} = 0$ *and* $\mathfrak{G}_{m,n} = 0$ *have the form:*

$$\begin{aligned}
u_{1,1} &= a_1u_{0,0} + a_2u_{0,1} + a_3u_{1,0} + a_4x_{0,0} + a_5x_{0,1} + a_6x_{1,0} + a_7t_{0,0}\\
&\quad + a_8t_{0,1} + a_9t_{1,0},\\
x_{1,1} &= b_1u_{0,0} + b_2u_{0,1} + b_3u_{1,0} + b_4x_{0,0} + b_5x_{0,1} + b_6x_{1,0} + b_7t_{0,0}\\
&\quad + b_8t_{0,1} + b_9t_{1,0},\\
t_{1,1} &= c_1u_{0,0} + c_2u_{0,1} + c_3u_{1,0} + c_4x_{0,0} + c_5x_{0,1} + c_6x_{1,0} + c_7t_{0,0}\\
&\quad + c_8t_{0,1} + c_9t_{1,0}.
\end{aligned} \tag{2.4.425}$$

PROOF. A generic PΔE (2.4.404), depending on $u_{m,n}(x_{m,n}, t_{m,n})$ and the lattice variables $x_{m,n}$ and $t_{m,n}$ in the four points $\{(m,n),(m+1,n),(m,n+1),(m+1,n+1)\}$, will depend on 12 variables. We require that (2.4.404) be invariant under the prolongation of (2.4.423), as given by

$$pr\,\hat{X}_{m,n} = \textstyle\sum_{i,j} \hat{X}_{m+i,n+j}. \tag{2.4.426}$$

The invariance condition is

$$\mathrm{pr}\hat{X}\, E_a\Big|_{E_c=0} = 0 \quad 1 \le a, c \le 3, \tag{2.4.427}$$

where E_a, $a = 1, 2, 3$ are the three equations (2.4.425) introduced in Theorem 20. For $\hat{X}$ given in (2.4.423) it follows that the generic linear equations $\mathfrak{E}_{m,n} = 0$, $\mathfrak{F}_{m,n} = 0$ and $\mathfrak{G}_{m,n} = 0$ should depend on a set of 11 independent invariants depending on $v_{m,n}$, $\sigma_{m,n}$ and $\eta_{m,n}$:

$$
\begin{aligned}
&L_1 = v_{0,0}u_{0,1} - v_{0,1}u_{0,0}, && L_2 = v_{0,0}u_{1,0} - v_{1,0}u_{0,0},\\
&L_3 = v_{0,0}u_{1,1} - \mathfrak{e}_{0,0}u_{0,0}, && L_4 = v_{0,0}x_{0,1} - \sigma_{0,1}u_{0,0},\\
&L_5 = v_{0,0}x_{1,0} - \sigma_{1,0}u_{0,0}, && L_6 = v_{0,0}x_{1,1} - \mathfrak{f}_{0,0}u_{0,0},\\
&L_7 = v_{0,0}t_{0,1} - \eta_{0,1}u_{0,0}, && L_8 = v_{0,0}t_{1,0} - \eta_{1,0}u_{0,0},\\
&L_9 = v_{0,0}t_{1,1} - \mathfrak{g}_{0,0}u_{0,0}, && L_{10} = v_{0,0}x_{0,0} - \sigma_{0,0}u_{0,0},\\
&L_{11} = v_{0,0}t_{0,0} - \eta_{0,0}u_{0,0}.
\end{aligned}
\tag{2.4.428}
$$

As the linear equations, here indicated as $F_{m,n}$, should not depend on the functions $(v_{m,n}, \sigma_{m,n}, \eta_{m,n})$ in the points (m, n), $(m + 1, n)$ and $(m, n + 1)$ we have nine constraints given by

$$
\frac{\partial F_{m,n}}{\partial v_{m+i,n+j}} = 0, \quad \frac{\partial F_{m,n}}{\partial \sigma_{m+i,n+j}} = 0, \quad \frac{\partial F_{m,n}}{\partial \tau_{m+i,n+j}} = 0, \quad (i, j) = (0, 0), (0, 1), (1, 0).
\tag{2.4.429}
$$

Eq. (2.4.429) are first order PDEs for the function $F_{m,n}$ with respect to the 11 invariants. Eq. (2.4.429) can be solved on the characteristics to define three invariants

$$
\begin{aligned}
K_1 = {}& v_{0,0}\{u_{1,1} - [\mathfrak{e}_{0,0,v_{0,1}}u_{0,1} + \mathfrak{e}_{0,0,v_{1,0}}u_{1,0} + \mathfrak{e}_{0,0,\sigma_{0,0}}x_{0,0} + \mathfrak{e}_{0,0,\sigma_{0,1}}x_{0,1}\\
&+ \mathfrak{e}_{0,0,\sigma_{1,0}}x_{1,0} + \mathfrak{e}_{0,0,\eta_{0,0}}t_{0,0} + \mathfrak{e}_{0,0,\eta_{0,1}}t_{0,1} + \mathfrak{e}_{0,0,\eta_{1,0}}t_{1,0}]\} - u_{0,0}\{\mathfrak{e}_{0,0}\\
&- [\mathfrak{e}_{0,0,v_{0,1}}v_{0,1} + \mathfrak{e}_{0,0,v_{1,0}}v_{1,0} + \mathfrak{e}_{0,0,\sigma_{0,0}}\sigma_{0,0} + \mathfrak{e}_{0,0,\sigma_{0,1}}\sigma_{0,1} + \mathfrak{e}_{0,0,\sigma_{1,0}}\sigma_{1,0}\\
&+ \mathfrak{e}_{0,0,\eta_{0,0}}\eta_{0,0} + \mathfrak{e}_{0,0,\eta_{0,1}}\eta_{0,1} + \mathfrak{e}_{0,0,\eta_{1,0}}\eta_{1,0}]\},\\
K_2 = {}& v_{0,0}\{u_{1,1} - [\mathfrak{f}_{0,0,v_{0,1}}u_{0,1} + \mathfrak{f}_{0,0,v_{1,0}}u_{1,0} + \mathfrak{f}_{0,0,\sigma_{0,0}}x_{0,0} + \mathfrak{f}_{0,0,\sigma_{0,1}}x_{0,1}\\
&+ \mathfrak{f}_{0,0,\sigma_{1,0}}x_{1,0} + \mathfrak{f}_{0,0,\eta_{0,0}}t_{0,0} + \mathfrak{f}_{0,0,\eta_{0,1}}t_{0,1} + \mathfrak{f}_{0,0,\eta_{1,0}}t_{1,0}]\} - u_{0,0}\{\mathfrak{f}_{0,0}\\
&- [\mathfrak{f}_{0,0,v_{0,1}}v_{0,1} + \mathfrak{f}_{0,0,v_{1,0}}v_{1,0} + \mathfrak{f}_{0,0,\sigma_{0,0}}\sigma_{0,0} + \mathfrak{f}_{0,0,\sigma_{0,1}}\sigma_{0,1} + \mathfrak{f}_{0,0,\sigma_{1,0}}\sigma_{1,0}\\
&+ \mathfrak{f}_{0,0,\eta_{0,0}}\eta_{0,0} + \mathfrak{f}_{0,0,\eta_{0,1}}\eta_{0,1} + \mathfrak{f}_{0,0,\eta_{1,0}}\eta_{1,0}]\},\\
K_3 = {}& v_{0,0}\{u_{1,1} - [\mathfrak{g}_{0,0,v_{0,1}}u_{0,1} + \mathfrak{g}_{0,0,v_{1,0}}u_{1,0} + \mathfrak{g}_{0,0,\sigma_{0,0}}x_{0,0} + \mathfrak{g}_{0,0,\sigma_{0,1}}x_{0,1}\\
&+ \mathfrak{g}_{0,0,\sigma_{1,0}}x_{1,0} + \mathfrak{g}_{0,0,\eta_{0,0}}t_{0,0} + \mathfrak{g}_{0,0,\eta_{0,1}}t_{0,1} + \mathfrak{g}_{0,0,\eta_{1,0}}t_{1,0}]\} - u_{0,0}\{\mathfrak{g}_{0,0}\\
&- [\mathfrak{g}_{0,0,v_{0,1}}v_{0,1} + \mathfrak{g}_{0,0,v_{1,0}}v_{1,0} + \mathfrak{g}_{0,0,\sigma_{0,0}}\sigma_{0,0} + \mathfrak{g}_{0,0,\sigma_{0,1}}\sigma_{0,1} + \mathfrak{g}_{0,0,\sigma_{1,0}}\sigma_{1,0}\\
&+ \mathfrak{g}_{0,0,\eta_{0,0}}\eta_{0,0} + \mathfrak{g}_{0,0,\eta_{0,1}}\eta_{0,1} + \mathfrak{g}_{0,0,\eta_{1,0}}\eta_{1,0}]\}.
\end{aligned}
\tag{2.4.430}
$$

By construction the three invariants K_i, $i = 1, 2, 3$ are independent and the three equations $\mathfrak{E}_{m,n} = 0$, $\mathfrak{F}_{m,n} = 0$ and $\mathfrak{G}_{m,n} = 0$ will be defined in terms of them. The three invariants K_1, K_2 and K_3 still depend on the functions $(v_{m,n}, \sigma_{m,n}, \eta_{m,n})$ in the points (m, n), $(m+1, n)$ and $(m, n+1)$ while they should depend just on the variables $(u_{m,n}, x_{m,n}, t_{m,n})$ in the points (m, n), $(m+1, n)$, $(m, n+1)$ and $(m+1, n+1)$. The derivatives F_{m,n,K_i}, $i = 1, 2, 3$ will satisfy a set of nine linear equations whose coefficients will form a matrix $\mathfrak{A}$ 9x3. The matrix $\mathfrak{A}$ can have rank 3, 2 or 1. In the case of rank 3 we have $F_{m,n,K_i} = 0$, $i = 1, 2, 3$ i.e. the function $F_{m,n}$ does not depend on the 3 invariants. If the rank of $\mathfrak{A}$ is 2 or 1 we can have at most two independent invariants. If we want to have three invariants we need to require that the coefficients of the matrix $\mathfrak{A}$ be zero, i.e. defining $\alpha_1 = v_{0,0}$, $\alpha_2 = v_{0,1}$, $\alpha_3 = v_{1,0}$,

$\cdots, \alpha_9 = \eta_{1,0}$ we have

$$\frac{\partial K_p}{\partial \alpha_q} = 0, \qquad p = 1, 2, 3,\ q = 1, \cdots, 9. \tag{2.4.431}$$

Eqs. (2.4.431) are linear homogeneous expressions in $u_{i,j}$, $x_{i,j}$ and $t_{i,j}$ with coefficients depending on $v_{i,j}$, $\sigma_{i,j}$ and $\eta_{i,j}$, for appropriate values of i and j. Consequently we have (2.4.425). Then (2.4.431) turn out to be a set of 159 overdetermined PDEs for the functions $\mathfrak{E}_{m,n}$, $\mathfrak{F}_{m,n}$ and $\mathfrak{G}_{m,n}$ whose solution (2.4.424) is obtained using Maple. It depends on 27 integration constants which must be set equal zero if (2.4.425) does not depend on $v_{i,j}$, $\sigma_{i,j}$ and $\eta_{i,j}$. □

A few remarks can be derived from Theorem 20 and must be stressed.

Remark 7. *The equation for $u_{m,n}$ and those for the lattice variables $x_{m,n}$ and $t_{m,n}$ are independent, however the functions appearing in the symmetry* (2.4.423) *do not satisfy equations independent from those satisfied by the lattice scheme. In fact these symmetries correspond to independent superposition laws for the equation and the lattice.*

Remark 8. *If the linear equation for $u_{m,n}$ is autonomous then the coefficients $\{a_4, \cdots, a_9\}$ are zero. The variable $v_{m,n}$ will satisfy a similar equation but the lattice equations can depend linearly on $u_{m,n}$.*

Remark 9. *The proof of Theorem* 20 *does not depend on the position of the four lattice points considered, i.e.* $\{(m,n), (m+1,n), (m,n+1), (m+1,n+1)\}$. *The same result is also valid if the four points are put on the triangle shown in Fig.* 2.6, *i.e.* $\{(m,n), (m+1,n), (m-1,n), (m,n+1)\}$.

Linearizable non linear schemes. Each equation of a difference scheme depends from the continuous variable $u_{m,n}$, $x_{m,n}$ and $t_{m,n}$. If the equations for the lattice variables, $x_{m,n}$ and $t_{m,n}$, are solvable we get

$$x_{m,n} = \mathcal{X}(m, n, c_0, c_1, \cdots), \qquad t_{m,n} = \mathcal{T}(m, n, d_0, d_1, \cdots), \tag{2.4.432}$$

and then the remaining equation for the variable $u_{m,n}$ depends explicitly on n, m and on the integration constants $(c_0, c_1, \cdots, d_0, d_1, \cdots)$ contained in (2.4.432). It will be an algebraic, maybe transcendental, equation of $u_{m,n}$ in the various lattice points involved in the equation. So the difference scheme reduce to a non autonomous equation on a fixed lattice and for its linearization we can apply the result presented in Section 2.4.11 [518].

If the equations for the lattice are not solvable the difference scheme can be thought as a system of coupled equations for the variables $u_{m,n}$, $x_{m,n}$ and $t_{m,n}$ on a fixed lattice. In this way taking into account the results of the previous Section, we can propose the following linearizability theorem:

Theorem 21. *A non linear difference scheme* (1.3.7) *involving $i_1 + i_2$ different points in the m index and $j_1 + j_2$ in the n index for a scalar function $u_{m,n}$ of a 2–dimensional space of coordinates $x_{m,n}$ and $t_{m,n}$ will be linearizable by point transformations*

$$\begin{aligned} w_{m,n}(y_{m,n}, z_{m,n}) = \ & f(x_{m,n}, t_{m,n}, u_{m,n}), \qquad & y_{m,n} &= g(x_{m,n}, t_{m,n}, u_{m,n}), \\ & & z_{m,n} &= k(x_{m,n}, t_{m,n}, u_{m,n}), \end{aligned} \tag{2.4.433}$$

to a linear difference scheme (2.4.425) *for $w_{m,n}$, $y_{m,n}$ and $z_{m,n}$ if it possesses a symmetry generator*

$$\begin{gathered} \hat{X} = \xi(x,t,u)\partial_x + \phi(x,t,u)\partial_t + \psi(x,t,u)\partial_u, \\ \xi(x,u) = \alpha(x,t,u)y, \quad \phi(x,t,u) = \beta(x,t,u)z, \quad \psi(x,t,u) = \gamma(x,t,u)w \end{gathered} \tag{2.4.434}$$

with α, β and γ given functions of their arguments and y, z and w an arbitrary solution of (2.4.424).

Application. We consider here the discretization of the potential Burgers [659] presented by Dorodnitsyn [223] and show that, even if it is reducible by a point transformation to the discrete scheme of the heat equation, is not linearizable by a point transformation. As a consequence we have that also the symmetry preserving discretization of the heat equation presented in [223] is not a linear difference scheme.

The symmetry preserving discretization of the potential Burgers [223] is given by the following scheme written on the stencil defined in terms of (τ, x, Δx, h^+, h^-, w, $\hat{w}$, w_+, w_-)

(2.4.435) $$\frac{\Delta x}{\tau} = \frac{1}{h^+ + h^-}\Big[\frac{h^-}{h^+}(w_+ - w) + \frac{h^+}{h^-}(w - w_-)\Big]$$

(2.4.436) $$e^{\hat{w} - w - \frac{\Delta^2 x}{2\tau}} = 1 + \frac{2\tau}{(h^+)^2}\Big[\frac{w_+ - w}{h^+} - \frac{w - w_-}{h^-}\Big]$$

(2.4.437) $$\tau = t_{m,n+1} - t_{m,n}, \qquad t_{m+1,n} = t_{m-1,n} = t_{m,n} = t.$$

In (2.4.435, 2.4.436) τ is the lattice spacing in the t direction, i.e. $t = t_0 + \tau n$ and

$$w = w_{m,n}(x_{m,n}, t_{m,n}), \ \hat{w} = w_{m,n+1}, \ w_- = w_{m-1,n}, \ w_+ - w_{m+1,n},$$

$$\Delta x = x_{m,n+1} - x_{m,n}, \quad h^+ = x_{m+1,n} - x_{m,n}, \quad h^- = x_{m,n} - x_{m-1,n}.$$

The continuous potential Burgers equation

(2.4.438) $$w_t = w_{xx} - \frac{1}{2}w_x^2,$$

has the following algebra of point symmetries

(2.4.439) $$\hat{X}_1 = \partial_t, \quad \hat{X}_2 = \partial_x, \quad \hat{X}_3 = t\partial_x + x\partial_t, \quad \hat{X}_4 = 2t\partial_t + x\partial_x,$$
$$\hat{X}_5 = \partial_w, \quad \hat{X}_6 = t^2\partial_t + tx\partial_x + \Big(\frac{1}{2}x^2 + t\Big)\partial_w.$$

The discrete invariants of (2.4.439) are

(2.4.440) $$\mathcal{I}_1 = \frac{h^+}{h^-}, \quad \mathcal{I}_2 = \frac{\tau^{1/2}}{h^+}e^{\frac{1}{2}(w - \hat{w}) + \frac{\Delta^2 x}{4\tau}},$$
$$\mathcal{I}_3 = \frac{1}{4}\frac{h^{+2}}{\tau} + \frac{h^{+2}}{h^+ + h^-}\Big[\frac{w_+ - w}{h^+} - \frac{w - w_-}{h^-}\Big],$$
$$\mathcal{I}_4 = \Delta x\frac{h^+}{\tau} - \frac{2h^+}{h^+ + h^-}\Big[\frac{h^-}{h^+}(w_+ - w) + \frac{h^-}{h^+}(w - w_-)\Big],$$

and the discretization of the Burgers (2.4.435,-2.4.437) is written in term of them.

We can apply on the lattice scheme (2.4.435, 2.4.436, 2.4.437) the symmetry generator

(2.4.441) $$\hat{X} = \psi(x, t, w)u\partial_w + \phi(x, t, w)s\partial_t + \xi(x, t, w)y\partial_x,$$

with (x, t, w) satisfying (2.4.435, 2.4.436, 2.4.437) while (y, s, u) are solutions of the linear scheme prescribed by Theorem 20

$$\begin{aligned}
u_{m,n+1} &= a_1 u_{m,n} + a_2 u_{m-1,n} + a_3 u_{m+1,n} + a_4 y_{m,n} + a_5 y_{m-1,n} + a_6 y_{m+1,n} \\
&\quad + a_7 s_{m,n} + a_8 s_{m-1,n} + a_9 s_{m+1,n}, \\
y_{m,n+1} &= c_1 u_{m,n} + c_2 u_{m-1,n} + c_3 u_{m+1,n} + c_4 y_{m,n} + c_5 y_{m-1,n} + c_6 y_{m+1,n} \\
&\quad + c_7 s_{m,n} + c_8 s_{m-1,n} + c_9 s_{m+1,n}, \\
s_{m,n+1} &= b_1 u_{m,n} + b_2 u_{m-1,n} + b_3 u_{m+1,n} + b_4 y_{m,n} + b_5 y_{m-1,n} + b_6 y_{m+1,n} \\
&\quad + b_7 s_{m,n} + b_8 s_{m-1,n} + b_9 s_{m+1,n},
\end{aligned} \tag{2.4.442}$$

where $(a_j,\ b_j,\ c_j,\ j = 1, \cdots, 9)$ are parameters at most depending on n and m. By a long and tedious calculation carried out using a symbolic calculation program we get that

$$\begin{aligned}
\psi(x, t, w) &= \psi_0(t) + \psi_1(t)x + \psi_2(t)x^2, \\
\phi(x, t, w) &= \phi_0(t) + \phi_1(t)x + \phi_2(t)x^2, \\
\xi(x, t, w) &= \xi_0(t) + \xi_1(t)x.
\end{aligned} \tag{2.4.443}$$

Introducing (2.4.443) into the determining equations for the symmetries of the discrete potential Burgers scheme (2.4.435, 2.4.436, 2.4.437) we get 1672 equations for the functions $(\psi_j(t),\ \phi_j(t),\ \xi_j(t),\ j = 0, 1, 2)$ depending on the coefficients $(a_j,\ b_j,\ c_j,\ j = 1, \cdots, 9)$. 168 of those equations do not depend on the coefficients $(a_j,\ b_j,\ c_j,\ j = 1, \cdots, 9)$ and on $(\psi_j(t + \tau),\ \phi_j(t + \tau),\ \xi_j(t + \tau),\ j = 0, 1, 2)$; solving them imposing that $\tau \neq 0$ we get $\psi_j(t) = 0$ for $j = 0, 1, 2$, $\phi_k = 0$ for $k = 1, 2$ and $\xi_k = 0$ for $k = 0, 1$. Introducing this result in the remaining 1508 equations, we get the following 9 equations

$$\begin{aligned}
&b_1\phi_0(t + \tau) = b_2\phi_0(t + \tau) = b_3\phi_0(t + \tau) = b_4\phi_0(t + \tau) = b_5\phi_0(t + \tau) \\
&= b_6\phi_0(t + \tau) = \phi_0(t) - b_7\phi_0(t + \tau) = b_8\phi_0(t + \tau) = b_9\phi_0(t + \tau) = 0.
\end{aligned}$$

If we require $\phi_0(t) \neq 0$, the coefficients b_j, $j = 1, \cdots 6, 8, 9$ must be all zero and $b_7 \neq 0$. As a consequence $\phi_0(t) = b_7^{-n}\bar{\phi}$, with $\bar{\phi}$ an arbitrary constant. In this case we have a symmetry generator $\hat{X} = b_7^{-n} s\partial_t$ which is a consequence of the linearity of (2.4.437). So we can conclude that the potential Burgers scheme (2.4.435, 2.4.436) is not linearizable and that the corresponding discretization of the heat equation [223] is not given by a linear scheme.

CHAPTER 3

Symmetries as integrability criteria

1. Introduction

The generalized symmetry approach to the classification of integrable equations has mainly been developed by a group of researchers belonging to the scientific school of A.B. Shabat in Ufa, Russia (see e.g. the review articles [27, 349, 604, 606–608, 762, 764, 850]). Its discrete version, considered here, is discussed in the papers [12, 33, 548, 549, 552, 755, 841, 842, 845, 851, 852] and in the surveys [27, 349, 764, 850].

Our purpose is to provide a review of the progress made in this field during the last 25 years, and mainly to discuss the discrete version of the generalized symmetry method and the corresponding classification results. We will consider at first DΔEs which belong to the three most important classes of equations: Volterra, Toda and relativistic Toda type equations. Then we will consider PΔEs mainly of the quad-graph form.

The generalized symmetry approach is the only method, as far as we know which enables one not only to test equations for integrability but also to classify integrable equations in classes characterized by arbitrary functions of many variables. Using this method, the classification problem has been solved for classes which include such well-known and important PDEs as the Burgers, Korteweg-de Vries and NLS or, in the differential difference case, equations and systems defined on three lattice points like the Volterra, Toda and relativistic Toda lattice equations. Together with exhaustive lists of integrable equations, a number of essentially new equations have been obtained as a result of this classification.

Some recent results on DΔEs defined on five lattice points have been obtained. However they will not be considered here and we refer the interested reader to the original literature [302, 310–312].

As it is known, equations integrable by the IST have infinitely many generalized symmetries and conservation laws. The generalized symmetry approach enables one to recognize equations possessing these properties. The existence of infinite hierarchies of generalized symmetries and/or conservation laws is used by this method as an integrability criteria. Recent results based the existence of recursion operators for the construction of generalized symmetries and conserved densities of PΔEs have been considered in [611–613] but they will not be discussed here.

Before going over to the discrete case let us first briefly review the situation for PDEs, as presented in the surveys [27, 606–608, 762, 764]. We discuss equations of the form:

$$u_t = f(u, u_1, u_2, u_3)\,, \tag{3.1.1}$$

where $u_t = \frac{\partial u}{\partial t}$ and $u_j = \frac{\partial^j u}{\partial x^j}$ for any $j > 0$. The Korteweg-de Vries equation (2.2.1) is of this class. Here we write it in the form

$$u_t = u_3 + 6uu_1. \tag{3.1.2}$$

All functions (right hand sides of generalized symmetries, conserved densities, coefficients of formal series) will have the form:

(3.1.3) $$\phi = \phi(u, u_1, u_2, \dots, u_k)$$

(here $0 \le k \le \infty$, $u_0 = u$), and the number k is called the order of the function.

A generalized symmetry of order m of (3.1.1) is an equation of the form:

(3.1.4) $$u_\epsilon = g(u, u_1, u_2, \dots, u_m)$$

compatible with (3.1.1), where by ϵ we denote the group parameter. The compatibility condition between (3.1.1) and (3.1.4) implies for the functions f and g:

(3.1.5) $$\frac{\partial^2 u}{\partial t \partial \epsilon} - \frac{\partial^2 u}{\partial \epsilon \partial t} = D_t g - D_\epsilon f = 0\,,$$

where D_t, D_ϵ are the operators of total differentiation (see (1.1.10) in Section 1.1) corresponding to (3.1.1, 3.1.4), defined, together with the operator of total x-derivative D, by

$$D = \frac{\partial}{\partial x} + \sum_{i \ge 0} u_{i+1} \frac{\partial}{\partial u_i}\,, \quad D_t = \frac{\partial}{\partial t} + \sum_{i \ge 0} D^i f \frac{\partial}{\partial u_i}\,, \quad D_\epsilon = \frac{\partial}{\partial \epsilon} + \sum_{i \ge 0} D^i g \frac{\partial}{\partial u_i}\,.$$

A conservation law of (3.1.1) is given by the equation

(3.1.6) $$D_t p = D q$$

for some local functions p and q of the form (3.1.3). The function p is called a conserved density. p is such that $D_t p \in \mathrm{Im} D$, i.e. is represented as the total x-derivative of a function q.

A conserved density of the form $p = c + D\tilde{p}$, where c is a constant, is trivial, as in such a case one can always find a function q: $q = \tilde{c} + D_t \tilde{p}$, where $\tilde{c}$ is another constant. One can easily check if p is trivial by using a formal variational derivative operator. A formal variational derivative of a function ϕ of order k given by (3.1.3) is defined as:

(3.1.7) $$\frac{\delta \phi}{\delta u} = \sum_{i=0}^{k} (-D)^i \frac{\partial \phi}{\partial u_i}\,.$$

This notion is closely related to the notion of the variational derivative given in [659]. The following result is very useful:

(3.1.8) $$\frac{\delta \phi}{\delta u} = 0$$

iff ϕ can be written as:

(3.1.9) $$\phi = c + D\psi\,,$$

where c is a constant and ψ a function of the form (3.1.3). As a consequence of this result, a conserved density p is trivial if $\frac{\delta p}{\delta u} = 0$. The order of a nontrivial conserved density p is the same as the order of the function $\frac{\delta p}{\delta u}$.

Generalized symmetries and conservation laws lead to the so-called integrability conditions. Before introducing them we need to define formal series written in terms of powers of D. A formal series of order k, A_k (see (1.1.26) Section 1.1), is given by

(3.1.10) $$A_k = a_k D^k + a_{k-1} D^{k-1} + \cdots + a_0 + a_{-1} D^{-1} + \cdots\,,$$

where a_j are functions of the form (3.1.3). The product of two formal series is uniquely defined by

$$A_k \circ B_m = \sum_{i \le k, j \le m} a_i D^i \circ b_j D^j,$$

$$a_i D^i \circ b_j D^j = a_i b_j D^{i+j} + \sum_{n \ge 1} \binom{i}{n} a_i D^n(b_j) D^{i+j-n},$$

where

$$\binom{i}{n} = \frac{i(i-1)(i-2)\dots(i-n+1)}{n!}$$

is the standard binomial coefficient, and by $\circ$ we mean the multiplication of operators. For any function ϕ of the form (3.1.3), we define the Fréchet derivative ϕ_* and its corresponding adjoint operator $\phi_*^\dagger$ as

$$\phi_* = \sum_{i=0}^{k} \frac{\partial \phi}{\partial u_i} D^i, \qquad \phi_*^\dagger = \sum_{i=0}^{k} (-D^i) \circ \frac{\partial \phi}{\partial u_i}, \tag{3.1.11}$$

which are particular formal series (3.1.10).

One can rewrite the compatibility condition (3.1.5) as:

$$(D_t - f_*) g = 0 \tag{3.1.12}$$

and the relation $\frac{\delta}{\delta u} D_t p = 0$, which follows from (3.1.6), as:

$$(D_t + f_*^\dagger) \varrho = 0, \qquad \varrho = \frac{\delta p}{\delta u}. \tag{3.1.13}$$

One can obtain from (3.1.4) the formal series of order m:

$$L = g_* + 0D^{-1} + 0D^{-2} + \cdots, \tag{3.1.14}$$

which will be an approximate solution of length m of the equation:

$$L_t = [f_*, L] = f_* L - L f_*, \tag{3.1.15}$$

where L_t is obtained from L by differentiating its coefficients with respect to t. The series $L_t - [f_*, L]$ has the form:

$$L_t - [f_*, L] = b_{m+3} D^{m+3} + b_{m+2} D^{m+2} + b_{m+1} D^{m+1} + \cdots. \tag{3.1.16}$$

L is called an approximate solution of (3.1.15) of length l if the first l coefficients of (3.1.16) vanish, i.e.

$$L_t - [f_*, L] = b_{m+3-l} D^{m+3-l} + b_{m+2-l} D^{m+2-l} + \cdots.$$

By applying the Fréchet derivative to both sides of (3.1.12) one can prove that $l = m$. One obtains in fact

$$g_{*,t} - [f_*, g_*] = f_{*,t}.$$

In a similar way, if a conservation law has order $m > 3$, we can apply the Fréchet derivative to (3.1.13) and prove the following result: the formal series

$$\mathfrak{S} = \varrho_* + 0D^{-1} + 0D^{-2} + \dots \tag{3.1.17}$$

of order m is an approximate solution of length $m - 3$ of the equation

$$\mathfrak{S}_t + \mathfrak{S} f_* + f_*^\dagger \mathfrak{S} = 0. \tag{3.1.18}$$

Approximate solutions of length l are defined in this case in a quite similar way as for the approximate symmetries. Such approximate solutions of (3.1.15) and (3.1.18) are called formal symmetry and formal conserved density, respectively.

We can not only multiply formal series (3.1.10) but also obtain its inverse A_k^{-1} and its k-order root $A_k^{1/k}$, using the standard definitions: $A_k^{-1}A_k = A_kA_k^{-1} = 1$, $(A_k^{1/k})^k = A_k$. So, the fractional powers $A_k^{i/k}$, where i and k are an arbitrary integers, are well defined. Let us note that if one starts from the series (3.1.14) or (3.1.17), one can obtain infinitely many nonzero coefficients of the resulting series.

Two formal symmetries of orders m and $\hat{m}$, L and $\hat{L}$, by multiplication provide new formal symmetries: $L\hat{L}$ and $L^{i/m}$. Two formal conserved densities $\mathfrak{S}$ and $\hat{\mathfrak{S}}$ generate a formal symmetry: $L = \mathfrak{S}^{-1}\hat{\mathfrak{S}}$. The following formula is valid for constructing a new formal conserved density: $\mathfrak{S}L = \hat{\mathfrak{S}}$. I.e. given a formal symmetry L and a formal conserved density $\mathfrak{S}$ we can construct by multiplication a new formal conserved density $\hat{\mathfrak{S}}$. Using these properties, we can simplify the problem and consider just formal symmetries and conserved densities of order 1 and of an arbitrarily big length k.

If (3.1.1) has generalized symmetries and conservation laws of arbitrarily high orders, we can calculate arbitrarily many coefficients of the formal symmetries and formal conserved densities of first order using (3.1.15) and (3.1.18). In doing so, integrability conditions will appear which will have the form: $\mathscr{H} \in \operatorname{Im} D$, where the function $\mathscr{H}$ does not depend on the form and orders of the symmetries and conservation laws, but it is expressed only in terms of the right hand side f of (3.1.1).

As an example, let us write down the integrability conditions in the following particular case:

$$u_t = u_3 + F(u, u_1)\,. \tag{3.1.19}$$

The integrability conditions, which come from the existence of generalized symmetries are derived from (3.1.15). They are of the form:

$$D_t p_i = D q_i\,, \qquad i \ge 1\,, \tag{3.1.20}$$

i.e. have the form of conservation laws. The first three conserved densities read:

$$p_1 = \frac{\partial F}{\partial u_1}\,, \qquad p_2 = \frac{\partial F}{\partial u}\,, \qquad p_3 = q_1\,. \tag{3.1.21}$$

The conditions which come from the existence of conservation laws are obtained from (3.1.18). They are of the form:

$$p_{2j} = D\sigma_{2j}\,, \qquad j \ge 1\,. \tag{3.1.22}$$

The conditions (3.1.22) mean that the even conserved densities are trivial. The functions q_{2j} are easily expressed in terms of the functions σ_{2j}: $q_{2j} = c_{2j} + D_t\sigma_{2j}$, where c_{2j} are some constants.

The integrability conditions (3.1.20-3.1.22) can be formulated in the alternative way. We can write

$$D_t\frac{\partial F}{\partial u_1}\,, \quad \frac{\partial F}{\partial u}\,, \quad D_t q_1 \ \in \operatorname{Im} D\,,$$

which imply that there exist some functions q_1, q_3 which satisfy relations (3.1.20) with $i = 1$, $i = 3$ and a function σ_2 which satisfies relation (3.1.22) with $j = 1$. The other conserved densities p_i are similar to p_3 and have a dependence on the functions q_i defined by the previous conditions.

One has to check the integrability conditions step by step. At first we check (3.1.20) with $i = 1$ and find the function q_1. To check the condition (3.1.20) with $i = 1$, we use the equivalence between (3.1.8) and (3.1.9). We check (3.1.8) for $\phi = D_t p_1$ and then, if it is satisfied, represent ϕ in the form (3.1.9) and verify whether $c = 0$. Then we pass to (3.1.20) with $i = 3$.

Integrability conditions allow one to check whether a given equation is integrable. Moreover, in many cases these conditions enable us to classify equations, i.e. to obtain complete lists of integrable equations. As integrability conditions are only necessary conditions for the existence of generalized symmetries and/or conservation laws, we then have to prove that equations of the resulting list really possess generalized symmetries and conservation laws. To do so we mainly construct generalized symmetries using Miura type transformations and master symmetries.

The use of Miura type transformations in the classification problems is discussed in [607–610,756,762,782,784,844,846,847,850]. The original Miura transformation (2.2.3) [617] brings any solution v of the modified KdV (2.2.2) into a solution of the KdV (3.1.2). In the case of (3.1.1), a Miura type transformation has the form:

$$u = s(v, v_1, \dots v_k) , \quad k > 0 , \tag{3.1.23}$$

and transforms an equation of the form

$$v_t = \hat{f}(v, v_1, v_2, v_3) \tag{3.1.24}$$

into (3.1.1). If a conservation law (3.1.6) of (3.1.1) defined by the functions

$$p = p(u, u_1, \dots u_{k_1}) , \qquad q = q(u, u_1, \dots u_{k_2})$$

is known, one obtains a conservation law $D_t \hat{p} = D\hat{q}$ for (3.1.24), defining

$$\hat{p} = p(s, Ds, \dots D^{k_1} s) , \qquad \hat{q} = q(s, Ds, \dots D^{k_2} s) .$$

The notion of master symmetry has been introduced in [264] and later discussed in [262,284,285,652] and has been considered before in (2.2.113) when discussing the KdV in Section 2.2.2. The master symmetry of the KdV (3.1.2) is

$$u_\tau = xu_t + 4(u_2 + 2u^2) + 2u_1 D^{-1} u , \tag{3.1.25}$$

where D^{-1} is the inverse of the operator D, or an x-integral, as shown in [262,651]. In the previous Chapter we constructed master symmetries that are also symmetries of integrable equations, either PDEs or DΔEs or PΔEs. If $u_\tau = h$ is a master symmetry of (3.1.1), and p is its conserved density, then the new conserved densities p_i are obtained by total τ-differentiation: $p_i = D_\tau^i p$, $i \geq 1$. Generalized symmetries $u_{\epsilon_i} = g_i$ of (3.1.1) are given by (cf. (3.1.5)):

$$g_1 = D_\epsilon h - D_\tau f , \quad g_2 = D_{\epsilon_1} h - D_\tau g_1 , \quad \dots .$$

Let us now go back to the problem of symmetries as integrability criteria for DΔEs. In Section 3.2 the general theory of formal symmetries for DΔEs will be given in the simple case of scalar equations depending just on nearest neighboring lattice points. In doing so we introduce the notions for DΔEs of generalized symmetry, conservation law, formal symmetry, formal conserved density, Miura transformation and master symmetry. We then discuss the integrability conditions which do not depend on the form and order of generalized symmetries and conservation laws and are expressed only in terms of the equation at study. It will be explained how to derive the integrability conditions and how to use them for testing and classifying the equations. At the end we discuss briefly the case of systems of DΔEs which are necessary for studying Toda type and relativistic Toda type equations.

In Section 3.3 we present the classification results for the Volterra, Toda and relativistic Toda type DΔEs. We give some integrability conditions and the complete lists of integrable equations. Those lists of equations are obtained using the integrability conditions up to point transformations even though those conditions are only necessary conditions for the integrability. This is the reason why we need to show how to construct hierarchies of generalized symmetries and conservation laws. We also discuss non point transformations of Miura type or non point invertible transformations which relate different equations of the same list or equations belonging to two different lists.

An extension of the method to the case of inhomogeneous DΔEs with an explicit dependence on the discrete spatial variable and continuous time is discussed in Section 3.4. We mainly pay our attention to Volterra and Toda type equations, but an example of the relativistic Toda type will be presented as well.

The standard scheme of the generalized symmetry method provides results also in the case of scalar evolutionary DΔEs of low orders or systems of few equations. Different classes of equations mentioned in Section 3.5 require modification of the scheme or even the use of different methods. In Section 3.5.1 we consider scalar evolutionary DΔEs of arbitrary order and derive for such equations a few integrability conditions. Then, in Section 3.5.2, we discuss results concerning multi-component DΔEs (vector and matrix ones, in particular). We mainly pay attention in this case to multi-component generalizations of the potential Volterra equation.

In Section 3.6, we consider the generalized symmetry method for PΔEs on the square lattice. Those equations are discrete analogs of the hyperbolic equations including the well-known sine-Gordon, Tzitzèika and Liouville equations.

At the end in Section 3.7 we apply some of the methods introduced in this Chapter to the case of C-integrable or linearizable PΔEs on three-point or four-point lattices and carry out the classification in the multilinear case.

2. The generalized symmetry method for DΔEs

This Section is devoted to the general theory of the generalized symmetry method in the DΔE case. It will be discussed in the simple case of Volterra type equations defined by one arbitrary function of three variables. At the end we extend the discussion to systems of lattice equations. Further details on the theory can be found in the papers [21,27,549,552, 764,845,847,850,852].

In Section 3.2.1 we discuss generalized symmetries and conservation laws and in Section 3.2.2 we derive an integrability condition, using the existence of one generalized symmetry. The notion of formal symmetry is introduced in Section 3.2.3, and two more integrability conditions are derived, using this notion. In Section 3.2.4 we introduce a formal conserved density and then obtain two additional integrability conditions. Properties of all five integrability conditions are discussed in Section 3.2.5. The use of the Hamiltonian structure, Miura transformations and master symmetries for the construction of generalized symmetries and conservation laws is discussed in Sections 3.2.6 and 3.2.7. The case of systems of lattice equations is considered in Section 3.2.8 in the example of Toda type equations. In Section 3.2.9 we derive the integrability conditions in the more difficult case of relativistic Toda type equations.

2.1. Generalized symmetries and conservation laws. Let us consider the class of lattice equations

$$\dot{u}_n = f(u_{n+1}, u_n, u_{n-1}) \equiv f_n\,, \qquad \frac{\partial f_n}{\partial u_{n+1}} \neq 0\,, \quad \frac{\partial f_n}{\partial u_{n-1}} \neq 0\,, \tag{3.2.1}$$

where $u_n = u_n(t)$, the dot denotes the derivative with respect to the continuous time variable t, and the index n is an arbitrary integer. We can think of $u_n(t)$ as an infinite set of functions of one continuous variable: $\{u_n(t) \,:\, n \in \mathbb{Z}\}$, and (3.2.1) as an infinite system of ODEs defined by one arbitrary function of three variables: $f(z_1, z_2, z_3)$. The well-known Volterra equation considered in Section 2.3.3

$$\dot{u}_n = u_n(u_{n+1} - u_{n-1}) \tag{3.2.2}$$

belongs to this class.

The time t of (3.2.1) may be complex and one may, when necessary, use the transformation $\tilde{t} = it$; the functions $u_n(t)$ and $f(z_1, z_2, z_3)$ are complex-valued functions of complex variables. By considering complex functions we avoid looking at many particular cases and simplify the calculation when we derive the integrability conditions or solve the classification problem. If, as a result of classification, we obtain an integrable equation, it will be also integrable in the real case, when we can pass to the real variables t, u_n and to the real function f. For instance, the Volterra equation (3.2.2) possesses the same infinite hierarchies of generalized symmetries and conservation laws in both cases: when t and u_n are complex or real.

Definition 1. *Let us consider complex-valued functions of N complex variables which are analytic on an open and connected subset of $\mathbb{C}^N$. We consider only single-valued functions and, in the case of multi-valued ones (like $\sqrt{z}$ and $\log z$), we choose a single-valued branch, the principal branch. We will call such functions locally analytic functions.*

By reducing, if necessary, the domain of definition of the function one can apply any arithmetical operation to a locally analytic function, compose them and compute their inverses $\varphi^{-1}(z)$, or find implicitly defined functions:

$$w = \varphi(z_1, z_2) \quad \Rightarrow \quad z_1 = \psi(w, z_2)\,.$$

Any problem under consideration (such as the classification, derivation of the integrability conditions or testing an equation for integrability) deals with a finite number of such functions and is solved in a finite number of steps.

Definition 2. *By the classification problem we mean looking for an unknown function of many variables, such as the function $f(z_1, z_2, z_3)$ which appears at the right hand side of* (3.2.1) *in such a way that for it we can find generalized symmetries.*

From the defining equations for the existence of generalized symmetries, we get some differential-functional relations for the unknown function which must be satisfied identically. To find these identities we use the following two properties.

Property 1: For any locally analytic function φ by reducing, if necessary, the domain of definition we have only two possible cases: either $\varphi \neq 0$ everywhere in the domain or $\varphi \equiv 0$.

Property 2: There are no divisors of zero, i.e.

$$\varphi_1 \varphi_2 \equiv 0 \qquad \Rightarrow \qquad \varphi_1 \equiv 0 \quad \text{or} \quad \varphi_2 \equiv 0\,.$$

We can differentiate functions as many times as necessary and solve differential equations. As a result of the classification, we will obtain in the right hand side of an equation such functions as $\sqrt{z}$, $\log z$, hyperbolic and elliptic functions, but not $\bar{z}$ and $|z|$. The obtained integrable equations will be expressed in terms of analytic functions defined in a domain which may be very small. But those equations remain integrable if one passes to globally defined analytic complex functions or to real functions and real time.

S-integrable equations [141] are known to have infinitely many generalized symmetries. Also C-integrable equations have this property. Generalized symmetries of (3.2.1) will be equations of the form

$$u_{n,\epsilon} = g(u_{n+m}, u_{n+m-1}, \dots u_{n+m'+1}, u_{n+m'}) \equiv g_n\,, \qquad \frac{\partial g_n}{\partial u_{n+m}} \frac{\partial g_n}{\partial u_{n+m'}} \neq 0\,, \tag{3.2.3}$$

where $u_n = u_n(t, \epsilon)$ and ϵ is the symmetry variable. In (3.2.3) by the index ϵ we denote the ϵ-derivative of $u_n(t, \epsilon)$, and $m \geq m'$ are two finite fixed integers. The symmetry of a DΔE (3.2.1) is defined by one locally analytic function of many variables:

$$g(z_1, z_2, z_3, \dots z_{1+m-m'})\,. \tag{3.2.4}$$

We will call (3.2.3) a local generalized symmetry of (3.2.1) as in the right hand side it does not contain integrals or summations. Moreover, we choose this symmetry to have no explicit dependence on the discrete spatial variable n and on the time of (3.2.1) t. It is known that local and n and t independent S-integrable equations like (3.2.1) may possess n and t dependent generalized symmetries. Moreover 1+1 dimensional S-integrable equations have infinitely many n and t independent local generalized symmetries. This property is also true for many C-integrable equations. So, the existence of an infinite hierarchy of local, n and t independent generalized symmetries of the form (3.2.3) is a natural requirement for the integrability (3.2.1).

Lie point symmetries of (3.2.2) have been presented in Section 2.3.3. They are of the form:

$$u_{n,\epsilon} = a(t)\dot{u}_n + b_n(t, u_n) \tag{3.2.5}$$

and are a subcase of the generalized symmetries. We will be interested in symmetries (3.2.3) with $m > 1$ and $m' < -1$ which are not Lie point symmetries, more precisely, with $m = -m' > 1$. For an explanation of the last requirement, see Section 3.2.4.1.

A generalized symmetry of (3.2.1) is an equation of the form (3.2.3) compatible with (3.2.1), i.e. such that they have a common set of solutions. Before giving a precise definition of generalized symmetries, we derive and discuss the conditions necessary for their existence. If $u_n(t, \epsilon)$ is a common solution of (3.2.1, 3.2.3), we have (3.1.5), where

$$D_t g_n = \sum_{j=m'}^{m} \frac{\partial g_n}{\partial u_{n+j}} f_{n+j}\,, \qquad D_\epsilon f_n = \sum_{j=-1}^{1} \frac{\partial f_n}{\partial u_{n+j}} g_{n+j}\,. \tag{3.2.6}$$

By f_{n+j}, g_{n+j} we mean

$$f_{n+j} = f(u_{n+j+1}, u_{n+j}, u_{n+j-1})\,, \quad g_{n+j} = g(u_{n+j+m}, u_{n+j+m-1}, \dots u_{n+j+m'})\,.$$

In this case the compatibility condition (3.1.5, 3.2.6) reads:

$$D_t g_n = \frac{\partial f_n}{\partial u_{n+1}} g_{n+1} + \frac{\partial f_n}{\partial u_n} g_n + \frac{\partial f_n}{\partial u_{n-1}} g_{n-1}. \tag{3.2.7}$$

Eq. (3.2.7) has to be satisfied for any common solution of (3.2.1, 3.2.3) and given f_n. It turns out to be an equation for the function g_n if f_n is known.

In the generalized symmetry method for DΔEs, we assume that (3.2.7) must be identically satisfied for all values of the variables

$$u_0, u_1, u_{-1}, u_2, u_{-2}, \dots \tag{3.2.8}$$

which are considered independent, and for all $n \in \mathbb{Z}$.

From (3.2.7) we often will get OΔEs of the form

$$\phi_{n+1} - \phi_n = 0 . \tag{3.2.9}$$

which must be satisfied identically for all values of the variables (3.2.8). We are looking for solutions of (3.2.9), such that ϕ_n is a function defined on a finite interval of the lattice:

$$\phi_n = \phi(u_{n+k}, u_{n+k-1}, \dots u_{n+k'}) , \qquad k \geq k' , \tag{3.2.10}$$

where

$$\frac{\partial \phi_n}{\partial u_{n+k}} \neq 0 , \quad \frac{\partial \phi_n}{\partial u_{n+k'}} \neq 0, \tag{3.2.11}$$

if ϕ_n is a not a constant function.

Let us analyze the consequences of (3.2.9). Let us assume that there exists a nonconstant solution ϕ_n of (3.2.9) given by (3.2.10) which satisfies (3.2.11). If we differentiate (3.2.9) with respect to $u_{n+k'}$, one can see that $\partial\phi_n/\partial u_{n+k'} = 0$ identically. Consequently we have a contradiction with (3.2.11), and thus the relation (3.2.9) implies that ϕ_n must be a constant function. Introducing the standard shift operator $S_n = S$, defined in (1.2.14) in Section 1.2.3, such that for any integer power j we have:

$$S^j \phi_n = \phi_{n+j} = \phi(u_{n+j+k}, u_{n+j+k-1}, \dots u_{n+j+k'}) , \tag{3.2.12}$$

we can rewrite (3.2.9) as $(S - 1)\phi_n = 0$ and thus we get

$$\ker(S - 1) = \mathbb{C} . \tag{3.2.13}$$

Definition 3. *Eq. (3.2.3) is called a* generalized symmetry *of (3.2.1) if the compatibility condition (3.2.7) is identically satisfied for all values of the independent variables (3.2.8). The numbers m and m' are called respectively the left order (or the order) and the right order of the generalized symmetry (3.2.3).*

For any generalized symmetry (3.2.3), the integers m and m' are fixed and define essentially different cases. In order to derive integrability conditions, we will only use the left order m and, for this reason, sometimes we will call m for simplicity the order of the generalized symmetry. The case of the right order m' will be discussed in Section 3.2.5.4.

Definition 3 is constructive. For any given (3.2.1) and any given orders m and m', with $m \geq m'$, one is able either to find a generalized symmetry (3.2.3) or to prove that it does not exist. In Section 3.3.1.1 we will show how we can construct generalized symmetries of the Volterra equation (3.2.2) with $m = 2$ and $m' = -2$. The resulting generalized symmetry (see also (2.3.184)) is:

$$u_{n,\epsilon} = u_n(u_{n+1}(u_{n+2} + u_{n+1} + u_n) - u_{n-1}(u_n + u_{n-1} + u_{n-2})) . \tag{3.2.14}$$

It is a general property of S-integrable equations by the IST in the 1+1 dimensional case that evolution local equations like (3.2.1), which have no explicit dependence on n and t, possess infinitely many n and t independent local conservation laws. We will assume that this is true for (3.2.1) and that also their conservation laws have no explicit n and t dependence.

Definition 4. *A relation of the form*

$$D_t p_n = (S-1) q_n \,, \tag{3.2.15}$$

where p_n and q_n are functions of the form (3.2.10), and D_t is a differentiation operator corresponding to (3.2.1) (see (3.2.6)), is called a local conservation law of (3.2.1). The relation (3.2.15) must be satisfied identically for all values of the independent variables (3.2.8). The function p_n is called a conserved density of (3.2.1).

It can be easily proved, as we did in the case of (3.2.13), that if the conserved density p_n has the form

$$p_n = p(u_{n+m_1}, u_{n+m_1-1}, \dots u_{n+m_2}) \,, \qquad m_1 \geq m_2 \,, \qquad \frac{\partial p_n}{\partial u_{n+m_1}} \frac{\partial p_n}{\partial u_{n+m_2}} \neq 0 \,, \tag{3.2.16}$$

then q_n must be a function of the form

$$q_n = q(u_{n+m_1}, u_{n+m_1-1}, \dots u_{n+m_2-1}) \,, \qquad \frac{\partial q_n}{\partial u_{n+m_1}} \frac{\partial q_n}{\partial u_{n+m_2-1}} \neq 0 \,. \tag{3.2.17}$$

It is obvious that if p_n cannot be expressed in the form (3.2.16), then it is a constant function, as well as q_n, and this is a trivial case. The two simplest conservation laws of the Volterra equation (3.2.2), with $m_1 = m_2 = 0$, are:

$$D_t u_n = (S-1)(u_n u_{n-1}) \,, \qquad D_t \log u_n = (S-1)(u_n + u_{n-1}) \,. \tag{3.2.18}$$

Local conservation laws, as well as generalized symmetries, can be used to solve (3.2.1). A conservation law (3.2.15) can be used to construct constants of motion (or conserved quantities, or first integrals).

Let us consider the periodic closure of (3.2.1) of period $N \geq 1$, an equation such that $u_n = u_{n+N}$ for any n. Then we can write (3.2.1) as a system of N ODEs for N functions $u_1(t), u_2(t), \dots u_N(t)$. A constant of motion of this system is a function $I = I(u_1, u_2, \dots u_N)$ such that $\frac{dI}{dt} = 0$. Any conservation law (3.2.15) of (3.2.1) generates a constant of motion $I = \sum_{n=1}^{N} p_n$ for this finite system. In fact,

$$\frac{dI}{dt} = \sum_{n=1}^{N} D_t p_n = \sum_{n=1}^{N} (q_{n+1} - q_n) = q_{N+1} - q_1 = 0 \,.$$

In the case of the periodically closed Volterra equation (3.2.2), we get from (3.2.18) two constants of motion I_1, I_2. We have:

$$I_1 = \sum_{n=1}^{N} u_n \,, \qquad \tilde{I}_2 = e^{I_2} = \prod_{n=1}^{N} u_n \,,$$

and I_1, $\tilde{I}_2$ are arbitrary t independent constants.

Any generalized symmetry (3.2.3) is a non linear DΔE which has common solutions $u_n(t, \epsilon)$ with (3.2.1). As in the case of Lie point symmetries, for generalized symmetries we can perform a symmetry reduction [543] by considering solutions such that $\frac{\partial u_n}{\partial \epsilon} = 0$, i.e. stationary solutions of (3.2.3) which satisfy the following OΔE:

$$g(u_{n+m}, u_{n+m-1}, \dots u_{n+m'}) = 0 \,.$$

This is the analog of the reduced ODE we obtain in the case of PDEs in two variables. If we solve this equation we obtain a function $u_n(t)$ which depends on arbitrary functions of t. These arbitrary functions can be obtained by introducing $u_n(t)$ into (3.2.1). In such a way we can, by symmetry reduction, construct particular solutions of (3.2.1) such as, for example, its soliton solutions.

Conservation laws, conserved densities and generalized symmetries generate linear spaces. In fact, the operators D_t and $S - 1$ are linear. For any pair of conservation laws of (2.4.8), with p_n, q_n and $\hat{p}_n$, $\hat{q}_n$ such that (3.2.15) and $D_t\hat{p}_n = (S-1)\hat{q}_n$ are satisfied, we have the following conservation law:

$$D_t(\alpha p_n + \beta\hat{p}_n) = (S-1)(\alpha q_n + \beta\hat{q}_n) ,$$

where α, β are arbitrary constants. In the case of two generalized symmetries $u_{n,\epsilon} = g_n$ and $u_{n,\hat{\epsilon}} = \hat{g}_n$ of (3.2.1), the functions g_n and $\hat{g}_n$ satisfy the linear equation (3.2.7), as well as any their linear combination $\alpha g_n + \beta\hat{g}_n$. Hence the equation

$$u_{n,\epsilon'} = \alpha g_n + \beta\hat{g}_n$$

will be a generalized symmetry of (3.2.1). The function f_n, given in (3.2.1), satisfies the compatibility condition (3.2.7) for any (3.2.1), i.e. the equation $u_{n,\epsilon'} = f_n$ is a trivial generalized symmetry of (3.2.1) and can be used in linear combinations with any other generalized symmetries to simplify them.

Conserved densities possess an important additional property: total differences can be added to them. Given any conservation law (3.2.15) and any function (3.2.10), we can construct for (3.2.1) the following conservation law:

$$D_t(p_n + (S-1)\phi_n) = (S-1)(q_n + D_t\phi_n) . \tag{3.2.19}$$

The conservation laws (3.2.15, 3.2.19) do not essentially differ one from the other and will be considered as equivalent.

Definition 5. *Two functions a_n and b_n of the form* (3.2.10) *are said to be equivalent, and we will write $a_n \sim b_n$, if the difference $a_n - b_n$ is given by the following equation:*

$$a_n - b_n = (S-1)c_n ,$$

where c_n also is a function of the form (3.2.10).

This equivalence relation allows us to split conserved densities and conservation laws into equivalence classes. Using it, we are able to transform any conserved density into a simplified reduced form and to define the order of a conservation law.

In particular, $a_n \sim 0$ iff $a_n = (S-1)c_n$, where a_n, c_n are of the form (3.2.10). For example, in the case of conservation law (3.2.15), one can write $D_t p_n \sim 0$. It follows from formulas (3.2.15, 3.2.19) that if $a_n \sim b_n$ and a_n is a conserved density of (3.2.1), then b_n also is a conserved density. So conservation laws with equivalent conserved densities are equivalent. The equivalence relation introduced by Definition 5 has the following three properties:

$$\begin{aligned} &a_n \sim a_n , \\ &a_n \sim b_n \quad \Rightarrow \quad b_n \sim a_n , \\ &a_n \sim b_n , \ b_n \sim c_n \quad \Rightarrow \quad a_n \sim c_n . \end{aligned}$$

Moreover, it is easy to see that

$$a_n \sim b_n , \ c_n \sim d_n \quad \Rightarrow \quad \alpha a_n + \beta c_n \sim \alpha b_n + \beta d_n \tag{3.2.20}$$

for any complex constants α and β. One also has

$$a_n = a_{n+1} - (S-1)a_n \sim a_{n+1} , \qquad a_n = a_{n-1} + (S-1)a_{n-1} \sim a_{n-1} ,$$

hence

$$a_n \sim a_{n+i} \qquad \text{for all} \qquad i \in \mathbb{Z} . \tag{3.2.21}$$

It follows from (3.2.20, 3.2.21) that

(3.2.22) $$a_n + b_n \sim a_{n+i} + b_{n+j} \qquad \text{for all} \qquad i, j \in \mathbb{Z}\,.$$

We can prove, in the same way as we did for the property (3.2.13), that if (3.2.10) is a total difference, i.e. $\phi_n = \phi(u_{n+k}, u_{n+k-1}, \dots u_{n+k'}) \sim 0$, then

(3.2.23) $$\phi_n = \text{constant} \quad \text{or} \quad k = k' \qquad \Rightarrow \qquad \phi_n = 0\,,$$

(3.2.24) $$\text{if } k > k' \qquad \Rightarrow \qquad \frac{\partial^2 \phi_n}{\partial u_{n+k} \partial u_{n+k'}} = 0\,.$$

We present now a theorem which helps us to simplify functions of the form (3.2.10), remaining inside a class of equivalence.

Theorem 22. *Let us consider any function of the form*

(3.2.25) $$a_n = a(u_{n+k_1}, u_{n+k_1-1}, \dots u_{n+k_2})\,, \qquad k_1 \geq k_2\,,$$

where a_n is not a constant function, such that $\frac{\partial a_n}{\partial u_{n+k_1}} \frac{\partial a_n}{\partial u_{n+k_2}} \neq 0$. Eq. (3.2.25) can be expressed in the form

(3.2.26) $$a_n = b_n + (S-1)c_n\,,$$

where c_n is a function of the form (3.2.10). The function b_n is such that:

(3.2.27) $$b_n = b(u_{n+k_3}, u_{n+k_3-1}, \dots u_{n+k_4})\,, \qquad k_1 \geq k_3 \geq k_4 \geq k_2\,,$$

where only one of the following two possibilities takes place:

(3.2.28) $$b_n = \textit{constant} \quad \textit{or} \quad k_3 = k_4\,,$$

(3.2.29) $$\textit{if } k_3 > k_4\,, \quad \Rightarrow \quad \frac{\partial^2 b_n}{\partial u_{n+k_3} \partial u_{n+k_4}} \neq 0\,.$$

The relation $a_n \sim 0$ is possible only in the case (3.2.28) if $b_n = 0$.

PROOF. Let us show how to construct the functions b_n and c_n. We will do so, using a trick which can be applied as many times as necessary.

In the case of (3.2.25) with $a_n = \text{constant}$, and $k_1 = k_2$, we choose in (3.2.26) $b_n = a_n$, $c_n = 0$ and have the required result. The only other remaining possibility is if

(3.2.30) $$k_1 > k_2 \quad \text{and} \quad \frac{\partial^2 a_n}{\partial u_{n+k_1} \partial u_{n+k_2}} = 0\,.$$

In this case one splits the function a_n in two components:

(3.2.31) $$\begin{aligned} a_n &= a_n^1 + a_n^2\,, \\ a_n^1 &= a^1(u_{n+k_1}, \dots u_{n+k_2+1})\,, \\ a_n^2 &= a^2(u_{n+k_1-1}, \dots u_{n+k_2})\,. \end{aligned}$$

Then one can rewrite (3.2.31) as:

(3.2.32) $$\begin{aligned} a_n &= a_n^3 + (S-1)a_{n-1}^1\,, \\ a_n^3 &= a_{n-1}^1 + a_n^2 = a^3(u_{n+k_1-1}, \dots u_{n+k_2})\,. \end{aligned}$$

If a_n^3 is nonconstant, then there exist two numbers $\hat{k}_1$, $\hat{k}_2$, such that $k_1 > \hat{k}_1 \geq \hat{k}_2 \geq k_2$, $a_n^3 = \hat{a}^3(u_{n+\hat{k}_1}, \dots u_{n+\hat{k}_2})$, $\frac{\partial a_n^3}{\partial u_{n+\hat{k}_1}} \frac{\partial a_n^3}{\partial u_{n+\hat{k}_2}} \neq 0$. If $a_n^3 = \text{constant}$, either $\hat{k}_1 = \hat{k}_2$, or $\hat{k}_1 > \hat{k}_2$

and then $\frac{\partial^2 a_n^3}{\partial u_{n+\hat{k}_1}\partial u_{n+\hat{k}_2}} \neq 0$. So we have the required result. If $\frac{\partial^2 a_n^3}{\partial u_{n+\hat{k}_1}\partial u_{n+\hat{k}_2}} = 0$, we can simplify a_n^3 by applying the procedure again. As $\hat{k}_1 < k_1$, this procedure will be applied only a finite number of times. It is clear that one is led at the end to formulas (3.2.26, 3.2.27) corresponding to the case (3.2.28) or (3.2.29).

If $a_n \sim 0$, then one has $b_n \sim 0$. From (3.2.29) we arrive at a contradiction with (3.2.24). In the case (3.2.28) it follows from (3.2.23) that $b_n = 0$. □

Theorem 22 enables us to verify if a function ϕ_n of the form (3.2.10) is a total difference, i.e. $\phi_n \sim 0$. Thus, we can check whether a function p_n is a conserved density of (3.2.1) and, in the case of positive answer, find the corresponding function q_n appearing in (3.2.15). In order to do so, one applies Theorem 22 to the function $a_n = D_t p_n$ and verifies if $b_n = 0$ (3.2.26). If this it is so, then q_n is given by: $q_n = c_n + \alpha$, where α is an arbitrary constant, as it follows from (3.2.13).

As an example of the proof of a conserved density, let us consider the case of $p_n = \log(u_{n+1}u_n)$ for the Volterra equation (3.2.2). In this case

$$a_n = D_t p_n = \frac{\dot{u}_{n+1}}{u_{n+1}} + \frac{\dot{u}_n}{u_n} = u_{n+2} - u_n + u_{n+1} - u_{n-1}$$

with $k_1 = 2$, $k_2 = -1$. We are in the case (3.2.30) and from (3.2.31) one can take $a_n^1 = u_{n+2}$. Then from (3.2.32), one obtains

$$a_n = 2u_{n+1} - u_n - u_{n-1} + (S-1)u_{n+1} .$$

Applying again the scheme used in the proof of Theorem 22, one gets:

$$a_n = u_n - u_{n-1} + (S-1)(u_{n+1} + 2u_n) ,$$

and we are led on the next step to the following conservation law:

$$D_t \log(u_{n+1}u_n) = (S-1)(u_{n+1} + 2u_n + u_{n-1}) . \tag{3.2.33}$$

Eq. (3.2.33) is equivalent to the second conservation laws (3.2.18), as from property (3.2.22) we have: $\log(u_{n+1}u_n) \sim 2\log u_n$.

The conserved density p_n plays a leading role in the conservation law present in (3.2.15). If p_n is known, the function q_n can be easily found, using Theorem 22. For this reason, we will mainly work with conserved densities without writing down the whole conservation law.

Now we have all the notions necessary to define the order of a conserved density and of a conservation law. The notion of order will allow us to distinguish essentially different cases. In accordance with Theorem 22, any conserved density p_n is equivalent to a function $\hat{p}_n$ of the following form:

$$p_n \sim \hat{p}_n = \mathcal{P}(u_{n+\hat{m}_1}, u_{n+\hat{m}_1-1}, \dots u_{n+\hat{m}_2}) . \tag{3.2.34}$$

If $\hat{p}_n$ is not a constant function, one has

$$\text{either} \quad \hat{m}_1 = \hat{m}_2, \quad \text{or} \quad \hat{m}_1 > \hat{m}_2 \text{ with } \frac{\partial^2 \hat{p}_n}{\partial u_{n+\hat{m}_1}\partial u_{n+\hat{m}_2}} \neq 0.$$

Introducing the function $\mathcal{P}_n = \hat{p}_{n-\hat{m}_2}$ together with number $m = \hat{m}_1 - \hat{m}_2$ and using the property (3.2.21), we obtain that any conserved density p_n is equivalent to a conserved density $\mathscr{P}_n$,

$$p_n \sim \mathscr{P}_n . \tag{3.2.35}$$

Eq. (3.2.35) has three possible forms given in the following definition:

Definition 6. *A conserved density p_n and the corresponding conservation law (3.2.15) are called trivial if*

$$\mathscr{P}_n = c \in \mathbb{C}\,, \tag{3.2.36}$$

and nontrivial if

$$\mathscr{P}_n = \mathscr{P}(u_n)\,, \qquad \mathscr{P}'(u_n) \neq 0\,; \tag{3.2.37}$$

or

$$\mathscr{P}_n = \mathscr{P}(u_{n+m}, u_{n+m-1}, \dots u_n)\,, \qquad m > 0\,, \qquad \frac{\partial^2 \mathscr{P}_n}{\partial u_{n+m} \partial u_n} \neq 0\,. \tag{3.2.38}$$

The order of a nontrivial conserved density p_n and of the corresponding conservation law (3.2.15) is given by the number 0 or m, depending if (3.2.37) or (3.2.38) takes place.

Properties (3.2.23, 3.2.24) imply that the conserved densities corresponding to cases (3.2.36-3.2.38) cannot be equivalent to each other. Formulae (3.2.35-3.2.38) define the special form of a conserved density necessary to define its order.

The following four functions:

$$\begin{aligned} &p_n^1 = \log u_n\,, \quad p_n^2 = u_n\,, \quad p_n^3 = u_{n+1}u_n + \tfrac{1}{2}u_n^2\,, \\ &p_n^4 = u_{n+2}u_{n+1}u_n + u_{n+1}^2 u_n + u_{n+1}u_n^2 + \tfrac{1}{3}u_n^3 \end{aligned} \tag{3.2.39}$$

exemplify the simplest nontrivial conserved densities of the Volterra equation (3.2.2). The densities p_n^1 and p_n^2 are taken from (3.2.18). The reader easily can check that $D_t p_n^3 \sim D_t p_n^4 \sim 0$. The orders of the conserved densities (3.2.39) are 0, 0, 1 and 2, respectively.

Let us define the formal variational derivative of a function ϕ_n (3.2.10) as

$$\frac{\delta \phi_n}{\delta u_n} = \sum_{i=k'}^{k} T^{-i} \frac{\partial \phi_n}{\partial u_{n+i}} = \sum_{j=-k}^{-k'} \frac{\partial \phi_{n+j}}{\partial u_n} \tag{3.2.40}$$

(see e.g. [219, 222, 755, 764, 840, 843]). The operator $\frac{\delta}{\delta u_n}$ is the discrete analog of (3.1.7) and possesses similar properties (see Theorem 24 in Section 3.2.2 below). Sometimes it is called Euler operator [409, 587], but we use the name formal variational derivative, following the continuous case. We will use it to calculate the order of a conserved density as from (3.1.8) it follows (3.1.9).

Let us show that, for the discrete variational derivative (3.2.40) the following result holds

$$\phi_n = (S-1)\psi_n \qquad \Rightarrow \qquad \frac{\delta \phi_n}{\delta u_n} = 0\,, \tag{3.2.41}$$

where ϕ_n, ψ_n are functions of the form (3.2.10). From (3.2.23) we see that we need to consider only the nontrivial case when $k > k'$ and $\frac{\partial \phi_n}{\partial u_{n+k}} \frac{\partial \phi_n}{\partial u_{n+k'}} \neq 0$. In this case $\psi_n = \psi(u_{n+k-1}, \dots u_{n+k'})$ and

$$\begin{aligned} \frac{\delta \phi_n}{\delta u_n} &= \sum_{j=-k}^{-k'} \frac{\partial}{\partial u_n}(\psi_{n+j+1} - \psi_{n+j}) = \frac{\partial}{\partial u_n}(\psi_{n-k'+1} - \psi_{n-k}) \\ &= \frac{\partial}{\partial u_n}(\psi(u_{n+k-k'}, \dots u_{n+1}) - \psi(u_{n-1}, \dots u_{n+k'-k})) = 0. \end{aligned} \tag{3.2.42}$$

In (3.2.42) $k - k' \geq 1$ and thus ϕ_n does not depend on the variable u_n.

To find the order of a conserved density p_n, we can calculate its formal variational derivative $\delta p_n/\delta u_n$. It follows from (3.2.41) and the fact that $\delta/\delta u_n$ is a linear operator that the application of the formal variational derivative operator to equivalent conserved densities gives the same result. For this reason $\frac{\delta p_n}{\delta u_n} = \frac{\delta \mathscr{P}_n}{\delta u_n}$, where $\mathscr{P}_n$ is a conserved density defined by (3.2.35-3.2.38). The form of $\delta \mathscr{P}_n/\delta u_n$ is obvious in the cases (3.2.36) and (3.2.37), while in the case (3.2.38) we have:

$$\frac{\delta \mathscr{P}_n}{\delta u_n} = \varrho_n = \varrho(u_{n+m}, u_{n+m-1}, \dots u_{n-m}),$$

$$\frac{\partial \varrho_n}{\partial u_{n+m}} = \frac{\partial^2 \mathscr{P}_n}{\partial u_{n+m} \partial u_n} \neq 0, \qquad \frac{\partial \varrho_n}{\partial u_{n-m}} = \frac{\partial^2 \mathscr{P}_{n-m}}{\partial u_n \partial u_{n-m}} = T^{-m} \frac{\partial \varrho_n}{\partial u_{n+m}} \neq 0.$$

Let us introduce the function:

$$\varrho_n = \frac{\delta p_n}{\delta u_n}. \tag{3.2.43}$$

Eq. (3.2.43) can have one of the following three essentially different forms:

$$\varrho_n = 0, \tag{3.2.44}$$

$$\varrho_n = \varrho(u_n) \neq 0, \tag{3.2.45}$$

$$\varrho_n = \varrho(u_{n+m}, u_{n+m-1}, \dots u_{n-m}), \quad m > 0, \quad \frac{\partial \varrho_n}{\partial u_{n+m}} \frac{\partial \varrho_n}{\partial u_{n-m}} \neq 0. \tag{3.2.46}$$

In the case (3.2.44), the conserved density p_n and the corresponding conservation law (3.2.15) are trivial. In the cases (3.2.45) and (3.2.46), the conserved density are nontrivial, and the corresponding orders are either 0 or $m > 0$, respectively.

As an example, let us consider the conserved densities (3.2.39). We have

$$\frac{\delta p_n^1}{\delta u_n} = \frac{1}{u_n}, \qquad \frac{\delta p_n^2}{\delta u_n} = 1, \qquad \frac{\delta p_n^3}{\delta u_n} = u_{n+1} + u_n + u_{n-1},$$

$$\frac{\delta p_n^4}{\delta u_n} = u_{n+2}u_{n+1} + u_{n+1}^2 + 2u_{n+1}u_n + u_n^2 + u_{n+1}u_{n-1} + 2u_n u_{n-1} + u_{n-1}^2 + u_{n-1}u_{n-2}.$$

Hence the orders of p_n^1, p_n^2, p_n^3, p_n^4 are respectively equal to 0, 0, 1, 2.

The Definition 6 of local conservation law and of conserved density is as constructive as that of generalized symmetry. For any given equation of the form (3.2.1) and any order, it is possible to find all conservation laws of that order or to prove that no conservation law exists. In Section 3.3.1.1, in the example of the Volterra equation, we will construct all conservation laws of the first order, using Definition 6.

2.2. First integrability condition. We discuss in this Section how to obtain the generalized symmetry (3.2.3) of an equation of the form (3.2.1) using the compatibility condition (3.2.7). In this way the first integrability condition for (3.2.1) will arise. At the end we briefly describe the general scheme of the generalized symmetry method for DΔEs. From the first integrability condition presented here Adler in [21] derived integrability conditions for higher order evolutionary DΔEs.

Given an equation characterized by the function f_n, one looks for a symmetry characterized by the function g_n with $m > 0$, $m' < 0$. For convenience we introduce for the derivatives of f_n and g_n the following notation:

$$f_n^{(i)} = \frac{\partial f_n}{\partial u_{n+i}}\ , i = 1, 0, 1 \quad g_n^{(j)} = \frac{\partial g_n}{\partial u_{n+j}},\ j = m, m-1, \cdots, m'+1, m'. \tag{3.2.47}$$

Then the compatibility condition reads:

$$D_t g_n = \sum_{i=m'}^{m} g_n^{(i)} f_{n+i} = f_n^{(1)} g_{n+1} + f_n^{(0)} g_n + f_n^{(-1)} g_{n-1}\ . \tag{3.2.48}$$

If $m \geq 1$, applying the operator $\dfrac{\partial}{\partial u_{n+m+1}}$ to (3.2.48), one obtains the following relation:

$$g_n^{(m)} f_{n+m}^{(1)} = g_{n+1}^{(m)} f_n^{(1)}\ , \qquad m \geq 1\ . \tag{3.2.49}$$

Applying $\dfrac{\partial}{\partial u_{n+m}}$ and $\dfrac{\partial}{\partial u_{n+m-1}}$ to (3.2.48), two other analogous relations can be derived:

$$D_t g_n^{(m)} + g_n^{(m)} f_{n+m}^{(0)} + g_n^{(m-1)} f_{n+m-1}^{(1)} = g_n^{(m)} f_n^{(0)} + g_{n+1}^{(m-1)} f_n^{(1)}\ , \quad m \geq 2\ , \tag{3.2.50}$$

$$\begin{aligned} D_t g_n^{(m-1)} &+ g_n^{(m)} f_{n+m}^{(-1)} + g_n^{(m-1)} f_{n+m-1}^{(0)} + g_n^{(m-2)} f_{n+m-2}^{(1)} \\ &= g_{n-1}^{(m)} f_n^{(-1)} + g_n^{(m-1)} f_n^{(0)} + g_{n+1}^{(m-2)} f_n^{(1)}\ , \quad m \geq 3\ . \end{aligned} \tag{3.2.51}$$

Let us define for any $N \geq 0$ the function:

$$\Phi_n^{(N)} = f_n^{(1)} f_{n+1}^{(1)} \cdots f_{n+N}^{(1)}\ , \tag{3.2.52}$$

where $f_n^{(1)} \neq 0$ (see (3.2.1)). If we divide (3.2.49-3.2.51) by $\Phi_n^{(m)}$, $\Phi_n^{(m-1)}$, $\Phi_n^{(m-2)}$ respectively, we obtain:

$$(S-1)\frac{g_n^{(m)}}{\Phi_n^{(m-1)}} = 0\ , \tag{3.2.53}$$

$$(S-1)\frac{g_n^{(m-1)}}{\Phi_n^{(m-2)}} = \Theta_n^{(1)}(g_n^{(m)})\ , \tag{3.2.54}$$

$$(S-1)\frac{g_n^{(m-2)}}{\Phi_n^{(m-3)}} = \Theta_n^{(2)}(g_n^{(m-1)}, g_n^{(m)})\ . \tag{3.2.55}$$

Here the left hand sides are total differences, and the functions $\Theta_n^{(i)}$ $(i = 1, 2)$ depend on the partial derivatives of $g_n^{(j)}$ defined in the previous equation.

Due to the property (3.2.13), (3.2.53) can be easily solved, and $g_n^{(m)}$ can be found. Hence the right hand side of (3.2.54) is known and the following condition appears: the function $\Theta_n^{(1)}(g_n^{(m)})$ must be a total difference. If this condition is satisfied, one finds $g_n^{(m-1)}$. Then the function $g_n^{(m-2)}$ can be found from (3.2.55) if an analogous condition is satisfied. In a quite similar way, we can write down equations for the other partial derivatives $g_n^{(i)}$, for $0 < i \leq m$ or $m' \leq i < 0$. Those equations have the same structure and lead to analogous conditions. In the integrable cases, i.e. if all such conditions are satisfied, we can define the function g_n up to arbitrary constants and an arbitrary function of u_n which can be specified easily, using (3.2.48). This will be done in Section 3.3.1.1 in the example of the Volterra equation for a generalized symmetry of the orders $m = 2$, $m' = -2$.

Let us pass now to the case when not only the function g_n but also f_n are unknown, i.e. when we consider the problem of classifying equations of the form (3.2.1) which have generalized symmetries. It turns out that in this case some conditions can be derived from equations of the type (3.2.54, 3.2.55), and those conditions do not depend on g_n and are expressed only in terms of f_n, i.e. of (3.2.1) itself. One obtains the same conditions, using any generalized symmetry of any high enough order m. Those integrability conditions will be necessary for the existence of high enough order generalized symmetries. The first of them is given by the following theorem:

Theorem 23. *If an equation of the form* (3.2.1) *possesses a generalized symmetry of the form* (3.2.3) *of order* $m \geq 2$, *then there must exist a function* $q_n^{(1)}$ *of the form* (3.2.10), *such that*

(3.2.56) $$\dot{p}_n^{(1)} = (S-1)q_n^{(1)} \quad \textit{with} \quad p_n^{(1)} = \log \frac{\partial f_n}{\partial u_{n+1}} ,$$

where $\dot{p}_n^{(1)} = D_t p_n^{(1)}$ *and* f_n *is given by* (3.2.1).

PROOF. In the case of generalized symmetries of the order $m \geq 2$, we can use (3.2.49, 3.2.50) and then (3.2.53, 3.2.54). From (3.2.53) it follows that

(3.2.57) $$g_n^{(m)} = \alpha \Phi_n^{(m-1)} ,$$

where the constant α does not vanish due to (3.2.3). It is easy to see that the right hand side of (3.2.54) has the form:

(3.2.58) $$\Theta_n^{(1)}(g_n^{(m)}) = \alpha D_t \log \Phi_n^{(m-1)} + \alpha (f_{n+m}^{(0)} - f_n^{(0)}) .$$

From (3.2.54) it follows that $\Theta_n^{(1)}(g_n^{(m)})$ must be equivalent to zero, and (3.2.22) implies that $f_{n+m}^{(0)} - f_n^{(0)} \sim 0$. The same (3.2.22) together with $p_n^{(1)} = \log f_n^{(1)}$ and (3.2.52) provide the following result:

(3.2.59) $$D_t \log \Phi_n^{(m-1)} = \dot{p}_n^{(1)} + \dot{p}_{n+1}^{(1)} + \cdots + \dot{p}_{n+m-1}^{(1)} \sim m\dot{p}_n^{(1)} .$$

Dividing (3.2.58) by αm and using the equivalence relations discussed above, we can see that $\dot{p}_n^{(1)} \sim 0$. This shows, in accordance with Definition 5, that the function $\dot{p}_n^{(1)}$ can be expressed in the form (3.2.56). □

Condition (3.2.56) has the form of a local conservation law, and Theorem 23 tells us that if there is a generalized symmetry of order $m \geq 2$, then (3.2.1) must have a conservation law with conserved density $p_n^{(1)}$ defined by (3.2.1). If an equation satisfies (3.2.56), then one automatically obtains for that equation a conserved density. A priori this conserved density may be trivial or its order may be equal to 0, 1 or 2. Any of these possibilities is realized in the examples we present in Section 3.3.1.2. In the case of the Volterra equation, for instance, $p_n^{(1)} = \log u_n$, and this is nothing but the first conserved density (3.2.39) of order 0.

In order to check the first integrability condition (3.2.56) for a given equation, one can use Theorem 22. Such checking can be simplified as we did in (3.1.8, 3.1.9) for PDEs.

Theorem 24. *A variational derivative* (3.2.40) *has the following property:*

(3.2.60) $$\frac{\delta \phi_n}{\delta u_n} = 0 \quad \textit{iff} \quad \phi_n = \sigma + (S-1)\psi_n ,$$

where σ *is a constant,* ϕ_n *and* ψ_n *are two functions of the form* (3.2.10).

PROOF. One part of the proof follows from (3.2.41), as $\frac{\delta\sigma}{\delta u_n} = 0$. Let us discuss the other part and suppose that $\frac{\delta\phi_n}{\delta u_n} = 0$. According to Theorem 22, for $a_n = \phi_n$ we have (3.2.26, 3.2.27) with (3.2.28) or (3.2.29). In the second case (3.2.29),

$$\frac{\delta\phi_n}{\delta u_n} = \frac{\delta a_n}{\delta u_n} = \frac{\delta b_n}{\delta u_n} = B_n = B(u_{n+K}, u_{n+K-1}, \dots u_{n-K}),$$

where $K = k_3 - k_4 > 0$ and

$$\frac{\partial B_n}{\partial u_{n+K}} = \frac{\partial^2 b_{n-k_4}}{\partial u_{n+K}\partial u_n} \neq 0 .$$

This contradicts the request that $\frac{\delta\phi_n}{\delta u_n} = 0$. In the case (3.2.28), the function b_n can be written as $b_n = b(u_{n+k_4})$. Hence

$$\frac{\delta\phi_n}{\delta u_n} = \frac{\delta b_n}{\delta u_n} = \frac{\partial b_{n-k_4}}{\partial u_n} = b'(u_n) = 0 ,$$

i.e. b_n is a constant. □

To check the first integrability condition (3.2.56), we can use property (3.2.60) with $\phi_n = \dot{p}_n^{(1)}$. At first we check if

$$\frac{\delta}{\delta u_n}\dot{p}_n^{(1)} = 0 . \tag{3.2.61}$$

Then, if this is true, using Theorem 22 we represent $\dot{p}_n^{(1)}$ as:

$$\dot{p}_n^{(1)} = \sigma + (S-1)q_n^{(1)} . \tag{3.2.62}$$

Here σ is a constant, and $p_n^{(1)}$ will be a conserved density only if $\sigma = 0$.

Having introduced all the necessary notions, we can briefly describe here the standard scheme of the generalized symmetry method. At first we choose a class of equations, such as (3.2.1), with an unknown right hand side. Then, assuming the existence of generalized symmetries and/or conservation laws of a high enough order, we derive a few integrability conditions like (3.2.56). These conditions have no dependence on the generalized symmetries and conservation laws and are expressed in terms of the function f_n. Then we try to describe the class of equations satisfying the integrability conditions. The aim is to obtain a list of equations, in which there are no arbitrary functions, but only arbitrary constants. If this is impossible, we have to obtain more integrability conditions. We will show in the following that one can derive as many conditions as necessary to characterize any class of equations. We usually choose as starting point of the classification a class of equations which is invariant under point transformations. In the case of (3.2.1) point transformations have the form: $\tilde{u}_n = s(u_n)$, $\tilde{t} = \mu t$, where μ is a constant. In this way a complete, up to point transformations, list of equations is obtained. Equations of this list will satisfy a finite number of integrability conditions and will contain no arbitrary functions.

As it will be shown, the integrability conditions are necessary conditions for the existence of generalized symmetries and conservation laws. For this reason we will prove, using Miura transformations, master symmetries and Hamiltonian structures, the existence of infinite hierarchies of generalized symmetries and conservation laws (see Sections 3.2.6 and 3.2.7). By doing so we obtain an exhaustive list of integrable equations of the given form. An example of such exhaustive classification will be given in Section 3.3.1.1.

2.3. Formal symmetries and further integrability conditions. From the compatibility condition (3.2.7), in addition to (3.2.56), we can derive more integrability conditions. However, the calculation will become more and more complicate on each step. These calculations can be drastically simplified, using the notion offormal symmetry. In this Section we introduce and discuss formal symmetries and then derive a second and a third integrability conditions.

Let us introduce, in analogy to the continuous case, the discrete Fréchet derivative of a function ϕ_n of the form (3.2.10) as the following operator

$$\phi_n^* = \sum_{i=k'}^{k} \frac{\partial \phi_n}{\partial u_{n+i}} S^i . \tag{3.2.63}$$

This is the discrete analog of the operator ϕ_* given by (3.1.11). For a function f_n given by (3.2.1), f_n^* is the operator

$$f_n^* = f_n^{(1)} S + f_n^{(0)} + f_n^{(-1)} S^{-1}, \tag{3.2.64}$$

with coefficients defined according to (3.2.47).

Formal symmetries are closely related to the following Lax equation

$$\dot{L}_n = [f_n^*, L_n] , \tag{3.2.65}$$

where $[f_n^*, L_n] = f_n^* L_n - L_n f_n^*$ is the standard commutator. The solutions of (3.2.65) will be formal series in powers of the shift operator S and will have the following form

$$L_n = \sum_{i=-\infty}^{N} l_n^{(i)} S^i , \qquad l_n^{(N)} \neq 0 , \tag{3.2.66}$$

where the coefficients $l_n^{(i)}$ are functions of the form (3.2.10). The number N will be called the order of L_n, and we write $\operatorname{ord} L_n = N$. The series $\dot{L}_n$ is obtained by applying the operator D_t given by (3.2.6) to the coefficients of L_n

$$\dot{L}_n = \dot{l}_n^{(N)} S^N + \dot{l}_n^{(N-1)} S^{N-1} + \cdots .$$

The Fréchet derivative operator f_n^* is a particular case of the series (3.2.66). In the case of f_n^*, $N = 1$ and $l_n^{(i)} = 0$ for all $i \leq -2$. The set of series (3.2.66) forms a linear space. Such series can be multiplied according to the rule: $l_n S^i \circ \hat{l}_n S^j = l_n \hat{l}_{n+i} S^{i+j}$, where $S^0 = 1$. The inverse of (3.2.66)

$$L_n^{-1} = \sum_{i=-\infty}^{\hat{N}} \hat{l}_n^{(i)} S^i, \qquad \hat{l}_n^{(\hat{N})} \neq 0 , \tag{3.2.67}$$

is uniquely defined by the equations: $L_n^{-1} L_n = L_n L_n^{-1} = 1$. In fact,

$$L_n^{-1} L_n = \hat{l}_n^{(\hat{N})} l_{n+\hat{N}}^{(N)} S^{\hat{N}+N} + (\hat{l}_n^{(\hat{N})} l_{n+\hat{N}}^{(N-1)} + \hat{l}_n^{(\hat{N}-1)} l_{n+\hat{N}-1}^{(N)}) S^{\hat{N}+N-1} + \dots ,$$

where the first coefficient cannot vanish. Hence $\hat{N} = -N$, and the first coefficients of (3.2.67) are defined by

$$\hat{l}_n^{(-N)} = (l_{n-N}^{(N)})^{-1}, \qquad \hat{l}_n^{(-N-1)} = -(l_{n-N}^{(N)})^{-1} l_{n-N}^{(N-1)} (l_{n-N-1}^{(N)})^{-1}, \quad \dots . \tag{3.2.68}$$

Let us introduce the operator A such that

$$A(L_n) = \dot{L}_n - [f_n^*, L_n] . \tag{3.2.69}$$

We can easily check the following two general formulas

(3.2.70) $$A(L_n^{-1}) = -L_n^{-1} A(L_n) L_n^{-1},$$

(3.2.71) $$A(L_n \tilde{L}_n) = A(L_n)\tilde{L}_n + L_n A(\tilde{L}_n).$$

These formulas show that given any two solutions L_n and $\tilde{L}_n$ of (3.2.65), their product $L_n\tilde{L}_n$ and the inverse L_n^{-1} satisfy the same equation. So any integer power L_n^i, where $L_n^0 = 1$, will also satisfy (3.2.65).

The solution L_n of (3.2.65) is nothing but the recursion operator of the integrable hierarchy of equations because it transforms the right hand side g_n of a generalized symmetry (2.4.10) into the right hand side $L_n g_n$ of a new generalized symmetry. The compatibility condition (3.2.7) can be written in terms of the Fréchet derivative f_n^* as:

(3.2.72) $$(D_t - f_n^*)g_n = 0.$$

Then, using (3.2.65, 3.2.72), one can easily check that

$$D_t L_n g_n = \dot{L}_n g_n + L_n \dot{g}_n = (f_n^* L_n - L_n f_n^*)g_n + L_n f_n^* g_n = f_n^* L_n g_n,$$

i.e. $u_{n,\epsilon'} = L_n g_n$ is a new generalized symmetry, maybe non-local. Taking into account that any integer power L_n^i satisfies (3.2.65), as well as the fact that f_n is a trivial solution of (3.2.72), we obtain infinitely many generalized symmetries of (3.2.1)

(3.2.73) $$u_{n,\epsilon_i} = L_n^i f_n,$$

where $i \in \mathbb{Z}$, $t_0 = t$.

As it was shown in Section 2.3.3 and we will prove again in Section 3.2.6,

(3.2.74) $$\tilde{\mathscr{L}} = u_n + u_n(u_{n+1}S - u_{n-1}S^{-2})(1 - S^{-1})^{-1}u_n^{-1}$$

is the recursion operator of the Volterra equation (3.2.2), i.e. it satisfies (3.2.65). Using the formula

$$(1 - S^{-1})^{-1} = 1 + S^{-1} + S^{-2} + \cdots = \sum_{i=-\infty}^{0} S^i,$$

one can rewrite $\tilde{\mathscr{L}}$ as

(3.2.75) $$\tilde{\mathscr{L}} = u_n S + u_{n+1} + u_n + \frac{u_{n+1}u_n}{u_{n-1}} S^{-1} + \dot{u}_n \left(\sum_{i=-\infty}^{-2} S^i \right) u_n^{-1}.$$

In this way we can write down the coefficients $l_n^{(i)}$ of the representation (3.2.66). It can be proved that (3.2.73) provides a local generalized symmetry for any $i \geq 1$. The orders m and m' of this symmetry, defined in (2.4.10), are such that $m = -m' = i + 1$. As (3.2.2) can be written as

$$\dot{u}_n = u_n(1 - S^{-1})(u_{n+1} + u_n),$$

from (3.2.74) we obtain that

$$u_{n,\epsilon} = \tilde{\mathscr{L}}\dot{u}_n = u_n\dot{u}_n + u_n(u_{n+1}S - u_{n-1}S^{-2})(u_{n+1} + u_n)$$

is the generalized symmetry (3.2.14). So, in the case of $i = 1$, formula (3.2.73) provides the generalized symmetry (3.2.14).

The recursion operator allows one to construct not only generalized symmetries but also conserved densities. This will be demonstrated by Theorem 25 which will allow us to derive some new integrability conditions. Let us define the residue of a series (3.2.66) as the coefficient at S^0, i.e.

(3.2.76) $$\operatorname{res} L_n = l_n^{(0)}.$$

From (3.2.76) it follows that if $N < 0$, then $\mathrm{res}L_n = 0$.

Theorem 25. *Let a series* L_n *(3.2.66) with* $N > 0$ *satisfy (3.2.65). Then*

(3.2.77) $$\log l_n^{(N)},$$

(3.2.78) $$resL_n^i, \quad i \geq 1,$$

are conserved densities of (3.2.1).

PROOF. First of all we prove that

(3.2.79) $$\mathrm{res}[\tilde{L}_n, \hat{L}_n] \sim 0$$

for any formal series $\tilde{L}_n$ and $\hat{L}_n$ of the form (3.2.66). Let $\mathrm{ord}\tilde{L}_n = N_1$ and $\mathrm{ord}\hat{L}_n = N_2$. If $\mathrm{ord}[\tilde{L}_n, \hat{L}_n] = N_1 + N_2 \geq 0$, then

(3.2.80) $$\mathrm{res}[\tilde{L}_n, \hat{L}_n] = \mathrm{res}[\sum_{i=-\infty}^{N_1} \tilde{l}_n^{(i)} S^i, \sum_{j=-\infty}^{N_2} \hat{l}_n^{(j)} S^j] = \sum_{i=-N_2}^{N_1} \mathrm{res}[\tilde{l}_n^{(i)} S^i, \hat{l}_n^{(-i)} S^{-i}].$$

The last sum in (3.2.80) is a total difference, as

$$\mathrm{res}[\tilde{l}_n^{(i)} S^i, \hat{l}_n^{(-i)} S^{-i}] = \tilde{l}_n^{(i)} \hat{l}_{n+i}^{(-i)} - \hat{l}_n^{(-i)} \tilde{l}_{n-i}^{(i)} \sim 0$$

due to the property (3.2.22). If $N_1 + N_2 < 0$, then $\mathrm{res}[\tilde{L}_n, \hat{L}_n] = 0$.

As any integer power L_n^i satisfies (3.2.65), one has

$$D_t \mathrm{res} L_n^i = \mathrm{res} D_t(L_n^i) = \mathrm{res}[f_n^*, L_n^i] \sim 0,$$

i.e. the functions $\mathrm{res}L_n^i$ are conserved densities. The series L_n given by (3.2.66) is such that $N > 0$, hence the functions $\mathrm{res}L_n^i$ are equal to 0 or to 1 if $i \leq 0$, and these densities are trivial. These conserved densities can be nontrivial only if $i \geq 1$. In (3.2.78) we have exactly this case.

Eq. (3.2.65) implies

(3.2.81) $$\dot{L}_n L_n^{-1} = [f_n^* L_n^{-1}, L_n].$$

It follows from (3.2.79, 3.2.81) that

$$\mathrm{res}(\dot{L}_n L_n^{-1}) = \dot{l}_n^{(N)} (l_n^{(N)})^{-1} = D_t \log l_n^{(N)} \sim 0.$$

This is the reason why (3.2.77) is another conserved density. □

A priori, we do not know whether the conserved densities (3.2.77, 3.2.78) are nontrivial, and which are their orders. Almost all such conserved densities can be trivial in the case of a linearizable equation. In the case of known S-integrable equations the recursion operator provides an infinite hierarchy of conserved densities of arbitrarily high order. This is the case of the Volterra equation (3.2.2). Using formula (3.2.75) for its recursion operator, one easily checks that

$$\log u_n = p_n^1, \qquad \mathrm{res}\mathscr{L} = u_{n+1} + u_n \sim 2p_n^2,$$
$$\mathrm{res}\mathscr{L}^2 = u_{n+2}u_{n+1} + 3u_{n+1}u_n + u_{n+1}^2 + u_n^2 \sim 4p_n^3,$$

where p_n^i are conserved densities (3.2.39) of (3.2.2). Moreover, it is possible to prove that the conserved densities $\mathrm{res}\mathscr{L}^i$ have the order $i - 1$ for any $i \geq 1$.

Remark 10. *We can also construct conserved densities of the Volterra equation, using the alternative Lax pair*

$$\dot{L}_n = [A_n, L_n]\,, \tag{3.2.82}$$

$$L_n = u_{n+1}^{1/2} S + u_n^{1/2} S^{-1}, \quad A_n = \tfrac{1}{2} u_{n+2}^{1/2} u_{n+1}^{1/2} S^2 - \tfrac{1}{2} u_n^{1/2} u_{n-1}^{1/2} S^{-2}. \tag{3.2.83}$$

We can prove here an analog of Theorem 25, as in its proof we do use neither the precise formula (3.2.64) for f_n^* nor the fact that L_n is an infinite series. One can check, for example, that

$$\log u_{n+1}^{1/2} \sim \tfrac{1}{2} p_n^1\,, \qquad res L_n = 0\,, \qquad res L_n^2 \sim 2p_n^2\,,$$
$$res L_n^3 = 0\,, \qquad res L_n^4 \sim 4p_n^3\,,$$

i.e. we find the densities (3.2.39). Moreover, we have: $res L_n^i = 0$ for all odd positive i, and $res L_n^{2j}$ is a conserved density of order $j-1$ for all $j \geq 1$.

In practice, it is difficult to construct a recursion operator and difficult to prove that a generalized symmetry (3.2.73) is local, i.e. its right hand side is of the form (3.2.10). We shall be interested below in approximate solutions of (3.2.65). They are easy to construct and can be used for deriving the integrability conditions. These solutions can be called approximate recursion operators, but we prefer to use the name formal symmetry because of its close connection with generalized symmetry (see Theorem 26 below).

Let us notice that for any series L_n of order N given by (3.2.66), the series $A(L_n)$ given by (3.2.69) can be expressed as

$$A(L_n) = a_n^{(N+1)} S^{N+1} + a_n^{(N)} S^N + a_n^{(N-1)} S^{N-1} + \dots\,. \tag{3.2.84}$$

Definition 7. *The series (3.2.66) is called a formal symmetry of (3.2.1) of length l (we write $lgt L_n = l$) if the first l coefficients of the series $A(L_n)$ (3.2.84) vanish:*

$$a_n^{(i)} = 0\,, \qquad N+1 \geq i \geq N+2-l\,. \tag{3.2.85}$$

We assume, moreover, that $l \geq 1$ and $a_n^{(N+1-l)} \neq 0$.

The recursion operator L_n of order N is such that all coefficients $a_n^{(i)}$ of $A(L_n)$ vanish. The equations $a_n^{(N+1-j)} = 0$, $0 \leq j \leq l-1$, define the coefficients $l_n^{(N-j)}$ of the recursion operator L_n and of a formal symmetry L_n, such that $\mathrm{lgt} L_n = l$, $\mathrm{ord} L_n = N$. So, the first l coefficients of such formal symmetry and of the N-th order recursion operator are defined by the same equations.

In order to find the length l of a given formal symmetry L_n of order N, we have to specify (3.2.84). If

$$A(L_n) = \sum_{i=-\infty}^{k} a_n^{(i)} S^i\,, \qquad a_n^{(k)} \neq 0\,, \tag{3.2.86}$$

then we have: $l = N+1-k$. So, for any formal symmetry, we have:

$$\mathrm{lgt} L_n = \mathrm{ord} L_n + 1 - \mathrm{ord} A(L_n)\,. \tag{3.2.87}$$

Recalling that the Fréchet derivative of the right hand side g_n of a generalized symmetry (3.2.3) is the following operator

$$g_n^* = \sum_{i=m'}^{m} \frac{\partial g_n}{\partial u_{n+i}} S^i = \sum_{i=m'}^{m} g_n^{(i)} S^i \tag{3.2.88}$$

(see (3.2.47, 3.2.63)), we state the following theorem:

Theorem 26. *If* (3.2.1) *has a generalized symmetry* (3.2.3) *of order* $m \geq 1$*, then it has a formal symmetry* L_n *with ord*$L_n = m$*, lgt*$L_n \geq m$ *defined by*

$$L_n = g_n^* + \sum_{i=-\infty}^{m'-1} 0\, S^i. \tag{3.2.89}$$

I.e all the terms with powers of S *lower than* m' *are zero* (3.2.3).

PROOF. Let us apply the Fréchet derivative to both sides of the defining equation for the generalized symmetry (3.2.72). We see that

$$\begin{aligned}(D_t g_n)^* &= \left(\sum_{i=m'}^{m} g_n^{(i)} f_{n+i}\right)^* = \sum_{i,j} \frac{\partial^2 g_n}{\partial u_{n+i}\partial u_{n+j}} f_{n+i} S^j + \sum_{i,j} \frac{\partial g_n}{\partial u_{n+i}} \frac{\partial f_{n+i}}{\partial u_{n+j}} S^j \\ &= \sum_{j=m'}^{m} \dot{g}_n^{(j)} S^j + \left(\sum_{i=m'}^{m} g_n^{(i)} S^i\right)\left(\sum_{\sigma=-1}^{1} f_n^{(\sigma)} S^\sigma\right),\end{aligned}$$

where $\sigma = j - i$ and the functions $f_n^{(\sigma)}$ are defined by (3.2.47). We can express the result in terms of the Fréchet derivatives g_n^* (3.2.88) and f_n^* (3.2.64):

$$(D_t g_n)^* = \dot{g}_n^* + g_n^* f_n^* \,. \tag{3.2.90}$$

As $f_n^* g_n = D_\epsilon f_n$ (see (3.2.6)), we can write the following analog of (3.2.90)

$$(f_n^* g_n)^* = (D_\epsilon f_n)^* = f_{n,\epsilon}^* + f_n^* g_n^* \,. \tag{3.2.91}$$

Using (3.2.69, 3.2.90, 3.2.91), from (3.2.72) we obtain the relation

$$A(g_n^*) = f_{n,\epsilon}^* = f_{n,\epsilon}^{(1)} S + f_{n,\epsilon}^{(0)} + f_{n,\epsilon}^{(-1)} S^{-1}, \tag{3.2.92}$$

where $f_{n,\epsilon}^{(i)}$ are the ϵ-derivatives of $f_n^{(i)}$.

Introducing the series (3.2.89), we see that $\mathrm{ord} L_n = m$, as $g_n^{(m)} \neq 0$ (see (2.4.10)), and from (3.2.92) we get

$$\mathrm{ord} A(L_n) = \mathrm{ord} f_{n,\epsilon}^* \leq 1 \,,$$

as $f_{n,\epsilon}^{(i)}$ may vanish. Formula (3.2.87) implies that $\mathrm{lgt} L_n \geq m \geq 1$, i.e. this series L_n is a formal symmetry. □

Theorem 26 shows how to obtain a formal symmetry from the generalized symmetry. To derive the integrability conditions, we need to use these formal symmetries. The coefficients of these formal symmetries have, due to (3.2.89), the same structure as the right hand side of a generalized symmetry. This is the reason why the coefficients $l_n^{(i)}$ of a formal symmetry, which is a series of the form (3.2.66), have no explicit dependence on n and t and are functions of the form (3.2.10).

As we have shown before, formal series can be multiplied and inverted. The same is also true for formal symmetries. Using relations (3.2.70, 3.2.71) together with (3.2.87), we can check that, if L_n and $\tilde{L}_n$ are formal symmetries, then the series L_n^{-1} and $L_n \tilde{L}_n$ also are formal symmetries, and we can find their orders and lengths. In fact, we always have $\mathrm{ord}(L_n \tilde{L}_n) = \mathrm{ord} L_n + \mathrm{ord} \tilde{L}_n$ and

$$\begin{aligned}\mathrm{ord} A(L_n \tilde{L}_n) &\leq \max(\mathrm{ord}(A(L_n)\tilde{L}_n)\,, \\ \mathrm{ord}(L_n A(\tilde{L}_n))) &= \max(\mathrm{ord} A(L_n) + \mathrm{ord} \tilde{L}_n\,, \\ \mathrm{ord} L_n + \mathrm{ord} A(\tilde{L}_n)) &= \max(\mathrm{ord} L_n + 1 - \mathrm{lgt} L_n + \mathrm{ord} \tilde{L}_n\,, \\ \mathrm{ord} L_n + \mathrm{ord} \tilde{L}_n + 1 - \mathrm{lgt} \tilde{L}_n) &= \mathrm{ord} L_n + \mathrm{ord} \tilde{L}_n + 1 - \min(\mathrm{lgt} L_n, \mathrm{lgt} \tilde{L}_n)\,.\end{aligned}$$

Formula (3.2.87) implies:

$$\mathrm{lgt}(L_n\tilde{L}_n) \geq \mathrm{ord}(L_n\tilde{L}_n) + 1 - (\mathrm{ord}L_n + \mathrm{ord}\tilde{L}_n + 1) + \min(\mathrm{lgt}L_n, \mathrm{lgt}\tilde{L}_n)\,.$$

Thus for any formal symmetries L_n and $\tilde{L}_n$ we get

$$\begin{aligned} &\mathrm{ord}(L_n\tilde{L}_n) = \mathrm{ord}L_n + \mathrm{ord}\tilde{L}_n\,, \\ &\mathrm{lgt}(L_n\tilde{L}_n) \geq \min(\mathrm{lgt}L_n, \mathrm{lgt}\tilde{L}_n)\,. \end{aligned} \qquad (3.2.93)$$

In a quite similar way, one can derive from (3.2.70, 3.2.87) for any formal symmetry L_n that,

$$\mathrm{ord}L_n^{-1} = -\mathrm{ord}L_n\,, \qquad \mathrm{lgt}L_n^{-1} = \mathrm{lgt}L_n\,. \qquad (3.2.94)$$

Let us take into account (3.2.93, 3.2.94) and that $L_n^0 = 1$ is the solution of (3.2.65) of order 0. Then, for any integer power i of the formal symmetry L_n, we get the following result

$$\mathrm{ord}L_n^i = i\,\mathrm{ord}L_n\,, \qquad \mathrm{lgt}L_n^i \geq \mathrm{lgt}L_n\,. \qquad (3.2.95)$$

In the following we will need formal symmetries of order 1. We will make the ansatz, valid in the case of the Volterra equation, that (3.2.1) has two generalized symmetries of the left orders m and $m+1$, where m is a high enough number. So, let g_n and $\hat{g}_n$ be the right hand sides of two generalized symmetries with the left orders $m \geq 1$ and $m+1$, respectively. Theorem 26 shows that the Fréchet derivative g_n^* is a formal symmetry of order m and $\mathrm{lgt}g_n^* \geq m$, and $\hat{g}_n^*$ is such that $\mathrm{ord}\hat{g}_n^* = m+1$, $\mathrm{lgt}\hat{g}_n^* \geq m+1$.

We can construct the following series

$$L_n = \hat{g}_n^*(g_n^*)^{-1}. \qquad (3.2.96)$$

As it follows from (3.2.93, 3.2.94), this series will be a formal symmetry of (3.2.1) of order 1 and $\mathrm{lgt}L_n \geq m$. This result is formulated in the following theorem:

Theorem 27. *If* (3.2.1) *possesses two generalized symmetries* $u_{n,\epsilon} = g_n$ *and* $u_{n,\hat{\epsilon}} = \hat{g}_n$ *of left orders* $m \geq 1$ *and* $m+1$, *then it possesses a formal symmetry* L_n *given by formula* (3.2.96), *such that* $\mathrm{ord}L_n = 1$ *and* $\mathrm{lgt}L_n \geq m$.

A formal symmetry (3.2.96) of the first order can be written as

$$L_n = l_n^{(1)}S + l_n^{(0)} + l_n^{(-1)}S^{-1} + l_n^{(-2)}S^{-2} + \dots\,, \qquad l_n^{(1)} \neq 0\,. \qquad (3.2.97)$$

Its length $l = \mathrm{lgt}L_n$ can be as high as necessary. This formal symmetry generates a number of conserved densities for (3.2.1) as well as a recursion operator given in Theorem 25. In fact, if $l \geq 3$, then the highest three coefficients of the series (3.2.84) with $N = 1$ vanish: $a_n^{(2)} = a_n^{(1)} = a_n^{(0)} = 0$. One can show, following Theorem 25, that $l_n^{(0)} = \mathrm{res}L_n$ is a conserved density. In the case $l \geq 4$, it follows from (3.2.95) that $\mathrm{ord}L_n^2 = 2$ and $\mathrm{lgt}L_n^2 \geq 4$. This means that the coefficients at S^i $(3 \geq i \geq 0)$ of the series $A(L_n^2)$ are equal to zero, and thus also the function $\mathrm{res}L_n^2$ is a conserved density. In this way we prove the following general statement: if $\mathrm{lgt}L_n \geq 3$, then the functions

$$\mathrm{res}L_n^i\,, \qquad 1 \leq i \leq \mathrm{lgt}L_n - 2\,, \qquad (3.2.98)$$

are conserved densities of (3.2.1).

Taking into account (3.2.69), (3.2.81) can be written as: $A(L_n)L_n^{-1} = 0$. In the case $l \geq 2$, it follows from (3.2.87) that $\mathrm{ord}A(L_n) \leq 0$, hence $\mathrm{ord}(A(L_n)L_n^{-1}) < 0$. For this reason one can show, following the proof of Theorem 25, that the function

$$\log l_n^{(1)} \qquad (3.2.99)$$

is a conserved density. Consequently, we can state the following theorem, which is the analog of Theorem 25:

Theorem 28. *If* (3.2.1) *has a formal symmetry* (3.2.97) *of the first order and if* $\operatorname{lgt} L_n \geq 2$, *then the function* (3.2.99) *is one of its conserved densities. If* $\operatorname{lgt} L_n \geq 3$, *the functions* (3.2.98) *are also conserved densities of* (3.2.1).

In particular, starting from two generalized symmetries of orders $m \geq 2$ and $m+1$ and using Theorems 27 and 28, we can construct $m-1$ conserved densities which, however, may be trivial.

In the following one considers two new theorems, based on Theorems 27 and 28, where further integrability conditions are written down. Here, instead of considering the existence of generalized symmetries and the compatibility condition (3.2.7), we use the Lax equation (3.2.65) and the existence of a formal symmetry of first order and of high enough length (3.2.97). Such formal symmetry not only makes the calculation much simpler but also provides us with integrability conditions which a priori have no dependence on the order m of the generalized symmetry (cf. Theorem 23 and its proof).

Theorem 29. *If* (3.2.1) *has a formal symmetry* (3.2.97) *of first order and if* $\operatorname{lgt} L_n \geq 3$, *then it satisfies condition* (3.2.56). *Let* $q_n^{(1)}$ *be a function obtained from* (3.2.56). *Then there exists a function* $q_n^{(2)}$ *of the form* (3.2.10), *such that*

(3.2.100) $$\dot{p}_n^{(2)} = (S-1)q_n^{(2)} \quad \textit{with} \quad p_n^{(2)} = q_n^{(1)} + \frac{\partial f_n}{\partial u_n} .$$

PROOF. In the case when $\operatorname{lgt} L_n \geq 3$, the first three coefficients of the series $A(L_n)$ (3.2.69, 3.2.84) with $N = 1$ must be equal to zero: $a_n^{(2)} = a_n^{(1)} = a_n^{(0)} = 0$. This request will give us some equations for the first three coefficients of the formal symmetry L_n: $l_n^{(1)}, l_n^{(0)}, l_n^{(-1)}$. From Theorem 28, we also have two conserved densities given by (3.2.99, 3.2.98) with $i = 1$. These three equations for the coefficients of L_n are the direct analogs of (3.2.49-3.2.51). The first of them, $a_n^{(2)} = 0$, can be written in the form

(3.2.101) $$l_n^{(1)} f_{n+1}^{(1)} = f_n^{(1)} l_{n+1}^{(1)}$$

(see (3.2.47, 3.2.64) for the used notation). Dividing (3.2.101) by $f_n^{(1)} f_{n+1}^{(1)}$ and using (3.2.13), we are led to the following formula: $l_n^{(1)} = c f_n^{(1)}$, $c \neq 0 \in \mathbb{C}$.

As the operator A defined by (3.2.69) is linear, a formal symmetry can be multiplied by any nonzero constant, and the length is not changed. Dividing L_n by c, we obtain, without loss of generality, the following formula for $l_n^{(1)}$

(3.2.102) $$l_n^{(1)} = f_n^{(1)}.$$

Theorem 28 guarantees that the function

$$\log l_n^{(1)} = \log f_n^{(1)} = \log \frac{\partial f_n}{\partial u_{n+1}} = p_n^{(1)}$$

is a conserved density of (3.2.1), i.e. condition (3.2.56) is satisfied.

The second equation, $a_n^{(1)} = 0$, reads

(3.2.103) $$\dot{l}_n^{(1)} + l_n^{(1)} f_{n+1}^{(0)} + l_n^{(0)} f_n^{(1)} = f_n^{(0)} l_n^{(1)} + f_n^{(1)} l_{n+1}^{(0)} .$$

Dividing (3.2.103) by $f_n^{(1)}$ and using (3.2.102), one obtains

(3.2.104) $$\dot{p}_n^{(1)} = (S-1)(l_n^{(0)} - f_n^{(0)}) .$$

Comparing (3.2.104) with (3.2.56) and using property (3.2.13), one gets: $q_n^{(1)} = l_n^{(0)} - f_n^{(0)} + \alpha$, where α is a constant. Then one has

$$l_n^{(0)} + \alpha = q_n^{(1)} + f_n^{(0)} = p_n^{(2)}, \tag{3.2.105}$$

where $p_n^{(2)}$ is the function given by (3.2.100). Theorem 28 guarantees that the function $\mathrm{res} L_n = l_n^{(0)}$ is a conserved density of (3.2.1), i.e. the function $p_n^{(2)}$ is also a conserved density. □

Theorem 30. *Let* (3.2.1) *have a formal symmetry* (3.2.97) *with length* $lgt L_n \geq 4$. *Let* $q_n^{(1)}$ *be a function defined by* (3.2.56), *while* $p_n^{(2)}$, $q_n^{(2)}$ *be functions given by* (3.2.100). *Then there exists a function* $q_n^{(3)}$ *of the form* (3.2.10), *such that*

$$\dot p_n^{(3)} = (S-1) q_n^{(3)} \quad \text{with} \quad p_n^{(3)} = q_n^{(2)} + \frac{1}{2}(p_n^{(2)})^2 + \frac{\partial f_n}{\partial u_{n+1}} \frac{\partial f_{n+1}}{\partial u_n}. \tag{3.2.106}$$

PROOF. This proof is a direct continuation of the calculations we did to prove Theorem 29. We will need to compute the coefficient $l_n^{(-2)}$ of the formal symmetry L_n. We could consider the equation $a_n^{(-1)} = 0$, however we prefer to use the conserved density $\mathrm{res} L_n^2$ provided by Theorem 28.

Let us write down $a_n^{(0)} = 0$ explicitly

$$\dot l_n^{(0)} + l_n^{(1)} f_{n+1}^{(-1)} + l_n^{(0)} f_n^{(0)} + l_n^{(-1)} f_{n-1}^{(1)} = f_n^{(-1)} l_{n-1}^{(1)} + f_n^{(0)} l_n^{(0)} + f_n^{(1)} l_{n+1}^{(-1)}. \tag{3.2.107}$$

Using (3.2.102, 3.2.105), (3.2.107) can be rewritten as

$$\dot l_n^{(0)} = \dot p_n^{(2)} = (S-1)(\, f_{n-1}^{(1)} (l_n^{(-1)} - f_n^{(-1)})\,).$$

As in the case of (3.2.104), we can now express $l_n^{(-1)}$ in terms of $q_n^{(2)}$ defined by (3.2.100)

$$l_n^{(-1)} = (q_n^{(2)} + \beta)/f_{n-1}^{(1)} + f_n^{(-1)}, \tag{3.2.108}$$

where β is a constant.

Let us write down the formula for the conserved density $\mathrm{res} L_n^2$. Using the equivalence relation (3.2.22), we get

$$\mathrm{res} L_n^2 = l_n^{(1)} l_{n+1}^{(-1)} + (l_n^{(0)})^2 + l_n^{(-1)} l_{n-1}^{(1)} \sim 2\, l_n^{(1)} l_{n+1}^{(-1)} + (l_n^{(0)})^2.$$

The densities can be multiplied by a nonzero constant and, taking into account (3.2.102, 3.2.105, 3.2.108) and (3.2.106) for $p_n^{(3)}$, one has

$$\begin{aligned} \frac{1}{2}\mathrm{res} L_n^2 &\sim q_{n+1}^{(2)} + \beta + f_n^{(1)} f_{n+1}^{(-1)} + \frac{1}{2}(p_n^{(2)} - \alpha)^2 \\ &\sim q_n^{(2)} + \frac{1}{2}(p_n^{(2)})^2 + f_n^{(1)} f_{n+1}^{(-1)} + \beta - \alpha p_n^{(2)} + \frac{1}{2}\alpha^2 = p_n^{(3)} - \alpha p_n^{(2)} + \frac{1}{2}\alpha^2 + \beta. \end{aligned}$$

A constant is a trivial conserved density, and conserved densities of (3.2.1) generate a linear space. This is the reason why the function $p_n^{(3)}$ must be a conserved density of (3.2.1), and hence (3.2.106) is satisfied. □

As one can see from (3.2.13), the functions $q_n^{(1)}$ and $q_n^{(2)}$ of the conditions (3.2.56, 3.2.100) are defined up to arbitrary constants. Therefore, the functions $p_n^{(2)}$ and $p_n^{(3)}$, given by (3.2.100, 3.2.106), may depend in general on those constants. As we have shown in the proofs of Theorems 29 and 30, $p_n^{(2)}$ and $p_n^{(3)}$ are conserved densities for all values of those constants, and one has no need to take into account those arbitrary constants, when checking the integrability conditions. In another words, when checking the integrability conditions (3.2.56, 3.2.100, 3.2.106), any choice of the functions $q_n^{(1)}$, $q_n^{(2)}$ will give the same result.

When proving Theorems 29 and 30, we have given a scheme for deriving the integrability conditions. This scheme corresponds to finding the coefficients of the first order formal symmetry and to the application of Theorem 28. Starting from a formal symmetry of a sufficiently big length, we can obtain as many integrability conditions as necessary, and all those conditions will have the form of local conservation laws.

As in the case of condition (3.2.56), Theorem 24 is essential for checking the integrability conditions (3.2.100, 3.2.106) as we did in (3.2.61, 3.2.62). If, given an equation (3.2.1), all these integrability conditions are satisfied, we obtain three conserved densities of low orders. In the case of the Volterra equation (3.2.2), for example, one gets

$$p_n^{(1)} = p_n^1 , \qquad p_n^{(2)} \sim 2p_n^2 + c_1 , \qquad p_n^{(3)} \sim 2p_n^3 + 2c_1 p_n^2 + \tfrac{1}{2}c_1^2 + c_2 ,$$

where p_n^1, p_n^2, p_n^3 are the conserved densities given in the list (3.2.39), with c_1, c_2 arbitrary constants. The conserved densities $p_n^{(1)}$, $p_n^{(2)}$ and $p_n^{(3)}$ have the orders 0, 0 and 1, respectively.

For any integrable equation of the Volterra type we get from Theorem 27 an arbitrarily long formal symmetry of the first order. The coefficients of such formal symmetry and Theorem 28 provide us with as many conserved densities as we need. Formulae (3.2.102, 3.2.105, 3.2.108) give the first three coefficients of such formal symmetry in terms of $q_n^{(1)}$ and $q_n^{(2)}$. In the case of the Volterra equation, we have

$$\begin{aligned} f_n^{(1)} &= u_n , \qquad f_n^{(0)} = u_{n+1} - u_{n-1} , \qquad f_n^{(-1)} = -u_n , \\ q_n^{(1)} &= u_n + u_{n-1} + c_1 , \qquad q_n^{(2)} = u_{n+1}u_n + u_n u_{n-1} + c_2 , \end{aligned}$$

where c_1, c_2 are constants, and denoting $c_3 = c_1 - \alpha$, $c_4 = c_2 + \beta$, we can write down explicitly the first three terms of the formal symmetry:

$$L_n = u_n S + u_{n+1} + u_n + c_3 + \frac{u_{n+1}u_n + c_4}{u_{n-1}} S^{-1} + \dots$$

(cf. this result with that obtained using the recursion operator (3.2.75)).

2.4. Formal conserved density. We have obtained in Sections 3.2.2 and 3.2.3 the integrability conditions (3.2.56, 3.2.100, 3.2.106) which follow from the existence of generalized symmetries. However, in order to carry out the exhaustive classification of integrable equations of the form (3.2.1), we need some additional integrability conditions which come from the conservation laws.[1] So, starting from the conservation laws, we introduce and discuss in this section the formal conserved densities, in analogy with the formal symmetries, and then derive two new integrability conditions. At the end we will prove a general statement which explains why the shape of an equation, possessing a higher order local conservation law, must have some symmetry.

Given a conserved density p_n of (3.2.1) we introduce and discuss an equation for its variational derivative ϱ_n, defined by (3.2.40, 3.2.43)

$$(D_t + f_n^{*\dagger})\varrho_n = 0 . \qquad (3.2.109)$$

[1] In the case of (3.1.1), the classification problem can be solved without using this kind of integrability conditions (see e.g. the review [607]). Such conditions help to make the problem easier and lead to a shorter list of equations. In the case of the lattice equations (3.2.1), these additional integrability conditions seem to be necessary to solve the problem.

This is the analog of (3.2.72). If f_n^*, given by (3.2.64), is the Fréchet derivative of f_n, the operator $f_n^{*\dagger}$ is its adjoint operator defined by

$$(a_n S^i)^\dagger = S^{-i} \circ a_n = a_{n-i} S^{-i}. \tag{3.2.110}$$

Then

$$f_n^{*\dagger} = S^{-1} \circ f_n^{(1)} + f_n^{(0)} + S \circ f_n^{(-1)} = f_{n+1}^{(-1)} S + f_n^{(0)} + f_{n-1}^{(1)} S^{-1}. \tag{3.2.111}$$

Taking into account (3.2.47), (3.2.111) can be rewritten as

$$f_n^{*\dagger} = \sum_{i=-1}^{1} f_{n+i}^{(-i)} S^i = \sum_{i=-1}^{1} \frac{\partial f_{n+i}}{\partial u_n} S^i. \tag{3.2.112}$$

Then we can prove the following theorem:

Theorem 31. *For any conserved density p_n of (3.2.1), its variational derivative ϱ_n satisfies (3.2.109).*

PROOF. Let p_n be a conserved density of (3.2.1), then

$$\dot{p}_n = \sum_i \frac{\partial p_n}{\partial u_{n+i}} f_{n+i} \sim \sum_i \frac{\partial p_{n-i}}{\partial u_n} f_n = \varrho_n f_n \sim 0\,, \tag{3.2.113}$$

where we have used the definition (3.2.40). On the other hand,

$$\frac{\partial \varrho_n}{\partial u_{n+i}} = \frac{\partial}{\partial u_{n+i}} \sum_j \frac{\partial p_{n+j}}{\partial u_n} = \frac{\partial}{\partial u_n} \sum_j \frac{\partial p_{n+j}}{\partial u_{n+i}} = \frac{\partial}{\partial u_n} T^i \sum_\sigma \frac{\partial p_{n+\sigma}}{\partial u_n}\,,$$

i.e. we have

$$\frac{\partial \varrho_n}{\partial u_{n+i}} = \frac{\partial \varrho_{n+i}}{\partial u_n} \qquad \text{for any} \qquad i \in \mathbb{Z}\,. \tag{3.2.114}$$

Using Theorem 24 together with relation (3.2.113), we get

$$\frac{\delta \dot{p}_n}{\delta u_n} = \frac{\delta(\varrho_n f_n)}{\delta u_n} = 0\,. \tag{3.2.115}$$

Moreover, using (3.2.112, 3.2.114), we obtain:

$$\begin{aligned} \frac{\delta(\varrho_n f_n)}{\delta u_n} &= \sum_i \frac{\partial \varrho_{n+i}}{\partial u_n} f_{n+i} + \sum_i \varrho_{n+i} \frac{\partial f_{n+i}}{\partial u_n} \\ &= \sum_i \frac{\partial \varrho_n}{\partial u_{n+i}} f_{n+i} + \sum_i \frac{\partial f_{n+i}}{\partial u_n} \varrho_{n+i} = (D_t + f_n^{*\dagger}) \varrho_n\,. \end{aligned}$$

This formula together with (3.2.115) imply (3.2.109). □

The following equation, analogous to (3.2.65), plays the main role in this section

$$\dot{\mathcal{S}}_n + \mathcal{S}_n f_n^* + f_n^{*\dagger} \mathcal{S}_n = 0\,. \tag{3.2.116}$$

Here $\mathcal{S}_n$ is a formal series of the same type as L_n (3.2.66)

$$\mathcal{S}_n = \sum_{i=-\infty}^{M} s_n^{(i)} S^i, \qquad s_n^{(M)} \neq 0\,, \tag{3.2.117}$$

and $s_n^{(i)}$ are functions of the form (3.2.10). Examples of exact solutions of (3.2.116) will be given in Section 3.2.6. We know that the exact solution L_n of (3.2.65) is the recursion operator for (3.2.1). The solution $\mathcal{S}_n$ of (3.2.116) is the inverse of a Noether or Hamiltonian operator. The details of this statement will be discussed in Section 3.2.6.

Eqs. (3.2.65, 3.2.116) are closely related. Let us introduce an operator B, such that

$$B(\mathscr{S}_n) = \dot{\mathscr{S}}_n + \mathscr{S}_n f_n^* + f_n^{*\dagger}\mathscr{S}_n\,. \tag{3.2.118}$$

For any formal series L_n, $\mathscr{S}_n$, $\tilde{\mathscr{S}}_n$ of the form (3.2.66, 3.2.117), the following identities take place

$$B(\mathscr{S}_n L_n) = B(\mathscr{S}_n)L_n + \mathscr{S}_n A(L_n)\,, \tag{3.2.119}$$

$$A(\mathscr{S}_n^{-1}\tilde{\mathscr{S}}_n) = \mathscr{S}_n^{-1}B(\tilde{\mathscr{S}}_n) - \mathscr{S}_n^{-1}B(\mathscr{S}_n)\mathscr{S}_n^{-1}\tilde{\mathscr{S}}_n\,. \tag{3.2.120}$$

The identity (3.2.119) shows that, for any solutions $\mathscr{S}_n$ of (3.2.116) and L_n of (3.2.65), the series $\mathscr{S}_n L_n$ will be a new solution of (3.2.116). On the other hand, as it follows from. (3.2.120), if $\mathscr{S}_n$ and $\tilde{\mathscr{S}}_n$ are any two solutions of (3.2.116), then the series $\mathscr{S}_n^{-1}\tilde{\mathscr{S}}_n$ satisfies (3.2.65).

As in the case of (3.2.65), we are interested here in approximate solutions of (3.2.116). Such solutions will be called formal conserved densities (see Definition 8 below) because of their close connection with conserved densities, as shown in Theorem 32 below.

For the formal series $B(\mathscr{S}_n)$, defined by (3.2.118), we obtain in general

$$B(\mathscr{S}_n) = b_n^{(M+1)}S^{M+1} + b_n^{(M)}S^M + b_n^{(M-1)}S^{M-1} + \dots, \tag{3.2.121}$$

where $\mathscr{S}_n$ is a series of the form (3.2.117) and thus has the order M.

Definition 8. *If a series $\mathscr{S}_n$ (3.2.117) is such that the first $l \geq 1$ coefficients of the series $B(\mathscr{S}_n)$ (3.2.121) vanish, i.e.*

$$B(\mathscr{S}_n) = b_n^{(M+1-l)}S^{M+1-l} + b_n^{(M-l)}S^{M-l} + \dots, \qquad b_n^{(M+1-l)} \neq 0\,, \tag{3.2.122}$$

then $\mathscr{S}_n$ is called a formal conserved density *of (3.2.1) of the order M and the length l, and we will write: $\mathrm{ord}\mathscr{S}_n = M$, $\mathrm{lgt}\mathscr{S}_n = l$.*

Comparing (3.2.121, 3.2.122), we easily obtain the following formula

$$\mathrm{lgt}\mathscr{S}_n = \mathrm{ord}\mathscr{S}_n + 1 - \mathrm{ord}B(\mathscr{S}_n) \tag{3.2.123}$$

relating the length and the order of formal conserved density $\mathscr{S}_n$. This is the analog of (3.2.87) which we obtained in the case of formal symmetries.

Theorem 32. *If (3.2.1) possesses a conserved density p_n of order $m \geq 2$, then it has a formal conserved density $\mathscr{S}_n$, such that $\mathrm{ord}\mathscr{S}_n = m$ and $\mathrm{lgt}\mathscr{S}_n \geq m-1$. This formal conserved density $\mathscr{S}_n$ is given by the formula*

$$\mathscr{S}_n = \varrho_n^* + \sum_{i=-\infty}^{-m-1} 0\,S^i, \qquad \varrho_n = \frac{\delta p_n}{\delta u_n}\,, \tag{3.2.124}$$

where ϱ_n has the form (3.2.46), and thus ϱ_n^* is given by

$$\varrho_n^* = \sum_{i=-m}^{m} \frac{\partial \varrho_n}{\partial u_{n+i}} S^i. \tag{3.2.125}$$

PROOF. Theorem 31 allows us to pass from a conserved density p_n to (3.2.109). Let us apply the Fréchet derivative to both sides of this equation. Using (3.2.112) and (3.2.63), we check that

$$(f_n^{*\dagger}\varrho_n)^* = \left(\sum_{i=-1}^{1} \frac{\partial f_{n+i}}{\partial u_n}\varrho_{n+i}\right)^* = \sum_{i,j} \frac{\partial^2 f_{n+i}}{\partial u_n \partial u_{n+j}}\varrho_{n+i}S^j + \sum_{i,j} \frac{\partial f_{n+i}}{\partial u_n}\frac{\partial \varrho_{n+i}}{\partial u_{n+j}}S^j$$

$$= \sum_j \Big(\sum_i \frac{\partial^2 f_{n+i}}{\partial u_n \partial u_{n+j}} \varrho_{n+i} \Big) S^j + \Big(\sum_{i=-1}^{1} \frac{\partial f_{n+i}}{\partial u_n} S^i \Big) \Big(\sum_{\sigma=-m}^{m} \frac{\partial \varrho_n}{\partial u_{n+\sigma}} S^\sigma \Big),$$

where $\sigma = j - i$. The first term has the form: $\sum_{j=-2}^{2} c_n^{(j)} S^j$, as $\frac{\partial^2 f_{n+i}}{\partial u_n \partial u_{n+j}} = 0$ for $j > 2$ and $j < -2$ (see (3.2.1)). The second term in the last expression is equal to $f_n^{*\dagger} \varrho_n^*$ (see (3.2.125)). Hence one is led to the following result

(3.2.126) $$(f_n^{*\dagger} \varrho_n)^* = f_n^{*\dagger} \varrho_n^* + \sum_{j=-2}^{2} c_n^{(j)} S^j .$$

On the other hand,

(3.2.127) $$(D_t \varrho_n)^* = \dot{\varrho}_n^* + \varrho_n^* f_n^*$$

(cf. (3.2.90)). Now (3.2.126, 3.2.127) together with (3.2.109, 3.2.118) imply

(3.2.128) $$B(\varrho_n^*) = - \sum_{j=-2}^{2} c_n^{(j)} S^j .$$

Introducing the series (3.2.124), i.e. $\mathscr{S}_n = \varrho_n^*$, we see that $\mathrm{ord}\mathscr{S}_n = m$ due to (3.2.46, 3.2.125). Formula (3.2.128) provides the inequality $\mathrm{ord} B(\mathscr{S}_n) \le 2$, and (3.2.123) implies that $\mathrm{lgt}\mathscr{S}_n \ge m-1$. This means that the formal series $\mathscr{S}_n$ is a formal conserved density. □

From (3.2.119, 3.2.120) it follows that there is the same connection between formal symmetries and formal conserved densities of (3.2.1) as in the case of exact solutions of the Lax equation (3.2.65) and (3.2.116). Two formal conserved densities $\mathscr{S}_n$ and $\tilde{\mathscr{S}}_n$ give a formal symmetry $L_n = \mathscr{S}_n^{-1} \tilde{\mathscr{S}}_n$. Formal conserved density $\mathscr{S}_n$ together with formal symmetry L_n generate another formal conserved density $\tilde{\mathscr{S}}_n = \mathscr{S}_n L_n$. The orders and lengths of the resulting formal symmetries and conserved densities can easily be found, using (3.2.119, 3.2.120, 3.2.87, 3.2.123). For example, the following analogs of (3.2.93) take place

(3.2.129) $$\begin{aligned} &\mathrm{ord}(\mathscr{S}_n L_n) = \mathrm{ord}\mathscr{S}_n + \mathrm{ord} L_n , \\ &\mathrm{lgt}(\mathscr{S}_n L_n) \ge \min(\mathrm{lgt}\mathscr{S}_n, \mathrm{lgt} L_n) . \end{aligned}$$

Let us consider a formal conserved density $\mathscr{S}_n$ (3.2.124) and a first order formal symmetry L_n (3.2.97) such that $\mathrm{lgt} L_n \ge \mathrm{lgt}\mathscr{S}_n$. We can consider a new formal conserved density $\hat{\mathscr{S}}_n = \mathscr{S}_n L_n^i$. Its length will satisfy the inequality $\mathrm{lgt}\hat{\mathscr{S}}_n \ge \mathrm{lgt}\mathscr{S}_n$, as it follows from (3.2.95, 3.2.129). In this way we can obtain a formal conserved density $\mathscr{S}_n$ which has order 1 or 0 and an arbitrarily big length. This provides a simple calculation of the coefficients of $\mathscr{S}_n$ and an easy derivation of additional integrability conditions (cf. Theorems 29, 30).

However, for the classification of (3.2.1), we need only two additional integrability conditions. These can be obtained, using just one conservation law of order $m \ge 3$. More precisely, we can and shall derive those integrability conditions, using (3.2.116) and one formal conserved density $\mathscr{S}_n$ of order m and $\mathrm{lgt}\mathscr{S}_n \ge m - 1$ obtained from Theorem 32.

Theorem 33. *Let* (3.2.1) *have a conservation law of order* $m \ge 3$ *and a generalized symmetry of order* $m \ge 2$. *Then there will exist functions* $\sigma_n^{(1)}$ *and* $\sigma_n^{(2)}$ *of the form* (3.2.10)

which satisfy the following relations

$$r_n^{(1)} = (S-1)\sigma_n^{(1)}, \qquad r_n^{(1)} = \log\left(-\frac{\partial f_n}{\partial u_{n+1}} / \frac{\partial f_n}{\partial u_{n-1}}\right), \tag{3.2.130}$$

$$r_n^{(2)} = (S-1)\sigma_n^{(2)}, \qquad r_n^{(2)} = \dot{\sigma}_n^{(1)} + 2\,\frac{\partial f_n}{\partial u_n}\,. \tag{3.2.131}$$

PROOF. Theorem 23 has shown that the existence of a generalized symmetry of order $m \geq 2$ guarantees that (3.2.56) is satisfied, i.e. $\dot{p}_n^{(1)} \sim 0$. According to Theorem 32, a conservation law of order $m \geq 3$ implies the existence of a formal conserved density $\mathscr{S}_n$ of the form (3.2.117) of order $M = m$ and length l, such that $l \geq m-1 \geq 2$. From (3.2.121) and Definition 8 it follows that we have: $b_n^{(m+1)} = b_n^{(m)} = 0$.

Using (3.2.64, 3.2.111, 3.2.118), $b_n^{(m+1)} = 0$ is written as

$$s_n^{(m)} f_{n+m}^{(1)} + f_{n+1}^{(-1)} s_{n+1}^{(m)} = 0\,, \tag{3.2.132}$$

where $f_n^{(1)} f_n^{(-1)} s_n^{(m)} \neq 0$ due to (3.2.1, 3.2.117). Applying the operator S^{-1} and then dividing by $s_{n-1}^{(m)} f_n^{(-1)}$, one gets:

$$\frac{s_n^{(m)}}{s_{n-1}^{(m)}} = -\frac{f_{n+m-1}^{(1)}}{f_n^{(-1)}} = -\frac{f_n^{(1)}}{f_n^{(-1)}} \frac{\Phi_{n+1}^{(m-2)}}{\Phi_n^{(m-2)}}\,,$$

where $\Phi_n^{(N)}$ is given by (3.2.52). Applying the logarithm to both sides of this relation, one obtains the condition

$$r_n^{(1)} = \log(-f_n^{(1)}/f_n^{(-1)}) = (S-1)(\log s_{n-1}^{(m)} - \log \Phi_n^{(m-2)})\,.$$

It is easy to see now that there will exist a function $\sigma_n^{(1)}$ satisfying (3.2.130), of thel form $\sigma_n^{(1)} = c + \log(s_{n-1}^{(m)}/\Phi_n^{(m-2)})$, where c is a constant. Then $s_n^{(m)}$ can be expressed in terms of $\sigma_n^{(1)}$

$$s_n^{(m)} = \Phi_{n+1}^{(m-2)} e^{\sigma_{n+1}^{(1)} - c}\,. \tag{3.2.133}$$

Eq. $b_n^{(m)} = 0$ reads

$$\dot{s}_n^{(m)} + s_n^{(m)} f_{n+m}^{(0)} + s_n^{(m-1)} f_{n+m-1}^{(1)} + f_{n+1}^{(-1)} s_{n+1}^{(m-1)} + f_n^{(0)} s_n^{(m)} = 0\,.$$

We exclude $f_{n+1}^{(-1)}$, using (3.2.132), and then divide the result by $s_n^{(m)}$. So we get

$$D_t \log s_n^{(m)} + f_{n+m}^{(0)} + f_n^{(0)} + (1-S)\frac{s_n^{(m-1)} f_{n+m-1}^{(1)}}{s_n^{(m)}} = 0\,. \tag{3.2.134}$$

Formulae (3.2.52, 3.2.133) and (3.2.22) allow one to check that

$$D_t \log s_n^{(m)} = D_t \log \Phi_{n+1}^{(m-2)} + \dot{\sigma}_{n+1}^{(1)} = \sum_{i=1}^{m-1} D_t \log f_{n+i}^{(1)} + \dot{\sigma}_{n+1}^{(1)} \sim$$

$$(m-1) D_t \log f_n^{(1)} + \dot{\sigma}_n^{(1)} = (m-1)\dot{p}_n^{(1)} + \dot{\sigma}_n^{(1)},$$

where $p_n^{(1)}$ is defined in (3.2.56). As $\dot{p}_n^{(1)} \sim 0$, then $D_t \log s_n^{(m)} \sim \dot{\sigma}_n^{(1)}$, and hence (3.2.134) can be rewritten as: $\dot{\sigma}_n^{(1)} + 2 f_n^{(0)} \sim 0$. Consequently the second part of Theorem 33 is proven. □

Using (3.2.116), we can derive arbitrarily many integrability conditions analogous to (3.2.130, 3.2.131). In the general case, it is easier to check these integrability conditions, applying Theorem 24. However, we do not need this theorem in the simple case of the Volterra equation (3.2.2). In fact, when $f_n = u_n(u_{n+1} - u_{n-1})$, we have $r_n^{(1)} = 0$, i.e. (3.2.130) is trivially satisfied. Moreover, $\sigma_n^{(1)}$ is a constant function, and thus $r_n^{(2)} = 2(u_{n+1} - u_{n-1}) \sim 0$.

2.4.1. *Why the shape of scalar S-integrable evolutionary DΔEs are symmetric.* From the results presented up to now, we can obtain a theorem which explains why only symmetrical evolutionary DΔEs may possess higher order conservation laws. We will illustrate the result by the example of the discrete Burgers equation (see Section 2.3.5). Then we discuss the implication of this theorem for other classes of equations [513].

Here we will consider n and t independent evolutionary DΔEs of the following very general form

$$\dot{u}_n = f_n = f(u_{n+N}, u_{n+N-1}, \dots u_{n+M}), \tag{3.2.135}$$

$$N \geq M, \qquad \frac{\partial f_n}{\partial u_{n+N}} \frac{\partial f_n}{\partial u_{n+M}} \neq 0. \tag{3.2.136}$$

For a given equation N and M are fixed integers. The definitions of conservation laws, conserved densities and their orders are given by Definitions 4 and 6. Eqs. (3.2.43-3.2.46) will give us the orders also in this case.

In a quite similar way, we can prove the following analog of Theorem 31. If p_n is a conserved density of (3.2.135), then its variational derivative ϱ_n satisfies (3.2.109), where

$$D_t = \sum_i f_{n+i} \frac{\partial}{\partial u_{n+i}}, \qquad f_n^{*\dagger} = \sum_i \frac{\partial f_{n+i}}{\partial u_n} S^i.$$

When considering a conserved density p_n of order $m > 0$, (3.2.46) for ϱ_n is valid, and we can rewrite (3.2.109) as

$$\sum_{i=-m}^{m} \frac{\partial \varrho_n}{\partial u_{n+i}} f_{n+i} + \sum_{i=-N}^{-M} \frac{\partial f_{n+i}}{\partial u_n} \varrho_{n+i} = 0. \tag{3.2.137}$$

We can now formulate and prove the following theorem:

Theorem 34. *If an equation of the form* (3.2.135), (3.2.136) *possesses a conservation law of order m, such that*

$$m > \min(|N|, |M|), \tag{3.2.138}$$

then $N = -M$ *and* $N \geq 0$.

PROOF. If $m > 0$, we can define the variational derivative ϱ_n of the conserved density p_n which will satisfy (3.2.46, 3.2.137).

The following table will be helpful:

$$\begin{array}{llll}
\frac{\partial}{\partial u_{n+j}} \frac{\partial \varrho_n}{\partial u_{n+i}} = 0 & \text{for} & j < -m & \text{or} \quad j > m \\
\frac{\partial}{\partial u_{n+j}} \frac{\partial f_{n+i}}{\partial u_n} = 0 & \text{for} & j < M - N & \text{or} \quad j > N - M \\
\frac{\partial \varrho_n}{\partial u_{n+i}} \frac{\partial f_{n+i}}{\partial u_{n+j}} = 0 & \text{for} & j < M - m & \text{or} \quad j > N + m \\
\frac{\partial f_{n+i}}{\partial u_n} \frac{\partial \varrho_{n+i}}{\partial u_{n+j}} = 0 & \text{for} & j < -m - N & \text{or} \quad j > m - M
\end{array} \tag{3.2.139}$$

This result, obtained using only (3.2.46, 3.2.135)), does not depend on the number i. We will need to take into account the table when we will differentiate (3.2.137) with respect to u_{n+j}.

The proof of this theorem is based on the formulation of two conditions which will lead to a contradiction. The first condition means that m, N, M satisfy the following inequalities

$$N > 0\,, \qquad m > -M\,, \qquad N > -M\,. \tag{3.2.140}$$

In this case, differentiating (3.2.137) with respect to u_{n+N+m} and using (3.2.139), we obtain:

$$\frac{\partial}{\partial u_{n+N+m}}\left(\frac{\partial \varrho_n}{\partial u_{n+m}} f_{n+m}\right) = \frac{\partial \varrho_n}{\partial u_{n+m}}\frac{\partial f_{n+m}}{\partial u_{n+N+m}} = \frac{\partial \varrho_n}{\partial u_{n+m}} T^m \frac{\partial f_n}{\partial u_{n+N}} = 0\,.$$

This result is in contradiction with conditions (3.2.46, 3.2.136). The situation is quite similar in the second case, when

$$-M > 0\,, \qquad -M > N\,, \qquad m > N\,. \tag{3.2.141}$$

Here we can differentiate (3.2.137) with respect to u_{n+m-M} and are led to

$$\frac{\partial}{\partial u_{n+m-M}}\left(\frac{\partial f_{n-M}}{\partial u_n}\varrho_{n-M}\right) = \frac{\partial f_{n-M}}{\partial u_n}\frac{\partial \varrho_{n-M}}{\partial u_{n+m-M}} = T^{-M}\left(\frac{\partial f_n}{\partial u_{n+M}}\frac{\partial \varrho_n}{\partial u_{n+m}}\right) = 0\,.$$

In this case the result is again in contradiction with (3.2.46, 3.2.136).

Now we are going to prove that we must have $N - -M$. Let us considering the following two possible cases:

$$\text{Case 1}: \qquad N > -M$$
$$\text{Case 2}: \qquad -M > N\,.$$

Using the results just obtained, Case 1 is compatible with (3.2.140). As $N \geq M$, we have

$$N > -M \geq -N\,, \qquad N \geq M > -N\,, \tag{3.2.142}$$

and hence $N > 0$. This result together with (3.2.142) imply: $|N| = N \geq |M| \geq -M$. Now, using (3.2.138), we obtain that: $m > -M$. So, (3.2.140) must take place, i.e. Case 1 is impossible.

Case 2 is dealt with in a very similar way. As $N \geq M$, then

$$-M > N \geq M\,, \qquad -M \geq -N > M\,,$$

hence $-M > 0$. Now $|M| = -M \geq |N| \geq N$, and therefore $m > N$ due to (3.2.138). So, (3.2.141) has been obtained, and thus Case 2 is also impossible. Consequently $N = -M$. As $N \geq M$, we also have $N \geq 0$. □

In all known S-integrable DΔEs and generalized symmetries of the form (3.2.1) we find that they possess infinite hierarchy of conservation laws. That is why generalized symmetries of (3.2.1) are symmetrical in the sense of Theorem 34.

Let us discuss the example of the discrete Burgers equation (whose properties are discussed at lenght in Section 2.3.5), a C-integrable linearizable DΔE which is not symmetrical (2.3.329)

$$\dot{u}_n = u_n(u_n - u_{n+1})\,. \tag{3.2.143}$$

This equation has an infinite hierarchy of generalized symmetries, but no local conservation laws of a positive order.

As we saw in Section 2.3.5.2 we can construct symmetries of the discrete Burgers equation like

(3.2.144) $$u_{n,\epsilon_2} = u_n u_{n+1}(u_{n+2} - u_n)\,, \qquad u_{n,\epsilon_{-1}} = 1 - u_n/u_{n-1}\,.$$

According to Theorem 34, (3.2.143), as well as the generalized symmetries (3.2.144), cannot have conservation laws of the order $m > 0$. The function $\log u_n$ is a conserved density of order $m = 0$ of (3.2.143, 3.2.144), e.g.

$$(\log u_n)_{\epsilon_{-1}} = 1/u_n - 1/u_{n-1} \sim 0\,.$$

In the case of the linear equations

(3.2.145) $$v_{n,\epsilon_k} = v_{n+k}$$

with $k \neq 0$, Theorem 34 guarantees that there is no conservation law of order $m > |k|$.

2.4.2. *Discussion of PDEs from the point of view of Theorem* 34. From Theorem 34 if (3.2.135) is an S-integrable DΔE and if the function f_n contains the highest shift u_{n+N} then it should also contain as lowest shift u_{n-N}. Defining the symmetric differences

$$v_k^{(\pm)} = u_{n+k} \pm u_{n-k},$$

we can rewrite (3.2.135) in the S-integrable case as

$$\dot{u}_n = f(v_N^{(+)}, v_N^{(-)}, v_{N-1}^{(+)}, v_{N-1}^{(-)}, \cdots, v_1^{(+)}, v_1^{(-)}, u_n).$$

To perform the continuous limit we define $x = nh$, $u_n = w(x)$ and, when $h \to 0$, we have the following Taylor expansion

$$u_{n\pm j} = w(x \pm jh) = w(x) \pm j\,h\,w_x + \frac{1}{2} j^2\,h^2 w_{2x} \pm \frac{1}{3!} j^3\,h^3 w_{3x} + \frac{1}{4!} j^4\,h^4 w_{4x} + \cdots\,,$$

where w_{nx} is the n-th derivative of $w(x)$ with respect to x. So we can write a list of Taylor expansions for $v_j^{(\pm)}$. For the lowest values of j we have:

$$\begin{aligned}
v_1^{(+)} &= 2w(x) + h^2 w_{2x} + \frac{h^4}{12} w_{4x} + \cdots,\\
v_1^{(-)} &= 2hw_x + \frac{1}{3} h^3 w_{3x} + \cdots,\\
v_2^{(+)} &= 2w(x) + 4h^2 w_{2x} + \frac{4}{3} h^4 w_{4x} + \cdots,\\
v_2^{(-)} &= 4hw_x + \frac{8}{3} h^3 w_{3x} + \cdots,\\
v_3^{(+)} &= 2w(x) + 9h^2 w_{2x} + \frac{27}{4} h^4 w_{4x} + \cdots.
\end{aligned}$$

Consequently we can express all derivatives of the function $w(x)$ in terms of the Taylor expansions of $v_j^{(\pm)}$ up to order h^2. Here we present the lowest terms:

$$\begin{aligned}
w_x &= \frac{1}{2h} v_1^{(-)} + \mathcal{O}(h^2),\\
w_{2x} &= \frac{1}{3h^2}[v_2^{(+)} - v_1^{(+)}] + \mathcal{O}(h^2),\\
w_{3x} &= \frac{1}{2h^3}[v_2^{(-)} - 2v_1^{(-)}] + \mathcal{O}(h^2),\\
w_{4x} &= \frac{1}{10h^4}[3v_3^{(+)} - 8v_2^{(+)} + 5v_1^{(+)}] + \mathcal{O}(h^2),
\end{aligned}$$

As all continuous derivatives can be expressed in terms of symmetric differences, we do have from Theorem 34 **no constraint** on the form of an S-integrable PDE. This result is coherent from what we know about S-integrable PDEs, i.e. as far as we know there is no special form for S-integrable PDEs.

2.4.3. *Discussion of PΔEs from the point of view of Theorem* 34. In this section we discuss by examples the semi continuous limit of S-integrable PΔEs.

One first example already treated in section 2.4.4 is the lpKdV (2.4.56), or the equation H_1 of the ABS classification, whose straight semi-continuous limit is given by (2.4.61) and its skew semi continuous limit (2.4.63) is a scalar evolutionary DΔE, involving the points $k-1$, k and $k+1$. Thus the lpkdv satisfies Yamilov's condition for S-integrability.

An integrable map obtained by factorization. Let us consider the non linear PΔE presented by Hietarinta and Viallet [387]

$$\begin{aligned} &c_1 u_{n,m} u_{n+1,m} + c_2(u_{n,m} u_{n+1,m+1} + u_{n+1,m} u_{n,m+1}) + c_3 u_{n,m+1} u_{n+1,m+1} \\ &+ c_5 u_{n,m} u_{n,m+1} + c_6 u_{n+1,m} u_{n+1,m+1} = 0. \end{aligned} \tag{3.2.146}$$

Eq. (3.2.146) has been proven to be integrable for all values of the constants c_i by checking its algebraic entropy. The equation is invariant under the exchange of n into m when c_1 goes into c_5 and c_3 into c_6.

Let us introduce a parameter ε so that we can carry out a continuous limit in the discrete variable $t = \varepsilon m$ and $u_{n,m} = v_n(t)$. We get

$$\begin{aligned} &c_5 v_n^2 + c_6 v_{n+1}^2 + (c_1 + 2c_2 + c_3) v_n v_{n+1} \\ &+ \varepsilon(c_5 v_n \dot{v}_n + c_6 v_{n+1} \dot{v}_{n+1} + (c_2 + c_3)(v_{n+1} \dot{v}_n + v_n \dot{v}_{n+1})) + \mathcal{O}(\varepsilon^2) = 0. \end{aligned} \tag{3.2.147}$$

The terms of order zero in ε do not contain the t derivatives. So to get a DΔE we have to require that the coefficients c_i depend on ε. We have the following possibilities:

(1) $c_5 = \alpha_5 \varepsilon,\ c_6 = \alpha_6 \varepsilon,\ c_1 + 2c_2 + c_3 = 0$
and we then have

$$(c_1 + c_2)\frac{\mathrm{d}}{\mathrm{d}t}(v_n v_{n+1}) = \alpha_5 v_n^2 + \alpha_6 v_{n+1}^2 = 0, \tag{3.2.148}$$

(2) $c_5 = c_6 = 0,\ c_1 + 2c_2 + c_3 = \alpha_{123}\varepsilon$
and we then have

$$(c_1 + c_2)\frac{\mathrm{d}}{\mathrm{d}t}(v_n v_{n+1}) = \alpha_{123} v_n v_{n+1}. \tag{3.2.149}$$

Eq. (3.2.148) is non local. Eq. (3.2.149) a linear ODE for $v_n v_{n+1}$.

As in the case of lpKdV equation, we do a skew change of variables

$$k = n + m + 1, \qquad u_{n+i,m+j} = w_{k+i+j,m+j}, \tag{3.2.150}$$

i.e. mixing the lattice indexes as in [384], (3.2.146) becomes

$$\begin{aligned} &c_1 w_{k-1,m} w_{k,m} + c_2(w_{k,m} w_{k,m+1} + w_{k-1,m} w_{k+1,m+1}) + c_3 w_{k,m+1} w_{k+1,m+1} \\ &+ c_5 w_{k-1,m} w_{k,m+1} + c_6 w_{k,m} w_{k+1,m+1} = 0. \end{aligned} \tag{3.2.151}$$

Introducing a small parameter ε and sending m to infinity so that

$$t = m\varepsilon, \qquad w_{k,m} = U_k(t),$$

(3.2.151) becomes, at the lowest orders in ε

$$\begin{aligned} &(c_1 + c_5) U_{k-1} U_k + (c_3 + c_6) U_k U_{k+1} + c_2(U_{k-1} U_{k+1} + U_k^2) \\ &+ \varepsilon((c_2 U_{k-1} + (c_3 + c_6) U_k)\dot{U}_{k+1} + (c_5 U_{k-1} + c_2 U_k + c_3 U_{k+1})\dot{U}_k) + \mathcal{O}(\varepsilon^2) = 0. \end{aligned}$$

To get a DΔE we need to require

$$c_1 + c_5 = \varepsilon\alpha, \quad c_3 + c_6 = \varepsilon\beta, \quad c_2 = \varepsilon\gamma,$$

and choose c_1 and c_6 of order one. In this way we get at lowest order in ϵ

$$(c_6 U_{k+1} + c_1 U_{k-1})\dot{U}_k - (\beta U_{k+1} + \gamma U_k + \alpha U_{k-1})U_k - \gamma U_{k-1} U_{k+1} = 0. \tag{3.2.152}$$

Eq. (3.2.152) satisfies Yamilov S-integrability theorem. So for all values of c_i (3.2.146) has the form of an S-integrable equation and the result obtained by the algebraic entropy is confirmed.

The $\mathbf{H_2}$ equation of the ABS classification. The H_2 equation is presented in (2.4.131) and we repeat it here for the convenience of the reader

$$\begin{aligned} &(u_{n,m} - u_{n+1,m+1})(u_{n+1,m} - u_{n,m+1}) \\ &+ (\beta - \alpha)(u_{n,m} + u_{n+1,m} + u_{n,m+1} + u_{n+1,m+1}) - \alpha^2 + \beta^2 = 0. \end{aligned} \tag{3.2.153}$$

Eq. (3.2.153) is one of the discrete integrable equations of the ABS list (see [22]).

As in the other cases, we admit that the constants α and β will depend on the small parameter ϵ, parameter in which we will carry out the limiting process.

Following [389], we redefine the parameters of H_2 (3.2.153)

$$p = r - a^2, \quad q = r - b^2,$$

and (3.2.153) becomes

$$\begin{aligned} &(u_{n,m} - u_{n+1,m+1})(u_{n+1,m} - u_{n,m+1}) \\ &+ (a^2 - b^2)(u_{n,m} + u_{n+1,m} + u_{n,m+1} + u_{n+1,m+1} + 2r - a^2 - b^2) = 0. \end{aligned} \tag{3.2.154}$$

Eq. (3.2.154) has the exact solution

$$u_0 = (an + bm + \gamma)^2 - \frac{1}{2}r. \tag{3.2.155}$$

Introducing (3.2.155) in (2.4.57) we can rewrite (3.2.154) as

$$\begin{aligned} &(v_{n,m} - v_{n+1,m+1})(v_{n+1,m} - v_{n,m+1}) \\ &+ 2(a - b)\{[a(n+1) + b(m+1) + \gamma]v_{n,m} - (an + bm + \gamma)v_{n+1,m+1}\} \\ &- 2(a + b)\{[an + b(m+1) + \gamma]v_{n+1,m} - (a(n+1) + bm + \gamma)v_{n,m+1}\} = 0 \end{aligned} \tag{3.2.156}$$

which has the solution $v_0 = 0$.

The standard straight limit gives a non local DΔE. To avoid it we take the skew limit in one of the indexes as in (3.2.150). Then, (3.2.156) is transformed into

$$\begin{aligned} &(w_{k,m} - w_{k+2,m+1})(w_{k+1,m} - v_{k+1,m+1}) \\ &+2(a - b)((a(k - m + 1) + b(m+1) + \gamma)w_{k,m} \\ &-(a(k - m) + bm + \gamma)w_{k+2,m+1}) \\ &-2(a + b)((a(k - m) + b(m+1) + \gamma)w_{k+1,m} \\ &-(a(k - m + 1) + bm + \gamma)w_{k+1,m+1}) = 0. \end{aligned} \tag{3.2.157}$$

In order to take the continuous limit in the index m, we substitute:

$$a \to \varepsilon\tilde{a}, \quad b \to \varepsilon\tilde{b}$$

and define

$$w_{k+i,m} = U_{k+i}(t), \quad w_{k+i,m+j} \to U_{k+i}(t) + \epsilon\, j\, \dot{U}_{k+i}(t) + \mathcal{O}(\epsilon^2).$$

Then, the lower order terms of (3.2.157) are

$$(3.2.158)\qquad \begin{aligned}&(U_k - U_{k+2} - \varepsilon\,\dot U_{k+2})(U_{k+1} - U_{k+1} - \varepsilon\,\dot U_{k+1})\\ &+ 2\epsilon(\tilde a - \tilde b)((\varepsilon\,\tilde a(k-m+1) + \varepsilon\,\tilde b(m+1) + \gamma)U_k\\ &- (\varepsilon\,\tilde a(k-m) + \varepsilon\,\tilde b m + \gamma)(U_{k+2} + \varepsilon\,\dot U_{k+2}))\\ &- 2\varepsilon(\tilde a + \tilde b)((\varepsilon\,\tilde a(k-m) + \varepsilon\,\tilde b(m+1) + \gamma)U_{k+1}\\ &- (\varepsilon\,\tilde a(k-m+1) + \varepsilon\,\tilde b m + \gamma)U_{k+1} + \varepsilon\,\dot U_{k+1})) + \cdots = 0.\end{aligned}$$

The lowest order term in ϵ of (3.2.158) is

$$\dot U_k = 2\gamma(\tilde a - \tilde b)(U_{k-1} - U_{k+1}),$$

an equation which satisfies Yamilov's theorem.

The equation ${}_r\mathbf{H}_1^\pi$ from ABS extended classification. Let us consider the following equation presented in Section 2.4.7, which we repeat here for the convenience of the reader

$$(3.2.159)\qquad \begin{aligned}{}_rH_1^\pi =&(\alpha-\beta)\left(\pi^2\chi_{m+n}u_{n+1,m}u_{n,m+1} + \pi^2\chi_{m+n+1}u_{n+1,m+1}u_{n,m} - 1\right)\\ &+ (u_{n,m} - u_{n+1,m+1})(u_{n+1,m} - u_{n,m+1}) = 0,\end{aligned}$$

where α, β, π are three parameters, which could depend on the steps of the lattice. The function χ_m is given in (2.4.148) with $k = m$. When $\pi = 0$ we have the H_1 equation or lpKdV. Eq. (3.2.159) is an S-integrable equation of Boll classification [112], appearing in the ΛBS list [23].

As in the previous case, we will transform the variable $u_{n,m} \to v_{n,m}$ in such a way that the resulting equation will have $v_0 = 0$ as a particular solution. Let us set

$$u_{n,m} = v_{n,m} + f,$$

where f, a constant, must satisfy the relation

$$(3.2.160)\qquad (\alpha-\beta)(\pi^2 f^2 - 1) = 0.$$

Choosing one of the two signs for f in (3.2.160) (with $\pi \neq 0$) we get:

$$(3.2.161)\qquad u_{n,m} = v_{n,m} + \frac{1}{\pi}.$$

The equation for $v_{n,m}$ is:

$$(3.2.162)\qquad \begin{aligned}&(\alpha-\beta)(\chi_{n+m+1}(1+\pi v_{n,m})(1+\pi v_{n+1,m+1})\\ &+\chi_{n+m}(1+\pi v_{n,m+1})(1+\pi v_{n+1,m}) - 1)\\ &+(v_{n+1,m} - v_{n,m+1})(v_{n,m} - v_{n+1,m+1}) = 0\end{aligned}$$

and $v_0 = 0$ is a solution of (3.2.162).

We take a semi continuous skew limit introducing the definition (3.2.150). The equation, with $w_{k,m}$ defined as in (3.2.150), is now

$$(3.2.163)\qquad \begin{aligned}&(\alpha-\beta)(\chi_{k+1}(1+\pi w_{k,m})(1+\pi w_{k+2,m+1})\\ &+ \chi_k(1+\pi w_{k+1,m+1})(1+\pi w_{k+1,m}) - 1)\\ &+ (w_{k+1,m} - w_{k+1,m+1})(w_{k,m} - w_{k+2,m+1}) = 0.\end{aligned}$$

Let us introduce an order parameter ε such that $t = \varepsilon m$ and take the continuous limit in the m direction. So we have

$$(3.2.164)\qquad \begin{aligned}w_{k+i,m} =&U_{k+i}(t),\\ w_{k+i,m+1} =&U_{k+i}(t) + \varepsilon\dot U_{k+i} + \mathcal{O}(\mu^2).\end{aligned}$$

Substituting (3.2.164) in (3.2.163), we get

$$(\alpha-\beta)(\chi_{k+1}(1+\pi U_k)(1+\pi(U_{k+2}+\varepsilon\dot{U}_{k+2}+\mathcal{O}(\varepsilon^2))) + \chi_k(1+\pi(U_{k+1}+\varepsilon\dot{U}_{k+1}+\mathcal{O}(\varepsilon^2)))(1+\epsilon U_{k+1})-1) + (U_{k+1}-U_{k+1}-\varepsilon\dot{U}_{k+1}+\mathcal{O}(\varepsilon^2))(U_k-U_{k+2}-\varepsilon\dot{U}_{k+2}+\mathcal{O}(\varepsilon^2))=0. \tag{3.2.165}$$

Choosing α and β functions of ε so that $\alpha-\beta=\gamma\varepsilon$, the first order term gives a DΔE satisfying Yamilov's theorem

$$\dot{U}_k=\gamma\frac{\chi_{k-1}(1+\pi U_k)^2+\chi_k(1+\pi U_{k-1})(1+\pi U_{k+1})}{U_{k-1}-U_{k+1}}. \tag{3.2.166}$$

Eq. (3.2.166) is a subcase of the k dependent generalization of the Yamilov discretization of the Krichever-Novikov equation postulated in [549] presented in (2.4.198, 2.4.199).

So S-integrable PΔE, both autonomous and non autonomous, satisfy Yamilov's theorem on the symmetric form of S-integrable DΔEs.

2.5. Discussion of the integrability conditions. Here in the following we will discuss some properties of the integrability conditions (3.2.56, 3.2.100, 3.2.106, 3.2.130, 3.2.131). More precisely, we will see:

- How to derive the integrability conditions, starting from the existence of two conservation laws without using the generalized symmetries.
- How to obtain an explicit form of the integrability conditions convenient for testing the integrability of a given equation.
- When the integrability conditions (3.2.56, 3.2.100, 3.2.106) allow one to construct nontrivial conservation laws.
- The problem of the left and right orders of the generalized symmetry, and one more set of integrability conditions.

2.5.1. *Derivation of integrability conditions from the existence of conservation laws.* As it will be explained in Section 3.3.1, (3.2.56, 3.2.130, 3.2.131) are sufficient to provide an exhaustive classification of integrable equations of the form (3.2.1). The other conditions (3.2.100, 3.2.106) are automatically satisfied by all the equations of the resulting list and will be used for constructing of conservation laws for the equations of the list.

Let us consider the three integrability conditions (3.2.56, 3.2.130, 3.2.131) derived in Theorems 23 and 33, starting from the existence of one generalized symmetry of the order $m\geq 2$ and one conservation law of the order $m\geq 3$. Now, instead, we require the existence of two conservation laws of orders m_1 and m_2: $m_1>m_2\geq 3$. In accordance with Theorem 32, from these conservation laws we can obtain two formal conserved densities $\mathscr{S}_n$ and $\tilde{\mathscr{S}}_n$, such that

$$\operatorname{ord}\mathscr{S}_n<\operatorname{ord}\tilde{\mathscr{S}}_n\,,\qquad \operatorname{lgt}\mathscr{S}_n\geq 2\,,\qquad \operatorname{lgt}\tilde{\mathscr{S}}_n\geq 2\,.$$

Using (3.2.71) and (3.2.120), we can pass from these formal conserved densities $\mathscr{S}_n$ and $\tilde{\mathscr{S}}_n$ to the following formal symmetry:

$$L_n=(\mathscr{S}_n^{-1}\tilde{\mathscr{S}}_n)^2. \tag{3.2.167}$$

The formal symmetry L_n given in (3.2.167) will be such that $\operatorname{ord}L_n\geq 2$ and $\operatorname{lgt}L_n\geq 2$, as we have (cf. (3.2.93, 3.2.95, 3.2.129)):

$$\operatorname{ord}L_n=2(\operatorname{ord}\tilde{\mathscr{S}}_n-\operatorname{ord}\mathscr{S}_n)\,,\qquad \operatorname{lgt}L_n\geq\min(\operatorname{lgt}\mathscr{S}_n,\operatorname{lgt}\tilde{\mathscr{S}}_n)\,.$$

The series L_n has the form (3.2.66) with $N \geq 2$. As $\mathrm{lgt}L_n \geq 2$, we obtain from (3.2.65) the following system of equations for the coefficients $l_n^{(N)}, l_n^{(N-1)}$:

$$(3.2.168)\qquad \begin{aligned} &l_n^{(N)} f_{n+N}^{(1)} = l_{n+1}^{(N)} f_n^{(1)}, \\ &\dot{l}_n^{(N)} + l_n^{(N)} f_{n+N}^{(0)} + l_n^{(N-1)} f_{n+N-1}^{(1)} = l_n^{(N)} f_n^{(0)} + l_{n+1}^{(N-1)} f_n^{(1)}. \end{aligned}$$

The integrability condition (3.2.56) has been derived in Section 3.2.2 from (3.2.48) for the generalized symmetry (3.2.3) of (3.2.1). However, the proof of Theorem 23 uses only the system of equations (3.2.49, 3.2.50) for the functions $g_n^{(m)}$ and $g_n^{(m-1)}$ which follows from the compatibility conditions (3.2.48). As the structure of two systems (3.2.168) and (3.2.49, 3.2.50) is the same, one can derive the integrability condition (3.2.56) from the system (3.2.168). The proof of Theorem 33 uses condition (3.2.56) instead of the existence of a generalized symmetry of the order $m \geq 2$. This means that we can write down an obvious modification of Theorem 33 in order to obtain conditions (3.2.130, 3.2.131). So, we are led to the following result:

Theorem 35. *If* (3.2.1) *has two conservation laws with orders $m_1 > m_2 \geq 3$, then it satisfies the integrability conditions* (3.2.56, 3.2.130, 3.2.131).

One further result of this kind can be obtained with a reasoning similar to that used to prove Theorem 35. Let us make the ansatz, valid in the case of the Volterra equation, that there are two conservation laws of orders m and $m+1$, with $m \geq 5$. In this case, in accordance with Theorem 32, we can pass to a pair of formal conserved densities $\mathcal{S}_n$ and $\tilde{\mathcal{S}}_n$:

$$\mathrm{ord}\mathcal{S}_n = m\,, \quad \mathrm{ord}\tilde{\mathcal{S}}_n = m+1\,, \quad \mathrm{lgt}\mathcal{S}_n \geq 4\,, \quad \mathrm{lgt}\tilde{\mathcal{S}}_n \geq 5\,.$$

Therefore a formal symmetry L_n, given by $L_n = \mathcal{S}_n^{-1}\tilde{\mathcal{S}}_n$, is such that $\mathrm{ord}L_n = 1$ and $\mathrm{lgt}L_n \geq 4$. Theorems 29 and 30 provide us with three integrability conditions. Using Theorem 33, we can derive two other conditions. More precisely, the following result takes place:

Theorem 36. *If* (3.2.1) *possesses two conservation laws of the orders $m \geq 5$ and $m+1$, then it satisfies the five integrability conditions* (3.2.56, 3.2.100, 3.2.106, 3.2.130, 3.2.131).

2.5.2. *Explicit form of the integrability conditions.* It will be proved in Section 3.3.1 that the three integrability conditions (3.2.56, 3.2.130, 3.2.131), which can be written in the form

$$(3.2.169)\qquad \dot{p}_n^{(1)} \sim 0\,, \qquad r_n^{(1)} \sim 0\,, \qquad r_n^{(2)} \sim 0\,,$$

are not only necessary but also sufficient for the integrability of an equation of the form (3.2.1). For this reason, the conditions (3.2.169) can be used for testing the integrability of a given equation. To be able to do so, we rewrite these conditions in an explicit form.

The first two conditions (3.2.169) are explicit, as the functions $\dot{p}_n^{(1)}$ and $r_n^{(1)}$, given by (3.2.56, 3.2.130), are explicitly defined in terms of the right hand side of (3.2.1). One can easily rewrite $r_n^{(2)}$, given by (3.2.131), in an explicit form. In fact, as it follows from (3.2.130), the function $\sigma_n^{(1)}$ may only depend on the variables u_n and u_{n-1}, as it is defined by the relation $\sigma_{n+1}^{(1)} - \sigma_n^{(1)} = r_n^{(1)}$. Differentiating it with respect to u_{n+1} and u_{n-1}, we can find the partial derivatives $\frac{\partial \sigma_n^{(1)}}{\partial u_n}, \frac{\partial \sigma_n^{(1)}}{\partial u_{n-1}}$ and then rewrite the time derivative

$$\dot{\sigma}_n^{(1)} = \frac{\partial \sigma_n^{(1)}}{\partial u_n} f_n + \frac{\partial \sigma_n^{(1)}}{\partial u_{n-1}} f_{n-1}$$

in the definition of $r_n^{(2)}$. Thus we get the following explicit expression for $r_n^{(2)}$

$$r_n^{(2)} = \frac{\partial r_{n-1}^{(1)}}{\partial u_n} f_n - \frac{\partial r_n^{(1)}}{\partial u_{n-1}} f_{n-1} + 2\,\frac{\partial f_n}{\partial u_n}\,. \tag{3.2.170}$$

One can also use the following form of conditions (3.2.169): $\frac{\delta \dot{p}_n^{(1)}}{\delta u_n} = 0$, $\frac{\delta r_n^{(1)}}{\delta u_n} = 0$, $\frac{\delta r_n^{(2)}}{\delta u_n} = 0$, as it has been proved in Theorem 24 that these two conditions are equivalent up to some integration constants.

2.5.3. *Construction of conservation laws from the integrability conditions.* The integrability conditions (3.2.56, 3.2.100, 3.2.106) provide an easy way for constructing conservation laws for integrable equations (3.2.1), whose complete list will be presented in Section 3.3.1.2. We explain in this section why, for most of those equations, these three conservation laws are nontrivial.

In the case of the Volterra equation (3.2.2), such conservation laws have low orders 0, 0, 1, as it has been shown in Section 3.2.3. In the case of the equation

$$\dot{u}_n = (u_{n+1} - u_n)^{1/2}(u_n - u_{n-1})^{1/2}, \tag{3.2.171}$$

all these three conservation laws are trivial. In fact, introducing the function $w_n = u_n - u_{n-1}$, we obtain for instance that

$$p_n^{(1)} = \log\tfrac{1}{2} - \tfrac{1}{2}(S-1)\log w_n\,, \qquad p_n^{(2)} = c - \tfrac{1}{2}(S-1)\frac{w_{n-1}^{1/2}}{w_n^{1/2}}\,,$$

where c is a constant, $p_n^{(1)}$ and $p_n^{(2)}$ are given by (3.2.56, 3.2.100).

However, if an equation satisfies the condition

$$\Theta_n = \frac{\partial^2 p_n^{(1)}}{\partial u_{n+1}\partial u_{n-1}} \neq 0\,, \tag{3.2.172}$$

then we prove in Theorem 37 that the conserved densities $p_n^{(1)}, p_n^{(2)}, p_n^{(3)}$ defined by (3.2.56, 3.2.100, 3.2.106) are nontrivial and have rather high orders 2, 3, 4. One can easily check that for most integrable equations, presented in Section 3.3.1.2, the condition (3.2.172) is satisfied. This condition means that the conserved density $p_n^{(1)}$ is of order 2, as $p_n^{(1)} \sim p_{n+1}^{(1)} = \mathcal{P}(u_{n+2}, u_{n+1}, u_n)$.

Theorem 37. *Let us assume that an equation of the form* (3.2.1) *satisfies the integrability conditions* (3.2.56, 3.2.100, 3.2.106)*, and that the conserved density $p_n^{(1)}$ is of order 2, i.e.* (3.2.172) *is satisfied. Then the conserved densities $p_n^{(2)}$ and $p_n^{(3)}$ are of orders 3 and 4, respectively.*

PROOF. We shall use here, in addition to (3.2.172), the following condition: $f_n^{(-1)} \neq 0$, obtained from (3.2.1, 3.2.47). At first we obtain some information on $q_n^{(1)}$ and $q_n^{(2)}$, using (3.2.56, 3.2.100). The function $q_n^{(1)}$ may depend on the variables $u_{n+1}, u_n, u_{n-1}, u_{n-2}$ only, and one has

$$\frac{\partial q_n^{(1)}}{\partial u_{n-2}} = -\frac{\partial p_n^{(1)}}{\partial u_{n-1}} f_{n-1}^{(-1)}\,. \tag{3.2.173}$$

It is easy to see that $p_n^{(2)}$ must depend on the same variables. Then $q_n^{(2)}$ may only depend on $u_{n+1}, u_n, \dots u_{n-3}$, and due to (3.2.100, 3.2.173) we have

(3.2.174)
$$\frac{\partial q_n^{(2)}}{\partial u_{n-3}} = -\frac{\partial p_n^{(2)}}{\partial u_{n-2}} f_{n-2}^{(-1)} = -\frac{\partial q_n^{(1)}}{\partial u_{n-2}} f_{n-2}^{(-1)} = \frac{\partial p_n^{(1)}}{\partial u_{n-1}} f_{n-1}^{(-1)} f_{n-2}^{(-1)} .$$

Using (3.2.100, 3.2.172, 3.2.173), one can show that

$$\frac{\partial^2 p_n^{(2)}}{\partial u_{n+1} \partial u_{n-2}} = \frac{\partial^2 q_n^{(1)}}{\partial u_{n+1} \partial u_{n-2}} = -\Theta_n f_{n-1}^{(-1)} \neq 0 ,$$

i.e. the density $p_n^{(2)}$ has order 3. The conserved density $p_n^{(3)}$, given by (3.2.106), has the following structure

$$p_n^{(3)} \sim q_n^{(2)} + \frac{1}{2}(p_n^{(2)})^2 + f_{n-1}^{(1)} f_n^{(-1)} = \mathscr{P}_n = \mathscr{P}(u_{n+1}, u_n, \dots u_{n-3}) .$$

Moreover, we derive from (3.2.172, 3.2.174) that

$$\frac{\partial^2 \mathscr{P}_n}{\partial u_{n+1} \partial u_{n-3}} = \frac{\partial^2 q_n^{(2)}}{\partial u_{n+1} \partial u_{n-3}} = \Theta_n f_{n-1}^{(-1)} f_{n-2}^{(-1)} \neq 0 ,$$

and therefore $p_n^{(3)}$ has order 4. □

We illustrate Theorem 37, considering the equation

(3.2.175)
$$\dot{u}_n = (u_{n+1} - u_{n-1})^{-1}$$

which satisfies all five integrability conditions. Introducing the function $w_n = u_{n+1} - u_{n-1}$, one can check that $p_n^{(1)} = \log(-w_n^{-2})$, $\Theta_n = -2w_n^{-2}$, and the condition (3.2.172) is satisfied. We then easily find that

$$p_n^{(2)} \sim c_1 - 2w_{n+1}^{-1} w_n^{-1},$$

$$p_n^{(3)} \sim c_2 - 2c_1 w_{n+1}^{-1} w_n^{-1} + w_{n+1}^{-2} w_n^{-2} + 2w_{n+2}^{-1} w_{n+1}^{-2} w_n^{-1},$$

where c_1, c_2 are arbitrary constants. So, these conserved densities have orders 3 and 4, respectively.

2.5.4. *Left and right order of generalized symmetries.* In deriving the integrability conditions (3.2.56, 3.2.100, 3.2.106), we have considered just the left order m of a generalized symmetry (3.2.3). According to Theorems 23, 27, 29 and 30, m was required to be sufficiently high. On the other hand, we used only the first two conditions given by (3.2.1), namely $f_n^{(1)} \neq 0$.

One can assume that the right order m' of a generalized symmetry (3.2.3) is sufficiently low and use the condition $f_n^{(-1)} \neq 0$. Following the proof of Theorem 23, one differentiates compatibility condition (3.2.48) with respect to

$$u_{n+m'-1} \ , \ u_{n+m'} \ , \ u_{n+m'+1} \ , \ u_{n+m'+2} \ , \ \dots \ .$$

Then, following the proofs of Theorems 29 and 30, one considers instead of (3.2.97) a formal symmetry of the form

$$L_n = l_n^{(-1)} S^{-1} + l_n^{(0)} + l_n^{(1)} S + l_n^{(2)} S^2 + \dots , \qquad l_n^{(-1)} \neq 0 ,$$

which is a formal series in positive powers of the shift operator S.

One more set of integrability conditions can be obtained in this way which have the form

(3.2.176)
$$D_t \hat{p}_n^{(i)} = (S-1)\hat{q}_n^{(i)}, \qquad i = 1, 2, 3, \dots ,$$

where, for example,

$$\hat{p}_n^{(1)} = \log f_n^{(-1)}, \qquad \hat{p}_n^{(2)} = \hat{q}_n^{(1)} - f_n^{(0)}. \tag{3.2.177}$$

The conditions (3.2.176) are the analogous of (3.2.56, 3.2.100, 3.2.106), and the functions (3.2.177) are similar to the conserved densities $p_n^{(1)}$, $p_n^{(2)}$ which can be written as

$$p_n^{(1)} = \log f_n^{(1)}, \qquad p_n^{(2)} = q_n^{(1)} + f_n^{(0)}. \tag{3.2.178}$$

Let us explain why the integrability conditions (3.2.176) are not important. In fact, such conditions can be obtained as corollaries of the integrability conditions (3.2.56, 3.2.100, 3.2.106, 3.2.130, 3.2.131). This will be demonstrated, considering as an example the conditions (3.2.176) with $i = 1, 2$ and using in addition to (3.2.177, 3.2.178) the formulas for the functions $r_n^{(1)}$ and $r_n^{(2)}$, given by (3.2.130, 3.2.131)

$$r_n^{(1)} = \log(-f_n^{(1)}/f_n^{(-1)}), \qquad r_n^{(2)} = \dot{\sigma}_n^{(1)} + 2f_n^{(0)}.$$

If the integrability conditions (3.2.56, 3.2.130) are satisfied, then

$$D_t\hat{p}_n^{(1)} = D_t(p_n^{(1)} - r_n^{(1)}) = (S-1)(q_n^{(1)} - \dot{\sigma}_n^{(1)}).$$

This means that the condition (3.2.176) with $i = 1$ is satisfied too. A function $\hat{q}_n^{(1)}$ exists, and its general form is: $\hat{q}_n^{(1)} = q_n^{(1)} - \dot{\sigma}_n^{(1)} + c$, where c is an arbitrary integration constant. From (3.2.100, 3.2.131) we obtain

$$D_t\hat{p}_n^{(2)} = D_t(q_n^{(1)} - \dot{\sigma}_n^{(1)} - f_n^{(0)}) = D_t(p_n^{(2)} - r_n^{(2)}) \sim 0,$$

i.e. the integrability condition (3.2.176) with $i = 2$ is obtained as a corollary of (3.2.100, 3.2.131).

2.6. Hamiltonian equations and their properties. We discuss here Hamiltonian lattice equations of the form (3.2.1) and explain why such equations are useful in the generalized symmetry method.

Let us consider the anti-symmetric operator K_n, such that $K_n^\dagger = -K_n$, where the definition of an adjoint operator is given by (3.2.110). The operator K_n has the form

$$K_n = \sum_{j=1}^{\nu}(k_n^{(j)}S^j - k_{n-j}^{(j)}S^{-j}). \tag{3.2.179}$$

Eq. (3.2.179) will satisfy an equation similar to that for the formal conservation densities (3.2.116)

$$\dot{K}_n = f_n^* K_n + K_n f_n^{*\dagger}. \tag{3.2.180}$$

The Hamiltonian equation is given by

$$\dot{u}_n = f_n, \qquad f_n = K_n \frac{\delta h_n}{\delta u_n}, \tag{3.2.181}$$

where h_n is any function of the form (3.2.10). It is more convenient in the case of the generalized symmetry method to introduce the following definition for Hamiltonian equations and Hamiltonian operators:

Definition 9. *An equation* (3.2.1) *is called Hamiltonian if it can be written as* (3.2.181), *where* K_n *is an anti-symmetric operator of the form* (3.2.179) *satisfying* (3.2.180). *The operator* K_n *and function* h_n *are called* Hamiltonian operator *and* Hamiltonian density, *respectively.*

The name Hamiltonian density is due to the fact that h_n is a conserved density of (3.2.181), as it follows from (3.2.179). In fact,

$$D_t h_n = \sum_i \frac{\partial h_n}{\partial u_{n+i}} f_{n+i} \sim \frac{\delta h_n}{\delta u_n} f_n = \frac{\delta h_n}{\delta u_n} K_n \frac{\delta h_n}{\delta u_n}$$

$$= \sum_{j=1}^{\nu} (1 - S^{-j}) \left(\frac{\delta h_n}{\delta u_n} k_n^{(j)} \frac{\delta h_{n+j}}{\delta u_{n+j}} \right) \sim 0 .$$

As the operator K_n satisfies (3.2.180), we can construct, starting from any conserved density, generalized symmetries of (3.2.181).

Theorem 38. *If p_n is a conserved density of (3.2.181), then the equation*

$$u_{n,\epsilon} = g_n \;\; with \;\; g_n = K_n \frac{\delta p_n}{\delta u_n} \qquad (3.2.182)$$

is its generalized symmetry.

PROOF. If p_n is a conserved density of (3.2.181), then its variational derivative $\varrho_n = \frac{\delta p_n}{\delta u_n}$ solves, according to Theorem 31, (3.2.109). From (3.2.109, 3.2.180) it follows that the function $g_n = K_n \varrho_n$ satisfies (3.2.72)

$$D_t g_n = \dot{K}_n \varrho_n + K_n \dot{\varrho}_n = (f_n^* K_n + K_n f_n^{*\dagger}) \varrho_n - K_n f_n^{*\dagger} \varrho_n = f_n^* K_n \varrho_n = f_n^* g_n .$$

This means that (3.2.182) is a generalized symmetry of (3.2.181). □

It is obvious that Theorem 38 is also true if K_n is an infinite formal series of the form (3.2.66) and satisfies (3.2.180). However in this case, one has to check that (3.2.181) and its symmetry (3.2.182) are local, i.e. have the form (3.2.1) and (3.2.3), respectively. So, an infinite formal series satisfying (3.2.180) may also map conserved densities into generalized symmetries. Sometimes, such formal series is called Noether operator [286].

Let us consider, for example, equations of the form

$$\dot{u}_n = P(u_n)(u_{n+1} - u_{n-1}) , \qquad (3.2.183)$$

where P is any non-zeroth function. Any (3.2.183) is an Hamiltonian equation. The Hamiltonian density h_n is given by: $h_n = \int \frac{u_n}{P(u_n)} du_n$. The operator

$$K_n = P(u_n)(S - S^{-1})P(u_n) = P(u_n)P(u_{n+1})S - P(u_n)P(u_{n-1})S^{-1} \qquad (3.2.184)$$

is Hamiltonian as one can prove, checking (3.2.180) by direct calculation.

There are two non linear integrable equations in this class. These are the Volterra equation (3.2.2) and the modified Volterra equation

$$\dot{u}_n = (c^2 - u_n^2)(u_{n+1} - u_{n-1}) , \qquad (3.2.185)$$

where c is an arbitrary constant. Omitting constants of integration, which play no role in this case, we can write down the Hamiltonian densities as

$$h_n = u_n , \qquad h_n = -\tfrac{1}{2} \log(c^2 - u_n^2) , \qquad (3.2.186)$$

respectively. In the case of the Volterra equation, if we use formula (3.2.182) and the conserved densities (3.2.39), we obtain the trivial symmetry $u_{n,\epsilon'} = 0$ from p_n^1 and the generalized symmetry (3.2.14) from p_n^3. Let us notice that the case of $p_n^2 = h_n$ is not interesting here. The conserved density p_n^4 is transformed into a generalized symmetry (3.2.3) with $m = 3$, $m' = -3$.

Theorem 39. *If K_n is a Hamiltonian operator of* (3.2.181)*, then its inverse $S_n = K_n^{-1}$ is a solution of the equation* (3.2.116)*.*

PROOF. This follows immediately from (3.2.180). In fact,

$$\dot{S}_n + S_n f_n^* + f_n^{*\dagger} S_n = -K_n^{-1}\dot{K}_n K_n^{-1} + K_n^{-1} f_n^* + f_n^{*\dagger} K_n^{-1}$$

$$= K_n^{-1}(-\dot{K}_n + f_n^* K_n + K_n f_n^{*\dagger})K_n^{-1} = 0 . \quad \square$$

The same is true for any formal series of the form (3.2.66) satisfying (3.2.180), i.e. for Noether operators. As we showed in Section 3.2.3, the exact solution of (3.2.65) is the recursion operator. Theorem 39 states that the exact solutions of (3.2.116) are the inverses of Hamiltonian and Noether operators.

Let us now consider an important application of Theorem 39. If an Hamiltonian equation (3.2.181) possesses generalized symmetries of high enough orders, then it has a formal symmetry L_n of the first order and of length l as big as necessary (see Theorem 27 of Section 3.2.3). In this case, we have not only the exact solution $\mathcal{S}_n = K_n^{-1}$ of (3.2.116) but also formal conserved densities $K_n^{-1} L_n^i$ of any order and of length l (Section 3.2.4). Additional integrability conditions of the form of (3.2.130, 3.2.131), which come from the existence of conservation laws, are automatically satisfied in this case.

So, when studying Hamiltonian equations like (3.2.183), we can use the following results:

- Generalized symmetries can be constructed, starting from conservation laws. Moreover, as it will be shown in Section 3.2.7, conservation laws can be obtained using Miura type transformations. The Hamiltonian structure provides the equations with generalized symmetries.
- Additional integrability conditions of the form of (3.2.130, 3.2.131) are automatically satisfied. If we classify integrable Hamiltonian equations or test a given Hamiltonian equation for integrability, we can use only integrability conditions of the form of (3.2.56, 3.2.100, 3.2.106) which come from the existence of generalized symmetries.

In the case of Toda and relativistic Toda type lattice equations, there are many classes of Hamiltonian equations. Then these two properties will be very useful.

Bi-Hamiltonian equations, i.e. equations possessing two compatible Hamiltonian structures, are known to be integrable [579]. Let us briefly provide an explanation of this fact. If an equation has two representations (3.2.181) with Hamiltonian operators K_n and $\hat{K}_n$ of different orders ($\operatorname{ord} K_n > \operatorname{ord} \hat{K}_n$), we can introduce the formal series $\mathcal{S}_n = K_n^{-1}$, $\hat{\mathcal{S}}_n = \hat{K}_n^{-1}$ and then, due to (3.2.120), the series $L_n = \mathcal{S}_n^{-1}\hat{\mathcal{S}}_n = K_n \hat{K}_n^{-1}$. Eqs. (3.2.70, 3.2.71, 3.2.119) imply that L_n^i and $\mathcal{S}_n L_n^i$, where i is an arbitrary integer, are the exact solutions of (3.2.65) and (3.2.116), respectively. This means that all integrability conditions are satisfied, as those conditions are derived from (3.2.65, 3.2.116). Moreover, L_n is the recursion operator, and one can construct, using it, an infinite number of conserved densities and generalized symmetries.

The Volterra equation (3.2.2) exemplifies a bi-Hamiltonian equation [654]. It is easy to check that both operators

$$K_n = u_n(S - S^{-1})u_n ,$$
$$\hat{K}_n = u_n(u_{n+1}S^2 + (u_{n+1} + u_n)S - (u_n + u_{n-1})S^{-1} - u_{n-1}S^{-2})u_n ,$$

together with the functions $h_n = u_n$ and $\hat{h}_n = \frac{1}{2}\log u_n$, define Hamiltonian representations (3.2.181) for the Volterra equation. As we shall show, $\tilde{\mathscr{L}} = \hat{K}_n K_n^{-1}$ is the formal series (3.2.74). This will prove that (3.2.74) is an exact solution of (3.2.65), i.e. it is the recursion operator. Moreover, formula $K_n^{-1}\tilde{\mathscr{L}}^i$ $(i \in \mathbb{Z})$ will give for the Volterra equation exact solutions of (3.2.116).

In fact, the Hamiltonian operators K_n and $\hat{K}_n$ can be rewritten as:

$$K_n = u_n(1 - S^{-1})(S + 1)u_n\,,$$
$$\hat{K}_n = u_n[u_n(1 - S^{-1}) + u_{n+1}S - u_{n-1}S^{-2}](S + 1)u_n\,,$$

and hence the inverse of K_n is given by:

$$K_n^{-1} = u_n^{-1}(S + 1)^{-1}(1 - S^{-1})^{-1}u_n^{-1}.$$

In fact

$$\hat{K}_n K_n^{-1} = u_n(u_n(1 - S^{-1}) + u_{n+1}S - u_{n-1}S^{-2})(1 - S^{-1})^{-1}u_n^{-1} = \tilde{\mathscr{L}}\,,$$

where $\tilde{\mathscr{L}}$ was introduced in (3.2.74).

2.7. Discrete Miura transformations and master symmetries. Using integrability conditions, necessary conditions for the existence of generalized symmetries and conservation laws, we obtain a list of equations. To prove their integrability we need to construct for the resulting equations higher order generalized symmetries and conservation laws.

Here we show some ways which one can prove integrability. One can construct a few conservation laws, using the integrability conditions (3.2.56, 3.2.100, 3.2.106) as presented in Sections 3.2.2 and 3.2.3. One can find coefficients of the formal series L_n (3.2.97) of the first order and then, using Theorem 28 contained in Section 3.2.3, obtain conserved densities. One more way is obtained by using the recursion operator considered in Section 3.2.3, which generates the infinite hierarchies of conservation laws and generalized symmetries. If an equation is Hamiltonian, one can, using the results presented in Section 3.2.6, constructs generalized symmetries starting from the conserved densities.

We are going to provide equations with infinite hierarchies of conservation laws and generalized symmetries and we will do that using Miura type transformations and local master symmetries together with Hamiltonian and Lagrangian structures.

The discrete analogue of the Miura transformation (3.1.23) is given by the following definition:

Definition 10. *The equation*

$$\dot{v}_n = \hat{f}_n = \hat{f}(v_{n+1}, v_n, v_{n-1}) \tag{3.2.187}$$

is transformed into (3.2.1) *by the transformation*

$$u_n = s_n = s(v_n, v_{n+1}, \ldots v_{n+k})\,, \quad k > 0\,, \quad \frac{\partial s_n}{\partial v_n}\frac{\partial s_n}{\partial v_{n+k}} \neq 0\,, \tag{3.2.188}$$

if s_n satisfies:

$$D_t s_n = \sum_{j=0}^{k} \frac{\partial s_n}{\partial v_{n+j}} \hat{f}_{n+j} = f(s_{n+1}, s_n, s_{n-1})\,. \tag{3.2.189}$$

The transformation (3.2.188) *is called a* Miura type transformation.

The name Miura type transformation is due to the fact that such transformations are similar to the original Miura transformation (2.2.3). As an example, let us present the transformation:

$$\tilde{u}_n = (c + u_n)(c - u_{n+1}) \tag{3.2.190}$$

which brings solutions u_n of the modified Volterra equation (3.2.185) into solutions $\tilde{u}_n$ of the Volterra equation (3.2.2). See in Section 2.4.5.1 for other examples of discrete Miura type transformations for the lSKdV.

Miura transformations, unlike point transformations which have the form: $u_n = s(v_n)$, are not invertible. Definition 10 is constructive because, for any given pair of equations (3.2.1, 3.2.187) and for any number k, we can find a Miura type transformation (3.2.188) or prove that it does not exist.

If (3.2.1) possesses a conservation law, it can be rewritten easily as a conservation law of (3.2.187). To do so, one has to replace the dependent variables u_{n+i} by the functions s_{n+i}. It is important that nontrivial conservation laws of positive order remain nontrivial. This will be shown in detail for the conservation laws (3.2.35-3.2.38).

Let the relation

$$\begin{aligned} D_t p_n &= (S-1)q_n\,, \\ p_n &= p(u_n, u_{n+1}, \dots u_{n+m})\,, \\ q_n &= q(u_{n-1}, u_n, \dots u_{n+m})\,, \end{aligned} \tag{3.2.191}$$

be a conservation law of (3.2.1) which has positive order m. This means that

$$\frac{\partial^2 p_n}{\partial u_n \partial u_{n+m}} \neq 0\,. \tag{3.2.192}$$

Then the following theorem will provide a conservation law for (3.2.187):

Theorem 40. *Let the Miura transformation* (3.2.188) *transform* (3.2.187) *into* (3.2.1). *Let us assume that there exists a conservation law* (3.2.191) *of* (3.2.1) *of positive order m. Then* (3.2.187) *possesses a conservation law*

$$\begin{aligned} D_t \hat{p}_n &= (S-1)\hat{q}_n\,, \\ \hat{p}_n &= \hat{p}(v_n, v_{n+1}, \dots v_{n+k+m}) = p(s_n, s_{n+1}, \dots s_{n+m})\,, \\ \hat{q}_n &= \hat{q}(v_{n-1}, v_n, \dots v_{n+k+m}) = q(s_{n-1}, s_n, \dots s_{n+m})\,, \end{aligned} \tag{3.2.193}$$

and its order is equal to $k+m$.

PROOF. Let us prove the first part of the statement, namely, that (3.2.193) is a conservation law of (3.2.187). We use the formula $\frac{\partial \hat{p}_n}{\partial v_{n+i}} = \sum_j \frac{\partial p_n}{\partial u_{n+j}} \frac{\partial s_{n+j}}{\partial v_{n+i}}$. As

$$\begin{aligned} D_t \hat{p}_n &= \sum_i \frac{\partial \hat{p}_n}{\partial v_{n+i}} \hat{f}_{n+i} = \sum_i \left(\sum_j \frac{\partial p_n}{\partial u_{n+j}} \frac{\partial s_{n+j}}{\partial v_{n+i}} \right) \hat{f}_{n+i} \\ &= \sum_j \frac{\partial p_n}{\partial u_{n+j}} \left(\sum_i \frac{\partial s_{n+j}}{\partial v_{n+i}} \hat{f}_{n+i} \right) = \sum_j \frac{\partial p_n}{\partial u_{n+j}} D_t s_{n+j}\,, \end{aligned}$$

using (3.2.1, 3.2.188, 3.2.189, 3.2.191) together with (3.2.193), we obtain

$$\begin{aligned} D_t \hat{p}_n &= \sum_j \frac{\partial p_n}{\partial u_{n+j}} f(s_{n+j+1}, s_{n+j}, s_{n+j-1}) = \sum_j \frac{\partial p_n}{\partial u_{n+j}} f(u_{n+j+1}, u_{n+j}, u_{n+j-1}) \\ &= \sum_j \frac{\partial p_n}{\partial u_{n+j}} \dot{u}_{n+j} = D_t p_n = (S-1)q_n = (S-1)\hat{q}_n\,. \end{aligned}$$

Let us now prove the second assertion. As $m > 0$, using (3.2.188, 3.2.192), we easily check that

$$\frac{\partial^2 \hat{p}_n}{\partial v_n \partial v_{n+k+m}} = \frac{\partial}{\partial v_n}\left(\frac{\partial p_n}{\partial u_{n+m}} \frac{\partial s_{n+m}}{\partial v_{n+k+m}}\right) = \frac{\partial^2 p_n}{\partial u_n \partial u_{n+m}} \frac{\partial s_n}{\partial v_n} T^m \frac{\partial s_n}{\partial v_{n+k}} \neq 0\,,$$

i.e. this is a conservation law of order $k + m$. □

Theorem 40 provides a way of constructing conservation laws and it shows that if (3.2.1) is integrable, in the sense that it possesses conservation laws of arbitrarily high orders, and (3.2.187) is transformed into it by a Miura type transformation, then (3.2.187) is integrable in the same sense.

Let us give a few examples. As we showed, the modified Volterra equation (3.2.185) is transformed into the Volterra equation (3.2.2) by transformation (3.2.190). Let us now use Theorem 40 and the conserved densities (3.2.39) of (3.2.2) to construct two conserved densities for (3.2.185). Starting from p_n^1, we obtain the following conserved density

$$\log(c + u_n) + \log(c - u_{n+1}) \sim \log(c^2 - u_n^2) = \hat{p}_n^1\,.$$

Starting from p_n^2, we are led to

$$(c + u_n)(c - u_{n+1}) = -u_n u_{n+1} + c^2 - c(S - 1)u_n\,.$$

Omitting the trivial terms and multiplying the result by -1, we obtain for (3.2.185) the density $\hat{p}_n^2 = u_n u_{n+1}$.

The modified Volterra equation is Hamiltonian (see (3.2.183, 3.2.184, 3.2.186)), and one can use Theorem 38 to construct a generalized symmetry. The case of $\hat{p}_n^1 = -2h_n$ (see (3.2.186)) is trivial. In the case of $\hat{p}_n^2$, one obtains $\frac{\delta \hat{p}_n^2}{\delta u_n} = u_{n+1} + u_{n-1}$ and is led, using formula (3.2.182), to the following generalized symmetry of (3.2.185)

(3.2.194) $$u_{n,\epsilon} = (c^2 - u_n^2)((c^2 - u_{n+1}^2)(u_{n+2} + u_n) - (c^2 - u_{n-1}^2)(u_n + u_{n-2}))\,.$$

Another example is given by (3.2.175) discussed in Section 3.2.5. It can be transformed, by the Miura transformation $\tilde{u}_n = (u_{n+1} - u_{n-1})^{-1}$, into the modified Volterra equation (3.2.185) with $c = 0$. This proves that (3.2.175) is integrable.

Let us consider the non linear DΔE (3.2.171). It can be shown that in this case the three conserved densities $p_n^{(1)}$, $p_n^{(2)}$, $p_n^{(3)}$ given by (3.2.56, 3.2.100, 3.2.106) are trivial. Moreover any conserved density which can be obtained, using (3.2.65) and (3.2.77, 3.2.78) is trivial too. Nevertheless, (3.2.171) has an infinite hierarchy of nontrivial conserved densities. In fact, if we introduce the function

(3.2.195) $$w_n = (u_{n+1} - u_n)^{1/2},$$

(3.2.171) is the transformed into

(3.2.196) $$2\dot{w}_n = w_{n+1} - w_{n-1}\,,$$

i.e. (3.2.171) is linearizable. It is easy to check that the functions $p_n^m = w_n w_{n+m}$ $(m \geq 0)$ are conserved densities of this linear equation, as

$$2D_t p_n^m = (S - 1)(w_{n-1} w_{n+m} + w_n w_{n+m-1})\,.$$

The order of such density p_n^m equals m. Using transformation (3.2.195), we obtain from p_n^m the following conserved density for (3.2.171)

(3.2.197) $$(u_{n+1} - u_n)^{1/2}(u_{n+m+1} - u_{n+m})^{1/2}.$$

If $m = 0$, such density is trivial. If $m > 0$, one can see that the conserved density (3.2.197) is nontrivial and has order $m + 1$, as Theorem 40 guarantees.

Master Symmetries. The notion of master symmetry has been introduced in Section 2.2.2.5 in the case of KdV and later in Section 2.3.2.2 for the Toda lattice. We consider here only local master symmetries, i.e. such master symmetries whose right hand side, unlike (3.1.25), contains no operators like $(S - 1)^{-1}$. Such master symmetry has been found for the first time in [285] for the Landau-Lifshitz equation. It is known that there are many local master symmetries in the case of DΔEs [27, 30, 169, 170, 653, 654, 866].

The master symmetry is an equation which depends explicitly on the spatial variable and may depend on its time. In the case of (3.2.1), we consider local master symmetries of the form

(3.2.198) $$u_{n,\tau} = \phi_n(\tau, u_{n+1}, u_n, u_{n-1}) .$$

If in the master symmetry there is an essential dependence on τ, then the corresponding equation (3.2.1) and its generalized symmetries (3.2.3) will also depend on τ which, for these equations, is an outer parameter. More details on this will be given at the end of this Section. This is the reason why the evolution differentiation D_τ corresponding to (3.2.198) is defined by

(3.2.199) $$D_\tau = \frac{\partial}{\partial \tau} + \sum_j \phi_{n+j} \frac{\partial}{\partial u_{n+j}}$$

(cf. (3.2.6)). An example of a master symmetry is given by:

(3.2.200) $$u_{n,\tau} = u_n((n+2)u_{n+1} + u_n - (n-1)u_{n-1}) .$$

This equation, introduced in [494], is a master symmetry of the Volterra equation (3.2.2) (see [654]), as it can be checked, using the following Definition 11. It is a subcase of the non isospectral symmetry of the Volterra equation presented in (2.3.186).

Let us define a Lie algebra structure as we did in the Introduction on the set of functions ϕ_n of the form (3.2.10) and $\phi_n(\tau, u_{n+1}, u_n, u_{n-1})$ defined in (3.2.198). For any functions ϕ_n and $\hat{\phi}_n$, we introduce the equations $u_{n,\tau} = \phi_n$ and $u_{n,\hat{\tau}} = \hat{\phi}_n$ and corresponding evolution differentiations D_τ and $D_{\hat{\tau}}$. A new function is defined as follows

(3.2.201) $$[\phi_n, \hat{\phi}_n] \equiv D_\tau \hat{\phi}_n - D_{\hat{\tau}} \phi_n .$$

Here [,] is a Lie bracket. It is obviously anti-symmetric: $[\phi_n, \hat{\phi}_n] = -[\hat{\phi}_n, \phi_n]$, and one can check by a direct calculation that it satisfies the Jacobi identity:

(3.2.202) $$[[\phi_n, \hat{\phi}_n], \tilde{\phi}_n] = [[\phi_n, \tilde{\phi}_n], \hat{\phi}_n] + [\phi_n, [\hat{\phi}_n, \tilde{\phi}_n]] .$$

The right hand side g_n of a generalized symmetry (3.2.3) of (3.2.1) satisfies (3.1.5), i.e. $[g_n, f_n] = 0$. In the case of the master symmetry (3.2.198), the function

(3.2.203) $$g_n = [\phi_n, f_n]$$

is the right hand side of a generalized symmetry. This generalized symmetry must be nontrivial, i.e. in (3.2.3) $m > 1$ and $m' < -1$. The function ϕ_n satisfies the following equation

(3.2.204) $$[[\phi_n, f_n], f_n] = 0 .$$

Any generalized symmetry (3.2.3) has the trivial solution: $\phi_n = g_n$. The master symmetry corresponds to a nontrivial solution of (3.2.204).

Definition 11. *Eq.* (3.2.198) *is a* master symmetry *of* (3.2.1) *if the function* ϕ_n *satisfies* (3.2.204)*, and the function* (3.2.203) *is the right hand side of a generalized symmetry* (3.2.3) *with orders* $m > 1$ *and* $m' < -1$.

In the case of a local master symmetry, this definition is constructive because, for any given (3.2.1), one can find a master symmetry (3.2.198) or prove that it does not exist.

The master symmetry is closely related to a t dependent generalized symmetry of (3.2.1), where t is the time of (3.2.1) as one can see in the case of non isospectral symmetries presented in Sections 2.3 and 2.4. This generalized symmetry is of the form: $u_{n,\hat{\epsilon}} = \hat{g}_n = tg_n + \phi_n$, where g_n given by (3.2.203) is the right hand side of a generalized symmetry $u_{n,\epsilon'} = g_n$. One easily checks that

$$[\hat{g}_n, f_n] = D_{\hat{\epsilon}} f_n - D_t \hat{g}_n = \sum_j \frac{\partial f_n}{\partial u_{n+j}} (tg_{n+j} + \phi_{n+j}) - D_t(tg_n + \phi_n)$$
$$= tD_{\tau'} f_n + D_\tau f_n - g_n - tD_t g_n - D_t \phi_n = t[g_n, f_n] + [\phi_n, f_n] - g_n = 0 \,.$$

Master symmetries enable one to construct infinite hierarchies of generalized symmetries. Let us introduce the adjoint action operator ad_{ϕ_n} corresponding to the master symmetry (3.2.198)

$$\text{(3.2.205)} \qquad \mathrm{ad}_{\phi_n} \hat{\phi}_n = [\phi_n, \hat{\phi}_n] \,.$$

Then, in terms of its powers $ad^i_{\phi_n}$, we can construct generalized symmetries for any $i \geq 1$

$$\text{(3.2.206)} \qquad u_{n,\epsilon_i} = g_n^{(i)} = \mathrm{ad}^i_{\phi_n} f_n \,.$$

In spite of the fact that (3.2.198) has an explicit dependence on the variable n, resulting generalized symmetries (3.2.206) do not depend on n. It is clear that the way of constructing generalized symmetries is simpler, using local master symmetry, than in the case of non local master symmetries or recursion operators, as non local functions can never appear when one applies the adjoint action operator ad_{ϕ_n} (3.2.205).

In the generic case it is not easy to prove that (3.2.206) are generalized symmetries and do not depend explicitly on n. This can be proved only for some integrable equations, using specific additional properties (see e.g. [170], where the Volterra equation is discussed). Definition 11 implies that (3.2.206) with $i = 1$ is a generalized symmetry of (3.2.1). We only prove here that also (3.2.206) with $i = 2$ is a generalized symmetry (see e.g. [262]).

Theorem 41. *If* (3.2.198) *is the master symmetry of* (3.2.1), *then* (3.2.206) *with* $i = 2$ *is a generalized symmetry of this equation.*

PROOF. Introducing the notation $g_n^{(0)} = f_n$, we obtain from (3.2.205, 3.2.206) the following result for all $i \geq 0$

$$\text{(3.2.207)} \qquad g_n^{(i+1)} = \mathrm{ad}_{\phi_n} g_n^{(i)} = [\phi_n, g_n^{(i)}] \,.$$

Then, using the Jacobi identity (3.2.202) and the fact that (3.2.206) with $i = 1$ is a generalized symmetry, we have

$$[g_n^{(2)}, f_n] = [[\phi_n, g_n^{(1)}], f_n] = [[\phi_n, f_n], g_n^{(1)}] + [\phi_n, [g_n^{(1)}, f_n]] =$$
$$[g_n^{(1)}, g_n^{(1)}] + [\phi_n, 0] = 0 \,,$$

i.e (3.2.206) with $i = 2$ is a generalized symmetry of (3.2.1). □

It can be checked, using Definition 11, that (3.2.200) and

$$\text{(3.2.208)} \qquad u_{n,\tau} = (c^2 - u_n^2)[(n+1)u_{n+1} - (n-1)u_{n-1}]$$

are master symmetries of the Volterra equation (3.2.2) and of the modified Volterra equation (3.2.185) (see [866] in the second case). Formula (3.2.206) with $i = 1$ gives the generalized symmetry (3.2.14) in the first case and (3.2.194) in the second one. Formula (3.2.206) for $i = 2$ provides in both cases a generalized symmetry (3.2.3) of orders $m = 3$, $m' = -3$.

Using local master symmetries, one easily can construct not only generalized symmetries but also conserved densities. Indeed, if p_n is a conserved density of (3.2.1), then the functions

$$D_\tau^j p_n\,, \qquad j \geq 1\,, \tag{3.2.209}$$

are also conserved densities. Here D_τ^j are powers of the operator D_τ (3.2.199). Adding total differences to the functions (3.2.209), one can not only simplify the obtained conserved densities but also remove the explicit dependence on the variable n.

We cannot prove formula (3.2.209) in the general case. For a given equation, such proof requires using additional properties, see e.g. [27, 169, 170], where a proof is given for the Volterra equation, using the Lax pair. It is a general property of integrable equations that the equation and its generalized symmetries possess common conserved densities. We consider below the case when a function p_n is the common conserved density for (3.2.1) and the corresponding generalized symmetries (3.2.206). In this case we can prove formula (3.2.209), which is a corollary of the following theorem:

Theorem 42. *Let* (3.2.198) *be a master symmetry of* (3.2.1), *and let p_n be the common conserved density of* (3.2.1) *and its generalized symmetries* (3.2.206). *Then the function $D_\tau p_n$ is also a common conserved density of* (3.2.1, 3.2.206).

PROOF. Introducing the notations $\epsilon_0 = t$ and $g_n^{(0)} = f_n$, using the derivative operator $D_{\epsilon_i} = \sum_j g_{n+j}^{(i)} \frac{\partial}{\partial u_{n+j}}$ for all $i \geq 0$, and taking into account (3.2.206) together with (3.2.199, 3.2.201, 3.2.207), we obtain

$$\begin{aligned} D_\tau D_{\epsilon_i} - D_{\epsilon_i} D_\tau &= \textstyle\sum_j (D_\tau g_{n+j}^{(i)} - D_{\epsilon_i} \phi_{n+j}) \frac{\partial}{\partial u_{n+j}} \\ &= \textstyle\sum_j [\phi_{n+j}, g_{n+j}^{(i)}] \frac{\partial}{\partial u_{n+j}} = \sum_j g_{n+j}^{(i+1)} \frac{\partial}{\partial u_{n+j}}\,. \end{aligned}$$

So, for any $i \geq 0$, we have the following general formula

$$[D_\tau, D_{\epsilon_i}] = D_\tau D_{\epsilon_i} - D_{\epsilon_i} D_\tau = D_{\epsilon_{i+1}}\,. \tag{3.2.210}$$

If a function p_n is the common conserved density of (3.2.1, 3.2.206), then one has the set of conservation laws

$$D_{\epsilon_i} p_n = (S - 1)\omega_n^{(i)}\,, \qquad i \geq 0\,. \tag{3.2.211}$$

Relations (3.2.210, 3.2.211) imply

$$D_{\epsilon_i} D_\tau p_n = D_\tau D_{\epsilon_i} p_n - D_{\epsilon_{i+1}} p_n = (S - 1)(D_\tau \omega_n^{(i)} - \omega_n^{(i+1)})\,,$$

i.e. $D_\tau p_n$ is also a conserved density of (3.2.1, 3.2.206). □

Let us consider, as an example, the master symmetry (3.2.200) of the Volterra equation (3.2.2) and the conserved densities (3.2.39). We see that

$$\begin{aligned} D_\tau p_n^1 &= (n+2)u_{n+1} + u_n - (n-1)u_{n-1} = (n+1)u_n \\ &+ (S-1)((n+1)u_n) + u_n - nu_n + (S-1)((n-1)u_{n-1}) \sim 2u_n = 2p_n^2\,. \end{aligned}$$

In particular, the explicit dependence on n disappears after passing to an equivalent density. Moreover, one can check that $D_\tau p_n^2 \sim 2p_n^3$, $D_\tau p_n^3 \sim 3p_n^4$, i.e. master symmetry (3.2.200) allows one to construct, starting from p_n^1, the conserved densities p_n^2, p_n^3, p_n^4.

The explicit dependence of the master symmetry on its time is more difficult to understand and will be discussed in the following. We do that by considering the following equation as an example, subcase of the YdKN (2.4.129) whose coefficients different from zero are $B_0 = 1$, $C_3 = 2$ and $C_6 = c$

(3.2.212) $$\dot{u}_n = \frac{u_{n+1} + u_{n-1} + 2u_n + c}{u_{n+1} - u_{n-1}},$$

where c is a constant. If we try to find a master symmetry (3.2.198) for (3.2.212), using Definition 11, we fail. However, if we consider the equation

(3.2.213) $$\dot{u}_n = \frac{u_{n+1} + u_{n-1} + 2u_n + a(\tau)}{u_{n+1} - u_{n-1}},$$

where $a(\tau)$ is an unknown function of the time of the master symmetry τ, we find a master symmetry if $a'(\tau) = -2$. Let us choose a solution of this ODE, satisfying the initial condition $a(0) = c$, and let us write down the master symmetry of (3.2.213) as

(3.2.214) $$u_{n,\tau} = n\dot{u}_n, \qquad a(\tau) = -2\tau + c.$$

Generalized symmetries and conserved densities, generated by (3.2.214) for (3.2.213), explicitly depend on τ and remain generalized symmetries and conserved densities for any value of the parameter τ (unlike the master symmetry). Putting $\tau = 0$, we obtain generalized symmetries and conserved densities for (3.2.212) with any given number c. So, a master symmetry is constructed for the generalization (3.2.213) of (3.2.212) depending on τ, and that master symmetry provides generalized symmetries and conserved densities for both (3.2.212, 3.2.213).

2.8. Generalized symmetries for systems of lattice equations: Toda type equations. Here we discuss the generalized symmetry method in the case of systems of lattice equations. We do that by the example of Toda type equations. The Toda lattice will be written as a systems of two evolution equations on the lattice. We will see that the general theory in this case is quite similar to the scalar one. The main difference is that the coefficients of formal symmetries and conserved densities are 2×2 matrices.

Let us consider the following class of equations

(3.2.215) $$\ddot{u}_n = f_n = f(\dot{u}_n, u_{n+1}, u_n, u_{n-1}), \qquad \frac{\partial f_n}{\partial u_{n+1}} \frac{\partial f_n}{\partial u_{n-1}} \neq 0,$$

which includes the well-known Toda lattice (1.4.16) [794–796] which we repeat here for the convenience of the reader

(3.2.216) $$\ddot{u}_n = e^{u_{n+1}-u_n} - e^{u_n - u_{n-1}}.$$

This class will be discussed in Section 3.3.2.

Local conservation laws of (3.2.215) have the form (3.2.15), where the scalar functions p_n, q_n are analogous to (3.2.10), but depend on a finite number of the variables u_{n+j}, $\dot{u}_{n+j}$. The time derivatives $d^i u_{n+j}/dt^i$ with $i \geq 2$ are expressed in terms of these variables in virtue of (3.2.215). The differentiation D_t is defined in this case as

(3.2.217) $$D_t = \sum_j \dot{u}_{n+j} \frac{\partial}{\partial u_{n+j}} + \sum_j f_{n+j} \frac{\partial}{\partial \dot{u}_{n+j}}.$$

Generalized symmetries of (3.2.215) are equations of the form

(3.2.218) $$u_{n,\epsilon} = \varphi_n = \varphi(u_{n+m}, \dot{u}_{n+m}, u_{n+m-1}, \dot{u}_{n+m-1}, \dots u_{n+m'}, \dot{u}_{n+m'}),$$

where $m \geq m'$. Using the fact that (3.2.215, 3.2.218) have common solutions $u_n(t, \epsilon)$ and applying the operator D_ϵ to (3.2.215), we obtain the compatibility condition for (3.2.215, 3.2.218), i.e. the following equation for the function φ_n

$$D_t^2 \varphi_n = D_\epsilon f_n . \tag{3.2.219}$$

From the point of view of the generalized symmetry method, it is more convenient to rewrite (3.2.215) in the form of a system of two equations. Let us introduce the function $v_n = \dot{u}_n$ and rewrite (3.2.215) as the system

$$\dot{u}_n = v_n , \qquad \dot{v}_n = f_n = f(v_n, u_{n+1}, u_n, u_{n-1}) \tag{3.2.220}$$

which is equivalent to (3.2.215) from the point of view of the definitions of generalized symmetries and conservation laws. The following formula

$$\phi_n = \phi(u_{n+k_1}, v_{n+k_2}, u_{n+k_1-1}, v_{n+k_2-1}, \dots u_{n+k_1'}, v_{n+k_2'}) , \tag{3.2.221}$$

with finite $k_1 \geq k_1'$, $k_2 \geq k_2'$, expresses the most general function which will enter into the symmetries and conserved densities. Conservation laws for (3.2.220) have the same form (3.2.15), p_n and q_n are functions of the form (3.2.221), and a formula for the operator D_t is obviously rewritten from (3.2.217).

The system (3.2.220) can be written in vector form

$$\dot{U}_n = F_n = F(U_{n+1}, U_n, U_{n-1}) , \quad U_n = \begin{pmatrix} u_n \\ v_n \end{pmatrix} , \quad F_n = \begin{pmatrix} v_n \\ f_n \end{pmatrix} . \tag{3.2.222}$$

Then, its generalized symmetry reads

$$U_{n,\epsilon} = G_n = G(U_{n+m}, U_{n+m-1}, \dots U_{n+m'}) , \quad G_n = \begin{pmatrix} \varphi_n \\ \psi_n \end{pmatrix} , \tag{3.2.223}$$

where $m \geq m'$ and φ_n, ψ_n are functions of the form (3.2.221). The standard compatibility condition $D_t G_n = D_\epsilon F_n$ implies the relations $D_t \varphi_n = v_{n,\epsilon}$ and $D_t \psi_n = D_\epsilon f_n$. Thus, we see that ψ_n is expressed via φ_n: $\psi_n = D_t \varphi_n$, and the function φ_n satisfies condition (3.2.219).

Main formulas, notions, definitions and theorems presented in this section are very similar to ones presented in Sections 3.2.1, 3.2.3, 3.2.4, 3.2.6. Let us introduce the following notation for the vector-function G_n defined by (3.2.223)

$$\frac{\partial G_n}{\partial U_{n+j}} = \begin{pmatrix} \partial \varphi_n / \partial u_{n+j} & \partial \varphi_n / \partial v_{n+j} \\ \partial \psi_n / \partial u_{n+j} & \partial \psi_n / \partial v_{n+j} \end{pmatrix} . \tag{3.2.224}$$

We assume that (3.2.223) is such that $\frac{\partial G_n}{\partial U_{n+m}} \neq 0$, $\frac{\partial G_n}{\partial U_{n+m'}} \neq 0$, i.e. G_n really depends on U_{n+m}, $U_{n+m'}$. We call the numbers m and m' the left and right orders of this symmetry. The order of the conserved densities and conservation laws is also defined as in the scalar case. As in Definition 6, we say that a conserved density p_n (which is a scalar function) is called trivial if it is equivalent to a constant. If $p_n \sim \mathcal{P}(U_n)$, where $\mathcal{P}$ is not a constant function, then p_n is of order 0. A conserved density p_n has order $m > 0$ if

$$p_n \sim \mathcal{P}_n = \mathcal{P}(U_n, U_{n+1}, \dots U_{n+m}) , \quad \begin{pmatrix} \frac{\partial^2 \wp_n}{\partial u_n \partial u_{n+m}} & \frac{\partial^2 \wp_n}{\partial u_n \partial v_{n+m}} \\ \frac{\partial^2 \wp_n}{\partial v_n \partial u_{n+m}} & \frac{\partial^2 \wp_n}{\partial v_n \partial v_{n+m}} \end{pmatrix} \neq 0 .$$

We can easily calculate the order of a conserved density, using the formal variational derivative. Let us introduce, as in (3.2.40), the variational derivatives with respect to u_n

and v_n for any function ϕ_n (3.2.221) as

$$\frac{\delta\phi_n}{\delta u_n} = \sum_{j=-k_1}^{-k_1'} \frac{\partial\phi_{n+j}}{\partial u_n}, \qquad \frac{\delta\phi_n}{\delta v_n} = \sum_{j=-k_2}^{-k_2'} \frac{\partial\phi_{n+j}}{\partial v_n}, \tag{3.2.225}$$

and then define for the conserved density p_n a variational derivative with respect to U_n

$$\varrho_n = \frac{\delta p_n}{\delta U_n} = \begin{pmatrix} \delta p_n/\delta u_n \\ \delta p_n/\delta v_n \end{pmatrix}. \tag{3.2.226}$$

As in (3.2.43-3.2.46), we have $\varrho_n = 0$ if p_n is a trivial conserved density, $\varrho_n = \varrho(U_n) \neq 0$ if ip_n is of order 0, and

$$\varrho_n = \varrho(U_{n+m}, U_{n+m-1}, \dots U_{n-m}), \qquad \frac{\partial\varrho_n}{\partial U_{n+m}} \frac{\partial\varrho_n}{\partial U_{n-m}} \neq 0,$$

if p_n has the order $m > 0$.

Following (3.2.63, 3.2.110), the Fréchet derivative G_n^* of the vector-function G_n (3.2.223) and its adjoint operator $G_n^{*\dagger}$ are defined in this case as

$$G_n^* = \sum_{j=m'}^{m} \frac{\partial G_n}{\partial U_{n+j}} S^j, \qquad G_n^{*\dagger} = \sum_{j=m'}^{m} \left(\frac{\partial G_{n-j}}{\partial U_n}\right)^{\dagger} S^{-j}, \tag{3.2.227}$$

where the coefficients of $G_n^{*\dagger}$ are the transposed matrices of those of G_n^*. We see that the coefficients of the operators are matrices and thus do not commute. In the same way and using the compact notation

$$f_n^{(j)} = \frac{\partial f_n}{\partial u_{n+j}}, \qquad f_n^{(v)} = \frac{\partial f_n}{\partial v_n} \tag{3.2.228}$$

(see (3.2.220)), we obtain the following formulas for the operators F_n^*, $F_n^{*\dagger}$ in the case of F_n given by (3.2.222)

$$F_n^* = \begin{pmatrix} 0 & 0 \\ f_n^{(1)} & 0 \end{pmatrix} S + \begin{pmatrix} 0 & 1 \\ f_n^{(0)} & f_n^{(v)} \end{pmatrix} + \begin{pmatrix} 0 & 0 \\ f_n^{(-1)} & 0 \end{pmatrix} S^{-1}, \tag{3.2.229}$$

$$F_n^{*\dagger} = \begin{pmatrix} 0 & f_{n+1}^{(-1)} \\ 0 & 0 \end{pmatrix} S + \begin{pmatrix} 0 & f_n^{(0)} \\ 1 & f_n^{(v)} \end{pmatrix} + \begin{pmatrix} 0 & f_{n-1}^{(1)} \\ 0 & 0 \end{pmatrix} ST^{-1}. \tag{3.2.230}$$

As in Sections 3.2.3 and 3.2.4, we can derive for the right hand side of a generalized symmetry (3.2.223) and for a variational derivative (3.2.226) in place of (3.2.72, 3.2.109), the following equations

$$(D_t - F_n^*)G_n = 0, \qquad (D_t + F_n^{*\dagger})\varrho_n = 0. \tag{3.2.231}$$

Here $D_t\Phi_n = \sum_j \frac{\partial\Phi_n}{\partial U_{n+j}} F_{n+j}$ for any vector-function Φ_n.

As in the scalar case given by (3.2.65, 3.2.116), (3.2.231) are connected to

$$\dot{L}_n = [F_n^*, L_n], \tag{3.2.232}$$

$$\dot{\mathscr{S}}_n + \mathscr{S}_n F_n^* + F_n^{*\dagger}\mathscr{S}_n = 0, \tag{3.2.233}$$

but L_n, $\mathscr{S}_n$ are now formal series of the form (3.2.66, 3.2.117) with 2×2 matrix coefficients. Applying the Fréchet derivative to (3.2.231), one can show that

$$L_n = G_n^*, \qquad \mathscr{S}_n = \varrho_n^* \tag{3.2.234}$$

provide the corresponding approximate solutions of (3.2.232, 3.2.233). In this way we can construct formal symmetries and conserved densities, and the definition of orders and lengths will be the same as before. As in Theorems 26 and 32, using formulas (3.2.234), we obtain from a generalized symmetry of order $m \geq 1$ a formal symmetry L_n, such that $\operatorname{ord} L_n = m$, $\operatorname{lgt} L_n \geq m$. From the conserved density of order $m \geq 2$ we derive a formal conserved density $\mathcal{S}_n$, such that $\operatorname{ord}\mathcal{S}_n = m$ and $\operatorname{lgt}\mathcal{S}_n \geq m-1$.

Formal symmetries and conserved densities in the case of the systems under consideration can be of two types. Namely, formal symmetries and conserved densities (3.2.66) and (3.2.117) can be such that $\det l_n^{(N)} = \det s_n^{(M)} = 0$, and then they are called degenerate. If $\det l_n^{(N)} \det s_n^{(M)} \neq 0$ they are non degenerate. The non degenerate case is equivalent to the scalar one and formal series (3.2.66, 3.2.117) can easily be inverted. The degenerate case is new and needs a detailed discussion. Let us introduce compact notations for the operator F_n^* (3.2.229) (3.2.66)

$$F_n^* = F_n^{(1)} S + F_n^{(0)} + F_n^{(-1)} S^{-1}, \qquad F_n^{(i)} = \begin{pmatrix} \alpha_n^{(i)} & \beta_n^{(i)} \\ \gamma_n^{(i)} & \delta_n^{(i)} \end{pmatrix}. \tag{3.2.235}$$

Then we can present the following theorem:

Theorem 43. *If the formal series L_n (3.2.66) is a degenerate formal symmetry of (3.2.222) of length $l \geq 2$, then it can be written as*

$$L_n = \begin{pmatrix} 0 & 0 \\ \gamma_n^{(N)} & 0 \end{pmatrix} S^N + \begin{pmatrix} \alpha_n^{(N-1)} & \beta_n^{(N-1)} \\ \gamma_n^{(N-1)} & \delta_n^{(N-1)} \end{pmatrix} S^{N-1} + \dots, \tag{3.2.236}$$

where $\gamma_n^{(N)} \beta_n^{(N-1)} \neq 0$.

Proof. The condition $l \geq 2$ means that we can set to zero coefficients at S^{N+1}, S^N in (3.2.232). Taking into account (3.2.229, 3.2.235) and collecting coefficients at S^{N+1}, we obtain the condition $F_n^{(1)} l_{n+1}^{(N)} = l_n^{(N)} F_{n+N}^{(1)}$ which is equivalent to the following ones

$$f_n^{(1)} \beta_{n+1}^{(N)} = \beta_n^{(N)} f_{n+N}^{(1)} = 0\,, \tag{3.2.237}$$

$$f_n^{(1)} \alpha_{n+1}^{(N)} = \delta_n^{(N)} f_{n+N}^{(1)}\,. \tag{3.2.238}$$

As from (3.2.215, 3.2.228) $f_n^{(1)} \neq 0$, (3.2.237) implies $\beta_n^{(N)} = 0$. Using (3.2.238) and the condition $\det l_n^{(N)} = \alpha_n^{(N)} \delta_n^{(N)} = 0$, we obtain $\alpha_n^{(N)} = \delta_n^{(N)} = 0$.

Taking from (3.2.66) the condition $l_n^{(N)} \neq 0$, one can rewrite it in the form: $\gamma_n^{(N)} \neq 0$. Coefficients at S^N give the following matrix equation

$$\dot{l}_n^{(N)} = F_n^{(1)} l_{n+1}^{(N-1)} + F_n^{(0)} l_n^{(N)} - l_n^{(N)} F_{n+N}^{(0)} - l_n^{(N-1)} F_{n+N-1}^{(1)}\,.$$

The left upper corner element is $\gamma_n^{(N)} = \beta_n^{(N-1)} f_{n+N-1}^{(1)}$, thus $\beta_n^{(N-1)} \neq 0$. □

One can prove in a quite similar way that the degenerate formal conserved density (3.2.117) of length $l \geq 2$ has the form

$$\mathcal{S}_n = \begin{pmatrix} a_n^{(M)} & 0 \\ 0 & 0 \end{pmatrix} S^M + \begin{pmatrix} a_n^{(M-1)} & b_n^{(M-1)} \\ c_n^{(M-1)} & d_n^{(M-1)} \end{pmatrix} S^{M-1} + \dots, \tag{3.2.239}$$

where $a_n^{(M)} d_n^{(M-1)} \neq 0$.

Theorem 43 enables us to obtain the following important result: the second power L_n^2 of a formal symmetry L_n of (3.2.222) is non degenerate if $\operatorname{lgt} L_n \geq 2$. In fact, let us consider the series L_n (3.2.236) and, taking into account (3.2.66), write down the first two coefficients of L_n^2

$$L_n^2 = l_n^{(N)} l_{n+N}^{(N)} S^{2N} + (l_n^{(N)} l_{n+N}^{(N-1)} + l_n^{(N-1)} l_{n+N-1}^{(N)}) S^{2N-1} + \dots .$$

The first coefficient of this formal series is a zero matrix, and the second one reads

$$\begin{pmatrix} \beta_n^{(N-1)} \gamma_{n+N-1}^{(N)} & 0 \\ \gamma_n^{(N)} \alpha_{n+N}^{(N-1)} + \delta_n^{(N-1)} \gamma_{n+N-1}^{(N)} & \gamma_n^{(N)} \beta_{n+N}^{(N-1)} \end{pmatrix} . \tag{3.2.240}$$

This matrix is non degenerate, then the formal symmetry L_n^2 is also non degenerate and has order $2N-1$ and $\mathrm{lgt} L_n^2 \geq \mathrm{lgt} L_n - 1$. This is the reason why, when deriving the integrability conditions from (3.2.232), we can consider only non degenerate formal symmetries.

The easiest way to derive integrability conditions is to use the following property of the Toda lattice (3.2.216). The Toda lattice has, for any order $m \geq 1$, two generalized symmetries and two conserved densities of order m, such that one of the corresponding formal symmetries and conserved densities is degenerate, while the other one is non degenerate. In such a case, one can avoid considering degenerate formal conserved densities. In fact, for any degenerate formal conserved densities $\mathscr{S}_n$, such that $\mathrm{lgt}\mathscr{S}_n \geq 2$, we can obtain a degenerate formal symmetry L_n, such that $\mathrm{lgt} L_n > \mathrm{lgt}\mathscr{S}_n$. Then, using (3.2.236, 3.2.239), we easily prove that the formal conserved density $\mathscr{S}_n L_n$ is non degenerate and

$$\mathrm{ord}(\mathscr{S}_n L_n) = \mathrm{ord}\mathscr{S}_n + \mathrm{ord} L_n - 1 , \qquad \mathrm{lgt}(\mathscr{S}_n L_n) \geq \mathrm{lgt}\mathscr{S}_n - 1 .$$

In accordance with what we have said above, one can start from the non degenerate formal symmetries L_n, $\hat{L}_n$ and conserved density $\mathscr{S}_n$

$$\begin{aligned} &\mathrm{ord} L_n = m , \quad \mathrm{ord}\hat{L}_n = m+1 , \quad \mathrm{ord}\mathscr{S}_n = m+1 , \\ &\mathrm{lgt} L_n \geq m , \quad \mathrm{lgt}\hat{L}_n \geq m+1 , \quad \mathrm{lgt}\mathscr{S}_n \geq m , \end{aligned}$$

where $m \geq 1$. Then the non degenerate formal symmetry $\tilde{L}_n = \hat{L}_n L_n^{-1}$ and conserved density $\tilde{\mathscr{S}}_n = \mathscr{S}_n L_n^{-1}$, such that

$$\mathrm{ord}\tilde{L}_n = \mathrm{ord}\tilde{\mathscr{S}}_n = 1 , \quad \mathrm{lgt}\tilde{L}_n \geq m , \quad \mathrm{lgt}\tilde{\mathscr{S}}_n \geq m ,$$

can be used for deriving integrability conditions. For the formal symmetry

$$\tilde{L}_n = \tilde{l}_n^{(1)} S + \tilde{l}_n^{(0)} + \tilde{l}_n^{(-1)} S^{-1} + \dots ,$$

we can write down some useful formulas for the conserved densities in terms of the matrix trace and determinant (cf. Theorem 28)

$$\begin{aligned} &D_t \log\det \tilde{l}_n^{(1)} \sim 0 , \\ &D_t \mathrm{tr}\, \mathrm{res} \tilde{L}_n^i \sim 0 , \qquad 1 \leq i \leq \mathrm{lgt}\tilde{L}_n - 2 . \end{aligned}$$

These formulas are valid if $\mathrm{lgt}\tilde{L}_n \geq 2$ and $\mathrm{lgt}\tilde{L}_n \geq 3$, respectively.

The integrability conditions can also be derived, starting from the existence of two conservation laws or one generalized symmetry and one conservation law. A precise statement is given in Section 3.3.2, but the corresponding calculation would be quite long in this case. One has to consider both non degenerate and degenerate formal conserved densities. One also has to use the fact that the degenerate formal conserved densities of (3.2.222) are invertible (as well as the degenerate formal symmetries!). Indeed, it is easy to check that the inverse $\mathscr{S}_n^{-1}$ of the formal series $\mathscr{S}_n$ (3.2.239) exists and is unique. First coefficients of $\mathscr{S}_n^{-1}$ have the following form

$$\mathscr{S}_n^{-1} = \begin{pmatrix} 0 & 0 \\ 0 & \tilde{d}_n^{(1-M)} \end{pmatrix} S^{1-M} + \begin{pmatrix} \tilde{a}_n^{(-M)} & \tilde{b}_n^{(-M)} \\ \tilde{c}_n^{(-M)} & \tilde{d}_n^{(-M)} \end{pmatrix} ST^{-M} + \dots ,$$

$$\tilde{d}_n^{(1-M)} = \frac{1}{d_{n-M+1}^{(M-1)}} , \qquad \tilde{a}_n^{(-M)} = \frac{1}{a_{n-M}^{(M)}} ,$$

$$\tilde{b}_n^{(-M)} = -\frac{b_{n-M}^{(M-1)}}{a_{n-M}^{(M)} d_{n-M}^{(M-1)}} , \qquad \tilde{c}_n^{(-M)} = -\frac{c_{n-M+1}^{(M-1)}}{a_{n-M}^{(M)} d_{n-M+1}^{(M-1)}} , \qquad \dots .$$

This allows us, starting from the existence of two conservation laws, to pass to formal conserved densities $\mathscr{S}_n$, $\hat{\mathscr{S}}_n$ and then to obtain the formal symmetry $L_n = \mathscr{S}_n^{-1}\hat{\mathscr{S}}_n$ even if $\mathscr{S}_n$ is degenerate.

As a result, we can derive for the systems (3.2.220) integrability conditions which have the same structure and meaning as the conditions (3.2.56, 3.2.100, 3.2.106, 3.2.130, 3.2.131). Similarly to (3.2.56), the first of them is

$$D_t \log \frac{\partial f_n}{\partial u_{n+1}} = (S-1)q_n^{(1)} ,$$

the other ones will be presented in Section 3.3.2. In this case, integrability conditions can be checked, using the following property formulated for functions ϕ_n of the form (3.2.221)

$$\frac{\delta \phi_n}{\delta u_n} = \frac{\delta \phi_n}{\delta v_n} = 0 \qquad \Leftrightarrow \qquad \phi_n = c + (S-1)\psi_n$$

(cf. Theorem 24). Here the formal variational derivatives are defined by (3.2.225), c is a constant, and ψ_n is another function of the form (3.2.221).

Let us discuss in conclusion the Hamiltonian systems

$$\text{(3.2.241)} \qquad \dot{u}_n = \varphi_n \frac{\delta h_n}{\delta v_n} , \qquad \dot{v}_n = -\varphi_n \frac{\delta h_n}{\delta u_n} , \qquad \varphi_n = \varphi(u_n, v_n) ,$$

where h_n is a function of the form (3.2.221). For example, if

$$\text{(3.2.242)} \qquad \varphi_n = 1 , \qquad h_n = e^{u_{n+1}-u_n} + \tfrac{1}{2}v_n^2 ,$$

we obtain the Toda lattice (3.2.216). Almost all integrable equations in Sections 3.3.2 and 3.3.3.2 are Hamiltonian with respect to this Hamiltonian structure.

In order to check that (3.2.241) is Hamiltonian, one has to rewrite it in vector form:

$$\text{(3.2.243)} \qquad \dot{U}_n = F_n = K_n \frac{\delta h_n}{\delta U_n} , \qquad K_n = \varphi_n \begin{pmatrix} 0 & 1 \\ -1 & 0 \end{pmatrix} .$$

The vector U_n is given in (3.2.222), and the operator $\frac{\delta}{\delta U_n}$ is defined by (3.2.226). One can see that K_n is a Hamiltonian operator, as it is obviously anti-symmetric (i.e $K_n^\dagger = -K_n$) and satisfies the equation

$$\text{(3.2.244)} \qquad \dot{K}_n = F_n^* K_n + K_n F_n^{*\dagger}$$

for any functions h_n, φ_n. The condition (3.2.244) is checked by a straightforward, but rather long calculation. It is easier to prove that the function h_n is the conserved density of system

(3.2.241). In fact,

$$D_t h_n = \sum_i \frac{\partial h_n}{\partial u_{n+i}} \varphi_{n+i} \frac{\delta h_{n+i}}{\delta v_{n+i}} - \sum_i \frac{\partial h_n}{\partial v_{n+i}} \varphi_{n+i} \frac{\delta h_{n+i}}{\delta u_{n+i}} \sim \left(\sum_j \frac{\partial h_{n+j}}{\partial u_n} \right) \varphi_n \frac{\delta h_n}{\delta v_n}$$
$$- \left(\sum_j \frac{\partial h_{n+j}}{\partial v_n} \right) \varphi_n \frac{\delta h_n}{\delta u_n} = \frac{\delta h_n}{\delta u_n} \varphi_n \frac{\delta h_n}{\delta v_n} - \frac{\delta h_n}{\delta v_n} \varphi_n \frac{\delta h_n}{\delta u_n} = 0 .$$

In Section 3.3.3.2, where all equations have the Hamiltonian structure (3.2.241), we write down only those integrability conditions which come from the existence of generalized symmetries, i.e. from (3.2.232), as explained in Section 3.2.6. For all Hamiltonian equations of Sections 3.3.2 and 3.3.3.2, we take into account that if p_n is a conserved density of (3.2.243), then the equation $U_{n,\epsilon} = K_n \frac{\delta p_n}{\delta U_n}$ is its generalized symmetry. In the case of system (3.2.241), such formula for the generalized symmetry takes the form

$$u_{n,\epsilon} = \varphi_n \frac{\delta p_n}{\delta v_n} , \qquad v_{n,\epsilon} = -\varphi_n \frac{\delta p_n}{\delta u_n} , \tag{3.2.245}$$

where p_n is its conserved density.

2.9. Integrability conditions for relativistic Toda type equations. In this section we consider two classes of lattice equations

$$\ddot{u}_n = f_n = f(u_{n+1}, u_n, u_{n-1}, \dot{u}_{n+1}, \dot{u}_n, \dot{u}_{n-1}) , \qquad \frac{\partial f_n}{\partial \dot{u}_{n+1}} \neq 0 , \tag{3.2.246}$$

and

$$\dot{u}_n = f_n = f(u_{n+1}, u_n, v_n) , \quad \dot{v}_n = g_n = g(v_{n-1}, v_n, u_n) , \qquad \frac{\partial f_n}{\partial u_{n+1}} \neq 0 . \tag{3.2.247}$$

These classes include two different forms of relativistic Toda type equations which are discussed in Section 3.3.3. The relativistic Toda lattice equation [721, 849]

$$\ddot{u}_n = \frac{\dot{u}_{n+1}\dot{u}_n}{1 + e^{u_n - u_{n+1}}} - \frac{\dot{u}_n \dot{u}_{n-1}}{1 + e^{u_{n-1} - u_n}} \tag{3.2.248}$$

is of the form (3.2.246).

Our purpose is to derive integrability conditions for such equations as (3.2.246, 3.2.247). The cases of (3.2.246) and (3.2.247) are more difficult than the case of (3.2.215) from the theoretical point of view, and the standard scheme of the generalized symmetry method does not give results. For this reason, we present here the new Theorems 45 and 47. We follow in this section the paper [851].

Let us at first discuss the class (3.2.246). It is convenient to rewrite (3.2.246) in the form

$$\dot{u}_n = v_n , \qquad \dot{v}_n = f_n = f(u_{n+1}, u_n, u_{n-1}, v_{n+1}, v_n, v_{n-1}) , \tag{3.2.249}$$

introducing the function $v_n = \dot{u}_n$. In such notation, the most general form of a scalar function will be given by (3.2.221), as was in Section 3.2.8. The vector form of the system (3.2.249) is given by (3.2.222) and generalized symmetries of (3.2.222) have the form (3.2.223). As before, starting from a generalized symmetry of order $m \geq 1$ and using the standard formula $L_n = G_n^*$ (see (3.2.227)), we obtain an approximate solution of the equation

$$\dot{L}_n = [F_n^*, L_n] , \tag{3.2.250}$$

i.e. a formal symmetry L_n, such that $\mathrm{ord} L_n = m$, $\mathrm{lgt} L_n \geq m$. The Fréchet derivative F_n^* takes in this case the form

$$F_n^* = \begin{pmatrix} 0 & 0 \\ f_n^{(1)} & g_n^{(1)} \end{pmatrix} S + \begin{pmatrix} 0 & 1 \\ f_n^{(0)} & g_n^{(0)} \end{pmatrix} + \begin{pmatrix} 0 & 0 \\ f_n^{(-1)} & g_n^{(-1)} \end{pmatrix} S^{-1}, \tag{3.2.251}$$

where

$$f_n^{(i)} = \frac{\partial f_n}{\partial u_{n+i}}, \qquad g_n^{(i)} = \frac{\partial f_n}{\partial v_{n+i}}. \tag{3.2.252}$$

As in the case of Toda type equations, the generalized symmetry (3.2.223) and formal symmetry $L_n = \sum_{i=-\infty}^{m} l_n^{(i)} S^i$ may be degenerate, i.e. such that $\det \frac{\partial G_n}{\partial U_{n+m}} = 0$ and $\det l_n^{(m)} = 0$, and non degenerate ($\det \frac{\partial G_n}{\partial U_{n+m}} \neq 0$, $\det l_n^{(m)} \neq 0$). For example, the relativistic Toda lattice (3.2.248) has for any order $m \geq 1$ both degenerate and non degenerate symmetries (see Section 3.3.3.5, where the construction of generalized symmetries for this equation is discussed). However, the degenerate formal symmetry cannot be inverted in the case of (3.2.246). For this reason, we consider here equations of the relativistic Toda type which possess two non degenerate generalized symmetries of orders $m \geq 1$ and $m+1$ (the corresponding formal symmetries of the form $L_n = G_n^*$ will be non degenerate as well). For equations of this kind, we can construct in the standard way (see e.g. Theorem 27, Section 3.2.3) a formal symmetry of the first order

$$L_n = \sum_{i \leq 1} l_n^{(i)} S^i, \qquad l_n^{(i)} = \begin{pmatrix} a_n^{(i)} & b_n^{(i)} \\ c_n^{(i)} & d_n^{(i)} \end{pmatrix}. \tag{3.2.253}$$

Theorem 44. *If (3.2.249) has two non degenerate generalized symmetries $U_{n,\epsilon} = G_n$ and $U_{n,\hat{\epsilon}} = \hat{G}_n$ of orders $m \geq 1$ and $m+1$, then it possesses a non degenerate formal symmetry L_n, such that $\mathrm{ord} L_n = 1$ and $\mathrm{lgt} L_n \geq m$, given by $L_n = \hat{G}_n^* (G_n^*)^{-1}$.*

Now let us write down the first two integrability conditions for systems of the form (3.2.249). Then we will be able to discuss the main theoretical problem in the case of the relativistic Toda type equations.

Note that we use here and below only the restriction f_n: $g_n^{(1)} = \frac{\partial f_n}{\partial v_{n+1}} \neq 0$ (see (3.2.246, 3.2.252)). The symmetrical case $g_n^{(-1)} \neq 0$ is reduced to this one by the change of variables $\tilde{u}_n = u_{-n}$, $\tilde{v}_n = v_{-n}$. It transforms the generalized symmetries into generalized symmetries, and so an integrable equation remains integrable.

Let us consider a formal symmetry of the form (3.2.253) such that $\mathrm{lgt} L_n \geq 2$. Using the condition $\det l_n^{(1)} \neq 0$ and multiplying, if necessary, L_n by a constant, we easily obtain from (3.2.250)

$$\begin{gathered} b_n^{(1)} = 0, \qquad a_n^{(1)} \neq 0, \qquad d_n^{(1)} = g_n^{(1)}, \\ c_n^{(1)} = f_n^{(1)} - a_n^{(1)} \varrho_n, \qquad b_n^{(0)} = 1 - a_n^{(1)} / g_n^{(1)}, \end{gathered} \tag{3.2.254}$$

where

$$\varrho_n = f_{n-1}^{(1)} / g_{n-1}^{(1)}. \tag{3.2.255}$$

On the next step we are led to the relations

$$D_t \log a_n^{(1)} = (S-1)\varrho_n, \tag{3.2.256}$$

$$D_t \log g_n^{(1)} = (S-1)(d_n^{(0)} - a_n^{(1)} \varrho_n / g_n^{(1)} - g_n^{(0)}) \tag{3.2.257}$$

which define the functions $a_n^{(1)}$, $d_n^{(0)}$ in an implicit way.

The first two integrability conditions for (3.2.249) require the existence of functions $a_n^{(1)}$, $d_n^{(0)}$ of the form (3.2.221) satisfying (3.2.256, 3.2.257). The condition (3.2.257) is analogous to the integrability conditions we have considered above. We can easily check it and find the unknown function $d_n^{(0)} - a_n^{(1)}\varrho_n/g_n^{(1)}$. The other condition is of a different sort: it has the form of a local conservation law with an unknown conserved density $\log a_n^{(1)}$. Here $D_t = \sum_i v_{n+i}\frac{\partial}{\partial u_{n+i}} + \sum_i f_{n+i}\frac{\partial}{\partial v_{n+i}}$, and it is unclear how to check this condition and how to find the density. The equations for $a_n^{(i)}$ are similar to (3.2.256), i.e. every new $a_n^{(i)}$ is defined by the conserved density. The relations for the functions $d_n^{(i)}$ depend on $a_n^{(i)}$, and consequently we cannot use these integrability conditions. This is a main theoretical problem. It will be solved by using a special exact solution $L_n = \Lambda_n$ of (3.2.250) which will be presented in Theorem 45. Such solution exists for any system of the form (3.2.249), i.e. is in some sense a trivial solution. There is an obvious trivial solution of (3.2.250): $L_n = I = IS^0$ which is the operator of multiplication by the unit matrix I. It turns out that Λ_n is one of the square roots of this operator.

Theorem 45. *There exists a unique solution Λ_n of (3.2.250),*

$$\Lambda_n = \sum_{i\le 0} \lambda_n^{(i)} S^i, \quad \lambda_n^{(i)} = \begin{pmatrix} \alpha_n^{(i)} & \beta_n^{(i)} \\ \gamma_n^{(i)} & \delta_n^{(i)} \end{pmatrix}, \quad \lambda_n^{(0)} = \begin{pmatrix} 1 & 0 \\ \gamma_n^{(0)} & -1 \end{pmatrix}, \tag{3.2.258}$$

such that $\Lambda_n^2 = I$.

PROOF. Let us introduce for the formal series (3.2.258) the following representation:

$$\Lambda_n = \sigma + \mathbb{R}_n + \mathbb{S}_n\,, \qquad \mathbb{R}_n = \sum_{i\le -1} r_n^{(i)} S^i, \qquad \mathbb{S}_n = \sum_{i\le 0} s_n^{(i)} S^i, \tag{3.2.259}$$

$$\sigma = \begin{pmatrix} 1 & 0 \\ 0 & -1 \end{pmatrix}, \quad r_n^{(i)} = \begin{pmatrix} \alpha_n^{(i)} & 0 \\ 0 & \delta_n^{(i)} \end{pmatrix}, \quad s_n^{(i)} = \begin{pmatrix} 0 & \beta_n^{(i)} \\ \gamma_n^{(i)} & 0 \end{pmatrix}, \tag{3.2.260}$$

where $\beta_n^{(0)} = 0$. Then we consider the equation $\Lambda_n^2 = E$ which is equivalent to the system

$$2\sigma\mathbb{R}_n + \mathbb{R}_n^2 + \mathbb{S}_n^2 = 0\,, \tag{3.2.261}$$

$$\mathbb{R}_n\mathbb{S}_n + \mathbb{S}_n\mathbb{R}_n = 0\,. \tag{3.2.262}$$

Using (3.2.261), we can express the series $\mathbb{R}_n$ via $\mathbb{S}_n$. Denoting

$$\chi_n^{(-1)} = 0\,, \qquad \chi_n^{(k)} = \sum_{k+1\le i\le -1} (r_n^{(i)} r_{n+i}^{(k-i)} + s_n^{(i)} s_{n+i}^{(k-i)})\,, \quad k \le -2\,,$$

and collecting all coefficients at the same powers of S, we obtain from (3.2.261) the following recurrent formulas for $r_n^{(k)}$

$$2\sigma r_n^{(k)} + \chi_n^{(k)} + s_n^{(0)} s_n^{(k)} + s_n^{(k)} s_{n+k}^{(0)} = 0\,, \quad k \le -1\,. \tag{3.2.263}$$

Eq. (3.2.263) gives the coefficients of $\mathbb{R}_n$ in an explicit and unique way.

Using again (3.2.261), we construct for $\mathbb{R}_n$ a further representation of the form: $\mathbb{R}_n = \sigma\sum_{i\ge 1} c_i \mathbb{S}_n^{2i}$ with constant coefficients c_i. It is well-defined, as $(s_n^{(0)})^2 = 0$ and thus $\mathbb{S}_n^{2i} = \sum_{j\le -i} \tilde{s}_n^{(j)} S^j$. Collecting in (3.2.261) coefficients at the same powers of $\mathbb{S}_n$, we find for the constants c_i the following recursion relation

$$2c_1 + 1 = 0\,, \qquad 2c_k + \sum_{1\le i\le k-1} c_i c_{k-i} = 0\,, \quad k \ge 2\,.$$

As the solution $\mathbb{R}_n$ of (3.2.261) is unique, we have in (3.2.263) another representation for the formal series $\mathbb{R}_n$. Now one can see that

$$\mathbb{R}_n\mathbb{S}_n + \mathbb{S}_n\mathbb{R}_n = \sigma[\sum_{i\geq 1} c_i\mathbb{S}_n^{2i}, \mathbb{S}_n]\,,$$

i.e. the series $\mathbb{R}_n$ satisfies also (3.2.262).

Let us consider (3.2.250). As in the scalar case (3.2.69), introducing the operator $A(L_n) = \dot{L}_n - [F_n^*, L_n]$, we obtain $A(\Lambda_n) = 0$. Let us consider separately its diagonal and anti-diagonal parts: $A(\Lambda_n)^{\|} = 0$ and $A(\Lambda_n)^{\perp} = 0$, respectively. Introducing the following scalar operators and formal series

$$f_n^{*,u} = f_n^{(1)}S + f_n^{(0)} + f_n^{(-1)}S^{-1}, \qquad f_n^{*,v} = g_n^{(1)}S + g_n^{(0)} + g_n^{(-1)}S^{-1},$$

$$A_n = \sum_{i\leq -1}\alpha_n^{(i)}S^i, \quad B_n = \sum_{i\leq -1}\beta_n^{(i)}S^i, \quad C_n = \sum_{i\leq 0}\gamma_n^{(i)}S^i, \quad D_n = \sum_{i\leq -1}\delta_n^{(i)}S^i,$$

one can rewrite the equation $A(\Lambda_n)^{\perp} = 0$ as:

(3.2.264)
$$\begin{aligned} &B_n f_n^{*,v} + \dot{B}_n + A_n - D_n + 2 = 0\,,\\ &f_n^{*,v}C_n - \dot{C}_n + f_n^{*,u}A_n - D_n f_n^{*,u} + 2f_n^{*,u} = 0\,, \end{aligned}$$

where $\alpha_n^{(i)}, \delta_n^{(i)}$ are expressed in terms of $\beta_n^{(i)}, \gamma_n^{(i)}$ by (3.2.263). From the different powers of S in the system (3.2.264) we obtain some recurrent and explicit formulas for the functions $\beta_n^{(i)}, \gamma_n^{(i)}$ which define the series $\mathbb{S}_n$ in a unique way.

Let us show that the formal series Λ_n defined by (3.2.263, 3.2.264) satisfies (3.2.250). It follows from the relations $A(\Lambda_n)^{\perp} = 0$, $\Lambda_n^2 = E$ and the fact that E is a solution of (3.2.250) that

(3.2.265)
$$A(E) = A(\Lambda_n^2) = \Lambda_n A(\Lambda_n) + A(\Lambda_n)\Lambda_n = \Lambda_n A(\Lambda_n)^{\|} + A(\Lambda_n)^{\|}\Lambda_n = 0$$

(see (3.2.71)). Here $B^{\|}$ and $B^{\perp}$ are the diagonal and antidiagonal parts of a matrix B. If $A(\Lambda_n)^{\|} \neq 0$, then this series is expressed as: $A(\Lambda_n)^{\|} = \sum_{i\leq l}\omega_n^{(i)}S^i$, where $\omega_n^{(i)}$ are the diagonal matrices and $\omega_n^{(l)} \neq 0$. Then, collecting coefficients at S^l in (3.2.265) and considering the diagonal part of the result, we are led to a contradiction: so $2\sigma\omega_n^{(l)} = 0$ (see (3.2.260)). □

All coefficients of the formal series (3.2.258) can be found explicitly using (3.2.263, 3.2.264). As an example let us write down a few formulas. Eq. (3.2.263) allow us to express $\alpha_n^{(i)}, \delta_n^{(i)}$ in terms of $\beta_n^{(i)}, \gamma_n^{(i)}$

$$\alpha_n^{(-1)} = -\tfrac{1}{2}\beta_n^{(-1)}\gamma_{n-1}^{(0)}\,, \qquad \delta_n^{(-1)} = \tfrac{1}{2}\gamma_n^{(0)}\beta_n^{(-1)},$$

$$\alpha_n^{(-2)} = -\tfrac{1}{2}(\alpha_n^{(-1)}\alpha_{n-1}^{(-1)} + \beta_n^{(-1)}\gamma_{n-1}^{(-1)} + \beta_n^{(-2)}\gamma_{n-2}^{(0)})\,,$$

$$\delta_n^{(-2)} = \tfrac{1}{2}(\delta_n^{(-1)}\delta_{n-1}^{(-1)} + \gamma_n^{(-1)}\beta_{n-1}^{(-1)} + \gamma_n^{(0)}\beta_n^{(-2)})\,.$$

Eq. (3.2.264) enable us to express $\beta_n^{(i)}, \gamma_n^{(i)}$ via the partial derivatives of f_n given by (3.2.252)

$$\beta_n^{(-1)} = -2/g_{n-1}^{(1)}\,, \qquad \gamma_n^{(0)} = -2f_{n-1}^{(1)}/g_{n-1}^{(1)}\,,$$

$$\beta_n^{(-2)} = -(\beta_n^{(-1)}g_{n-1}^{(0)} + \dot{\beta}_n^{(-1)} + \alpha_n^{(-1)} - \delta_n^{(-1)})/g_{n-2}^{(1)}\,,$$

$$\gamma_n^{(-1)} = -(g_{n-1}^{(0)}\gamma_{n-1}^{(0)} - \dot{\gamma}_{n-1}^{(0)} + f_{n-1}^{(1)}\alpha_n^{(-1)} - \delta_{n-1}^{(-1)}f_{n-2}^{(1)} + 2f_{n-1}^{(0)})/g_{n-1}^{(1)}\,.$$

Now we can derive the standard integrability conditions by constructing a convenient formal symmetry. Using the formal series Λ_n and L_n, presented in Theorems 44 and 45, we get

$$\Lambda_n^+ = \tfrac{1}{2}(E + \Lambda_n)\,, \qquad \Lambda_n^- = \tfrac{1}{2}(E - \Lambda_n)\,, \tag{3.2.266}$$

$$L_n^+ = \Lambda_n^+ L_n \Lambda_n^+\,, \qquad L_n^- = \Lambda_n^- L_n \Lambda_n^-\,. \tag{3.2.267}$$

It is clear that Λ_n^+ and Λ_n^- are the exact solutions of (3.2.250), while L_n^+ and L_n^- are the formal symmetries which have the same order and length as L_n.

Let us consider the formal symmetry L_n^- and use the same notations for its coefficients as for (3.2.253). It follows from (3.2.267) and (3.2.254) that $b_n^{(1)} = a_n^{(1)} = 0$, $d_n^{(1)} = g_n^{(1)}$. It is convenient to construct the other coefficients of L_n^- by using the relation $\Lambda_n^- L_n^- \Lambda_n^- = L_n^-$ which follows from the property $(\Lambda_n^-)^2 = \Lambda_n^-$. So we get

$$c_n^{(1)} = f_n^{(1)}, \qquad b_n^{(0)} = 1\,, \qquad a_n^{(0)} = \varrho_n\,,$$

with ϱ_n given in (3.2.255). In this way we can express any function of $a_n^{(i)}$, $b_n^{(i)}$, $c_n^{(i)}$ in terms of the functions (3.2.252) and $d_n^{(i)}$ with $i \le 0$.

The functions $d_n^{(i)}$ have to be found from (3.2.250). Equations for these functions will give the standard integrability conditions. In order to write those conditions in the form of conservation laws

$$D_t p_n^{(i)} = (S - 1) q_n^{(i)}, \tag{3.2.268}$$

we use, as in Section 3.2.8, the following relations

$$D_t \operatorname{tr} \operatorname{res} L_n^i \sim 0\,, \qquad 1 \le i \le \operatorname{lgt} L_n - 2\,, \tag{3.2.269}$$

valid for formal symmetries with $\operatorname{ord} L_n = 1$ and $\operatorname{lgt} L_n \ge 3$. Using the formal symmetry L_n^- with $\operatorname{lgt} L_n^- \ge 4$, we obtain three integrability conditions of the form (3.2.268), with $i = 1, 2, 3$.

Let us write down the resulting formulas in terms of the functions $q_n^{(i)}$ instead of $d_n^{(i)}$. We replace the function v_n by $\dot u_n$ in order to obtain integrability conditions for equations of the form (3.2.246). The conserved densities now read

$$p_n^{(1)} = \log \frac{\partial f_n}{\partial \dot u_{n+1}}\,, \qquad p_n^{(2)} = q_n^{(1)} + \frac{\partial f_n}{\partial \dot u_n} + \varrho_n\,, \tag{3.2.270}$$

$$p_n^{(3)} = q_n^{(2)} + \frac{1}{2}(p_n^{(2)})^2 + \frac{\partial f_n}{\partial \dot u_{n-1}} \frac{\partial f_{n-1}}{\partial \dot u_n} + \omega_n\,, \tag{3.2.271}$$

where

$$\varrho_n = \frac{\partial f_{n-1}}{\partial u_n} \left(\frac{\partial f_{n-1}}{\partial \dot u_n} \right)^{-1}, \qquad \omega_n = \frac{\partial f_n}{\partial u_n} - \frac{\partial f_n}{\partial \dot u_n} \varrho_n - \varrho_n^2 + D_t \varrho_n\,. \tag{3.2.272}$$

We have three standard necessary conditions for the integrability of (3.2.246) which require the existence of functions $q_n^{(1)}$, $q_n^{(2)}$, $q_n^{(3)}$ of the form

$$\varphi_n = \varphi(u_{n+k_1}, \dot u_{n+k_2}, u_{n+k_1-1}, \dot u_{n+k_2-1}, \dots u_{n+k_1'}, \dot u_{n+k_2'})\,, \tag{3.2.273}$$

where $k_1 \ge k_1'$ and $k_2 \ge k_2'$, such that (3.2.268, 3.2.270, 3.2.271) are satisfied.

These integrability conditions can be checked for functions of the form (3.2.273) by using the following statements

$$\frac{\delta \varphi_n}{\delta u_n} = \sum_i \frac{\partial \varphi_{n+i}}{\partial u_n} = 0\,, \qquad \frac{\delta \varphi_n}{\delta \dot u_n} = \sum_i \frac{\partial \varphi_{n+i}}{\partial \dot u_n} = 0$$

So φ_n is expressed as: $\varphi_n = c + (S-1)\psi_n$, where c is a constant, while ψ_n is another function of the form (3.2.273) with possibly different k_i, k_i'. The functions $q_n^{(i)}$ are defined in (3.2.268) up to arbitrary constants and one can chose those constants arbitrarily when testing the integrability of an equation. Calculating further coefficients of the formal symmetry L_n^-, we can continue checking a given equation for integrability and we can construct for an equation more conservation laws.

The case of the formal symmetry L_n^+, given in (3.2.267), is similar. However, it leads to integrability conditions of a different type. The relation

$$\Lambda_n^+ L_n^+ \Lambda_n^+ = L_n^+ \tag{3.2.274}$$

helps us to express $b_n^{(i)}$, $c_n^{(i)}$, $d_n^{(i)}$ via $a_n^{(i)}$, where $a_n^{(1)} \neq 0$. A few of the simplest resulting formulas read

$$b_n^{(1)} = d_n^{(1)} = 0\,, \quad c_n^{(1)} = -a_n^{(1)}\varrho_n\,, \quad b_n^{(0)} = -a_n^{(1)}/g_n^{(1)}, \quad d_n^{(0)} = a_n^{(1)}\varrho_n/g_n^{(1)},$$

with ϱ_n given in (3.2.255). Assuming that $\operatorname{lgt} L_n^+ \geq 3$, we obtain from (3.2.250) some equations for $a_n^{(1)}$ and $a_n^{(0)}$. In this way we derive two integrability conditions which will be rewritten in terms of (3.2.246)

$$D_t \hat p_n^{(i)} = (S-1)\hat q_n^{(i)}, \qquad i = 1, 2, \tag{3.2.275}$$

$$\hat q_n^{(1)} = \varrho_n\,, \qquad \hat q_n^{(2)} = \omega_n \left(\frac{\partial f_n}{\partial \dot u_{n+1}}\right)^{-1} \exp \hat p_n^{(1)}, \tag{3.2.276}$$

where we use the notations (3.2.272).

In this case we require the existence of functions $\hat p_n^{(1)}$ and $\hat p_n^{(2)}$ of the form (3.2.273), satisfying (3.2.275, 3.2.276). These integrability conditions are nonstandard, and it is an open problem how to use them. We have written down these conditions as an example of formulas which could be helpful in future studies.

Taking into account Theorem 44, we can formulate the following result:

Theorem 46. *If an equation of the form* (3.2.246) *has two non degenerate generalized symmetries of orders $m \geq 4$ and $m+1$, there exist functions $q_n^{(i)}$ with $i = 1, 2, 3$ and $\hat p_n^{(i)}$ with $i = 1, 2$ of the form* (3.2.273)*, which satisfy the conditions* (3.2.268, 3.2.270, 3.2.271, 3.2.275, 3.2.276)*.*

Let us consider as an example the relativistic Toda lattice (3.2.248). The integrability conditions (3.2.268, 3.2.270, 3.2.271) can easily be checked and, introducing the new functions

$$\phi(z) = 1/(1+e^{-z})\,, \qquad w_n = u_{n+1} - u_n\,, \tag{3.2.277}$$

we find

$$p_n^{(1)} = \log \dot u_n + \log \phi(w_n)\,, \qquad q_n^{(1)} = \dot u_n + \dot u_{n-1}\phi(w_{n-1})\,. \tag{3.2.278}$$

The second of these conditions is,

$$p_n^{(2)} = 2\dot u_n + (S-1) r_n\,, \qquad q_n^{(2)} = 2\dot u_{n-1} r_n + D_t r_n\,, \tag{3.2.279}$$

where $r_n = \dot{u}_n \phi(w_{n-1})$. It turns out that we can also solve in this case (3.2.275) for the functions $\hat{q}_n^{(i)}$ given by (3.2.276). We use below the function

$$\check{p}_n = \frac{1}{\dot{u}_n}(e^{w_n} + 1)(e^{w_{n-1}} + 1) \tag{3.2.280}$$

which is a conserved density of (3.2.248), as $D_t \check{p}_n = (1 - S)e^{w_{n-1}}$. There are the following solutions of (3.2.275)

$$\hat{p}_n^{(1)} = \log \check{p}_n + \log \phi(w_n), \qquad \hat{p}_n^{(2)} = 2\check{p}_n + (S-1)(\check{p}_n(2 - \phi(w_n))).$$

It should be remarked that $p_n^{(2)} \sim 2\dot{u}_n$, $\hat{p}_n^{(2)} \sim 2\check{p}_n$, while $\hat{p}_n^{(1)} \sim -p_n^{(1)}$, as

$$\hat{p}_n^{(1)} + p_n^{(1)} = (S-1)(u_n + u_{n-1} + \log \phi(w_{n-1})).$$

So, considering four integrability conditions, we have obtained three essentially different and nontrivial conserved densities: $p_n^{(1)}$, $\dot{u}_n$, $\check{p}_n$.

Let us briefly discuss the derivation of the integrability conditions for systems of the form (3.2.247). This class is quite similar to the class (3.2.249), only some formulas are different. For example, in the case of vector form (3.2.222), one has $F_n = \begin{pmatrix} f_n \\ g_n \end{pmatrix}$, hence

$$F_n^* = \begin{pmatrix} f_n^{(1)} & 0 \\ 0 & 0 \end{pmatrix} S + \begin{pmatrix} f_n^{(0)} & f_n^{(v)} \\ g_n^{(u)} & g_n^{(0)} \end{pmatrix} + \begin{pmatrix} 0 & 0 \\ 0 & g_n^{(-1)} \end{pmatrix} S^{-1},$$

where

$$f_n^{(i)} = \frac{\partial f_n}{\partial u_{n+i}}, \quad f_n^{(v)} = \frac{\partial f_n}{\partial v_n}, \qquad g_n^{(u)} = \frac{\partial g_n}{\partial u_n}, \quad g_n^{(i)} = \frac{\partial g_n}{\partial v_{n+i}}.$$

The Hamiltonian form (3.3.61) (see Section 3.3.3.1) of the relativistic Toda lattice (3.2.248), which belongs to the class (3.2.247), also possesses for any order $m \geq 1$ both degenerate and non degenerate generalized symmetries. That is why we also can suppose the existence of two non degenerate generalized symmetries of orders $m \geq 1$ and $m+1$ for relativistic Toda type equations. Then we obtain the complete analog of Theorem 44 in which we only replace the system (3.2.249) by (3.2.247).

For the non degenerate formal symmetry (3.2.253) with $\mathrm{lgt} L_n \geq 2$, one easily finds from (3.2.250)

$$b_n^{(1)} = c_n^{(1)} = 0, \qquad a_n^{(1)} = f_n^{(1)}, \qquad d_n^{(1)} \neq 0, \tag{3.2.281}$$

$$D_t \log d_n^{(1)} = (1-S) g_n^{(0)}, \qquad D_t \log f_n^{(1)} = (S-1)(a_n^{(0)} - f_n^{(0)}). \tag{3.2.282}$$

Eq. (3.2.282) provide us integrability conditions of two different types, where the first one is nonstandard. As in the previous case, the new conditions are unusable. An analog of Theorem 45 reads:

Theorem 47. *For any system (3.2.247), there exists a unique solution Λ_n of (3.2.250), such that*

$$\Lambda_n = \sum_{i \leq 0} \lambda_n^{(i)} S^i, \qquad \lambda_n^{(i)} = \begin{pmatrix} \alpha_n^{(i)} & \beta_n^{(i)} \\ \gamma_n^{(i)} & \delta_n^{(i)} \end{pmatrix}, \qquad \lambda_n^{(0)} = \begin{pmatrix} 1 & 0 \\ 0 & -1 \end{pmatrix},$$

and $\Lambda_n^2 = E$.

The proof of Theorems 45 and 47 are similar as well as the construction of the coefficients of Λ_n. For example, the first coefficients are given by

$$\alpha_n^{(-1)} = \delta_n^{(-1)} = 0\,, \qquad \alpha_n^{(-2)} = -\tfrac{1}{2}\beta_n^{(-1)}\gamma_{n-1}^{(-1)}, \qquad \delta_n^{(-2)} = \tfrac{1}{2}\gamma_n^{(-1)}\beta_{n-1}^{(-1)},$$
$$\beta_n^{(-1)} = 2f_{n-1}^{(v)}/f_{n-1}^{(1)}\,, \qquad \gamma_n^{(-1)} = 2g_n^{(u)}/f_{n-1}^{(1)}\,.$$

In order to obtain standard integrability conditions, like the second of conditions (3.2.282), we use in this case the formal symmetry L_n^+ given by (3.2.266, 3.2.267). Taking into account (3.2.281) and (3.2.274), we get

$$(3.2.283) \qquad l_n^{(1)} = \begin{pmatrix} f_n^{(1)} & 0 \\ 0 & 0 \end{pmatrix}, \qquad l_n^{(0)} = \begin{pmatrix} a_n^{(0)} & f_n^{(v)} \\ g_n^{(u)} & 0 \end{pmatrix},$$

where we used (3.2.253) for the coefficients of L_n^+. In this way one can express $b_n^{(i)}$, $c_n^{(i)}$, $d_n^{(i)}$ in terms of $a_n^{(i)}$ and f_n, g_n defining the system (3.2.247).

As in the previous case, we write down the equations for $a_n^{(i)}$, using (3.2.250) and (3.2.269). If $\operatorname{lgt} L_n^+ \geq 4$, such equations provide three standard integrability conditions, which will be presented in Section 3.3.3.2. It is interesting that formulas (3.2.283) are sufficient in this simple case to obtain all three integrability conditions.

3. Classification results

Here we present the classification results for lattice equations of the Volterra, Toda and relativistic Toda types. The classification theorems will be given together with the integrability conditions and the complete lists of integrable equations. The involved theorems are presented with no proof, but we refer to the original literature for them.

In the case of the classification of Volterra type equations we present in Appendix D a partial proof of the classification Theorem 49 contained in [841].

When necessary, Miura type transformations and master symmetries are written down in order to explain why those equations possess infinite hierarchies of generalized symmetries and conservation laws.

3.1. Volterra type equations. As proofs of the classification theorems are not included, we show at first some examples of simple classification problems in order to show how to carry out the calculations in this way.

3.1.1. *Examples of classification.* The following three problems are discussed here:

- How to find all generalized symmetries of given orders in the case of the Volterra equation.
- How to find for this equation all conservation laws of a given order.
- Classification problem for a simple class of equations including the Volterra and modified Volterra equations.

The problem of finding generalized symmetries.

Let us find for the Volterra equation (3.2.2) all generalized symmetries (2.4.10) of orders $m = 2$ and $m' = -2$. Here the equation is given by

$$(3.3.1) \qquad f_n = u_n(u_{n+1} - u_{n-1})\,,$$

and the right hand sides of symmetries are of the form

$$(3.3.2) \qquad g_n = g(u_{n+2}, u_{n+1}, u_n, u_{n-1}, u_{n-2})\,, \qquad \frac{\partial g_n}{\partial u_{n+2}}\frac{\partial g_n}{\partial u_{n-2}} \neq 0\,.$$

In order to find the function g_n, we are going to use the compatibility condition (3.2.7) and take into account the property (3.2.13).

The functions in (3.2.7) may depend only on the variables u_{n+j} with $-3 \le j \le 3$. Differentiating (3.2.7) with respect to u_{n+3}, one is led to the equation

$$\frac{\partial g_n}{\partial u_{n+2}} \frac{\partial f_{n+2}}{\partial u_{n+3}} = \frac{\partial f_n}{\partial u_{n+1}} \frac{\partial g_{n+1}}{\partial u_{n+3}}$$

which, after dividing by $u_n u_{n+1} u_{n+2}$, turns into

$$(S-1)\left(\frac{1}{u_n u_{n+1}} \frac{\partial g_n}{\partial u_{n+2}}\right) = 0 .$$

Using (3.2.13), we obtain $\frac{\partial g_n}{\partial u_{n+2}} = \alpha u_n u_{n+1}$, where α is a nonzero constant due to (3.3.2). So, the dependence of g_n on u_{n+2} is specified by

$$g_n = \alpha u_n u_{n+1} u_{n+2} + a_n , \qquad a_n = a(u_{n+1}, u_n, u_{n-1}, u_{n-2}) . \tag{3.3.3}$$

In the next step, we differentiate (3.2.7) with respect to u_{n+2}

$$D_t \frac{\partial g_n}{\partial u_{n+2}} + \frac{\partial g_n}{\partial u_{n+1}} \frac{\partial f_{n+1}}{\partial u_{n+2}} + \frac{\partial g_n}{\partial u_{n+2}} \frac{\partial f_{n+2}}{\partial u_{n+2}} = \frac{\partial f_n}{\partial u_{n+1}} \frac{\partial g_{n+1}}{\partial u_{n+2}} + \frac{\partial f_n}{\partial u_n} \frac{\partial g_n}{\partial u_{n+2}} . \tag{3.3.4}$$

Then we divide this relation by $u_n u_{n+1}$ and, using (3.3.3), obtain

$$(S-1)\left(\frac{1}{u_n} \frac{\partial a_n}{\partial u_{n+1}} - \alpha(2u_{n+1} + u_n)\right) = 0 .$$

Eq. (3.2.13) implies $\frac{\partial a_n}{\partial u_{n+1}} = \alpha(2u_n u_{n+1} + u_n^2) + \beta u_n$, where β is another constant. In such a way we specify the dependence of a_n on u_{n+1}

$$a_n = \alpha(u_n u_{n+1}^2 + u_n^2 u_{n+1}) + \beta u_n u_{n+1} + b_n , \qquad b_n = b(u_n, u_{n-1}, u_{n-2}) . \tag{3.3.5}$$

Compatibility condition (3.2.7) is symmetrical. In a quite similar way, differentiating (3.2.7) with respect to u_{n-3} and u_{n-2}, one can find the dependence on u_{n-2} and u_{n-1} of the functions appearing in (3.3.3, 3.3.5). Consequently one obtains the following formula for the function g_n

$$\begin{aligned} g_n = \alpha u_n u_{n+1}(u_{n+2} + u_{n+1} + u_n) + \gamma u_n u_{n-1}(u_n + u_{n-1} + u_{n-2}) + \\ \beta u_n u_{n+1} + \delta u_n u_{n-1} + c(u_n) \end{aligned} \tag{3.3.6}$$

with four arbitrary constants and one arbitrary function $c(u_n)$. Compatibility condition (3.2.7) takes now the form

$$\begin{aligned} &(\alpha + \gamma)u_n^2(u_{n+1}^2 - u_{n-1}^2 + f_n) + (\beta + \delta)u_n f_n + \\ &u_n(c(u_{n+1}) - c(u_{n-1})) + (u_{n+1} - u_{n-1})c(u_n) = f_n c'(u_n) , \end{aligned} \tag{3.3.7}$$

with f_n given by (3.3.1).

Applying to both sides of (3.3.7) the operator $\frac{\partial^4}{\partial u_n^2 \partial u_{n+1}^2}$, we obtain the restriction $\alpha + \gamma = 0$. Then, applying $\frac{\partial^3}{\partial u_n \partial u_{n+1}^2}$, we are led to the condition $c''(u_n) = 0$, i.e. $c(u_n) = c_1 u_n + c_2$. Dividing (3.3.7) by $u_{n+1} - u_{n-1}$, we obtain

$$(\beta + \delta)u_n^2 + c_1 u_n + c_2 = 0 .$$

Setting equal to zero all coefficients of this polynomial, we get g_n of the form (3.3.6) with

$$\gamma = -\alpha , \qquad \delta = -\beta , \qquad c(u_n) = 0 .$$

In the particular case $\alpha = 1$ and $\beta = 0$, one has the generalized symmetry (3.2.14). The general formula for the symmetry (3.2.3) of orders $m = 2$ and $m' = -2$ of the Volterra equation (3.2.2) is

$$u_{n,\epsilon'} = \alpha u_{n,\epsilon} + \beta \dot{u}_n\,, \qquad \alpha \neq 0\,.$$

Here α and β are arbitrary constants, $u_{n,\epsilon}$ is defined by generalized symmetry (3.2.14), and $\dot{u}_n$ is given by the Volterra equation itself.

First order conservation laws of the Volterra equation.

We look for conservation laws (3.2.15) of the first order with conserved densities p_n of the special form (3.2.38), i.e. such that

$$p_n = p(u_{n+1}, u_n)\,, \qquad \frac{\partial^2 p_n}{\partial u_{n+1}\partial u_n} \neq 0\,. \tag{3.3.8}$$

Let us introduce the function $a_n = D_t p_n$, then $a_n \sim 0$. Using the scheme of the proof of Theorem 22, we must be able to express this function in the form (3.2.26) with $b_n = 0$. In doing so, we will get a restriction on p_n. In fact,

$$a_n = \frac{\partial p_n}{\partial u_{n+1}}(u_{n+1}u_{n+2} - u_{n+1}u_n) + \frac{\partial p_n}{\partial u_n}(u_n u_{n+1} - u_n u_{n-1})\,,$$

and condition (3.2.30) is satisfied. According to Theorem 22 we can choose, for instance, $a_n^1 = \frac{\partial p_n}{\partial u_{n+1}} u_{n+1}u_{n+2}$. Then the function a_n^3 of (3.2.32) takes the form

$$a_n^3 = \left(\frac{\partial p_{n-1}}{\partial u_n} - \frac{\partial p_n}{\partial u_{n+1}} + \frac{\partial p_n}{\partial u_n}\right) u_n u_{n+1} - \frac{\partial p_n}{\partial u_n} u_n u_{n-1} = a^3(u_{n+1}, u_n, u_{n-1})\,.$$

It follows from (3.2.32) that $a_n \sim a_n^3 \sim 0$. Thus, due to (3.2.24), the following condition must be satisfied

$$\frac{\partial^2 a_n^3}{\partial u_{n+1}\partial u_{n-1}} = u_n(\Phi_{n-1} - \Phi_n) = 0\,, \qquad \Phi_n = \frac{\partial^2 p_n}{\partial u_{n+1}\partial u_n}\,. \tag{3.3.9}$$

This is a first restriction for the density p_n.

Now one can, applying the operator $-S\frac{1}{u_n}$ to (3.3.9), rewrite it as $(S-1)\Phi_n = 0$. This means that $\Phi_n = \alpha \in \mathbb{C}$, where $\alpha \neq 0$ due to (3.3.8). So, the conserved density p_n can be expressed as: $p_n = \alpha u_{n+1}u_n + \varphi(u_{n+1}) + \omega(u_n)$ or equivalently as

$$p_n = \alpha u_{n+1}u_n + \psi(u_n) + (S-1)\varphi(u_n)\,, \tag{3.3.10}$$

where $\psi(z) = \varphi(z) + \omega(z)$. The function ψ, which defines the nontrivial part of this conserved density, will be specified in the following.

The function $D_t p_n$ reads

$$D_t p_n = \Omega_n + (S-1)(\alpha u_{n+1}u_n u_{n-1} + \psi'(u_n)u_n u_{n-1} + D_t\varphi(u_n))\,, \tag{3.3.11}$$

where

$$\Omega_n = u_{n+1}u_n(\alpha u_{n+1} - \alpha u_n - \psi'(u_{n+1}) + \psi'(u_n))\,.$$

As $D_t p_n \sim \Omega_n \sim 0$, (3.2.24) implies

$$\frac{\partial^2 \Omega_n}{\partial u_{n+1}\partial u_n} = 2\alpha(u_{n+1} - u_n) - \Psi(u_{n+1}) + \Psi(u_n) = (S-1)(2\alpha u_n - \Psi(u_n)) = 0\,,$$

where $\Psi(z) = (z\psi'(z))'$. We are led to the following ODE for the function ψ: $\Psi(z) = (z\psi'(z))' = 2\alpha z + \beta$, where β is an arbitrary constant. Solving this ODE, we obtain a formula for ψ which depends on two other arbitrary constants

(3.3.12) $$\psi(z) = \frac{\alpha}{2}z^2 + \beta z + \gamma \log z + \delta\,.$$

The function Ω_n reads

(3.3.13) $$\Omega_n = (S-1)(\gamma u_n)\,.$$

One can see from (3.3.11, 3.3.13) that $D_t p_n \sim 0$, i.e. no more restriction for the density p_n will appear. Using (3.3.10-3.3.13), one obtains the following formulas for p_n and q_n defining conservation law (3.2.15)

(3.3.14) $$\begin{aligned} p_n &= \alpha p_n^3 + \beta p_n^2 + \gamma p_n^1 + \delta + (S-1)\varphi(u_n)\,, \qquad \alpha \neq 0\,, \\ q_n &= \alpha(u_{n+1}u_n u_{n-1} + u_n^2 u_{n-1}) + \beta u_n u_{n-1} + \\ &\quad \gamma(u_n + u_{n-1}) + \sigma + D_t\varphi(u_n)\,. \end{aligned}$$

Here p_n^j are given by (3.2.39), and σ is an arbitrary integration constant.

The functions p_n and q_n depend on five arbitrary constants and one arbitrary function φ, and the conserved density p_n is nothing but the linear combination of known conserved densities p_n^j and a trivial one $\delta + (T-1)\varphi$. Formulae (3.3.14) give the most general form of a conservation law with density (3.3.8) of the Volterra equation.

An example of classification problem.

The class of equations considered here is very simple, but it includes an integrable case apart from the Volterra equation. The classification problem is solved, using only integrability condition (3.2.56) and its corollary (3.2.61).

The starting point of our classification is the following class of lattice equations

(3.3.15) $$\dot{u}_n = P(u_n)(u_{n+1} - u_{n-1})\,,$$

where $P'(u_n) \neq 0$, as we are interested in the non linear equations. There is only one unknown function P here, and the aim is to find all (3.3.15) satisfying integrability condition (3.2.56).

Using the corollary (3.2.61) of (3.2.56), we find:

(3.3.16) $$p_n^{(1)} = \log P(u_n)\,, \qquad \dot{p}_n^{(1)} = P'(u_n)(u_{n+1} - u_{n-1})\,.$$

Now we can rewrite (3.2.61) as the relation

(3.3.17) $$P''(u_n)(u_{n+1} - u_{n-1}) + P'(u_{n-1}) - P'(u_{n+1}) = 0$$

which must be identically satisfied for all values of three independent variables u_{n+1}, u_n, u_{n-1}. Applying the operator $\frac{\partial^2}{\partial u_n \partial u_{n+1}}$, we see that $P'''(u_n) = 0$, i.e. P is the quadratic polynomial with arbitrary constant coefficients

(3.3.18) $$P(u_n) = \alpha u_n^2 + \beta u_n + \gamma\,.$$

With P given by (3.3.18), (3.3.17) and (3.2.61) are satisfied. Moreover, $\sigma = 0$ in the representation (3.2.62), i.e. (3.2.56) is also satisfied. This follows from the formula

$$\dot{p}_n^{(1)} = (-1)(2\alpha u_n u_{n-1} + \beta u_n + \beta u_{n-1})\,,$$

see (3.3.16, 3.3.18). So, the polynomial (3.3.18) describes all equations of the form (3.3.15) satisfying integrability condition (3.2.56).

Using the following linear point transformations $\tilde{u}_n = c_1 u_n + c_2$, where $c_1 \neq 0$ and c_2 are constants, one can transform any equation of the form (3.3.15, 3.3.18) into the Volterra equation (3.2.2) or into the modified Volterra equation (3.2.185). This means that, up to point transformations, the resulting list of integrable equations of the form (3.3.15) consists of the Volterra and modified Volterra equations. As it is known, (3.2.185) is transformed into (3.2.2) by the discrete Miura transformation (3.2.190), i.e. the list of integrable equations (3.3.15), up to Miura type transformations, is given by the Volterra equation only.

The classification has been finished in this simple case because we already know that the Volterra and modified Volterra equations are integrable and have infinite hierarchies of generalized symmetries and conservation laws.

3.1.2. *Lists of equations, transformations and master symmetries.* Let us discuss the classification of equations of the form (3.2.1). In this case only three integrability conditions (3.2.56, 3.2.130, 3.2.131) are used for the classification. The other two conditions (3.2.100, 3.2.106) are used for constructing simple conservation laws. For the sake of convenience, we present some of the above results, contained in Theorems 23, 33, 35, in the following summarizing theorem:

Theorem 48. *If (3.2.1) has one generalized symmetry of order $m_1 \geq 2$ and one conservation law of order $m_2 \geq 3$ or it possesses two conservation laws of orders $m_1 > m_2 \geq 3$, then it must satisfy the following three conditions:*

$$D_t \log \frac{\partial f_n}{\partial u_{n+1}} \sim 0 \,, \tag{3.3.19}$$

$$r_n = \log\left(-\frac{\partial f_n}{\partial u_{n+1}} \Big/ \frac{\partial f_n}{\partial u_{n-1}}\right) \sim 0 \,, \qquad D_t \sigma_n + 2\frac{\partial f_n}{\partial u_n} \sim 0 \,,$$

where $(S-1)\sigma_n = r_n$.

A list of equations satisfying these integrability conditions is written down below with no index n as (3.2.1) have no explicit dependence on the variable n. The Volterra equation (3.2.2) takes in such case the form: $\dot{u} = u(u_1 - u_{-1})$.

List of Volterra type equations

$$(\mathrm{V}_1) \qquad \dot{u} = P(u)(u_1 - u_{-1}),$$

$$(\mathrm{V}_2) \qquad \dot{u} = P(u^2)\left(\frac{1}{u_1 + u} - \frac{1}{u + u_{-1}}\right),$$

$$(\mathrm{V}_3) \qquad \dot{u} = Q(u)\left(\frac{1}{u_1 - u} + \frac{1}{u - u_{-1}}\right),$$

$$(\mathrm{V}_4) \qquad \dot{u} = \frac{R(u_1, u, u_{-1}) + \nu R(u_1, u, u_1)^{1/2} R(u_{-1}, u, u_{-1})^{1/2}}{u_1 - u_{-1}},$$

$$(\mathrm{V}_5) \qquad \dot{u} = y(u_1 - u) + y(u - u_{-1}) \,, \quad y' = P(y),$$

$$(\mathrm{V}_6) \qquad \dot{u} = y(u_1 - u)y(u - u_{-1}) + \mu \,, \quad y' = P(y)/y$$

$$(\mathrm{V}_7) \qquad \dot{u} = (y(u_1 - u) + y(u - u_{-1}))^{-1} + \mu \,, \quad y' = P(y^2),$$

$$(\mathrm{V}_8) \qquad \dot{u} = (y(u_1 + u) - y(u + u_{-1}))^{-1} \,, \quad y' = Q(y),$$

$$(\mathrm{V}_9) \qquad \dot{u} = \frac{y(u_1 + u) - y(u + u_{-1})}{y(u_1 + u) + y(u + u_{-1})} \,, \quad y' = P(y^2)/y,$$

$$(\mathrm{V}_{10}) \qquad \dot u = \frac{y(u_1+u)+y(u+u_{-1})}{y(u_1+u)-y(u+u_{-1})}\,, \quad y' = Q(y)/y,$$

$$(\mathrm{V}_{11}) \qquad \dot u = \frac{(1-y(u_1-u))(1-y(u-u_{-1}))}{y(u_1-u)+y(u-u_{-1})} + \mu\,, \quad y' = \frac{P(y^2)}{1-y^2}.$$

Here $\nu \in \{0, \pm 1\}$, the functions $P(u)$ and $Q(u)$ are polynomials of the form:

$$P(u) = \alpha u^2 + \beta u + \gamma\,, \tag{3.3.20}$$

$$Q(u) = \alpha u^4 + \beta u^3 + \gamma u^2 + \delta u + \pi\,, \tag{3.3.21}$$

while R is the following polynomial of three variables

$$R(u, v, w) = (\alpha v^2 + 2\beta v + \gamma)uw + (\beta v^2 + \lambda v + \delta)(u+w) + \gamma v^2 + 2\delta v + \pi. \tag{3.3.22}$$

Coefficients of P, Q, R and the number μ are arbitrary constants, the functions y are given by ODEs.

It should be remarked that, using n and t dependent transformations, one can reduce this list. For example, using the transformation $\tilde u_n = (-1)^n u_n$, one can rewrite (V_2) in the form (V_3), as

$$P(u^2) = \alpha u^4 + \beta u^2 + \gamma\,. \tag{3.3.23}$$

However, we do not discuss here transformations of this kind.

The form of (V_3) and (V_4) is invariant under the linear-fractional transformations

$$\tilde u_n = \frac{c_1 u_n + c_2}{c_3 u_n + c_4} \tag{3.3.24}$$

with constant coefficients. Only the coefficients of Q, R are changed, while the number ν remains unchanged. Eq. (V_4) when $\nu = 0$ is nothing but the YdKN equation. The particular case of (V_4) with $\nu = 0$

$$\dot u_n = \frac{(u_{n+1}-u_n)(u_n-u_{n-1})}{u_{n+1}-u_{n-1}}$$

is completely invariant under the action of transformations (3.3.24).

The classification is carried out up to point transformations of the form

$$\tilde u_n = s(u_n)\,, \qquad \tilde t = ct\,, \tag{3.3.25}$$

where $s' \neq 0$, and $c \neq 0$ is a constant. It can be shown that the integrability conditions (3.3.19) are invariant under (3.3.25). Generalized symmetries and conservation laws are transformed into generalized symmetries and conservation laws of the same order. The following theorem has been formulated in [842], and its complete proof can be found in [843] (see also the review articles [27, 850])

Theorem 49. *Eq. (3.2.1) satisfies (3.3.19) if and only if it can be written, using (3.3.25), as one of the equations* $(\mathrm{V}_1$-$\mathrm{V}_{11})$*.*

A partial proof of this theorem is contained in [841] whose English translation is presented in Appendix D.

The integrability of almost all equations of the list, except for (V_4) with $\nu = 0$, can be shown, using Miura type transformations. Eq. (V_6) with $P = \gamma \neq 0$ should be considered separately because it is linearizable. In this case the function y satisfies the ODE $y' = \gamma/y$ whose solution is: $y(z) = (2\gamma z + c)^{1/2}$, where c is an integration constant. An obvious point transformation (3.3.25) allows one to set $\gamma = 1/2$. Then the transformation $w_n = y(u_{n+1} - u_n)$ transforms (V_6) with $y' = \gamma/y$ into the linear equation (3.2.196).The conserved densities for (V_6) are constructed in the same way as we did for (3.2.171) in Section 3.2.7 (see formula (3.2.197)).

Many equations of the list are transformed into the Volterra equation (3.2.2) by non invertible Miura type transformations of the form

$$\tilde{u}_n = s(u_{n+k_1}, u_{n+k_1-1}, \dots u_{n+k_2}) , \tag{3.3.26}$$

where $k_1 > k_2$. Transformations (3.3.26) are very similar to the transformation (2.4.42) of Section 3.2.7. After the change of variables $\hat{u}_n = \tilde{u}_{n-k_2}$, (2.2.19) takes the form (3.2.188) with $k = k_1 - k_2$. However, one can prove that some of the equations, like for example (V_3), cannot be transformed into the Volterra equation in this way. In such case we need to use transformations of the form

$$\begin{aligned} \tilde{u}_n &= U(u_{n+k_1}, u_{n+k_1-1}, \dots u_{n+k_2}) , \\ \tilde{v}_n &= V(u_{n+k_3}, u_{n+k_3-1}, \dots u_{n+k_4}) , \end{aligned} \tag{3.3.27}$$

with $k_1 > k_2$ and $k_3 > k_4$, in order to reduce (V_3) to the system

$$\dot{u}_n = u_n(v_{n+1} - v_{n-1}) , \qquad \dot{v}_n = u_{n+1} - u_{n-1} . \tag{3.3.28}$$

Using (3.3.26, 3.3.27), one can construct the conserved densities in the same way as we did in Section 3.2.7, using (3.2.188).

Let us explain how one can obtain conserved densities for the system (3.3.28). This system is nothing but one of the forms of the Toda lattice. In fact, denoting $\tilde{u}_n = u_{2n}$, $\tilde{v}_n = v_{2n-1}$ or $\tilde{u}_n = u_{2n+1}$, $\tilde{v}_n = v_{2n}$, one obtains from (3.3.28) the following system

$$\dot{u}_n = u_n(v_{n+1} - v_n) , \qquad \dot{v}_n = u_n - u_{n-1} . \tag{3.3.29}$$

One can see that (3.3.28) consists of two copies of systems (3.3.29). On the other hand this is nothing else but Flaschka representation of Toda lattice (3.2.216,1.4.16) using (2.3.9) which, for the convenience of the reader we repeat here

$$\tilde{u}_n = e^{u_{n+1}-u_n} , \qquad \tilde{v}_n = \dot{u}_n . \tag{3.3.30}$$

So we can call (3.3.29) the Toda system as we did in Section 2.3.2 (2.3.8).

The Lax pair of the Toda system (3.3.28) can be rewritten for as, analogosly as (2.3.10, 2.3.11)

$$\begin{aligned} \dot{L}_n &= [A_n, L_n] , \\ L_n &= u_{n+1}^{1/2} S^2 + v_n + u_{n-1}^{1/2} S^{-2}, \qquad A_n = \frac{1}{2} u_{n+1}^{1/2} S^2 - \frac{1}{2} u_{n-1}^{1/2} S^{-2}. \end{aligned}$$

Conserved densities for (3.3.28) are obtained, using formulas (3.2.77, 3.2.78) of Section 3.2.3, as in the case of the Lax pair (3.2.82, 3.2.83).

It turns out that also the Volterra equation can be transformed into (3.3.28). We have the following general theorem:

Theorem 50. *Any non linear equation of the form* (V_1-V_{11})*, except for* (V_4) *with* $\nu = 0$ *and* (V_6) *with* $y' = \gamma / y$*, can be transformed into the system* (3.3.28) *by a transformation of the form* (3.3.27)*.*

PROOF. The Volterra equation (3.2.2) is transformed into (3.3.28) by: $\tilde{u}_n = u_{n+1} u_n$, $\tilde{v}_n = u_{n+1} + u_n$. As pointed out at the end of Section 3.3.1.1, the non linear equations (V_1) split into the Volterra equation (3.2.2) and the modified Volterra equation (3.2.185), using point transformations. Eq. (3.2.185) is transformed into the Volterra equation (3.2.2) by the discrete Miura transformation (3.2.190).

It is easy to check that transformations of the form $\tilde{u}_n = y(u_{n+1} + u_n)$ transform (V_9) into (V_2) and (V_8, V_{10}) into (V_3). The transformations $\tilde{u}_n = y(u_{n+1} - u_n)$ transform (V_5, V_6) into (V_1) and (V_7, V_{11}) into (V_2). So, we have to discuss now only three equations: (V_2), (V_3) and (V_4) with $\nu \neq 0$.

Equations of the form (V_2) defined by (3.3.23) split into two cases: $\alpha = 0$ and $\alpha \neq 0$. If $\alpha \neq 0$, then the polynomial (3.3.23) can be written as

$$P(u^2) = \alpha(u^2 - a^2)(u^2 - b^2)\,.$$

We can transform (V_2) into the Volterra equation in both cases

$$\tilde{u}_n = -\frac{P(u_n^2)}{(u_{n+1} + u_n)(u_n + u_{n-1})}\,, \qquad \tilde{u}_n = -\alpha\frac{(u_{n+1} + a)(u_n^2 - b^2)(u_{n-1} - a)}{(u_{n+1} + u_n)(u_n + u_{n-1})}\,.$$

The linear-fractional transformations (3.3.24) can be used to simplify (V_3, V_4). In the case of (V_3), one can obtain in this way $\alpha = 0$ in the polynomial (3.3.21). If $\beta = 0$, then we can transformation (V_3) into (3.2.2) by

$$\tilde{u}_n = \Omega_n\,, \qquad \Omega_n = -\frac{Q(u_n)}{(u_{n+1} - u_n)(u_n - u_{n-1})}\,.$$

In the case $\beta \neq 0$, (V_3) is transformed into the system (3.3.28) by the transformation

$$\tilde{u}_n = \Omega_{n+1}\Omega_n\,, \qquad \tilde{v}_n = \Omega_{n+1} + \Omega_n - \beta(u_{n+1} + u_n)\,.$$

In the case of (V_4) with $\nu \neq 0$, we introduce the function

$$\Delta_n = \frac{R(u_{n+1}, u_n, u_{n+1})^{1/2} + \nu R(u_{n-1}, u_n, u_{n-1})^{1/2}}{u_{n+1} - u_{n-1}}\,.$$

Using transformations (3.3.24), we obtain two cases for the polynomial (3.3.22): $\alpha = \beta = 0$ and $\alpha = 1$, $\beta = \gamma = 0$. In the first case, we can transform (V_4) into the Volterra equation by

$$\tilde{u}_n = -\tfrac{\nu}{2}(\Delta_{n+1} + \gamma^{1/2})(\Delta_n - \gamma^{1/2})\,.$$

In the second case, we have to consider two subcases: $\nu = -1$ and $\nu = 1$. In the first of them, we transform (V_4) into the Volterra equation, using

$$\tilde{u}_n = \tfrac{1}{2}(\Delta_{n+1} + u_{n+1})(\Delta_n - u_n)\,.$$

In the last case,

$$\tilde{u}_n = \tfrac{1}{4}(\Delta_{n+1} + u_{n+1})(\Delta_n^2 - u_n^2)(\Delta_{n-1} - u_{n-1})\,,$$

$$\tilde{v}_n = -\tfrac{1}{2}((\Delta_{n+1} - u_{n+1})(\Delta_n - u_n) + (\Delta_n + u_n)(\Delta_{n-1} + u_{n-1}))$$

enable us to transform (V_4) with $\nu = \alpha = 1$, $\beta = \gamma = 0$ into (3.3.28). □

The transformations presented above are given in their complete form in references [843, 850], some of them can also be found in [169, 170, 842, 847]. As for (V_4) with $\nu = 0$, we can state the following two results [843, 850]:

- Eq. (V_4) with $\nu = 0$ and R given by (3.3.22) is transformed into the Volterra equation (3.2.2) by a transformation of the form (3.3.26) if and only if it can be reduced, using the linear-fractional transformations (3.3.24), to the case $\alpha = \beta = 0$. Then the transformation is given by

$$\tilde{u}_n = A_n\,, \qquad A_n = -\frac{R(u_{n+1}, u_n, u_{n+1})}{(u_{n+2} - u_n)(u_{n+1} - u_{n-1})}\,.$$

- Eq. (V_4) is transformed into the system (3.3.28) by the Miura transformation (3.3.27) iff it can be reduced by (3.3.24) to the case $\alpha\gamma = \beta^2$, $\alpha\delta = \beta(\lambda - \gamma)$. In this case the transformation into (3.3.28) is given by

$$\tilde{u}_n = A_{n+1}A_n\,, \quad \tilde{v}_n = B_{n+1} + B_n\,, \quad B_n = A_n - \alpha u_{n+1}u_n - \beta(u_{n+1} + u_n)\,.$$

We see that, in general, (V_4) cannot be transformed into (3.3.28). So, up to Miura type transformations, we have in this section three non linear cases: the system (3.3.28), (V_4) with $\nu = 0$ and (V_6) with $y' = \gamma/y$.

The integrability of all (V_1-V_{11}) has been shown, using Miura type transformations, except for (V_4) with $\nu = 0$. In this case we can construct a local master symmetry [27]. The form of the master symmetry is simple

$$u_{n,\tau} = n\dot{u}_n\,, \tag{3.3.31}$$

where $\dot{u}_n$ is given by (V_4) with $\nu = 0$. However, we need to introduce an explicit dependence on the time τ of (3.3.31) into the master symmetry and equation itself, and we do that, following the example of (3.2.212) presented at the end of Section 3.2.7, which corresponds to (V_4) with $\nu = 0$, $\alpha = \beta = \gamma = \lambda = 0$, $\delta = 1$ and $\pi = c$.

In the case of (V_4) let the coefficients of the polynomial R (3.3.22) be functions of τ. Then we define a new polynomial ρ as

$$\begin{aligned}\rho(u,v) &= R(u,v,u)\\ &= \alpha u^2v^2 + 2\beta uv(u+v) + \gamma(u^2+v^2) + 2\lambda uv + 2\delta(u+v) + \sigma\,.\end{aligned} \tag{3.3.32}$$

The dependence on τ in (3.3.31) and in (V_4) with $\nu = 0$ is obtained by solving the PDE

$$2\frac{\partial\rho}{\partial\tau} = \rho\frac{\partial^2\rho}{\partial u\partial v} - \frac{\partial\rho}{\partial u}\frac{\partial\rho}{\partial v}\,. \tag{3.3.33}$$

In the left hand side of (3.3.33), we only differentiate the coefficients of ρ with respect to τ. The polynomial in the right hand side has the same form as ρ, but with different coefficients. Collecting coefficients corresponding to the same powers u^iv^j, we obtain from (3.3.33) a system of six ODEs for six coefficients of the polynomial ρ. That system has solutions $\alpha(\tau), \beta(\tau), \ldots$ for any initial conditions $\alpha(0) = \alpha_0, \beta(0) = \beta_0, \ldots$. Therefore, as in the case of (3.2.212), we can construct conservation laws and generalized symmetries of (V_4) with $\nu = 0$ for any given constant coefficient of the polynomial (3.3.22).

A more detailed discussion of the master symmetry of (V_4) with $\nu = 0$ as well as a general formula for the simplest generalized symmetry, constructed with the help of this master symmetry, can be found in [492].

One can see that the existence of a pair of conservation laws or of one generalized symmetry and one conservation law, with orders given as in Theorem 48, implies the existence of an infinite hierarchy of conservation laws. It turns out that the integrability conditions (3.3.19) are not only necessary but also sufficient for the integrability of (3.2.1). That is why these conditions can be used for testing a given equation for integrability. It is convenient to use for such testing an explicit form of the last of (3.3.19) which is given by (3.2.170).

The problem of constructing the generalized symmetries for all equations (V_1-V_{11}) remains open. However, many equations of the form (V_1-V_3) and (V_4) with $\nu \neq 0$ have local master symmetries (see [169, 170]) and therefore generalized symmetries. For example, (V_2) defined by the polynomial

$$P(u^2) = (1-u^2)(a^2 - b^2u^2)$$

has the following master symmetry

$$u_{n,\tau} = P(u_n^2)\left(\frac{n}{u_{n+1}+u_n} - \frac{n-1}{u_n+u_{n-1}}\right) + b^2 u_n(1-u_n^2)\,.$$

The τ dependence of the coefficients a and b is given by

$$a(\tau) = \lambda_1(\tau) - \lambda_2(\tau)\,, \qquad b(\tau) = \lambda_1(\tau) + \lambda_2(\tau)\,,$$

where both functions λ_j satisfy the same ODE, namely $\lambda_j' = \frac{1}{2}\lambda_j^3$.

The Lax pairs also can be constructed, if needed, for all integrable equations obtained by the generalized symmetry method. For instance, a Lax pair for (V_2, is given by 3.3.23) and can be found in [847].

3.2. Toda type equations. We discuss here the class of lattice equations (3.2.215) including the well-known Toda lattice (3.2.216,1.4.16). It should be remarked that a more narrow class, such that $\frac{\partial f_n}{\partial \dot{u}_n} = 0$, also contains (3.2.216), but has only one more integrable example that is a trivial modification of (3.2.216, 1.4.16), given by (T_3) below, and thus it is not interesting. The class (3.2.215) is the most simple nontrivial class of equations, including the Toda lattice, which is invariant under (3.3.25).

The integrability conditions [27, 845] necessary and sufficient for the exhaustive classification are

(3.3.34)
$$\begin{gathered} D_t p_n^{(i)} = (S-1)q_n^{(i)}, \qquad i = 1,2,3\,, \\ p_n^{(1)} = \log \frac{\partial f_n}{\partial u_{n+1}}\,, \qquad p_n^{(2)} = 2q_n^{(1)} + \frac{\partial f_n}{\partial \dot{u}_n}\,, \\ p_n^{(3)} = 2q_n^{(2)} - \frac{1}{2}D_t\frac{\partial f_n}{\partial \dot{u}_n} + \frac{1}{4}\left(\frac{\partial f_n}{\partial \dot{u}_n}\right)^2 + \frac{1}{4}(p_n^{(2)})^2 + \frac{\partial f_n}{\partial u_n}\,; \end{gathered}$$

(3.3.35)
$$\begin{gathered} r_n^{(j)} = (S-1)\sigma_n^{(j)}, \qquad j = 1,2\,, \\ r_n^{(1)} = \log\left(\frac{\partial f_n}{\partial u_{n+1}} \Big/ \frac{\partial f_n}{\partial u_{n-1}}\right), \qquad r_n^{(2)} = D_t\sigma_n^{(1)} + \frac{\partial f_n}{\partial \dot{u}_n}\,. \end{gathered}$$

In (3.3.34, 3.3.35) we require the existence of functions $q_n^{(i)}$, $\sigma_n^{(j)}$ depending on a finite number of independent variables u_{n+k}, $\dot{u}_{n+k}$. Using (3.3.34), one can construct for a given integrable equation also low order conservation laws.

Theorem 51. *Let us assume that* (3.2.215) *has one generalized symmetry of order* $m_1 \geq 5$ *and one conservation law of order* $m_2 \geq 4$ *or possesses two conservation laws of orders* $m_1 > m_2 \geq 7$*. Then this equation satisfies the conditions* (3.3.34, 3.3.35)*.*

Let us write down a complete list of lattice equations of the form (3.2.215) satisfying the conditions (3.3.34, 3.3.35). For simplicity, we write them down, using the notations: $u = u_n$, $u_1 = u_{n+1}$ and $u_{-1} = u_{n-1}$.

List of Toda type equations

(T_1)
$$\ddot{u} = P(\dot{u})(y(u_1 - u) - y(u - u_{-1}))\,, \qquad y' = Q(y)$$

(T_2)
$$\ddot{u} = (X(u) - \dot{u}^2)\left(\frac{1}{u_1 - u} - \frac{1}{u - u_{-1}}\right) + \frac{X'(u)}{2},$$

(T_3)
$$\ddot{u} = e^{u_1 - 2u + u_{-1}} + \mu,$$

(T_4)
$$\ddot{u} = (\dot{u}^2 - s(u))\left(\frac{1}{u_1 + u} + \frac{1}{u + u_{-1}}\right) + \frac{s'(u)}{2}.$$

Here P, Q, X and s are the following polynomials:

$$P(z) = \pi z^2 + az + b\,, \qquad Q(z) = \pi z^2 + cz + d\,, \tag{3.3.36}$$

$$X(z) = c_4 z^4 + c_3 z^3 + c_2 z^2 + c_1 z + c_0\,, \quad s(z) = c_4 z^4 + c_2 z^2 + c_0\,. \tag{3.3.37}$$

Coefficients of these polynomials and the number μ are arbitrary constants.

Eq. (T_2) with $X = 0$ has been studied in [143], and the same equation with a particular case of $X \neq 0$ has been considered in [453]. The list of equations (T_1-T_4) has been presented in [27, 755, 845, 850]. The following theorem and its proof can be found in [845, 848].

Theorem 52. *Eq.* (3.2.215) *satisfies the conditions* (3.3.34, 3.3.35) *if and only if it can be rewritten, up to point transformations* (3.3.25), *as one of the equations* (T_1-T_4).

If, instead of (3.3.25), we consider simple point transformations depending explicitly on n and t

$$\tilde{u}_n = s_n(t, u_n)\,, \qquad \tilde{t} = \Theta(t)\,, \tag{3.3.38}$$

we can reduce the number of arbitrary constants in the above list of equations and rewrite, for example, (T_1) in an explicit form. Namely, one can transform any equation of the class (T_1) into one of the following lattice equations:

Explicit form of (T_1)

$$\ddot{u} = e^{u_1 - u} - e^{u - u_{-1}}, \tag{Td$_1$}$$

$$\ddot{u} = \dot{u}(u_1 - 2u + u_{-1}), \tag{Td$_2$}$$

$$\ddot{u} = \dot{u}(e^{u_1 - u} - e^{u - u_{-1}}), \tag{Td$_3$}$$

$$\ddot{u} = (\alpha^2 - \dot{u}^2)\left(\frac{1}{u_1 - u} - \frac{1}{u - u_{-1}}\right), \tag{Td$_4$}$$

$$\ddot{u} = (\alpha^2 - \dot{u}^2)(\tanh(u_1 - u) - \tanh(u - u_{-1})). \tag{Td$_5$}$$

Here α is an arbitrary constant, (Td_1) is the Toda lattice (3.2.216, 1.4.16).

Moreover, the change of variables $\tilde{u}_n = (-1)^n u_n$ allows one to transform (T_4) into an equation of the form (T_2). The transformations (3.3.38), we use here, are invertible and enable us to rewrite solutions and, if necessary, generalized symmetries and conservation laws. Eq. (3.3.38) do not introduce any explicit dependence on the variables n and t into the generalized symmetries and conservation laws, but that will not be proved. For this reason, the integrability will be shown below only for (T_2, T_3) and (Td_1-Td_5).

Eqs. (T_1, T_2), and therefore (Td_1-Td_5), can be expressed in Hamiltonian and Lagrangian forms. This is useful from the viewpoint of physical applications. The Hamiltonian form is given by (3.2.241) (see also (3.2.225)), where $v_n = \dot{u}_n$. Eq. (3.2.245) can be used for constructing generalized symmetries. The function φ_n and Hamiltonian density h_n are given for (T_1) by

$$\varphi_n = P(v_n)\,, \qquad h_n = Y(u_{n+1} - u_n) + Z(v_n)\,, \tag{3.3.39}$$

$$Y'(z) = y(z)\,, \qquad Z'(z) = z/P(z)\,, \tag{3.3.40}$$

and in the case of (T_2) by

$$\varphi_n = v_n^2 - X(u_n)\,, \qquad h_n = \tfrac{1}{2}\log\varphi_n - \log(u_{n+1} - u_n)\,. \tag{3.3.41}$$

Eqs. (3.3.39-3.3.41) have been taken from [845].

Some of Toda type equations have Lagrangian forms. The Lagrangian form, discussed in details in Section 3.3.3.1, will be defined by (3.3.54, 3.3.55). In particular we have formulas (3.3.75, 3.3.76) for constructing two extra conservation laws. Here we only write down Lagrangians [27] of the form

$$L = R(\dot{u}_n, u_n) - Y(u_{n+1} - u_n) . \tag{3.3.42}$$

In the case of (T_1), the functions R and Y are defined by

$$R = R(\dot{u}_n) , \qquad R''(z) = 1/P(z) , \qquad Y'(z) = y(z) , \tag{3.3.43}$$

while in the case of (T_2), defining

$$X_+ = \sqrt{X(u_n)} + \dot{u}_n , \qquad X_- = \sqrt{X(u_n)} - \dot{u}_n ,$$

one has for $X \neq 0$:

$$R = \frac{X_+ \log X_+ + X_- \log X_-}{2\sqrt{X(u_n)}} , \qquad Y = \log(u_{n+1} - u_n) .$$

(T_2) with $X = 0$ is nothing but (Td_4) with $\alpha = 0$, and thus we can use (3.3.43).

Theorem 53. *Any equation of the form* (T_3) *or* (Td_1-Td_5) *can be transformed into the Toda system* (3.3.29) *by a Miura type transformation.*

PROOF. Eq. (T_3) is transformed into the Toda lattice (Td_1) by the following transformation: $\tilde{u}_n = u_{n+1} - u_n$. All ($Td_1$-$Td_5$) can be rewritten as systems of the form:

$$\begin{aligned} \dot{u}_n &= A(u_n)(v_{n+1} - v_n) , \\ \dot{v}_n &= B(v_n)(u_n - u_{n-1}) , \end{aligned} \tag{3.3.44}$$

with the following three possibilities:

$$\begin{array}{lll} \text{Case 1}: & A(z) = z, & B(z) = 1, \\ \text{Case 2}: & A(z) = z, & B(z) = z, \\ \text{Case 3}: & A(z) = z^2 - \alpha^2, & B(z) = z^2 - \beta^2. \end{array}$$

In fact, (Td_1) is transformed into Case 1 by (3.3.30). Transformations of (Td_2) into the same Case 1 and of (Td_3) into Case 2 are given by the Miura.

$$\tilde{u}_n = \dot{u}_n, \quad \tilde{v}_n = y(u_n - u_{n-1}),$$

where the function y is defined by (T_1). The transformation $\tilde{u}_n = \dot{u}_n$, $\tilde{v}_n = -y(u_n - u_{n-1})$ brings (Td_4, Td_5) into Case 3, where $\beta = 0$ for (Td_4) and $\beta = 1$ for (Td_5).

Moreover, Case 3 is transformed into Case 2 by the Miura

$$\tilde{u}_n = (u_n + \alpha)(v_{n+1} + \beta) , \qquad \tilde{v}_n = (u_n - \alpha)(v_n - \beta) ,$$

and Case 2 is transformed into Case 1 by the transformation $\tilde{u}_n = u_n v_{n+1}$, $\tilde{v}_n = u_n + v_n$. As (3.3.29) corresponds to Case 1, the Theorem is proved. □

All transformations contained in the proof of Theorem 53 can be found in [756]. It turns out that we can write down local master symmetries for all three cases associated to system (3.3.44) (see [495, 654]). Those master symmetries are of the form

$$\begin{aligned} u_{n,\tau} &= A(u_n)((2n + k)v_{n+1} - 2nv_n) + \gamma u_n^2 , \\ v_{n,\tau} &= B(v_n)((2n - 1 + k)u_n - (2n - 1)u_{n-1}) + \delta v_n^2 , \end{aligned} \tag{3.3.45}$$

where

$$\begin{array}{llll} \text{Case 1}: & k=4, & \gamma=0, & \delta=1, \\ \text{Case 2}: & k=3, & \gamma=1, & \delta=1, \\ \text{Case 3}: & k=2, & \gamma=0, & \delta=0. \end{array}$$

In order to construct new conserved densities, using (3.2.209), we need a starting conserved density. The systems (3.3.44) have two obvious densities

$$p_n^+ = \int \frac{du_n}{A(u_n)}, \qquad p_n^- = \int \frac{dv_n}{B(v_n)}.$$

In Cases 1 and 2, both can be used as starting points. In Case 1, a master symmetry (3.3.45) provides the system (3.3.44) with the following conserved density

$$D_\tau p_n^+ \sim 2p_n^-, \quad D_\tau p_n^- \sim 2u_n + v_n^2,$$

and in Case 2, we are led to

$$D_\tau p_n^+ \sim D_\tau p_n^- \sim u_n + v_n .$$

As $D_\tau p_n^+ \sim D_\tau p_n^- \sim 0$, in Case 3 one needs to start from a different density

$$p_n^0 = \log(u_n+\alpha) + \log(v_n+\beta), \qquad D_\tau p_n^0 \sim u_n(v_{n+1}+v_n).$$

We can construct the conserved densities not only for (3.3.44) but also for (T_3, Td_1-Td_5), using the transformations given in the proof of Theorem 53. (Td_1-Td_5) are Hamiltonian, therefore we can obtain for them also generalized symmetries.

Let us now discuss (T_2). We do not know how to transform it into the Toda lattice. However, there is a connection with (V_4), with $\nu = 0$, which has been considered in Section 3.3.1.2. In fact, introducing $\tilde{u}_n = u_{2n}$, $\tilde{v}_n = u_{2n-1}$ and $\tilde{t} = t/2$, we can pass from (V_4) with $\nu = 0$ to the system

$$\dot{u}_n = \frac{2\rho}{v_{n+1}-v_n} + \rho_{v_n}, \quad \dot{v}_n = \frac{2\rho}{u_n - u_{n-1}} - \rho_{u_n}, \quad \rho = \rho(u_n, v_n). \tag{3.3.46}$$

In (3.3.46) we denote by indexes partial derivatives of ρ. ρ is defined by (3.3.32, 3.3.22). The function ρ is a quadratic polynomial of each variable, and if one considers the discriminator $\mathcal{D}$ with respect to v_n

$$\mathcal{D} = (\rho_{v_n})^2 - 2\rho\rho_{v_n v_n}, \tag{3.3.47}$$

one obtains a function of just u_n. Besides, $\mathcal{D}$, given by (3.3.47), is a fourth degree polynomial. Differentiating the first of (3.3.46) with respect to t and then using the second one, it is possible to express the result only in terms of u_{n+j}, $\dot{u}_{n+j}$ and to rewrite it as (T_2) with $\mathcal{D}$ given by (3.3.47). This is why for any solution (u_n, v_n) of the system (3.3.46), the function u_n satisfies an equation of the form (T_2), with $\mathcal{S}$ given by (3.3.37). However, this connection cannot be used for constructing local conservation laws and generalized symmetries of (T_2).

We can construct local conservation laws and generalized symmetries of (T_2) using a master symmetry [27]. Passing from (T_2) to an equivalent system for the functions u_n and $v_n = \dot{u}_n$, we can write the master symmetry as

$$u_{n,\tau} = (\lambda + 2n)v_n, \qquad v_{n,\tau} = (\sigma + 2n)\dot{v}_n + u_{n,t'}. \tag{3.3.48}$$

Here σ is an arbitrary constant, $\dot{v}_n$ is given by (T_2), while $u_{n,t'}$ by

$$u_{n,t'} = (X(u_n) - v_n^2)\left(\frac{1}{u_{n+1}-u_n} + \frac{1}{u_n - u_{n-1}}\right).$$

This equation with $v_n = \dot{u}_n$ is nothing but the generalized symmetry of (T_2). Conserved densities can be constructed, starting for instance from the Hamiltonian density h_n given by (3.3.41).

If the number σ in (3.3.48) is not an even integer, then we can exclude the function v_n from the second of (3.3.48), using the first one. In this way one obtains from the master symmetry an n-dependent second order DΔE

$$u_{n,\tau\tau} = \left(\eta X(u_n) - \frac{u_{n,\tau}^2}{\eta}\right)\left(\frac{\eta+1}{u_{n+1}-u_n} - \frac{\eta-1}{u_n-u_{n-1}}\right) + \frac{\eta^2}{2}X'(u_n), \tag{3.3.49}$$

where $\eta = \sigma + 2n$. Master symmetries are known to be integrable in some sense (see, in the case of lattice equations, e.g. [106, 494, 495, 795]), and (3.3.49) exemplifies a nice equation of this kind.

Eqs. (3.3.34, 3.3.35) are necessary and sufficient conditions for the integrability and can be used as a testing tool for (3.2.215). As in Section 3.2.5.2, in the case of Volterra type equations, all five conditions (3.3.34, 3.3.35) can be rewritten in an explicit form convenient for such testing [852].

As in the case of (V_3, V_4) considered in Section 3.3.1.2, the form of (T_2) is invariant under the linear-fractional transformations (3.3.24). Only the coefficients of S (3.3.37) are changed by these transformations. Equations of this kind can appear in practice and are expressed in terms of elliptic functions. Let us consider an interesting example of equations of this kind of Toda type [453]

$$\begin{aligned} \ddot{u}_n &= (\dot{u}_n^2 - 1)(\zeta(u_n + u_{n+1}) + \zeta(u_n - u_{n+1}) \\ &\quad + \zeta(u_n + u_{n-1}) + \zeta(u_n - u_{n-1}) - 2\zeta(2u_n)), \end{aligned} \tag{3.3.50}$$

where by ζ we mean the ζ-function of Weierstrass. Recall that $\zeta'(z) = -\wp(z)$, where the $\wp$-function of Weierstrass is defined by the ODE $\wp'^2(z) = 4\wp^3(z) + \alpha\wp(z) + \beta$ with constants coefficients. Standard formulas for elliptic functions allow us to rewrite (3.3.50) as

$$\ddot{u}_n = (\dot{u}_n^2 - 1)\left(\frac{\wp'(u_n)}{\wp(u_n) - \wp(u_{n+1})} + \frac{\wp'(u_n)}{\wp(u_n) - \wp(u_{n-1})} - \frac{\wp''(u_n)}{\wp'(u_n)}\right).$$

Introducing $\tilde{u}_n = \wp(u_n)$, we can transform this equation into (T_2) with $S(z) = 4z^3 + \alpha z + \beta$.

3.3. Relativistic Toda type equations. We discuss in this section lattice equations of the following two classes

$$\begin{gathered} \ddot{u}_n = f(u_{n+1}, u_n, \dot{u}_{n+1}, \dot{u}_n) - g(u_n, u_{n-1}, \dot{u}_n, \dot{u}_{n-1}), \\ f_{\dot{u}_{n+1}} g_{\dot{u}_{n-1}} \neq 0, \end{gathered} \tag{3.3.51}$$

$$\begin{gathered} \dot{u}_n = f(u_{n+1}, u_n, v_n), \qquad \dot{v}_n = g(v_{n-1}, v_n, u_n), \\ f_{u_{n+1}} f_{v_n} g_{v_{n-1}} g_{u_n} \neq 0, \end{gathered} \tag{3.3.52}$$

where partial derivatives of f, g are denoted by indexes. Each of these classes is important in itself and has its own applications [27]. As it will be shown below, there is a nontrivial connection between them [27]. All (3.3.51) will be Lagrangian, while systems of the form (3.3.52) will correspond to Hamiltonian systems. The relativistic Toda lattice (3.2.248) is of the form (3.3.51). Other integrable equations of the form (3.3.51, 3.3.52) have analogous algebraic properties and are called relativistic Toda type equations for this reason. The Lagrangian and Hamiltonian equations of this kind have been discussed in [25, 27, 219, 222, 409, 587, 755, 845].

In Section 3.3.3.1 we discuss a non point transformation connection between the Lagrangian equations (see [27, 849]) and at the end give some useful remarks about generalized symmetries and conservation laws of the Lagrangian equations [849]. Then, in Sections 3.3.3.2 and 3.3.3.3, we separately describe these Hamiltonian and Lagrangian forms and give two lists of integrable equations (H_1-H_3) and (L_1, L_2) together with some classification theorems and integrability conditions. In Section 3.3.3.4 we point out the exact correspondence between equations of two lists and show in Section 3.3.3.5, by constructing the master symmetries, that all of them possess generalized symmetries and conservation laws.

3.3.1. *Non point connection between Lagrangian and Hamiltonian equations, and properties of Lagrangian equations.* At first, let us recall some well-known facts of classical mechanics. Given a Lagrangian function $L = L(u, \dot{u})$, the Euler-Lagrange equation is

$$\frac{d}{dt}\frac{\partial L}{\partial \dot{u}} = \frac{\partial L}{\partial u}\,, \qquad \frac{\partial^2 L}{\partial \dot{u}^2} \neq 0\,. \tag{3.3.53}$$

If we introduce the function $v = L_{\dot{u}}$, we can express $\dot{u}$ in terms of u and v, i.e. we have an invertible transformation: $(u, \dot{u}) \leftrightarrow (u, v)$. The Legendre transformation $H = v\dot{u} - L$ defines a relation between the Lagrangian L and Hamilton's function $H = H(u, v)$ and leads to the equations of Hamilton

$$\dot{u} = \frac{\partial H}{\partial v}\,, \qquad \dot{v} = -\frac{\partial H}{\partial u}\,.$$

It can be easily proved that the Euler-Lagrange equation and Hamilton equations are equivalent. We give such proof below in a more general case.

Let us now consider Lagrangians, depending on a field u_n living on the lattice, of the form

$$L = L(\dot{u}_n, u_{n+1}, u_n)\,, \qquad \frac{\partial^2 L}{\partial \dot{u}_n^2} \neq 0\,. \tag{3.3.54}$$

The Euler-Lagrange equation is then defined as

$$\frac{d}{dt}\frac{\partial L}{\partial \dot{u}_n} = \frac{\partial}{\partial u_n}(1 + S^{-1})L\,. \tag{3.3.55}$$

On the right hand side we have the formal variational derivative $\frac{\delta L}{\delta u_n}$ (see (3.2.225)). The relativistic Toda lattice (3.2.248) is an Euler-Lagrange equation of the form (3.3.54, 3.3.55) [25] with the Lagrangian

$$L = \dot{u}_n \log \frac{\dot{u}_n}{e^{u_{n+1}-u_n} + 1}\,. \tag{3.3.56}$$

The Legendre transformation

$$H = v_n\dot{u}_n - L\,, \qquad v_n = L_{\dot{u}_n} \tag{3.3.57}$$

leads in this case to an invertible change of variables between the two sets of variables: $\{u_n, \dot{u}_n\}$ and $\{u_n, v_n\}$. The formula for v_n has the form $v_n = s(\dot{u}_n, u_{n+1}, u_n)$, while u_n remains unchanged. As $s_{\dot{u}_n} \neq 0$ due to (3.3.54), then $\dot{u}_n$ easily can be expressed via v_n, u_{n+1}, u_n. This is a non point transformation because the function s depends also on u_{n+1}.

It can be easily proved that the Legendre transformation (3.3.57) gives the following Hamiltonian system

$$\dot{u}_n = \frac{\delta H}{\delta v_n}\,, \qquad \dot{v}_n = -\frac{\delta H}{\delta u_n}\,, \qquad H = H(v_n, u_{n+1}, u_n)\,, \tag{3.3.58}$$

$$\frac{\delta H}{\delta v_n} = \frac{\partial H}{\partial v_n}\,, \qquad \frac{\delta H}{\delta u_n} = \frac{\partial}{\partial u_n}(1+S^{-1})H \tag{3.3.59}$$

[see (3.2.225) for the definition of the variational derivative contained in (3.3.59, 3.3.58)]. Comparing (3.3.58) with (3.2.241), one can see that we have substituted h_n by H to be more closed to the classical formulas.

All known integrable non linear equations (3.3.51) have the Lagrangian structure (3.3.54, 3.3.55). In those cases, it is possible not only to pass to the Hamiltonian system (3.3.58) but also to transform (3.3.58) into the simpler form (3.3.52). This can be done, using an additional point transformation of the form: $\hat{u}_n = a(u_n)$, $\hat{v}_n = b(u_n, v_n)$. The Hamiltonian H (or the Hamiltonian density) is simplified, and the Hamiltonian system now reads

$$\dot{u}_n = \varphi(u_n, v_n)\frac{\delta H}{\delta v_n}\,, \qquad \dot{v}_n = -\varphi(u_n, v_n)\frac{\delta H}{\delta u_n}\,, \qquad H = \Phi(u_n, v_n) + \Psi(u_{n+1}, v_n)\,. \tag{3.3.60}$$

For example, in the case of the relativistic Toda lattice equation (3.2.248), the additional point transformation is: $\hat{u}_n = e^{u_n}$, $\hat{v}_n = e^{v_n - u_{n-1}}$. The resulting system of the form (3.3.52) is

$$\dot{u}_n = u_n v_n(u_{n+1} + u_n)\,, \qquad \dot{v}_n = -u_n v_n(v_n + v_{n-1})\,. \tag{3.3.61}$$

This system, equivalent to (3.2.248), has the Hamiltonian structure (3.3.60) with $\varphi = u_n v_n$ and $H = u_n v_n + u_{n+1} v_n$. The invertible transformation from (3.2.248) into (3.3.61) is given by

$$\hat{u}_n = e^{u_n}\,, \qquad \hat{v}_n = \frac{\dot{u}_n}{e^{u_{n+1}} + e^{u_n}}\,. \tag{3.3.62}$$

Let us discuss in the following theorem how to pass from the Hamiltonian system (3.3.60) to the Lagrangian equation (3.3.54, 3.3.55).

Theorem 54. *If* (u_n, v_n) *is a solution of* (3.3.52, 3.3.60) *with* $f_{v_n} \neq 0$ *and with Hamiltonian* H *of the general form* $H(u_{n+1}, u_n, v_n)$*, then the functions* y_n, z_n *are given by*

$$y_n = u_n\,, \qquad z_n = \dot{u}_n = f(u_{n+1}, u_n, v_n) \tag{3.3.63}$$

satisfy the equation

$$\frac{d}{dt}\frac{\partial L}{\partial z_n} = \frac{\partial}{\partial y_n}(1+S^{-1})L\,, \tag{3.3.64}$$

with Lagrangian $L(z_n, y_{n+1}, y_n)$ *defined by*

$$L = \psi(u_n, v_n)\dot{u}_n - H\,, \qquad \psi_{v_n} = 1/\varphi\,. \tag{3.3.65}$$

PROOF. The invertible transformation (3.3.63) implies the following relations for the partial derivatives

$$\frac{\partial}{\partial v_n} = f_{v_n}\frac{\partial}{\partial z_n}\,, \qquad \frac{\partial}{\partial u_n} = \frac{\partial}{\partial y_n} + f_{u_n}\frac{\partial}{\partial z_n} + S^{-1}(f_{u_{n+1}})\frac{\partial}{\partial z_{n-1}}\,. \tag{3.3.66}$$

Eqs. (3.3.65, 3.3.66) allow us to find

$$L_{z_n} = \psi\,, \qquad L_{y_n} = \psi_{u_n} f - H_{u_n}\,, \qquad L_{y_{n+1}} = -H_{u_{n+1}}\,, \tag{3.3.67}$$

where we have used the relation $f = \varphi H_{v_n}$ which comes from (3.3.52, 3.3.59, 3.3.60). Then (3.3.64) follows from (3.3.52, 3.3.59, 3.3.60, 3.3.67) and $\psi_{v_n} = 1/\varphi$ from

$$D_t L_{z_n} = D_t\psi = \psi_{u_n} f + \psi_{v_n} g = \psi_{u_n} f - (H_{u_n} + S^{-1}H_{u_{n+1}}) = L_{y_n} + S^{-1}L_{y_{n+1}}\,.$$

We can see that

$$L_{z_n z_n} = (\varphi f_{v_n})^{-1} \neq 0 \,, \tag{3.3.68}$$

and, due to (3.3.63), the equivalence of (3.3.64) and (3.3.54, 3.3.55) is obvious. □

By a point transformation

$$\tilde{u}_n = s(u_n), \tag{3.3.69}$$

one can change an equation and its Lagrangian, but the form of Euler-Lagrange equation (3.3.54, 3.3.55) remains unchanged. On the other hand, one can introduce a new Lagrangian

$$\tilde{L} = \alpha L + \beta + \sigma(u_n)\dot{u}_n + (S-1)\omega(u_n) \,, \tag{3.3.70}$$

where $\alpha \neq 0$ and β are constants, while σ and ω are arbitrary functions. In this case, not only the Lagrangian structure (3.3.54, 3.3.55) but also the corresponding lattice equation is not changed by (3.3.69).

Using Theorem 54, we pass from the system (3.3.52, 3.3.60) to the Lagrangian equation (3.3.54, 3.3.55). We obtain in general an equation of the form (3.2.246). However, in known integrable cases, the resulting form of the Lagrangian is

$$L = L_0(\dot{u}_n, u_n) + \dot{u}_n V(u_{n+1}, u_n) + U(u_{n+1}, u_n) \,. \tag{3.3.71}$$

An equation with such Lagrangian is of the class (3.3.51). Such form of equations and Lagrangians is invariant under the transformation (3.3.69).

Let us consider as an example the well-known lattice system

$$\dot{u}_n = u_{n+1} + u_n^2 v_n \,, \qquad \dot{v}_n = -v_{n-1} - v_n^2 u_n \,. \tag{3.3.72}$$

This system together with the Lax pair and Hamiltonian structure can be found in the papers [601,755,866] and preprints [685,753]. Its Hamiltonian structure (3.3.60) is defined by the functions $\varphi = 1$, $H = u_{n+1}v_n + \frac{1}{2}u_n^2 v_n^2$. Using Theorem 54, we obtain a Lagrangian equation which, by the point transformation $\tilde{u}_n = \log u_n$, can be rewritten as

$$\ddot{u}_n = \dot{u}_{n+1} e^{u_{n+1}-u_n} - \dot{u}_{n-1} e^{u_n - u_{n-1}} - e^{2(u_{n+1}-u_n)} + e^{2(u_n - u_{n-1})}. \tag{3.3.73}$$

Eq. (3.3.73) corresponds to the Lagrangian $L = (\dot{u}_n - e^{u_{n+1}-u_n})^2$. The invertible transformation of the system (3.3.72) into (3.3.73) is

$$\tilde{u}_n = \log u_n \,, \qquad \tilde{u}_{n,t} = \frac{u_{n+1}}{u_n} + u_n v_n \,. \tag{3.3.74}$$

Let discuss now the conservation laws of Lagrangian equations. In the classical case given by (3.3.53), one has the constant of motion $I_1 = \dot{u} L_{\dot{u}} - L$. Indeed, using (3.3.53), one can easily prove that $dI_1/dt = 0$. If $L_u = 0$, $I_2 = L_{\dot{u}}$ is another constant of motion. Passing to lattice equations (3.3.54, 3.3.55), we have local conservation laws instead of constants of motion.

The Hamiltonian H is always a conserved density for the system (3.3.60), as we have shown at the very end of Section 3.2.8. Rewriting H in terms of the variables (3.3.63), one is lead to a conserved density for the Lagrangian equation. Using (3.3.65, 3.3.67), we obtain

$$H = \psi \dot{u}_n - L = \dot{u}_n L_{\dot{u}_n} - L,$$

and this is the conserved density of the Lagrangian equation. Indeed, one can easily check that for (3.3.55) the following conservation law takes place:

$$D_t(\dot{u}_n L_{\dot{u}_n} - L) = (S^{-1} - 1)(\dot{u}_{n+1} L_{u_{n+1}}) \,. \tag{3.3.75}$$

If the Lagrangian has the form $L = L(\dot{u}_n, u_{n+1} - u_n)$, we have another conservation law for this equation given by

$$D_t L_{\dot{u}_n} = (1 - S^{-1}) L_{u_n} . \tag{3.3.76}$$

As in the classical case, there is for the Lagrangian equations (3.3.54, 3.3.55) the standard Noether's connection between conservation laws and symmetries. The construction of a local conservation law, starting from a generalized symmetry, is discussed in [25, 219]. However, we are more interested here in the passage from conservation laws to symmetries [849]. So in the following, using the equivalence of Lagrangian and Hamiltonian equations, we write down a simple formula for constructing generalized symmetries.

Let us consider the Hamiltonian systems (3.3.60) and the Euler-Lagrange equations (3.3.54), (3.3.55) related by Theorem 54. This is the general case, as the Hamiltonian H has the general form $H(u_{n+1}, u_n, v_n)$. Let p be a conserved density of the Euler-Lagrange equation of the form

$$p = p(\dot{u}_{n+i_1}, \dot{u}_{n+i_1-1}, \dots \dot{u}_{n+i_2}, u_{n+j_1}, u_{n+j_1-1}, \dots u_{n+j_2}) , \tag{3.3.77}$$

where $i_1 \geq i_2$, $j_1 \geq j_2$. Using the invertible transformation (3.3.63), we can pass to a density $\hat{p}$ of the corresponding Hamiltonian system which depends on the variables u_{n+i}, v_{n+i}. By changing p, we only replace $\dot{u}_{n+i}$ by the functions $f(u_{n+1+i}, u_{n+i}, v_{n+i})$. As it has been shown in Section 3.2.8 by (3.2.245), the generalized symmetry of the Hamiltonian system can be obtained as

$$u_{n,\epsilon} = \varphi \frac{\delta \hat{p}}{\delta v_n} , \qquad v_{n,\epsilon} = -\varphi \frac{\delta \hat{p}}{\delta u_n} . \tag{3.3.78}$$

If we return to the variables u_{n+i} and $\dot{u}_{n+i}$, i.e. to the Euler-Lagrange equation (3.3.54, 3.3.55), we will have instead of (3.3.78) two formulas of the form $u_{n,\epsilon} = G$, $\dot{u}_{n,\epsilon} = \hat{G}$ expressed in terms of the conserved density (3.3.77) and Lagrangian L. The second equation follows from the first one, as $\hat{G} = D_t G$, and we only rewrite the first equation in order to obtain a generalized symmetry of the Lagrangian equation. Using (3.3.66, 3.3.68), we have in terms of the variables (3.3.63)

$$y_{n,\epsilon} = u_{n,\epsilon} = \varphi \frac{\partial}{\partial v_n} \sum_i S^i \hat{p} = \varphi f_{v_n} \frac{\partial}{\partial z_n} \sum_i S^i p = (L_{z_n z_n})^{-1} \frac{\delta p}{\delta z_n} .$$

So, we are led to the following generalized symmetry

$$u_{n,\epsilon} = \frac{1}{L_{\dot{u}_n \dot{u}_n}} \frac{\delta p}{\delta \dot{u}_n} , \qquad \frac{\delta p}{\delta \dot{u}_n} = \frac{\partial}{\partial \dot{u}_n} \sum_{i=-i_1}^{-i_2} S^i p . \tag{3.3.79}$$

We can formulate the obtained results in the following theorem:

Theorem 55. *The Euler-Lagrange equation* (3.3.54, 3.3.55) *always possesses the local conservation law* (3.3.75)*. If* $L = L(\dot{u}_n, u_{n+1} - u_n)$*, this equation also has the conservation law* (3.3.76)*. If the function* (3.3.77) *is a conserved density of this Lagrangian equation, then* (3.3.79) *is the generalized symmetry.*

In the case of the standard conservation laws (3.3.75, 3.3.76), formula (3.3.79) gives the trivial Lie point symmetries: $u_{n,\epsilon} = \dot{u}_n$ and $u_{n,\epsilon} = 1$. Nontrivial examples will be presented in the case of the relativistic Toda lattice (3.2.248). The relativistic Toda has the following conserved densities (3.2.280) and

$$\hat{p}_n = \dot{u}_{n+1} \dot{u}_n \phi(w_n) + \tfrac{1}{2} \dot{u}_n^2 , \tag{3.3.80}$$

with $\phi(z)$, w_n given by (3.2.277). In this case we have $L_{\dot u_n \dot u_n} = 1/\dot u_n$, for the Lagrangian (3.3.56), and easily obtain two generalized symmetries

$$u_{n,\epsilon_1} = \dot u_{n+1}\dot u_n \phi(w_n) + \dot u_n \dot u_{n-1}\phi(w_{n-1}) + \dot u_n^2 \,, \tag{3.3.81}$$

$$u_{n,\epsilon_2} = -\hat p_n \,. \tag{3.3.82}$$

3.3.2. *Hamiltonian form of relativistic lattice equations.* Let us consider lattice systems of the form (3.3.52). It is explained at the end of Section 3.2.9 how to derive the integrability conditions in this case. The following theorem takes place:

Theorem 56. *If a system of the form (3.2.247) has two non degenerate generalized symmetries of orders $m \ge 4$ and $m+1$, then there exist functions $q_n^{(1)}$, $q_n^{(2)}$, $q_n^{(3)}$ of the form (3.2.221), which satisfy the following integrability conditions*

$$\begin{aligned} D_t p_n^{(i)} &= (S-1) q_n^{(i)}, \qquad i = 1,2,3\,, \\ p_n^{(1)} &= \log f_{u_{n+1}}\,, \qquad p_n^{(2)} = q_n^{(1)} + f_{u_n}\,, \\ p_n^{(3)} &= q_n^{(2)} + \tfrac{1}{2}(p_n^{(2)})^2 + f_{v_n} g_{u_n}\,. \end{aligned} \tag{3.3.83}$$

As usually, the functions $q_n^{(i)}$ can be chosen arbitrarily when checking these integrability conditions. Unlike the case of Volterra and Toda type equations, we restrict ourselves here by considering systems with a special Hamiltonian structure. Namely, we study systems (3.3.52) which have the following structure

$$\dot u_n = \varphi(u_n, v_n)\frac{\delta H}{\delta v_n}\,, \qquad \dot v_n = -\varphi(u_n, v_n)\frac{\delta H}{\delta u_n}\,, \tag{3.3.84}$$

with the Hamiltonian H of the form (3.2.221) (recall also the definition (3.2.225)). It is easy to see that H can be expressed as

$$H = \Phi(u_n, v_n) + \Psi(u_{n+1}, v_n)\,, \qquad \frac{\partial^2 \Psi}{\partial u_{n+1}\partial v_n} \ne 0\,. \tag{3.3.85}$$

In the case of (3.3.85), it is sufficient to use the existence of generalized symmetries (i.e. conditions (3.3.83)). The existence of higher order conservation laws implies no additional integrability conditions, as one can see in Sections 3.2.6 and 3.2.8. On the other hand, due to this Hamiltonian structure, the conditions (3.3.83) provide a system with both low order conservation laws and generalized symmetries (see (3.2.245) and (3.3.78)).

Let us discuss a classification result for systems of the form (3.3.52) with the Hamiltonian structure (3.3.84, 3.3.85) satisfying the conditions (3.3.83). The classification is carried out up to Lie point transformations of the form

$$\tilde u_n = \nu(u_n)\,, \qquad \tilde v_n = \eta(v_n)\,, \qquad \tilde t = ct\,, \tag{3.3.86}$$

where $c \ne 0$ is a constant, while ν and η are nonconstant functions. Such transformations do not change (3.3.52, 3.3.84, 3.3.85). The complete list consists of (H_1-H_3). Here we omit the index n for simplicity. The coefficients c_j of the polynomials $r(u,v)$, as well as α and β, are arbitrary constants.

List of Hamiltonian relativistic lattice equations

$$(\mathrm{H}_1) \qquad \dot{u} = u_1 + u^2 v + \alpha u\,, \qquad -\dot{v} = v_{-1} + v^2 u + \alpha v,$$

$$(\mathrm{H}_2) \qquad \dot{u} = r(u_1 - u + \alpha r_v) + \beta r_v\,, \qquad -\dot{v} = r(v_{-1} - v + \alpha r_u) + \beta r_u,$$
$$r = c_1 uv + c_2 u + c_3 v + c_4\,, \quad r_u r_v \neq 0,$$

$$(\mathrm{H}_3) \qquad \dot{u} = \frac{2r}{u_1 - v} + r_v + \alpha u + \beta\,, \qquad -\dot{v} = \frac{2r}{v_{-1} - u} + r_u - \alpha v - \beta,$$
$$\text{Case 1}: \quad \alpha = 0\,, \quad r = c_1(u-v)^2 + c_2(u-v) + c_3,$$
$$\text{Case 2}: \quad \beta = 0\,, \quad r = c_1 u^2 + c_2 v^2 + c_3 uv,$$
$$\text{Case 3}: \quad \alpha = \beta = 0\,,$$
$$r = c_1 u^2 v^2 + c_2 uv(u+v) + c_3(u^2+v^2) + c_4 uv + c_5(u+v) + c_6.$$

Let us write down for all systems of the list above the functions φ, H defining the Hamiltonian structure (3.3.84). The system (H_1) corresponds to the functions

$$\varphi = 1\,, \qquad H = u_{n+1}v_n + \tfrac{1}{2}u_n^2 v_n^2 + \alpha u_n v_n\,.$$

In the case of (H_2), the Hamiltonian structure is given by

$$\varphi = r\,, \qquad H = (u_{n+1} - u_n)v_n + \alpha r + \beta \log r\,,$$

with $r(u_n, v_n)$ specified above in (H_2). For (H_3) one has

$$\varphi = r\,, \qquad H = \log r - 2\log(u_{n+1} - v_n) + \sigma(u_n, v_n)\,,$$

where $r = r(u_n, v_n)$ is given above in (H_3), and $\sigma = 0$ in Case 3. In Case 2 the function σ is defined by the two compatible PDEs

$$\sigma_{v_n} = \alpha u_n / r\,, \qquad \sigma_{u_n} = -\alpha v_n / r\,.$$

In Case 1 both functions r and σ depend on $z = u_n - v_n$, and σ is given by a solution of the ODE $\sigma'(z) = -\beta / r(z)$.

Theorem 57. *A system* (3.3.52) *with the Hamiltonian structure* (3.3.84, 3.3.85) *satisfies the conditions* (3.3.83) *if and only if it can be transformed by a Lie point transformation* (3.3.86) *into one of the systems* $(\mathrm{H}_1$-$\mathrm{H}_3)$.

PROOF. Theorem 57 and the list of integrable systems $(\mathrm{H}_1$-$\mathrm{H}_3)$ can be found in [848] and the review articles [27, 850]. The integrability conditions (3.3.83) are presented also in those references, but the analog of Theorem 56 has unnatural assumptions there. Part of the list $(\mathrm{H}_1$-$\mathrm{H}_3)$ has been published earlier in [755]. The master symmetries for systems of the form (H_3) can be found in [27]. The Bäcklund auto-transformations and Lax pairs for some systems of the list have been constructed in [33]. Schlesinger type auto-transformations are presented in [849].

All equations of the Volterra, Toda and relativistic Toda type, considered in Section 3.3, generate Bäcklund auto-transformations for NLS type equations [24, 33, 562, 754, 755]. The systems $(\mathrm{H}_1$-$\mathrm{H}_3)$ are closely connected with such well-known equations as the Ablowitz-Ladik and Sklyanin lattices and allow one to construct a list of integrable systems of hyperbolic equations similar to the Pohlmeyer-Lund-Regge system [27]. On the other hand, the systems $(\mathrm{H}_1, \mathrm{H}_2)$ give a simple polynomial representation for some well-known relativistic

Toda type equations, presented in Sections 3.3.3.3 and 3.3.3.4, including the relativistic Toda lattice itself (see [850] and the end of Section 3.3.3.4).

3.3.3. *Lagrangian form of relativistic lattice equations.* Here we discuss the class (3.3.51). In this case we have the integrability conditions (3.2.268, 3.2.270, 3.2.271), given by Theorem 46 for lattice equations of the form (3.2.246). Eq. (3.2.246) is more general than (3.3.51) and from Theorem 46 we get conditions which can be used for checking a given equation for integrability. However, the list (L_1, L_2) presented below has been obtained in the paper [24] (see also [776]) by a simpler scheme than the generalized symmetry method, without using the integrability conditions presented in Theorem 46.

Let us briefly discuss the simpler scheme. If we use the existence of only one generalized symmetry of a simple fixed form, we also can obtain, in principle, a list of integrable equations. It is assumed in [24] that (3.3.51) possess symmetries of the form

$$u_{n,\epsilon} = f(u_{n+1}, u_n, \dot{u}_{n+1}, \dot{u}_n) + g(u_n, u_{n-1}, \dot{u}_n, \dot{u}_{n-1}) \,, \tag{3.3.87}$$

with the same functions f, g as in (3.3.51). The relativistic Toda lattice (3.2.248) has also a symmetry of this kind, namely (3.3.81). Such symmetry can be expressed always as a NLS type system in terms of $u = u_{n+1}$ and $v = u_n$

$$u_\epsilon = u_{tt} + 2g(u, v, u_t, v_t) \,, \qquad v_\epsilon = -v_{tt} + 2f(u, v, u_t, v_t) \,. \tag{3.3.88}$$

In order to do so, one rewrites, on their common solutions, the symmetry (3.3.87), using (3.3.51). One uses the additional condition that the system (3.3.88) must be integrable. As it is known from [608], if the system (3.3.88) possess a higher order conservation law, it must satisfy the following integrability condition

$$g_{u_t} - f_{v_t} \in \operatorname{Im} D_t \,. \tag{3.3.89}$$

Here D_t is the total derivative with respect to t, and thus (3.3.89) reads

$$g_{u_t} - f_{v_t} = D_t s(u, v) = s_u u_t + s_v v_t \,.$$

Returning to the variables u_{n+j}, $\dot{u}_{n+j}$, we pass to the relation

$$S\frac{\partial g}{\partial \dot{u}_n} - \frac{\partial f}{\partial \dot{u}_n} = \frac{\partial s}{\partial u_{n+1}}\dot{u}_{n+1} + \frac{\partial s}{\partial u_n}\dot{u}_n \,, \qquad s = s(u_{n+1}, u_n) \,, \tag{3.3.90}$$

in terms of the functions f, g given in (3.3.51). So, apart from the existence of a symmetry of the form (3.3.87), we obtain the condition that there must exist a function s satisfying (3.3.90).

Using these two conditions, we can write down a list of two equations, L_1 and L_2, with many arbitrary constants. Here coefficients of the polynomials P, Q, r are arbitrary constants, and the functions $a(z)$, $b(z)$ are defined by a system of ODE. The function $\mathbb{S}$ in (L_2) is a 4th degree polynomial, as it is the discriminator of r. The index n is omitted.

List of Lagrangian relativistic lattice equations

$$(\mathrm{L}_1) \qquad \ddot u = P(\dot u)(\dot u_1 a(u_1 - u) - \dot u_{-1} a(u - u_{-1}) + b(u_1 - u) - b(u - u_{-1})),$$

$$P(z) = \epsilon z^2 + \alpha z + \beta\,, \qquad Q(z) = \epsilon z^2 + \gamma z + \delta,$$

$$a' = aQ'(b) - \alpha a^2\,, \qquad b' = Q(b) - \beta a^2,$$

$$(\mathrm{L}_2) \qquad 2\ddot u = (\dot u^2 - \mathbb{S}(u))\left(\frac{\partial s_1/\partial u - \dot u_1}{s_1} + \frac{\partial s/\partial u + \dot u_{-1}}{s}\right) + \mathbb{S}'(u),$$

$$r(x,y) = c_1 x^2 y^2 + c_2 xy(x+y) + c_3(x^2+y^2) + c_4 xy + c_5(x+y) + c_6,$$

$$\mathbb{S}(x) = r_y^2 - 2rr_{yy}\,, \qquad s_1 = r(u_1, u)\,, \qquad s = r(u, u_{-1}).$$

As $f_{\ddot u_{n+1}} \neq 0$ from (3.3.51) one can easily check that $Pa \neq 0$. It is not easy to solve the system of ODE for the functions a, b. That is why we write down in Section 3.3.3.4 (L_1) in a more explicit and detailed form. To solve that system for $a(z)$ and $b(z)$, we use the invariant function $I = Q(b)/a + \beta a - \alpha b$. By direct calculation one can check that $\frac{dI}{dz} = 0$.

Theorem 58. *The complete list of equations of the form* (3.3.51)*, possessing a generalized symmetry of the form* (3.3.87) *and satisfying the condition* (3.3.90)*, consists, up to point transformations* (3.3.25)*, of the equations* $(\mathrm{L}_1, \mathrm{L}_2)$*.*

So, we obtain in this case a list of integrable equations (the integrability will be shown in Section 3.3.3.5), but not an exhaustive classification compared with the other classes and, especially, compared with Volterra and Toda type equations. The reason is that the form of generalized symmetry has been fixed. However, integrable equations are known to have infinitely many generalized symmetries, and a different form of symmetry (see e.g. (3.3.82, 3.2.280) in the case of the relativistic Toda lattice) may lead, in principle, to a quite different resulting list of equations.

Any equation of the complete list can be expressed in the Lagrangian form (3.3.54, 3.3.55) [25]. In the case of (L_1), the Lagrangian is given by

$$L = R(\dot u_n) - \dot u_n A(u_{n+1} - u_n) - B(u_{n+1} - u_n)\,, \tag{3.3.91}$$

$$R''(z) = 1/P(z)\,, \qquad A'(z) = a(z)\,, \qquad B'(z) = b(z)\,. \tag{3.3.92}$$

The functions R, A, B are not completely defined. However, (L_1) is well-defined by this Lagrangian because, changing a Lagrangian according to the formula (3.3.70), we do not change the corresponding lattice equation.

The Lagrangian for (L_2) reads:

$$L = \log \frac{r(u_{n+1}, u_n)}{\dot u_n^2 - S(u_n)} + \dot u_n(A(u_{n+1}, u_n) + B(\dot u_n, u_n))\,, \tag{3.3.93}$$

$$A_{u_{n+1}} = \frac{1}{r(u_{n+1}, u_n)}\,, \qquad B_{\dot u_n} = \frac{2}{\dot u_n^2 - S(u_n)}\,. \tag{3.3.94}$$

The functions A, B are defined up to arbitrary functions of u_n. Those arbitrary functions do not arise in the equation due to (3.3.70), so that the equation with such Lagrangian is also well-defined.

The auto-Bäcklund transformations for equations of the form (L_1) can be found in [19,25], and Lax pairs for (L_1) are presented in [26]. Schlesinger type auto-transformations for all $(\mathrm{L}_1, \mathrm{L}_2)$ have been constructed in [849]. The local master symmetries for $(\mathrm{L}_1, \mathrm{L}_2)$ can be found in [27,850].

3.3.4. *Relations between the presented lists of relativistic equations.* Here we discuss the precise correspondence between the relativistic lattice equations of the two lists (H_1-H_3) and (L_1, L_2). They are related by non point, but invertible transformations. Such transformations allow one to rewrite generalized symmetries, conservation laws and solutions. These two lists have been obtained independently, and later an equivalence between them has been observed in [27]. This is probably the first nontrivial application of non point invertible transformations to integrable equations.

First of all we write down a detailed and explicit form of (L_1). The index n is omitted here again. The coefficients μ and ν are arbitrary constants. The exact correspondence and detailed lists of equations are given below in accordance with [849].

Detailed form of Lagrangian equations ($\mathbf{L}_1$)

$$(\mathrm{Ld}_1)\qquad \ddot{u} = \dot{u}_1 e^{u_1-u} - \dot{u}_{-1} e^{u-u_{-1}} - e^{2(u_1-u)} + e^{2(u-u_{-1})},$$

$$(\mathrm{Ld}_2)\qquad \ddot{u} = \dot{u}\left(\frac{\dot{u}_1}{u_1-u} - \frac{\dot{u}_{-1}}{u-u_{-1}} + u_1 - 2u + u_{-1}\right),$$

$$(\mathrm{Ld}_3)\qquad \ddot{u} = \dot{u}\left(\frac{\dot{u}_1}{1+\mu e^{u-u_1}} - \frac{\dot{u}_{-1}}{1+\mu e^{u_{-1}-u}} - \nu(e^{u_1-u} - e^{u-u_{-1}})\right),$$

$$(\mathrm{Ld}_4)\qquad \ddot{u} = \dot{u}(1-\dot{u})\left(\frac{\dot{u}_1}{u_1-u} - \frac{\dot{u}_{-1}}{u-u_{-1}}\right),$$

$$(\mathrm{Ld}_5)\qquad \ddot{u} = \dot{u}(\dot{u}-\mu)\left(\frac{\dot{u}_1}{e^{u_1-u}+\mu} - \frac{\dot{u}_{-1}}{e^{u-u_{-1}}+\mu}\right).$$

Eqs. (Ld_1-Ld_5) are obtained by solving a system of ODE for the functions a, b given in (L_1) and by applying simple n and t dependent point transformations. Using point transformations of the form (3.3.38), one can transform any (L_1) into the form (L_2), or into one of (Ld_1-Ld_5), or into the equation

$$(3.3.95)\qquad \ddot{u}_n = \frac{\dot{u}_{n+1}\dot{u}_n}{u_{n+1}-u_n} - \frac{\dot{u}_n\dot{u}_{n-1}}{u_n-u_{n-1}}.$$

Eq. (3.3.95) is trivial (cf. (Ld_2)). Indeed, the invertible transformation $w_n = \frac{\dot{u}_n}{u_n-u_{n-1}}$, with u_n unchanged, allows one to rewrite (3.3.95) as the system

$$(3.3.96)\qquad \dot{w}_n = w_n(w_{n+1} - w_n), \qquad \dot{u}_n = w_n(u_n - u_{n-1}).$$

The first of (3.3.96) is the discrete Burgers equation (see (2.4.31-2.3.319)), while the second one is a linear equation for u_n. Generalized symmetries for (3.3.95) can be constructed, if necessary, using the master symmetry of (Ld_2) (3.3.108, 3.3.109) with $b_n = 0$, which will be presented in Section 3.3.3.5.

We consider below (L_2) and (Ld_1-Ld_5) only. The derivation of the Lagrangians for (Ld_1-Ld_5) will be left to the reader as they can be easily obtained, using formulas (3.3.91, 3.3.92) together with (L_1).

Using n and t dependent point transformations of the form

$$(3.3.97)\qquad \tilde{u}_n = \kappa_n(t, u_n), \qquad \tilde{v}_n = \eta_n(t, v_n), \qquad \tilde{t} = \Theta(t),$$

we can not only reduce the number of arbitrary constants in the systems (H_1-H_3) but also rewrite these systems so that the equivalence relation between the Lagrangian and Hamiltonian forms will become simpler and clearer. Here we present a list of systems corresponding to (Ld_1-Ld_5). The index n is omitted, μ and ν are arbitrary constants. We have in the list particular modified cases of (H_1-H_3), that is why we write down also the functions φ and H defining the Hamiltonian structure (3.3.84).

Hamiltonian systems corresponding to (Ld_1 - Ld_5)

(Hd_1)
$$\dot u = e^{u_1-u} + e^{u-v}, \qquad \dot v = e^{v-v_{-1}} + e^{u-v},$$
$$\varphi = -e^{v-u}, \quad H = e^{u_1-v} + \frac{1}{2}e^{2(u-v)},$$

(Hd_2)
$$\dot u = (u_1 - u)(u - v), \qquad \dot v = (v - v_{-1})(u - v),$$
$$\varphi = v - u, \quad H = v(u - u_1),$$

(Hd_3)
$$\dot u = (e^{u_1-u} + \mu)(e^{u-v} + \nu), \quad \dot v = (e^{v-v_{-1}} + \mu)(e^{u-v} + \nu),$$
$$\varphi = -(1 + \nu e^{v-u}), \quad H = e^{u_1-v} + \mu e^{u-v},$$

(Hd_4)
$$\dot u = \frac{u_1 - u}{u_1 - v}, \qquad \dot v = \frac{v - v_{-1}}{u - v_{-1}},$$
$$\varphi = v - u, \quad H = \log\frac{u - v}{u_1 - v},$$

(Hd_5)
$$\dot u = \frac{e^{u_1-u} + \mu}{e^{u_1-v} + 1}, \qquad \dot v = \frac{e^{v-v_{-1}} + \mu}{e^{u-v_{-1}} + 1},$$
$$\varphi = \mu - e^{v-u}, \quad H = \log\frac{1 + e^{u_1-v}}{1 - \mu e^{u-v}}.$$

It can be proved using (3.3.97) that one can transform any of (H_1, H_2) and (H_3), Cases 1,2 into (H_3), Case 3 or into one of (Hd_1-Hd_5). So, instead of (H_1-H_3), we consider below (Hd_1-Hd_5) and (H_3), Case 3. The precise correspondence between the Lagrangian equations and Hamiltonian systems is

(3.3.98)
$$(Hd_i) \sim (Ld_i), \qquad i = 1, 2, 3, 4, 5,$$
$$(H_3), \text{Case } 3 \sim (L_2).$$

If one starts from any of the Hamiltonian systems (Hd_1-Hd_5) and (H_3), Case 3 and passes to its Lagrangian formulation in accordance with Theorem 54 one obtains the equations shown in (3.3.98), and no additional point transformation is necessary.

According to Theorem 54 u_n remains unchanged, and a relation between $\dot u_n$ and v_n is given by the first equation of the Hamiltonian system. All those relations can be easily inverted except for the first equation of the system (H_3), Case 3. However, this equation can be rewritten as
$$\dot u_n = \frac{2r(u_{n+1}, u_n)}{u_{n+1} - v_n} - \frac{\partial r(u_{n+1}, u_n)}{\partial u_{n+1}},$$
and then the function v_n can be easily expressed in terms of $\dot u_n$, u_n, u_{n+1}.

One can pass from the Hamiltonian systems (H_3), Case 3 and (Hd_1-Hd_5) to the corresponding Lagrangian equation. One differentiates the first equation of the Hamiltonian system with respect to the time t, then excludes $\dot v_n$, using the second equation, and eliminates v_{n+j}, using the first equation. This gives for the function u_n the Lagrangian equation shown in (3.3.98). A Lagrangian constructed by (3.3.65) will coincide with the Lagrangian, given above for all (Ld_1-Ld_5, L_2), up to the formula (3.3.70).

It should be remarked that the correspondence (3.3.98) between the two lists of equations provides a simple polynomial representation of the relativistic Toda type equations

(Ld_1-Ld_3), including the relativistic Toda lattice (3.2.248) itself. In fact, using the transformation $\hat{u}_n = e^{u_n}$, $\hat{v}_n = e^{-v_n}$, one transforms the system (Hd_1) into (3.3.72) and (Hd_3) into the following one

$$\begin{aligned} \dot{u}_n &= (u_{n+1} + \mu u_n)(u_n v_n + \nu)\,, \\ \dot{v}_n &= -(v_{n-1} + \mu v_n)(u_n v_n + \nu)\,. \end{aligned} \tag{3.3.99}$$

The systems (3.3.72, 3.3.99) have a polynomial form as well as (Hd_2). The invertible transformations of (Ld_1-Ld_3) into the polynomial systems (3.3.72, Hd_2, 3.3.99) respectively read

$$\begin{aligned} \hat{u}_n &= e^{u_n}, & \hat{v}_n &= \frac{\dot{u}_n - e^{u_{n+1}-u_n}}{e^{u_n}}\,, \\ \hat{u}_n &= u_n\,, & \hat{v}_n &= u_n + \frac{\dot{u}_n}{u_n - u_{n+1}}\,, \\ \hat{u}_n &= e^{u_n}, & \hat{v}_n &= \frac{\dot{u}_n}{e^{u_{n+1}} + \mu e^{u_n}} - \nu e^{-u_n}\,. \end{aligned}$$

3.3.5. *Master symmetries for the relativistic lattice equations.* Here we show the integrability of relativistic Toda type equations, using master symmetries and simple non invertible transformations. We have proved in the previous sections the equivalence of the Hamiltonian and Lagrangian relativistic lattice equations. Generalized symmetries, conservation laws and master symmetries can be transformed from one equation into another. So, we can explain why all relativistic equations are integrable, considering the Hamiltonian or Lagrangian form only. According to (3.3.98), we will construct generalized symmetries and conservation laws only for the Lagrangian equations (Ld_1-Ld_5) and the Hamiltonian system (H_3), Case 3. All master symmetries, presented in this section, have been taken from the papers [27, 850] (see also [653, 866]).

First of all we write down the master symmetries for (Ld_1)-Ld_5). Eqs. (Ld_4, Ld_5) are of the form (L_1) with $b = 0$ and can be expressed as systems in terms of u_n and $v_n = \dot{u}_n$ they are

$$\begin{aligned} \dot{u}_n &= v_n\,, \\ \dot{v}_n &= P(v_n)(v_{n+1}a_n - v_{n-1}a_{n-1})\,, \qquad a_n = a(u_{n+1} - u_n)\,, \end{aligned} \tag{3.3.100}$$

where the functions $P(z)$, $a(z)$ are given by the table

$$\begin{aligned} P &= z(1-z) & a &= \tfrac{1}{z} & &\text{for} & &(Ld_4) \\ P &= z(z-\mu) & a &= \tfrac{1}{e^z+\mu} & &\text{for} & &(Ld_5) \end{aligned} \tag{3.3.101}$$

The local master symmetries for these systems are

$$\begin{aligned} u_{n,\tau} &= n v_n\,, \\ v_{n,\tau} &= P(v_n)((n+1)v_{n+1}a_n - (n-1)v_{n-1}a_{n-1} + \sigma)\,, \end{aligned} \tag{3.3.102}$$

where

$$\begin{aligned} \sigma &= 0 & &\text{for} & &(Ld_4) \\ \sigma &= -1 & &\text{for} & &(Ld_5) \end{aligned} \tag{3.3.103}$$

In the case of (Ld_1-Ld_3), we use the non invertible transformation

$$w_n = u_{n+1} - u_n\,, \qquad v_n = \dot{u}_n \tag{3.3.104}$$

to rewrite them in the form

$$\begin{aligned} \dot{w}_n &= v_{n+1} - v_n\,, \\ \dot{v}_n &= P(v_n)(v_{n+1}a_n - v_{n-1}a_{n-1} + b_n - b_{n-1})\,, \end{aligned} \tag{3.3.105}$$

where

(3.3.106)
$$a_n = a(w_n)\,, \qquad b_n = b(w_n)$$

(see (L_1)). The functions $P(z)$, $a(z)$, $b(z)$ are defined by

(3.3.107)
$$\begin{array}{lllll} P = 1, & a = e^z, & b = -e^{2z}, & \text{for} & (\mathrm{Ld}_1), \\ P = z, & a = z^{-1}, & b = z, & \text{for} & (\mathrm{Ld}_2), \\ P = z, & a = \frac{1}{1+\mu e^{-z}}, & b = -\nu e^z, & \text{for} & (\mathrm{Ld}_3). \end{array}$$

The master symmetries for the systems (3.3.105-3.3.107) are given by

(3.3.108)
$$\begin{aligned} w_{n,\tau} &= (n+k)v_{n+1} - (n-k+1)v_n + c_n\,, \\ v_{n,\tau} &= P(v_n)[(n+k)(v_{n+1}a_n + b_n) - (n-k)(v_{n-1}a_{n-1} \\ &\quad + b_{n-1})] + \sigma v_n^2\,, \end{aligned}$$

with the constants k, σ and function c_n defined in the following table

(3.3.109)
$$\begin{array}{llll} (\mathrm{Ld}_1): & k = 2, & \sigma = 1, & c_n = -2a_n, \\ (\mathrm{Ld}_2): & k = 2, & \sigma = 0, & c_n = b_n^2, \\ (\mathrm{Ld}_3): & k = \frac{3}{2}, & \sigma = 1, & c_n = b_n - \mu\nu. \end{array}$$

In the case of (Ld_3), this is a master symmetry only if $\mu = 0$ or $\nu = 0$. If $\mu\nu \neq 0$, we introduce into both the system (3.3.105) and its master symmetry (3.3.108) a dependence on the variable τ of the master symmetry, so that

(3.3.110)
$$\nu = \nu(\tau)\,, \qquad \frac{d\nu}{d\tau} = \mu\nu^2$$

(see a discussion about τ dependent master symmetries at the end of Section 3.2.7).

We can construct the conserved densities for the systems (3.3.100, 3.3.105), using formula (3.2.209), but need a starting density. Let us write down three simple densities for (Ld_1-Ld_5),

(3.3.111)
$$p_n^- = \int \frac{dv_n}{P(v_n)} - \int a_n dw_n\,, \qquad p_n^+ = \int \frac{v_n dv_n}{P(v_n)} + \int b_n dw_n\,,$$

(3.3.112)
$$p_n^0 = \log P(v_n) + \log a_n\,.$$

They are written in terms of the functions v_n, w_n, a_n, b_n defined by (3.3.104, 3.3.106). The functions P, a, b are given by (3.3.101) with $b = 0$ and by (3.3.107). These formulas also provide the conserved densities of (L_1).

In fact, (L_1) are Lagrangian and have, according to (3.3.75, 3.3.76), the conserved densities $p_n^- = L_{\dot u_n}$, $p_n^+ = \dot u_n L_{\dot u_n} - L$ in terms of the functions (3.3.91, 3.3.92). Taking into account that $(zR'(z) - R(z))' = z/P(z)$, we obtain (3.3.111). The function (3.3.112) is nothing but the conserved density $p_n^{(1)}$ given by (3.2.270). This is another form of the densities (3.3.111), which is used below, because it can be obtained as the linear combination

$$p_n^0 = c_1 p_n^- + c_2 p_n^+ + c_3(u_{n+1} - u_n) + c_4$$

with some constant coefficients c_j.

Eqs. (Ld_4, Ld_5) are equivalent to (3.3.100, 3.3.101). Eqs. (3.3.111, 3.3.112), written in terms of v_n and $w_n = u_{n+1} - u_n$, give the conserved densities for these systems. The function p_n^- cannot be a starting density for the master symmetry (3.3.102, 3.3.103), as $D_\tau p_n^- \sim 0$. The next density p_n^+ can be taken as a starting one, and on the first step we obtain:

(3.3.113)
$$D_\tau p_n^+ \sim v_n(v_{n+1}a_n + \pi)\,.$$

This function with $v_n = \dot{u}_n$ is a new conserved density for (Ld_4, Ld_5).

In the case of (Ld_1-Ld_3), the transformation (3.3.104) is non invertible, and we have to check directly that (3.3.111, 3.3.112), considered as the functions of v_n and w_n, are also conserved densities for (3.3.105, 3.3.106). The density p_n^+ can be used again as a starting point. Using the master symmetry (3.3.108, 3.3.109), one obtains a new conserved density for the systems (3.3.105-3.3.107)

$$D_\tau p_n^+ \sim (2k-1)(v_n v_{n+1} a_n + (v_n + v_{n+1}) b_n) + \frac{\sigma v_n^3}{P(v_n)} + b_n c_n + \frac{\partial b_n}{\partial \tau}\,. \tag{3.3.114}$$

Here $\frac{\partial b_n}{\partial \tau} = 0$ in all cases, except for (Ld_3) with $\mu\nu \neq 0$, when

$$b_n = -\nu e^{w_n}, \qquad \frac{\partial b_n}{\partial \tau} = -\mu \nu^2 e^{w_n},$$

due to (3.3.110). Now we can construct a conserved density for (Ld_1-Ld_3), using the transformation (3.3.104), namely, replacing v_n, w_n in (3.3.114) by the functions $\dot{u}_n$, $u_{n+1} - u_n$.

So, using the master symmetries (3.3.102, 3.3.103) and (3.3.108, 3.3.109), we construct conserved densities for the systems (3.3.100, 3.3.101) and (3.3.105-3.3.107) and then rewrite those conserved densities for (Ld_1-Ld_5). As these equations are Lagrangian, we can then use (3.3.79) to obtain generalized symmetries. The Lagrangians for (Ld_1-Ld_5) have the form (3.3.91, 3.3.92), hence $L_{\dot{u}_n \dot{u}_n} = 1/P(\dot{u}_n)$. For example, starting from the density (3.3.113), we obtain for (Ld_4, Ld_5) the following symmetry

$$u_{n,\epsilon} = P(\dot{u}_n)(\dot{u}_{n+1} a(u_{n+1} - u_n) + \dot{u}_{n-1} a(u_n - u_{n-1}) + \pi)\,.$$

Example of the relativistic Toda lattice (3.2.248).

Let us discuss the relativistic Toda lattice in details. The relativistic Toda lattice equation is included in (Ld_3) by choosing $\mu = 1$, $\nu = 0$. Its representation (L_1) is defined, according to (3.3.107), by $P = z$, $b = 0$, $a = \phi(z) = \frac{1}{1+e^{-z}}$. The corresponding system of the form (3.3.105, 3.3.106) is

$$\begin{aligned} \dot{w}_n &= v_{n+1} - v_n\,, \\ \dot{v}_n &= v_n v_{n+1} \phi(w_n) - v_n v_{n-1} \phi(w_{n-1})\,, \end{aligned} \tag{3.3.115}$$

and the master symmetry (3.3.108, 3.3.109) takes the form

$$\begin{aligned} w_{n,\tau} &= (n + \tfrac{3}{2}) v_{n+1} - (n - \tfrac{1}{2}) v_n\,, \\ v_{n,\tau} &= (n + \tfrac{3}{2}) v_n v_{n+1} \phi(w_n) - (n - \tfrac{3}{2}) v_n v_{n-1} \phi(w_{n-1}) + v_n^2\,. \end{aligned} \tag{3.3.116}$$

We will discuss here only conserved densities for (3.3.115). Conserved densities for the relativistic Toda lattice are constructed from (3.3.104), and the corresponding generalized symmetries are obtained, using (3.3.79) with $L_{\dot{u}_n \dot{u}_n} = 1/\dot{u}_n$ (see (3.3.56)).

By direct calculation one can check that the following functions are the conserved densities for (3.3.115):

(P_1) $$p_n^- = p_n^0 - w_n\,,$$

(P_2) $$p_n^0 = \log v_n + \log \phi(w_n)\,,$$

(P_3) $$p_n^+ = v_n\,,$$

(P_4) $$\hat{p}_n = v_n v_{n+1} \phi(w_n) + \frac{1}{2} v_n^2\,,$$

(P$_5$) $$\hat{p}_n^{\diamond} = v_n v_{n+1} v_{n+2}\phi(w_n)\phi(w_{n+1}) + (v_n + v_{n+1})v_n v_{n+1}\phi(w_n) + \frac{1}{3}v_n^3 ,$$

(P$_6$) $$\check{p}_n = \frac{1}{v_n}(1 + e^{w_n})(1 + e^{w_{n-1}}) ,$$

(P$_7$) $$\check{p}_n^{\diamond} = \check{p}_n\check{p}_{n+1}\phi(w_n) + \frac{1}{2}\check{p}_n^2 .$$

The conserved densities (P$_1$-P$_3$) come from (3.3.111, 3.3.112) (see also $p_n^{(1)}$, $p_n^{(2)}$ given by (3.2.278, 3.2.279)), while (P$_4$, P$_6$) are taken from (3.2.280, 3.3.80). The densities (P$_1$, P$_3$) are written down here up to some nonessential constants of integration. The density (P$_1$) becomes trivial when we pass to (3.2.248) as $w_n = (S-1)u_n \sim 0$, but it is not trivial at the level of the system (3.3.115) and helps us to show the complete picture.

Using the ODE $\phi' = \phi - \phi^2$, it is easy to check that the operator D_τ corresponding to the master symmetry (3.3.116) acts on (P$_1$-P$_4$) giving

$$D_\tau w_n \sim p_n^+ , \qquad D_\tau p_n^0 \sim 2p_n^+ , \qquad D_\tau p_n^+ \sim 2\hat{p}_n , \qquad D_\tau \hat{p}_n \sim 3\hat{p}_n^{\diamond} .$$

So, starting from (P$_1$) or (P$_2$), one can construct a new density (P$_5$) as well as infinitely many densities depending on the variables v_{n+j}. The system (3.3.115) also has an infinite hierarchy of conserved densities with rational dependence on v_{n+j} which are similar to (P$_6$, P$_7$). However, the master symmetry (3.3.116) does not help us to construct them. In fact,

$$D_\tau \check{p}_n = (S^{-1} - 1)\left(\tfrac{2n+1}{2}(1 + e^{w_n})\right) \sim 0 , \qquad D_\tau \check{p}_n^{\diamond} \sim -\check{p}_n .$$

All relativistic lattice equations have a second hierarchy of conservation laws and generalized symmetries. It is an open problem how to construct them.

Let us discuss now the Hamiltonian system (H$_3$), Case 3, namely

(3.3.117) $$\dot{u}_n = \frac{2r}{u_{n+1} - v_n} + r_{v_n} , \qquad \dot{v}_n = \frac{2r}{u_n - v_{n-1}} - r_{u_n} ,$$

where r can be defined as

(3.3.118) $$r = r(u_n, v_n) = r(v_n, u_n) , \qquad \frac{\partial^3 r}{\partial u_n^3} = 0 .$$

The local master symmetry in this case has the very simple form

(3.3.119) $$u_{n,\tau} = n\dot{u}_n , \qquad v_{n,\tau} = (n-1)\dot{v}_n ,$$

but the coefficients of the polynomial r depend on τ. The dependence on τ is given by

(3.3.120) $$r_\tau = r r_{u_n v_n} - r_{u_n} r_{v_n} ,$$

exactly as in the case of (3.3.32, 3.3.33). This master symmetry can be rewritten as a master symmetry for the Lagrangian equation (L$_2$). It is equivalent to the system (3.3.117, 3.3.118), and, as such, it has been presented in [27].

Generalized symmetries are obtained by the master symmetry (3.3.119, 3.3.120) or, as (3.3.117, 3.3.118) is Hamiltonian, using (3.3.78). One can construct a hierarchy of conserved densities, starting from the density $p_n^{(1)}$ given in (3.3.83). It is not difficult to check that, up to some nonessential constants,

(3.3.121) $$p_n^{(1)} = \log r - 2\log(u_{n+1} - v_n) ,$$

(3.3.122) $$p_n^{(2)} = -\frac{4r}{(u_{n+1} - v_n)(u_n - v_{n-1})} + \frac{2r_{u_n}}{u_{n+1} - v_n} - \frac{2r_{v_n}}{u_n - v_{n-1}} + r_{u_n v_n} ,$$

see (3.3.83). Applying the operator D_τ, we obtain

$$D_\tau p_n^{(1)} = p_n^{(2)} + (1-S)\frac{2n\dot{u}_n}{u_n - v_{n-1}} + \frac{r_\tau - r r_{u_n v_n} + r_{u_n} r_{v_n}}{r},$$

i.e $D_\tau p_n^{(1)} \sim p_n^{(2)}$, only if (3.3.120) is satisfied. The two simplest conserved densities of the second hierarchy, symmetrical to (3.3.121, 3.3.122), are

$$\tilde{p}_n^{(1)} = \log r - 2\log(v_{n+1} - u_n),$$

$$\tilde{p}_n^{(2)} = -\frac{4r}{(v_{n+1}-u_n)(v_n - u_{n-1})} + \frac{2r_{v_n}}{v_{n+1}-u_n} - \frac{2r_{u_n}}{v_n - u_{n-1}} + r_{u_n v_n}.$$

4. Explicit dependence on the discrete spatial variable n and time t

Here we consider lattice equations which explicitly depend on the discrete spatial variable and continuous time. In the case of integrable DΔEs in Section 2.3.2 we showed that we can construct Toda, Volterra, Burgers and dNLS like equations with variable coefficients.

In particular inhomogeneous Toda equations have also been presented in Section 2.3.2.7, inhomogeneous Volterra equations in Section 2.3.3.5, inhomogeneous Burgers in Section 2.3.5 and inhomogeneous dNLS in Section 2.3.4.1.

First, in Sections 3.4.1 and 3.4.2, we will discuss Volterra and Toda type equations, as the generalized symmetry method has been developed for these classes. Then we present, in Section 3.4.3, an n-dependent integrable example of relativistic Toda type equations. We will consider, among others, results published in the papers [26, 27, 33, 549, 755, 852].

4.1. Dependence on n in Volterra type equations. Let us consider the n-dependent generalizations of equations of the form (3.2.1). In Section 3.4.1.1 the differences between n-dependent and n-independent equations will be discussed, and the integrability conditions for these new type equations will be given. In the next Section 3.4.1.2 we will show a few integrable examples obtained with the help of those conditions. The theory and examples, we use here, can be found in [549, 552].

4.1.1. *Discussion of the general theory.* We consider here the class of equations

$$\dot{u}_n = f_n(u_{n+1}, u_n, u_{n-1}), \tag{3.4.1}$$

where

$$\frac{\partial f_n}{\partial u_{n+1}} \neq 0, \quad \frac{\partial f_n}{\partial u_{n-1}} \neq 0 \quad \text{for all } n. \tag{3.4.2}$$

The most general form of functions in this case is

$$\phi_n(u_{n+k}, u_{n+k-1}, \dots u_{n+k'}), \qquad k \geq k', \tag{3.4.3}$$

where

$$\frac{\partial \phi_n}{\partial u_{n+k}} \neq 0, \quad \frac{\partial \phi_n}{\partial u_{n+k'}} \neq 0 \quad \text{for some } n \tag{3.4.4}$$

if ϕ_n is not an n-dependent constant. The words "for some n" in the condition (3.4.4) mean that there exist numbers n_1 and n_2 such that $\frac{\partial \phi_{n_1}}{\partial u_{n_1+k}} \neq 0$, $\frac{\partial \phi_{n_2}}{\partial u_{n_2+k'}} \neq 0$. A function ϕ_n will be called an n-dependent constant if it depends only on n.

The n-dependent function (3.4.3) is defined by an infinite set of functions of $k-k'+1$ variables: $\phi_n(z_1, z_2, \dots z_{k-k'+1})$. Such functions may be *a priori* quite different, and we only know that ϕ_{n_1} and ϕ_{n_2} must depend on z_1 and $z_{k-k'+1}$, respectively. Analogously, the

n-dependent equation (3.4.1) is defined, unlike (3.2.1), by an infinite set of *a priori* quite different functions $f_n(z_1, z_2, z_3)$, all of which depend on the variables z_1, z_3.

The shift operator S acts on the function (3.4.3) as follows:

$$S\phi_n = \phi_{n+1}(u_{n+1+k}, u_{n+k}, \dots u_{n+1+k'})$$

(cf. (3.2.12)). One can prove in a way similar to what we did in Section 3.2.1 the property (3.2.13) for functions of the form (3.4.3). More precisely, the statement now reads: if $\phi_n \in \ker(S-1)$ i.e. $\phi_{n+1} = \phi_n$ for all n, then ϕ_n is a standard (i.e. n-independent) constant.

The generalized symmetry of (3.4.1) is an equation of the form

$$u_{n,\epsilon} = g_n(u_{n+m}, u_{n+m-1}, \dots u_{n+m'}) , \tag{3.4.5}$$

where $m \geq m'$ and

$$\frac{\partial g_n}{\partial u_{n+m}} \neq 0 , \quad \frac{\partial g_n}{\partial u_{n+m'}} \neq 0 \quad \text{for some } n , \tag{3.4.6}$$

which satisfies the compatibility conditions (3.2.7) for all n. The numbers m and m' are called the left and right orders of this symmetry. The form (3.4.5, 3.4.6) of the symmetry seems to be very general, but it really is more restrictive due to the condition (3.4.2).

In fact, as in Section 3.2.2, if $m \geq 1$ we obtain (3.2.49) from the compatibility condition. As $f_n^{(1)} \neq 0$ for all n due to (3.4.2), we can pass to (3.2.53) which must take place for any n. In this way we are led to (3.2.57), where α is an n-independent constant. It follows from (3.4.6) that $\alpha \neq 0$, thus $g_n^{(m)} \neq 0$ for all n. See (3.2.47) for the definition of $g_n^{(j)}$. The case $m' \leq -1$ is quite similar, so we can state the following theorem:

Theorem 59. *If a generalized symmetry* (3.4.5) *is such that* $m \geq 1$ *(or* $m' \leq -1$*), then* $\frac{\partial g_n}{\partial u_{n+m}} \neq 0$ *(correspondingly* $\frac{\partial g_n}{\partial u_{n+m'}} \neq 0$*) for any value of* n.

Two functions a_n, b_n of the form (3.4.3) are said to be equivalent in this case ($a_n \sim b_n$) if there exists one function c_n such that $a_n - b_n = (S-1)c_n$ for all n. Now, however, $a \sim 0$ for any constant a, if $a = (S-1)(na)$. Moreover, $a_n \sim 0$ for any n-dependent constant a_n, as the equation $a_n = c_{n+1} - c_n$ for an n-dependent constant c_n always has a solution. We choose c_0 arbitrarily and then we can find the constants c_n recursively.

The definition of the formal variational derivative for functions of the form (3.4.3) remains the same, i.e. (3.2.40). Theorem 24 can obviously be modified, and the proof will be quite similar. A statement in the n-dependent case reads:

Proposition 3. *For any function* (3.4.3),

$$\phi_n \sim 0 \qquad \textit{iff} \qquad \frac{\delta\phi_n}{\delta u_n} = 0 \tag{3.4.7}$$

for all n.

This property will be used for checking the integrability conditions.

The local conservation law of (3.4.1) is a relation of the form (3.2.15), valid for all n, where p_n, q_n are functions of the form (3.4.3). For the Volterra equation (3.2.2), there is the following example of conservation law with an explicit dependence on n (cf. (3.2.18)). We have

$$D_t((-1)^n \log u_n) = (S-1)((-1)^{n+1}u_n + (-1)^n u_{n-1}) .$$

Definition 6 must be slightly modified. A conservation law (3.2.15) is trivial if $p_n \sim 0$. For a nontrivial conservation law, the first of two possibilities is: $p_n \sim \mathcal{P}_n(u_n)$, where $\mathcal{P}_n'(u_n) \neq 0$ for some n. Such conservation law has the order 0. The second case is such that

$$p_n \sim \mathcal{P}_n(u_{n+m}, u_{n+m-1}, \dots u_n)\,,$$

where $m > 0$ and $\dfrac{\partial^2 \mathcal{P}_n}{\partial u_{n+m} \partial u_n} \neq 0$ for some n. This is a conservation law of order $m > 0$.

Using the property (3.4.7) and the fact that $\dfrac{\delta p_n}{\delta u_n} = \dfrac{\delta \mathcal{P}_n}{\delta u_n}$, one can find the order of a local conservation law with the help of (3.2.43) (cf. (3.2.44-3.2.46)). The case $\varrho_n = 0$, for any n, corresponds to a trivial conservation law. In the case $\varrho_n(u_n) \neq 0$ for some n, the order is equal to 0. If

$$\varrho_n = \varrho_n(u_{n+m}, u_{n+m-1}, \dots u_{n-m})\,,$$

where $m > 0$ and $\dfrac{\partial \varrho_n}{\partial u_{n+m}} \neq 0$, $\dfrac{\partial \varrho_n}{\partial u_{n-m}} \neq 0$ for some n, then the order is equal to m.

It is interesting that, in the n-dependent case, almost any difference disappears between lattice equations and systems of equations. Let us consider in detail the following class of systems

$$\dot{U}_k = F_k(V_{k+1}, U_k, V_k)\,, \qquad \dot{V}_k = G_k(U_k, V_k, U_{k-1})\,, \tag{3.4.8}$$

which contains systems of the form (3.3.44) and (3.3.46) as well as the Toda system (3.3.29). Eqs. (3.4.8) are related by (3.3.30) to the classical Toda model (3.2.216). Introducing the new dependent variables U_k, V_k and the functions F_k, G_k in such a way that

$$U_k = u_{2k+1}\,, \quad V_k = u_{2k}\,, \quad F_k = f_{2k+1}\,, \quad G_k = f_{2k}\,, \tag{3.4.9}$$

one can rewrite (3.4.1) as a system (3.4.8). For example, the Volterra equation (3.2.2) takes the form

$$\dot{U}_k = U_k(V_{k+1} - V_k)\,, \qquad \dot{V}_k = V_k(U_k - U_{k-1})\,, \tag{3.4.10}$$

and this is nothing but (3.3.44), Case 2.

Before going to the inverse transformation, let us discuss the n-dependent constant χ_n, often used below and introduced in (2.4.148) when dealing with Boll equations in Section 2.4.7. It is clear that $\chi_{2k} = 1$, $\chi_{2k+1} = 0$, and, for any n, we have

$$\chi_n = \chi_{n+2}\,, \quad \chi_n + \chi_{n+1} = 1\,, \quad \chi_n \chi_{n+1} = 0\,, \quad \chi_n^2 = \chi_n. \tag{3.4.11}$$

The inverse transformation is given by the same formulas (3.4.9), i.e. u_n, f_n are defined in accordance with (3.4.9). For instance, given the Toda system (3.3.29), written in the form (3.4.8), we are led to

$$\dot{u}_{2k+1} = u_{2k+1}(u_{2k+2} - u_{2k})\,, \qquad \dot{u}_{2k} = u_{2k+1} - u_{2k-1}\,.$$

Using χ_n (2.4.148), we can express this system as the following n-dependent equation

$$\dot{u}_n = (\chi_{n+1} u_n + \chi_n)(u_{n+1} - u_{n-1})\,. \tag{3.4.12}$$

The generalized symmetries are transformed in the same way as the corresponding equations. Let us discuss in detail the conservation laws. Given a conservation law (3.2.15) of (3.4.1), we have the relation $D_t(p_{n+1} + p_n) = q_{n+2} - q_n$ and therefore can introduce the functions $P_k = p_{2k+1} + p_{2k}$, $Q_k = q_{2k}$ in order to obtain for the corresponding system (3.4.8) a conservation law

$$\dot{P}_k = Q_{k+1} - Q_k\,. \tag{3.4.13}$$

In the case of the Volterra equation (3.2.2) when considering the first of its conservation laws (3.2.18) we obtain for the system (3.4.10)

$$P_k = u_{2k+1} + u_{2k} = U_k + V_k\,, \qquad Q_k = u_{2k}u_{2k-1} = V_k U_{k-1}\,.$$

If instead we have (3.4.8) and (3.4.13), we define the functions p_n, q_n for a conservation law of (3.4.1) as

$$p_{2k} = P_k\,, \quad p_{2k+1} = 0\,, \quad q_{2k} = q_{2k-1} = Q_k\,.$$

Let us consider the example of the Toda system (3.3.29), also written as (3.3.44), Case 1. As it has been written in Section 3.3.2, its conserved densities can be constructed by using the master symmetry (3.3.45). One of those conserved densities is $P_k = 2U_k + V_k^2$, and thus we have: $p_{2k} = 2u_{2k+1} + u_{2k}^2$. Using the function χ_n (2.4.148), we can rewrite this formula in a more convenient form. In this case,

$$p_n = \chi_n(2u_{n+1} + u_n^2) \sim \hat{p}_n = 2\chi_{n+1}u_n + \chi_n u_n^2\,, \tag{3.4.14}$$

and it is easy to check for (3.4.12), using the relations (3.4.11), that $D_t\hat{p}_n = (S-1)(2u_n u_{n-1})$. In the same way, we can write down for (3.4.12) the other conserved densities. The three simplest conserved densities are

$$\chi_{n+1}\log u_n\,, \qquad \chi_n u_n\,, \qquad 3u_{n+1}u_n + \chi_n u_n^3\,. \tag{3.4.15}$$

The systems (3.4.8) and (3.4.1) can be considered equivalent as the transformation (3.4.9) is nothing but a change of notation. The integrability is preserved in the sense that there exist infinitely many local conservation laws and generalized symmetries. In a quite similar way, we can pass from (3.4.1) to systems of 3, 4 and more lattice equations. Introducing e.g.

$$U_k = u_{3k+2}\,, \quad V_k = u_{3k+1}\,, \quad W_k = u_{3k}\,,$$

we obtain from the Volterra equation (3.2.2) the following system

$$\begin{aligned} \dot{U}_k &= U_k(W_{k+1} - V_k)\,, \\ \dot{V}_k &= V_k(U_k - W_k)\,, \\ \dot{W}_k &= W_k(V_k - U_{k-1})\,. \end{aligned} \tag{3.4.16}$$

So, the class (3.4.1) includes not only the Volterra equation but also the Toda lattice written in the form (3.4.12). Studying (3.4.1), we also study many systems of equations like (3.4.8, 3.4.16).

If we derive integrability conditions for (3.4.1), using Lax equation (3.2.65), we have to do the same calculations as in the n-independent case and we get the same results. Namely, the following statement can be obtained:

Theorem 60. *If an equation* (3.4.1, 3.4.2) *has two generalized symmetries* (3.4.5, 3.4.6) *of left orders $m \geq 4$ and $m+1$, then there exist functions $q_n^{(1)}$, $q_n^{(2)}$, $q_n^{(3)}$ of the form* (3.4.3)*, such that the integrability conditions* (3.2.56, 3.2.100, 3.2.106) *are satisfied. If this equation also possesses a conservation law of order $m \geq 3$, then there exist functions $\sigma_n^{(1)}$, $\sigma_n^{(2)}$ of the form* (3.4.3) *satisfying the conditions* (3.2.130, 3.2.131)*.*

The proof can be found in reference [549]. The only difference, compared with the n-independent case, is that all functions in the integrability conditions are now of the form (3.4.3). For this reason, the classification problem is more difficult in the n-dependent case. For example, we have now divisors of zero, i.e. nontrivial functions φ_n, ψ_n: $\varphi_n\psi_n = 0$ such that for all n (nontrivial means that $\varphi_n \neq 0$, $\psi_n \neq 0$ for some n). On the other hand, in the n-independent case, one always can split the problem into two subcases: $c = 0$ and $c \neq 0$, using a constant c. Now we often cannot do so, as instead of a constant c we have to do with an n-dependent constant c_n. Nevertheless, using the integrability conditions, we can test any given equation (3.4.1) for integrability as well as can find new integrable examples, as it will be shown in the next section.

4.1.2. *Examples.* Here we consider some classification problems and examples of n-dependent integrable equations. All equations in Examples 1, 2 will be two-periodic in the sense that $f_n = f_{n+2}$, for all n, for the function $f_n(z_1, z_2, z_3)$ defining (3.4.1). A lattice equation of Example 3 will not have such periodicity.

Example 1 The starting point of this classification is

$$\dot{u}_n = P_n(u_{n+1} - u_{n-1}), \qquad P_n = P_n(u_n), \tag{3.4.17}$$

which includes the Volterra and Toda equations in the form (3.2.2) and (3.4.12). It is interesting to compare this example with the last example of Section 3.3.1.1, as such comparison will show the difference between the n-dependent and n-independent cases. We assume that

$$P_n \neq 0 \quad \text{for all } n, \tag{3.4.18}$$

$$P_n' \neq 0 \quad \text{for some } n. \tag{3.4.19}$$

The first requirement corresponds to the condition (3.4.2). The second one means that we are interested in non linear equations.

We check the integrability conditions using the property (3.4.7). As

$$\dot{p}_n^{(1)} = P_n'(u_{n+1} - u_{n-1}),$$

we can rewrite the condition (3.2.56), Section 3.2.2, as

$$\frac{\delta \dot{p}_n^{(1)}}{\delta u_n} = P_n''(u_{n+1} - u_{n-1}) + P_{n-1}' - P_{n+1}' = 0. \tag{3.4.20}$$

Applying the operator $\frac{\partial}{\partial u_{n+1}}$, we obtain $(S-1)P_n'' = 0$ for all n, and this implies that P_n'' is a constant. Thus P_n is a polynomial of the form

$$P_n = au_n^2 + b_n u_n + c_n, \tag{3.4.21}$$

where a is a constant, while b_n, c_n are n-dependent functions. Inserting this result into (3.4.20), we see that b_n is two-periodic, i.e.

$$b_n = b_{n+2} \tag{3.4.22}$$

for any value of n.

In this case, unlike (3.3.15), one integrability condition (3.2.56) is not enough to describe all integrable cases. As the conditions (3.2.130, 3.2.131) are satisfied due to (3.4.21, 3.4.22), let us consider the conditions (3.2.100, 3.2.106). Constants of integration for the functions $q_n^{(1)}$, $q_n^{(2)}$ given by (3.2.56, 3.2.100) can be chosen as in the n-independent case. We choose

$$q_n^{(1)} = 2au_n u_{n-1} + b_{n+1}u_n + b_n u_{n-1}$$

and obtain instead of (3.2.100) the following condition

$$\frac{\delta \dot{p}_n^{(2)}}{\delta u_n} = -2P_n'(c_{n+1} - c_{n-1}) = 0.$$

This gives us two restrictions for c_n

$$a(c_{n+1} - c_{n-1}) = b_n(c_{n+1} - c_{n-1}) = 0. \tag{3.4.23}$$

In an analogous way, we write down the integrability condition $\frac{\delta \dot{p}_n^{(3)}}{\delta u_n} = 0$, corresponding to (3.2.106), and see that up to (3.4.22, 3.4.23) we get the new restriction

$$b_{n+1}c_n(c_{n+1} - c_{n-1}) = 0\,. \tag{3.4.24}$$

Now, using the conditions (3.4.18, 3.4.19, 3.4.21-3.4.24), we can prove that

$$c_n = c_{n+2} \tag{3.4.25}$$

for any n. Let us assume that $\hat{c}_n = c_{n+1} - c_{n-1} \neq 0$ for $n = 0$. We can do so without loss of generality, as for any $n = n_0$ the results will be the same. Eqs. (3.4.23, 3.4.24) imply:

$$a = b_0 = b_1 c_0 = 0\,.$$

If $c_0 = 0$, then $P_0 = 0$, but this is in contradiction with (3.4.18). If $b_1 = 0$, then for all k

$$b_{2k} = b_0 = 0\,, \quad b_{2k+1} = b_1 = 0$$

due to (3.4.22). This means that $P_n' = 0$ for any n and is in contradiction with the second restriction (3.4.19). So, $\hat{c}_n$ vanishes for all values of n, i.e. c_n is two-periodic.

So we have proved the following proposition:

Proposition 4. *An equation of the form* (3.4.17) *with the restrictions* (3.4.18, 3.4.19) *satisfies the integrability conditions* (3.2.56, 3.2.100, 3.2.106, 3.2.130, 3.2.131) *iff P_n is of the form* (3.4.21)*, where a is a constant, while b_n, c_n are n-dependent two-periodic constants* (3.4.22, 3.4.25)*. All such equations are integrable. In fact, using the change of variables* (3.4.9)*, we can transform any such equation into a system* (3.3.44) *of Section* 3.3.2 *with five arbitrary constants*

$$A(z) = az^2 + b_1 z + c_1\,, \quad B(z) = az^2 + b_0 z + c_0\,.$$

Due to (3.4.18, 3.4.19) one has: $AB \neq 0$, besides $A' \neq 0$ or $B' \neq 0$. Moreover, using the following linear transformations with constant coefficients

$$\tilde{u}_n = \kappa_1 u_n + \kappa_2\,, \quad \tilde{v}_n = \kappa_3 v_n + \kappa_4\,, \quad \tilde{t} = \kappa_5 t\,,$$

any system of this kind can be transformed into a system (3.3.44) corresponding to Cases 1, 2 or 3, considered in Section 3.3.2.

Example 2 Let us discuss a generalization of (V_4) with $\nu = 0$ presented in Section 3.3.1.2. Let us recall that this is the only equation in the list (V_1-V_{11}) which cannot be transformed into the Volterra and Toda lattice equations. In the continuous limit, one can obtain from (V_4) with $\nu = 0$ the well-known Krichever-Novikov equation [549].

Let us replace the coefficients of the polynomial (3.3.22) by nine arbitrary n-dependent constants, as one can see in (2.4.198, 2.4.199) and test such equation for integrability. The identity

$$2f_n = 2\frac{\rho_n}{v_n} - \frac{\partial \rho_n}{\partial u_{n+1}} = 2\frac{\tilde{\rho}_n}{v_n} + \frac{\partial \tilde{\rho}_n}{\partial u_{n-1}}\,, \tag{3.4.26}$$

where we use the notations $v_n = u_{n+1} - u_{n-1}$,

$$\rho_n = p_n u_{n+1}^2 + 2q_n u_{n+1} + r_n\,, \qquad \tilde{\rho}_n = p_n u_{n-1}^2 + 2q_n u_{n-1} + r_n\,,$$

is very helpful in the following calculations.

We can express the integrability condition (3.2.56) in the form

$$\dot{p}_n^{(1)} \sim \Phi_n = 2\frac{\rho_n - \tilde{\rho}_{n+1}}{v_{n+1}v_n} + \varphi_n(u_{n+1}, u_n, u_{n-1}) \sim 0\,,$$

where the precise formula for φ_n is not necessary. As in the n-independent case (see the property (3.2.24)), we have the following consequence of such condition: $\frac{\partial^2 \Phi_n}{\partial u_{n+2} \partial u_{n-1}} = 0$ for any n. Hence $\rho_n = \tilde{\rho}_{n+1}$, i.e. the following nine equations:

$$\begin{array}{lll} \alpha_n = \alpha_{n+1} = \alpha\,, & \sigma_n = \sigma_{n+1} = \sigma\,, & \pi_n = \pi_{n+1} = \pi\,, \\ \beta_n = \beta_{n+2}\,, & \gamma_n = \gamma_{n+2}\,, & \delta_n = \delta_{n+2}\,, \\ \tilde{\beta}_n = \beta_{n+1}\,, & \tilde{\gamma}_n = \gamma_{n+1}\,, & \tilde{\delta}_n = \delta_{n+1}, \end{array} \tag{3.4.27}$$

take place for all n. In another words, α_n, σ_n, π_n must be constant, while β_n, γ_n, δ_n must be two-periodic.

The equations of the form (2.4.198, 2.4.199), with n-dependent constants given by (3.4.27), satisfy all the five integrability conditions. For example, the conditions (3.2.56, 3.2.100) provide us with conserved densities of orders 2 and 3

$$p_n^{(1)} \sim \log \rho_n - 2 \log v_n\,, \qquad -2p_n^{(2)} \sim 4\frac{\rho_n}{v_{n+1} v_n} + \frac{\partial^2 \rho_n}{\partial u_{n+1} \partial u_n}\,.$$

The new n-dependent equation has the same master symmetry (3.3.31) as in the case of (V_4) with $\nu = 0$ [27]. Introducing a dependence on τ into the coefficients of the polynomial $\rho_n(u_{n+1}, u_n)$, we obtain a function $\rho_n(\tau, u_{n+1}, u_n)$ such that the τ-dependence is given by the following PDE

$$2\frac{\partial \rho_n}{\partial \tau} = \rho_n \frac{\partial^2 \rho_n}{\partial u_{n+1} \partial u_n} - \frac{\partial \rho_n}{\partial u_{n+1}} \frac{\partial \rho_n}{\partial u_n}\,,$$

which must be satisfied for any n (cf. (3.3.33)).

Eq. (2.4.198, 2.4.199) is the non autonomous generalization of the YdKN (V_4) with $\nu = 0$. This equation has been obtained for the first time in [548, 549] and has been discussed at length in Section 2.4.7.5. Its integrability has been also proved by studying its algebraic entropy. A linear-fractional transformation (3.3.24), with the coefficients c_i replaced by two-periodic n-dependent constants, does not change the form (2.4.198, 2.4.199) of the equation and the conditions (3.4.27). In general, such transformation allows one to remove only one of three two-periodic constants β_n, γ_n, δ_n, i.e. we have obtained a non-trivial integrable generalization of (V_4) with $\nu = 0$. As the generalization is two-periodic, it can be rewritten as an n-independent system. In fact, using the transformation (3.4.9) together with $\tilde{t} = t/2$ and the identity (3.4.26), we can express the equation as a system (3.3.46) with the function $\rho_{2n}(u_n, v_n)$ instead of $\rho(u_n, v_n)$. More precisely, we have now the polynomial

$$\begin{aligned} \rho_{2n}(u_n, v_n) \;\; = \;\; & \alpha u_n^2 v_n^2 + 2\beta_0 u_n^2 v_n + 2\beta_1 u_n v_n^2 + 2\lambda u_n v_n + \\ & \gamma_0 u_n^2 + \gamma_1 v_n^2 + 2\delta_1 u_n + 2\delta_0 v_n + \pi \end{aligned}$$

with nine arbitrary constants (cf. (3.3.32)).

Example 3 Let us consider the class of equations

$$\dot{u}_n = \varphi_n(u_{n+1} - u_{n-1}) \tag{3.4.28}$$

and discuss integrable examples of this form with no classification results. In Example 1 we have obtained a class of integrable equations of the form (3.4.17) defined by a polynomial P_n, which satisfies the conditions (3.4.18, 3.4.19, 3.4.21, 3.4.22, 3.4.25). If we define the function $\varphi_n(z)$ by the following ordinary differential equation: $\varphi_n' = P_n(\varphi_n)$, we can transform (3.4.28) into an integrable equation (3.4.17) by the transformation

$$\tilde{u}_n = \varphi_n(u_{n+1} - u_{n-1})\,. \tag{3.4.29}$$

Eq. (3.4.28) with the function $\varphi_n(z)$ is integrable too, and the transformation (3.4.29) allows us to construct a hierarchy of local conservation laws, as it has been done in Section 3.2.7.

In this way we obtain, for example, the following n-independent functions $\varphi_n(z)$: e^z, z^{-1}, $\tanh z$ defining integrable equations of the form (3.4.28). These equations, of course, can be found in the list (V_1-V_{11}) of Section 3.3.1.2 as particular cases of (V_6, V_7, V_9). There are the following two-periodic examples of $\varphi_n(z)$:

(3.4.30) $$\phi_n = \chi_{n+1} e^z + \chi_n z\,, \qquad \phi_n = \chi_{n+1} \tanh z + \chi_n z^{-1},$$

with χ_n given by (2.4.148). These functions are such that

$$\varphi_n' = \chi_{n+1}\varphi_n + \chi_n\,, \qquad \varphi_n' = \chi_{n+1} - \varphi_n^2\,.$$

This is the reason why the corresponding equations are integrable. We see, in particular, that the first of (3.4.28, 3.4.30) is transformed by (3.4.29) exactly into the Toda lattice in the form (3.4.12). It turns out that this equation admits an integrable generalization with no n-periodicity.

The generalization reads

(3.4.31) $$\dot u_n = \chi_{n+1}\left(\exp\frac{v_n}{\pi_n} + \frac{\sigma_n}{\pi_n}\right) + \chi_n \pi_{n+1}\pi_{n-1} v_n\,,$$

where $v_n = u_{n+1} - u_{n-1}$, and π_n, σ_n are n-dependent constants with linear dependence on n

(3.4.32) $$\pi_n = \alpha n + \beta \neq 0\,, \qquad \sigma_n = \gamma n + \delta,\ \forall n.$$

for all n. Using the point transformations $\tilde u_n = s_n(u_n)$, we cannot make this equation neither n-independent nor two-periodic. All the five integrability conditions are satisfied, and the conditions (3.2.56, 3.2.100, 3.2.106) provide us with two nontrivial conserved densities

(3.4.33) $$p_n^{(1)} \sim 2\alpha\rho_n^{(1)}, \qquad \rho_n^{(1)} = \frac{\chi_n u_n}{\pi_{n+1}\pi_{n-1}}\,, \qquad p_n^{(2)} \sim 0\,,$$

(3.4.34) $$p_n^{(3)} \sim \rho_n^{(2)} = 2\chi_{n+1}\pi_n \exp\frac{v_n}{\pi_n} + \chi_n(\pi_{n+1}u_{n+1} - \pi_{n-1}u_{n-1})^2,$$

where $\rho_n^{(1)}$ and $\rho_n^{(2)}$ are the conserved densities of orders 0 and 2.

The standard transformation (3.4.29) gives in this case (3.4.17) with

$$P_n = \chi_{n+1}\pi_n^{-2}(\pi_n u_n - \sigma_n) + \chi_n \pi_{n+1}\epsilon_{n-1}\,,$$

which is integrable only if $\alpha = \gamma = 0$. It turns out that there exists the following complicate transformation

(3.4.35) $$\tilde u_n = \chi_{n+1}\pi_n \exp\frac{v_n}{\pi_n} + \chi_n(\pi_{n+1}u_{n+1} - \pi_{n-1}u_{n-1} - 2\gamma t)$$

of (3.4.31, 3.4.32) into the Toda lattice (3.4.12). This allows one to construct a hierarchy of conservation laws for the new equation. For example, three of the conserved densities (3.4.14, 3.4.15) give nothing new with respect to (3.4.33, 3.4.34), while the last of densities (3.4.15) provides the equation with a conserved density of order 3. It is interesting that if $\gamma \neq 0$, then the transformation (3.4.35) depends on time, and the conservation laws for (3.4.31, 3.4.32) become t-dependent. A simplest nontrivial example of t-dependent conserved density can be obtained from the last of the densities (3.4.15).

4.2. Toda type equations with an explicit n and t dependence. Following [852], we consider here n, t-dependent equations of the form

$$\ddot{u}_n = f_n(t, \dot{u}_n, u_{n+1}, u_n, u_{n-1}), \tag{3.4.36}$$

where

$$\frac{\partial f_n}{\partial u_{n+1}} \neq 0, \quad \frac{\partial f_n}{\partial u_{n-1}} \neq 0 \quad \text{for all } n, \tag{3.4.37}$$

possessing n, t-dependent generalized symmetries and conservation laws. In Section 2.3.2.7 we constructed equations of the class (3.4.36) by considering non isospectral deformations of the discrete Schrödinger spectral problem associated to the Toda lattice. Recall that, in the case of Volterra type equations, we have discussed examples of such generalized symmetries in Section 3.2.7 (just after Definition 11) and of conservation laws at the end of Section 3.4.1.2. There are in this case no new examples of integrable equations with local symmetry structure, but equations with n-dependent coefficients. However, we derive here the integrability conditions which illustrate the specific features of the t-dependent case and allow one to check any given equation for integrability.

We will mainly pay attention to the difference between the class (3.4.36) and the two classes discussed in Sections 3.2.8 and 3.4.1.1. As usually, let us pass from (3.4.36) to systems of the form

$$\dot{u}_n = v_n, \quad \dot{v}_n = f_n(t, v_n, u_{n+1}, u_n, u_{n-1}). \tag{3.4.38}$$

We will consider also generalizations of f_n given by

$$\phi_n(t, u_{n+k_1}, v_{n+k_2}, u_{n+k_1-1}, v_{n+k_2-1}, \dots u_{n+k_1'}, v_{n+k_2'}), \tag{3.4.39}$$

where $k_1 \geq k_1'$, $k_2 \geq k_2'$. Here we assume, in general, that

$$\frac{\partial \phi_n}{\partial u_{n+k_1}} \neq 0, \quad \frac{\partial \phi_n}{\partial u_{n+k_1'}} \neq 0, \quad \frac{\partial \phi_n}{\partial v_{n+k_2}} \neq 0, \quad \frac{\partial \phi_n}{\partial v_{n+k_2'}} \neq 0$$

for at least some n. Of course, simpler cases are possible: with no dependence on u_{n+i} (the derivatives $\frac{\partial \phi_n}{\partial v_{n+k_2}}, \frac{\partial \phi_n}{\partial v_{n+k_2'}}$ satisfy the same condition), with no dependence on v_{n+i}, and the case when $\phi_n = \phi_n(t)$.

We can express the system (3.4.38) and any of its generalized symmetries in vector form, in terms of the vector $U_n = \begin{pmatrix} u_n \\ v_n \end{pmatrix}$

$$\dot{U}_n = F_n(t, U_{n+1}, U_n, U_{n-1}) = \begin{pmatrix} v_n \\ f_n \end{pmatrix}, \tag{3.4.40}$$

$$U_{n,\tau} = G_n(t, U_{n+m}, U_{n+m-1}, \dots U_{n+m'}) = \begin{pmatrix} \varphi_n \\ \psi_n \end{pmatrix}, \tag{3.4.41}$$

where $m \geq m'$ and φ_n, ψ_n are functions of the form (3.4.39). We suppose that $\frac{\partial G_n}{\partial U_{n+m}} \neq 0$, $\frac{\partial G_n}{\partial U_{n+m'}} \neq 0$ for some n (see the definition (3.2.224)), and the numbers m and m' are called left and right orders of the symmetry. The compatibility condition for (3.4.40, 3.4.41) is standard

$$D_t G_n = D_\tau F_n = \sum_i \frac{\partial F_n}{\partial U_{n+i}} G_{n+i},$$

but the operator D_t acts on the vector function G_n and any scalar function (3.4.39) in the following way

$$D_t G_n = \frac{\partial G_n}{\partial t} + \sum_i \frac{\partial G_n}{\partial U_{n+i}} F_{n+i} ,$$
$$D_t \phi_n = \frac{\partial \phi_n}{\partial t} + \sum_i \left(\frac{\partial \phi_n}{\partial u_{n+i}} v_{n+i} + \frac{\partial \phi_n}{\partial v_{n+i}} f_{n+i} \right) .$$

As in Section 3.4.1.1, we say that two scalar functions a_n, b_n of the form (3.4.39) are equivalent if there exists another function c_n of this class: $a_n - b_n = (S-1)c_n$ for all n. There is the following criteria for the scalar functions (3.4.39)

(3.4.42) $$\phi_n \sim 0 \quad \text{iff} \quad \frac{\delta \phi_n}{\delta u_n} = \frac{\delta \phi_n}{\delta v_n} = 0 \quad \text{for all } n .$$

The formal variational derivatives $\frac{\delta}{\delta u_n}$, $\frac{\delta}{\delta v_n}$ are given by (3.2.225). In particular, $\phi_n(t) \sim 0$.

A conservation law of the system (3.4.38) is a relation (3.2.15), valid for all n, where scalar functions p_n, q_n are of the form (3.4.39). The conservation law is trivial if $p_n \sim 0$. The order of a nontrivial conservation law and of its conserved density p_n can be defined, using the vector function (3.2.226). The order equals 0 if $\varrho_n = \varrho_n(t, U_n) \neq 0$ for some n (the case $\frac{\partial \varrho_n}{\partial U_n} = 0$ is possible too). The order is equal to $m > 0$ if

$$\varrho_n = \varrho_n(t, U_{n+m}, U_{n+m-1}, \dots U_{n-m})$$

and $\frac{\partial \varrho_n}{\partial U_{n+m}} \neq 0$, $\frac{\partial \varrho_n}{\partial U_{n-m}} \neq 0$ for some n.

The other main formulas, definitions and statements are the same as in Section 3.2.8. Below we write down five integrability conditions for (3.4.36, 3.4.37) analogous to (3.3.34, 3.3.35) of Section 3.3.2. Two of them will be the same as (3.3.35). The other three will differ from (3.3.34).

Let us explain why there is a difference by the following calculation. We consider the standard equation of formal symmetries of (3.4.38)

(3.4.43) $$\dot{L}_n = [F_n^*, L_n]$$

(cf. (3.2.232,3.2.233) of Section 3.2.8). Let us recall that $L_n = \sum_{i=-\infty}^{N} l_n^{(i)} S^i$ with 2×2 matrix coefficients $l_n^{(i)}$, while F_n^* is given by (3.2.228, 3.2.229). As in Section 3.2.8, in order to derive the integrability conditions, we can use a formal symmetry L_n of an arbitrarily big length, which is such that $N = 1$ and $\det l_n^{(1)} \neq 0$ for any n.

Collecting the coefficients of the powers S^2, S in (3.4.43), one obtains the equations

(3.4.44) $$\beta_n^{(1)} f_{n+1}^{(1)} = 0 ,$$
(3.4.45) $$\delta_n^{(1)} - \alpha_n^{(1)} = \beta_n^{(1)} f_{n+1}^{(v)} + \dot{\beta}_n^{(1)} ,$$
(3.4.46) $$f_n^{(1)} \alpha_{n+1}^{(1)} = \delta_n^{(1)} f_{n+1}^{(1)}$$

for the elements of $l_n^{(i)}$ given by (3.2.235). As $f_n^{(1)} \neq 0$ for any n due to the condition (3.4.37), then (3.4.44, 3.4.45) imply: $\beta_n^{(1)} = 0$, $\delta_n^{(1)} = \alpha_n^{(1)}$ for all n. For this reason, (3.4.46) can be rewritten as: $(S-1)(\alpha_n^{(1)}/f_n^{(1)}) = 0$.

The $\ker(S-1)$ consists of functions of the form $c(t)$. That is why we can write: $\alpha_n^{(1)} = \mu^2(t) f_n^{(1)}$, where $\mu \neq 0$ because $\det l_n^{(1)} \neq 0$. Passing, as usually, to a new formal symmetry $\tilde{L}_n = \mu^{-2} L_n$, we obtain: $\alpha_n^{(1)} = f_n^{(1)}$. However, unlike the t-independent case, (3.4.43) changes its form

$$\dot{L}_n + \frac{2\mu'}{\mu} L_n = [F_n^*, L_n] \,. \tag{3.4.47}$$

Now we find further coefficients of L_n from (3.4.47) and derive integrability conditions in the standard way, i.e. as in the case of (3.4.43) (see details in [852]). The resulting conditions, rewritten in terms of (3.4.36, 3.4.37), read

$$\begin{gathered} D_t p_n^{(i)} = (S-1) q_n^{(i)}, \qquad i = 1,2,3\,, \\ p_n^{(1)} = \log \frac{\partial f_n}{\partial u_{n+1}}\,, \qquad p_n^{(2)} = \mu\left(2q_n^{(1)} + \frac{\partial f_n}{\partial \dot{u}_n}\right), \qquad p_n^{(3)} = \\ 2(\mu q_n^{(2)} + n\mu' p_n^{(2)}) + \tfrac{1}{4}(p_n^{(2)})^2 + \mu^2\left(\frac{1}{4}\left(\frac{\partial f_n}{\partial \dot{u}_n}\right)^2 - \frac{1}{2} D_t \frac{\partial f_n}{\partial \dot{u}_n} + \frac{\partial f_n}{\partial u_n}\right), \end{gathered} \tag{3.4.48}$$

where $q_n^{(i)}$ are functions of the form (3.4.39) with $v_{n+j} = \dot{u}_{n+j}$. In the derivation of the two additional integrability conditions, which come from the existence of conservation laws, there is nothing new. Let us formulate a precise statement similar to Theorem 51 on the necessary conditions of integrability:

Theorem 61. *Let* (3.4.36, 3.4.37) *have one generalized symmetry of the left order* $m_1 \geq 5$ *and one conservation law of order* $m_2 \geq 4$. *Then this equation satisfies* (3.4.48) *and* (3.3.35) *of Section* 3.3.2. *These conditions require the existence of a function* $\mu = \mu(t) \neq 0$ *and functions* $q_n^{(i)}$, $\sigma_n^{(j)}$ *of the form* (3.4.39) *with* $v_{n+k} = \dot{u}_{n+k}$, *which are the solutions of* (3.4.48, 3.3.35).

Point transformations, which can be used in this case, are of the form (3.3.38), where $\Theta' \neq 0$ and $\frac{\partial s_n}{\partial u_n} \neq 0$ for any n. The functions $q_n^{(i)}$, $\sigma_n^{(j)}$ are defined up to five t-dependent constants of integration: $c_i(t)$, $\hat{c}_j(t)$. Unlike the t-independent case, we cannot remove the functions μ, c_i, $\hat{c}_j$ and must consider all possibilities when checking the integrability conditions.

Let us consider the equation

$$\ddot{u}_n = e^{u_{n+1}-u_n} - e^{u_n - u_{n-1}} - \alpha \dot{u}_n - 2\alpha^2 n\,, \tag{3.4.49}$$

where $\alpha \neq 0$ is a constant, which will give us an example with $\mu' \neq 0$. Checking its integrability, we will use the conditions (3.4.48) with $i = 1,2$. The density $p_n^{(1)}$ has the form $p_n^{(1)} = (S-1)u_n$, hence $q_n^{(1)} = \dot{u}_n + \beta(t)$, where β is an arbitrary integration function. Then $p_n^{(2)} \sim 2\mu\dot{u}_n$, and we obtain for its time derivative

$$D_t p_n^{(2)} \sim 2(\mu'\dot{u}_n + \mu\ddot{u}_n) \sim 2(\mu' - \alpha\mu)\dot{u}_n \sim 0\,.$$

Now the criteria (3.4.42) with $v_n = \dot{u}_n$ gives: $\mu' = \alpha\mu$. We see that $\mu' \neq 0$ because $\alpha\mu \neq 0$. The equation with such μ satisfies all five integrability conditions. This is not surprising, as the change of variables

$$\tilde{u}_n = u_n + 2n(\alpha t - \log\alpha)\,, \qquad \tilde{t} = e^{-\alpha t}$$

transforms (3.4.49) into the Toda lattice (3.2.216) (see also [685]).

There is no classification result in the case of (3.4.36, 3.4.37). For this reason, it is useful to prove the following very simple and general necessary condition of integrability, which require that an integrable equation may have only quadratic dependence on $\dot{u}_n$.

Theorem 62. *The conditions* (3.3.35) *imply the following necessary condition of integrability for* (3.4.36, 3.4.37)

$$\partial^3 f_n/\partial \dot{u}_n^3 = 0 \quad \text{for all } n\,. \tag{3.4.50}$$

PROOF. Taking into account the criteria (3.4.42) with $v_n = \dot{u}_n$, we obtain from the first of conditions (3.3.35): $\frac{\delta r_n^{(1)}}{\delta \dot{u}_n} = \frac{\partial r_n^{(1)}}{\partial \dot{u}_n} = 0$. Then it can be proved that

$$\sigma_n^{(1)} = \sigma_n^{(1)}(t, u_n, u_{n-1})\,, \tag{3.4.51}$$

and consequently

$$r_n^{(2)} = \frac{\partial \sigma_n^{(1)}}{\partial t} + \frac{\partial \sigma_n^{(1)}}{\partial u_n}\dot{u}_n + \frac{\partial \sigma_n^{(1)}}{\partial u_{n-1}}\dot{u}_{n-1} + \frac{\partial f_n}{\partial \dot{u}_n}\,.$$

Using the same criteria (3.4.42), we get:

$$\frac{\delta r_n^{(2)}}{\delta \dot{u}_n} = \frac{\partial(\sigma_n^{(1)} + \sigma_{n+1}^{(1)})}{\partial u_n} + \frac{\partial^2 f_n}{\partial \dot{u}_n^2} = 0, \ \forall n. \tag{3.4.52}$$

Now the condition (3.4.50) follows from (3.4.51, 3.4.52). □

In Section 2.3.2.7 [497] one can find a sequence of inhomogeneous Toda lattice equations with n-dependent coefficients associated to discrete Schrödinger type spectral problems. All the resulting equations satisfy Theorem 62.

As we mentioned above, there is in this case no new integrable example with the local symmetry structure. In particular, the following negative result takes place (its proof can be found in [852]):

Theorem 63. *Any integrable equation, belonging to the subclass*

$$\ddot{u}_n = \varphi_n(t, u_{n+1}, u_n) + \psi_n(t, u_n, u_{n-1})$$

of the class (3.4.36, 3.4.37)*, is transformed by a point transformation of the form* (3.3.38) *into the Toda lattice* (3.2.216) *or into an equation linear in u_{n+1}, u_n and u_{n-1}.*

There is, however, an interesting different example, very similar to the Toda lattice [549]. The equation is

$$\frac{\ddot{u}_n}{c_{n+1}c_n} = \exp\frac{u_{n+1} - u_n}{c_{n+1}} - \exp\frac{u_n - u_{n-1}}{c_n}\,, \qquad c_n = an + b\,, \tag{3.4.53}$$

where a, b are constants, such that $ab \neq 0$. It is related by a non point and non invertible change of variables to the integrable (T_3) of Section 3.3.2 which is integrable, as it is shown in Theorem 53. The transformation of solutions u_n of (T_3) with $\mu = 0$ into solutions $\tilde{u}_n$ of (3.4.53) reads:

$$\tilde{u}_n = c_{n+1}c_n(S-1)\frac{u_n + d_n}{c_n}\,, \tag{3.4.54}$$

where d_n is given by: $d_{n+1} - 2d_n + d_{n-1} = \log c_n^{-1}$. Eq. (3.4.53) is integrable in the sense that we can construct its solutions with the help of the transformation (3.4.54). However, this transformation provides (3.4.53) with non local symmetries and conservation laws, which cannot be expressed in terms of functions of the form (3.4.39). We can show that (3.4.53) does not have the local symmetry structure by using the conditions (3.4.48) with $i = 1$. In fact,

$$D_t p_n^{(1)} = \frac{\dot{u}_{n+1} - \dot{u}_n}{c_{n+1}} \sim \left(\frac{1}{c_n} - \frac{1}{c_{n+1}}\right)\dot{u}_n \sim 0.$$

The criteria (3.4.42) with $v_n = \dot{u}_n$ gives: $c_{n+1} - c_n = a = 0$, i.e. we are led to a contradiction.

4.3. Example of relativistic Toda type. The lattice equations we discuss in this section have a dependence on the discrete variable n of a quite different type compared with sections 3.4.1 and 3.4.2. These equations have arbitrary n-dependent parameters. As it is shown in [24, 755], relativistic Toda type equations generate auto-Bäcklund transformations for NLS type equations. Such a Bäcklund transformation admits a generalization with an arbitrary constant parameter. When we pass to a chain of Bäcklund transformations, we obtain an integrable lattice equation with an arbitrary n-dependent parameter.

Such n-dependent relativistic Toda type equations together with Lax pairs are presented in [26] (in the Lagrangian form) and in [33] (in the Hamiltonian form). The simplest examples are:

$$\ddot{u}_n = \dot{u}_n \left(\frac{\dot{u}_{n+1}}{1 + e^{u_n - u_{n+1}}} - \frac{\dot{u}_{n-1}}{1 + e^{u_{n-1} - u_n}} - \nu_n e^{u_{n+1} - u_n} + \nu_{n-1} e^{u_n - u_{n-1}} \right), \tag{3.4.55}$$

$$\dot{u}_n = u_{n+1} + \nu_n u_n + u_n^2 v_n, \qquad \dot{v}_n = -v_{n-1} - \nu_n v_n - v_n^2 u_n, \tag{3.4.56}$$

where ν_n is an arbitrary n-dependent coefficient. Eq. (3.4.55) generalizes the relativistic Toda equation (3.2.248), cf. also (Ld_3) of Section 3.3.3.4. The system (3.4.56) generalizes (3.3.72). The system (3.4.56) together with its Lax pair has been considered even earlier in [755], where it is shown that this system is an auto-Bäcklund transformation for the NLS.

We, however, discuss here in more details a different, more complicate, example. Following [27], we consider a generalization of (H_3) with $\alpha = \beta = 0$ considered in Section 3.3.3.2. This generalization has the same form as (3.3.117)

$$\dot{u}_n = \frac{2r_n}{u_{n+1} - v_n} + \frac{\partial r_n}{\partial v_n}, \qquad \dot{v}_n = \frac{2r_n}{u_n - v_{n-1}} - \frac{\partial r_n}{\partial u_n}, \tag{3.4.57}$$

however in this case r_n is not defined as in (3.3.119) but by

$$r_n = r_n(u_n, v_n), \qquad \frac{\partial^3 r_n}{\partial u_n^3} = \frac{\partial^3 r_n}{\partial v_n^3} = 0. \tag{3.4.58}$$

r_n is an arbitrary bi-quadratic polynomial, which may be asymmetric, with 9 arbitrary n-dependent coefficients

$$r_n = \alpha_n u_n^2 v_n^2 + \beta_n u_n^2 v_n + \hat{\beta}_n u_n v_n^2 + \gamma_n u_n v_n + \delta_n u_n^2 + \hat{\delta}_n v_n^2 + \pi_n u_n + \hat{\pi}_n v_n + \mu_n.$$

In the integrable cases, these coefficients must satisfy one more condition. Let us introduce the discriminators of r_n:

$$d_n^{(u)}(v_n) = \left(\frac{\partial r_n}{\partial u_n} \right)^2 - 2r_n \frac{\partial^2 r_n}{\partial u_n^2}, \qquad d_n^{(v)}(u_n) = \left(\frac{\partial r_n}{\partial v_n} \right)^2 - 2r_n \frac{\partial^2 r_n}{\partial v_n^2}, \tag{3.4.59}$$

cf. (3.3.47). The functions $d_n^{(u)}$ and $d_n^{(v)}$ depend only on v_n and u_n, respectively, and are 4th degree polynomials. The condition for coefficients of r_n is that

$$d_n^{(u)}(z) = d_{n+1}^{(v)}(z) \tag{3.4.60}$$

for any values of n, z.

The form of (3.4.57, 3.4.58), as well as condition (3.4.60), is invariant under the linear-fractional transformation

$$\tilde{u}_n = \frac{\nu_n^1 u_n + \nu_n^2}{\nu_n^3 u_n + \nu_n^4}, \qquad \tilde{v}_n = \frac{\nu_{n+1}^1 v_n + \nu_{n+1}^2}{\nu_{n+1}^3 v_n + \nu_{n+1}^4}, \qquad \tilde{t} = \eta t, \tag{3.4.61}$$

with n-dependent coefficients v_n^i and constant η. Only the coefficients of r_n are changed. This system is Hamiltonian, as

$$\dot{u}_n = r_n \frac{\delta h_n}{\delta v_n}, \qquad \dot{v}_n = -r_n \frac{\delta h_n}{\delta u_n},$$

where the formal variational derivatives are defined by (3.2.225), and the Hamiltonian density is $h_n = \log \frac{r_n}{(u_{n+1}-v_n)^2}$.

System (3.4.57) is integrable, as it has a master symmetry. As in the case of (3.3.117), the master symmetry has the form (3.3.119), and a dependence on τ is introduced in the coefficients of r_n by (3.3.120), i.e. by

$$\frac{\partial r_n}{\partial \tau} = r_n \frac{\partial^2 r_n}{\partial u_n \partial v_n} - \frac{\partial r_n}{\partial u_n}\frac{\partial r_n}{\partial v_n}, \tag{3.4.62}$$

which is to hold for all n. Conserved densities of the system (3.4.57, 3.4.58, 3.4.60) are constructed, starting from the Hamiltonian density h_n. It can be checked by direct calculation that $D_\tau h_n \sim \rho_n$, where

$$\rho_n = -\frac{4r_n}{(u_{n+1}-v_n)(u_n-v_{n-1})} + \frac{2\partial r_n/\partial u_n}{u_{n+1}-v_n} - \frac{2\partial r_n/\partial v_n}{u_n - v_{n-1}} + \frac{\partial^2 r_n}{\partial u_n \partial v_n},$$

and this function is a conserved density for the system (3.4.57, 3.4.58, 3.4.60). As system (3.4.57) is Hamiltonian, the following system

$$u_{n,\epsilon'} = r_n \frac{\delta \rho_n}{\delta v_n}, \qquad v_{n,\epsilon'} = -r_n \frac{\delta \rho_n}{\delta u_n}$$

is its simplest generalized symmetry.

So, the integrable cases are defined by (3.4.60), equivalent to a system of 5 discrete equations for the 9 coefficients of the polynomial r_n. It is a difficult open problem to describe in a more clear and explicit form all particular cases of the system (3.4.57, 3.4.58, 3.4.60) up to transformations (3.4.61). Below we discuss a few integrable examples of this kind.

First we consider a generalization of (H_3), Case 1 with $\beta = 0$, of Section 3.3.3.2, such that

$$r_n = \delta_n(u_n - v_n)^2 + \sigma_n(u_n - v_n) + \mu_n,$$

where $\delta_n \neq 0$ for all n. A change of variables $\tilde{u}_n = u_n + \nu_n$, $\tilde{v}_n = v_n + \nu_{n+1}$, where $\nu_n - \nu_{n+1} = \sigma_n/(2\delta_n)$, allows us to reduce r_n to $r_n = \delta_n(u_n - v_n)^2 + \mu_n$. Introducing the function $c_n = \delta_n \mu_n$, we see that $d_n^{(u)}(v_n) = d_n^{(v)}(u_n) = -4c_n$. It follows from (3.4.60) that $c_n = c_{n+1} = c$ with no dependence on n. We are led to the polynomial

$$r_n = \delta_n(u_n - v_n)^2 + c/\delta_n$$

corresponding to an integrable case. Eq. (3.4.62) implies that c does not depend on τ, and $\delta_n = \delta_n(\tau)$ satisfies the ODE $\delta_n' = 2\delta_n^2$ for all n.

Another example is given by

$$r_n = \gamma_n u_n v_n + \delta_n(u_n^2 + v_n^2). \tag{3.4.63}$$

From (3.4.60) it follows that the function $\gamma_n^2 - 4\delta_n^2$ does not depend on n, and we can define δ_n as:

$$4\delta_n^2 = \gamma_n^2 - c, \tag{3.4.64}$$

where c is an arbitrary constant. An integrable system is defined by (3.4.63, 3.4.64). Eq. (3.4.62) implies that c does not depend on τ, and $\gamma_n(\tau)$ satisfies the ODE $\gamma_n' = c - \gamma_n^2$.

The last example is when one of the discriminators (3.4.59), e.g. $d_n^{(u)}$, has no multiple roots for any n, i.e. the curve $w^2 = d_n^{(u)}(z)$ is elliptic. As it is shown in [27], the transformation (3.4.61) allows us to transform both discriminators into the Weierstrass form:

$$d_n^{(u)}(z) = d_n^{(v)}(z) = P(z) = 4z^3 - g_2 z - g_3,$$

with n-independent coefficients g_2, g_3. For such discriminators,

$$r_n = \frac{1}{\sqrt{P(\pi_n)}}\left(\left(u_n v_n + (u_n + v_n)\pi_n + \frac{g_2}{4}\right)^2 - (u_n + v_n + \pi_n)(4u_n v_n \pi_n - g_3)\right),$$

with one n-dependent parameter π_n. This polynomial r_n satisfies the condition (3.4.60), i.e. defines an integrable system. Eq. (3.4.62) gives the following ODE for $\pi_n(\tau)$: $\pi_n' = \sqrt{P(\pi_n)}$; g_2, g_3 have no dependence on τ.

The Lax pairs for the three systems discussed above can be found in [33]. It is also shown in [33] that these integrable systems generate auto-Bäcklund transformations for the stereographic projections of the Heisenberg and Landau-Lifshitz equations.

5. Other types of lattice equations

Here we discuss some other types of lattice equations to which the generalized symmetry method can be applied. In Section 3.5.1 we consider scalar evolutionary DΔEs of an arbitrary order. Then, in Section 3.5.2, we present some results concerning the multi-component DΔEs, i.e. systems of arbitrarily many equations.

5.1. Scalar evolutionary DΔEs of an arbitrary order. Let us consider n, t–independent equations of order k

(3.5.1) $$\dot{u}_n = f(u_{n+k}, u_{n+k-1}, \dots u_{n-k}) \equiv f_n, \qquad \frac{\partial f_n}{\partial u_{n+k}} \frac{\partial f_n}{\partial u_{n-k}} \neq 0,$$

where $k > 0$. Such equations are symmetric in the sense of Section 3.2.4.1. So they satisfy the necessary conditions for the integrability and may have not only higher order generalized symmetries but also conservation laws.

Integrable examples of this kind are given by Volterra type equations, presented in Section 3.3.1.2, and by their generalized symmetries, such as (3.2.14). A quite different type of examples is provided by the equations

(3.5.2) $$\dot{u}_n = u_n \sum_{i=1}^{k} (u_{n+i} - u_{n-i})$$

integrable for any $k > 0$. Eq. (3.5.2) can be called the Narita-Itoh-Bogoyavlensky equation [105, 420, 627]; if $k = 1$, we have the Volterra equation. Some integrable modifications of these equations are also known, see e.g. [105, 106, 308].

We are going to derive for the class (3.5.1) the following integrability conditions

(3.5.3) $$\dot{p}_n^{(1)} = (S - 1)q_n^{(1)}, \qquad p_n^{(1)} = \log \frac{\partial f_n}{\partial u_{n+k}},$$

(3.5.4) $$\dot{p}_n^{(2)} = (S - 1)q_n^{(2)}, \qquad p_n^{(2)} = q_n^{(1)} + k\frac{\partial f_n}{\partial u_n},$$

$$r_n^{(1)} = (S-1)\sigma_n^{(1)}, \qquad r_n^{(1)} = \log\left(-\frac{\partial f_n}{\partial u_{n+k}} / \frac{\partial f_n}{\partial u_{n-k}}\right), \tag{3.5.5}$$

$$r_n^{(2)} = (S-1)\sigma_n^{(2)}, \qquad r_n^{(2)} = \dot{\sigma}_n^{(1)} + 2k\frac{\partial f_n}{\partial u_n}. \tag{3.5.6}$$

Here, as usually, the functions $q_n^{(i)}, \sigma_n^{(i)}$ are of the form (3.2.10), i.e. are functions of finite range.

Eqs. (3.5.3-3.5.6) are generalizations of the integrability conditions presented in Section 3.2.2–3.2.4. The conditions (3.5.3, 3.5.5, 3.5.6) will be derived for any $k > 0$, and proofs will be carried out under such general assumptions as in Theorems 23, 33 where these conditions have been obtained for the case $\nu = 1$. The integrability conditions (3.5.3, 3.5.5) can be found in [764, 843], but in the case of (3.5.3) we give here a simpler proof. The condition (3.5.4) will be proved only for $k = 1, 2$. In the case $k = 1$, following [549], we provide a more general result with respect to Theorem 29.

Let us consider two examples. It is easy to check that (3.5.2) for any k satisfies all the four integrability conditions (3.5.3–3.5.6), while (3.5.3, 3.5.4) provide the equation with two obvious conserved densities $\log u_n$ and u_n. The equation

$$\dot{u}_n = u_n^2(u_{n+2} + u_{n+1} - u_{n-1} - u_{n-2}) \tag{3.5.7}$$

is similar to (3.5.2) with $k = 2$ and is a second order analogue of the modified Volterra equation (3.2.185) with $c = 0$. The conditions (3.5.3, 3.5.5, 3.5.6) are satisfied, and the first of them gives the conserved density $\log u_n$. It follows from (3.5.4) that the function $2u_{n+2}u_n + u_{n+1}u_n$ must be a conserved density. It is easy to see that this is not true, thus (3.5.7) is not integrable.

We shall use in the proofs we present below the standard Lax–like equations for the formal symmetry L_n and the formal conserved density $\mathscr{S}_n$ of (3.5.1). They are

$$\dot{L}_n = [f_n^*, L_n], \tag{3.5.8}$$

$$\dot{\mathscr{S}}_n + \mathscr{S}_n f_n^* + f_n^{*\dagger}\mathscr{S}_n = 0, \tag{3.5.9}$$

where

$$f_n^* = \sum_{i=-k}^{k} f_n^{(i)} S^i, \qquad f_n^{*\dagger} = \sum_{i=-k}^{k} f_{n+i}^{(-i)} S^i, \qquad f_n^{(i)} = \frac{\partial f_n}{\partial u_{n+i}}.$$

Besides,

$$L_n = \sum_{i=-\infty}^{m} l_n^{(i)} S^i, \qquad \mathscr{S}_n = \sum_{i=-\infty}^{m} s_n^{(i)} S^i,$$

where the coefficients $l_n^{(i)}, s_n^{(i)}$ are n, t–independent functions of finite range, and $l_n^{(m)} s_n^{(m)} \neq 0$.

By analogy with Theorems 26 and 32, we can prove the following properties. If (3.5.1) has a generalized symmetry of left order $m > 0$, then it possesses a formal symmetry L_n of order m and length $\operatorname{lgt} L_n \geq m$. If there exists a conservation law of order $m > k$, then there is a formal conserved density $\mathscr{S}_n$, such that $\operatorname{ord}\mathscr{S}_n = m$ and $\operatorname{lgt}\mathscr{S}_n \geq m - k$.

Theorem 64. *If an equation of the form* (3.5.1) *possesses a generalized symmetry of left order $m > k$, then there must exist a function $q_n^{(1)}$ of finite range, such that the condition* (3.5.3) *is satisfied.*

PROOF. We have a formal symmetry with $\mathrm{ord} L_n = m$ and $\mathrm{lgt} L_n > k$. Such length allows us to collect in (3.5.8) coefficients at S^m that gives an equation containing the time–derivative $\dot{l}_n^{(m)}$. A form of the same equation can be obtained, collecting coefficients at S^0 in the equation

$$\dot{L}_n L_n^{-1} = [f_n^* L_n^{-1}, L_n] \tag{3.5.10}$$

which is equivalent to (3.5.8). Moreover, we can apply the residue (see (3.2.76) of Section 3.2.3) to (3.5.10) and obtain the following condition

$$\mathrm{res}(\dot{L}_n L_n^{-1}) = D_t \log l_n^{(m)} \sim 0. \tag{3.5.11}$$

To find $l_n^{(m)}$, we consider coefficients of S^{m+k} in (3.5.8): $f_n^{(k)} l_{n+k}^{(m)} = l_n^{(m)} f_{n+m}^{(k)}$. Applying the logarithm, we get

$$(S^k - 1) \log l_n^{(m)} = (S^m - 1) \log f_n^{(k)},$$

where $m, k > 0$. Then, applying the operator $(S-1)^{-1}$, we obtain

$$(1 + S + \cdots + S^{k-1}) \log l_n^{(m)} = c + (1 + S + \cdots + S^{m-1}) \log f_n^{(k)}, \tag{3.5.12}$$

where c is an integration constant. Applying the operator D_t to the last relation and passing to equivalent functions, we are led to: $k D_t \log l_n^{(m)} \sim m D_t \log f_n^{(k)}$. From the relation (3.5.11) it follows that $D_t \log f_n^{(k)} \sim 0$, i.e. the condition (3.5.3) is satisfied. □

Theorem 65. *If, in addition to a generalized symmetry of the left order $m > k$, an equation of the form (3.5.1) possesses a conservation law of order $m > 2k$, then there must exist functions $\sigma_n^{(1)}, \sigma_n^{(2)}$ of finite range, such that the conditions (3.5.5, 3.5.6) are satisfied.*

PROOF. From Theorem 64 it follows that we can use the condition (3.5.3). Besides we have a formal conserved density with $\mathrm{ord} \mathscr{S}_n = m > 2k$ and $\mathrm{lgt} S_n > k$.

Collecting coefficients of S^{m+k} in (3.5.9) and applying S^{-k}, we get

$$s_{n-k}^{(m)} f_{n+m-k}^{(k)} = -f_n^{(-k)} s_n^{(m)}.$$

An equivalent form of this equation is

$$(1 - S^{-k}) \log s_n^{(m)} = (S^{m-k} - 1) \log f_n^{(k)} + \log(-f_n^{(k)} / f_n^{(-k)}), \tag{3.5.13}$$

where $k > 0$, $m - k > 0$. We see that $\log(-f_n^{(k)} / f_n^{(-k)}) \sim 0$, i.e. the condition (3.5.5) is satisfied.

Moreover, applying to (3.5.13) the operator $(S-1)^{-1}$ and using the condition (3.5.5), we obtain

$$(S^{-1} + S^{-2} + \cdots + S^{-k}) \log s_n^{(m)} = (1 + S + \cdots + S^{m-k-1}) \log f_n^{(k)} + \sigma_n^{(1)} + c,$$

where c is an integration constant. Then, differentiating this equation with respect to t, using the condition (3.5.3), and passing to equivalent functions, we are led to the relation

$$k D_t \log s_n^{(m)} \sim D_t \sigma_n^{(1)}. \tag{3.5.14}$$

Eq. (3.5.9) is equivalent to: $\dot{\mathscr{S}}_n \mathscr{S}_n^{-1} + \mathscr{S}_n f_n^* \mathscr{S}_n^{-1} + f_n^{*\dagger} = 0$. The length $\mathrm{lgt} \mathscr{S}_n > k$ allows us to consider in this equation coefficients at S^0 and thus to apply the residue. Using the relation (3.5.14), we get

$$k \,\mathrm{res}(\dot{\mathscr{S}}_n \mathscr{S}_n^{-1} + f_n^* + f_n^{*\dagger}) = k(D_t \log s_n^{(m)} + 2 f_n^{(0)}) \sim D_t \sigma_n^{(1)} + 2k f_n^{(0)} \sim 0,$$

i.e. the condition (3.5.6) has been derived. □

Theorem 66. *If* (3.5.1) *with* $k = 1$, *possess a generalized symmetry of the left order* $m > 2$, *than it must satisfy* (3.5.4).

PROOF. We have a formal symmetry: $\mathrm{ord} L_n = m > 2$, $\mathrm{lgt} L_n > 2$. Let us introduce a notation for any $\ell \geq 0$: $\Phi_n^{(\ell)} = \prod_{i=0}^{\ell} f_{n+i}^{(1)}$. Collecting coefficients of S^{m+1} and then dividing by $\Phi_n^{(m)}$, we get: $(S-1)(l_n^{(m)}/\Phi_n^{(m-1)}) = 0$. Multiplying L_n by a constant, we obtain the following formula: $l_n^{(m)} = \Phi_n^{(m-1)}$.

Let us consider coefficients at S^m and divide the resulting equation by $\Phi_n^{(m-1)}$

$$D_t \log \Phi_n^{(m-1)} = (S-1)\left(\frac{l_n^{(m-1)}}{\Phi_n^{(m-2)}} - \sum_{i=0}^{m-1} f_{n+i}^{(0)}\right). \tag{3.5.15}$$

From Theorem 64 it follows that we can use (3.5.3). Due to (3.5.3) we have the following general formula

$$D_t \log \Phi_n^{(\ell)} = \sum_{i=0}^{\ell} \dot{p}_{n+i}^{(1)} = (S-1)\sum_{i=0}^{\ell} q_{n+i}^{(1)} = q_{n+\ell+1}^{(1)} - q_n^{(1)}. \tag{3.5.16}$$

By using (3.5.15, 3.5.16) we find

$$\frac{l_n^{(m-1)}}{\Phi_n^{(m-2)}} = c + \sum_{i=0}^{m-1} p_{n+i}^{(2)}, \tag{3.5.17}$$

where c is a constant, while $p_n^{(2)}$ is given by the second of (3.5.4).

Collecting coefficients of S^{m-1} and dividing by $\Phi_n^{(m-2)}$, we get

$$D_t \frac{l_n^{(m-1)}}{\Phi_n^{(m-2)}} + \frac{l_n^{(m-1)}}{\Phi_n^{(m-2)}}(f_{n+m-1}^{(0)} - f_n^{(0)} + D_t \log \Phi_n^{(m-2)}) =$$

$$f_n^{(-1)} f_{n-1}^{(1)} - f_{n+m}^{(-1)} f_{n+m-1}^{(1)} + (S-1)\frac{l_n^{(m-2)}}{\Phi_n^{(m-3)}} \sim 0.$$

The left hand side of this relation can be rewritten due to (3.5.16, 3.5.17) as

$$\sum_{i=0}^{m-1} \dot{p}_{n+i}^{(2)} + \left(c + \sum_{i=0}^{m-1} p_{n+i}^{(2)}\right)(p_{n+m-1}^{(2)} - p_n^{(2)}) \sim m\dot{p}_n^{(2)}.$$

So, $\dot{p}_n^{(2)} \sim 0$, i.e. the condition (3.5.4) is satisfied. □

In the case of (3.5.1) with $k = 2$, it is difficult to derive the condition (3.5.4) on the same level of generality as before, and an additional requirement will be introduced. For this reason, let us first discuss the symmetry structure of some known equations of the form (3.5.1).

Volterra type equations and their symmetries have generalized symmetries of any left order $m > 0$. Eq. (3.5.2) and their modifications have generalized symmetries of any left order $m = jk \geq k$, see e.g. [866]. For instance, the simplest generalized symmetry of (3.5.2) with $k = 2$ is

$$\begin{aligned} u_{n,\epsilon} = u_n(&u_{n+2}(u_{n+4} + u_{n+3} + u_{n+2} + u_{n+1} + u_n) \\ &+ u_{n+1}(u_{n+3} + u_{n+2} + u_{n+1} + u_n) \\ &- u_{n-1}(u_n + u_{n-1} + u_{n-2} + u_{n-3}) \\ &- u_{n-2}(u_n + u_{n-1} + u_{n-2} + u_{n-3} + u_{n-4})). \end{aligned} \tag{3.5.18}$$

It is not difficult to explain why (3.5.2) has no generalized symmetry of odd order.

In fact, if there were such symmetries, we would have a formal symmetry: $\operatorname{ord} L_n = m = 2\ell + 1 > 0$, $\operatorname{lgt} L_n > 0$. As it follows from the proof of Theorem 64, such length of the formal symmetry is sufficient to derive (3.5.12) which reads

$$(1+S)\log l_n^{(m)} = c + (1 + S + \cdots + S^{2\ell})\log u_n.$$

This condition can be rewritten as

$$\log u_n = (1+S)\Theta_n, \tag{3.5.19}$$

where $\Theta_n = \log l_n^{(m)} - c/2$ if $\ell = 0$ and

$$\Theta_n = \log l_n^{(m)} - \frac{c}{2} - \sum_{i=1}^{\ell} \log u_{n+2i-1}$$

if $\ell > 0$. It is easy to prove that (3.5.19) has no solution Θ_n of finite range (cf. (3.2.13) of Section 3.2.1 and its proof), i.e. we have a contradiction.

So, all known integrable equations of the form (3.5.1), namely, Volterra type equations together with their symmetries, Narita-Itoh-Bogoyavlensky type equations, and their symmetries like (3.5.18) have generalized symmetries of any left order $jk \geq k$. We may require the existence of two generalized symmetries of orders jk and $(j+1)k$. Using corresponding formal symmetries $\tilde{L}_n$ and $\hat{L}_n$, we construct a new formal symmetry $L_n = \hat{L}_n \tilde{L}_n^{-1}$, such that $\operatorname{ord} L_n = k$, $\operatorname{lgt} L_n \geq jk$ (see (3.2.93, 3.2.94) in Section 3.2.3).

Theorem 67. *Let (3.5.1) with $k = 2$ have two generalized symmetries of the left orders $2j > 5$ and $2(j+1)$. Then it must satisfy (3.5.4).*

PROOF. From Theorem 64 it follows that (3.5.3) is satisfied. Besides we have a formal symmetry with $\operatorname{ord} L_n = 2$ and $\operatorname{lgt} L_n > 5$. Applying the residue to (3.5.8), we get the condition $\dot{l}_n^{(0)} \sim 0$ which will be used below.

Collecting in (3.5.8) coefficients of S^4 and then dividing by $f_n^{(2)} f_{n+2}^{(2)}$, we get: $(S^2 - 1)(l_n^{(2)}/f_n^{(2)}) = 0$. Multiplying L_n by a constant, we are led to the formula $l_n^{(2)} = f_n^{(2)}$.

Collecting coefficients of S^3 and then multiplying by the function $(l_{n+1}^{(1)} - f_{n+1}^{(1)})/(f_n^{(2)} f_{n+1}^{(2)})$, we can express the result as $(S-1)\frac{(l_{n+1}^{(1)} - f_{n+1}^{(1)})(l_n^{(1)} - f_n^{(1)})}{f_n^{(2)}} = 0$, i.e.

$$(l_{n+1}^{(1)} - f_{n+1}^{(1)})(l_n^{(1)} - f_n^{(1)}) = \alpha^2 f_n^{(2)},$$

with α a constant. By using this equation, we get two possible cases:

$$\text{Case 1}: \quad \alpha = 0 \Rightarrow l_n^{(1)} = f_n^{(1)};$$

$$\text{Case 2}: \quad \alpha \neq 0 \Rightarrow l_n^{(1)} = f_n^{(1)} + \alpha\phi_n, \quad \phi_{n+1}\phi_n = f_n^{(2)}. \tag{3.5.20}$$

Case 1 is obvious. If $\alpha \neq 0$, we denote $\phi_n = (l_n^{(1)} - f_n^{(1)})/\alpha$ and show that there exists a function ϕ_n satisfying (3.5.20).

Using the coefficient of S^2 in (3.5.8) and dividing by $f_n^{(2)}$, we express the result in the form

$$D_t \log f_n^{(2)} = (S^2 - 1)(l_n^{(0)} - f_n^{(0)}) + \Phi_n. \tag{3.5.21}$$

Here

$$\Phi_n = \frac{f_n^{(1)} l_{n+1}^{(1)} - l_n^{(1)} f_{n+1}^{(1)}}{f_n^{(2)}} = (1-S)\varrho_n,$$

where $\varrho_n = 0$ in Case 1 and $\varrho_n = \alpha f_n^{(1)}/\phi_n$ in Case 2. By using (3.5.3) together with (3.5.21), we find $q_n^{(1)}$. It is

$$q_n^{(1)} = \beta - \varrho_n + (S+1)(l_n^{(0)} - f_n^{(0)}),$$

with β a constant.

Taking into account the last equation for $q_n^{(1)}$ and the condition $\dot{l}_n^{(0)} \sim 0$ we get for the function $p_n^{(2)}$ of (3.5.4):

$$\dot{p}_n^{(2)} = \dot{q}_n^{(1)} + 2\dot{f}_n^{(0)} \sim -\dot{\varrho}_n. \tag{3.5.22}$$

We see that (3.5.4) is satisfied in Case 1, as $\varrho_n = 0$. To finish the proof in Case 2, we need to show that the function ϱ_n is a conserved density.

First of all we need a formula for ϱ_n. From (3.5.20) it follows that $\log \phi_n \sim \frac{1}{2} \log f_n^{(2)}$ is a conserved density, i.e. we can define a function ψ_n

$$D_t \log \phi_n = (S-1)\psi_n. \tag{3.5.23}$$

We express $D_t \log f_n^{(2)} = (S^2 - 1)\psi_n$ and find from (3.5.21)

$$\varrho_n = \gamma + (S+1)(l_n^{(0)} - f_n^{(0)} - \psi_n), \tag{3.5.24}$$

where γ is an integration constant.

Collecting in (3.5.8) the coefficients of S, then multiplying by α/ϕ_n and using (3.5.20), we get an equation of the form

$$\dot{\varrho}_n + (\varrho_n + \alpha^2)D_t \log \phi_n = (S^2-1)[\alpha\phi_{n-1}(l_n^{(-1)} - f_n^{(-1)})] + \\ \varrho_n(S-1)l_n^{(0)} + (\varrho_n + \alpha^2)(1-S)f_n^{(0)}.$$

By using (3.5.23) and passing to equivalent functions, we simplify the last equation to

$$\dot{\varrho}_n \sim \varrho_n(S-1)(l_n^{(0)} - f_n^{(0)} - \psi_n).$$

Now we replace ϱ_n in the right hand side of the previous relation by (3.5.24) and see that $\dot{\varrho}_n \sim 0$, i.e. (3.5.4) is satisfied due to (3.5.22). □

5.2. Multi-component DΔEs. In this Section we consider the multi–component DΔEs which are systems of arbitrarily many equations. We discuss an approach which is closely related to the generalized symmetry method and to non associative algebraic structures. It allows one to construct integrable examples of multi–component equations and even to carry out the exhaustive symmetry classification of such equations. This approach has been developed mainly by Svinolupov, see e.g. the articles [779–781] and review [349]. It should be remarked that there is an alternative and older method for the construction of integrable multi–component equations, also connected to algebraic structures, which is presented e.g. in [57, 58, 272, 670]. However, we will not discuss it here.

Let us explain the method for PDEs without going into details. Instead of the KdV (2.4.56), we consider the following system of N equations [780]

$$u_t^i = u_3^i + c_{jk}^i u^j u_1^k, \tag{3.5.25}$$

where $i = 1, 2, \dots N$, by c_{jk}^i we denote some constant coefficients, and the summation over repeated indexes is assumed. The constants c_{jk}^i are interpreted as the structure constants of the multiplication in an algebra J. A priori, such algebra is not assumed to be commutative or associative. Then we suppose that a system like (3.5.25) possesses a generalized symmetry or a conservation law of a certain form and of fixed low order. Such conditions lead to some conditions for the algebra J. In the case of (3.5.25), J must be a Jordan algebra. In the other cases we obtain some other algebraic structures, such as the Jordan triple systems, Jordan pairs, and left–symmetric algebras. Such structures are well–known in algebra and, using known algebraic examples, we can construct integrable examples of multi–component equations.

The exhaustive symmetry classification problem can be solved for a class like (3.5.25). This is the case of e.g. [123–125, 129, 146, 147, 493, 502, 758, 780, 781, 783], where the multi–component generalizations of the KdV, mKdV,NLS, Toda and Burgers are considered. All integrable cases are described, using the existence of sufficiently high order generalized symmetries and conservation laws of the most general form.

The DΔEs have been studied in the articles [16,30,785,786], see also the review [349]. The multi–component analogs of the relativistic type lattice system (3.3.72), closely related to the NLS, are considered in [785]. Generalizations of the n–dependent form (3.4.56) of (3.3.72) are discussed in [16]. In [786] integrable generalizations of the lattice equation

$$\ddot{v}_n - v_n^{-1}\dot{v}_n^2 = v_{n+1} - v_n^2 v_{n-1}^{-1} \tag{3.5.26}$$

are studied. Eq. (3.5.26) is nothing but the Toda lattice (3.2.216) written in the rational form by the point transformation $v_n = \exp u_n$. Such form is used because the method under discussion is mainly suitable for studying the polynomial and rational equations.

In this Section, following [30], we are going to discuss the multi–component analogs of the lattice equation

$$u_{n,t} = u_n^2(u_{n+1} - u_{n-1}), \tag{3.5.27}$$

of Volterra type (V_1). The aim will be to construct integrable examples of this kind. Eq, (3.5.27) is more convenient for the method than the Volterra equation itself.

Eq. 3.5.27 is a degeneration of the modified Volterra equation (3.2.185), where we put $c = 0$ and change $t \to -t$. Moreover the simple transformation $\hat{u}_n = u_{n+1}u_n$ brings (3.5.27) into the Volterra equation. After the point transformations $\hat{u}_n = \exp \hat{u}_n^*$ and $u_n = \exp u_n^*$, we obtain $\hat{u}_n^* = u_{n+1}^* + u_n^*$ which is a discrete analog of the potentiation $\hat{u} = u_x$. For this reason, (3.5.27) can be called the potential Volterra equation.

As it has been shown in Section 3.2.7, (3.5.27) possesses the following master symmetry

$$u_{n,\tau} = u_n^2((n+1)u_{n+1} - (n-1)u_{n-1}), \tag{3.5.28}$$

generalized symmetry

$$u_{n,\epsilon'} = u_n^2(u_{n+1}^2(u_{n+2} + u_n) - u_{n-1}^2(u_n + u_{n-2})),$$

and local conservation law

$$D_t(u_{n+1}u_n) = (S-1)(u_{n+1}u_n^2 u_{n-1}).$$

All these objects will be constructed in the multi–component case too.

Let us first introduce the necessary algebraic structures. The finite–dimensional linear space J with a triple product $\{\,,\,,\,\} : J^3 \to J$ will be called a ternary algebra $\tilde{J}$. A ternary algebra $\tilde{J}$ will be called the Jordan triple system if the following identities take place for arbitrary elements of $\tilde{J}$:

$$\{a, b, c\} = \{c, b, a\}, \tag{3.5.29}$$

$$\{a, b, \{c, d, e\}\} - \{c, d, \{a, b, e\}\} = \{\{c, b, a, \}d, e\} - \{c, \{b, a, d, \}e\}. \tag{3.5.30}$$

Details and bibliography can be found, for example, in [569, 628].

Now we can write down a multi–component analog of the potential Volterra equation. Let $\mathbf{u}_n$ belong to a ternary algebra J. Then for any such algebra we have the following multi–component equation

$$\mathbf{u}_{n,t} = \{\mathbf{u}_n, (\mathbf{u}_{n+1} - \mathbf{u}_{n-1}), \mathbf{u}_n\} \tag{3.5.31}$$

analogous to (3.5.27). If J is the Jordan triple system, we call (3.5.31) the Jordan–Volterra equation. Eq. (3.5.31) is integrable, as it will be shown below.

The lattice equation (3.5.31) can be rewritten more explicitly, using the expansion $\mathbf{u}_n = u_n^m e_m$ (where the summation on repeated indexes is assumed) over a basis $e_1, e_2, \dots e_N$ in J. The multiplication $\{,\}$ is uniquely defined by the formula $\{e_i, e_j, e_k\} = a_{ijk}^m e_m$ (the factors a_{ijk}^m are the structure constants of J), and (3.5.31) takes the form

$$u_{n,t}^m = a_{ijk}^m u_n^i u_n^k (u_{n+1}^j - u_{n-1}^j).$$

However, we prefer to use coordinate–free notations.

It is convenient to introduce a linear operator $P_a : J \to J$, so that $P_a(b) = \{a, b, a\}$. Then (3.5.31) can be written as

$$\mathbf{u}_{n,t} = P_{\mathbf{u}_n}(\mathbf{u}_{n+1} - \mathbf{u}_{n-1}). \tag{3.5.32}$$

Instead of the master symmetry (3.5.28) we introduce the following multi–component equation

$$\mathbf{u}_{n,\tau} = P_{\mathbf{u}_n}((n+1)\mathbf{u}_{n+1} - (n-1)\mathbf{u}_{n-1}). \tag{3.5.33}$$

By defining the new differentiation $D_{\epsilon'} = [D_\tau, D_\epsilon]$ in the standard way (cf. (3.2.210) of Section 3.2.7), we construct a candidate for the generalized symmetry

$$\mathbf{u}_{n,\epsilon'} = P_{\mathbf{u}_n}(P_{\mathbf{u}_{n+1}}(\mathbf{u}_{n+2} + \mathbf{u}_n) - P_{\mathbf{u}_{n-1}}(\mathbf{u}_n + \mathbf{u}_{n-2})). \tag{3.5.34}$$

The following result is proved by direct calculation.

Theorem 68. *Let (3.5.32) correspond to an arbitrary ternary algebra $\tilde{J}$. Then (3.5.34) is its generalized symmetry if and only if the identities (3.5.29, 3.5.30) take place for the multiplication of $\tilde{J}$, i.e. $\tilde{J}$ is a Jordan triple system.*

We see that any Jordan-Volterra equation has a master symmetry of the form (3.5.33) and is integrable in this sense. Now we are going to present the most important examples of Jordan triple systems and of corresponding Jordan-Volterra equations. It should be remarked that those examples cover all the simple Jordan triple systems aside from two exceptional ones, see e.g. [569].

Example 1. A linear space J of $N \times N$ matrices becomes the Jordan triple system if one defines the triple product, using the standard matrix multiplication, as follows

$$\{a, b, c\} = \frac{1}{2}(abc + cba).$$

The matrix equation, corresponding to (3.5.31), reads

$$\mathbf{u}_{n,t} = \mathbf{u}_n(\mathbf{u}_{n+1} - \mathbf{u}_{n-1})\mathbf{u}_n. \tag{3.5.35}$$

The subspaces of symmetric and skew–symmetric matrices with the same triple product are also Jordan triple systems.

Example 2. The previous example can be generalized if one defines the triple product of $N \times M$ matrices by

$$\{a, b, c\} = \frac{1}{2}(ab^\dagger c + cb^\dagger a),$$

where † denotes the transposition. An integrable matrix equation in this case will have the form

$$\mathbf{u}_{n,t} = \mathbf{u}_n(\mathbf{u}^\dagger_{n+1} - \mathbf{u}^\dagger_{n-1})\mathbf{u}_n. \tag{3.5.36}$$

If $N = M$, we easily can obtain (3.5.35) by: $u_{2k} \to u_{2k}$, $u_{2k+1} \to u^\dagger_{2k+1}$.

In the particular case $N = 1$, J turns into an M–dimensional vector space with the multiplication

$$\{a, b, c\} = \frac{1}{2}(\langle a, b\rangle c + \langle c, b\rangle a),$$

where $\langle , \rangle$ denotes the standard scalar product. This example of a Jordan triple system usually leads to a nontrivial vector equation. In this case, (3.5.36) takes the form

$$\mathbf{u}_{n,t} = \left\langle \mathbf{u}_n, \mathbf{u}_{n+1} - \mathbf{u}_{n-1}\right\rangle \mathbf{u}_n. \tag{3.5.37}$$

Eq. (3.5.37) is degenerate in some sense. In fact, all the coordinates of the vector $\mathbf{u}_n$ are proportional to each other, i.e. $\mathbf{u}_n = v_n\mathbf{c}_n$, where $\mathbf{c}_n$ is a constant vector, and $v_n = v_n(t)$ is a scalar function. Eq. (3.5.37) is equivalent to

$$v_{n,t} = v_n^2(\alpha_n v_{n+1} - \alpha_{n-1}v_{n-1}), \tag{3.5.38}$$

where $\alpha_n = \left\langle \mathbf{c}_n, \mathbf{c}_{n+1}\right\rangle$ are t-independent factors. So, we have the scalar lattice equation with n-dependent coefficients. If $\alpha_n \neq 0$ for all n, we introduce $v_n = \beta_n u_n$, where $\beta_n\beta_{n+1} = 1/\alpha_n$, and (3.5.38) is transformed into (3.5.27). In spite of such degeneracy, the generalized symmetries of (3.5.37), as well as one corresponding to (3.5.34), are nontrivial integrable vector equations.

Example 3. A nontrivial vector example generalizing the potential Volterra equation is obtained by introducing the rule

$$\{a, b.c\} = \langle a, b\rangle c + \langle c, b\rangle a - \langle a, c\rangle b,$$

which defines the structure of the Jordan triple system in an M–dimensional vector space. The corresponding vector equation in this case reads

$$\mathbf{u}_{n,t} = 2\left\langle \mathbf{u}_n, \mathbf{u}_{n+1} - \mathbf{u}_{n-1}\right\rangle \mathbf{u}_n - \langle \mathbf{u}_n, \mathbf{u}_n\rangle (\mathbf{u}_{n+1} - \mathbf{u}_{n-1}). \tag{3.5.39}$$

It is clear that, starting from (3.5.33, 3.5.34), we can construct for all the three examples (3.5.35, 3.5.36, 3.5.39) a generalized symmetry and a master symmetry. With the help of such master symmetry we can construct an infinity of generalized symmetries. We can also write down for these three examples simple conservation laws [30], for instance, (3.5.39) has the following one

$$D_t\left\langle \mathbf{u}_n, \mathbf{u}_{n+1}\right\rangle = (S-1)\left\langle \{\mathbf{u}_n\mathbf{u}_{\mathbf{n-1}}\mathbf{u}_n\}, \mathbf{u}_{n+1}\right\rangle.$$

Applying the master symmetry to such starting conservation law, we can get more conservation laws.

Moreover, we can show that these examples are Hamiltonian and we can construct for them Lax pairs.

Closely related to these multicomponent equations are the integrable multi–component Toda type lattice equations and the NLS type continuous coupled systems [30].

6. Completely discrete equations

The discovery of new integrable PΔEs (or $\mathbb{Z}^2$-lattice equations) is always a very challenging problem as, by proper continuous limits, many other results on DΔEs and PDEs may be obtained. Moreover, many physical and biological applications involve discrete systems, see for instance [295, 731] and references therein.

Application of the generalized symmetry method in the case of PΔEs is more difficult than in cases of DΔEs we considered in the previous Sections. For this reason, the first classification results for such equations have been obtained by different methods in [19] by Adler and in [22] by Adler, Bobenko and Suris. Results on the ABS, Boll and on examples of quad-graph equations of the form (2.4.404) not belonging to those classes have been presented in Sections 2.4.6, 2.4.7 and 2.4.8.

Here we consider PΔEs of the form (2.4.130).

Following [303, 555–558], we demonstrate in Section 3.6.1, 3.6.2 that the generalized symmetry method can be applied in case of PΔEs of the form (2.4.130) too. All integrable autonomous discrete equations (2.4.404) known in the literature possess generalized symmetries, see e.g. [492, 556, 699, 800, 837, 838]. Therefore the generalized symmetry method should have the most wide use compared to other known methods.

In Section 3.6.1 we use the existence of the three-point generalized symmetries and derive the simplest integrability conditions. Then we explain how to solve those integrability conditions by applying annihilation operators.

In Section 3.6.2 the integrability conditions are applied to some known examples and classes of PΔEs depending on a few arbitrary constants. In this way we show, on examples, how to test PΔEs for the integrability and how to solve simple classification problems.

In Section 3.7 C-integrable PΔEs are considered with methods similar to the one considered in the derivation of integrability conditions in Section 3.6.1.3 . They provide the complete classification for multilinear PΔEs of the form (3.7.1) and (2.4.130).

6.1. Generalized symmetries for PΔEs and integrability conditions. In this Section we introduce the generalized symmetry method for PΔEs showing similarity and differences with respect to the case of DΔEs introduced in Section 3.2. Partial results in this direction can be found also in [727].

6.1.1. *Preliminary definitions.* As (2.4.130) has no explicit dependence on the point (n, m) of the lattice, we assume that the same will be for the generalized symmetries and conservation laws we will be considering in the following. For this reason, without loss of generality, we write down symmetries and conservation laws at the point $(0, 0)$. Eq. (2.4.130) must satisfy the following conditions

$$(F_{u_{0,0}}, F_{u_{1,0}}, F_{u_{0,1}}, F_{u_{1,1}}) \neq 0, \tag{3.6.1}$$

where indexes in (3.6.1) denote partial derivatives. These conditions are not sufficient to rule out trivial equations. The equation

$$(u_{0,0} + u_{1,0})(u_{0,1} + u_{1,1} + 1) = 0$$

provide an example of equation which is degenerate but satisfies (3.6.1).

We require that (2.4.130) be rewritable in the form

$$u_{1,1} = f^{(1,1)}(u_{1,0}, u_{0,0}, u_{0,1}), \tag{3.6.2}$$

where

$$(f^{(1,1)}_{u_{1,0}}, f^{(1,1)}_{u_{0,0}}, f^{(1,1)}_{u_{0,1}}) \neq 0, \tag{3.6.3}$$

and the apex $^{(1,1)}$ indicate that the function f is obtained from (2.4.130) by explicitating the function u in the point $(1, 1)$. The conditions (3.6.3) are necessary but not sufficient conditions to prevent triviality of the equation (3.6.2). Moreover, to get a scheme which is invertible and to provide propagation in both discrete directions, we have to suppose that the function f depends on all its variables, i.e.

$$f^{(1,1)}_{u_{1,0}} \cdot f^{(1,1)}_{u_{0,0}} \cdot f^{(1,1)}_{u_{0,1}} \neq 0. \tag{3.6.4}$$

Whenever convenient we will express our formulas in terms of the two shift operators, S_n, S_m introduced in Section 2.4.1 with S_n defined in (1.2.13) and S_m in (1.2.14). To avoid the presence of many indexes we provide an alternative definition. To simplify the notation we will denote $S_n = T_1$ and $S_m = T_2$ whose action on a function $u_{0,0}$ is thus:

$$T_1 u_{0,0} = u_{1,0}, \quad T_2 u_{0,0} = u_{0,1}. \tag{3.6.5}$$

The functions $u_{n,m}$ are related among themselves by (3.6.2) and its shifted values

$$u_{n+1,m+1} = T_1^n T_2^m f^{(1,1)}(u_{1,0}, u_{0,0}, u_{0,1}) = f^{(1+n,1+m)}(u_{n+1,m}, u_{n,m}, u_{n,m+1}),$$

and it is easy to see that all of them can be expressed in terms of the functions

$$u_{n,0}, \quad u_{0,m}, \tag{3.6.6}$$

where n, m are arbitrary integers. This is not the only possible choice of independent variables [31], but, being the simplest, is the one we will use in the following. The functions (3.6.6) play the role of boundary–initial conditions for (3.6.2).

The evolutionary form of a generalized symmetry of (3.6.2) is given by the following equation

$$\frac{d}{d\epsilon} u_{0,0} = g_{0,0} = G(u_{\ell,0}, u_{\ell-1,0}, \ldots, u_{\ell',0}, u_{0,k}, u_{0,k-1}, \ldots, u_{0,k'}), \tag{3.6.7}$$

where $\ell \geq \ell'$, $k \geq k'$ and ϵ takes the role of the symmetry variable. The form of this equation at the various points of the lattice is obtained by applying the shift operators T_1 and T_2

$$\frac{d}{d\epsilon} u_{n,m} = T_1^n T_2^m g_{0,0} = g_{n,m} = G(u_{\ell+n,m}, \ldots, u_{\ell'+n,m}, u_{n,m+k}, \ldots, u_{n,m+k'}).$$

Eq. (3.6.7) is a generalized symmetry of (3.6.2) if the two equations (3.6.2, 3.6.7) are compatible for all independent variables (3.6.6), i.e.

$$\frac{du_{1,1}}{d\epsilon} - \left.\frac{df^{(1,1)}}{d\epsilon}\right|_{u_{1,1}=f^{(1,1)}} = 0. \tag{3.6.8}$$

In practice, (3.6.8) reads

$$g_{1,1} = \left. g_{1,0} f^{(1,1)}_{u_{1,0}} + g_{0,0} f^{(1,1)}_{u_{0,0}} + g_{0,1} f^{(1,1)}_{u_{0,1}} \right|_{u_{1,1}=f^{(1,1)}}. \tag{3.6.9}$$

Eq. (3.6.9) must be identically satisfied when all the variables $u_{n,m}$ contained in the functions $g_{n,m}$ and in the derivatives of $f^{(1,1)}$ are expressed in terms of the independent variables (3.6.6). This result provides strict conditions, given by a set of equations for the functions $f^{(1,1)}$ and G, often overdetermined.

Let us consider some autonomous functions $p_{0,0}, q_{0,0}$ which depend on a finite number of functions $u_{n,m}$ and have no explicit dependence on the point (n, m) of the lattice. The relation

$$\left.(T_1 - 1)p_{0,0} - (T_2 - 1)q_{0,0}\right|_{u_{1,1}=f^{(1,1)}} = 0 \tag{3.6.10}$$

is called a (local n, m-independent) conservation law of (3.6.2) if it is satisfied on the solutions set of this equation. To check it, we need to express all variables in terms of the independent variables (3.6.6) and require that it is identically satisfied.

Starting from the choice of the independent variables (3.6.6) and the class of autonomous PΔEs (3.6.2), we can prove a few useful statements which will be used for studying the compatibility condition (3.6.9). Let us consider the functions $u_{n,1}, u_{1,m}$ appearing in (3.6.9). We can prove the following theorem:

Theorem 69. *The transformation* $\mathcal{T}\,:\,\{u_{n,0}, u_{0,m}\} \rightarrow \{\tilde{u}_{n,0}, \tilde{u}_{0,m}\}$*, given by the shift operator* T_2

$$\tilde{u}_{0,m} = u_{0,m+1}, \qquad \tilde{u}_{n,0} = u_{n,1}, \quad n \neq 0, \tag{3.6.11}$$

is invertible under the equation (3.6.2)*. Moreover, if a function* ϕ *is non–zero, then* $T_2\phi \neq 0$ *too.*

PROOF. The invertibility of the transformation $\tilde{u}_{0,m} = u_{0,m+1}$ is obvious. Let us show by induction that for any $n \geq 1$ we have

$$\tilde{u}_{n,0} = \tilde{u}_{n,0}(u_{n,0}, u_{n-1,0}, \ldots, u_{1,0}, u_{0,0}, u_{0,1}), \quad \partial_{u_{n,0}}\tilde{u}_{n,0} \neq 0, \quad \partial_{u_{0,1}}\tilde{u}_{n,0} \neq 0. \tag{3.6.12}$$

It follows from (3.6.2) and (3.6.4) that the proposition is true for $\tilde{u}_{1,0} = u_{1,1}$. For $n \geq 1$, from (3.6.2) we get

$$\tilde{u}_{n+1,0} = u_{n+1,1} = f^{(1,1)}(u_{n+1,0}, u_{n,0}, \tilde{u}_{n,0}), \tag{3.6.13}$$

with $\tilde{u}_{n,0}$ given by (3.6.12). So $\tilde{u}_{n+1,0}$ has the same structure as $\tilde{u}_{n,0}$ and thus (3.6.12) is true. As $\tilde{u}_{n,0}$ depends on $u_{0,1}$, then the functions $u_{n+1,0}, u_{n,0}, \tilde{u}_{n,0}$ are functionally independent, i.e. $\partial_{u_{n+1,0}}\tilde{u}_{n+1,0} \neq 0$ and $\partial_{u_{0,1}}\tilde{u}_{n+1,0} \neq 0$. A similar analysis can be carried out in the case of the functions $\tilde{u}_{n,0}$ when $n \leq -1$. In this case we have

$$\tilde{u}_{n,0} = \tilde{u}_{n,0}(u_{n,0}, u_{n+1,0}, \ldots, u_{-1,0}, u_{0,0}, u_{0,1}), \qquad \partial_{u_{n,0}}\tilde{u}_{n,0} \neq 0, \quad \partial_{u_{0,1}}\tilde{u}_{n,0} \neq 0. \tag{3.6.14}$$

From (3.6.12, 3.6.14) it follows that (3.6.11) is invertible.

To prove the second part of this theorem, let us consider a non-constant function $\phi \neq 0$. Taking into account (3.6.2) and its shifted values, ϕ can always be expressed in terms of the independent variables as

$$\phi = \Phi(u_{N,0}, u_{N-1,0}, \ldots, u_{N',0}, u_{0,K}, u_{0,K-1}, \ldots, u_{0,K'}), \tag{3.6.15}$$

for some integer numbers N, N', K and K' such that $N \geq N'$, $K \geq K'$. Then we will have

$$T_2\phi = \Phi(\tilde{u}_{N,0}, \ldots, \tilde{u}_{N',0}, \tilde{u}_{0,K}, \ldots, \tilde{u}_{0,K'}). \tag{3.6.16}$$

If ϕ depends essentially on the variables $u_{i,0}$ with $i \neq 0$, then there must exist two numbers N and N' such that $\partial_{u_{N,0}}\phi \neq 0$ and $\partial_{u_{N',0}}\phi \neq 0$. When $N > 0$, from (3.6.12) it follows that only the function $\tilde{u}_{N,0}$ appearing in (3.6.16) depends on $u_{N,0}$. Hence $\partial_{u_{N,0}}T_2\phi \neq 0$, i.e. $T_2\phi \neq 0$. The case, when $N' < 0$, is analogous. If ϕ depends only on $u_{0,j}$, then $\partial_{u_{0,K}}\phi \neq 0$ and $\partial_{u_{0,K'}}\phi \neq 0$, and the proof is obvious. □

The operators T_1, T_1^{-1}, T_2^{-1} act on the variables (3.6.6) in an analogous way. Consequently they also define invertible transformations. As a result we can state the following Proposition:

Proposition 5. *For any non-zero function* ϕ, $T_1^{\ell}T_2^{k}\phi \neq 0$ *for any* $\ell, k \in \mathbb{Z}$.

From (3.6.11, 3.6.12, 3.6.14) we can derive the structure of some of the partial derivatives of the functions $u_{n,1}$. For convenience, from now on we will define

$$g_{u_{n,m}} = \partial_{u_{n,m}} g_{0,0} \tag{3.6.17}$$

for the derivatives of the functions $g_{0,0}$ appearing in (3.6.7). For $n > 0$, from (3.6.12, 3.6.13) it follows that $\partial_{u_{n+1,0}} u_{n+1,1} = T_1^n \partial_{u_{1,0}} u_{1,1}$. From (3.6.2) we can also get $u_{-1,1} = \hat{f}^{(-1,1)}(u_{-1,0}, u_{0,0}, u_{0,1})$ and then by differentiation

$$\partial_{u_{-1,0}} u_{-1,1} = -T_1^{-1} \frac{f^{(1,1)}_{u_{0,0}}}{f^{(1,1)}_{u_{0,1}}}. \tag{3.6.18}$$

Then, applying the operator T_1^{n+1}, with $n < 0$, to (3.6.18) it follows that $\partial_{u_{n,0}} u_{n,1} = -T_1^n \frac{f^{(1,1)}_{u_{0,0}}}{f^{(1,1)}_{u_{0,1}}}$. For the functions of the form $u_{1,m}$ we get similar results. So we can state the following Proposition:

Proposition 6. *The functions $u_{n,1}, u_{1,m}$ are such that*

$$\begin{aligned}
n > 0: \quad & u_{n,1} = u_{n,1}(u_{i,0}, u_{n-1,0}, \dots, u_{1,0}, u_{0,0}, u_{0,1}), \\
& \partial_{u_{n,0}} u_{n,1} = T_1^{n-1} f^{(1,1)}_{u_{1,0}}; \\
n < 0: \quad & u_{n,1} = u_{n,1}(u_{n,0}, u_{n+1,0}, \dots, u_{-1,0}, u_{0,0}, u_{0,1}), \\
& \partial_{u_{n,0}} u_{n,1} = -T_1^n \frac{f^{(1,1)}_{u_{0,0}}}{f^{(1,1)}_{u_{0,1}}}; \\
m > 0: \quad & u_{1,m} = u_{1,m}(u_{1,0}, u_{0,0}, u_{0,1}, \dots, u_{0,m-1}, u_{0,m}), \\
& \partial_{u_{0,m}} u_{1,m} = T_2^{m-1} f^{(1,1)}_{u_{0,1}}; \\
m < 0: \quad & u_{1,m} = u_{1,m}(u_{1,0}, u_{0,0}, u_{0,-1}, \dots, u_{0,m+1}, u_{0,m}), \\
& \partial_{u_{0,m}} u_{1,m} = -T_2^m \frac{f^{(1,1)}_{u_{0,0}}}{f^{(1,1)}_{u_{1,0}}}.
\end{aligned} \tag{3.6.19}$$

6.1.2. *Derivation of the first integrability conditions.* Following the standard scheme of the generalized symmetry method, we derive here from the compatibility condition (3.6.9) four conditions necessary for the integrability of (3.6.2).

For a generalized symmetry (3.6.7) we suppose that if $g_{0,0}$ depends on at least one variable of the form $u_{n,0}$, then $g_{u_{\ell,0}} \neq 0$ and $g_{u_{\ell',0}} \neq 0$, and the numbers ℓ, ℓ' are called orders of the symmetry. The same can be said about the variables $u_{0,m}$ and the corresponding numbers k, k' if $g_{u_{0,k}} \neq 0$ and $g_{u_{0,k'}} \neq 0$.

Theorem 70. *Let (3.6.2) possess a generalized symmetry (3.6.7) of orders ℓ, ℓ', k and k'. Then the following relations must take place*

$$\text{If} \quad \ell > 0 \quad \Longrightarrow \quad (T_1^{\ell} - 1)\log f^{(1,1)}_{u_{1,0}} = (1 - T_2)T_1 \log g_{u_{\ell,0}}; \tag{3.6.20}$$

$$\text{If} \quad \ell' < 0 \quad \Longrightarrow \quad (T_1^{\ell'} - 1)\log \frac{f^{(1,1)}_{u_{0,0}}}{f^{(1,1)}_{u_{0,1}}} = (1 - T_2)\log g_{u_{n',0}}; \tag{3.6.21}$$

(3.6.22) $$\text{If} \quad k > 0 \quad \Longrightarrow \quad (T_2^k - 1)\log f^{(1,1)}_{u_{0,1}} = (1 - T_1)T_2 \log g_{u_{0,k}};$$

(3.6.23) $$\text{If} \quad k' < 0 \quad \Longrightarrow \quad (T_2^{k'} - 1)\log \frac{f^{(1,1)}_{u_{0,0}}}{f^{(1,1)}_{u_{1,0}}} = (1 - T_1)\log g_{u_{0,k'}}.$$

Before going over to the proof of this theorem, let us clarify its meaning by noticing that in the case of a three point symmetry with $g_{0,0} = G(u_{1,0}, u_{0,0}, u_{-1,0})$, for which $\ell > 0$ and $\ell' < 0$, one can use both relations (3.6.20, 3.6.21).

PROOF. Let us consider the compatibility condition (3.6.9) expressed in terms of the independent variables (3.6.6). As $g_{0,0}$ depends on $u_{n,0}$ and $u_{0,m}$, the functions $(g_{1,1}, g_{1,0}, g_{0,1})$ depend on $(u_{n,1}, u_{1,m})$, whose form is given by Proposition 6. Moreover, (3.6.9) will contain $u_{n,0}$ with $\ell + 1 \geq n \geq \ell'$ and $u_{0,m}$ with $k + 1 \geq m \geq k'$.

If $\ell > 0$, applying to (3.6.9) the operator $\partial_{u_{\ell+1,0}}$ and using the results (3.6.19) contained in Proposition 6, we get:

$$T_1T_2(g_{u_{\ell,0}})T_1^{\ell} f_{u_{1,0}} = f_{u_{1,0}} T_1 g_{u_{\ell,0}}.$$

Applying the logarithm to both sides of the previous equation, we obtain (3.6.20). The other cases are obtained in a similar way by differentiating (3.6.9) with respect to $u_{\ell',0}$, $u_{0,k+1}$, and $u_{0,k'}$. □

Eqs. (3.6.20–3.6.23) can be expressed as a standard conservation law of the form (3.6.10), using the obvious well-known formulas

$$S_l^m - 1 = (S_l - 1)(1 + S_l + \cdots + S_l^{m-1}), \qquad m > 0,$$
$$S_l^m - 1 = (1 - S_l)(S_l^{-1} + S_l^{-2} + \cdots + S_l^{m}), \quad m < 0,$$

where $l = 1, 2$. This means that, from the existence of a generalized symmetry, one can construct some conservation laws.

Theorem 70 provides integrability conditions, i.e. the fact that for an integrable equation there must exist a function $g_{0,0}$ satisfying (3.6.20–3.6.23). The unknown function $g_{0,0}$ must depend on a finite number of independent variables. These integrability conditions turn out to be difficult to use for testing and classifying PΔEs.

In the case of the DΔEs of Volterra or Toda type [850], as we saw in Section 3.2, there are integrability conditions equivalent to (3.6.20–2.4.41). In order to check these integrability conditions one can use the formal variational derivatives [219, 409, 840, 850], defined as (3.2.40) for ϕ given by (3.6.15). Using such variational derivatives, for example the integrability conditions (3.6.20, 3.6.22) are reduced to

(3.6.24) $$\frac{\delta^{(2)}}{\delta u_{0,0}}(T_1^n - 1)\log f_{u_{1,0}} = 0, \qquad \frac{\delta^{(1)}}{\delta u_{0,0}}(T_2^k - 1)\log f_{u_{0,1}} = 0,$$

which do not involve any unknown function. This result is due to the fact that in this case all discrete variables are independent.

In the case of PΔEs the situation is essentially different. Some of the discrete variables are dependent and the variational derivatives must be calculated modulo the equation (3.6.2). So, (3.6.24) will not be anymore valid. If we apply here the variational derivatives, we will get, at most, some partial results depending on the choice of the independent variables introduced.

The conservation laws (3.6.20–3.6.23) depend on the order of the symmetries. These conservation laws can be simplified under some assumptions on the structure of the Lie algebra of the generalized symmetries. If we assume that for a given equation we are able to get generalized symmetries for any value of n and k, then we can derive order-independent conservation laws, using a trick standard in the generalized symmetry method [850]. This assumption implies that if, for example, we have a generalized symmetry of order ℓ then there must be also one of order $\ell + 1$. This is a very constraining assumption which is not always verified, as we know from the continuous case [608]. Here it is used just as an example for the construction of simplified formulas. In fact such simplified formulas can be obtained assuming any difference between the orders of two generalized symmetries, and in next Section we consider an example with difference 2.

So, in the following theorem, we will assume that in addition to (3.6.7) a second generalized symmetry

$$u_{0,0,\tilde{t}} = \tilde{g}_{0,0} = \tilde{G}(u_{\tilde{\ell},0}, u_{\tilde{\ell}-1,0}, \dots, u_{\tilde{\ell}',0}, u_{0,\tilde{k}}, u_{0,\tilde{k}-1}, \dots, u_{0,\tilde{k}'}) \tag{3.6.25}$$

of orders $\tilde{\ell}, \tilde{\ell}', \tilde{k}, \tilde{k}'$ will exist. With this assumption we shall obtain four conservation laws:

$$(T_1 - 1)p_{0,0}^{(j)} = (T_2 - 1)q_{0,0}^{(j)}, \qquad j = 1, 2, 3, 4, \tag{3.6.26}$$

with $p_{0,0}^{(j)}$ or $q_{0,0}^{(j)}$ expressed in terms of (3.6.2).

Theorem 71. *Let (3.6.2) possess two generalized symmetries (3.6.25) and (3.6.7). Then (3.6.2) admits the conservation laws (3.6.26)*

$$n > 0,\ \tilde{\ell} = \ell + 1 \quad \Longrightarrow \quad j = 1,\ p_{0,0}^{(1)} = \log f_{u_{1,0}}^{(1,1)}; \tag{3.6.27}$$

$$\ell' < 0,\ \tilde{\ell}' = \ell' - 1 \quad \Longrightarrow \quad j = 2,\ p_{0,0}^{(2)} = \log \frac{f_{u_{0,0}}^{(1,1)}}{f_{u_{0,1}}^{(1,1)}}; \tag{3.6.28}$$

$$k > 0,\ \tilde{k} = k + 1 \quad \Longrightarrow \quad j = 3,\ q_{0,0}^{(3)} = \log f_{u_{0,1}}^{(1,1)}; \tag{3.6.29}$$

$$k' < 0,\ \tilde{k}' = k' - 1 \quad \Longrightarrow \quad j = 4,\ q_{0,0}^{(4)} = \log \frac{f_{u_{0,0}}^{(1,1)}}{f_{u_{1,0}}^{(1,1)}}. \tag{3.6.30}$$

PROOF. Let us consider in detail just the case when $\ell > 0$, $\tilde{\ell} = \ell + 1$. Due to Theorem 70 (3.6.20) must be satisfied and consequently

$$(T_1^{n+1} - 1)p_{0,0}^{(1)} = (1 - T_2)T_1 \log \tilde{g}_{u_{\ell+1,0}}, \tag{3.6.31}$$

where $p_{0,0}^{(1)}$ is given by (3.6.27). Applying the operator $-T_1$ to (3.6.20) and adding the result to (3.6.31), we get the conservation law (3.6.26) with $j = 1$, where $q_{0,0}^{(1)}$ is given by

$$q_{0,0}^{(1)} = T_1^2 \log g_{u_{\ell,0}} - T_1 \log \tilde{g}_{u_{\ell+1,0}}.$$

The other cases are proved in an analogous way. □

So for (3.6.2) we have four necessary conditions of integrability: there must exist some functions of finite range $q_{0,0}^{(1)}, q_{0,0}^{(2)}, p_{0,0}^{(3)}, p_{0,0}^{(4)}$ of the form (3.6.15) satisfying the conservation laws (3.6.26) with $p_{0,0}^{(1)}, p_{0,0}^{(2)}, q_{0,0}^{(3)}, q_{0,0}^{(4)}$ defined by (3.6.27–3.6.30).

The following theorem will precise the structure of the unknown functions $q_{0,0}^{(1)}, q_{0,0}^{(2)}, p_{0,0}^{(3)}$, and $p_{0,0}^{(4)}$.

Theorem 72. *If the functions $q_{0,0}^{(1)}$, $q_{0,0}^{(2)}$, $p_{0,0}^{(3)}$, and $p_{0,0}^{(4)}$ satisfy (3.6.26), with $p_{0,0}^{(1)}$, $p_{0,0}^{(2)}$, $q_{0,0}^{(3)}$, and $q_{0,0}^{(4)}$ given by (3.6.27–3.6.30), and are written in the form (3.6.15), then $q_{0,0}^{(1)}$ and $q_{0,0}^{(2)}$ may depend only on the variables $u_{n,0}$, and $p_{0,0}^{(3)}$ and $p_{0,0}^{(4)}$ on $u_{0,m}$.*

PROOF. Let us consider (3.6.26) with $j = 1$. The functions therein involved have the form

$$
\begin{aligned}
p_{0,0}^{(1)} &= P^{(1)}(u_{1,0}, u_{0,0}, u_{0,1}), \qquad p_{1,0}^{(1)} = P^{(1)}(u_{2,0}, u_{1,0}, u_{1,1}),\\
q_{0,0}^{(1)} &= Q^{(1)}(u_{N,0}, \dots, u_{N',0}, u_{0,K}, \dots, u_{0,K'}),\\
q_{0,1}^{(1)} &= Q^{(1)}(u_{N,1}, \dots, u_{N',1}, u_{0,K+1}, \dots, u_{0,K'+1}).
\end{aligned}
\tag{3.6.32}
$$

Let us consider the function $q_{0,0}^{(1)}$ and let us study its dependence on the variables $u_{0,j}$ with $j \neq 0$. Using Proposition 6, we see that the functions $u_{n,1}$ in $p_{1,0}^{(1)}, q_{0,1}^{(1)}$ may depend only on $u_{0,1}$. If $K > 0$, we differentiate (3.6.26) with $j = 1$ with respect to $u_{0,K+1}$ and get: $\partial_{u_{0,K+1}} q_{0,1}^{(1)} = T_2 \partial_{u_{0,K}} q_{0,0}^{(1)} = 0$. Then, from Proposition 5, it follows that $q_{0,0}^{(1)}$ does not depend on $u_{0,K}$. If $K' < 0$, let us differentiate with respect to $u_{0,K'}$ and we get $\partial_{u_{0,K'}} q_{0,0}^{(1)} = 0$. This shows that the function $q_{0,0}^{(1)}$ cannot depend on $u_{0,j}$ with $j \neq 0$.

The proof for the other cases is quite similar. □

As we cannot use the formal variational derivative, we have to work directly with functions $q_{0,0}^{(1)}, q_{0,0}^{(2)}, p_{0,0}^{(3)}, p_{0,0}^{(4)}$ which have the structure

$$
\begin{aligned}
q_{0,0}^{(j)} &= Q^{(j)}(u_{N_j,0}, \dots, u_{N'_j,0}), \qquad j = 1, 2;\\
p_{0,0}^{(l)} &= P^{(l)}(u_{0,K_l}, \dots, u_{0,K'_l}), \qquad l = 3, 4.
\end{aligned}
$$

In Section 3.6.1.3 we are going to limit ourselves to just five-point symmetries. This will make the problem more definite in the sense that the numbers N_j, N'_j, K_l, K'_l will be specified and small.

6.1.3. *Integrability conditions for five point symmetries.* From the definition of Lie symmetry, we can construct a new symmetry by adding the right hand sides of two symmetries $u_{0,0,t} = g_{0,0}$ and $u_{0,0,\tilde{t}} = \tilde{g}_{0,0}$: $u_{0,0,\hat{t}} = \hat{g}_{0,0} = c_1 g_{0,0} + c_2 \tilde{g}_{0,0}$, where c_1, c_2 are arbitrary constants. Symmetries form a linear space. For example, (2.4.281) of Section 2.4.8 has two three point symmetries (2.4.282) and (2.4.283), therefore it has a five-point generalized symmetry

$$
u_{0,0,\epsilon} = g_{0,0} = G(u_{1,0}, u_{-1,0}, u_{0,0}, u_{0,1}, u_{0,-1}),
\tag{3.6.33}
$$

with

$$
(G_{u_{1,0}}, G_{u_{-1,0}}, G_{u_{0,1}}, G_{u_{0,-1}}) \neq 0.
\tag{3.6.34}
$$

One can also prove in all generality for a symmetry of the form (3.6.33) (see details and references in [555]) that its right hand side must be of the form

$$
G = \Phi(u_{1,0}, u_{0,0}, u_{-1,0}) + \Psi(u_{0,1}, u_{0,0}, u_{0,-1}).
\tag{3.6.35}
$$

So to get a coherent result we need to add to the generalized symmetry equation (3.6.33) the conditions (3.6.34), from now on called *non degeneracy conditions*.

All known integrable autonomous equations (3.6.2) have symmetries of the types

$$\Psi = 0 \quad \text{and} \quad \Psi_{u_{0,1}}\Psi_{u_{0,-1}} \neq 0; \tag{3.6.36}$$

$$\Phi = 0 \quad \text{and} \quad \Phi_{u_{1,0}}\Phi_{u_{-1,0}} \neq 0. \tag{3.6.37}$$

Thus any symmetry of the form (3.6.33, 3.6.35) is the linear combination of a symmetry (3.6.36) and (3.6.37). However, we cannot prove this property theoretically.

The other known integrable examples of the form (2.4.130) have also five point generalized symmetries. We are going to use the existence of a five point generalized symmetry of the form (3.6.33) as an integrability criteria. This may be a severe restriction, as there might be integrable equations with symmetries depending on more lattice points.

In the ABS classification in Section 2.4.6 all three point generalized symmetries turn out to be Miura transformations of the Volterra equation or of the YdKN (see Section 2.4.6.3) [492]. If we expect to find new type integrable discrete equations of the form (2.4.130) these should have as generalized symmetries some new type integrable equations. One example of such equation is given by the Narita-Itoh-Bogoyavlensky (NIB) [105, 420, 627] equation, a sub case of (3.5.2)

$$u_{0,0,\epsilon} = g_{0,0} = u_{0,0}(u_{2,0} + u_{1,0} - u_{-1,0} - u_{-2,0}). \tag{3.6.38}$$

We will prove in the Appendix E that no equation of the form (2.4.130) can have (3.6.38) as a symmetry.

We can then state the following theorem:

Theorem 73. *If* (3.6.2, 3.6.4) *possesses a generalized symmetry of the form* (3.6.33), *then the functions*

$$q_{0,0}^{(j)} = Q^{(j)}(u_{2,0}, u_{1,0}, u_{0,0}), \qquad j = 1, 2; \tag{3.6.39}$$

$$p_{0,0}^{(j)} = P^{(j)}(u_{0,2}, u_{0,1}, u_{0,0}), \qquad jm = 3, 4,$$

satisfy the conditions (3.6.26, 3.6.27, 3.6.28, 3.6.29, 3.6.30).

PROOF. From the relations (3.6.20–3.6.23), as $n = k = 1$ and $n' = k' = -1$, we are able to construct the functions

$$q_{0,0}^{(1)} = -T_1 \log g_{u_{1,0}}, \qquad q_{0,0}^{(2)} = T_1 \log g_{u_{-1,0}}, \tag{3.6.40}$$

$$p_{0,0}^{(3)} = -T_2 \log g_{u_{0,1}}, \qquad p_{0,0}^{(4)} = T_2 \log g_{u_{0,-1}},$$

satisfying conditions (3.6.26, 3.6.27–3.6.30). It follows from (3.6.19, 3.6.33) that the function $q_{0,0}^{(1)}$ has the structure

$$q_{0,0}^{(1)} = \hat{Q}^{(1)}(u_{2,0}, u_{1,0}, u_{0,0}, u_{1,1}, u_{1,-1}) = Q^{(1)}(u_{2,0}, u_{1,0}, u_{0,0}, u_{0,1}, u_{0,-1}).$$

In analogy to Theorem 72 we get that $Q^{(1)}$ cannot depend on $u_{0,1}, u_{0,-1}$. The proof for the other functions contained in (3.6.40) is obtained in the same way. □

Let us analyze in detail Theorem 73 when $k = 1$. In this case we just require $\mathcal{G}_{u_{1,0}} \neq 0$, while the dependence of $\mathcal{G}$ on the other variables is not important. In this case Theorem 73 tells us that there exists a conservation law with a function $p_{0,0}^{(1)}$ completely defined by (3.6.2) while $q_{0,0}^{(1)}$ is an arbitrary function of $u_{2,0}, u_{1,0}, u_{0,0}$. The other three cases are similar. Therefore Theorem 73 provides four *integrability conditions* in the form of conservation laws. In the case of a non degenerate symmetry (3.6.33, 3.6.34), all these integrability conditions must be satisfied.

Let us notice that, from the existence of generalized symmetries, we can easily derive many integrability conditions of this kind. Such conditions have been written down in [611]. However, the other conditions are in general more complicate, and more difficult to use in practice. Further integrability conditions can be obtained requiring the existence of a recursion operator for the symmetries. This would allow us to get new integrable equations of this class corresponding to the existence of higher symmetries [612].

If the integrability conditions given in Theorem 73 are satisfied, then there must exist functions $q^{(1)}_{0,0}$, $q^{(2)}_{0,0}$, $p^{(3)}_{0,0}$, $p^{(4)}_{0,0}$ local in their argument. Once we know all such functions, we can check if some autonomous five-point symmetries might exist. In fact in such a case we can construct the four partial derivatives of G [555]

$$G_{u_{1,0}} = \exp(-q^{(1)}_{-1,0}), \qquad G_{u_{-1,0}} = \exp(q^{(2)}_{-1,0}), \tag{3.6.41}$$

$$G_{u_{0,1}} = \exp(-p^{(3)}_{0,-1}), \qquad G_{u_{0,-1}} = \exp(p^{(4)}_{0,-1}). \tag{3.6.42}$$

These partial derivatives must be compatible

$$G_{u_{1,0},u_{-1,0}} = G_{u_{-1,0},u_{1,0}}, \qquad G_{u_{0,1},u_{0,-1}} = G_{u_{0,-1},u_{0,1}}. \tag{3.6.43}$$

Eq. (3.6.43) is an additional integrability condition. If this further integrability condition is satisfied we can construct G in the form

$$G = \Phi(u_{1,0}, u_{0,0}, u_{-1,0}) + \Psi(u_{0,1}, u_{0,0}, u_{0,-1}) + \nu(u_{0,0}), \tag{3.6.44}$$

where Φ and Ψ are known functions of their arguments while $\nu(u_{0,0})$ is an unknown arbitrary function which may correspond to a Lie point symmetry of the equation. The function ν can be specified by considering the compatibility condition (3.6.9), the last and most fundamental integrability condition.

The first problem is to check the integrability conditions given in Theorem 73. In the case of DΔEs we had a similar situation, i.e. the integrability conditions were given by conservation laws depending on arbitrary functions of a limited number of variables. However such problem was easier to solve as all discrete variables were independent and we could use the variational derivative to check them [555]. Here the calculation of the variational derivative is not sufficient to prove if a given expression is a conservation law.

So, in the following, we present a scheme for solving this problem for any given equation of form (3.6.2), i.e. we show how we can solve (3.6.39) to obtain $q^{(1)}_{0,0}$, $q^{(2)}_{0,0}$, $p^{(3)}_{0,0}$, $p^{(4)}_{0,0}$ as local functions of their argument. We will split the explanation into two steps.

Step 1. At first we consider the integrability conditions (3.6.26) corresponding to $j = 1, 2$. The unknown functions in the right hand side of (3.6.26) when $j = 1, 2$ contain the dependent variable $u_{2,1}$ which, from (3.6.2), depends on $u_{2,0}$, $u_{1,0}$, $u_{1,1}$ and thus it is not immediately expressed in terms of independent variables, but give rise to extremely complicated functional expressions of the independent variables. We can avoid this problem by applying the operators T_1^{-1}, T_2^{-1} to (3.6.26). In this case we have

$$\begin{aligned} &p^{(j)}_{0,0} - p^{(j)}_{-1,0} = q^{(j)}_{-1,1} - q^{(j)}_{-1,0} = \\ &Q^{(j)}(u_{1,1}, u_{0,1}, u_{-1,1}) - Q^{(j)}(u_{1,0}, u_{0,0}, u_{-1,0}), \end{aligned} \tag{3.6.45}$$

$$\begin{aligned} &p^{(j)}_{0,-1} - p^{(j)}_{-1,-1} = q^{(j)}_{-1,0} - q^{(j)}_{-1,-1} = \\ &Q^{(j)}(u_{1,0}, u_{0,0}, u_{-1,0}) - Q^{(j)}(u_{1,-1}, u_{0,-1}, u_{-1,-1}). \end{aligned} \tag{3.6.46}$$

Here $p^{(j)}_{n,m}$ are known functions expressed in term of (3.6.2). The functions $q^{(j)}_{n,m}$ are unknown, and $(q^{(j)}_{-1,1}, q^{(j)}_{-1,-1})$ contain the dependent variables $u_{1,1}, u_{-1,1}, u_{1,-1}, u_{-1,-1}$. Our aim is to derive from (3.6.45, 3.6.46) a set of equations for the unknown function, $q^{(j)}_{-1,0}$.

To do so let us extract from (2.4.404) three further expressions of the form of (3.6.2) for the dependent variables contained in (3.6.45, 3.6.46)

(3.6.47)
$$u_{-1,1} = f^{(-1,1)}(u_{-1,0}, u_{0,0}, u_{0,1}),$$
$$u_{1,-1} = f^{(1,-1)}(u_{1,0}, u_{0,0}, u_{0,-1}),$$
$$u_{-1,-1} = f^{(-1,-1)}(u_{-1,0}, u_{0,0}, u_{0,-1}).$$

All functions $f^{(n,m)}$ have a nontrivial dependence on all their variables, as it is the case of $f^{(1,1)}$, and are expressed in terms of independent variables. Let us introduce the two differential operators

(3.6.48)
$$\mathcal{A} = \partial_{u_{0,0}} - \frac{f^{(1,1)}_{u_{0,0}}}{f^{(1,1)}_{u_{1,0}}}\partial_{u_{1,0}} - \frac{f^{(-1,1)}_{u_{0,0}}}{f^{(-1,1)}_{u_{-1,0}}}\partial_{u_{-1,0}},$$

(3.6.49)
$$\mathcal{B} = \partial_{u_{0,0}} - \frac{f^{(1,-1)}_{u_{0,0}}}{f^{(1,-1)}_{u_{1,0}}}\partial_{u_{1,0}} - \frac{f^{(-1,-1)}_{u_{0,0}}}{f^{(-1,-1)}_{u_{-1,0}}}\partial_{u_{-1,0}},$$

chosen in such a way to annihilate the functions $q^{(j)}_{-1,1}$ and $q^{(j)}_{-1,-1}$, namely, $\mathcal{A}q^{(j)}_{-1,1} = 0$, $\mathcal{B}q^{(j)}_{-1,-1} = 0$. Applying $\mathcal{A}$ to (3.6.45) and $\mathcal{B}$ to (3.6.46), we obtain two equations for the unknown $q^{(j)}_{-1,0}$:

(3.6.50)
$$\mathcal{A}q^{(j)}_{-1,0} = r^{(j,1)}, \quad \mathcal{B}q^{(j)}_{-1,0} = r^{(j,2)},$$

where $r^{(j,1)}$, $r^{(j,2)}$ are some explicitly known functions of (3.6.2). Considering the standard commutator of $\mathcal{A}$ and $\mathcal{B}$, $[\mathcal{A},\mathcal{B}] = \mathcal{AB} - \mathcal{BA}$, we can add a further equation

(3.6.51)
$$[\mathcal{A},\mathcal{B}]q^{(j)}_{-1,0} = r^{(j,3)}.$$

Eqs. (3.6.50, 3.6.51) represent a linear partial differential system of three equations for the unknown $q^{(j)}_{-1,0} = Q^{(j)}(u_{1,0}, u_{0,0}, u_{-1,0})$. For the three partial derivatives of $q^{(j)}_{-1,0}$, this is just a linear algebraic system of three equations in three unknown. In most of the examples considered below, this system is non degenerate and thus it provides one and only one solution for the three derivatives of $q^{(j)}_{-1,0}$. In these cases we can find in unique way the partial derivatives of $q^{(j)}_{0,0}$. Then we can check the consistency of the partial derivatives and, if satisfied, find $q^{(j)}_{0,0}$ up to an arbitrary constant. Finally we check (3.6.26) with $j = 1, 2$ in any of the equivalent forms (3.6.45) or (3.6.46).

The non degeneracy of (3.6.50, 3.6.51) depends on (3.6.2) only. So, if we have checked the non degeneracy for $j = 1$, we know that this is also true for $j = 2$ and vice versa. So both functions $q^{(1)}_{0,0}$, $q^{(2)}_{0,0}$ are found in unique way up to a constant of integration.

If (3.6.50, 3.6.51) is degenerate, the functions $q^{(1)}_{0,0}$, $q^{(2)}_{0,0}$ are defined up to some arbitrary functions. In this case the checking of the integrability conditions (3.6.26) may be more difficult.

In principle the coefficients of (3.6.50, 3.6.51) may depend, in addition to the natural variables $u_{0,0}$, $u_{1,0}$, $u_{-1,0}$, entering in $q^{(j)}_{-1,0}$, on the independent variables $u_{0,1}$, $u_{0,-1}$. In such a case we have to require that the solution $q^{(j)}_{-1,0}$ does not depend on them. In this case we have to split the equations of the system (3.6.50, 3.6.51) with respect to the various powers of the independent variables $u_{0,1}$, $u_{0,-1}$, if (2.4.11) is rational, and obtain an overdetermined system of equations for $q^{(j)}_{-1,0}$. Moreover, overdetermined systems of equations are usually simpler to solve. There will be some examples of this kind in Section 3.6.2.

Step 2. Let us consider now the conditions (3.6.26) with $j = 3, 4$. In this case we have a similar situation. By appropriate shifts we rewrite (3.6.26) in the two equivalent forms

(3.6.52)
$$p^{(j)}_{1,-1} - p^{(j)}_{0,-1} = q^{(j)}_{0,0} - q^{(j)}_{0,-1} = P^{(j)}(u_{1,1}, u_{1,0}, u_{1,-1}) - P^{(j)}(u_{0,1}, u_{0,0}, u_{0,-1}),$$

(3.6.53)
$$p^{(j)}_{0,-1} - p^{(j)}_{-1,-1} = q^{(j)}_{-1,0} - q^{(j)}_{-1,-1} = P^{(j)}(u_{0,1}, u_{0,0}, u_{0,-1}) - P^{(j)}(u_{-1,1}, u_{-1,0}, u_{-1,-1}).$$

We can introduce the annihilation operators

(3.6.54)
$$\hat{\mathcal{A}} = \partial_{u_{0,0}} - \frac{f^{(1,1)}_{u_{0,0}}}{f^{(1,1)}_{u_{0,1}}}\partial_{u_{0,1}} - \frac{f^{(1,-1)}_{u_{0,0}}}{f^{(1,-1)}_{u_{0,-1}}}\partial_{u_{0,-1}},$$

(3.6.55)
$$\hat{\mathcal{B}} = \partial_{u_{0,0}} - \frac{f^{(-1,1)}_{u_{0,0}}}{f^{(-1,1)}_{u_{0,1}}}\partial_{u_{0,1}} - \frac{f^{(-1,-1)}_{u_{0,0}}}{f^{(-1,-1)}_{u_{0,-1}}}\partial_{u_{0,-1}},$$

such that $\hat{\mathcal{A}} p^{(j)}_{1,-1} = 0$ and $\hat{\mathcal{B}} p^{(j)}_{-1,-1} = 0$. Then we are led to the system

(3.6.56)
$$\hat{\mathcal{A}} p^{(j)}_{0,-1} = \hat{r}^{(j,1)}, \quad \hat{\mathcal{B}} p^{(j)}_{0,-1} = \hat{r}^{(j,2)}, \quad [\hat{\mathcal{A}}, \hat{\mathcal{B}}] p^{(j)}_{0,-1} = \hat{r}^{(j,3)}$$

for the function $p^{(j)}_{0,-1}$ depending on $u_{0,1}$, $u_{0,0}$, $u_{0,-1}$, where $\hat{r}^{(j,l)}$ are known functions expressed in terms of $f^{(n,m)}$.

After we have solved (3.6.39) we can construct a generalized symmetry. When both systems (3.6.50, 3.6.51) and (3.6.56) are non degenerate, we find Φ and Ψ, given by (3.6.44), up to at most four arbitrary constants. Two of them may be specified by the consistency conditions (3.6.43) while the remaining constants are specified, using the compatibility condition (3.6.9), together with the function ν. In practice we always look for symmetries of the form (3.6.35). Such symmetries are defined uniquely up to multiple factors and the addition of functions of the form $\nu(u_{0,0})$ corresponding to the right hand side of point symmetries $u_{0,0,\tau} = \nu(u_{0,0})$. We write down only the generalized symmetry as we are not interested in the point symmetries.

If one of the systems (3.6.50, 3.6.51) and (3.6.56) is degenerate, then Φ and Ψ in the right hand side of a generalized symmetry (3.6.44) may be found up to some arbitrary functions. Those arbitrary functions must be specified using the compatibility condition (3.6.9). However, in almost all degenerate examples considered below, (3.6.2) turns out to be trivial and it can be rewritten in one of the following four forms

(3.6.57)
$$(T_1 \pm 1)w(u_{0,0}, u_{0,1}) = 0, \qquad (T_2 \pm 1)w(u_{1,0}, u_{0,0}) = 0.$$

Eqs. (3.6.57) can be integrated once and give some equations depending on a reduced number of lattice variables.

In the next Section we test PΔEs which depend on arbitrary constants and thus we solve some simple classification problem. We look for such particular cases that satisfy our integrability test and are not of Klein type (2.4.137) or not transformable into Klein type equations by n, m-dependent Möbius transformations.

It is worthwhile to notice that integrability conditions analogous to (3.6.26, 3.6.27–3.6.30) have been derived for hyperbolic systems of the form (3.5.10) by Zhiber and Shabat in [870].

6.2. Testing PΔEs for the integrability and some classification results.

6.2.1. *A simple classification problem.* Here we apply the formulas introduced in the previous Section to study the class of PΔEs

$$u_{1,1} = f_{0,0} = u_{1,0} + u_{0,1} + \varphi(u_{0,0}). \tag{3.6.58}$$

This class is not empty, it contains trivial approximations of some well-known integrable equations, namely, the sine-Gordon, Tzitzèika and Liouville equations.

The class of equations (3.6.58) depends on an unknown function φ, and we require that (3.6.58) possess a generalized symmetry of the form (3.6.33). To do so it must satisfy the integrability conditions (3.6.26, 3.6.27–3.6.30, 3.6.39). If $\varphi'' = 0$, equation (3.6.58) is linear, and all the integrability conditions are satisfied trivially. So we require that $\varphi'' \neq 0$.

The proof that (3.6.26, 3.6.27–3.6.30, 3.6.39) are conservation laws for (3.6.58) is carried out by differentiating them in such a way to reduce them to simple differential equations, a scheme introduced in 1823 by Abel [2] (see [14] for a review) for solving functional equations. The applications of this scheme for PΔEs can be found in [407, 544, 700]. In [700] the scheme was used for finding conservation laws for known equations, i.e. when the dependence of the functions $p_{0,0}$ and $q_{0,0}$ on the symmetries was unknown while the difference equation (3.6.2) was given. In [727] the existence of a simple conservation law is used as an integrability condition.

Here we consider the case when either $p_{0,0}$ or $q_{0,0}$ is expressed in terms of the unknown right hand side of (3.6.2). The conservation laws are allowed to depend on arbitrary functions of the variables $u_{1,0}, u_{0,0}, u_{0,1}$. Moreover, as it will be shown at the end of this Section, the existence of simple conservation laws is not sufficient to prove integrability. One can have non linear PΔEs of this class (3.6.58) with two local conservation laws but with no generalized symmetry.

Let us study the class of PΔEs (3.6.58). For later use we can rewrite (3.6.58) in three equivalent forms, applying to it the operators T_1^{-1}, T_2^{-1}

$$\begin{aligned} u_{-1,1} &= u_{0,1} - u_{0,0} - \varphi(u_{-1,0}), \\ u_{1,-1} &= u_{1,0} - u_{0,0} - \varphi(u_{0,-1}), \\ u_{-1,-1} &= \varphi^{-1}(u_{0,0} - u_{-1,0} - u_{0,-1}). \end{aligned} \tag{3.6.59}$$

Let us consider condition (3.6.26) with $j = 2$. Applying the shift operators T_1^{-1}, T_2^{-1}, we rewrite it in two equivalent forms:

$$p^{(2)}_{0,0} - p^{(2)}_{-1,0} = q^{(2)}_{-1,1} - q^{(2)}_{-1,0}, \tag{3.6.60}$$

$$p^{(2)}_{0,-1} - p^{(2)}_{-1,-1} = q^{(2)}_{-1,0} - q^{(2)}_{-1,-1}, \tag{3.6.61}$$

where $p^{(2)}_{0,0} = \log \varphi'(u_{0,0})$ and $q^{(2)}_{0,0}$ is given by (3.6.39). Taking into account (3.6.58, 3.6.59), (3.6.60, 3.6.61) can be expressed in terms of the independent variables (3.6.6).

Eqs. (3.6.60, 3.6.61) are two functional equations for $q_{0,0}^{(2)}$. By applying the annihilation operators

$$\hat{\mathcal{A}} = \partial_{u_{0,0}} + \partial_{u_{1,0}} + \partial_{u_{-1,0}}, \qquad \hat{\mathcal{B}} = \partial_{u_{0,0}} - \varphi'(u_{0,0})\partial_{u_{1,0}} - \frac{1}{\varphi'(u_{-1,0})}\partial_{u_{-1,0}},$$

we reduce them to PDEs. Using (3.6.59), we can show that $\hat{\mathcal{A}}$ annihilates any function $\Phi(u_{1,-1}, u_{0,-1}, u_{-1,-1})$. So, applying $\hat{\mathcal{A}}$ to (3.6.61), we get

$$\hat{\mathcal{A}}\, q_{-1,0}^{(2)} = 0. \tag{3.6.62}$$

The operator $\hat{\mathcal{B}}$ annihilates $q_{-1,1}^{(2)}$. Thus applying the operator $\hat{\mathcal{B}}$ to (3.6.60) we get:

$$\hat{\mathcal{B}}\, q_{-1,0}^{(2)} = -\hat{\mathcal{B}}\,(p_{0,0}^{(2)} - p_{-1,0}^{(2)}).$$

If we introduce the difference operator $\hat{\mathcal{C}} = \hat{\mathcal{A}} - \hat{\mathcal{B}}$, we get

$$\hat{\mathcal{C}}\, q_{-1,0}^{(2)} = \hat{\mathcal{B}}\,(p_{0,0}^{(2)} - p_{-1,0}^{(2)}). \tag{3.6.63}$$

From (3.6.62, 3.6.63) we also get

$$[\hat{\mathcal{A}}, \hat{\mathcal{C}}]\, q_{-1,0}^{(2)} = \hat{\mathcal{A}}\hat{\mathcal{B}}\,(p_{0,0}^{(2)} - p_{-1,0}^{(2)}), \tag{3.6.64}$$

where $[\hat{\mathcal{A}}, \hat{\mathcal{C}}]$ is the standard commutator of two operators. So (3.6.62–3.6.64) can be rewritten as a PDE for the function $q = q_{-1,0}^{(2)}$, where, as before, by the indexes we denote partial derivatives and by apices derivatives with respect to the argument

$$\begin{aligned} q_{u_{0,0}} + q_{u_{1,0}} + q_{u_{-1,0}} &= 0, \\ a(u_{0,0})q_{u_{1,0}} + b(u_{-1,0})q_{u_{-1,0}} &= c(u_{0,0}) - b'(u_{-1,0}), \\ a'(u_{0,0})q_{u_{1,0}} + b'(u_{-1,0})q_{u_{-1,0}} &= c'(u_{0,0}) - b''(u_{-1,0}). \end{aligned} \tag{3.6.65}$$

The functions $a(z)$, $b(z)$ and $c(z)$ are given by

$$a(z) = \varphi'(z) + 1, \quad b(z) = \frac{1}{\varphi'(z)} + 1, \quad c(z) = \frac{\varphi''(z)}{\varphi'(z)},$$

where $a'(z)b'(z)c(z) \neq 0$, as $\varphi''(z) \neq 0$.

The solvability of the system (3.6.65) depends on the following determinant

$$\Delta = \begin{vmatrix} a(u_{0,0}) & b(u_{-1,0}) \\ a'(u_{0,0}) & b'(u_{-1,0}) \end{vmatrix}.$$

We must have $\Delta \neq 0$. If we have $\Delta = 0$, as $u_{0,0}$ and $u_{-1,0}$ are independent variables, we obtain the relations $\frac{a'(u_{0,0})}{a(u_{0,0})} = \frac{b'(u_{-1,0})}{b(u_{-1,0})} = \nu$, where ν is a constant. These relations are in contradiction with the condition that $\varphi'' \neq 0$.

If we differentiate the system (3.6.65) with respect to $u_{1,0}$, we easily deduce that $q_{u_{1,0}} = \alpha$, where α is a constant. Then from (3.6.65) we obtain two different expressions for $q_{u_{-1,0}}$

$$q_{u_{-1,0}} = \frac{d(u_{0,0}) - b'(u_{-1,0})}{b(u_{-1,0})} = \frac{d'(u_{0,0}) - b''(u_{-1,0})}{b'(u_{-1,0})}, \qquad d(z) = c(z) - \alpha a(z). \tag{3.6.66}$$

If $d' \neq 0$, differentiating (3.6.66) with respect to $u_{0,0}$, we get $\frac{d''(u_{0,0})}{d'(u_{0,0})} = \frac{b'(u_{-1,0})}{b(u_{-1,0})} = \sigma$, where σ is a constant. This result is again in contradiction with the condition $\varphi'' \neq 0$. So, $d = \beta$, a constant, and we get the following ODE for φ

$$\varphi''/\varphi' = \alpha\varphi' + \alpha + \beta. \tag{3.6.67}$$

If φ satisfies (3.6.67), (3.6.66) is satisfied, and $q_{u_{-1,0}} = \alpha + \beta$.

The system (3.6.65) provides us with another partial derivative of q

$$q_{u_{0,0}} = -2\alpha - \beta,$$

from which we deduce that

$$q = q^{(2)}_{-1,0} = \alpha u_{1,0} - (2\alpha + \beta)u_{0,0} + (\alpha + \beta)u_{-1,0} + \delta,$$

where δ is an arbitrary constant. The integration of (3.6.67) gives

$$\log \varphi'(z) = \alpha\varphi(z) + (\alpha + \beta)z + \gamma,$$

where γ is a further constant. If we introduce these last two equations into (3.6.60), we get $u_{0,0} + \beta\varphi(u_{-1,0}) = 0$, which implies $\beta = 0$.

Thus, we have proved that (3.6.58) satisfies the condition (3.6.26) with $j = 2$

$$\log \varphi'(z) = \alpha(\varphi(z) + z) + \gamma, \tag{3.6.68}$$

with $\alpha \neq 0$, as $\varphi'' \neq 0$. Consequently the solution of (3.6.68) is

$$\varphi = \frac{\gamma}{\alpha} - z - \frac{1}{\alpha}\log\left(e^{c-z} - 1\right), \tag{3.6.69}$$

with c an integration constant and $\alpha \neq 0$. If (3.6.68) is satisfied,

$$p^{(2)}_{0,0} = \log \varphi'(u_{0,0}), \qquad q^{(2)}_{0,0} = \alpha(u_{2,0} - 2u_{1,0} + u_{0,0}) + \delta, \tag{3.6.70}$$

and these functions define a nontrivial conservation law.

If (3.6.58), with φ given by (3.6.68), has a generalized symmetry of the form (3.6.33), the other conditions (3.6.26, 3.6.27–3.6.30, 3.6.39) must be satisfied. From (3.6.27) we get that the condition (3.6.26) with $j = 1$ becomes

$$(T_2 - 1)q^{(1)}_{0,0} = 0. \tag{3.6.71}$$

This equation has a trivial solution, $q^{(1)}_{0,0}$ a constant. We now look for a nontrivial solution. From (3.6.39) it follows that the functions $q^{(1)}_{0,0}$ and $q^{(2)}_{0,0}$ depend on the same set of variables. Hence $\tilde{q} = q^{(1)}_{-1,0}$ also satisfies (3.6.65), but with zeros on the right hand side. As $q_{u_{1,0}}$ is a constant, it follows that also $q^{(1)}_{0,0}$ must be a constant, i.e. the constant solution is the most general solution of (3.6.71).

From (3.6.40), we get the partial derivatives of the right hand side of the symmetry (3.6.33), $g_{u_{1,0}}$ and $g_{u_{-1,0}}$. It is easy to verify that the first of the conditions (3.6.43) is not satisfied. Consequently (3.6.58), with φ given by (3.6.68), has no generalized symmetry of the form (3.6.33).

In Section 3.6.1.3 we have considered the simpler symmetries (3.6.36, 3.6.37). Using the previous reasoning, we can prove that there is no symmetry defined by (3.6.35, 3.6.36). Eq. (3.6.58) is symmetric under the involution $u_{n,m} \to u_{m,n}$. Also (3.6.26) with $j = 3, 4$ are symmetric with respect to (3.6.26) with $j = 1, 2$. So these further conditions will provide a conservation law symmetric to the one defined by (3.6.70) and prove that there is no symmetry given by (3.6.35, 3.6.37).

Let us collect the results obtained so far in the following theorem, where the conservation laws will be written in a simplified form, omitting inessential constants.

Theorem 74. *Eq.* (3.6.58) *satisfies the integrability conditions* (3.6.26, 3.6.27–3.6.30, 3.6.39) *iff* φ *is a solution of* (3.6.68). *Eq.* (3.6.58), *when* φ *is given by* (3.6.68), *has two nontrivial conservation laws*

$$\begin{aligned}(T_1-1)(\varphi(u_{0,0})+u_{0,0})&=(T_2-1)(u_{2,0}-2u_{1,0}+u_{0,0}),\\(T_2-1)(\varphi(u_{0,0})+u_{0,0})&=(T_1-1)(u_{0,2}-2u_{0,1}+u_{0,0}).\end{aligned}\tag{3.6.72}$$

However, in this case, (3.6.58) *does not have a generalized symmetry of the form* (3.6.33) *or of the form given by* (3.6.35, 3.6.36) *or* (3.6.35, 3.6.37).

Let us notice that (3.6.58) possesses the conservation laws (3.6.72) for any φ, not only when φ satisfies (3.6.68). However, the integrability conditions are satisfied only if φ satisfies (3.6.68), but no generalized symmetry of the form mentioned in Theorem 74 exists.

6.2.2. *Further application of the method to examples and classes of equations.* Here we apply the test to prove the integrability of a number of non linear nontrivial PΔEs introduced by various authors using different approaches [23, 374, 387, 398, 410, 554]. It should be stressed that all PΔEs below are affine linear, i.e. they can be written in the form (3.6.2), (3.6.3), in such a way that (3.6.1) is satisfied.

Example 1. This will be a simple illustrative example (2.4.281) introduced before in Section 2.4.8 as a PΔE not in the ABS or Boll class. We will discuss it in all details. Up to a rotation (change of axes) (2.4.281) is equivalent to the equations presented in [399] and [637]. In Section 2.4.8 [410, 554, 555] its Lax pairs and some conservation laws are presented. Two generalized symmetries of the form (3.6.9) have been constructed in [554]. So (3.2.69) satisfies our integrability test, but nevertheless, it is instructive to try out the test with this equation.

The study of this equation splits into two different steps.

Step 1. Let us consider the integrability condition (3.6.26) with $j=1$. The corresponding system (3.6.50, 3.6.51) reads

$$\begin{aligned}q_{u_{0,0}}-\tfrac{u_{1,0}+1}{u_{0,0}-1}q_{u_{1,0}}-\tfrac{u_{-1,0}-1}{u_{0,0}+1}q_{u_{-1,0}}&=\tfrac{2u_{0,0}}{1-u_{0,0}^2},\\q_{u_{0,0}}-\tfrac{u_{1,0}-1}{u_{0,0}+1}q_{u_{1,0}}-\tfrac{u_{-1,0}+1}{u_{0,0}-1}q_{u_{-1,0}}&=\tfrac{2u_{0,0}}{1-u_{0,0}^2},\\(u_{1,0}u_{0,0}+1)q_{u_{1,0}}-(u_{-1,0}u_{0,0}+1)q_{u_{-1,0}}&=0,\end{aligned}$$

where $q=q_{-1,0}^{(1)}$ and by the index we denote the argument of the derivative. This system is non degenerate, and its solution is

$$q_{u_{1,0}}=q_{u_{-1,0}}=0,\quad q_{u_{0,0}}=\frac{2u_{0,0}}{1-u_{0,0}^2}.$$

Hence $q_{0,0}^{(1)}=-\log(u_{1,0}^2-1)+c_1$, where c_1 is an arbitrary constant. $q_{0,0}^{(1)}$ together with $p_{0,0}^{(1)}=\log\frac{u_{0,0}-1}{u_{0,1}+1}$ satisfy the relation (3.6.26), and provide a conservation law for (2.4.281).

Eq. (3.6.26) with $j=2$ can be solved simply. As $p_{0,0}^{(2)}=-p_{0,0}^{(1)}+\log(-1)$, then the solution of (3.6.26) with $j=2$ is given by $q_{0,0}^{(2)}=-q_{0,0}^{(1)}+c_2$, with c_2 another arbitrary constant. As the corresponding system (3.6.50, 3.6.51) is non degenerate, there is no other solution.

Now we look for G, the r.h.s. of the symmetry in (3.6.33). As follows from (3.6.41), a candidate for such a symmetry is given by

$$u_{0,0,\epsilon_1}=(u_{0,0}^2-1)(\alpha u_{1,0}+\beta u_{-1,0})+\nu(u_{0,0}).$$

Here α, β are nonzero arbitrary constants, and $\nu(u_{0,0})$ is an arbitrary function of its argument. Rescaling ϵ_1, we can set in all generality $\alpha = 1$. Substituting this into (3.6.9) we get $\beta = -1$, $\nu(u_{0,0}) \equiv 0$. This symmetry is nothing but the well-known modified Volterra equation (3.2.185, 2.4.278) [850] which here we write as

$$u_{0,0,\epsilon_1} = (u_{0,0}^2 - 1)(u_{1,0} - u_{-1,0}). \tag{3.6.73}$$

Step 2. Let us consider the integrability conditions (3.6.26) with $j = 3, 4$. In this case the corresponding system (3.6.56) is degenerate. So we have to modify the procedure presented in the previous Section and applied up above in the case when $j = 1, 2$. For $j = 3$ this system reads

$$(u_{0,1} + u_{0,0})p_{u_{0,1}} - (u_{0,0} + u_{0,-1})p_{u_{0,-1}} = 2,$$
$$(u_{0,0}^2 - 1)p_{u_{0,0}} + (u_{0,1}u_{0,0} + 1)p_{u_{0,1}} + (u_{0,0}u_{0,-1} + 1)p_{u_{0,-1}} = 0,$$

where $p = p_{0,-1}^{(3)}$. Its general solution is

$$p = \Omega(\omega) + \log \frac{u_{0,1} + u_{0,0}}{u_{0,0} + u_{0,-1}}, \quad \text{with} \quad \omega = \frac{u_{0,0}^2 - 1}{(u_{0,1} + u_{0,0})(u_{0,0} + u_{0,-1})}. \tag{3.6.74}$$

The integrability condition (3.6.26) with $j = 3$ is satisfied iff $\Omega(\omega) = \gamma - \log \omega$, where γ is an arbitrary constant. The case $j = 4$ is quite similar, and we easily find the second generalized symmetry of (2.4.281)

$$u_{0,0,t_2} = (u_{0,0}^2 - 1)\left(\frac{1}{u_{0,1} + u_{0,0}} - \frac{1}{u_{0,0} + u_{0,-1}}\right). \tag{3.6.75}$$

Eq. (3.6.75) is a subcase of (V_2) in Section 3.3.1.2.

Step 2 of this example is not standard. In all the following discrete equations, either the systems (3.6.50, 3.6.51) and (3.6.56) are non degenerate or the equations themselves are trivial. As a result we have proved the following theorem:

Theorem 75. *Eq.* (2.4.281) *satisfies the generalized symmetry test and possesses the symmetries* (3.6.73) *and* (3.6.75).

Example 2. Let us consider a known PΔE closely related to (2.4.281) and studied in [398, 410, 598, 635]

$$u_{1,1}(u_{0,1} + c)(u_{1,0} - 1) = u_{0,0}(u_{1,0} + c)(u_{0,1} - 1), \tag{3.6.76}$$

where $c \neq -1, 0$. In particular, its Lax pair can be found in [635]. If $c = -1$, it is trivial. If $c = 0$, using the point transformation

$$u_{n,m} = \frac{2}{1 - \hat{u}_{n,m}}, \tag{3.6.77}$$

we can reduce it to (2.4.281) for $\hat{u}_{n,m}$. So, (3.6.76) generalizes (2.4.281). Thus, it will not be surprising that this equation satisfies our test. The calculation is quite similar to the one shown in Example 1, Step 1. We easily find two generalized symmetries of the form (3.6.33)

$$\frac{u_{0,0,t_1}}{u_{0,0}(u_{0,0} - 1)} = (T_1 - 1)\frac{1}{u_{0,0}u_{-1,0} + c(u_{0,0} + u_{-1,0} - 1)}, \tag{3.6.78}$$

$$\frac{u_{0,0,t_2}}{u_{0,0}(u_{0,0} + c)} = (T_2 - 1)\frac{1}{u_{0,0}u_{0,-1} - (u_{0,0} + u_{0,-1} + c)}. \tag{3.6.79}$$

As a result we get the theorem:

Theorem 76. *Eq.* (3.6.76) *satisfies the generalized symmetry test and possesses the symmetries* (3.6.78) *and* (3.6.79)*.*

Example 3. Another example is given by the equation

$$u_{1,1}u_{0,0}(u_{1,0}-1)(u_{0,1}+1)+(u_{1,0}+1)(u_{0,1}-1)=0 \tag{3.6.80}$$

taken from [410]. It is an equation which possesses five non autonomous conservation laws of the form

$$(T_1-1)p_{n,m}(u_{n,m},u_{n,m+1})=(T_2-1)q_{n,m}(u_{n,m},u_{n+1,m}), \tag{3.6.81}$$

where $p_{n,m}$, $q_{n,m}$ depend explicitly on the discrete variables n, m. In [410] the authors also calculated the algebraic entropy for (3.6.80), and demonstrated in this way that the equation should be integrable.

This example does not satisfy our test. The system (3.6.50, 3.6.51), corresponding to the first of the integrability conditions (3.6.26), is non degenerate, and we find from it $q_{0,0}^{(1)}$ in a unique way. However this function does not satisfy the condition (3.6.26). The same is true for all 4 integrability conditions. This means that all four assumptions of Theorem 73 are not satisfied. We can state the following theorem:

Theorem 77. *Eq.* (3.6.80) *does not satisfy any of four integrability conditions* (3.2.60)*-*(3.3.18)*. This equation cannot have an autonomous nontrivial generalized symmetry of the form* (2.4.10)*.*

Eq. (3.6.80) might have, however, non autonomous generalized symmetries.

Example 4. Let us consider the PΔE

$$(1+u_{0,0}u_{1,0})(\nu u_{1,1}+u_{0,1})=(1+u_{0,1}u_{1,1})(\nu u_{0,0}+u_{1,0}), \tag{3.6.82}$$

where the constant ν is such that $\nu^2 \neq 1$. When $\nu = \pm 1$ the equations are trivial, as they are equivalent to (3.6.57). Eq. (3.6.82) has been obtained in [554] by combining Miura type transformations relating DΔEs of the Volterra type. In [691] Miura type transformations have been found relating this equation to integrable equations of the form (3.6.2). Eq. (3.6.82) satisfies our test, and we find 2 generalized symmetries

$$u_{0,0,\epsilon_1}=\frac{(u_{0,0}^2-\nu)(\nu u_{0,0}^2-1)}{u_{0,0}}\left(\frac{1}{u_{1,0}u_{0,0}+1}-\frac{1}{u_{0,0}u_{-1,0}+1}\right), \tag{3.6.83}$$

$$u_{0,0,\epsilon_2}=\frac{(u_{0,0}^2-\nu)(\nu u_{0,0}^2-1)}{u_{0,0}}\left(\frac{1}{u_{0,1}u_{0,0}-1}-\frac{1}{u_{0,0}u_{0,-1}-1}\right). \tag{3.6.84}$$

In the particular case $\nu = 0$ (3.6.82) reduces to

$$u_{1,1}-u_{0,0}=\frac{1}{u_{1,0}}-\frac{1}{u_{0,1}}, \tag{3.6.85}$$

and (3.6.83, 3.6.84) to its generalized symmetries. Eq. (3.6.85), up to point transformations, can be found in [331, 399, 410]. As a result of this example we can state the following theorem:

Theorem 78. *Eq.* (3.6.82) *satisfies the generalized symmetry test and possesses the symmetries* (3.6.83) *and* (3.6.84)*.*

Example 5. A further interesting example is provided by the PΔE [374]

$$
\begin{aligned}
&2(u_{0,0}+u_{1,1})+u_{1,0}+u_{0,1}+\\
&\gamma(4u_{0,0}u_{1,1}+2u_{1,0}u_{0,1}+3(u_{0,0}+u_{1,1})(u_{1,0}+u_{0,1}))+\\
&(\xi_2+\xi_4)u_{0,0}u_{1,1}(u_{1,0}+u_{0,1})+(\xi_2-\xi_4)u_{1,0}u_{0,1}(u_{0,0}+u_{1,1})+\\
&\zeta u_{0,0}u_{1,1}u_{1,0}u_{0,1}=0,
\end{aligned}
\tag{3.6.86}
$$

where γ, ξ_2, ξ_4 and ζ are constant coefficients. This equation is obtained as a subclass of the most general multilinear dispersive equation on the square lattice, (2.4.133), which depends on 16 parameters (see on its integrability also [375] and the references therein contained). Eq. (3.6.86) is contained in the intersection of 5 of the 6 classes of equations belonging to

$$
\begin{aligned}
\mathcal{Q}_+ &= a_1\,(u_{0,0}+u_{1,1})+a_2\,(u_{1,0}+u_{0,1})+(\alpha_1-\alpha_2)\,u_{0,0}u_{1,0}\\
&+(\alpha_1+\alpha_2)\,u_{0,1}u_{1,1}+(\beta_1-\beta_2)\,u_{0,0}u_{0,1}+(\beta_1+\beta_2)\,u_{1,0}u_{1,1}+\gamma_1\,u_{0,0}u_{1,1}\\
&+\gamma_2\,u_{1,0}u_{0,1}+(\xi_1-\xi_3)\,u_{0,0}u_{1,0}u_{0,1}+(\xi_1-\xi_3)\,u_{0,0}u_{1,0}u_{1,1}\\
&+(\xi_2-\xi_4)\,u_{1,0}u_{0,1}u_{1,1}+(\xi_2+\xi_4)\,u_{0,0}u_{0,1}u_{1,1}+\zeta\,u_{0,0}u_{0,1}u_{1,0}u_{1,1},
\end{aligned}
\tag{3.6.87}
$$

which are reduced to an integrable NLS under a multiple scale reduction [375].

Using the transformation $u_{n,m}=1/(\hat{u}_{n,m}-\gamma)$ and redefining the original constants entered in (3.6.86)

$$\alpha=\xi_2+\xi_4-5\gamma^2,\qquad \beta=\xi_2-\xi_4-4\gamma^2,\qquad \delta=\zeta+12\gamma^3-4\gamma\xi_2,$$

we obtain for $\hat{u}_{n,m}$ a simpler equation depending on just 3 free parameters:

$$(\hat{u}_{0,0}\hat{u}_{1,1}+\alpha)(\hat{u}_{1,0}+\hat{u}_{0,1})+(2\hat{u}_{1,0}\hat{u}_{0,1}+\beta)(\hat{u}_{0,0}+\hat{u}_{1,1})+\delta=0. \tag{3.6.88}$$

If the three parameters are null, (3.6.88) is a linear equation in $\tilde{u}_{0,0}=1/\hat{u}_{0,0}$, and thus trivially integrable.

For (3.6.88) the test is complicate, as the system (3.6.50, 3.6.51) depends on the additional variables $\hat{u}_{0,1}$, $\hat{u}_{0,-1}$. It is written as a polynomial system and setting to zero the coefficients of the different powers of $\hat{u}_{0,1}$ and $\hat{u}_{0,-1}$ we obtain a simpler system of equations for $q^{(k)}_{-1,0}$.

Eq. (3.6.88) is a simple classification problem, as it depends on three arbitrary constants, and we search for all integrable cases, if any, contained in it. By looking at its generalized symmetries we find 2 integrable non linearizable cases:

(1) $\alpha=2\beta\neq 0$ and $\delta=0$, i.e. $\xi_2=3\xi_4+3\gamma^2$, $\zeta=12\gamma\xi_4$;
(2) $\beta=2\alpha\neq 0$ and $\delta=0$, i.e. $\xi_2=6\gamma^2-3\xi_4$, $\zeta=12\gamma(\gamma^2-\xi_4)$.

In Case 1, using the transformation $\hat{u}_{n,m}=u_{n,m}(-1)^m\beta^{1/2}$, we obtain (3.6.82) with $\nu=1/2$. In Case 2 we can always choose $\alpha=1$ and the equation reads

$$(u_{0,0}u_{1,1}+1)(u_{1,0}+u_{0,1})+2(u_{1,0}u_{0,1}+1)(u_{0,0}+u_{1,1})=0. \tag{3.6.89}$$

By applying the procedure presented in the previous section we find the generalized symmetries

$$u_{0,0,\epsilon_1}=(u_{0,0}^2-1)\frac{u_{1,0}-u_{-1,0}}{u_{1,0}u_{-1,0}-1},\qquad u_{0,0,\epsilon_2}=(u_{0,0}^2-1)\frac{u_{0,1}-u_{0,-1}}{u_{0,1}u_{0,-1}-1}. \tag{3.6.90}$$

This last example (3.6.89) seems to be a new integrable model. This result can be formulated as the following theorem:

Theorem 79. *There are 2 nontrivial cases when* (3.6.88) *satisfies the generalized symmetry test. The first one is given by the relations* $\alpha=2\beta\neq 0$, $\delta=0$, *and the equation is* *transformed into* (3.6.82)*. In the second case, an equation can be written as* (3.6.89) *which possesses the generalized symmetries* (3.6.90)*.*

Example 6. This example is also a PΔE with arbitrary constant coefficients, obtained by Hietarinta and Viallet [387] as an equation with good factorization properties and considered to be an equation worth further studying. It is

$$(u_{0,0} - u_{0,1})(u_{1,0} - u_{1,1}) + (u_{0,0} - u_{1,1})r_4 + (u_{0,1} - u_{1,0})r_3 + r = 0. \tag{3.6.91}$$

Hietarinta and Viallet claim that (3.6.91) is integrable for all values of the coefficients r, r_3 and r_4, as it has a quadratic growth of the iterations in the calculation of its algebraic entropy.

Here we see that if $r_4 = r_3 = r = 0$, the equation is trivial. So we consider only those cases when the triple of parameters r_4, r_3 and r is different from zero.

Let $r_4 + r_3 = v = 0$ in (3.6.91). We apply an n, m-dependent point transformation $u_{n,m} = \hat{u}_{n,m} + (n+m)r_4$ and obtain for $\hat{u}_{n,m}$ the Klein type equation

$$(u_{0,0} - u_{0,1})(u_{1,0} - u_{1,1}) + r - r_4^2 = 0,$$

more precisely, a particular case of the Q_V equation. However, it is obviously trivial whenever $r = r_4^2$. If $r \neq r_4^2$, we can rewrite it as

$$(T_1 + 1)[\log(u_{0,0} - u_{0,1}) - \frac{1}{2}\log(r_4^2 - r)] = 0,$$

i.e. the equation is trivial in this case too.

The other possible case is when $r_4 + r_3 = v \neq 0$. By the transformation $u_{n,m} = v\hat{u}_{n,m}$ we get the following two parameter equation

$$(u_{0,0} - u_{0,1} + a)(u_{1,0} - u_{1,1} + a) + u_{0,1} - u_{1,0} + b = 0, \tag{3.6.92}$$

where

$$r_4 = av, \qquad r_3 = (1-a)v, \qquad r = (b + a^2)v^2.$$

Eq. (3.6.92) has 2 generalized symmetries which we can construct using our procedure. They are

$$u_{0,0,\epsilon_1} = (u_{1,0} - u_{0,0} - a - b)(u_{0,0} - u_{-1,0} - a - b), \tag{3.6.93}$$

$$u_{0,0,\epsilon_2} = \frac{(1 - y_{0,0})(1 - y_{0,-1})}{y_{0,0} + y_{0,-1}} + 1, \quad y_{n,m} = 2(u_{n,m+1} - u_{n,m}) - 2a + 1, \tag{3.6.94}$$

thus showing its integrability. This result can be formulated in the theorem:

Theorem 80. *In the case $r_4 + r_3 = 0$,* (3.6.91) *is equivalent to a trivial equation. In the case $r_4 + r_3 \neq 0$, it can be rewritten in the form* (3.6.92)*. Eq.* (3.6.92) *satisfies the generalized symmetry test and possesses the generalized symmetries* (3.6.93) *and* (3.6.94)*.*

Example 7. The next example (3.2.146) is also taken from [387] and has been considered in Section 3.2.4.3. We repeat it here for the convenience of the reader

$$u_{0,0}u_{0,1}c_5 + u_{1,0}u_{1,1}c_6 + u_{0,0}u_{1,0}c_1 + u_{0,1}u_{1,1}c_3 + (u_{0,0}u_{1,1} + u_{1,0}u_{0,1})c_2 = 0. \tag{3.6.95}$$

This PΔE has been proven to be integrable for all values of constants c_i by checking its algebraic entropy. Also in this case we have a kind of classification problem once we exclude, up to some simple transformations, all Klein type and trivial subequations.

Let us observe at first that if $c_5 = c_6$ and $c_1 = c_3$, (3.6.95) is of Klein type, and if moreover $c_1 = c_5 = 0$, it is trivial. We can construct some point transformations which leave (3.6.95) invariant, but transform the coefficients among themselves. By the transformation

(3.6.96) $$u_{n,m} = \hat{u}_{m,n},$$

$c_5 \leftrightarrow c_1$, $c_6 \leftrightarrow c_3$, and by the transformation

(3.6.97) $$u_{n,m} = 1/\hat{u}_{n,m},$$

$c_5 \leftrightarrow c_6$, $c_1 \leftrightarrow c_3$. In both cases c_2 remains unchanged. Moreover, the n, m dependent transformation

(3.6.98) $$u_{n,m} = \hat{u}_{n,m}\kappa_1^n\kappa_2^m, \qquad \kappa_i \neq 0,\ i = 1, 2,$$

leaves the equation invariant with the following transformation of the coefficients

$$\hat{c}_5 = c_5/\kappa_1, \quad \hat{c}_6 = c_6\kappa_1, \quad \hat{c}_1 = c_1/\kappa_2, \quad \hat{c}_3 = c_3\kappa_2, \quad \hat{c}_2 = c_2.$$

So, if at least one of the coefficients c_i $(i \neq 2)$ is different from zero, using the transformations (3.6.96) and (3.6.97), we can make $c_5 \neq 0$. Let us assume that also $c_6 \neq 0$. If either c_1 or c_3 is equal to zero then, using the transformations (3.6.96) and (3.6.97), we can make $c_6 = 0$. If both c_1 and c_3 are either zero or different from zero, using the transformation (3.6.98), we can make $c_1 = c_3$ and $c_5 = c_6$, i.e. we obtain a Klein type equation. So the only possible remaining case is when $c_5 \neq 0$, $c_6 = 0$ and without loss of generality we can set

(3.6.99) $$c_5 = 1, \quad c_6 = 0.$$

The non degeneracy conditions (3.6.1) give 2 restrictions: $c_2 \neq 0$ or $c_2 = 0$, $c_1c_3 \neq 0$. In these 2 cases, the equation can be nontrivially rewritten in the form of (3.6.7). If $c_2 \neq 0$ and $c_1 = c_3 = 0$, (3.6.95) is trivial, as it is equivalent to

$$(T_2 + 1)\left(c_2\frac{u_{1,0}}{u_{0,0}} + \frac{1}{2}\right) = 0.$$

So, at the end we get 2 admissible cases:

(3.6.100) $$c_2 = 0 : \qquad c_1c_3 \neq 0,$$

(3.6.101) $$c_2 \neq 0 : \quad c_1 \text{ or } c_3 \neq 0.$$

The equation (3.6.95) with (3.6.99), satisfying conditions (3.6.100, 3.6.101), possesses 2 generalized symmetries. The first symmetry depends on the number $c_1c_3 - c_2^2$. If

(3.6.102) $$c_1c_3 - c_2^2 \neq 0,$$

the condition (3.6.100) is satisfied automatically. The symmetry reads

(3.6.103) $$u_{0,0,\epsilon_1} = (u_{1,0} - cu_{0,0})\left(\frac{u_{0,0}}{u_{-1,0}} - c\right), \qquad c = \frac{c_2}{c_1c_3 - c_2^2}.$$

In the case when $c_1c_3 = c_2^2$, as $c_2 \neq 0$ due to condition (3.6.100), the condition (3.6.101) is satisfied automatically. In this case

(3.6.104) $$c_1c_3 = c_2^2 \neq 0,$$

and the symmetry reads

(3.6.105) $$u_{0,0,\epsilon_1} = u_{1,0} + \frac{u_{0,0}^2}{u_{-1,0}}.$$

The form of the second symmetry depends on the number c_1c_3. If

$$c_1c_3 \neq 0, \tag{3.6.106}$$

then both non degeneracy conditions are satisfied, and we have the generalized symmetry

$$u_{0,0,\epsilon_2} = \frac{c_2c_3c_1(u_{0,1}u_{0,-1} + u_{0,0}^2) + \frac{1}{2}(c_2^2 + c_3c_1)u_{0,0}(u_{0,1}c_3 + u_{0,-1}c_1)}{u_{0,1}c_3 - u_{0,-1}c_1}. \tag{3.6.107}$$

If $c_1c_3 = 0$, then we cannot have $c_2 = 0$ due to condition (3.6.100). So, $c_2 \neq 0$, and we use condition (3.6.101). We have here 2 cases for which both non degeneracy conditions are satisfied. First of them is

$$c_3 = 0, \quad c_1c_2 \neq 0, \tag{3.6.108}$$

and the corresponding generalized symmetry has the form

$$u_{0,0,\epsilon_2} = \left(u_{0,1} + u_{0,0}\frac{c_1}{c_2}\right)\left(\frac{u_{0,0}}{u_{0,-1}} + \frac{c_1}{c_2}\right). \tag{3.6.109}$$

The second case is

$$c_1 = 0, \quad c_2c_3 \neq 0, \tag{3.6.110}$$

and the generalized symmetry reads

$$u_{0,0,\epsilon_2} = \left(\frac{u_{0,0}}{u_{0,1}} + \frac{c_3}{c_2}\right)\left(u_{0,-1} + u_{0,0}\frac{c_3}{c_2}\right). \tag{3.6.111}$$

Now we can state the following theorem:

Theorem 81. *For* (3.6.95) *we have:*

(1) *If it is not equivalent to a Klein type equation, then it can be rewritten in the form* (3.6.99) *by using transformations* (3.6.96), (3.6.97)*;*
(2) *The nontrivial equations* (3.2.109) *and* (3.6.99) *must satisfy the conditions* (3.6.100), (3.6.101)*;*
(3) *Eq.* (3.6.95), (3.6.99) *with the restriction* (3.6.100), (3.6.101) *satisfies the generalized symmetry test for any values of* c_1, c_2, c_3*;*
(4) *The first symmetry of this PΔE is of the form* (3.6.103) *in case* (3.6.102) *and of the form* (3.6.105) *in case* (3.6.104)*;*
(5) *The second symmetry is of the form* (3.6.107) *in case* (3.6.106)*, of the form* (3.6.109) *in case* (3.6.108) *and of the form* (3.6.111) *in case* (3.6.110)*.*

Example 8. The last example we will consider here is taken from an article by Adler, Bobenko and Suris [23], where an extended definition of 3D-consistency is discussed and the so-called deformations of H equations are presented. We consider in the following an equation related to (2.4.159a) presented in Section 2.4.7 which we repeat here for the convenience of the reader:

$$(u_{0,0} - u_{1,1})(u_{1,0} - u_{0,1}) = (\alpha - \beta)(1 - \pi u_{1,0}u_{0,1}), \tag{3.6.112}$$

where $\alpha \neq \beta$ and π are constants. Eq. (3.6.112) is a generalization, H_1^π, of the well-known discrete potential KdV or H_1 equation which is reobtained when $\pi = 0$.

Let us use the integrability condition (3.2.60) with $k = 1$ and obtain the system (3.6.50, 3.6.51). The first equation of this system depends on the additional variable $u_{0,1}$. We rewrite the equation in polynomial form and obtain a 4th degree polynomial in $u_{0,1}$. The coefficients of this polynomial provide us with 5 more equations for $q_{-1,0}^{(1)}$. Using these equations, we easily obtain as an integrability condition that $\pi = 0$. The other integrability conditions are similar, and none of them is satisfied if $\pi \neq 0$.

Theorem 82. *Eq.* (3.6.112) *with* $\pi \neq 0$ *satisfies none of the four integrability conditions* (3.2.60)*-*(3.3.18)*. This equation cannot have an autonomous nontrivial generalized symmetry of the form* (2.4.10)*.*

The result is not surprising, as (2.4.156a) is 3D-consistent on the so-called black-white lattice while (3.6.112) it is not. The integrable (2.4.159a) is an n-dependent PΔE. This means that to check 3D-consistency we have to use (3.6.112) together with another equation, i.e. (3.6.112) is conditionally 3D-consistent. Eq. (2.4.159a) is obtained in [839] instead of (3.6.112). Eq. (3.6.112) is obtained when $n+m$ is even, while if $n+m$ is odd we have a different equation. An n, m-dependent Lax pair can be found in [839] and n, m-dependent generalized symmetries for (2.4.159a) have been presented in Section 2.4.7.5.

7. Linearizability through change of variables in PΔEs

Here we will study the C-integrability conditions for PΔEs using, at difference from Section 2.4.11, techniques presented in Section 3.6.1.3 of this Chapter.

We consider three-point and four-point non linear PΔEs where one-point symmetries, two-point symmetries or Cole–Hopf transformations relate them to linear equations. Then we use the obtained linearizability conditions for the classifications of linearizable equations and to prove when specific equations are linearizable.

Three-point PΔEs. As far as we know C-integrable PΔEs defined on three points of the plane have not been studied up to now in details. Triangular lattices are important as on them we can define many discrete models [18, 622, 739]. A generic equation on three lattice points can be written as

$$\mathcal{E}_{m,n}(u_{m,n}, u_{m,n+1}, u_{m+1,n}) = 0. \tag{3.7.1}$$

This class of equations is not empty. An example of a PΔE belonging to (3.7.1) is the completely discrete Burgers given by (2.4.292) in Section 2.4.9. By a proper definition of the two independent continuous variables x and t in terms of the discrete indexes n and m and of some small parameters ϵ and τ we can get from (3.7.1) continuous equations of the form

$$u_t = f(x,t,u)u_x + g(x,t,u)u_{xx}, \qquad u = u(x,t). \tag{3.7.2}$$

This class of equations includes the continuous Burgers equation [502] which we studied in Section 2.2.5.

Four possible boundary value problems can be defined for equations involving just three lattice points. In Fig. 3.1 and 3.2 we present them. In Fig. 3.1a we consider an equation which relate the values of the field at the points, (m,n), $(m,n+1)$, and $(m+1,n)$. Then if we give the initial values along the line r_1 the solution of the equation will give the values of the field in all points in the upper half plane. If we give the initial values along the line r_2 the solution of the equation will give the values of the field in all points in the right half plane while if we give the initial values along the line r_3 the solution of the equation will construct the solution along a left descending staircase. A similar situation is obtained in Fig. 3.1b and in Fig. 3.2. For more on initial–boundary value problems see [810].

If we restrict ourselves to multilinear equations we can have at most cubic nonlinearity. In such a case, as we will see in the following, the classification problem for linearizable PΔEs can be carried out up to the end.

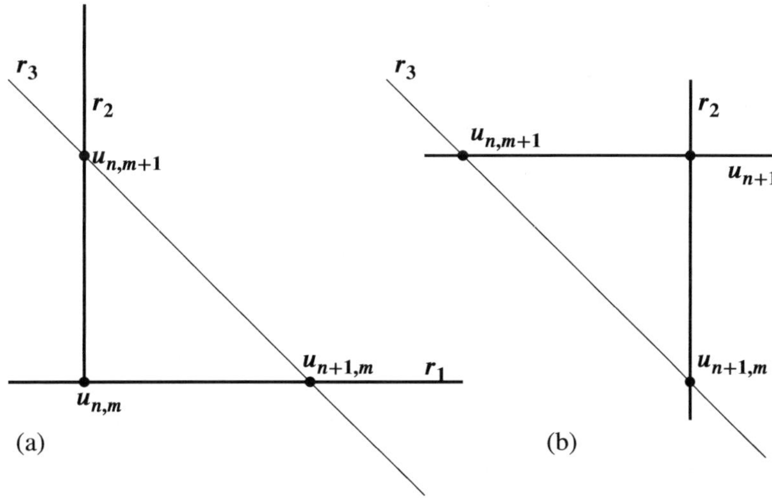

FIGURE 3.1. Points related by an equation defined on three points. In (a) giving the points on the line r_3 the equation constructs the staircase and propagates on the left. Given the points on r_1 the equation generate a propagation in the upper half plane while given the points on r_2 we will have propagation on the right. In (b) the situation is the opposite: given the points on r_3 it generate a staircase which propagates on the right while from r_1 it propagates in the lower half plane and from r_2 on the left [reprinted from [745]].

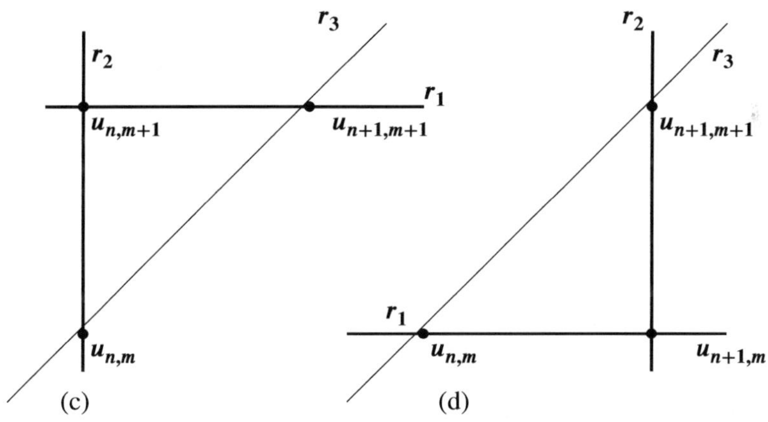

FIGURE 3.2. Points related by an equation defined on three points. The situation in the cases (c) and (d) is similar to the one of the cases (a) and (b) of the previous figure only the directions of propagation are different.

Four-point PΔEs. PΔEs on a quad-graph (see Fig. 2.3) are defined as (2.4.404). They have been considered on a fixed lattice by many authors [22, 29, 304, 305, 384, 518, 556] and many results on them can be found in Section 2.4.6. They are the simplest PΔEs which in the continuous limit turn into hyperbolic equations.

Let us define here one-point, two-point and Cole–Hopf transformations which transform (2.4.404) or (3.7.1) to a four-point or three-point linear equation with constant coefficients in the new field $\tilde{u}_{n,m}$.

Linearizing transformations. By a one-point transformation we mean a transformation

$$\tilde{u}_{n,m} = f_{n,m}(u_{n,m}) \tag{3.7.3}$$

characterized by a function depending just from the function $u_{n,m}$ and on the lattice point (n, m). It will be a Lie point symmetry transformation if $f = f_{n,m}$ satisfies all Lie group axioms which we considered in Section 1.1 of the Introduction. In the following we will only assume the differentiability of the function f up to at least second order.

A natural generalization is when one considers two-point transformations

$$\tilde{u}_{n,m} = g_{n,m}(u_{n,m}, u_{n,m+1}), \tag{3.7.4}$$

characterized by a function $g = g_{n,m}$ depending on the function u in two lattice points, say $u_{n,m}$ and $u_{n,m+1}$. This will correspond to a generalized symmetry. It is a non invertible transformation as the Miura (3.2.188) we considered in Section 3.2.7. The alternative choice when $g = g_{n,m}(u_{n,m}, u_{n+1,m})$ is to be considered separately if the function g depends on the lattice point (n, m). Otherwise it is recovered by a discrete symmetry transformation. Two lattice points is the minimum number of points necessary to provide in the continuous limit the first derivative and contact symmetries (see (1.1) for their definition) have been introduced by Lie as symmetries depending on first derivatives. However, as was shown in [521, 527] contact transformations on the lattice do not exist. Often contact symmetries are also called Miura transformations [617, 618] as R. Miura introduced them to transform the KdV into the mKdV equation and have played a very important role in the integrability of the KdV equation, see Section 2.2.2. Eq. (3.7.4) contains the transformation (3.7.3) as a subcase but here we will assume $\frac{\partial g_{n,m}}{\partial u_{n,m+1}} \neq 0$. Under this hypothesis the conditions for one-point transformations are not obtained as a limiting case of the two-point transformations. So one-point and two-point transformations will be treated as independent cases.

By a generalized Cole–Hopf transformation we mean a transformation [190, 400]

$$\tilde{u}_{n,m+1} = h_{n,m}(u_{n,m}, u_{n,m+1})\, \tilde{u}_{n,m}, \tag{3.7.5}$$

$$\tilde{u}_{n+1,m} = k_{n,m}(\{u_{n,m}\})\, \tilde{u}_{n,m}, \tag{3.7.6}$$

were the function $k_{n,m}$, depending on $u_{n,m}$ at various lattice points, is determined requiring that the compatibility between (3.7.5), (3.7.6) gives the expected PΔE. For example (3.7.5) reduces to the discrete Cole-Hopf transformation (2.3.319) when $h_{n,m}(u_{n,m}, u_{n,m+1}) = u_{n,m}$. Then $\tilde{u}_{n,m}$ will satisfy a linear discrete heat equation and $u_{n,m}$ will satisfy a discrete Burgers equation [502] as we saw in Section 2.4.9. In this case $k_{n,m}(\{u_{n,m}\}) = -u_{n,m}$.

In Section 3.7.1 we present the linearizability theorems for three-point PΔEs and the classification of three-point linearizable non linear PΔEs. In the subsequent Section 3.7.2 we deal with linearizable non linear quad-graph equations, presenting at the end a list of nontrivial examples. In Section 3.7.3 we present some results on the classification of quad-graph equations linearizable by one-point transformations.

7.1. Three-point PΔEs linearizable by local and non local transformations. To simplify the presentation we will limit ourselves to autonomous equations. So we will not need to index the complex dependent variable by its lattice point but just by its relative position with respect to the reference point m, n, i.e. $u_{m,n} \doteq u_{0,0}$. So equation (3.7.1) can be written as

$$\mathcal{E}\left(u_{0,0}, u_{1,0}, u_{0,1}\right) = 0. \tag{3.7.7}$$

This equation relates two points laying on a curve with a third laying in an independent direction. So, given a curve of points, we can extend it by this equation to all the points laying in the half plane above or below it (see Figure 3.1 or 3.2 in the case of a line and its perpendicular). In the following we will consider just the case of Fig. 3.1 when the evolution is in the upper half plane as by symmetry we can always reduce ourselves to this case.

7.1.1. *Linearizability conditions.* We will state in detail the procedure used, which will be applied later in all other cases. This procedure follows a similar one introduced in the case of the analysis of formal symmetries for integrable quad-graph equations [555, 556]. In particular we describe how we get on one side the determining equations which give the transformation and on the other side the conditions under which the equation (3.7.7) might be linearizable. The latter are necessary conditions which the given equation has to satisfy if a point transformation which linearizes the equation exists. If the conditions are satisfied then we can solve the PDE determining the transformation and get a first approximation to the point transformation. However only if also the initial determining equation is satisfied the system is linearizable.

We will assume that we can solve (3.7.7) with respect to any of the three variables

$$u_{0,0} = F\left(u_{1,0}, u_{0,1}\right), \quad F_{u_{1,0}} \neq 0, \quad F_{u_{0,1}} \neq 0, \tag{3.7.8a}$$

$$u_{1,0} = G\left(u_{0,0}, u_{0,1}\right), \quad G_{u_{0,0}} \neq 0, \quad G_{u_{0,1}} \neq 0, \tag{3.7.8b}$$

$$u_{0,1} = H\left(u_{0,0}, u_{1,0}\right), \quad H_{u_{0,0}} \neq 0, \quad H_{u_{1,0}} \neq 0. \tag{3.7.8c}$$

Moreover we assume the existence of a linearizing autonomous one-point transformation

$$\tilde{u}_{0,0} \doteq f_{0,0}\left(u_{0,0}\right), \quad \frac{df(x)}{dx} \neq 0, \tag{3.7.9}$$

which transforms (3.7.7) into the linear PΔE

$$a\tilde{u}_{0,0} + b\tilde{u}_{1,0} + c\tilde{u}_{0,1} + d = 0, \tag{3.7.10}$$

with a, b, c and d being complex constant coefficients. Hence, choosing for example as independent variables $u_{1,0}$ and $u_{0,1}$, we will have that

$$af_{0,0}|_{u_{0,0}=F} + bf_{1,0} + cf_{0,1} + d = 0, \tag{3.7.11}$$

must be identically satisfied for any choice of $u_{1,0}$ and $u_{0,1}$. Differentiating (3.7.11) with respect to the independent variables $u_{1,0}$ or $u_{0,1}$, we will respectively obtain

$$b\frac{df_{1,0}}{du_{1,0}} + a\left(\frac{df_{0,0}}{du_{0,0}}\Big|_{u_{0,0}=F}\right)F_{u_{1,0}} = 0, \tag{3.7.12a}$$

$$c\frac{df_{0,1}}{du_{0,1}} + a\left(\frac{df_{0,0}}{du_{0,0}}\Big|_{u_{0,0}=F}\right)F_{u_{0,1}} = 0, \tag{3.7.12b}$$

which have to be identically satisfied for all $u_{1,0}$ and $u_{0,1}$. From (3.7.12), as $\frac{df(x)}{dx} \neq 0$, if (3.7.11) is to be nontrivial, we derive that $a \neq 0$. Moreover, as $\frac{df(x)}{dx} \neq 0$, $F_{u_{1,0}} \neq 0$ and $F_{u_{0,1}} \neq 0$, and also $b \neq 0$ and $c \neq 0$. Then we can divide (3.7.10) by a and, introducing the new parameters $\beta \doteq \frac{b}{a} \neq 0$, $\gamma \doteq \frac{c}{a} \neq 0$ and $\delta \doteq \frac{d}{a}$, we can rewrite (3.7.10) as

$$\tilde{u}_{0,0} + \beta\tilde{u}_{1,0} + \gamma\tilde{u}_{0,1} + \delta = 0. \tag{3.7.13}$$

Taking the (principal value of the) logarithm of (3.7.12), we have

$$\log \frac{d f_{1,0}}{d u_{1,0}} - \log \frac{d f_{0,0}}{d u_{0,0}}\Big|_{u_{0,0}=F} = \log\left(-\frac{F_{u_{1,0}}}{\beta}\right) \pmod{2\pi \mathrm{i}}, \tag{3.7.14a}$$

$$\log \frac{d f_{0,1}}{d u_{0,1}} - \log \frac{d f_{0,0}}{d u_{0,0}}\Big|_{u_{0,0}=F} = \log\left(-\frac{F_{u_{0,1}}}{\gamma}\right) \pmod{2\pi \mathrm{i}}. \tag{3.7.14b}$$

Introducing the annihilation operator $\mathcal{A}$

$$\mathcal{A} \doteq \frac{\partial}{\partial u_{1,0}} - \frac{F_{u_{1,0}}}{F_{u_{0,1}}} \frac{\partial}{\partial u_{0,1}}, \tag{3.7.15}$$

such that $\mathcal{A}\phi(F) = 0$ for any arbitrary function $\phi(F)$ and applying it to (3.7.14) we obtain

$$\frac{d}{d u_{1,0}} \log \frac{d f_{1,0}}{d u_{1,0}} = \frac{W_{(u_{1,0})}\left[F_{u_{0,1}}; F_{u_{1,0}}\right]}{F_{u_{1,0}} F_{u_{0,1}}}, \tag{3.7.16a}$$

$$\frac{d}{d u_{0,1}} \log \frac{d f_{0,1}}{d u_{0,1}} = \frac{W_{(u_{0,1})}\left[F_{u_{1,0}}; F_{u_{0,1}}\right]}{F_{u_{1,0}} F_{u_{0,1}}}, \tag{3.7.16b}$$

where the Wronskian W is defined by

$$W_{(x)}[f; g] \doteq f g_x - g f_x. \tag{3.7.17}$$

The left hand side of (3.7.16a) depends only on $u_{1,0}$, while the right hand side depends on both $u_{1,0}$ and $u_{0,1}$. So, requiring that the right hand side be independent on $u_{0,1}$, we obtain the following **necessary condition for the linearizability**

$$\frac{\partial}{\partial u_{1,0}} \frac{\partial}{\partial u_{0,1}} \log\left(\frac{F_{,u_{1,0}}}{F_{,u_{0,1}}}\right) = 0. \tag{3.7.18}$$

This condition (and the two similar ones obtained choosing as independent variables $u_{0,0}$, $u_{1,0}$ or $u_{0,0}$, $u_{0,1}$) is identically satisfied when (3.7.7) is a multilinear equation, i.e. it has the form (3.7.33). Eq. (3.7.18) implies

$$F = \mathcal{F}\left(\mathcal{R}\left(u_{1,0}\right) + \tilde{\mathcal{R}}\left(u_{0,1}\right)\right), \tag{3.7.19}$$

where $\mathcal{F}$, $\mathcal{R}$ and $\tilde{\mathcal{R}}$ are arbitrary functions of their argument. Inserting (3.7.19) into (3.7.16a), integrating once with respect to $u_{1,0}$ and taking the exponential of both sides of the equation we get

$$\frac{d f_{1,0}}{d u_{1,0}} = \rho \frac{d\mathcal{R}\left(u_{1,0}\right)}{d u_{1,0}}, \tag{3.7.20}$$

where $\rho > 0$ is an arbitrary integration constant. Integrating again once with respect to $u_{1,0}$ we get an expression for the linearizing transformation

$$f_{0,0} = \rho \mathcal{R}\left(u_{0,0}\right) + \sigma, \tag{3.7.21}$$

where σ is an arbitrary integration constant. If instead we insert (3.7.19) into (3.7.16b) we get

$$f_{0,0} = \tilde{\rho} \tilde{\mathcal{R}}\left(u_{0,0}\right) + \tilde{\sigma}, \tag{3.7.22}$$

where $\tilde{\rho} > 0$ and $\tilde{\sigma}$ are two arbitrary integration constants a priori different from ρ and σ. Comparing (3.7.21) with (3.7.22) we find $\tilde{\mathcal{R}} = \tau\mathcal{R} + \pi$, $\tau \doteq \rho/\tilde{\rho} > 0$, $\pi \doteq (\sigma - \tilde{\sigma})/\tilde{\rho}$ and (3.7.19) reduces to

$$F = \mathcal{F}\left[\mathcal{R}\left(u_{1,0}\right) + \tau\mathcal{R}\left(u_{0,1}\right) + \pi\right]. \tag{3.7.23}$$

From (3.7.19) we get $F_{u_{1,0}} = \mathcal{F}'\mathcal{R}_{u_{1,0}}$ and $F_{u_{0,1}} = \mathcal{F}'\tilde{\mathcal{R}}_{u_{0,1}}$ and thus

$$\frac{F_{u_{1,0}}}{F_{u_{0,1}}} = \frac{\mathcal{R}_{u_{1,0}}}{\tilde{\mathcal{R}}_{u_{0,1}}}. \tag{3.7.24}$$

As $\mathcal{R}$ and $\tilde{\mathcal{R}}$ differ only by constants, if we set $u_{1,0} = u_{0,1} = x$ in (3.7.24) we get a constant.

By chosing as independent variables $u_{0,0}$, $u_{0,1}$ or $u_{0,0}$, $u_{1,0}$ and considering instead of the operator $\mathcal{A}$ the operators

$$\mathcal{B} \doteq \frac{\partial}{\partial u_{0,0}} - \frac{G_{u_{0,0}}}{G_{u_{0,1}}}\frac{\partial}{\partial u_{0,1}}, \tag{3.7.25a}$$

$$\mathcal{C} \doteq \frac{\partial}{\partial u_{0,0}} - \frac{H_{u_{0,0}}}{H_{u_{1,0}}}\frac{\partial}{\partial u_{1,0}}. \tag{3.7.25b}$$

we obtain

$$G = \mathcal{G}\left(\mathcal{J}\left(u_{0,0}\right) + \tilde{\mathcal{J}}\left(u_{0,1}\right)\right), \tag{3.7.26a}$$

$$H = \mathcal{H}\left(\mathcal{I}\left(u_{0,0}\right) + \tilde{\mathcal{I}}\left(u_{1,0}\right)\right). \tag{3.7.26b}$$

Moreover we have expressions similar to (3.7.21, 3.7.22) implying that $f_{0,0}$ is linear in $\mathcal{J}\left(u_{0,0}\right)$, $\tilde{\mathcal{J}}\left(u_{0,0}\right)$ and in $\mathcal{I}\left(u_{0,0}\right)$ and $\tilde{\mathcal{I}}\left(u_{0,0}\right)$. Consequently the ratio of the derivatives of $f_{0,0}$, expressed in terms of $\mathcal{R}$ and $\mathcal{J}$ or $\mathcal{I}$ and $\mathcal{R}$ will be a constant. From these expressions we can derive other necessary conditions for the linearizability which we can state as the following theorem:

Theorem 83. *A non linear PΔE defined on three points (3.7.7) will be linearizable by a point transformation only if the functions F, G or H defined in (3.7.8) satisfy for any x the following conditions:*

$$\frac{d}{dx}A(x) = 0, \quad A(x) \doteq \left.\frac{F_{u_{1,0}}}{F_{u_{0,1}}}\right|_{u_{1,0}=u_{0,1}=x} = \frac{\left.\frac{d\mathcal{R}(u_{1,0})}{du_{1,0}}\right|_{u_{1,0}=x}}{\left.\frac{d\tilde{\mathcal{R}}(u_{0,1})}{du_{0,1}}\right|_{u_{0,1}=x}} \neq 0, \tag{3.7.27a}$$

$$\frac{d}{dx}B(x) = 0, \quad B(x) \doteq \left.\frac{G_{u_{0,0}}}{G_{u_{0,1}}}\right|_{u_{0,0}=u_{0,1}=x} = \frac{\left.\frac{d\mathcal{J}(u_{0,0})}{du_{0,0}}\right|_{u_{0,0}=x}}{\left.\frac{d\tilde{\mathcal{J}}(u_{0,1})}{du_{0,1}}\right|_{u_{0,1}=x}} \neq 0, \tag{3.7.27b}$$

$$\frac{d}{dx}C(x) = 0, \quad C(x) \doteq \left.\frac{H_{u_{0,0}}}{H_{u_{1,0}}}\right|_{u_{0,0}=u_{1,0}=x} = \frac{\left.\frac{d\mathcal{I}(u_{0,0})}{du_{0,0}}\right|_{u_{0,0}=x}}{\left.\frac{d\tilde{\mathcal{I}}(u_{1,0})}{du_{1,0}}\right|_{u_{1,0}=x}} \neq 0, \tag{3.7.27c}$$

$$\frac{d}{dx}D(x) = 0, \quad D(x) \doteq \frac{\left.\frac{d\mathcal{I}(u_{0,0})}{du_{0,0}}\right|_{u_{0,0}=x}}{\left.\frac{d\mathcal{J}(u_{0,0})}{du_{0,0}}\right|_{u_{0,0}=x}} \neq 0, \tag{3.7.27d}$$

$$\frac{d}{dx}E(x)=0,\quad E(x)\doteq\frac{\left.\frac{d\tilde{\mathcal{I}}(u_{1,0})}{du_{1,0}}\right|_{u_{1,0}=x}}{\left.\frac{d\mathcal{R}(u_{1,0})}{du_{1,0}}\right|_{u_{1,0}=x}}\neq 0. \tag{3.7.27e}$$

The three conditions (3.7.27a, 3.7.27b, 3.7.27c), being expressed only in term of F, G and H, can always be used, while (3.7.27d, 3.7.27e) can be used only when one is able to calculate explicitly the functions $\frac{d\mathcal{I}}{dx}$, $\frac{d\tilde{\mathcal{I}}}{dx}$, $\frac{d\mathcal{J}}{dx}$ and $\frac{d\mathcal{R}}{dx}$.

Inserting (3.7.23, 3.7.21) into (3.7.12) and writing everything in term of $y\doteq\mathcal{R}\left(u_{1,0}\right)+\tau\mathcal{R}\left(u_{0,1}\right)+\pi$, we obtain

$$\beta+\frac{d\left(\mathcal{R}\circ\mathcal{F}\right)(y)}{dy}=0, \tag{3.7.28a}$$

$$\gamma+\tau\frac{d\left(\mathcal{R}\circ\mathcal{F}\right)(y)}{dy}=0, \tag{3.7.28b}$$

where $\mathcal{R}\circ\mathcal{F}$ stands for the compositions of the functions $\mathcal{R}$ and $\mathcal{F}$. Solving (3.7.28), we get

$$\mathcal{R}\circ\mathcal{F}(y)=-\beta y+\beta_0, \tag{3.7.29a}$$

$$\gamma=\beta\tau, \tag{3.7.29b}$$

where β_0 is an arbitrary integration constant. From (3.7.29a) we get $\mathcal{F}(y)=\mathcal{R}^{-1}\left(-\beta y+\beta_0\right)$, so that (3.7.23) becomes

$$F=\mathcal{R}^{-1}\left(-\beta\mathcal{R}\left(u_{1,0}\right)-\beta\tau\mathcal{R}\left(u_{0,1}\right)-\beta\pi+\beta_0\right). \tag{3.7.30}$$

Inserting (3.7.30), (3.7.21) and (3.7.29b) into (3.7.11), we have $\delta=\beta\pi\rho-\beta\sigma(\tau+1)-\rho\beta_0-\sigma$. Then we can state the following obvious theorem:

Theorem 84. *Necessary and sufficient condition for the linearizability by a point transformation of an equation belonging to the class (3.7.7) is that the equation can be written in the form*

$$u_{0,0}=\mathcal{R}^{-1}\left(-\beta\mathcal{R}\left(u_{1,0}\right)-\tilde{\tau}\mathcal{R}\left(u_{0,1}\right)-\tilde{\pi}\right), \tag{3.7.31}$$

with $\tilde{\tau}$, β and $\tilde{\pi}$ arbitrary integration constants. This equation is linearizable by the point transformation $\tilde{u}_{0,0}\doteq\rho\mathcal{R}\left(u_{0,0}\right)+\tilde{\sigma}\rho$ to the equation

$$\tilde{u}_{0,0}+\beta\tilde{u}_{1,0}+\tilde{\tau}u_{0,1}+\left[\tilde{\pi}-\tilde{\sigma}\left(\tilde{\tau}+\beta+1\right)\right]\rho=0, \tag{3.7.32}$$

where ρ and $\tilde{\theta}$ are arbitrary constants.

Let's note that, in all generality, we can assume $\rho=1$ as it can be always rescaled away by choosing $\tilde{u}_{0,0}\doteq\rho\tilde{\tilde{u}}_{0,0}$ so that $\tilde{\tilde{u}}_{0,0}=\mathcal{R}\left(u_{0,0}\right)+\tilde{\sigma}$. Moreover, by a proper choice of the parameters $\tilde{\sigma}$ e $\tilde{\tau}$ we can always reduce the linear equation to an homogeneous equation.

The main content of Theorem 84 is the assertion that there are no other equations defined on three points which can be linearized by a one-point transformation apart from those presented here.

7.1.2. *Classification of complex multilinear equations defined on a three-point lattice linearizable by one-point transformations.* Let us consider the multilinear equation

$$\begin{aligned}\tilde{a}u_{0,0}+\tilde{b}u_{0,1}+\tilde{c}u_{1,0}+\tilde{d}u_{0,0}u_{1,0}&\\ +\tilde{e}u_{0,0}u_{0,1}+\tilde{f}u_{1,0}u_{0,1}+\tilde{g}u_{0,0}u_{1,0}u_{0,1}+\tilde{h}&=0\end{aligned} \tag{3.7.33}$$

where $\tilde{a}$, $\tilde{b}$, $\tilde{c}$, $\tilde{d}$, $\tilde{e}$, $\tilde{f}$, $\tilde{g}$, $\tilde{h}$ are arbitrary complex parameters. From (3.7.27) we derive

(3.7.34a) $$\frac{d\mathcal{R}(x)}{dx} = \frac{\kappa_1}{\tilde{e}\tilde{h} - \tilde{a}\tilde{b} + \left(\tilde{c}\tilde{e} - \tilde{b}\tilde{d} - \tilde{a}\tilde{f} + \tilde{g}\tilde{h}\right)x + \left(\tilde{c}\tilde{g} - \tilde{d}\tilde{f}\right)x^2},$$

(3.7.34b) $$\frac{d\tilde{\mathcal{R}}(x)}{dx} = \frac{\kappa_1}{\tilde{d}\tilde{h} - \tilde{a}\tilde{c} + \left(\tilde{b}\tilde{d} - \tilde{c}\tilde{e} - \tilde{a}\tilde{f} + \tilde{g}\tilde{h}\right)x + \left(\tilde{b}\tilde{g} - \tilde{e}\tilde{f}\right)x^2},$$

(3.7.34c) $$\frac{d\mathcal{J}(x)}{dx} = \frac{\kappa_2}{\tilde{f}\tilde{h} - \tilde{b}\tilde{c} + \left(\tilde{g}\tilde{h} - \tilde{c}\tilde{e} - \tilde{b}\tilde{d} + \tilde{a}\tilde{f}\right)x + \left(\tilde{a}\tilde{g} - \tilde{d}\tilde{e}\right)x^2},$$

(3.7.34d) $$\frac{d\tilde{\mathcal{J}}(x)}{dx} = \frac{\kappa_2}{\kappa_1}\frac{d\tilde{\mathcal{H}}(x)}{dx}, \quad \frac{d\mathcal{I}(x)}{dx} = \frac{\kappa_3}{\kappa_2}\frac{d\mathcal{J}(x)}{dx},$$
$$\frac{d\tilde{\mathcal{I}}(x)}{dx} = \frac{\kappa_3}{\kappa_1}\frac{d\mathcal{H}(x)}{dx},$$

where κ_i, $i = 1,\ldots,3$ are arbitrary complex constants. Then the necessary conditions for the linearizability (3.7.27d, 3.7.27e) are automatically satisfied. Moreover, as from (3.7.34d) it follows for the functions $A(x)$, $B(x)$ and $C(x)$, defined in (3.7.27), we have the following equation $A(x) \cdot C(x) = B(x)$. Moreover if (3.7.27a, 3.7.27c) are satisfied then also (3.7.27b) will be satisfied. Hence, imposing the necessary conditions for the linearizability (3.7.27a, 3.7.27c), we obtain the following different cases:

(1) $\tilde{d} = \tilde{f} = \tilde{g} = 0$, $\tilde{b} = \tilde{a}$, $\tilde{e} = \tilde{a}(\tilde{a} + \tilde{c})$, $\tilde{h} = 1$, $\tilde{a} \neq 0$, $\tilde{c} \neq 0$, $\tilde{c} \neq -\tilde{a}$. Integrating the differential equation (3.7.16a) for $f_{1,0}\left(u_{1,0}\right)$ we get
$$f_{0,0}\left(u_{0,0}\right) = c_1\left\{\log\left[1 + (\tilde{a} + \tilde{c})u_{0,0}\right] + c_2\right\},$$
where $c_1 \neq 0$ and c_2 are arbitrary complex integration constants and we have chosen the principal branch of the logarithm. Inserting this result into (3.7.12) and requiring that they be identically satisfied for any $u_{1,0}$ and $u_{0,1}$, we get $a = c = -b$. Finally from (3.7.10) we have, where z an arbitrary entire parameter, $d = -c_1 b\left[c_2 - \log(-\tilde{c}/\tilde{a}) - 2\pi i z\right]$. In conclusion, choosing in all generality $c_1 = 1$, $c_2 = 0$, we find that the transformation $\tilde{u}_{0,0} = \log\left[1 + (\tilde{a} + \tilde{c})u_{0,0}\right]$ linearizes the non linear difference equation
$$u_{0,0} = -\frac{1}{\tilde{a}}\frac{1 + \tilde{c}u_{1,0} + \tilde{a}u_{0,1}}{1 + (\tilde{a} + \tilde{c})u_{0,1}},$$
into the linear equation
$$\tilde{u}_{0,0} - \tilde{u}_{1,0} + \tilde{u}_{0,1} - \log(-\tilde{c}/\tilde{a}) - 2\pi i z = 0.$$

(2) $\tilde{d} = \tilde{e} = \tilde{g} = 0$, $\tilde{c} = \tilde{b}$, $\tilde{f} = \tilde{b}\left(\tilde{a} + \tilde{b}\right)$, $\tilde{h} = 1$, $\tilde{a} \neq 0$, $\tilde{b} \neq 0$, $\tilde{b} \neq -\tilde{a}$. Integrating the differential equation (3.7.16a) for $f_{1,0}\left(u_{1,0}\right)$ we get
$$f_{0,0}\left(u_{0,0}\right) = c_1\left\{\log\left[1 + \left(\tilde{a} + \tilde{b}\right)u_{0,0}\right] + c_2\right\},$$
where $c_1 \neq 0$ and c_2 are arbitrary complex integration constants and with no loss of generality we can choose the principal branch of the logarithm function. Inserting this result into (3.7.12) and requiring that they be identically satisfied for any $u_{1,0}$ and $u_{0,1}$, we get $a = -b$, $c = b$ and from (3.7.10) $d = -c_1 b\left[c_2 + \log\left(-\tilde{a}/\tilde{b}\right) + 2\pi i z\right]$ with z an arbitrary entire parameter. In conclusion, choosing $c_1 = 1$, $c_2 = 0$, we have that the equation
$$u_{0,0} = -\frac{1 + \tilde{b}\left[\left(u_{1,0} + u_{0,1}\right) + \left(\tilde{a} + \tilde{b}\right)u_{1,0}u_{0,1}\right]}{\tilde{a}},$$

is linearizable by the transformation $\tilde{u}_{0,0} = \log\left[1 + \left(\tilde{a} + \tilde{b}\right) u_{0,0}\right]$ into the linear equation

$$\tilde{u}_{0,0} - \tilde{u}_{1,0} - \tilde{u}_{0,1} + \log\left(-\tilde{a}/\tilde{b}\right) + 2\pi \mathrm{i} z = 0.$$

(3) $\tilde{a} = \tilde{b} = \tilde{c} = \tilde{d} = \tilde{e} = \tilde{f} = 0$, $\tilde{g} = 1$, $\tilde{h} \neq 0$. Integrating the differential equation (3.7.16a) for $f_{1,0}\left(u_{1,0}\right)$ we get

$$f_{0,0}\left(u_{0,0}\right) = c_1\left(c_2 + \log u_{0,0}\right),$$

where $c_1 \neq 0$ and c_2 are arbitrary complex integration constants and we have chosen the principal branch of the logarithm function. Inserting this result into (3.7.12) and requiring them to be identically satisfied for any $u_{1,0}$ and $u_{0,1}$, we get $a = b = c$. Finally (3.7.10) gives, where z an arbitrary entire parameter, $d = -c_1 b\left[3c_2 + \log\left(-\tilde{h}\right) + 2\pi \mathrm{i} z\right]$. In conclusion, choosing $c_1 = 1$, $c_2 = 0$, we have that the equation

$$u_{0,0} = -\frac{\tilde{h}}{u_{1,0}u_{0,1}},$$

is linearizable by the transformation $\tilde{u}_{0,0} = \log u_{0,0}$ into the linear equation

$$\tilde{u}_{0,0} + \tilde{u}_{1,0} + \tilde{u}_{0,1} - \log\left(-\tilde{h}\right) - 2\pi \mathrm{i} z = 0.$$

All the other cases obtained imposing the necessary conditions for the linearizability (3.7.27a, 3.7.27c) either do not provide any solution or, if they do, these solutions are sub–cases of the cases just found or can be transformed into them or in a linear equation by a composition of a Möbius transformation of the dependent variable $u_{0,0}$ with an exchange of the independent variables $n \leftrightarrow m$. So we can state the following theorem:

Theorem 85. *Apart from the equations which are Möbius equivalent to a linear equation, as for example the non linear equation*

$$w_{1,0}w_{0,1} + \alpha w_{0,0}w_{0,1} + \beta w_{0,0}w_{1,0} + \delta w_{0,0}w_{1,0}w_{0,1} = 0, \tag{3.7.35}$$

where $\delta = 0, 1$ *and* $\alpha \neq 0$ *and* $\beta \neq 0$ *are otherwise arbitrary complex parameters which linearizes by the inversion* $\tilde{u}_{0,0} = 1/w_{0,0}$ *to the equation*

$$\tilde{u}_{0,0} + \alpha\tilde{u}_{1,0} + \beta\tilde{u}_{0,1} + \delta = 0,$$

the only equations belonging to the class (3.7.33) *which are linearizable by a point transformation are, up to a Möbius transformation of the dependent variable (eventually composed with an exchange of the independent variables* $n \leftrightarrow m$*), the following three*

$$w_{0,0}w_{1,0}w_{0,1} - 1 = 0, \tag{3.7.36a}$$

$$w_{0,0}w_{0,1} - w_{1,0} = 0, \tag{3.7.36b}$$

$$w_{1,0}w_{0,1} - w_{0,0} = 0. \tag{3.7.36c}$$

Eqs. (3.7.36a, 3.7.36b, 3.7.36c) *linearize by the transformation* $\tilde{u}_{0,0} = \log w_{0,0}$, *with* log *always standing for the principal branch of the complex logarithmic function, respectively to the equations*

$$\tilde{u}_{0,0} + \tilde{u}_{1,0} + \tilde{u}_{0,1} = 2\pi \mathrm{i} z, \tag{3.7.37}$$

$$\tilde{u}_{0,0} - \tilde{u}_{1,0} + \tilde{u}_{0,1} = 2\pi \mathrm{i} z, \tag{3.7.38}$$

$$-\tilde{u}_{0,0} + \tilde{u}_{1,0} + \tilde{u}_{0,1} = 2\pi \mathrm{i} z, \tag{3.7.39}$$

Let's note that, given a solution $w_{0,0}$ of (3.7.36a-3.7.36c), the choice of the principal branch of the complex logarithm function in the transformation $\tilde{u}_{0,0} = \log w_{0,0}$ implies that in the inhomogeneous terms of the linear eqs. (3.7.37-3.7.39) the parameter z can only take the values -1, 0 and 1. More on this point can be found in [745].

7.1.3. *Linearizability by a Cole–Hopf transformation.* Let us assume the existence of a linearizing Cole–Hopf transformation

$$\tilde{u}_{0,1} \doteq f_{0,0}\left(u_{0,0}\right)\tilde{u}_{0,0}, \quad \frac{df(x)}{dx} \neq 0, \tag{3.7.40}$$

which transforms the linear PΔE

$$a\tilde{u}_{0,0} + b\tilde{u}_{1,0} + c\tilde{u}_{0,1} = 0, \tag{3.7.41}$$

into (3.7.7), where a, b and c are complex coefficients with $(a,b,c) \neq (0,0,0)$. Hence, inserting (3.7.40) into (3.7.41), we get

$$-b\tilde{u}_{1,0} = \tilde{u}_{0,0}\left[a + cf_{0,0}\left(u_{0,0}\right)\right]. \tag{3.7.42}$$

As $b \neq 0$, otherwise the linear equation is trivial, we derive a non linear equation for $u_{0,0}$

$$f_{1,0}\left(a + cf_{0,0}\right) = f_{0,0}\left(a + cf_{0,1}\right), \tag{3.7.43}$$

which, as it should be, it depends essentially on the three points, implies $a \neq 0$ and $c \neq 0$. As a consequence we can divide (3.7.41, 3.7.43) by a. Then performing the non autonomous transformation $\tilde{u}_{0,0} \doteq (a/b)^n (a/c)^m w_{0,0}$ and setting $g_{0,0}\left(u_{0,0}\right) \doteq c/a\, f\left(u_{0,0}\right)$, so that $w_{0,1} \doteq g_{0,0}\left(u_{0,0}\right) w_{0,0}$, we can transform (3.7.41) into the linear equation

$$w_{0,0} + w_{1,0} + w_{0,1} = 0. \tag{3.7.44}$$

Then the non linear equation (3.7.43) becomes

$$\frac{1 + g_{0,0}}{g_{0,0}} = \frac{1 + g_{0,1}}{g_{1,0}}. \tag{3.7.45}$$

Taking the (principal value of the) logarithm of (3.7.45) and differentiating the result with respect to $u_{1,0}$ or $u_{0,1}$, we have

$$\frac{d}{du_{1,0}}\log g_{1,0} + \left(\frac{d}{du_{0,0}}\log\left(\frac{1+g_{0,0}}{g_{0,0}}\right)\Big|_{u_{0,0}=F}\right)F_{u_{1,0}} = 0. \tag{3.7.46a}$$

$$\frac{d}{du_{0,1}}\log\left(1+g_{0,1}\right) - \left(\frac{d}{du_{0,0}}\log\left(\frac{1+g_{0,0}}{g_{0,0}}\right)\Big|_{u_{0,0}=F}\right)F_{u_{0,1}} = 0. \tag{3.7.46b}$$

Finally, taking again the (principal value of the) logarithm of (3.7.46) and applying the annihilation operator $\mathcal{A}$ defined in (3.7.15), we obtain

$$\frac{d}{du_{1,0}}\log\frac{d}{du_{1,0}}\log g_{1,0} - \frac{d}{du_{1,0}}\log\frac{F_{u_{1,0}}}{F_{u_{0,1}}} = 0, \tag{3.7.47a}$$

$$\frac{d}{du_{0,1}}\log\frac{d}{du_{0,1}}\log\left(1+g_{0,1}\right) - \frac{d}{du_{0,1}}\log\frac{F_{u_{0,1}}}{F_{u_{1,0}}} = 0. \tag{3.7.47b}$$

If we had chosen as independent variables $u_{0,0}$, $u_{0,1}$ or $u_{0,0}$, $u_{1,0}$ and used the annihilation operators $\mathcal{B}$ or C defined by (3.7.25), (3.7.47) would have been replaced by

$$\frac{d}{du_{0,0}}\log\frac{d}{du_{0,0}}\log\left(\frac{1+g_{0,0}}{g_{0,0}}\right)-\frac{d}{du_{0,0}}\log\frac{G_{u_{0,0}}}{G_{u_{0,1}}}=0, \tag{3.7.48a}$$

$$\frac{d}{du_{0,1}}\log\frac{d}{du_{0,1}}\log\left(1+g_{0,1}\right)-\frac{d}{du_{0,1}}\log\frac{G_{u_{0,1}}}{G_{u_{0,0}}}=0, \tag{3.7.48b}$$

or

$$\frac{d}{du_{0,0}}\log\frac{d}{du_{0,0}}\log\left(\frac{1+g_{0,0}}{g_{0,0}}\right)-\frac{d}{du_{0,0}}\log\frac{H_{u_{0,0}}}{H_{u_{1,0}}}=0, \tag{3.7.49a}$$

$$\frac{d}{du_{1,0}}\log\frac{d}{du_{1,0}}\log g_{1,0}-\frac{d}{du_{1,0}}\log\frac{H_{u_{1,0}}}{H_{u_{0,0}}}=0. \tag{3.7.49b}$$

From (3.7.47, 3.7.48, 3.7.49) we have that (3.7.19, 3.7.26, 3.7.27a) remain valid. So, inserting (3.7.19, 3.7.26, 3.7.27a) into (3.7.47, 3.7.48, 3.7.49) after an integration and an exponentiation we obtain the following relations for the function $g\,(x)$

$$\frac{d}{dx}\log g=\alpha_1\frac{d}{dx}\mathcal{R}\,(x), \tag{3.7.50a}$$

$$\frac{d}{dx}\log\,(1+g)=\alpha_2\frac{d}{dx}\tilde{\mathcal{R}}\,(x), \tag{3.7.50b}$$

$$\frac{d}{dx}\log\left(\frac{1+g}{g}\right)=\alpha_3\frac{d}{dx}\mathcal{J}\,(x), \tag{3.7.50c}$$

$$\frac{d}{dx}\log\,(1+g)=\alpha_4\frac{d}{dx}\tilde{\mathcal{J}}\,(x), \tag{3.7.50d}$$

$$\frac{d}{dx}\log\left(\frac{1+g}{g}\right)=\alpha_5\frac{d}{dx}\mathcal{I}\,(x), \tag{3.7.50e}$$

$$\frac{d}{dx}\log g=\alpha_6\frac{d}{dx}\tilde{\mathcal{I}}\,(x), \tag{3.7.50f}$$

where α_i, $i=1,\ldots,6$ are six complex nonzero arbitrary integration constants. From system (3.7.50), after some slight manipulation, we have

$$\frac{d}{dx}\log g=\alpha_1\frac{d}{dx}\mathcal{R}\,(x),\quad \frac{d}{dx}\log\,(1+g)=\alpha_2\frac{d}{dx}\tilde{\mathcal{R}}\,(x), \tag{3.7.51a}$$

$$\frac{\frac{d\tilde{\mathcal{J}}(x)}{dx}}{\frac{d\tilde{\mathcal{R}}(x)}{dx}}=\frac{\alpha_2}{\alpha_4},\quad \frac{\frac{d\tilde{\mathcal{I}}(x)}{dx}}{\frac{d\mathcal{R}(x)}{dx}}=\frac{\alpha_1}{\alpha_6},\quad \frac{\frac{d\mathcal{I}(x)}{dx}}{\frac{d\mathcal{J}(x)}{dx}}=\frac{\alpha_3}{\alpha_5}, \tag{3.7.51b}$$

$$\frac{\alpha_2}{\alpha_1}-A\,(x)=\frac{\alpha_5}{\alpha_6}A\,(x)\cdot C\,(x), \tag{3.7.51c}$$

where the functions $A\,(x)$ and $C\,(x)$ are defined by (3.7.27a, 3.7.27c) as for one-point transformations. Eq. (3.7.51c) is a first integrability condition. Moreover, as a consequence of

(3.7.51b), we derive three additional integrability conditions

$$\frac{d}{dx} D(x) = 0, \quad D(x) \doteq \frac{\frac{d\mathcal{I}(x)}{dx}}{\frac{d\mathcal{J}(x)}{dx}} \neq 0, \tag{3.7.52a}$$

$$\frac{d}{dx} E(x) = 0, \quad E(x) \doteq \frac{\frac{d\tilde{\mathcal{I}}(x)}{dx}}{\frac{d\mathcal{R}(x)}{dx}} \neq 0, \tag{3.7.52b}$$

$$\frac{d}{dx} L(x) = 0, \quad L(x) \doteq \frac{\frac{d\tilde{\mathcal{J}}(x)}{dx}}{\frac{d\tilde{\mathcal{R}}(x)}{dx}} \neq 0, \tag{3.7.52c}$$

the first two being the same integrability conditions (3.7.27d, 3.7.27e) as for point transformations. From (3.7.51) we have that (3.7.51a), after some manipulation, give

$$\frac{d}{dx} A(x) = \alpha_5 A(x) \frac{d}{dx} \mathcal{I}(x), \tag{3.7.53a}$$

$$g(x) = \beta_1 e^{\alpha_1 \mathcal{R}(x)}, \tag{3.7.53b}$$

where $\beta_1 \neq 0$ is an arbitrary complex integration constant. Hence we can state the following theorem:

Theorem 86. *A non linear PΔE defined on three points* (3.7.7) *will be linearizable by a Cole–Hopf transformation only if the functions F, G or H defined in* (3.7.8) *satisfy for any x the C-integrability conditions* (3.7.51c, 3.7.52, 3.7.53a).

The condition (3.7.51c), being expressed only in term of F and H, can always be used, while (3.7.52, 3.7.53a) can be used only when one is able to calculate explicitly the functions $\frac{d\mathcal{I}}{dx}$, $\frac{d\tilde{\mathcal{I}}}{dx}$, $\frac{d\tilde{\mathcal{J}}}{dx}$ and $\frac{d\tilde{\mathcal{R}}}{dx}$. Finally (3.7.53b) gives the function $g(x)$ appearing in the linearizing transformation $w_{0,1} \doteq g_{0,0}\left(u_{0,0}\right) w_{0,0}$.

7.1.4. *Classification of complex multilinear equations defined on three points linearizable by Cole-Hopf transformation.* If we consider the class of complex multilinear equations defined by (3.7.33), we have that the three necessary linearizability conditions (3.7.52) are automatically satisfied. So, imposing the two remaining necessary linearizability conditions (3.7.51c, 3.7.53a), we always obtain for the function $g_{0,0}\left(u_{0,0}\right)$ a linear fractional function. As we classify up to a Möbius transformation, we can be always set $g_{0,0}\left(u_{0,0}\right) = u_{0,0}$. So in all generality the linearizing Cole–Hopf transformation is given by $w_{0,1} \doteq u_{0,0} w_{0,0}$. We can now state the following theorem:

Theorem 87. *The class of complex autonomous multilinear discrete equations defined on three-points which is linearizable to a homogeneous linear equation by a Cole–Hopf transformation $w_{0,1} = u_{0,0} w_{0,0}$, is given, up to a Möbius transformation of the dependent variable (eventually composed with an exchange of the independent variables $n \leftrightarrow m$), by the following equation*

$$u_{1,0}(1 + u_{0,0}) - u_{0,0}(1 + u_{0,1}) = 0. \tag{3.7.54a}$$

This equation linearizes to the equation

$$w_{0,0} + w_{1,0} + w_{0,1} = 0. \tag{3.7.54b}$$

Eq.(3.7.54a) is strictly related to the non linear wave equation (2.4.292) obtained as the first equation of the completely discrete Burgers hierarchy in Section 2.4.9.

7.2. Nonlinear equations on a quad-graph linearizable by one-point, two-point and generalized Cole–Hopf transformations.

7.2.1. *Linearization by one-point transformations.* Let us assume the existence of a point transformation (3.7.3) which linearizes (2.4.130) to the PΔE

$$\tilde{u}_{n,m} + a\tilde{u}_{n+1,m} + b\tilde{u}_{n,m+1} + c\tilde{u}_{n+1,m+1} = 0 \tag{3.7.55}$$

with a, b and c being (n, m)–independent arbitrary non all zero complex coefficients. The choice that (3.7.55) be autonomous is a restriction but it is also a natural simplifying ansatz when one is dealing with autonomous equations. Moreover, as (2.4.404, 3.7.55) have no n, m dependent coefficients, they are translationally invariant under shifts in n and m. So we can with no loss of generality choose as reference point $n = 0$ and $m = 0$. This will also be assumed to be true for the linearizing point tranformation. Then equation (3.7.55) reads

$$f_{0,0} + af_{1,0} + bf_{0,1} + cf_{1,1} = 0. \tag{3.7.56}$$

In (3.7.56) and in the following equations we assume u_{0j} and u_{i0} as independent variables and consequently the variable $u_{1,1}$ appearing in the last term is not independent but it can be written in term of independent variables using the equation (2.4.130) [555, 611]. If the equation relates four points, we assume that (2.4.130) is solvable with respect to $u_{1,1}$

$$u_{1,1} = F(u_{0,0}, u_{0,1}, u_{1,0}), \tag{3.7.57}$$

where, as (2.4.130) depends on all lattice points we must have

$$F_{,u_{0,1}} \neq 0, \qquad F_{,u_{1,0}} \neq 0, \qquad F_{,u_{0,0}} \neq 0. \tag{3.7.58}$$

To solve the functional equation (3.7.56) we apply the Abel technique [2, 14], i.e. we write its differential consequences and solve them. The solution of its differential consequences is a necessary condition for the functional equation to be satisfied.

Let us differentiate (3.7.56) with respect to $u_{0,1}$ and then apply the logarithmic function. We get:

$$(T_1 - 1)\log\frac{df_{0,1}}{du_{0,1}} = \log\tilde{F}, \tag{3.7.59}$$

where $\tilde{F}$ is given by

$$\tilde{F}\left(u_{0,0}, u_{0,1}, u_{1,0}\right) \doteq -\frac{b}{cF_{,u_{0,1}}}.$$

We can always introduce an annihilator operator $\mathcal{A}$ such that

$$\mathcal{A}\phi\left(u_{1,1}\right) = 0, \tag{3.7.60}$$

where ϕ is an arbitrary function of its argument. The most general annihilator operator of this form reads

$$\mathcal{A} = \frac{\partial}{\partial u_{0,1}} + S^{(1)}(u_{0,1}, u_{0,0}, u_{1,0})\frac{\partial}{\partial u_{1,0}} + S^{(2)}(u_{0,1}, u_{0,0}, u_{1,0})\frac{\partial}{\partial u_{0,0}},$$

where $S^{(i)}(u_{0,1}, u_{0,0}, u_{1,0})$, $i = 1, 2$ are arbitrary functions of their variables. Equation (3.7.60) is satisfied for any function ϕ if

$$S^{(1)} = -\frac{F_{u_{0,1}} + S^{(2)}F_{u_{0,0}}}{F_{u_{1,0}}}.$$

There is no further condition to fix $S^{(2)}(u_{0,1}, u_{0,0}, u_{1,0})$. Applying the annihilator operator $\mathcal{A}$ onto (3.7.59) we get

$$-\frac{d}{du_{0,1}} \log \frac{df_{0,1}}{du_{0,1}} = \mathcal{A} \log \tilde{F}$$

$$= -\frac{1}{F_{u_{1,0}} F_{u_{0,1}}} \big(W_{(u_{0,1})}[F_{u_{1,0}}; F_{u_{0,1}}] + S^{(2)} W_{(u_{0,1})}[F_{u_{1,0}}; F_{u_{0,0}}] \big), \tag{3.7.61}$$

which is a differential equation for $f_{0,1}(u_{0,1})$ where the Wronskian $W_{(x)}[f; g]$ is defined as in (3.7.17). The left hand side of (3.7.61) depends only on $u_{0,1}$ while the right hand side depends on $u_{0,0}$, $u_{1,0}$ and $u_{0,1}$ through the given non linear PΔE and the up to now arbitrary function $S^{(2)}(u_{0,1}, u_{0,0}, u_{1,0})$. So we must have

$$\frac{\partial}{\partial u_{0,0}} \mathcal{A} \log \tilde{F} = 0, \qquad \frac{\partial}{\partial u_{1,0}} \mathcal{A} \log \tilde{F} = 0,$$

i.e.

$$W_{(u_{0,1})}[F_{u_{1,0}}; F_{u_{0,0}}] S^{(2)}_{u_{0,0}} + \mathcal{K}^{(1)} S^{(2)} + \mathcal{K}^{(0)} = 0, \tag{3.7.62}$$

$$W_{(u_{0,1})}[F_{u_{1,0}}; F_{u_{0,0}}] S^{(2)}_{u_{1,0}} + \mathcal{H}^{(1)} S^{(2)} + \mathcal{H}^{(0)} = 0, \tag{3.7.63}$$

with

$$\mathcal{K}^{(0)}(u_{0,0}, u_{1,0}, u_{0,1}) \doteq \frac{W_{(u_{0,0})}[F_{u_{1,0}} F_{u_{0,1}}; W_{(u_{0,1})}[F_{u_{1,0}}; F_{u_{0,1}}]]}{F_{u_{1,0}} F_{u_{0,1}}},$$

$$\mathcal{K}^{(1)}(u_{0,0}, u_{1,0}, u_{0,1}) \doteq \frac{W_{(u_{0,0})}[F_{u_{1,0}} F_{u_{0,1}}; W_{(u_{0,1})}[F_{u_{1,0}}; F_{u_{0,0}}]]}{F_{u_{1,0}} F_{u_{0,1}}},$$

$$\mathcal{H}^{(0)}(u_{0,0}, u_{1,0}, u_{0,1}) \doteq \frac{W_{(u_{1,0})}[F_{u_{1,0}} F_{u_{0,1}}; W_{(u_{0,1})}[F_{u_{1,0}}; F_{u_{0,1}}]]}{F_{u_{1,0}} F_{u_{0,1}}},$$

$$\mathcal{H}^{(1)}(u_{0,0}, u_{1,0}, u_{0,1}) \doteq \frac{W_{(u_{1,0})}[F_{u_{1,0}} F_{u_{0,1}}; W_{(u_{0,1})}[F_{u_{1,0}}; F_{u_{0,0}}]]}{F_{u_{1,0}} F_{u_{0,1}}}.$$

As (3.7.62), (3.7.63) must be valid for any function $S^{(2)}(u_{0,1}, u_{0,0}, u_{1,0})$, it follows that

$$W_{(u_{0,1})}[F_{u_{1,0}}; F_{u_{0,0}}] = 0 \tag{3.7.64}$$

and

$$\mathcal{K}^{(0)} = 0, \qquad \mathcal{H}^{(0)} = 0. \tag{3.7.65}$$

When (3.7.64), (3.7.65) are satisfied also $\mathcal{K}^{(1)}$ and $\mathcal{H}^{(1)}$ are null. Equations (3.7.64), (3.7.65) are necessary conditions for the linearizability by point transformations. If (2.4.130) is multilinear (see (3.7.157)), the condition $\mathcal{H}^{(0)} = 0$ is always identically satisfied.

In conclusion, we can state the following theorems:

Theorem 88. *Given a PΔE* (2.4.404) *on a quad-graph, there exists a linearizing point transformation only if* (3.7.64) *and* (3.7.65) *are satisfied.*

Theorem 89. *Given a PΔE* (2.4.404) *on a quad-graph, if Theorem* 88 *is satisfied, then the solution of* (3.7.61) *will provide the possible point transformation up to an integration constant. If the obtained* $f(u_{0,0})$ *verifies* (3.7.56) *then the transformation* (3.7.3) *linearizes our PΔE.*

7.2.2. *Two-point transformations.* Introducing the two-point transformation (3.7.4) into the linear equation (3.7.55) we get

$$g_{0,0}(u_{0,0},u_{0,1})+ag_{1,0}(u_{1,0},u_{1,1})+bg_{0,1}(u_{0,1},u_{0,2})+cg_{1,1}(u_{1,1},u_{1,2})=0 \tag{3.7.66}$$

an equation which has to be identically satisfied for any value of the independent variables of our problem. In (3.7.66), apart from the independent functions $u_{0,0}$, $u_{1,0}$, $u_{0,2}$ and $u_{0,1}$, there appear the functions $u_{1,1}$ and $u_{1,2}$. The function $u_{1,1}$ is expressed in term of the independent variables taking into account the PΔE (2.4.130). An equation involving $u_{1,2}$ is obtained from (2.4.130) by shifting the second index by 1, i.e. $F\left(u_{0,1},u_{0,2},u_{1,1},u_{1,2}\right)=0$. However this equation does not express directly $u_{1,2}$ in terms of the independent variables but it involves $u_{1,1}$ which is expressed in term of the independent variables through the PΔE itself. This fact complicates the equations obtained for the differential consequences of (3.7.66). To avoid it we have two possibilities, either back-shifting (3.7.55) once with respect to the second index, i.e. considering, in place of (3.7.66), the following equation

$$g_{0,-1}(u_{0,-1},u_{0,0})+ag_{1,-1}(u_{1,-1},u_{1,0})+bg_{0,0}(u_{0,0},u_{0,1})+cg_{1,0}(u_{1,0},u_{1,1})=0 \tag{3.7.67}$$

or, back-shifting once with respect to the first index and once with respect to the second, i.e.

$$g_{-1,-1}(u_{-1,-1},u_{-1,0})+ag_{0,-1}(u_{0,-1},u_{0,0})+bg_{-1,0}(u_{-1,0},u_{-1,1})+cg_{0,0}(u_{0,0},u_{0,1})=0. \tag{3.7.68}$$

In (3.7.67), (3.7.68) the variables $u_{-1,-1}$, $u_{-1,1}$, $u_{1,-1}$ and $u_{1,1}$ are not independent and from (2.4.130) we have, apart from (3.7.57)

$$u_{-1,1}=G(u_{-1,0},u_{0,0},u_{0,1}),\qquad u_{-1,-1}=H(u_{0,0},u_{-1,0},u_{0,-1}),$$
$$u_{1,-1}=K(u_{0,-1},u_{0,0},u_{1,0}).$$

As (2.4.404) depends on all lattice points we must have, apart from (3.7.58)

$$\begin{array}{lll} G_{u_{0,1}}\neq 0, & G_{u_{-1,0}}\neq 0, & G_{u_{0,0}}\neq 0,\\ H_{u_{0,-1}}\neq 0, & H_{u_{-1,0}}\neq 0, & H_{u_{0,0}}\neq 0,\\ K_{u_{0,-1}}\neq 0, & K_{u_{1,0}}\neq 0, & K_{u_{0,0}}\neq 0. \end{array}$$

Moreover, introducing the operators T_1 and T_2 as in (3.6.5), we have the following relations between the derivatives of the functions F, G, H and K:

$$\begin{array}{ll} F_{u_{0,1}}[T_1G]_{u_{1,1}}\Big|_{u_{1,1}=F}=1, & F_{u_{1,0}}[T_2K]_{u_{1,1}}\Big|_{u_{1,1}=F}=1,\\ K_{u_{0,-1}}[T_1H]_{u_{1,-1}}\Big|_{u_{1,-1}=K}=1, & G_{u_{-1,0}}[T_2H]_{u_{-1,1}}\Big|_{u_{-1,1}=G}=1. \end{array} \tag{3.7.69}$$

Taking into account (3.7.69) it turns out that (3.7.67) and (3.7.68) give the same necessary conditions. So it is sufficient to consider one of them, say (3.7.67).

Differentiating (3.7.67) once with respect to $u_{0,1}$, we get

$$b\frac{\partial g_{0,0}}{\partial u_{0,1}}(u_{0,0},u_{0,1})+c\frac{\partial g_{1,0}}{\partial u_{1,1}}(u_{1,0},u_{1,1})\frac{\partial F}{\partial u_{0,1}}(u_{0,0},u_{0,1},u_{1,0})=0. \tag{3.7.70}$$

Applying the logarithmic function to (3.7.70) we get the DΔE for the function g

$$(T_1-1)\log\frac{\partial g_{0,0}}{\partial u_{0,1}}=\log\tilde{F},\qquad \tilde{F}(u_{0,0},u_{0,1},u_{1,0})\doteq-\frac{1}{\frac{c}{b}\frac{\partial F}{\partial u_{0,1}}}, \tag{3.7.71}$$

where $\tilde{F}$ is an explicit function given in term of the given quad-graph PΔE $F = 0$. A solution of (3.7.71) could be obtained by summing it up. However in this case the resulting solution g would not be of the required form (3.7.4).

So, to find a solution of (3.7.71) of the required form, we simplify (3.7.71) by introducing an annihilator operator $\mathcal{A}$ such that

$$\mathcal{A}\phi(u_{1,0}, u_{1,1}) = 0, \tag{3.7.72}$$

where ϕ is an arbitrary function of its arguments. The most general operator of this form reads

$$\mathcal{A} = \frac{\partial}{\partial u_{0,1}} + S_a^{(1)}(u_{0,0}, u_{0,1}, u_{1,0})\frac{\partial}{\partial u_{0,0}} + S_a^{(2)}(u_{0,0}, u_{0,1}, u_{1,0})\frac{\partial}{\partial u_{1,0}}, \tag{3.7.73}$$

where $S_a^{(i)}\left(u_{0,0}, u_{0,1}, u_{1,0}\right)$, $i = 1, 2$ are arbitrary functions of the independent variables to be determined. Equation (3.7.72) is satisfied for any function ϕ if

$$S_a^{(1)} = -\frac{F_{u_{0,1}}}{F_{u_{0,0}}}, \qquad S_a^{(2)} = 0. \tag{3.7.74}$$

Applying the operator $\mathcal{A}$ onto (3.7.71) and defining $\psi(u_{0,0}, u_{0,1}) \doteq \log \frac{\partial g_{0,0}}{\partial u_{0,1}}$, we get

$$\psi_{,u_{0,1}} - \frac{F_{u_{0,1}}}{F_{u_{0,0}}}\psi_{,u_{0,0}} = \mathcal{R}(u_{0,0}, u_{0,1}, u_{1,0}) \doteq \frac{1}{F_{u_{0,0}}F_{u_{0,1}}}W_{(u_{0,1})}[F_{u_{0,0}}; F_{u_{0,1}}]. \tag{3.7.75}$$

Equation (3.7.75) is a differential equation for the function $\psi(u_{0,0}, u_{0,1})$, i.e. for the function characterizing the two-point transformation whose coefficients depend on the given quad-graph PΔE $F = 0$ (2.4.130). In (3.7.75) the function ψ depends just on $u_{0,0}$, $u_{0,1}$ while the terms depending on the given quad-graph PΔE $F = 0$ depend on $u_{0,0}$, $u_{0,1}$, $u_{1,0}$. As the quad-graph equation $F = 0$ depends also on the variable $u_{1,0}$, (3.7.75) will be an equation determining the two-point transformation only if some further compatibility conditions are satisfied.

Differentiating (3.7.75) once with respect to $u_{1,0}$, we get the following alternatives:

1. If $W_{(u_{1,0})}[F_{u_{0,1}}; F_{u_{0,0}}] = 0$ identically, we must have

$$\frac{\partial}{\partial u_{1,0}}\mathcal{R} = 0, \tag{3.7.76}$$

which is a necessary condition for the linearizability of (2.4.404) through the two-point transformation (3.7.4).

2. If $W_{(u_{1,0})}[F_{u_{0,1}}; F_{u_{0,0}}] \neq 0$, we get

$$\psi_{u_{0,0}} = \mathcal{M}(u_{0,0}, u_{0,1}, u_{1,0}), \qquad \mathcal{M} \doteq \frac{F_{u_{0,0}}^2 \mathcal{R}_{u_{1,0}}}{W_{(u_{1,0})}[F_{u_{0,1}}; F_{u_{0,0}}]}. \tag{3.7.77}$$

Inserting (3.7.77) in (3.7.75), we get

$$\psi_{u_{0,1}} = \mathcal{N}(u_{0,0}, u_{0,1}, u_{1,0}), \tag{3.7.78}$$

where

$$\mathcal{N} \doteq \frac{F_{u_{0,1}}}{F_{u_{0,0}}}\mathcal{M} + \mathcal{R}.$$

As the left hand side of (3.7.77) is independent of $u_{1,0}$, we get the necessary condition

(3.7.79) $$\frac{\partial}{\partial u_{1,0}}\mathcal{M}=0.$$

It is straightforward to prove that, if (3.7.79) is satisfied, also $\frac{\partial}{\partial u_{1,0}}\mathcal{N}=0$ will be true. Moreover the compatibility of (3.7.77, 3.7.78) gives another necessary condition

(3.7.80) $$\frac{\partial}{\partial u_{0,1}}\mathcal{M}=\frac{\partial}{\partial u_{0,0}}\mathcal{N}.$$

We can differentiate (3.7.67) with respect to $u_{0,-1}$ and we get

(3.7.81) $$(T_1-1)\log\frac{\partial g_{0,-1}}{\partial u_{0,-1}}=\log\tilde{K},\quad \tilde{K}(u_{0,0},u_{1,0},u_{0,-1})\doteq-\frac{1}{a\frac{\partial K}{\partial u_{0,-1}}}.$$

We can introduce the annihilator operator $\mathcal{B}$

(3.7.82) $$\mathcal{B}=\frac{\partial}{\partial u_{0,-1}}+S_b^{(1)}(u_{0,0},u_{1,0},u_{0,-1})\frac{\partial}{\partial u_{0,0}}+S_b^{(2)}(u_{0,0},u_{1,0},u_{0,-1})\frac{\partial}{\partial u_{1,0}},$$

such that

(3.7.83) $$\mathcal{B}\xi(u_{1,0},u_{1,-1})=0,$$

where ξ is an arbitrary function of its arguments. As (3.7.83) has to be satisfied for any ξ, we get

$$S_b^{(1)}=-\frac{K_{u_{0,-1}}}{K_{u_{0,0}}},\qquad S_b^{(2)}=0.$$

Applying the annihilator operator $\mathcal{B}$ onto (3.7.81) and defining $\phi(u_{0,-1},u_{0,0})\doteq\log\frac{\partial g_{0,-1}}{\partial u_{0,-1}}$ we get the following differential equation for the function ϕ

(3.7.84) $$\phi_{u_{0,-1}}-\frac{K_{u_{0,-1}}}{K_{u_{0,0}}}\phi_{u_{0,0}}=\mathcal{T}(u_{0,0},u_{1,0},u_{0,-1})\doteq\frac{1}{K_{u_{0,0}}K_{u_{0,-1}}}W_{(u_{0,-1})}[K_{u_{0,0}};K_{u_{0,-1}}].$$

Differentiating equation (3.7.84) once with respect to $u_{1,0}$, we get the following alternatives:

1. $W_{(u_{1,0})}[K_{u_{0,-1}};K_{u_{0,0}}]=0$ identically, then we must have

(3.7.85) $$\frac{\partial}{\partial u_{1,0}}\mathcal{T}=0,$$

with $\mathcal{T}$ given by (3.7.84). Eq. (3.7.85) is a necessary condition for linearizability of (2.4.404) through the two-point transformation (3.7.4).

2. If $W_{(u_{1,0})}[K_{u_{0,-1}};K_{u_{0,0}}]\neq 0$, we get

(3.7.86) $$\phi_{u_{0,0}}=\mathcal{P}(u_{0,0},u_{1,0},u_{0,-1}),\qquad \mathcal{P}\doteq\frac{K_{u_{0,0}}^2\mathcal{T}_{u_{1,0}}}{W_{(u_{1,0})}[K_{u_{0,-1}};K_{u_{0,0}}]}.$$

Inserting (3.7.86) in (3.7.84), we get

(3.7.87) $$\phi_{u_{0,-1}}=\mathcal{Q}(u_{0,0},u_{1,0},u_{0,-1}),$$

where

$$\mathcal{Q}\doteq\frac{K_{u_{0,-1}}}{K_{u_{0,0}}}\mathcal{P}+\mathcal{T}.$$

As the left hand side of (3.7.86) is independent of $u_{1,0}$, we get the necessary condition

$$\frac{\partial}{\partial u_{1,0}}\mathcal{P} = 0. \tag{3.7.88}$$

It is straightforward to prove that, if (3.7.88) is satisfied, also $\frac{\partial}{\partial u_{1,0}}\mathcal{Q} = 0$ will be true. Moreover the compatibility of (3.7.86, 3.7.87) gives another necessary condition

$$\frac{\partial}{\partial u_{0,-1}}\mathcal{P} = \frac{\partial}{\partial u_{0,0}}\mathcal{Q}. \tag{3.7.89}$$

We summarize the results so far obtained in the following theorems:

Theorem 90. *Given a PΔE* (2.4.130) *on a quad-graph we can construct the Wronskian functions $\mathbb{F} = W_{(u_{1,0})}[F_{u_{0,1}}; F_{u_{0,0}}]$ and $\mathbb{K} = W_{(u_{1,0})}[K_{u_{0,-1}}; K_{u_{0,0}}]$. Depending on the values of $\mathbb{F}$ and $\mathbb{K}$ we have different necessary conditions for the existence of a linearizing two-point transformation* (3.7.4).

1. $\mathbb{F} \neq 0$, $\mathbb{K} \neq 0$. *We have different results according to the value of $\mathcal{M}$* (3.7.77).
   (*a*) *If $\mathcal{M} \neq 0$, apart from the linearizability conditions* (3.7.79, 3.7.80, 3.7.88, 3.7.89) *the following compatible conditions must be satisfied:*

$$\left(\log\frac{[T_2\mathcal{P}]}{\mathcal{M}}\right)_{u_{0,0}} = \mathcal{M} - [T_2\mathcal{Q}], \qquad \left(\log\frac{[T_2\mathcal{P}]}{\mathcal{M}}\right)_{u_{0,1}} = \mathcal{N} - [T_2\mathcal{P}]. \tag{3.7.90}$$

   (*b*) *If $\mathcal{M} = 0$, the linearizability conditions are* (3.7.80, 3.7.88, 3.7.89)*.*
2. $\mathbb{F} \neq 0$, $\mathbb{K} = 0$. *Apart from the linearizability conditions* (3.7.80, 3.7.79, 3.7.85) *we have different results according to the value of $\mathcal{T}_{u_{0,0}}$* (3.7.84).
   (*a*) *If $\mathcal{T}_{u_{0,0}} = 0$*

$$\mathcal{M}_{,u_{0,0}} = \left(\frac{[T_2K]_{u_{0,0}}}{[T_2K]_{u_{0,1}}}\mathcal{M}\right)_{u_{0,1}} + \left(\frac{[T_2K]_{u_{0,0}}}{[T_2K]_{u_{0,1}}}\mathcal{N} - \mathcal{M} + [T_2\mathcal{T}]\right)\mathcal{M}.$$

   (*b*) *If $\mathcal{T}_{u_{0,0}} \neq 0$, defining*

$$\mathcal{S} \doteq -\frac{\left(\frac{[T_2K]_{u_{0,0}}}{[T_2K]_{u_{0,1}}}\mathcal{M}\right)_{u_{0,1}} + \frac{[T_2K]_{u_{0,0}}}{[T_2K]_{u_{0,1}}}\mathcal{M}\mathcal{N} + [T_2\mathcal{T}]\mathcal{M} - \mathcal{M}^2 - \mathcal{M}_{u_{0,0}}}{[T_2\mathcal{T}]_{u_{0,1}}},$$

   we get the following linearizability conditions

$$\mathcal{S}_{u_{0,1}} = \mathcal{M} - \mathcal{S}\mathcal{N}, \qquad \mathcal{S}_{u_{0,0}} = \frac{[T_2K]_{u_{0,0}}}{[T_2K]_{u_{0,1}}}\mathcal{M} + [T_2\mathcal{T}]\mathcal{S} - \mathcal{S}\mathcal{M}.$$

3. $\mathbb{F} = 0$, $\mathbb{K} \neq 0$. *Apart from the linearizability conditions* (3.7.89, 3.7.88, 3.7.76) *we have different results according to the value of $\mathcal{R}_{u_{0,0}}$* (3.7.75).
   (*a*) *If $\mathcal{R}_{u_{0,0}} = 0$,*

$$\left([T_2\mathcal{P}]\right)_{,u_{0,1}} = \left(\frac{F_{u_{0,1}}}{F_{u_{0,0}}}[T_2\mathcal{P}]\right)_{u_{0,0}} + \left(\frac{F_{u_{0,1}}}{F_{u_{0,0}}}[T_2\mathcal{Q}] - [T_2\mathcal{P}] + \mathcal{R}\right)[T_2\mathcal{P}].$$

(*b*) *If* $\mathcal{R}_{u_{0,0}} \neq 0$, *defining*

$$\mathcal{U} \doteq -\frac{\left(\frac{F_{u_{0,1}}}{F_{u_{0,0}}}[T_2\mathcal{P}]\right)_{u_{0,0}} + \frac{F_{u_{0,1}}}{F_{u_{0,0}}}[T_2\mathcal{P}][T_2\mathcal{Q}] + \mathcal{R}[T_2\mathcal{P}] - ([T_2\mathcal{P}])^2 - ([T_2\mathcal{P}])_{u_{0,1}}}{\mathcal{R}_{u_{0,0}}} \tag{3.7.91}$$

we get the further linearizability conditions

$$\mathcal{U}_{u_{0,0}} = [T_2\mathcal{P}] - \mathcal{U}[T_2\mathcal{Q}], \qquad \mathcal{U}_{u_{0,1}} = \frac{F_{u_{0,1}}}{F_{u_{0,0}}}[T_2\mathcal{P}] + \mathcal{R}\mathcal{U} - \mathcal{U}[T_2\mathcal{P}].$$

4. $\mathbb{F} = 0$, $\mathbb{K} = 0$. *Apart from the linearizability conditions (*3.7.85, 3.7.76*) we have a set of conditions for the functions F and K involved, depending if* $\frac{[T_2K]_{u_{0,0}}}{[T_2K]_{u_{0,1}}}\frac{F_{u_{0,1}}}{F_{u_{0,0}}}$ *is equal to* 1 *or not. These conditions are obtained by requiring that the overdetermined system obtained by explicitating* (3.7.75), (3.7.84) *in term of* $g = g_{0,0}$ *and possibly shifting*

$$g_{u_{0,0}u_{0,0}} - \frac{[T_2K]_{u_{0,0}}}{[T_2K]_{u_{0,1}}}g_{u_{0,0},u_{0,1}} = [T_2\mathcal{T}]g_{u_{0,0}}, \tag{3.7.92}$$

$$g_{u_{0,1}u_{0,1}} - \frac{F_{u_{0,1}}}{F_{u_{0,0}}}g_{u_{0,0},u_{0,1}} = \mathcal{R}g_{u_{0,1}}, \tag{3.7.93}$$

be solvable for any $u_{0,0}$, $u_{0,1}$, $u_{1,0}$. *These equations are easy to derive by symbolic manipulation but too long to write down. So, for the sake of clarity, we do not write them down here.*

Theorem 91. *Given a P∆E* (2.4.130) *on a quad-graph, if Theorem* 90 *is satisfied, depending on the values of* $\mathbb{F}$ *and* $\mathbb{K}$, *we have different PDEs defining the two-point transformation.*

1. $\mathbb{F} \neq 0$, $\mathbb{K} \neq 0$. *We have different results according to the value of* $\mathcal{M}$ (3.7.77).
   (*a*) *If* $\mathcal{M} \neq 0$ *we have for* $g = g_{0,0}$

$$g_{u_{0,0}u_{0,0}} = [T_2\mathcal{Q}]g_{u_{0,0}}, \qquad g_{u_{0,1}} = \frac{[T_2\mathcal{P}]}{\mathcal{M}}g_{u_{0,0}}.$$

   (*b*) *If* $\mathcal{M} = 0$ *we have* $g = g^{(0)}(u_{0,0}) + g^{(1)}(u_{0,1})$ *and*

$$g^{(0)}_{u_{0,0}u_{0,0}} = [T_2\mathcal{Q}]g^{(0)}_{u_{0,0}}, \qquad g^{(1)}_{u_{0,1}u_{0,1}} = \mathcal{N}g^{(1)}_{u_{0,1}}.$$

2. $\mathbb{F} \neq 0$, $\mathbb{K} = 0$. *We have different results according to the value of* $\mathcal{T}_{u_{0,0}}$(3.7.84).
   (*a*) *If* $\mathcal{T}_{u_{0,0}} = 0$ *the two-point transformation is obtained by solving for* $g = g_{0,0}$ *the compatible system of PDEs* (3.7.77), (3.7.78) *and* (3.7.92);
   (*b*) *If* $\mathcal{T}_{u_{0,0}} \neq 0$ *the two-point transformation is obtained by solving the compatible PDEs* (3.7.78) *and*

$$g_{,u_{0,0}} = \mathcal{S}g_{u_{0,1}}.$$

3. $\mathbb{F} = 0$, $\mathbb{K} \neq 0$. *We have different results according to the value of* $\mathcal{R}_{,u_{0,0}}$ (3.7.75).
   (*a*) *If* $\mathcal{R}_{u_{0,0}} = 0$ *the two-point transformation is obtained by solving for* $g = g_{0,0}$ *the compatible system of PDEs (*3.7.86, 3.7.87*) and* (3.7.93);

(*b*) *If* $\mathcal{R}_{u_{0,0}} \neq 0$ *the two-point transformation is obtained by solving the compatible PDEs* (3.7.87) *and*

$$g_{u_{0,1}} = \mathcal{V} g_{u_{0,0}},$$

where $\mathcal{V}$ *is given by* (3.7.91).

4. $\mathbb{F} = 0$, $\mathbb{K} = 0$ *the two-point transformation is obtained by solving for* $g = g_{0,0}$ *the compatible system of PDEs* (3.7.93) *and* (3.7.92).

When we have solved the PDEs for the function $g(u_{0,0}, u_{0,1})$ we may still have arbitrary functions or arbitrary constants. These get fixed by inserting the function g into the equations (3.7.70), (3.7.81) and solving them and their consequences. At the end we need to verify that (3.7.66) or its shifted versions (3.7.67, 3.7.68) be satisfied.

7.2.3. *Linearization by a generalized Cole–Hopf transformation to an homogeneous linear equation.* Let us assume the existence of a generalized Cole–Hopf transformation (3.7.5) which linearizes (2.4.130) into (3.7.55), where a, b and c are arbitrary constants and the function $h = h_{0,0}(u_{0,0}, u_{0,1})$ of its two arguments is to be determined.

The non linear equation (2.4.130) can be written just in term of h and k as

$$h_{1,0} k_{0,0} = k_{0,1} h_{0,0} \tag{3.7.94}$$

and in terms of h alone as

$$\left(\frac{h_{0,0} + 1/b}{h_{0,0}} \right) = \left(\frac{h_{1,0} + a/c}{h_{1,0}} \right) \left(\frac{h_{0,1} + 1/b}{h_{1,1} + a/c} \right). \tag{3.7.95}$$

If we assume $h = h(u_{0,0})$, then, differentiating (3.7.95) with respect to $u_{0,1}$, we find

$$\frac{\partial}{\partial u_{0,1}} \log(h_{0,1} + 1/b) = \frac{\partial}{\partial u_{1,1}} \log(h_{1,1} + a/c) \frac{\partial}{\partial u_{0,1}} F, \tag{3.7.96}$$

and following Section 3.7.2.1, we find the same linearizability conditions (3.7.64, 3.7.65) as for one-point transformations, i.e. Theorem 88 will be valid. However the differential equation for the transformation is different and is given by

$$\frac{d}{du_{0,1}} \log \frac{d}{du_{0,1}} \log(h_{0,1} + 1/b) = \frac{1}{F_{,u_{1,0}} F_{,u_{0,1}}} W_{(u_{0,1})}[F_{,u_{1,0}}; F_{,u_{0,1}}]. \tag{3.7.97}$$

So, if an equation is linearizable by a one-point transformation it can also be linearizable by a Cole–Hopf transformation depending on one point only. However the effective linearizing transformation is different and thus one can find a nonlinear PΔE on the square which is linearizable by a Cole–Hopf transformation with $h = h(u_{0,0})$ but not by a point transformation (3.7.3).

If the left hand side of (3.7.95) depends on $u_{0,0}$ and $u_{0,1}$, the first left term in the right hand side depends on $u_{1,0}$ and $u_{1,1}$ and the second one on $u_{0,1}$, $u_{0,2}$, $u_{1,1}$ and $u_{1,2}$. The variable $u_{1,1}$ is given in terms of the independent variables by (3.7.57) while $u_{1,2}$ can be rewritten in term of the independent variables as

$$u_{1,2} = F(u_{0,1}, u_{0,2}, F(u_{0,0}, u_{0,1}, u_{1,0})). \tag{3.7.98}$$

From (3.7.98) the expression of $u_{1,2}$ in terms of the independent variables depends twice on the quad-graph equation (2.4.130). So we will consider in place of (3.7.94) the equation,

$$h_{1,-1} k_{0,-1} = k_{0,0} h_{0,-1}.$$

which in terms of h alone read

$$\left(\frac{h_{0,-1}+1/b}{h_{0,-1}}\right)=\left(\frac{h_{1,-1}+a/c}{h_{1,-1}}\right)\left(\frac{h_{0,0}+1/b}{h_{1,0}+a/c}\right), \tag{3.7.99}$$

The left hand side of (3.7.99) depends on $u_{0,0}$ and $u_{0,-1}$, the first left term in the right hand side depends on $u_{1,0}$ and $u_{1,-1} = K(u_{0,-1}, u_{1,0}, u_{0,0})$ i.e. on $u_{1,0}$, $u_{0,0}$ and $u_{0,-1}$ and the second one on $u_{0,1}$, $u_{0,0}$, $u_{1,1} = F(u_{0,1}, u_{1,0}, u_{0,0})$ i.e. on $u_{0,1}$, $u_{0,0}$ and $u_{1,0}$. Thus one can see that the three terms appearing in the equation (3.7.99) contain no overlapping set of variables. This is a condition necessary to get out of (3.7.99) some differential conditions for the functions F and K, i.e. for the equation (2.4.130) to be rewritable as the compatibility condition of (3.7.5) and (3.7.6).

Let us consider (3.7.99) and, as we have products, we reduce it to a sum of terms by applying to it the logarithmic function. Then we differentiate the resulting equation with respect to $u_{0,1}$. Only the second term on the r.h.s. of the equality depends on $u_{0,1}$ through the dependence of $h_{1,0}$ on $u_{1,1}$ and of $h_{0,0}$. So we get:

$$\frac{\partial}{\partial u_{0,1}}\log(h_{0,0}+1/b)=\frac{\partial}{\partial u_{1,1}}\log(h_{1,0}+a/c)\frac{\partial}{\partial u_{0,1}}F, \tag{3.7.100}$$

equivalent, in structure to (3.7.70). The term on the l.h.s. of (3.7.100) depends on $u_{0,0}$ and $u_{0,1}$ while the first factor on the r.h.s. depends on $u_{1,0}$ and $u_{1,1}$ and we can always introduce the differential operator $\mathcal{A}$ as given by (3.7.73). If we apply again the logarithmic function to equation (3.7.100) and then the annihilator operator $\mathcal{A}$ onto the resulting equation, setting $\psi(u_{0,0}, u_{0,1}) \doteq \log \frac{\partial}{\partial u_{0,1}} \log(h+1/b)$, we get the linear differential equation (3.7.75). It is worthwhile to notice that, even if the differential equation is the same when expressed in term of the variable ψ, its expression in term of f is different from the one in term of h.

Let us now differentiate (3.7.99) with respect to $u_{0,-1}$. Proceeding in an analogous way as we did before, we get

$$\frac{\partial}{\partial u_{0,-1}}\log\left(\frac{h_{0,-1}+1/b}{h_{0,-1}}\right)=\frac{\partial}{\partial u_{1,-1}}\log\left(\frac{h_{1,-1}+a/c}{h_{1,-1}}\right)\frac{\partial}{\partial u_{0,-1}}K. \tag{3.7.101}$$

The term on the l.h.s. of (3.7.101) depends on $u_{0,-1}$ and $u_{0,0}$ while the first factor on the r.h.s. depends on $u_{1,-1}$ and $u_{1,0}$. We can always introduce the annihilator operator $\mathcal{B}$ as given by (3.7.82). So if we apply again the logarithmic function to equation (3.7.101) and then the operator $\mathcal{B}$ onto the resulting equation, setting $\phi(u_{0,-1}, u_{0,0}) \doteq \log \frac{\partial}{\partial u_{0,-1}} \log \frac{h_{0,-1}+1/b}{h_{0,-1}}$, we get the linear differential equation (3.7.84) for ϕ. However, as before, even if the differential equation is the same when expressed in term of the variable ϕ, its expression in term of f is different from the one in term of h.

As the determining equations in terms of ψ and ϕ are exactly the same as those of contact transformations, the linearizability conditions are as presented in Theorem 90. However the function ψ and ϕ are defined here differently then in the case of two-point transformations. So the equations defining $h = h_{0,0}$ are different. In particular (3.7.92) and

(3.7.93) in this case have to be replaced by the non linear equations

$$\frac{h_{u_{0,1}u_{0,1}}}{h+1/b} - \left(\frac{h_{u_{0,1}}}{h+1/b}\right)^2 - \frac{F_{u_{0,1}}}{F_{u_{0,0}}}\left(\frac{h_{u_{0,1}u_{0,0}}}{h+1/b} - \frac{h_{u_{0,1}}h_{u_{0,0}}}{(h+1/b)^2}\right) = \mathcal{R},$$

$$\left[\left(\frac{h_{0,-1,u_{0,-1}}}{h_{0,-1}(h_{0,-1}+1/b)}\right)_{u_{0,-1}} - \frac{K_{u_{0,-1}}}{K_{u_{0,0}}}\left(\frac{h_{0,-1,u_{0,-1}}}{h_{0,-1}(h_{0,-1}+1/b)}\right)_{u_{0,0}}\right]\cdot$$

$$\cdot\frac{h_{0,-1}(h_{0,-1}+1/b)}{h_{0,-1,u_{0,-1}}} = \mathcal{T}.$$

A simpler and sometimes more useful non linear equation for $h_{0,0}$ can be obtained in the following way. Let us shift (3.7.101) by T_2. In such a way we get

$$\text{(3.7.102)} \qquad \frac{\partial}{\partial u_{0,0}}\log\left(\frac{h+1/b}{h}\right) = \frac{\partial}{\partial u_{1,0}}\log\left(\frac{h_{1,0}+a/c}{h_{1,0}}\right)\frac{\partial}{\partial u_{0,0}}[T_2K].$$

Then from (3.7.100), (3.7.102) we extract the partial derivatives of h with respect to $u_{0,0}$ and $u_{0,1}$,

$$\text{(3.7.103)} \qquad h_{u_{0,1}} = \frac{h+1/b}{h_{1,0}+a/c}h_{1,0,u_{1,1}}\frac{\partial F}{\partial u_{0,1}},$$

$$\text{(3.7.104)} \qquad h_{u_{0,0}} = \frac{a/ch}{1/bh_{1,0}}\frac{h+1/b}{h_{1,0}+a/c}h_{1,0,u_{1,0}}\frac{\partial[T_2K]}{\partial u_{0,0}}.$$

Dividing (3.7.103) by (3.7.104) we get the following equation:

$$\text{(3.7.105)} \qquad (T_1-1)\log\frac{hh_{u_{0,1}}}{h_{u_{0,0}}} + \log\frac{1/b\frac{\partial F}{\partial u_{0,1}}}{a/c\frac{\partial[T_2K]}{\partial u_{0,0}}} = 0.$$

Differentiating (3.7.105) with respect to $u_{1,0}$ we obtain a second order non linear PDE for the function $\chi(u_{0,0},u_{0,1}) = \log(h)$. We have:

$$\text{(3.7.106)} \qquad \left[\chi_{u_{0,0}} + \frac{\chi_{u_{0,0}u_{0,1}}}{\chi_{u_{0,1}}} - \frac{\chi_{u_{0,0}u_{0,0}}}{\chi_{u_{0,0}}}\right] + \mathcal{C}\left[\chi_{u_{0,1}} + \frac{\chi_{u_{0,1}u_{0,1}}}{\chi_{u_{0,1}}} - \frac{\chi_{u_{0,1}u_{0,0}}}{\chi_{u_{0,0}}}\right] + \mathcal{D} = 0,$$

where

$$\mathcal{C}(u_{-1,0},u_{0,0},u_{0,1}) \doteq \left\{T_1^{-1}\left[\frac{\partial F}{\partial u_{1,0}}\right]\right\}_{u_{-1,1}\to G(u_{-1,0},u_{0,0},u_{0,1})},$$

$$\mathcal{D}(u_{-1,0},u_{0,0},u_{0,1}) \doteq \left\{T_1^{-1}\left[\frac{\partial}{\partial u_{1,0}}\log\left(\frac{1}{\mathcal{H}}\frac{\partial F}{\partial u_{0,1}}\right)\right]\right\}_{u_{-1,1}\to G(u_{-1,0},u_{0,0},u_{0,1})},$$

$$\mathcal{H}(u_{0,0},u_{1,0},u_{0,1}) \doteq \left\{T_2\left[\frac{\partial K}{\partial u_{0,-1}}\right]\right\}_{u_{1,1}\to F(u_{0,0},u_{1,0},u_{0,1})}.$$

The first and second terms of (3.7.106) depend on derivatives of the unknown function $\chi(u_{0,0},u_{0,1})$ but the coefficient of the second term and the last one may contain also $u_{-1,0}$. So we have a further set of linearizability conditions. If $\frac{\partial}{\partial u_{-1,0}}\mathcal{C} = 0$, differentiating (3.7.106) with respect to $u_{-1,0}$ we have

$$\frac{\partial}{\partial u_{-1,0}}\mathcal{D} = 0,$$

while, if $\frac{\partial}{\partial u_{-1,0}} C \neq 0$, after a differentiation with respect to $u_{-1,0}$, we have the Wronskian

$$W_{(u_{-1,0})}[C_{u_{-1,0}}; D_{u_{-1,0}}] = 0. \tag{3.7.107}$$

In the first case, the solutions of (3.7.106) provides us with an ansatz of the function $h = h_{0,0}$, otherwise the function h is obtained by solving the following overdetermined system of non linear PDEs

$$\chi_{u_{0,0}} = \frac{\chi_{u_{0,0}u_{0,0}}}{\chi_{u_{0,0}}} - \frac{\chi_{u_{0,0}u_{0,1}}}{\chi_{u_{0,1}}} + \frac{W_{(u_{-1,0})}[C; D]}{C_{u_{-1,0}}}, \tag{3.7.108}$$
$$\chi_{u_{0,1}} = \frac{\chi_{u_{0,1}u_{0,0}}}{\chi_{u_{0,0}}} - \frac{\chi_{u_{0,1}u_{0,1}}}{\chi_{u_{0,1}}} - \frac{D_{u_{-1,0}}}{C_{u_{-1,0}}}.$$

If the condition (3.7.107) is satisfied, then $\frac{\partial}{\partial u_{-1,0}} \frac{W_{(u_{-1,0})}[C;D]}{C_{u_{-1,0}}} = 0$. The overdetermined system (3.7.108) is compatible iff

$$W_{(u_{0,1})}[W_{(u_{-1,0})}[C; D]; C_{u_{-1,0}}] = W_{(u_{0,0})}[C_{u_{-1,0}}; D_{u_{-1,0}}], \tag{3.7.109}$$

a further linearizability condition. Equations (3.7.106), (3.7.108) are non linear PΔEs which, introducing the function

$$\theta\left(u_{0,0}, u_{0,1}\right) \doteq \chi + \log\left(\frac{\chi_{u_{0,1}}}{\chi_{u_{0,0}}}\right), \tag{3.7.110}$$

can be linearized and read:

$$\theta_{u_{0,0}} + C\theta_{u_{0,1}} + D = 0, \tag{3.7.111}$$

$$\theta_{u_{0,0}} = \frac{W_{(u_{-1,0})}[C; D]}{C_{u_{-1,0}}}, \qquad \theta_{u_{0,1}} = -\frac{D_{u_{-1,0}}}{C_{u_{-1,0}}}. \tag{3.7.112}$$

Once the solution of the equations (3.7.111) or (3.7.112) has been obtained, the function h can be reconstructed. Starting from the definition (3.7.110) we get

$$e^{\chi}\frac{\partial}{\partial u_{0,1}} e^{\chi} = e^{\theta}\frac{\partial}{\partial u_{0,0}} e^{\chi},$$

or in terms of h

$$h_{u_{0,0}} = e^{-\theta} h h_{u_{0,1}}, \tag{3.7.113}$$

a *Hopf-like* equation whose solution can be obtained for example by separation of variables. Once we have a solution, we can introduce it into the lowest order differential equations and define the arbitrary functions or constant involved. The so obtained function h will provide us with a linearizing generalized Cole–Hopf transformation if the difference relation (3.7.95) is satisfied. Equation (3.7.113) can be introduced in (3.7.103), (3.7.104) and after some manipulations and the application of the annihilator operator $\mathcal{A}$ defined in (3.7.73), (3.7.74), we obtain a linear evolution equation for the function $\theta(u_{0,0}, u_{0,1})$

$$\theta_{u_{0,1}} - \frac{F_{u_{0,1}}}{F_{u_{0,0}}}\theta_{u_{0,0}} = \tilde{\mathcal{T}}(u_{0,0}, u_{1,0}, u_{0,1}) \doteq \mathcal{A} \log\left(\frac{1}{\mathcal{H}} F_{u_{0,1}}\right). \tag{3.7.114}$$

Differentiating equation (3.7.114) once with respect to $u_{1,0}$, we get the following alternatives according to the values of $\mathbb{F}$ defined in Theorem 90

1. $\mathbb{F} = 0$ identically, then we must have

$$\tilde{\mathcal{T}}_{u_{1,0}} = 0,$$

which is a necessary condition for linearizability through the Cole–Hopf transformation (3.7.5), (3.7.6).

2. If $\mathbb{F} \neq 0$, we get

$$\theta_{,u_{0,0}} = \tilde{\mathcal{P}}(u_{0,0}, u_{1,0}, u_{0,1}), \qquad \tilde{\mathcal{P}} \doteq \frac{F^2_{u_{0,0}} \tilde{\mathcal{T}}_{u_{1,0}}}{\mathbb{F}}. \tag{3.7.115}$$

Inserting (3.7.115) in (3.7.114), we get

$$\theta_{,u_{0,1}} = \tilde{\mathcal{Q}}(u_{0,0}, u_{1,0}, u_{0,1}), \tag{3.7.116}$$

where

$$\tilde{\mathcal{Q}} \doteq \frac{F_{u_{0,1}}}{F_{u_{0,0}}} \tilde{\mathcal{P}} + \tilde{\mathcal{T}}.$$

As the left hand side of (3.7.115) is independent of $u_{1,0}$, we get the necessary condition

$$\frac{\partial}{\partial u_{1,0}} \tilde{\mathcal{P}} = 0. \tag{3.7.117}$$

It is straightforward to prove that, if (3.7.117) is satisfied, also $\frac{\partial}{\partial u_{1,0}} \tilde{\mathcal{Q}} = 0$ will be true. Moreover the compatibility of (3.7.115), (3.7.116) gives another necessary condition

$$\frac{\partial}{\partial u_{0,1}} \tilde{\mathcal{P}} = \frac{\partial}{\partial u_{0,0}} \tilde{\mathcal{Q}}.$$

As in the case of contact transformations, the combination of the two cases defined by (3.7.111) or (3.7.112) and the two cases defined by (3.7.114) or (3.7.115, 3.7.116) gives a total of four subcases for the specification of the function θ. If (3.7.107, 3.7.109) are satisfied, the solution of the two equations (3.7.112) is given by

$$\theta = \int_{u'_{0,0}=\alpha}^{u'_{0,0}=u_{0,0}} \frac{W_{(u_{-1,0})}[\mathcal{C}; \mathcal{D}]}{\mathcal{C}_{u_{-1,0}}} (u'_{0,0}, u_{0,1}) du'_{0,0} - \int_{u'_{0,1}=\beta}^{u'_{0,1}=u_{0,1}} \frac{\mathcal{D}_{u_{-1,0}}}{\mathcal{C}_{u_{-1,0}}} (\alpha, u'_{0,1}) du'_{0,1} + \gamma,$$

where α and β are some fixed values of the variables $u_{0,0}$ and $u_{0,1}$ at which the integrals are well defined, while γ is an arbitrary integration constant.

7.2.4. *Examples.* Here we apply the linearizability conditions in the case of some interesting examples.

Liouville equation. Let us consider the discrete Liouville equation [767]

$$u_{1,1} = \frac{(u_{1,0} - 1)(u_{0,1} - 1)}{u_{0,0}} \doteq F(u_{0,0}, u_{0,1}, u_{1,0}). \tag{3.7.118}$$

In [767] it was shown that the transformation

$$u_{0,0} = \frac{\tilde{u}_{1,0}\tilde{u}_{0,1}}{(\tilde{u}_{1,0} - \tilde{u}_{0,0})(\tilde{u}_{0,1} - \tilde{u}_{0,0})},$$

maps solutions of the linear equation

$$\tilde{u}_{0,0} - \tilde{u}_{1,0} - \tilde{u}_{0,1} + \tilde{u}_{1,1} = 0,$$

into solutions of (3.7.118). This transformation is not of the form considered here as depends on three points. However to test the procedures introduced in this Section we can try to linearize this example by the transformations considered in the previous sections.

We can try to linearize this discrete Liouville equation by a point transformation. The necessary conditions (3.7.64, 3.7.65) are identically satisfied and the linearizing point transformation, obtained by integrating (3.7.61), is given by

$$f(u_{0,0}) = c_1[c_2 + \log(u_{0,0} - 1)], \tag{3.7.119}$$

where $c_1 \neq 0$ and c_2 are arbitrary constants. One can easily see that it does not exist any value of the constants c_2, $a \neq 0$, $b \neq 0$ and $c \neq 0$ of (3.7.56) for which the function $f(u_{0,0})$ given in (3.7.119) satisfies (3.7.59) identically modulo (3.7.118) (the multiplicative constant c_1 is inessential as it can be always rescaled away). Hence the Liouville equation (3.7.118) cannot be linearized by a one-point transformation.

We can try to linearize by a two-point transformation of the form (3.7.4). As $\mathbb{F} = \mathbb{K} = 0$ identically, we are in the fourth case. Moreover the two linearizability conditions (3.7.76, 3.7.85) are identically satisfied and the overdetermined system of differential equations (3.7.92) reads

$$\begin{aligned} g_{u_{0,0}u_{0,0}} - \frac{1-u_{0,1}}{u_{0,0}} g_{u_{0,0}u_{0,1}} + \frac{1}{u_{0,0}} g_{u_{0,0}} &= 0, \\ g_{u_{0,1}u_{0,1}} - \frac{u_{0,0}}{1-u_{0,1}} g_{u_{0,0}u_{0,1}} - \frac{1}{1-u_{0,1}} g_{u_{0,1}} &= 0, \end{aligned} \tag{3.7.120}$$

whose solution is given by

$$g\left(u_{0,0}, u_{0,1}\right) = \theta\left(\xi\right) + c_3 \log\left(u_{0,0}\right) + c_4, \qquad \xi = \frac{u_{0,1}-1}{u_{0,0}}, \tag{3.7.121}$$

where c_3 and c_4 are arbitrary constants and $\theta \neq 0$ is an arbitrary function of its arguments. As one can see, the system (3.7.120) does not specify the two-point transformation. To define it we need to introduce (3.7.121) into (3.7.70), (3.7.81). In this way we get a system of two first order DΔEs involving θ and $T_1\theta$. From them we can extract a first order ODE for θ which depends on ξ and $u_{1,0}$. As a consequence this equation splits into an overdetermined system of two first order ODEs for $\theta(\xi)$, whose solution is given by

$$\theta = c_3 \log(\xi + 1) + \alpha, \qquad b = -1, \qquad a = -c,$$

where α is an arbitrary constant and $c_3 \neq 0$. Hence, after a reparametrization of c_4,

$$g(u_{0,0}, u_{0,1}) = c_3[\log(u_{0,0} + u_{0,1} - 1) + c_4]. \tag{3.7.122}$$

A necessary condition for (3.7.122) to be a linearizing transformation, is that (3.7.66) be identically satisfied modulo (3.7.118). It is easy to show that it is not possible to find a value of c_4 and $c_3 \neq 0$ such that this condition is satisfied. In conclusion the equation (3.7.118) cannot be linearized by a two-point transformation.

If we consider the linearization through a Cole–Hopf transformation, we are in the case when $\mathbb{F} = \mathbb{K} = 0$ and the linearizability conditions $\mathcal{R}_{u_{1,0}} = \mathcal{U}_{,u_{1,0}} = 0$ are also satisfied. The equations for the functions ψ and ϕ read

$$\phi_{u_{0,1}} + \frac{u_{0,0}}{u_{0,1}-1}\phi_{u_{0,0}} + \frac{1}{u_{0,1}-1} = 0, \qquad \psi_{u_{0,-1}} + \frac{u_{0,0}-1}{u_{0,-1}}\psi_{u_{0,0}} + \frac{1}{u_{0,0}} = 0,$$

and their solution imply

$$h + 1/b = \sigma(\tilde{\xi})\rho(u_{0,0}), \qquad \frac{h+1/b}{h} = \kappa(\tilde{\xi})\tau(u_{0,1}), \qquad \tilde{\xi} \doteq \frac{u_{0,1}-1}{u_{0,0}}, \tag{3.7.123}$$

where σ, ρ, κ and τ are arbitrary nonzero functions of their argument. The function $\theta(u_{0,0}, u_{0,1})$ defined in (3.7.110), is specified by the conditions $C_{u_{-1,0}} = \mathbb{F} = 0$. The necessary conditions $\mathcal{D}_{u_{-1,0}} = \tilde{\mathcal{T}}_{u_{1,0}} = 0$ are satisfied and the two equations (3.7.111), (3.7.114) respectively read

$$\theta_{u_{0,0}} + \frac{u_{0,1}}{u_{0,0}-1}\theta_{u_{0,1}} = 0, \qquad \theta_{u_{0,1}} + \frac{u_{0,0}}{u_{0,1}-1}\theta_{u_{0,0}} = 0.$$

The only admissible solution of this system is a constant. The solution of the overdetermined system of the two functional equations (3.7.123) and of the Hopf-like PDE (3.7.113), after a reparametrization of the constant θ, is given by

$$h = -\frac{\gamma(u_{0,1}-1)+\delta}{b\delta + \tilde{\theta} u_{0,0}},$$

where $\gamma \neq 0$, δ and $\tilde{\theta} \neq 0$ are arbitrary constants. A necessary condition to obtain a linearizing transformation is that (3.7.94) be identically satisfied modulo (3.7.118). No nonzero value of a, b, c, γ, $\tilde{\theta}$ and δ can satisfy this condition, hence (3.7.118) cannot be linearized by a Cole–Hopf transformation too.

Multilinear quad-graph equation linearizable up to 5th order by a multiple scale expansion [374]. Let us consider the multilinear equation

$$\begin{aligned}
&\zeta u_{0,0}u_{1,0}u_{0,1}u_{1,1} + a_1(u_{0,0}+u_{1,1}) + a_2(u_{1,0}+u_{0,1}) + \gamma_1 u_{0,0}u_{1,1} \\
&+ \frac{a_2\gamma_1}{a_1}u_{1,0}u_{0,1} + \frac{(a_1+a_2)\gamma_1}{2a_1}(u_{0,0}u_{0,1}+u_{1,0}u_{1,1}+u_{0,0}u_{1,0}+u_{0,1}u_{1,1}) \\
&+ \frac{(a_1+2a_2)\gamma_1^2}{4a_1^2}(u_{0,0}+u_{1,1})u_{1,0}u_{0,1} + \frac{(2a_1+a_2)\gamma_1^2}{4a_1^2}(u_{1,0}+u_{0,1})u_{0,0}u_{1,1},
\end{aligned} \tag{3.7.124}$$

where a_1, a_2, γ_1, ζ are arbitrary real parameters with $|a_1| \neq |a_2| \neq 0$. In [374] it has been shown that this equation passes a linearizability test based on multiscale analysis up to fifth order in the perturbation parameter for small u.

Let us search for the possibility to linearize (3.7.124) by a one-point transformation. Of the necessary conditions (3.7.65), $\mathcal{H}^{(0)} = 0$ is automatically satisfied while $\mathcal{K}^{(0)} = 0$ and (3.7.64) can be satisfied if and only if

$$\zeta = \frac{(a_1+a_2)\,\gamma_1^3}{4a_1^3}. \tag{3.7.125}$$

In this case the linearizing point transformation obtained integrating (3.7.61) is given by

$$f(u_{0,0}) = c_1\left[\frac{1}{2a_1+\gamma_1 u_{0,0}} + c_2\right], \tag{3.7.126}$$

where $c_1 \neq 0$ and c_2 are constants. Inserting $f(u_{0,0})$ into (3.7.59), one finds that this relation is identically satisfied modulo (3.7.124), (3.7.125) when $c = ba_1/a_2$. Finally, inserting $f(u_{0,0})$ and c into (3.7.56), it is straightforward to see that this relation results identically satisfied modulo (3.7.124), (3.7.125) when $c_2 = -1/(2a_1)$, $c = 1$ and $a = b = a_2/a_1$. Equation (3.7.126) is the linearizing point transformation.

Let us consider the case $\zeta \neq (a_1+a_2)\gamma_1^3/(4a_1^3)$. If we search for a linearizing two-point transformation, as $a_j \neq 0$, $j = 1, 2$, we are always in the subcase (a). Then the necessary condition (3.7.79) cannot be satisfied. So, if the condition (3.7.125) is not satisfied, (3.7.124) cannot be linearized by a two-point transformation.

We can try to linearize (3.7.124) by a Cole–Hopf transformation. We are in the case where $\mathbb{F} \neq 0$ and the equation (3.7.100) splits into two equations. As the necessary condition $\mathcal{T}_{u_{1,0}} = 0$ cannot be satisfied, (3.7.124) cannot be linearized by a Cole-Hopf transformation. We can conclude that, if the condition (3.7.125) is not satisfied, (3.7.124) cannot be linearized by neither a point, nor contact or Cole–Hopf transformation.

Hietarinta equation. Let us consider the Hietarinta equation [338, 381, 693]

$$\frac{u_{0,0} + e_2}{u_{0,0} + e_1} \frac{u_{1,1} + o_2}{u_{1,1} + o_1} = \frac{u_{1,0} + e_2}{u_{1,0} + o_1} \frac{u_{0,1} + o_2}{u_{0,1} + e_1}, \tag{3.7.127}$$

where e_j and o_j, $j = 1, 2$ are arbitrary parameters.

The necessary conditions (3.7.64), (3.7.65) of linearizability through a one-point transformation are identically satisfied and the integration of equation (3.7.61) gives

$$f(u_{0,0}) = c_1 \left[\log\left(\frac{u_{0,0} + o_1}{u_{0,0} + o_2}\right) + c_2 \right],$$

where $c_1 \neq 0$ and c_2 are arbitrary constants. One can easily see that no values of the constants c_2, $a \neq 0$, $b \neq 0$ and $c \neq 0$ of (3.7.56) exist for which the function $f(u_{0,0})$ satisfies (3.7.59) identically modulo the Hietarinta equation. As a consequence (3.7.127) cannot be linearized by a one-point transformation.

Let us look for a linearizing two-point transformation. As $\mathbb{F} = \mathbb{K} = 0$, we are in the case (4). Moreover the two linearizability conditions (3.7.76, 3.7.85) are identically satisfied and the overdetermined system of differential equations (3.7.92) reads

$$g_{u_{0,0}u_{0,0}} + \frac{(e_2 - e_1)(u_{0,1} + o_1)(u_{0,1} + o_2)}{(o_2 - o_1)(u_{0,0} + e_1)(u_{0,0} + e_2)} g_{u_{0,0}u_{0,1}} + \frac{2u_{0,0} + e_1 + e_2}{(u_{0,0} + e_1)(u_{0,0} + e_2)} g_{u_{0,0}} = 0, \tag{3.7.128}$$

$$g_{u_{0,1}u_{0,1}} + \frac{(o_2 - o_1)(u_{0,0} + e_1)(u_{0,0} + e_2)}{(e_2 - e_1)(u_{0,1} + o_1)(u_{0,1} + o_2)} g_{u_{0,0}u_{0,1}} + \frac{2u_{0,1} + o_1 + o_2}{(u_{0,1} + o_1)(u_{0,1} + o_2)} g_{u_{0,1}} = 0.$$

The solution of (3.7.128) is given by

$$g(u_{0,0}, u_{0,1}) = \theta(\xi) + c_1 \log\left(\frac{u_{0,1} + o_2}{u_{0,1} + o_1}\right) + c_2, \qquad \xi = \frac{(u_{0,0} + e_2)(u_{0,1} + o_1)}{(u_{0,0} + e_1)(u_{0,1} + o_2)}, \tag{3.7.129}$$

where c_1 and c_2 are arbitrary constants and $\theta \neq 0$ is an arbitrary function of its argument. As one can see, the system (3.7.128) is not sufficient to specify the eventual two-point transformation. We need to introduce (3.7.129) into (3.7.70, 3.7.81). In this way we get a system of first order differential equations involving θ and $T_1\theta$. From them we can extract a first order ordinary differential equation for θ which depends on ξ, $u_{0,0}$ and $u_{0,1}$. As a consequence this equation splits into an overdetermined system of four ordinary differential equations for $\theta(\xi)$. This system has no solution for generic e_j, o_j, $j = 1, 2$. As a consequence the Hietarinta equation cannot be linearized by a two-point transformation.

When we look for a linearizing one point Cole-Hopf transformation we are in the case when (3.7.64), (3.7.65) are satisfied, and the integration of (3.7.97) gives

$$h(u_{0,0}) = \frac{1}{b} \left[c_1 \left(\frac{u_{0,0} + e_1}{u_{0,0} + o_2}\right)^{c_2} - 1 \right],$$

where $c_1 \neq 0$ and $c_2 \neq 0$ are arbitrary integration constants. We have that (3.7.96) can be identically satisfied modulo the Hietarinta equation if and only if

$$c_1 = \left(1 - \frac{ab}{c}\right)\left(\frac{o_2 - o_1}{e_1 - o_1}\right), \qquad c_2 = 1.$$

Finally (3.7.95) is identically satisfied modulo the Hietarinta equation if and only if

$$c_1 = -\frac{o_2 - e_2}{e_2 - e_1}, \qquad a = \frac{c(e_2 - o_1)(o_2 - e_1)}{b(e_2 - e_1)(o_2 - o_1)},$$

so that

$$\tilde{u}_{0,1} = -\frac{1}{b}\frac{(o_2 - e_1)(u_{0,0} + e_2)}{(e_2 - e_1)(u_{0,0} + o_2)}\tilde{u}_{0,0}.$$

Through the gauge transformation $\tilde{u}_{0,0} \doteq (-b/c)^n(-b)^{-m}w_{0,0}$ we get a simplified linearizing transformation

$$w_{0,1} = \frac{(e_1 - o_2)(u_{0,0} + e_2)}{(e_1 - e_2)(u_{0,0} + o_2)}w_{0,0}, \tag{3.7.130}$$

$$w_{0,0} + aw_{1,0} - w_{0,1} + w_{1,1} = 0, \qquad a = -\frac{(o_1 - e_2)(e_1 - o_2)}{(e_1 - e_2)(o_1 - o_2)}. \tag{3.7.131}$$

Is is moreover straightforward to demonstrate that if (3.7.130), (3.7.131) are satisfied, then also the Hietarinta equation is satisfied.

When we look for a linearizing two point Cole-Hopf transformation we are in the case defined by the conditions $\mathcal{C}_{,u_{-1,0}} = \mathcal{D}_{,u_{-1,0}} = \mathcal{T}_{,u_{1,0}} = \mathbb{F} = 0$ and thus the linearizing function is defined by the equations (3.7.111), (3.7.114) which read

$$\theta_{u_{0,0}} + \frac{(e_2 - o_1)(u_{0,1} + o_1)(u_{0,1} + o_2)}{(o_2 - o_1)(u_{0,0} + o_1)(u_{0,0} + e_2)}\theta_{u_{0,1}} = 2\frac{u_{0,0}(o_2 - o_1) - u_{0,1}(e_2 - o_1) + o_1(o_2 - e_2)}{(o_2 - o_1)(u_{0,0} + o_1)(u_{0,0} + e_2)},$$

$$\theta_{u_{0,1}} + \frac{(o_2 - e_1)(u_{0,0} + e_1)(u_{0,0} + e_2)}{(e_2 - e_1)(u_{0,1} + e_1)(u_{0,1} + o_2)}\theta_{u_{0,0}} = 2\frac{u_{0,0}(o_2 - e_1) - u_{0,1}(e_2 - e_1) + e_1(o_2 - e_2)}{(e_2 - e_1)(u_{0,1} + e_1)(u_{0,1} + o_2)},$$

whose solution is given by

$$\theta(u_{0,0}, u_{0,1}) = \log\left[\left(\frac{u_{0,0} + e_2}{u_{0,1} + o_2}\right)^2\right] + \alpha,$$

where α is an arbitrary integration constant. Then we can solve the Hopf-like equation (3.7.113) by separation of variables, $h = A(u_{0,0})B(u_{0,1})$, obtaining

$$h(u_{0,0}, u_{0,1}) = \frac{e^\alpha(u_{0,0} + e_2)}{\delta(u_{0,0} + e_2) + \beta}\frac{\gamma(u_{0,1} + o_2) - \beta}{u_{0,1} + o_2}, \tag{3.7.132}$$

where β, γ and δ are arbitrary integration constants. A necessary condition to obtain the linearization is that (3.7.94) be identically satisfied for all $u_{0,0}, u_{1,0}, u_{0,1}, u_{0,2}, e_j, o_j, j = 1, 2$ modulo the Hietarinta equation, from which we get

$$e^\alpha = \frac{a\beta(o_2 - o_1)}{c(e_2 - o_1)[\beta + \gamma(e_2 - o_2)]}, \qquad \delta = -\frac{\beta\gamma}{\beta + \gamma(e_2 - o_2)},$$
$$b = \frac{c(o_2 - e_1)(e_2 - o_1)}{a(e_2 - e_1)(o_2 - o_1)}.$$

When we insert the obtained values of e^α, δ and b into the transformation (3.7.132), the two equations for ψ and ϕ are identically satisfied. As h depends on $u_{0,0}$ and $u_{0,1}$, it is necessary that $\beta \neq 0$. By redefining $\gamma \doteq \beta\pi$ we can eliminate the parameter β from the transformation. The transformation as well as the coefficient b of the linear equation so far obtained depend in a multiplicative way from the ratio a/c. Hence, performing the transformation $\tilde{u}_{n,m} \doteq \chi^m v_{n,m}$, where χ is a constant, the linear equation (3.7.146) and the Cole–Hopf transformation (3.7.4) respectively read

$$v_{n,m} + a v_{n+1,m} + b\chi v_{n,m+1} + c\chi v_{n+1,m+1} = 0,$$
$$v_{n,m+1} = \frac{1}{\chi} h(u_{n,m}, u_{n,m+1}) v_{n,m},$$

and choosing $\chi = -a/c$ we can remove the ratio a/c from the expressions of h and b. In other words we can always choose a "gauge" for the linearizing transformation in which $a = -c$. In conclusion we have

$$h = \frac{(o_2 - o_1)(u_{0,0} + e_2)[1 - \pi(u_{0,1} + o_2)]}{(e_2 - o_1)[1 - \pi(u_{0,0} + o_2)](u_{0,1} + o_2)}, \qquad b = -\frac{(o_2 - e_1)(e_2 - o_1)}{(e_2 - e_1)(o_2 - o_1)}. \tag{3.7.133}$$

To be able to solve the Hietarinta equation explicitly we look here for the inverse formula, i.e. the formula which provide the solution of the Hietarinta equation in terms of those of the linear equation (3.7.55).

Given a solution of the Hietarinta equation, let us perform a Cole–Hopf transformation (3.7.5) with h given in (3.7.133). Let us extract from the Cole–Hopf transformation $u_{0,1}$ as a function of $u_{0,0}$ and $\eta_{0,0} \doteq \tilde{u}_{0,1}/\tilde{u}_{0,0}$ and insert it, together with its consequences, in the Hietarinta equation. If $\pi \neq \frac{1}{(o_2-o_1)}$, we obtain $u_{1,0}$ as a function of $u_{0,0}$, $\eta_{0,0}$ and $\eta_{1,0}$. The compatibility between the functions $u_{0,1}(u_{0,0}, \eta_{0,0})$ and $u_{1,0}(u_{0,0}, \eta_{0,0}, \eta_{1,0})$ gives a second degree polynomial equation in $u_{0,0}$ which must be satisfied for all $u_{0,0}$. Taking into account that the ratio $\eta_{0,0}$ cannot in general be a constant when $u_{0,0}$ satisfies the Hietarinta equation, we have that, if $\pi \neq \frac{1}{(o_2-o_1)}$, $\tilde{u}_{0,0}$ will satisfy the linear and in general non autonomous equation

$$(1 - \alpha_n)\tilde{u}_{0,0} - c\tilde{u}_{1,0} + b(1 - \alpha_n)\tilde{u}_{0,1} + c\tilde{u}_{1,1} = 0, \qquad c = \frac{(e_1 - o_1)(e_2 - o_1)}{(e_2 - e_1)(o_2 - o_1)}, \tag{3.7.134}$$

where α_n is an n-dependent integration function depending on the initial values $u_{n,0}$ and $\tilde{u}_{n,0}$ given by

$$\alpha_n = 1 - \frac{(e_1 - o_1)[1 - \pi(u_{n,0} + o_2)](u_{n+1,0} + o_1)\tilde{u}_{n+1,0}}{(o_2 - o_1)(u_{n,0} + e_1)[1 - \pi(u_{n+1,0} + o_2)]\tilde{u}_{n,0}}. \tag{3.7.135}$$

When $\alpha_n \neq 1$, performing the "gauge" transformation $\tilde{u}_{n,m} \doteq \tau_n v_{n,m}$, with $\tau_{n+1} = (1+\alpha_n)\tau_n$, the Cole–Hopf transformation is invariant while the function $v_{n,m}$ will satisfy the linear autonomous equation

$$v_{0,0} - c v_{1,0} + b v_{0,1} + c v_{1,1} = 0. \tag{3.7.136}$$

When $\pi = \frac{1}{(o_2-o_1)}$, the function h becomes

$$h = \frac{(o_2 - o_1)(u_{0,0} + e_2)(u_{0,1} + o_1)}{(e_2 - o_1)(u_{0,0} + e_1)(u_{0,1} + o_2)}. \tag{3.7.137}$$

Inserting the corresponding Cole–Hopf transformation into the Hietarinta equation we get $u_{0,0}$ in terms of $\eta_{0,0}$ and $\eta_{1,0}$

(3.7.138)

$$u_{0,0} = -\frac{e_2(e_1 - o_1)(o_2 - o_1) + o_1(e_2 - o_1)(o_2 - e_1)\eta_{0,0} - e_1(e_2 - o_1)(o_2 - o_1)\eta_{1,0}}{(e_1 - o_1)(o_2 - o_1) + (e_2 - o_1)(o_2 - e_1)\eta_{0,0} - (e_2 - o_1)(o_2 - o_1)\eta_{1,0}}.$$

The insertion of (3.7.138) with its consequences into $u_{0,1}(u_{0,0}, \eta_{0,0})$ implies for the function $\tilde{u}_{0,0}$ the same evolution (3.7.134) with α_n given by the limit as $\pi \to \frac{1}{(o_2-o_1)}$ of (3.7.135) and hence (3.7.136) for $v_{0,0}$, just as was shown in [693]. As a consequence $u_{0,0}$, given by (3.7.138), becomes

$$u_{0,0} = -\frac{e_1(o_2 - o_1)v_{0,0} - o_1(e_1 - o_1)v_{1,0}}{(o_2 - o_1)v_{0,0} - (e_1 - o_1)v_{1,0}}. \tag{3.7.139}$$

Conversely, if $v_{0,0}$ satisfies (3.7.136), then it is possible to demonstrate that $u_{0,0}$ given by (3.7.139) satisfies the Hietarinta equation.

In the relations (3.7.138), (3.7.139) the role of the variables $u_{n,m}$ and $v_{n,m}$ appears inverted with respect to those given by the Cole–Hopf transformation (3.7.5), (3.7.6).

The inverted transformation, consequently, appears only when $\pi = \frac{1}{(o_2-o_1)}$. The relation (3.7.139) is the inverse, in the space of the solutions of the Hietarinta equation, of the relation (3.7.6) corresponding to an h given in (3.7.137). Eq. (3.7.139) restrict the result to the space of the solutions of the Hietarinta equation, that is

$$\frac{v_{1,0}}{v_{0,0}} = k = \frac{(o_2 - o_1)(u_{0,0} + e_1)}{(e_1 - o_1)(u_{0,0} + o_1)}. \tag{3.7.140}$$

Finally, inserting the initial value at $m = 0$ of (3.7.140) into (3.7.135), we get $\alpha_n = 0$. By the transformation $v_{0,0} \doteq c^{-n} w_{0,0}$, we can simplify further (3.7.136) which together with (3.7.140) becomes

$$\tilde{w}_{0,0} - \tilde{w}_{1,0} + b\tilde{w}_{0,1} + \tilde{w}_{1,1} = 0, \tag{3.7.141}$$

$$\tilde{w}_{1,0} = \frac{(e_2 - o_1)(u_{0,0} + e_1)}{(e_2 - e_1)(u_{0,0} + o_1)}\tilde{w}_{0,0}. \tag{3.7.142}$$

The relation (3.7.142) represents another linearizing one point Cole–Hopf transformation $\tilde{w}_{1,0} = \tilde{h}(u_{0,0})\tilde{w}_{0,0}$.

Then the general integral of the Hietarinta equation is obtained inserting in the inverse of (3.7.142) the solution of the initial-boundary value problem for the linear equation given in (3.7.141). This solution is given by

- if $b = -1$ by $\tilde{w}_{n,m} = \tilde{w}_{n,0} + \tilde{w}_{0,m} - \tilde{w}_{0,0}$,
- if $b \neq -1$ by $\tilde{w}_{n,m} = \frac{1}{2\pi i}\oint_{\mathbb{C}} \zeta_m(z) z^{n-1} dz$, where

$$(1 - z)\zeta_m + (b + z)\zeta_{m+1} = z(\tilde{w}_{0,m+1} - \tilde{w}_{0,m}), \qquad \zeta_0(z) = \sum_{j=0}^{+\infty} \tilde{w}_{j,0} z^{-j},$$

and $\mathbb{C}$ represents a counterclockwise circumference in the complex z−plane, centered in $z = 0$ and lying inside the region of convergence of the series $\sum_{j=0}^{+\infty} \tilde{w}_{j,m} z^{-j}$.

For example, when $b \neq -1$, an explicit solution in the plane $n \geq 0$ of the initial-boundary value problemcharacterized by $\tilde{w}_{0,m} = 1$, $\tilde{w}_{n,0} = 2^{-n}$, $n \geq 0$ ($\tilde{w}_{n,0} = 0$, $n < 0$), so that $\zeta_0 = 2z/(2z-1)$, $|z| > 1/2$, is given, if $b = -\frac{1}{2}$, by

$$\tilde{\omega}_{n,m} = \sum_{\rho=\max\{n,m\}}^{n+m} \binom{m}{\rho-n}\binom{\rho}{m}\frac{(-1)^{n+m-\rho}}{2^{\rho-m}}, \qquad m \geq 0,$$

$$\tilde{\omega}_{n,m} = \sum_{\rho=\max\{n,|m|-1\}}^{n+|m|-1} \binom{|m|-1}{\rho-n}\binom{\rho}{|m|-1}\left(-\frac{1}{2}\right)^{n+|m|-\rho-1}, \qquad m \leq -1,$$

and, if $b \neq -\frac{1}{2}$, by

$$\tilde{\omega}_{n,m} = \frac{1}{2^n(-1-2b)^m} - \sum_{\kappa=0}^{m-1}\sum_{\rho=\max\{n,\kappa\}}^{n+m} \binom{m}{\rho-n}\binom{\rho}{\kappa}\frac{(-1)^{n+m-\kappa}b^{\rho-\kappa}}{(b+1/2)^{m-\kappa}}, \qquad m \geq 1,$$

$$\tilde{\omega}_{n,m} = \frac{(-1-2b)^{|m|}}{2^n} - \sum_{\kappa=0}^{|m|-1}\sum_{\rho=\max\{n,\kappa\}}^{n+|m|} \binom{|m|}{\rho-n}\binom{\rho}{\kappa}(-2)^{|m|-\kappa}b^{n+|m|-\rho}, \qquad m \leq -1.$$

A linearizable non linear quad-graph equation. As a further example let us consider the simple multilinear equation

(3.7.143) $$u_{0,0} + \alpha u_{1,0} + \pi[u_{0,0}u_{0,1} + \alpha u_{1,0}u_{1,1}] = 0.$$

As in the case of the discrete Liouville equation the necessary conditions for the linearizability by point transformations are identically satisfied and the linearizing one-point transformation, obtained by integrating (3.7.61), is given by

$$f(u_{0,0}) = c_1\left[c_2 + \log\left(u_{0,0} + \frac{1}{\epsilon}\right)\right],$$

where $c_1 \neq 0$ and c_2 are arbitrary constants. Inserting $f(u_{0,0})$ into (3.7.59), one can easily see that this relation can be identically satisfied modulo (3.7.143) only if $c = -b$. Differentiating (3.7.56) with respect to $u_{1,0}$ and inserting $f(u_{0,0})$ and $c = -b$ into the resulting relation, we have that no values of the constants c_2, $a \neq 0$, $b \neq 0$ and $c \neq 0$ of (3.7.56) exist for which this can be satisfied identically modulo (3.7.143). Hence this equation (3.7.143) cannot be linearized by a one-point transformation.

We can try to linearize it by a two-point transformation of the form (3.7.4). As $\mathbb{F} = \mathbb{K} = 0$ identically, we are in the fourth case. Moreover the two linearizability conditions (3.7.76, 3.7.85) are identically satisfied and the overdetermined system of differential equations (3.7.92), (3.7.93) reads

$$g_{u_{0,0}u_{0,0}} - \frac{1+\pi u_{0,1}}{\pi u_{0,0}} g_{u_{0,0},u_{0,1}} + \frac{1}{u_{0,0}} g_{u_{0,0}} = 0,$$

$$g_{u_{0,1}u_{0,1}} - \frac{\pi u_{0,0}}{1+u\pi_{0,1}} g_{u_{0,0},u_{0,1}} + \frac{\epsilon}{1+\epsilon u_{0,1}} g_{u_{0,1}} = 0,$$

whose solution is given by

(3.7.144) $$g(u_{0,0}, u_{0,1}) = \theta(\xi) + c_3\log(u_{0,0}) + c_4, \qquad \xi = u_{0,0}\left(u_{0,1} + \frac{1}{\pi}\right),$$

where c_3 and c_4 are arbitrary constants and $\theta \neq 0$ is an arbitrary function of its arguments. Introducing (3.7.144) in (3.7.70, 3.7.81) we get $b = \frac{c}{\alpha}$. The final determining equation (3.7.67) implies $c_3 = c_4 = 0$, $a = \alpha$ and $\theta = ku_{0,0}(1 + \pi u_{0,1})$. So, in conclusion (3.7.143) is linearizable by the transformation

$$g(u_{0,0}, u_{0,1}) = u_{0,0}(1 + \pi u_{0,1})$$

into the linear equation

$$\tilde{u}_{0,0} + \alpha\tilde{u}_{1,0} + b\tilde{u}_{0,1} + \alpha b\tilde{u}_{1,1} = 0.$$

Four QRT-type linearizable equations. As a final example we consider the four QRT-type linearizable non linear PΔEs recently presented in [807]

$$u_{1,1} = \frac{u_{1,0} + u_{0,1} - (1 - u_{1,0}u_{0,1})u_{0,0}}{1 - u_{1,0}u_{0,1} + (u_{1,0} + u_{0,1})u_{0,0}},$$

$$u_{1,1} = \frac{u_{1,0} - u_{0,1} + (1 + u_{1,0}u_{0,1})u_{0,0}}{1 + u_{1,0}u_{0,1} - (u_{1,0} - u_{0,1})u_{0,0}},$$

$$u_{1,1} = \frac{f_2(n,m) + f_1(n,m)u_{0,0}}{f_1(n,m) - f_2(n,m)u_{0,0}},$$

$$u_{1,1} = \frac{2u_{1,0} - u_{0,1} + u_{0,1}u_{1,0}^2 - \left(1 + 2u_{1,0}u_{0,1} - u_{1,0}^2\right)u_{0,0}}{1 + 2u_{1,0}u_{0,1} - u_{1,0}^2 + \left(2u_{1,0} - u_{0,1} + u_{0,1}u_{1,0}^2\right)u_{0,0}},$$

where

$$f_1(n,m) \doteq (1 + u_{1,0}u_{0,1})\left(u_{1,0}^2u_{0,1}^2 - 3u_{1,0}^2 - 3u_{0,1}^2 + 8u_{1,0}u_{0,1} + 1\right),$$
$$f_2(n,m) \doteq (u_{0,1} - u_{1,0})\left(3u_{1,0}^2u_{0,1}^2 - u_{1,0}^2 - u_{0,1}^2 + 8u_{1,0}u_{0,1} + 3\right).$$

All these equations are linearizable through the transformation $v_{0,0} = \arctan(u_{0,0})$ and give the following linear equations

$$\begin{aligned} v_{0,0} - v_{1,0} - v_{0,1} + v_{1,1} &= p\pi, \\ v_{0,0} + v_{1,0} - v_{0,1} - v_{1,1} &= p\pi, \\ v_{0,0} - 3v_{1,0} + 3v_{0,1} - v_{1,1} &= p\pi, \\ v_{0,0} - 2v_{1,0} + v_{0,1} + v_{1,1} &= p\pi, \end{aligned} \tag{3.7.145}$$

where $p \in \mathcal{Z}$. The p-dependent right-hand side is a consequence of the multi-valuedness of the arctan function. We can choose $p = 0$ if we choose the principle branch of the arctan function, i.e. $\arctan(0) = 0$.

Applying the formulas contained in Section 3.7.2.1 it is immediate to show that the necessary conditions (3.7.64, 3.7.65) of linearizability through a point transformation are identically satisfied. The integration of equation (3.7.61) gives

$$f(u_{0,0}) = c_1[\arctan(u_{0,0}) + c_2],$$

where $c_1 \neq 0$ and c_2 are arbitrary constants. The differential consequences of equation (3.7.56) (which in this case will have a constant right-hand side not necessarily equal to zero) can be identically satisfied modulo the QRT-type equations if and only if

$$\begin{aligned} &a = -1, && b = -1, && c = 1, \\ &a = 1, && b = -1, && c = -1, \\ &a = -3, && b = 3, && c = -1, \\ &a = -2, && b = 1, && c = 1. \end{aligned}$$

With these values of the coefficients (a, b, c), (3.7.56) is identically satisfied modulo the corresponding equations. In the first three cases its right hand side is equal to $p\pi$ for arbitrary c_2 while in the fourth case it is equal to $c_2 + p\pi$. As c_2 is arbitrary, for the sake of simplicity we will choose $c_2 = 0$ in all cases.

7.3. Results on the classification of multilinear PΔEs linearizable by point transformation on a square lattice. In previous Sections one has provided necessary conditions for the linearizability of real dispersive multilinear PΔEs defined on a quad–graph (see Fig. 2.3) and on three points (see Fig.3.1a).

These conditions, obtained by considering the existence of point transformations have been sufficient to classify the multilinear equations defined on three points but not those defined on four points. Here, starting from the results presented up to now we construct the largest possible set of linearizability conditions and, through them, we classify the multilinear equations on a square lattice. We assume a PΔE on a quad–graph to be given by (2.4.404) for a field $u_{n,m}$ which linearizes into a linear autonomous equation for $\tilde{u}_{n,m}$

$$a\tilde{u}_{0,0} + b\tilde{u}_{1,0} + c\tilde{u}_{0,1} + d\tilde{u}_{1,1} + e = 0 \tag{3.7.146}$$

with a, b, c, d and e being (n, m)–independent arbitrary non zero complex coefficients.

In Section 3.7.3.1 we discuss point transformations, present further linearizability conditions which ensure that the given equation is linearizable and the differential equations which define the transformation f. In Section 3.7.3.2 we classify all multilinear quad-graph equations which belong to the class (2.4.404) up to a Möbius transformation.

7.3.1. *Quad-graph PΔEs linearizable by a point transformation.* We will assume that we can solve (2.4.130) with conditions (3.6.1) with respect to each one of the four variables in its argument

$$u_{1,1} = F\left(u_{0,0}, u_{1,0}, u_{0,1}\right), \quad F_{u_{0,0}} \neq 0 \quad F_{u_{1,0}} \neq 0, \quad F_{u_{0,1}} \neq 0, \tag{3.7.147a}$$

$$u_{1,0} = G\left(u_{0,0}, u_{0,1}, u_{1,1}\right), \quad G_{u_{0,0}} \neq 0, \quad G_{u_{0,1}} \neq 0, \quad G_{u_{1,1}} \neq 0, \tag{3.7.147b}$$

$$u_{0,1} = H\left(u_{0,0}, u_{1,0}, u_{1,1}\right), \quad H_{u_{0,0}} \neq 0, \quad H_{u_{1,0}} \neq 0 \quad H_{u_{1,1}} \neq 0, \tag{3.7.147c}$$

$$u_{0,0} = T\left(u_{1,0}, u_{0,1}, u_{1,1}\right), \quad T_{u_{1,0}} \neq 0, \quad T_{u_{0,1}} \neq 0 \quad T_{u_{1,1}} \neq 0. \tag{3.7.147d}$$

Moreover that (2.4.130, 3.6.1) can be linearized by the linearizing autonomous point transformation (3.7.3) into the linear equation (3.7.146). Hence, assuming we can solve (2.4.130, 3.6.1) with respect to $u_{1,1}$, we can choose as independent variables $u_{0,0}$, $u_{1,0}$ and $u_{0,1}$ and we will have that

$$af_{0,0} + bf_{1,0} + cf_{0,1} + df_{1,1}\Big|_{u_{1,1}=F} + e = 0, \tag{3.7.148}$$

must be identically satisfied for any $u_{0,0}$, $u_{1,0}$ and $u_{0,1}$. Differentiating (3.7.148) with respect to $u_{0,0}$, $u_{1,0}$ or $u_{0,1}$, we obtain

$$a\frac{df_{0,0}}{du_{0,0}} + d\frac{df_{1,1}}{du_{1,1}}\Big|_{u_{1,1}=F} F_{u_{0,0}} = 0, \tag{3.7.149a}$$

$$b\frac{df_{1,0}}{du_{1,0}} + d\frac{df_{1,1}}{du_{1,1}}\Big|_{u_{1,1}=F} F_{u_{1,0}} = 0, \tag{3.7.149b}$$

$$c\frac{df_{0,1}}{du_{0,1}} + d\frac{df_{1,1}}{du_{1,1}}\Big|_{u_{1,1}=F} F_{u_{0,1}} = 0, \tag{3.7.149c}$$

which have to be identically satisfied for any $u_{0,0}$, $u_{1,0}$ and $u_{0,1}$. From them, considering that $\frac{df(x)}{dx} \neq 0$, we derive that $d \neq 0$, otherwise $a = b = c = e = 0$. As a consequence, considering that $\frac{df(x)}{dx} \neq 0$, $F_{u_{0,0}} \neq 0$, $F_{u_{1,0}} \neq 0$ and $F_{u_{0,1}} \neq 0$, we have also $a \neq 0$, $b \neq 0$ and $c \neq 0$. Then in all generality we can divide (3.7.10) by d and, introducing the new parameters $\alpha \doteq a/d \neq 0$, $\beta \doteq b/d \neq 0$, $\gamma \doteq c/d \neq 0$ and $\pi \doteq e/a$, (3.7.10) can be rewritten as

$$\alpha\tilde{u}_{0,0} + \beta\tilde{u}_{1,0} + \gamma\tilde{u}_{0,1} + \tilde{u}_{1,1} + \pi = 0. \tag{3.7.150}$$

Defining $\frac{df(x)}{dx} \doteq R(x)$, from (3.7.149) we obtain

$$\frac{F_{u_{0,0}}}{F_{u_{1,0}}} = \frac{\alpha R(u_{0,0})}{\beta R(u_{1,0})}, \tag{3.7.151a}$$

$$\frac{F_{u_{0,0}}}{F_{u_{0,1}}} = \frac{\alpha R(u_{0,0})}{\gamma R(u_{0,1})}. \tag{3.7.151b}$$

From (3.7.151) we get the following linearizability conditions

$$A(x, u_{0,1}) \doteq \frac{F_{u_{0,0}}}{F_{u_{1,0}}}\Big|_{u_{0,0}=u_{1,0}=x} = \frac{\alpha}{\beta}, \quad \forall x,\ u_{0,1}, \tag{3.7.152a}$$

$$B(x, u_{1,0}) \doteq \frac{F_{u_{0,0}}}{F_{u_{0,1}}}\Big|_{u_{0,0}=u_{0,1}=x} = \frac{\alpha}{\gamma}, \quad \forall x,\ u_{1,0}, \tag{3.7.152b}$$

$$C(x, u_{0,0}) \doteq \frac{F_{u_{0,1}}}{F_{u_{1,0}}}\Big|_{u_{1,0}=u_{0,1}=x} = \frac{\gamma}{\beta}, \quad \forall x,\ u_{0,0}, \tag{3.7.152c}$$

$$\frac{\partial}{\partial u_{0,1}}\frac{F_{u_{0,0}}}{F_{u_{1,0}}} = 0, \quad \forall u_{0,0},\ u_{1,0},\ u_{0,1}, \tag{3.7.152d}$$

$$\frac{\partial}{\partial u_{1,0}}\frac{F_{u_{0,0}}}{F_{u_{0,1}}} = 0, \quad \forall u_{0,0},\ u_{1,0},\ u_{0,1}, \tag{3.7.152e}$$

$$\frac{\partial}{\partial u_{0,0}}\frac{F_{u_{0,1}}}{F_{u_{1,0}}} = 0, \quad \forall u_{0,0},\ u_{1,0},\ u_{0,1}. \tag{3.7.152f}$$

Alternatively the conditions (3.7.152a-3.7.152c) can be substituted by the following ones

$$\frac{d}{dx}A(x, u_{0,1}) = 0, \quad \forall x,\ u_{0,1}, \tag{3.7.153a}$$

$$\frac{d}{dx}B(x, u_{1,0}) = 0, \quad \forall x,\ u_{1,0}, \tag{3.7.153b}$$

$$\frac{d}{dx}C(x, u_{0,0}) = 0, \quad \forall x,\ u_{0,0}. \tag{3.7.153c}$$

Taking the (principal value of the) logarithm of (3.7.149a), we have

$$\log\frac{df_{0,0}}{du_{0,0}} - \log\frac{df_{1,1}}{du_{1,1}}\Big|_{u_{1,1}=F} = \log\left(-\frac{F_{u_{0,0}}}{\alpha}\right) \pmod{2\pi i}. \tag{3.7.154}$$

Then, let us introduce the annihilation operator $\mathcal{B}$

$$\mathcal{B} \doteq \frac{\partial}{\partial u_{0,0}} - \frac{F_{u_{0,0}}}{F_{u_{1,0}}}\frac{\partial}{\partial u_{1,0}}, \tag{3.7.155}$$

such that and $B\phi = 0$, where $\phi(u_{0,0}, u_{1,0}, u_{0,1})$ is an arbitrary functions of its arguments. When we apply (3.7.155) to (3.7.154), we obtain an ordinary differential equation describing the evolution of the linearizing transformation

$$\frac{d}{du_{0,0}} \log \frac{df_{0,0}}{du_{0,0}} = \frac{1}{F_{u_{0,0}} F_{u_{1,0}}} W_{(u_{0,0})} \left[F_{u_{1,0}}; F_{u_{0,0}} \right], \tag{3.7.156}$$

where the Wronskian $W_{(x)}[f; g]$ is given by (3.7.17). Let's remark that the linearizability conditions (3.7.152d-3.7.152f) imply that the right hand member of the equation (3.7.156) does not depend on $u_{1,0}$ and $u_{0,1}$. These conditions were considered previously in subsection 3.7.2 and had not been sufficient to classify (2.4.404). Other similar conditions can be obtained starting from (3.7.149b) or (3.7.149c). The linearizability conditions presented here have been obtained starting from (3.7.147a). Similar results could be obtained starting from (3.7.147b), (3.7.147c) or (3.7.147d). However these results would not have provided any really new linearizability condition. So, here, in the next Section we start the classifying process from the more basic equations (3.7.152) as we did in the case of equations depending on just three points in subsection 3.7.1.

7.3.2. *Classification of complex autonomous multilinear quad-graph PΔEs linearizable by a point transformation.* Let us consider as $F\left(u_{0,0}, u_{1,0}, u_{0,1}, u_{1,1}\right)$ the complex multilinear equation defined by

$$\begin{aligned} &a_1 u_{0,0} + a_2 u_{1,0} + a_3 u_{0,1} + a_4 u_{1,1} + a_5 u_{0,0} u_{1,0} + a_6 u_{0,0} u_{0,1} + \\ &+ a_7 u_{0,0} u_{1,1} + a_8 u_{1,0} u_{0,1} + a_9 u_{1,0} u_{1,1} + a_{10} u_{0,1} u_{1,1} + \\ &+ a_{11} u_{0,0} u_{1,0} u_{0,1} + a_{12} u_{0,0} u_{1,0} u_{1,1} + a_{13} u_{0,0} u_{0,1} u_{1,1} + \\ &+ a_{14} u_{1,0} u_{0,1} u_{1,1} + a_{15} u_{0,0} u_{1,0} u_{0,1} u_{1,1} + a_{16} = 0, \end{aligned} \tag{3.7.157}$$

where a_j, $j = 1, \ldots, 16$, are arbitrary complex free parameters. This equation is invariant under a Möbius transformation of the dependent variable

$$u_{0,0} \doteq \frac{b_1 v_{0,0} + b_2}{b_3 v_{0,0} + b_4}, \tag{3.7.158}$$

where b_k, $k = 1, \ldots, 4$ are four arbitrary complex parameters such that $b_1 b_4 - b_2 b_3 \neq 0$. As we operate in the field of complex numbers and we classify up to Möbius transformations, using inversions, dilations and translations we can always simplify (3.7.157) by setting either $a_{15} = a_{16} = 0$ or $\sum_{k=1}^{4} a_k = \sum_{k=5}^{10} a_k = \sum_{k=11}^{14} a_k = a_{15} = 0$, $a_{16} = 1$. Let's now apply to these two cases the six necessary linearizability conditions (3.7.152). This amounts to solving a system of 96 algebraic in general non linear equations involving the coefficients a_j, $j = 1, \ldots, 15$. Their solution implies, through the integration of the differential equation (3.7.156), that the function $f(x)$ appearing in the linearizing transformation (3.7.146) can only be of the following two types:

(1) The fractional linear function

$$f(x) = \frac{c_1 x + c_2}{c_3 x + c_4}, \quad c_1 c_4 - c_2 c_3 \neq 0, \tag{3.7.159}$$

which, as the classification is up to Möbius transformations of $u_{0,0}$, can always be reduced to be $f(x) = 1/x$;

(2) The (principal branch) of the logarithmic function

$$f(x) = d_1 \log\left(\frac{d_2 x + d_3}{d_4 x + d_5}\right) + d_6 \quad d_1 \neq 0, \quad d_2 d_5 - d_3 d_4 \neq 0, \tag{3.7.160}$$

which, as the classification is up to Möbius transformations, can always be reduced to $f(x) = \log(x)$. Moreover in this case the ratios α/β and α/γ are always real and of modulus 1, so that the possible linear equations can be only of the following four types:

$$\alpha\left(\tilde{u}_{0,0}+\tilde{u}_{1,0}+\tilde{u}_{0,1}\right)+\tilde{u}_{1,1}+\pi=0; \tag{3.7.161a}$$

$$\alpha\left(\tilde{u}_{0,0}+\tilde{u}_{1,0}-\tilde{u}_{0,1}\right)+\tilde{u}_{1,1}+\pi=0; \tag{3.7.161b}$$

$$\alpha\left(\tilde{u}_{0,0}-\tilde{u}_{1,0}+\tilde{u}_{0,1}\right)+\tilde{u}_{1,1}+\pi=0; \tag{3.7.161c}$$

$$\alpha\left(\tilde{u}_{0,0}-\tilde{u}_{1,0}-\tilde{u}_{0,1}\right)+\tilde{u}_{1,1}+\pi=0. \tag{3.7.161d}$$

It is easy to prove that, if the transformation $\tilde{u}_{0,0} \doteq \log\left(u_{0,0}\right)$ has to produce a multilinear equation for $u_{0,0}$, we must have $\alpha = \pm 1$. In fact, as $F\left(u_{0,0}, u_{1,0}, u_{0,1}\right)$, given in (3.7.147a), should be a fractional linear function of $u_{0,0}$ with coefficients depending on $u_{1,0}$ and $u_{0,1}$, we have that the relations

$$u_{0,0}^{\alpha}\left(e_1 u_{0,0}+e_2\right)=e_3 u_{0,0}+e_4, \quad e_j=e_j\left(u_{1,0}, u_{0,1}\right), \quad j=1, \ldots, 4,$$

where $e_1 e_4 - e_2 e_3$ is not identically zero for all $u_{1,0}$ and $u_{0,1}$, must be identically satisfied for all $u_{0,0}$. Differentiating (3.7.162) twice with respect to $u_{0,0}$, we get $e_1(\alpha+1)u_{0,0} + e_2(\alpha-1) = 0$ identically for all $u_{0,0}$, so that

$$e_1(\alpha+1)=0, \quad e_2(\alpha-1)=0.$$

Considering that e_1 and e_2 cannot be simultaneously identically zero, it follows that $\alpha = \pm 1$. In this way we obtain a set of eight linear PΔEs corresponding to eight linearizable non linear PΔEs.

Hence we can summarize the results obtained in the following theorem:

Theorem 92. *Let us consider the class of multilinear equations* (3.7.157). *Apart from the class of equations linearizable by a Möbius transformation, which can be represented up to a Möbius transformation of the dependent variable (eventually composed with an exchange of the independent variables $n \leftrightarrow m$) by the equation*

$$w_{0,1}w_{1,1}\left(w_{1,0}+\beta w_{0,0}\right)+w_{0,0}w_{1,0}\left(\gamma w_{1,1}+\delta w_{0,1}\right)+\pi w_{0,0}w_{1,0}w_{0,1}w_{1,1}=0, \qquad \pi=0,1, \quad \beta,\gamma,\delta\neq 0, \tag{3.7.162}$$

linearizable by the inversion $\tilde{u}_{0,0} = 1/w_{0,0}$ to the equation

$$\tilde{u}_{0,0}+\beta\tilde{u}_{1,0}+\gamma\tilde{u}_{0,1}+\delta\tilde{u}_{1,1}+\pi=0, \tag{3.7.163}$$

the only other linearizable equations are up to a Möbius transformation of the dependent variable (eventually composed with an exchange of the independent variables $n \leftrightarrow m$), represented by the following six non linear equations

$$w_{0,0}w_{1,0}w_{0,1}w_{1,1}-1=0, \tag{3.7.164a}$$

$$w_{0,0}-w_{1,0}w_{0,1}w_{1,1}=0, \tag{3.7.164b}$$

$$w_{1,0}-w_{0,0}w_{0,1}w_{1,1}=0, \tag{3.7.164c}$$

$$w_{1,1}-w_{0,0}w_{1,0}w_{0,1}=0, \tag{3.7.164d}$$

$$w_{0,1}w_{1,1}-\theta w_{0,0}w_{1,0}=0, \tag{3.7.164e}$$

$$w_{0,0}w_{1,1}-\theta w_{1,0}w_{0,1}=0, \tag{3.7.164f}$$

where $\theta \neq 0$ is an otherwise arbitrary complex parameter. They are linearizable by the transformation $\tilde{u}_{0,0} = \log w_{0,0}$ to the equations

(3.7.164g) $$\tilde{u}_{0,0} + \tilde{u}_{1,0} + \tilde{u}_{0,1} + \tilde{u}_{1,1} = 2\pi \mathrm{i} z,$$

(3.7.164h) $$-\tilde{u}_{0,0} + \tilde{u}_{1,0} + \tilde{u}_{0,1} + \tilde{u}_{1,1} = 2\pi \mathrm{i} z,$$

(3.7.164i) $$\tilde{u}_{0,0} - \tilde{u}_{1,0} + \tilde{u}_{0,1} + \tilde{u}_{1,1} = 2\pi \mathrm{i} z,$$

(3.7.164j) $$-\tilde{u}_{0,0} - \tilde{u}_{1,0} - \tilde{u}_{0,1} + \tilde{u}_{1,1} = 2\pi \mathrm{i} z,$$

(3.7.164k) $$-\tilde{u}_{0,0} - \tilde{u}_{1,0} + \tilde{u}_{0,1} + \tilde{u}_{1,1} = \log \theta + 2\pi \mathrm{i} z,$$

(3.7.164l) $$\tilde{u}_{0,0} - \tilde{u}_{1,0} - \tilde{u}_{0,1} + \tilde{u}_{1,1} = \log \theta + 2\pi \mathrm{i} z,$$

where log *always stands for the principal branch of the complex logarithmic function and where* $\log \theta$ *stands for the principal branch of the complex logarithmic function of the parameter* θ.

APPENDIX A

Construction of lattice equations and their Lax pair

Let a consider a quad equation $Q = Q(x, x_1, x_2, x_{12}; \alpha_1, \alpha_2)$ of the ones introduced in **Theorem** 12 in Section 2.4.7. An equation of this kind, even if consistent around the cube, if not treated with care can turn into a non integrable equation. For example, let us consider the deformed H_1 equation, ${}_rH_1^\pi$ [23]:

$$(x - x_{12})(x_1 - x_2) - (\alpha_1 - \alpha_2)(1 + \pi^2 x_1 x_2) = 0, \tag{A.1}$$

and let us consider the trivial embedding of such equation into a lattice given by the identification[1]:

$$x \to u_{n,m}, \quad x_1 \to u_{n+1,m}, \quad x_2 \to u_{n,m+1}, \quad x_{12} \to u_{n+1,m+1}. \tag{A.2}$$

Eq. (A.1) with the identification (A.2) becomes the following lattice equation:

$$(u_{n,m} - u_{n+1,m+1})(u_{n+1,m} - u_{n,m+1}) - (\alpha_1 - \alpha_2)(1 + \pi^2 u_{n+1,m} u_{n,m+1}) = 0. \tag{A.3}$$

We apply the algebraic entropy test[2] to (A.3) and we find the following growth in the Nord-East $(-,+)$ direction of the degrees:

$$\{ d_{-,+} \} = \{ 1, 2, 4, 9, 21, 50, 120, 289 \dots \}. \tag{A.4}$$

This sequence has generating function

$$g_{-,+} = \frac{1 - s - s^2}{s^3 + s^2 - 3s + 1} \tag{A.5}$$

and therefore has a non-zero algebraic entropy given by:

$$\eta_{-,+} = \log\left(1 + \sqrt{2}\right), \tag{A.6}$$

corresponding to the entropy of a non-integrable lattice equation.

The identification (A.2) is not the only possible embedding. An embedding of (A.1) into a $\mathbb{Z}^2$ lattice is obtained by choosing an elementary cell of dimension greater than one as the one depicted in Fig. 2.5 in Section 2.4.7. In such a case we will get a different PΔE. Since *a priori* $Q \neq |Q \neq \underline{Q} \neq |\underline{Q}$ the obtained lattice will be a four color lattice, see Fig. 2.5. Choosing the origin on the $\mathbb{Z}^2$ lattice in the point x we obtain a lattice equation of the

[1]Generally from now we will call the field variables u to distinguish them from the "static" vertex indexes x

[2]For more details on degree of growth, algebraic entropy and related subjects see Appendix C.1 and references therein.

following form:

$$\widetilde{Q}[u] = \begin{cases} Q(u_{n,m}, u_{n+1,m}, u_{n,m+1}, u_{n+1,m+1}) & n = 2k, m = 2k, k \in \mathbb{Z}, \\ |Q(u_{n,m}, u_{n+1,m}, u_{n,m+1}, u_{n+1,m+1}) & n = 2k+1, m = 2k, k \in \mathbb{Z}, \\ \underline{Q}(u_{n,m}, u_{n+1,m}, u_{n,m+1}, u_{n+1,m+1}) & n = 2k, m = 2k+1, k \in \mathbb{Z}, \\ |\underline{Q}(u_{n,m}, u_{n+1,m}, u_{n,m+1}, u_{n+1,m+1}) & n = 2k+1, m = 2k+1, k \in \mathbb{Z}, \end{cases} \tag{A.7}$$

We could have constructed $\widetilde{Q}[u]$ starting from any other point in Fig. 2.5 as the origin, but such equations would differ from each other only by a translation, a rotation or a reflection. So in the sense of Theorem 12 they will be equivalent to (A.7).

In the case of ${}_rH_1^\pi$ we have:

$$ {}_r\widetilde{H}_1^\pi = \begin{cases} (u_{n,m} - u_{n+1,m+1})(u_{n+1,m} - u_{n,m+1}) & |n| + |m| = 2k, \\ \quad - (\alpha_1 - \alpha_2)(1 + \pi^2 u_{n+1,m} u_{n,m+1}), & k \in \mathbb{Z}, \\ (u_{n,m} - u_{n+1,m+1})(u_{n+1,m} - u_{n,m+1}) & |n| + |m| = 2k+1, \\ \quad - (\alpha_1 - \alpha_2)(1 + \pi^2 u_{n,m} u_{n+1,m+1}), & k \in \mathbb{Z}, \end{cases} \tag{A.8}$$

where we have used the symmetry properties of the equation ${}_rH_1^\pi$. This result can be written as one equation using the function χ_{n+m}. The resulting equation coincide with (2.4.159a).

We shall require that the quad equations satisfy CaC, since we are concerned about integrability. So let us consider the six-tuples of quad equations (2.4.145) assigned to the faces of the 3D cube displayed in Fig. 2.4 in Section 2.4.6.

First let us notice that, without loss of generality, we can assume that, if Q is the consistent quad equation we are interested in, then we may assume that Q is the bottom equation i.e. $Q = A$. Indeed if we are interested in an equation on the side of the cube given in Fig. 2.4 and this equation is different from A (once made the appropriate substitutions) we may just rotate it and re-label the vertices in an appropriate manner, so that our side equation will become the bottom equation. In this way following again [112] and taking into account the result stated above we may build an embedding in $\mathbb{Z}^3$, whose points we shall label as triples (n, m, p), of the consistency cube. To this end we reflect the consistency cube with respect to the normal of the back and the right side and then complete again with another reflection, just in the same way we did for the square. Using the same notations as in the planar case we see which are the proper equations which must be put on the sides of the "multicube". Their form can therefore be described as in (A.7). As a result we end up with Fig. A.1, where the functions appearing on the top and on the bottom can be defined as in (A.7)[3] while on the sides we shall have:

$$\underline{B}(x, x_2, x_3, x_{23}) = B(x_2, x, x_{23}, x_3), \tag{A.9a}$$

$$\underline{\bar{B}}(x_1, x_{12}, x_{13}, x_{123}) = \bar{B}(x_{12}, x_1, x_{123}, x_{13}), \tag{A.9b}$$

$$|C(x, x_1, x_3, x_{13}) = C(x_1, x, x_{13}, x_3), \tag{A.9c}$$

$$|\bar{C}(x_2, x_{12}, x_{23}, x_{123}) = \bar{C}(x_{12}, x_2, x_{123}, x_{23}). \tag{A.9d}$$

---

[3]Obviously in the case of $\bar{A}$ one should traslate every point by one in the p direction.

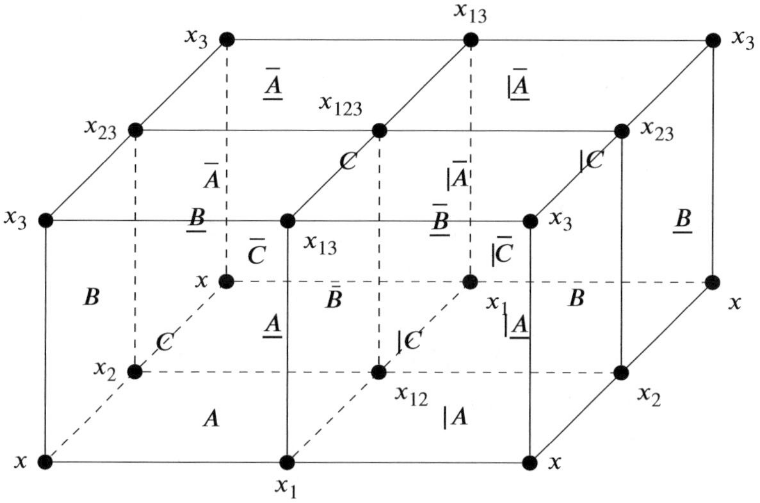

FIGURE A.1. The extension of the consistency cube [reprinted from [336]].

From (A.9) we obtain the analogous in $\mathbb{Z}^3$ of (A.7). We have a new consistency cube, see Fig. A.1, with equations given by[4]:

$$\text{(A.10a)} \qquad \tilde{\tilde{A}}[u] = \begin{cases} \bar{A}(u_{n,m,p+1}, u_{n+1,m,p+1}, u_{n,m+1,p+1}, u_{n+1,m+1,p+1}), \\ |\bar{A}(u_{n,m,p+1}, u_{n+1,m,p+1}, u_{n,m+1,p+1}, u_{n+1,m+1,p+1}), \\ \underline{\bar{A}}(u_{n,m,p+1}, u_{n+1,m,p+1}, u_{n,m+1,p+1}, u_{n+1,m+1,p+1}), \\ |\underline{A}(u_{n,m,p+1}, u_{n+1,m,p+1}, u_{n,m+1,p+1}, u_{n+1,m+1,p+1}), \end{cases}$$

$$\text{(A.10b)} \qquad \tilde{B}[u] = \begin{cases} B(u_{n,m,p}, u_{n,m+1,p}, u_{n,m,p+1}, u_{n,m+1,p+1}), \\ \bar{B}(u_{n,m,p}, u_{n,m+1,p}, u_{n,m,p+1}, u_{n,m+1,p+1}), \\ \underline{B}(u_{n,m,p}, u_{n,m+1,p}, u_{n,m,p+1}, u_{n,m+1,p+1}), \\ \underline{\bar{B}}(u_{n,m,p}, u_{n,m+1,p}, u_{n,m,p+1}, u_{n,m+1,p+1}), \end{cases}$$

$$\text{(A.10c)} \qquad \tilde{\tilde{B}}[u] = \begin{cases} \bar{B}(u_{n+1,m,p}, u_{n+1,m+1,p}, u_{n+1,m,p+1}, u_{n+1,m+1,p+1}), \\ B(u_{n+1,m,p}, u_{n+1,m+1,p}, u_{n+1,m,p+1}, u_{n+1,m+1,p+1}), \\ \underline{\bar{B}}(u_{n+1,m,p}, u_{n+1,m+1,p}, u_{n+1,m,p+1}, u_{n+1,m+1,p+1}), \\ \underline{B}(u_{n+1,m,p}, u_{n+1,m+1,p}, u_{n+1,m,p+1}, u_{n+1,m+1,p+1}), \end{cases}$$

[4]For the sake of the simplicity of the presentation we have left out if n and m are even or odd integers. They are recovered by comparing with (A.7).

$$\text{(A.10d)} \qquad \widetilde{C}\,[u] = \begin{cases} C(u_{n,m,p}, u_{n+1,m,p}, u_{n,m,p+1}, u_{n+1,m,p+1}), \\ |C(u_{n,m,p}, u_{n+1,m,p}, u_{n,m,p+1}, u_{n+1,m,p+1}), \\ \overline{C}(u_{n,m,p}, u_{n+1,m,p}, u_{n,m,p+1}, u_{n+1,m,p+1}), \\ |\overline{C}(u_{n,m,p}, u_{n+1,m,p}, u_{n,m,p+1}, u_{n+1,m,p+1}), \end{cases}$$

$$\text{(A.10e)} \qquad \widetilde{\widetilde{C}}\,[u] = \begin{cases} \overline{C}(u_{n,m+1,p}, u_{n+1,m+1,p}, u_{n,m+1,p+1}, u_{n+1,m+1,p+1}), \\ |\overline{C}(u_{n,m+1,p}, u_{n+1,m+1,p}, u_{n,m+1,p+1}, u_{n+1,m+1,p+1}), \\ C(u_{n,m,p}, u_{n+1,m+1,p}, u_{n,m+1,p+1}, u_{n+1,m+1,p+1}), \\ |\overline{C}(u_{n,m+1,p}, u_{n+1,m+1,p}, u_{n,m+1,p+1}, u_{n+1,m,+1p+1}). \end{cases}$$

This means that the "multicube" of Fig. A.1 appears as the usual consistency cube of Fig. 2.4 with the following identifications:

$$\text{(A.11)} \qquad A \rightsquigarrow \widetilde{A}, \quad \overline{A} \rightsquigarrow \widetilde{\widetilde{A}}, \quad B \rightsquigarrow \widetilde{B}, \quad \bar{B} \rightsquigarrow \widetilde{\widetilde{B}}, \quad C \rightsquigarrow \widetilde{C}, \quad \overline{C} \rightsquigarrow \widetilde{\widetilde{C}}.$$

As an example of this procedure, let us consider again the equation ${}_rH_1^{\pi}$. ${}_rH_1^{\pi}$ has the CaC equations:

$$\text{(A.12a)} \qquad A = (x - x_{12})(x_1 - x_2) + (\alpha_1 - \alpha_2)(1 + \pi x_1 x_2),$$

$$\text{(A.12b)} \qquad \overline{A} = (x_3 - x_{123})(x_{13} - x_{23}) + (\alpha_1 - \alpha_2)(1 + \pi x_3 x_{123}),$$

$$\text{(A.12c)} \qquad B = (x - x_{23})(x_2 - x_3) + (\alpha_2 - \alpha_3)(1 + \pi x_2 x_3),$$

$$\text{(A.12d)} \qquad \overline{B} = (x_1 - x_{123})(x_{12} - x_{13}) + (\alpha_2 - \alpha_3)(1 + \pi x_1 x_{123}),$$

$$\text{(A.12e)} \qquad C = (x - x_{13})(x_1 - x_3) + (\alpha_1 - \alpha_3)(1 + \pi x_1 x_3),$$

$$\text{(A.12f)} \qquad \overline{C} = (x_2 - x_{123})(x_{12} - x_{23}) + (\alpha_1 - \alpha_2)(1 + \pi x_2 x_{123}),$$

therefore from (A.10, A.11) we get the following consistency on the "multicube":

$$\text{(A.13a)} \quad A = \begin{cases} (u_{n,m,p} - u_{n+1,m+1,p})(u_{n+1,m,p} - u_{n,m+1,p}) \\ \quad -(\alpha_1 - \alpha_2)(1 + \pi^2 u_{n+1,m,p} u_{n,m+1,p}), & |n| + |m| = 2k,\ k \in \mathbb{Z}, \\ (u_{n,m,p} - u_{n+1,m+1,p})(u_{n+1,m,p} - u_{n,m+1,p}) \\ \quad -(\alpha_1 - \alpha_2)(1 + \pi^2 u_{n,m,p} u_{n+1,m+1,p}), & |n| + |m| = 2k + 1,\ k \in \mathbb{Z}, \end{cases}$$

$$\text{(A.13b)} \quad \overline{A} = \begin{cases} (u_{n,m,p+1} - u_{n+1,m+1,p+1})(u_{n+1,m,p+1} - u_{n,m+1,p+1}) & |n| + |m| = 2k, \\ \quad -(\alpha_1 - \alpha_2)(1 + \pi^2 u_{n,m,p+1} u_{n+1,m+1,p+1}), & k \in \mathbb{Z}, \\ (u_{n,m,p+1} - u_{n+1,m+1,p+1})(u_{n+1,m,p+1} - u_{n,m+1,p+1}) & |n| + |m| = 2k + 1, \\ \quad -(\alpha_1 - \alpha_2)(1 + \pi^2 u_{n+1,m,p+1} u_{n,m+1,p+1}), & k \in \mathbb{Z}, \end{cases}$$

$$\text{(A.13c)} \quad B = \begin{cases} (u_{n,m,p} - u_{n,m+1,p+1})(u_{n,m+1,p} - u_{n,m,p+1}) \\ \quad -(\alpha_2 - \alpha_3)(1 + \pi^2 u_{n,m+1,p} u_{n,m,p+1}), & |n| + |m| = 2k,\ k \in \mathbb{Z}, \\ (u_{n,m,p} - u_{n,m+1,p+1})(u_{n,m+1,p} - u_{n,m,p+1}) \\ \quad -(\alpha_2 - \alpha_3)(1 + \pi^2 u_{n,m,p} u_{n,m+1,p+1}), & |n| + |m| = 2k + 1,\ k \in \mathbb{Z}, \end{cases}$$

$$\text{(A.13d)}\quad \bar{B}=\begin{cases}(u_{n+1,m,p}-u_{n,m+1,p+1})(u_{n+1,m+1,p}-u_{n+1,m,p+1}) & |n|+|m|=2k,\\ \quad -(\alpha_2-\alpha_3)(1+\pi^2 u_{n+1,m,p}u_{n+1,m+1,p+1}, & k\in\mathbb{Z},\\ (u_{n+1,m,p}-u_{n+1,m+1,p+1})(u_{n+1,m+1,p}-u_{n+1,m,p+1}) & |n|+|m|=2k+1,\\ \quad -(\alpha_2-\alpha_3)(1+\pi^2 u_{n+1,m+1,p}u_{n+1,m,p+1}), & k\in\mathbb{Z},\end{cases}$$

$$\text{(A.13e)}\quad C=\begin{cases}(u_{n,m,p}-u_{n+1,m,p+1})(u_{n+1,m,p}-u_{n,m,p+1}) & \\ \quad -(\alpha_1-\alpha_3)(1+\pi^2 u_{n+1,m,p}u_{n,m,p+1}), & |n|+|m|=2k,\ k\in\mathbb{Z},\\ (u_{n,m,p}-u_{n+1,m,p+1})(u_{n+1,m,p}-u_{n,m,p+1}) & \\ \quad -(\alpha_1-\alpha_3)(1+\pi^2 u_{n,m,p}u_{n+1,m,p+1}), & |n|+|m|=2k+1,\ k\in\mathbb{Z},\end{cases}$$

$$\text{(A.13f)}\quad \bar{C}=\begin{cases}(u_{n,m+1,p}-u_{n+1,m+1,p+1})(u_{n+1,m+1,p}-u_{n,m+1,p+1}) & |n|+|m|=2k,\\ \quad -(\alpha_1-\alpha_3)(1+\pi^2 u_{n,m+1,p}u_{n+1,m+1,p+1}), & k\in\mathbb{Z},\\ (u_{n,m+1,p}-u_{n+1,m+1,p+1})(u_{n+1,m+1,p}-u_{n,m+1,p+1}) & |n|+|m|=2k+1,\\ \quad -(\alpha_1-\alpha_2)(1+\pi^2 u_{n+1,m+1,p}u_{n,m+1,p+1}), & k\in\mathbb{Z},\end{cases}$$

Up to now we showed how, given a CaC quad equation Q, it is possible to embed it into a PΔE in $\mathbb{Z}^2$ given by (A.7). Furthermore we showed that this procedure can be extended along the third dimension in such a way that the consistency is preserved. This have been done following [112] and filling the details (which are going to be important).

The PΔE (A.7) is not very manageable since we have to change equation according to the point of the lattice we are in. It will be more efficient to have an expression which "knows" by itself in which point we are. This can obtained by going over to non autonomous PΔEs as was done in the BW lattice case [839].

We shall present here briefly how from (A.7) it is possible to construct an equivalent non autonomous system, and moreover how to construct the non autonomous version of CaC (A.10).

We take an equation $\hat{Q}$ constructed by a linear combination of (2.4.146) with n and m depending coefficients:

$$\text{(A.14)}\qquad \begin{aligned}\hat{Q}=&\, f_{n,m}\,Q(u_{n,m},u_{n+1,m},u_{n,m+1},u_{n+1,m+1})+\\ &+|f_{n,m}\,|Q(u_{n,m},u_{n+1,m},u_{n,m+1},u_{n+1,m+1})+\\ &+\underline{f}_{n,m}\,\underline{Q}(u_{n,m},u_{n+1,m},u_{n,m+1},u_{n+1,m+1})+\\ &+|\underline{f}_{n,m}\,|\underline{Q}(u_{n,m},u_{n+1,m},u_{n,m+1},u_{n+1,m+1}).\end{aligned}$$

We require that it satisfies the following conditions:

(1) The coefficients are periodic of period 2 in both directions, since, in the $\mathbb{Z}^2$ embedding, the elementary cell is a 2×2 one.
(2) The coefficients are such that they produce the right equation in a given lattice point as specified in (A.7).

Condition 1 implies that any function $\tilde{f}_{n,m}$ in (A.14), i.e. either $f_{n,m}$ or $|f_{n,m}$ or $\underline{f}_{n,m}$ or $|\underline{f}_{n,m}$, solves the two ordinary difference equations:

$$\text{(A.15)}\qquad \tilde{f}_{n+2,m}-\tilde{f}_{n,m}=0,\quad \tilde{f}_{n,m+2}-\tilde{f}_{n,m}=0,$$

whose solution is:

$$\tilde{f}_{n,m} = c_0 + c_1 (-1)^n + c_2 (-1)^m + c_3 (-1)^{n+m} \tag{A.16}$$

with c_i constants to be determined.

The condition 2 depends on the choice of the equation in (A.7) and will give some "boundary conditions" for the function $\tilde{f}$, allowing us to fix the coefficients c_i. For $f_{n,m}$, for example, we have, substituting the appropriate lattice points, the following conditions:

$$f_{2k,2k} = 1, \quad f_{2k+1,2k} = f_{2k,2k+1} = f_{2k+1,2k+1} = 0, \tag{A.17}$$

which yield for $f_{n,m}$, $|f_{n,m}$, $\underline{f}_{n,m}$ and $|\underline{f}_{n,m}$ the definitions given in (2.4.158).: Then inserting (2.4.158) in (A.14) we obtain a non autonomous PΔE which corresponds to (A.7). If the quad-graph equation Q possesses some discrete symmetries, the expression (A.14) greatly simplify. If an equation Q is invariant under the discrete group D_4 we trivially have, using (2.4.149), $\hat{Q} = Q$. This result states that an equation with the symmetry (2.4.149) is defined on a monochromatic lattice, as expected since we are in the case of the ABS classification [22]. If the equation Q has the symmetries of the *rhombus*, namely (2.4.151), we get:

$$\hat{Q} = (f_{n,m} + |\underline{f}_{n,m})Q + (|f_{n,m} + \underline{f}_{n,m})|Q \tag{A.18}$$

and using (2.4.158):

$$f_{n,m} + |\underline{f}_{n,m} = \chi_{n+m}, \quad |f_{n,m} + \underline{f}_{n,m} = \chi_{n+m+1}, \tag{A.19}$$

where (2.4.148) is taken into account. This obviosly match with the results in [839].

In the case of *trapezoidal* symmetry (2.4.153) one obtains:

$$\hat{Q} = (f_{n,m} + |f_{n,m})Q + (\underline{f}_{n,m} + |\underline{f}_{n,m})\underline{Q}, \tag{A.20}$$

with

$$f_{n,m} + |f_{n,m} = \chi_m, \quad \underline{f}_{n,m} + |\underline{f}_{n,m} = \chi_{m+1}, \tag{A.21}$$

As an example of such construction let us consider again ${}_rH_1^{\pi}$ (A.1). Since we are in the rhombic case [839] we use formula (A.19) and get:

$$\begin{aligned} {}_r\hat{H}_1^{\pi} &= (u_{n,m} - u_{n+1,m+1})(u_{n+1,m} - u_{n,m+1}) - (\alpha_1 - \alpha_2) \\ &+ (\alpha_1 - \alpha_2)\pi^2 \left(\chi_{n+m}\, u_{n+1,m}u_{n,m+1} + \chi_{n+m+1}\, u_{n,m}u_{n+1,m+1}\right) = 0, \end{aligned} \tag{A.22}$$

which corresponds to the case $\sigma = 1$ of [839] (the discussion of the meaning of parameter σ in [839] is postponed to the end of this Appendix).

The consistency of a generic system of quad equations is obtained by considering the consistency of the tilded equations as displayed in (A.11). We now construct, starting from (A.13), the non autonomous PΔEs in the (n, m) variables using the weights $\tilde{f}_{n,m}$, as given in (A.14), applied to the relevant equations. Carrying out such construction, we end with the following sextuple of equations:

$$\widehat{A}(u_{n,m,p}, u_{n+1,m,p}, u_{n,m+1,p}, u_{n+1,m+1,p}) = f_{n,m}A + |f_{n,m}|A + \underline{f}_{n,m}\underline{A} + |\underline{f}_{n,m}|\underline{A} = 0, \tag{A.23a}$$

$$\widehat{\widehat{A}}(u_{n,m,p+1}, u_{n+1,m,p+1}, u_{n,m+1,p+1}, u_{n+1,m+1,p+1}) = f_{n,m}\bar{A} + |f_{n,m}|\bar{A} + \underline{f}_{n,m}\underline{\bar{A}} + |\underline{f}_{n,m}|\underline{\bar{A}} = 0, \tag{A.23b}$$

$$\widehat{B}(u_{n,m,p}, u_{n,m+1,p}, u_{n,m,p+1}, u_{n,m+1,p+1}) = f_{n,m}B + |f_{n,m}|\bar{B} + \underline{f}_{n,m}\underline{B} + |\underline{f}_{n,m}|\underline{\bar{B}} = 0, \tag{A.23c}$$

$$\widehat{\widehat{B}}(u_{n+1,m,p}, u_{n+1,m+1,p}, u_{n+1,m,p+1}, u_{n+1,m+1,p+1}) = f_{n,m}\bar{B} + |f_{n,m}|B + \underline{f}_{n,m}\underline{\bar{B}} + |\underline{f}_{n,m}|\underline{B} = 0, \tag{A.23d}$$

$$\widehat{C}(u_{n,m,p}, u_{n+1,m,p}, u_{n,m,p+1}, u_{n+1,m,p+1}) = f_{n,m}C + |f_{n,m}|C + \underline{f}_{n,m}\bar{C} + |\underline{f}_{n,m}|\bar{C} = 0, \tag{A.23e}$$

$$\widehat{\widehat{C}}(u_{n,m+1,p}, u_{n+1,m,p}, u_{n,m+1,p+1}, u_{n+1,m+1,p+1}) = f_{n,m}\bar{C} + |f_{n,m}|\bar{C} + \underline{f}_{n,m}C + |\underline{f}_{n,m}|C = 0, \tag{A.23f}$$

where all the functions on the right hand side of the equality sign are evaluated on the point indicated on the left hand side.

We note that a Lax pair obtained by making use of (A.23) will be effectively a pair, since $(\widehat{B}, \widehat{\widehat{B}})$ and $(\widehat{C}, \widehat{\widehat{C}})$ are related by translation so they are just two different solutions of the same equation. Indeed by using the properties of the functions $\tilde{f}_{n,m}$ we have:

$$\widehat{\widehat{B}} = S_n\widehat{B}, \quad \widehat{\widehat{C}} = S_m\widehat{C}, \tag{A.24}$$

where S_n is the operator of translation in the n direction, and S_m the operator of translation in the m direction. This allows us to construct Bäcklund transformations and Lax pair in the usual way[104, 644].

As a final example we shall derive the non autonomous side equations for H_1^π and its Lax pair. We will then confront the result with that obtained in [839]. Considering (A.12, A.23, A.24) and using the fact that the equation is rhombic (A.19) we get:

$$\widehat{A} = (u_{n,m,p} - u_{n+1,m+1,p})(u_{n+1,m,p} - u_{n,m+1,p}) - (\alpha_1 - \alpha_2)\cdot \cdot \left[1 + \pi^2\left(\chi_{n+m}\, u_{n+1,m,p}u_{n,m+1,p} + \chi_{n+m+1}\, u_{n,m,p}u_{n+1,m+1,p}\right)\right] = 0, \tag{A.25a}$$

$$\widehat{\widehat{A}} = (u_{n,m,p+1} - u_{n+1,m+1,p+1})(u_{n+1,m,p+1} - u_{n,m+1,p+1}) - (\alpha_1 - \alpha_2)\cdot \cdot \left[1 + \pi^2\left(\chi_{n+m+1}\, u_{n+1,m,p+1}u_{n,m+1,p+1} + \chi_{n+m}\, u_{n,m,p+1}u_{n+1,m+1,p+1}\right)\right] = 0, \tag{A.25b}$$

$$\widehat{B} = (u_{n,m,p} - u_{n,m+1,p+1})(u_{n,m+1,p} - u_{n,m,p+1}) - (\alpha_2 - \alpha_3)\cdot \cdot \left[1 + \pi^2\left(\chi_{n+m}\, u_{n,m+1,p}u_{n,m,p+1} + \chi_{n+m+1}\, u_{n,m,p}u_{n,m+1,p+1}\right)\right] = 0, \tag{A.25c}$$

(A.25d) $$\widehat{C} = (u_{n,m,p} - u_{n+1,m,p+1})(u_{n+1,m,p} - u_{n,m,p+1}) - (\alpha_1 - \alpha_3)\cdot \cdot \left[1 + \pi^2 \left(\chi_{n+m}\, u_{n+1,m,p}u_{n,m,p+1} + \chi_{n+m+1}\, u_{n,m,p}u_{n+1,m,p+1}\right)\right] = 0,$$

From the equations $\widehat{B}$ and $\widehat{C}$ we find, up to a sign and common factor, which we can eliminate since the Lax pair is defined from CaC only up to projective equivalence [119, 388], the following Lax pair:

(A.26a) $$L = \begin{pmatrix} u_{n,m} & \alpha_1 - \alpha_3 - u_{n,m}u_{n+1,m} \\ 1 & -u_{n+1,m} \end{pmatrix} + (\alpha_1 - \alpha_3)\pi^2 \begin{pmatrix} \chi_{n+m}u_{n+1,m} & 0 \\ 0 & -\chi_{n+m+1}u_{n,m} \end{pmatrix}$$

(A.26b) $$M = \begin{pmatrix} u_{n,m} & \alpha_2 - \alpha_3 - u_{n,m}u_{n,m+1} \\ 1 & -u_{n,m+1} \end{pmatrix} + \left(\alpha_2 - \alpha_3\right)\pi^2 \begin{pmatrix} \chi_{n+m}u_{n,m+1} & 0 \\ 0 & -\chi_{n+m+1}u_{n,m} \end{pmatrix},$$

(A.26c) $$\overline{L} = \begin{pmatrix} u_{n,m+1} & \alpha_1 - \alpha_3 - u_{n,m+1}u_{n+1,m+1} \\ 1 & -u_{n+1,m+1} \end{pmatrix} + (\alpha_1 - \alpha_3)\pi^2 \begin{pmatrix} \chi_{n+m+1}u_{n+1,m+1} & 0 \\ 0 & -\chi_{n+m}u_{n,m+1} \end{pmatrix}$$

(A.26d) $$\overline{M} = \begin{pmatrix} u_{n+1,m} & \alpha_2 - \alpha_3 - u_{n+1,m}u_{n+1,m+1} \\ 1 & -u_{n+1,m+1} \end{pmatrix} + \left(\alpha_2 - \alpha_3\right)\pi^2 \begin{pmatrix} \chi_{n+m+1}u_{n+1,m+1} & 0 \\ 0 & -\chi_{n+m}u_{n+1,m} \end{pmatrix}.$$

This is a Lax pair since $\overline{L} = S_m L$ and $\overline{M} = S_n M$. We find that such matrices give as compatibility (A.1) and correspond to the following proportionality factor τ [119]:

(A.27) $$\tau = \frac{1 + \pi^2 \left(\chi_{n+m+1}u_{n,m} + \chi_{n+m}u_{n+1,m}\right)}{1 + \pi^2 \left(\chi_{n+m+1}u_{n,m} + \chi_{n+m}u_{n,m+1}\right)}.$$

The Lax pair (A.26) is Gauge equivalent to that obtained in [839] with gauge:

(A.28) $$G = \begin{pmatrix} 0 & 1 \\ -1 & 0 \end{pmatrix}.$$

A similar calculation could be done for the other two rhombic equations, and, up to gauge transformations, will give the same result. Indeed the gauge transformations (A.28) is needed for H_1^π and H_2^π whereas for H_3^π we need the gauge:

(A.29) $$\tilde{G} = \begin{pmatrix} 0 & (-1)^{n+m} \\ -(-1)^{n+m} & 0 \end{pmatrix}.$$

Remark 11. *At the level of the non autonomous equations the choice of origin of $\mathbb{Z}^2$ in a point different from x in* (A.7) *would have led to different initial conditions in* (A.17), *which ultimately lead to the following form for the functions f:*

$$f_{n,m}^{\sigma_1,\sigma_2} = \frac{1+\sigma_1(-1)^n+\sigma_2(-1)^m+\sigma_1\sigma_2(-1)^{n+m}}{4}, \tag{A.30a}$$

$$|f_{n,m}^{\sigma_1,\sigma_2} = \frac{1-\sigma_1(-1)^n+\sigma_2(-1)^m-\sigma_1\sigma_2(-1)^{n+m}}{4}, \tag{A.30b}$$

$$\underline{f}_{n,m}^{\sigma_1,\sigma_2} = \frac{1+\sigma_1(-1)^n-\sigma_2(-1)^m-\sigma_1\sigma_2(-1)^{n+m}}{4}, \tag{A.30c}$$

$$|\underline{f}_{n,m}^{\sigma_1,\sigma_2} = \frac{1-\sigma_1(-1)^n-\sigma_2(-1)^m+\sigma_1\sigma_2(-1)^{n+m}}{4}, \tag{A.30d}$$

where the two constants $\sigma_i \in \{ \pm 1 \}$ depend on the point chosen. Indeed if the point is x we have $\sigma_1 = \sigma_2 = 1$, whereas if we choose x_1 we have $\sigma_1 = 1$, $\sigma_2 = -1$, if we choose x_2 then $\sigma_1 = -1$ and $\sigma_2 = 1$ and finally if we choose x_{12} we shall put $\sigma_1 = \sigma_2 = -1$. It is easy to see that in the rhombic and in the trapezoidal case (A.30) *collapse to the σ version of the functions χ_k as given by* (2.4.148)*:*

$$\chi_k^{(\sigma)} = \frac{1+\sigma(-1)^k}{2}. \tag{A.31}$$

The final equation will then depend on σ_1 or σ_2, only if rhombic or trapezoidal.

It was proved in [839] *that the transformations:*

$$v_{n,m} = u_{n+1,m}, \quad w_{n,m} = u_{n,m+1}, \tag{A.32}$$

map a rhombic equation with a certain σ into the same rhombic equation with $-\sigma$. In [839] *this fact was used to construct a Lax pair and Bäcklund transformations. An analogous result can be easily proven for trapezoidal equations: using the transformation $w_{n,m} = u_{n,m+1}$ we can send a trapezoidal equation with a certain σ into the same equation with $-\sigma$. A similar trasformation in the n direction would just trivally leave invariant the trapezoidal equation, since there is no explicit dependence on n. However in general, if an equation does not possess discrete symmetries, as it is the case for a H^6 equation, no trasformation like* (A.32) *would take the equation into itself with different coefficents. We can anyway construct a Lax pair with the procedure explained above, which is then slightly more general than the approach based on* (A.32).

APPENDIX B

Transformation groups for quad lattice equations.

Here in the following we present the extension of the Möbius transformations necessary to treat the equations we obtain when constructing quad graph equationsconsistent on an extended four color cube presented in Fig. A.1.

The classification of quad equations presented in Section 2.4.7.1, has been carried out up to a Möbius transformations in each vertex:

$$M:\ \left(x,x_1,x_2,x_{12}\right)\mapsto\left(\frac{a_0x+b_0}{c_0x+d_0},\frac{a_1x_1+b_1}{c_1x_1+d_1},\frac{a_2x_2+b_2}{c_2x_2+d_2},\frac{a_{12}x_{12}+b_{12}}{c_{12}x_{12}+d_{12}}\right). \tag{B.1}$$

As in the usual Möbius transformation we have here $(a_i,b_i,c_i,d_i)\in\mathbb{CP}^4\backslash V\left(a_id_i-b_ic_i\right)\simeq$ PGL$(2,\mathbb{C})$ with $i=0,1,2,12$, i.e. each set of parameters is defined up to to a multiplication by a number. Obviously as the usual Möbius transformations these transformations will form a group under composition and we shall call such group (Möb)^4.

On the other hand when dealing with equations defined on the lattice we have to follow the prescription of Appendix A and use the representation given by (A.14), i.e. we will have non autonomous lattice equations. Here we prove the following theorem which extends the group (Möb)^4 to the level of the transformations of the non autonomous lattice equations. We will call such group the "non autonomous lifting" of (Möb)^4, and denote it by $\widehat{(\text{Möb})}^4$.

Theorem 93. $\widehat{(\text{Möb})}^4$ *is the symmetry group of the non autonomous PΔEs obtained with the procedure presented in Appendix* A *by the "non autonomous lifting" of* (Möb)^4, *the symmetry group of the quad-graph equations presented in Section* 2.4.7.1.

This theorem states that the result that we get by acting with a group of transformations $\widehat{(\text{Möb})}^4$ is the same of what we obtain if we first transform the equations of Section

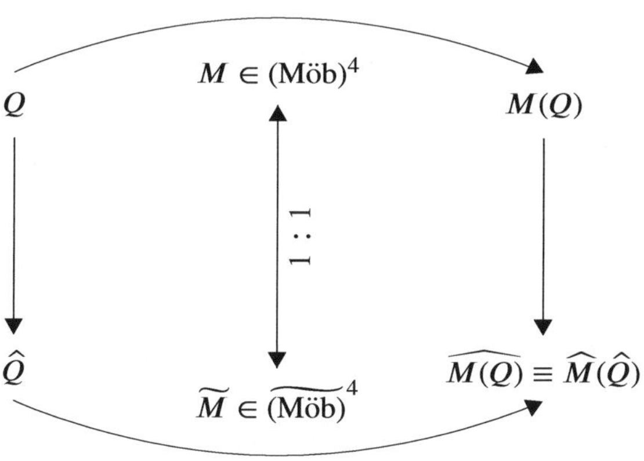

FIGURE B.1. The commutative diagram defining $\widehat{(\text{Möb})}^4$ [reprinted from [336]].

2.4.7.1 using $\boldsymbol{M} \in (\text{Möb})^4$ and then we construct the non autonomous quad-graph equation with the prescription of Appendix A or viceversa if we first construct the non autonomous equation and then we transform it using the "non autonomous lifting" of $\boldsymbol{M}$, $\widehat{\boldsymbol{M}}$, see Fig. B.1.

The Theorem 93 is made up of two parts, the proof that $\widehat{(\text{Möb})}^4$ is a group and that it is equivalent to the "non autonomous lifting" of $(\text{Möb})^4$.

Let us first construct, a transformation which will be the candidate to be the "non autonomous lifting" of $\boldsymbol{M} \in (\text{Möb})^4$. Given $\boldsymbol{M} \in (\text{Möb})^4$ using the same ideas of Appendix A we can construct the following transformation:

$$\text{(B.2)} \qquad \boldsymbol{M}_{n,m} \in \widehat{(\text{Möb})}^4 : u_{n,m} \mapsto f_{n,m}\frac{a_0 u_{n,m} + b_0}{c_0 u_{n,m} + d_0} + |f_{n,m}\frac{a_1 u_{n,m} + b_1}{c_1 u_{n,m} + d_1} + \underline{f}_{n,m}\frac{a_2 u_{n,m} + b_2}{c_2 u_{n,m} + d_2} + |\underline{f}_{n,m}\frac{a_{12} u_{n,m} + b_{12}}{c_{12} u_{n,m} + d_{12}}.$$

Eq. (B.2) give us a mapping Φ between the group $(\text{Möb})^4$ and a set of non autonomous transformations of the field $u_{n,m}$. Moreover there is a one to one correspondence between an element $\boldsymbol{M} \in (\text{Möb})^4$ (B.1) and one $\boldsymbol{M}_{n,m} \in \widehat{(\text{Möb})}^4$ (B.2). Φ is given by:

$$\text{(B.3a)} \qquad \Phi : \begin{pmatrix} \dfrac{ax+b}{cx+d} \\ \dfrac{a_1 x_1 + b_1}{c_1 x_1 + d_1} \\ \dfrac{a_2 x_2 + b_2}{c_2 x_2 + d_2} \\ \dfrac{a_{12} x_{12} + b_{12}}{c_{12} x_{12} + d_{12}} \end{pmatrix}^T \longmapsto \begin{matrix} f_{n,m}\dfrac{a u_{n,m} + b}{c u_{n,m} + d} + \\ |f_{n,m}\dfrac{a_1 u_{n,m} + b_1}{c_1 u_{n,m} + d_1} + \\ \underline{f}_{n,m}\dfrac{a_2 u_{n,m} + b_2}{c_2 u_{n,m} + d_2} + \\ |\underline{f}_{n,m}\dfrac{a_{12} u_{n,m} + b_{12}}{c_{12} u_{n,m} + d_{12}} \end{matrix},$$

and its inverse

(B.3b)

$$\text{(B.3c)} \qquad \Phi^{-1} : \begin{matrix} f_{n,m}\dfrac{\alpha^{(0)} u_{n,m} + \beta^{(0)}}{\gamma^{(0)} u_{n,m} + \delta^{(0)}} + \\ |f_{n,m}\dfrac{\alpha^{(1)} u_{n,m} + \beta^{(1)}}{\gamma^{(1)} u_{n,m} + \delta^{(1)}} + \\ \underline{f}_{n,m}\dfrac{\alpha^{(2)} u_{n,m} + \beta^{(2)}}{\gamma^{(2)} u_{n,m} + \delta^{(2)}} + \\ |\underline{f}_{n,m}\dfrac{\alpha^{(3)} u_{n,m} + \beta^{(3)}}{\gamma^{(3)} u_{n,m} + \delta^{(3)}} \end{matrix} \longmapsto \begin{pmatrix} \dfrac{\alpha^{(0)} x + \beta^{(0)}}{\gamma^{(0)} x + \delta^{(0)}} \\ \dfrac{\alpha^{(1)} x_1 + \beta^{(1)}}{\gamma^{(1)} x_1 + \delta^{(1)}} \\ \dfrac{\alpha^{(2)} x_2 + \beta^{(2)}}{\gamma^{(2)} x_2 + \delta^{(2)}} \\ \dfrac{\alpha^{(3)} x_{12} + \beta^{(3)}}{\gamma^{(3)} x_{12} + \delta^{(3)}} \end{pmatrix}^T.$$

We have now to prove that $\widehat{(\text{Möb})}^4$ is a group and that the mapping (B.3) is actually a group homomorphism.

$\widehat{(\text{Möb})}^4$ is a subset of the general non autonomous Möbius transformation:

$$\text{(B.4)} \qquad W_{n,m} : u_{n,m} \mapsto \frac{a_{n,m} u_{n,m} + b_{n,m}}{c_{n,m} u_{n,m} + d_{n,m}}.$$

From the general rule of composition of two Möbius transformations (B.4) we get:

$$W^1_{n,m}\left(W^0_{n,m}\left(u_{n,m}\right)\right) = \frac{(a^0_{n,m}a^1_{n,m} + b^1_{n,m}c^0_{n,m})u_{n,m} + a^1_{n,m}b^0_{n,m} + b^1_{n,m}d^0_{n,m}}{(a^0_{n,m}c^1_{n,m} + c^0_{n,m}d^1_{n,m})u_{n,m} + b^0_{n,m}c^1_{n,m} + d^0_{n,m}d^1_{n,m}}. \tag{B.5}$$

Its inverse is given by

$$W^{-1}_{n,m}(u_{n,m}) = \frac{d_{n,m}u_{n,m} - b_{n,m}}{-c_{n,m}u_{n,m} + a_{n,m}}, \tag{B.6}$$

Using the computational rules given in Table B.1 and in (B.5) we find that the composition of two elements $M^{(1)}_{n,m}, M^{(2)}_{n,m} \in \widehat{(\text{Möb})}^4$ with parameters $(a^{(i)}_j, b^{(i)}_j, c^{(i)}_j, d^{(i)}_j)$, $j = 0, 1, 2, 3$ and $i = 1, 2$, gives:

$$M^2_{n,m}\left(M^1_{n,m}\left(u_{n,m}\right)\right) = \frac{\left\{\begin{array}{r} \left[f_{n,m}(a^{(0)}_1 a^{(0)}_2 + b^{(0)}_2 c^{(0)}_1)+\right.\\ \mathrm{l}f_{n,m}(a^{(1)}_1 a^{(1)}_2 + b^{(1)}_2 c^{(1)}_1)+\\ \underline{f}_{n,m}(a^{(2)}_1 a^{(2)}_2 + b^{(2)}_2 c^{(2)}_1)+\\ \left.\mathrm{l}\underline{f}_{n,m}(a^{(3)}_1 a^{(3)}_2 + b^{(3)}_2 c^{(3)}_1)\right]u_{n,m}+\\ +\left[f_{n,m}(b^{(0)}_1 a^{(0)}_2 + d^{(0)}_1 b^{(0)}_2)+\right.\\ +\mathrm{l}f_{n,m}(b^{(1)}_1 a^{(1)}_2 + d^{(1)}_1 b^{(1)}_2)+\\ +\underline{f}_{n,m}(b^{(2)}_1 a^{(2)}_2 + d^{(2)}_1 b^{(2)}_2)+\\ \left.+\mathrm{l}\underline{f}_{n,m}(b^{(3)}_1 a^{(3)}_2 + d^{(3)}_1 b^{(3)}_2)\right] \end{array}\right\}}{\left\{\begin{array}{r} \left[f_{n,m}(a^{(0)}_1 c^{(0)}_2 + c^{(0)}_1 d^{(0)}_1)+\right.\\ \mathrm{l}f_{n,m}(a^{(1)}_1 c^{(1)}_2 + c^{(1)}_1 d^{(1)}_1)+\\ \underline{f}_{n,m}(a^{(2)}_1 c^{(2)}_2 + c^{(2)}_1 d^{(2)}_1)+\\ \left.\mathrm{l}\underline{f}_{n,m}(a^{(3)}_1 c^{(3)}_2 + c^{(3)}_1 d^{(3)}_1)\right]u_{n,m}+\\ \left[f_{n,m}(b^{(0)}_1 c^{(0)}_2 + d^{(0)}_1 d^{(0)}_2)+\right.\\ \mathrm{l}f_{n,m}(b^{(1)}_1 c^{(1)}_2 + d^{(1)}_1 d^{(1)}_2)+\\ \underline{f}_{n,m}(b^{(2)}_1 c^{(2)}_2 + d^{(2)}_1 d^{(2)}_2)+\\ \left.\mathrm{l}\underline{f}_{n,m}(b^{(3)}_1 c^{(3)}_2 + d^{(3)}_1 d^{(3)}_2)\right] \end{array}\right\}}. \tag{B.7}$$

Its inverse is given by

$$M^{-1}_{n,m}(u_{n,m}) = \frac{\left\{\begin{array}{l} (f_{n,m}d^{(0)} + \mathrm{l}f_{n,m}d_1 + \underline{f}_{n,m}d_2 + \mathrm{l}\underline{f}_{n,m}d_3)u_{n,m}-\\ (f_{n,m}b_0 + \mathrm{l}f_{n,m}b_1 + \underline{f}_{n,m}b_2 + \mathrm{l}\underline{f}_{n,m}b_3) \end{array}\right\}}{\left\{\begin{array}{l} -(f_{n,m}c_0 + \mathrm{l}f_{n,m}c_1 + \underline{f}_{n,m}c_2 + \mathrm{l}\underline{f}_{n,m}c_3)u_{n,m}+\\ (f_{n,m}a_0 + \mathrm{l}f_{n,m}a_1 + \underline{f}_{n,m}a_2 + \mathrm{l}\underline{f}_{n,m}a_3) \end{array}\right\}}. \tag{B.8}$$

Thus one has proven that $\widehat{(\text{Möb})}^4$ is a group.

$\cdot$	$f_{n,m}$	$\vert f_{n,m}$	$\underline{f}_{n,m}$	$\vert\underline{f}_{n,m}$
$f_{n,m}$	$f_{n,m}$	0	0	0
$\vert f_{n,m}$	0	$\vert f_{n,m}$	0	0
$\underline{f}_{n,m}$	0	0	$\underline{f}_{n,m}$	0
$\vert\underline{f}_{n,m}$	0	0	0	$\vert\underline{f}_{n,m}$

TABLE B.1. Multiplication rules for the functions $\tilde{f}_{n,m}$ as given by (2.4.158) [reprinted from [336]].

Let us show that the maps Φ and Φ^{-1} in (B.3) are group homomorphism. They preserve the identity, and from the formula of composition of Möbius transformations in $(\text{Möb})^4$ (B.7) we derive the required result.

Let us now check if the diagram of Fig. B.1 is satisfied. Let us consider (B.1) and (B.2) and a general multilinear quad-graph equation:

$$\begin{aligned} Q_{\text{gen}}\left(x, x_1, x_2, x_{12}\right) = {} & A_{0,1,2,12} x x_1 x_2 x_{12} + B_{0,1,2} x x_1 x_2 + \\ & + B_{0,1,12} x x_1 x_{12} + B_{0,2,12} x x_2 x_{12} + B_{1,2,12} x_1 x_2 x_{12} \\ & + C_{0,1} x x_1 + C_{0,2} x x_2 + C_{0,12} x x_{12} + C_{1,2} x_1 x_2 \\ & + C_{1,12} x_1 x_{12} + C_{2,12} x_2 x_{12} \\ & + D_0 x + D_1 x_1 + D_2 x_2 + D_{12} x_{12} + K \end{aligned} \tag{B.9}$$

where $A_{0,1,2,12}$, $B_{i,j,k}$, $C_{i,j}$, D_i and K with $i, j, k \in \{0, 1, 2, 12\}$ are arbitrary complex constants. The proof that $\widehat{Q(M(x, x_1, x_2, x_{12}))} = \widehat{Q}(M_{n,m}(u_{n,m}))$ where $M \in (\text{Möb})^4$ and $M_{n,m} = \Phi(M) \in \widehat{(\text{Möb})}^4$ is a very computationally heavy calculation due to the high number of parameters involved (twelve in the transformation[1] and fifteen in the equation (B.9), twenty-seven parameters in total) and to the fact that rational functions are involved. To simplify the problem it is sufficient to recall that every Möbius

$$m(z) = \frac{az + b}{cz + d}, \quad z \in \mathbb{C} \tag{B.10}$$

transformation can be obtained as a superposition of a translation:

$$T_a(z) = z + a, \tag{B.11a}$$

dilatation:

$$D_a(z) = az \tag{B.11b}$$

and *inversion*:

$$I(z) = \frac{1}{z}, \tag{B.11c}$$

i.e.

$$m(z) = \left(T_{a/c} \circ D_{(bc-ad)/c^2} \circ I \circ T_{d/c}\right)(z). \tag{B.12}$$

[1]Using that Möbius transformations are projectively defined one can lower the number of parameters from sixteen to twelve, but this implies to impose that some parameters are non-zero and thus all various different possibilities must be taken into account.

As the group $(\text{Möb})^4$ is obtained by four copies of the Möbius group each acting on a different variable we can decompose each entry $M \in (\text{Möb})^4$ as in (B.12). Therefore we need to check $3^4 = 81$ transformations, depending at most on four parameters. We can automatize such proof by making a specific computer program to generate all the possible fundamental transformations in $(\text{Möb})^4$ and then check them one by one reducing the computational effort. To this end we used the computer algebra system `SymPy`[787]. This ends the proof of the theorem. □

The details of the calculations are contained in [336].

APPENDIX C

Algebraic entropy of the non autonomous Boll equations

1. Algebraic entropy test for H^4 and H^6 trapezoidal equations

Algebraic entropy is used as a test of integrability for discrete systems. It measures the degree of growth of the iterates of a rational map [801, 814]. The classification of lattice equations based on the algebraic entropy test [387] is:

Linear growth: The equation is linearizable.

Polynomial growth: The equation is integrable.

Exponential growth: The equation is chaotic.

We have performed the algebraic entropy analysis in the principal growth directions [814] as shown in Figure C.1 on all non autonomous equations presented in Section 2.4.7.2 to identify their behaviour. To this end we used the SymPy [787] module ae2d.py [336]. We found that a non autonomous equation, at difference from what it is assumed in [814], may have non-constant degrees upon the diagonals. This fact implies that the equations of Section 2.4.7.2 do not have in general a single sequence of degrees.

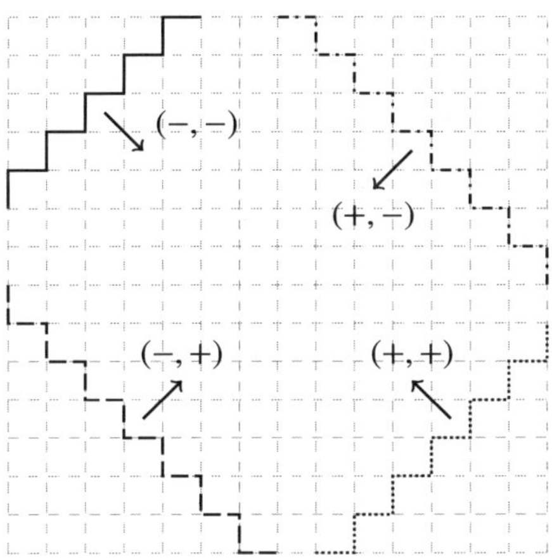

FIGURE C.1. Principal growth directions [reprinted from [336]].

Let us explain this result in details. Let us suppose that we wish to calculate the algebraic entropy in the North-East direction $(-,+)$. We should then solve our equation with respect to $u_{n+1,m+1}$ and then iteratively find the degree of the map in the $(-,+)$ direction. According to [814] one obtains a matrix, the evolution matrix, where in each element we put the degree d_k of the corresponding k lattice point. In the evolution matrix the degrees along the diagonals are assumed to be all equal. Then the evolution matrix will have the form:

$$
\begin{array}{cccccc}
1 & d_2 & d_3 & d_4 & d_5 & d_6 \\
1 & 1 & d_2 & d_3 & d_4 & d_5 \\
 & 1 & 1 & d_2 & d_3 & d_4 \\
 & & 1 & 1 & d_2 & d_3 \\
 & & & 1 & 1 & d_2 \\
 & & & & 1 & 1
\end{array}
\tag{C.1}
$$

The algebraic entropy η is given by (2.4.134). Here we obtain experimentally that the degrees along the diagonals are not the same. This means that the actual evolution matrix will have the form:

$$
\begin{array}{cccccc}
1 & d_2^{(5)} & d_3^{(4)} & d_4^{(3)} & d_5^{(2)} & d_6^{(1)} \\
1 & 1 & d_2^{(4)} & d_3^{(3)} & d_4^{(2)} & d_5^{(1)} \\
 & 1 & 1 & d_2^{(3)} & d_3^{(2)} & d_4^{(1)} \\
 & & 1 & 1 & d_2^{(2)} & d_3^{(1)} \\
 & & & 1 & 1 & d_2^{(1)} \\
 & & & & 1 & 1
\end{array}
\tag{C.2}
$$

where by $d_k^{(i)}$ we mean the degree of the iterate map proper to the i-th column of the evolution matrix. This is obviously a more general case than (C.1). In the framework of [814] the sequence of degrees will have the form (C.1) but in our case we get (C.2). We shall call the sequence

$$
1, d_2^{(1)}, d_3^{(1)}, d_4^{(1)}, d_5^{(1)}, \ldots
\tag{C.3}
$$

the principal sequence of growth. A sequence as

$$
1, d_2^{(i)}, d_3^{(i)}, d_4^{(i)}, d_5^{(i)}, \ldots
\tag{C.4}
$$

with $i = 2$ will be a secondary sequence of growth, for $i = 3$ a *third*, and so on. In principle any sequence (C.4) will define an entropy $\eta^{(i)}$:

$$
\eta^{(i)} = \lim_{k\to\infty} \frac{1}{k} \log d_k^{(i)}.
\tag{C.5}
$$

To any sequence (C.4) we can associate a generating function, i.e. the function $g^{(i)}(s)$ such that the coefficients of its Taylor expansion are as near as possible to the $d_k^{(i)}$:

$$
g^{(i)}(s) = \sum_{k=0}^{\infty} d_k^{(i)} s^i.
\tag{C.6}
$$

Usually we assume these generating functions to be rational and then the algebraic entropy can be calculated as the modulus of the smallest pole of the generating function:

$$
\eta^{(i)} = \min\left\{ |s| \;\middle|\; \lim_{\sigma\to s} \left|g^{(i)}(\sigma)\right| = \infty \right\}.
\tag{C.7}
$$

We built the program `ae2d.py` to analyze the evolution matrices and to search for recurring sequences of growth. It can extract from a (sufficiently big) evolution matrix the number of sequences of growth and analyze them. For further details see [336].

Before considering the algebraic entropy of the equations presented in Section 2.4.7.2 we discuss briefly the algebraic entropy for the rhombic H^4 equations, which are well

Equation	Growth direction			
	$-,+$	$+,+$	$+,-$	$-,-$
${}_tH_1^\varepsilon$	L_1, L_2	L_1, L_2	L_3, L_4	L_3, L_4
${}_tH_2^\varepsilon$	L_5, L_6	L_5, L_6	L_7, L_8	L_7, L_8
${}_tH_3^\varepsilon$	L_5, L_6	L_5, L_6	L_7, L_8	L_7, L_8
D_1	L_0, L_0	L_0, L_0	L_0, L_0	L_0, L_0
${}_1D_2$	L_9, L_{10}	L_9, L_{10}	L_9, L_{10}	L_9, L_{10}
D_3	L_{11}, L_{12}	L_{11}, L_{12}	L_{11}, L_{12}	L_{11}, L_{12}
${}_1D_4$	L_{13}, L_8	L_{13}, L_8	L_{13}, L_8	L_{13}, L_8
${}_2D_2$	L_{14}, L_{15}	L_{14}, L_{15}	L_{15}, L_{14}	L_{15}, L_{14}
${}_3D_2$	L_{16}, L_{17}	L_{16}, L_{17}	L_{17}, L_{16}	L_{17}, L_{16}
${}_2D_4$	L_{18}, L_8	L_{18}, L_8	L_{19}, L_{20}	L_{19}, L_{20}

TABLE C.1. Sequences of growth for the trapezoidal H^4 and H^6 equations. The first one is the principal sequence, while the second the secondary. All sequences L_j, $j = 0, \cdots, 20$ are presented in Table C.2. [reprinted from [336]].

known to be integrable [839]. Running ae2d.py on these equations one finds that they possess only a principal sequence with the following isotropic sequence of degrees:

$$1, 2, 4, 7, 11, 16, 22, 29, 37, \ldots. \tag{C.8}$$

To this sequence corresponds the generating function:

$$g = \frac{s^2 - s + 1}{(1 - s)^3}, \tag{C.9}$$

which gives, through the definition $g = \sum_k d_k s^k$, the asymptotic fit of the degrees:

$$d_k = \frac{k(k+1)}{2} + 1. \tag{C.10}$$

Since the growth is quadratic $\eta = 0$. This result is a confirmation by the algebraic entropy approach of the integrability of the rhombic H^4 equations. Let us note that the growth sequences (C.8) are the same in the rhombic H^4 equations also when $\pi = 0$, i.e. if we are in the case of the H equations of the ABS classification.

For the trapezoidal H^4 and H^6 equations the situation is a bit more complicate. Indeed those equations have in every direction two different sequence of growth, the principal and the secondary one, as the coefficients of the evolution matrix (C.2) are 2-periodic. The most surprising feature is however that all the sequences of growth are linear. This means that all such equations are not only integrable due to the CaC property, but also linearizable.

Instead of presenting the full evolution matrices (C.2), which would be very lengthly and obscure, we present two tables with the relevant properties. The interested reader will find the full matrices in [336]. In Table C.1 we present a summary of the sequences of growth of both trapezoidal $H^4$and H^6 equations. The explicit sequence of the degrees of growth with generating functions, asymptotic fit of the degrees of growth and entropy is given in Table C.2.

Observing Table C.1 and Table C.2 we may notice the following facts:

- The trapezoidal H^4 equations are not isotropic: the sequences in the $(-,+)$ and $(+,+)$ directions are different from those in the $(+,-)$ and $(-,-)$ directions. These results reflect the symmetry of the equations.
- The H^6 equations, except from ${}_2D_4$ which has the same behaviour as the trapezoidal H^4, are isotropic. Equations ${}_2D_2$ and ${}_3D_2$ exchange the principal and the secondary sequences from the $(-,+)$, $(+,+)$ directions and the $(+,-)$, $(-,-)$ directions.
- All growths, except L_0, L_3, L_4, L_7, L_8, L_{12} and L_{17}, exhibit a highly oscillatory behaviour. They have generating functions of the form:

(C.11)
$$g(s) = \frac{P(s)}{(s-1)^2(s+1)^2},$$

with the polynomial $P(s) \in \mathbb{Z}[s]$. We may write

(C.12)
$$g(s) = P_0(s) + \frac{P_1(s)}{(s-1)^2(s+1)^2},$$

with $P_0(s) \in \mathbb{Z}[s]$ of degree less than P and $P_1(s) = \alpha s^3+\beta s^2+\gamma s+\delta$. Expanding the second term in (C.12) in partial fractions we obtain:

(C.13)
$$\begin{aligned} g(s) = P_0(s) + \frac{1}{4}\Bigg[&\frac{-\alpha-\gamma+\beta+\delta}{(s+1)^2} + \frac{\alpha+\gamma+\beta+\delta}{(s-1)^2} \\ &+\frac{2\alpha+\beta-\delta}{s-1} + \frac{2\alpha-\beta+\delta}{s+1}\Bigg]. \end{aligned}$$

Expanding the term in square parentheses in Taylor series we find that:

(C.14)
$$g(s) = P_0(s) + \sum_{k=0}^{\infty}\left[A_0 + A_1(-1)^k + A_2 k + A_3(-1)^k k\right] s^k,$$

with $A_i = A_i(\alpha,\beta,\gamma,\delta)$ constants. This means the $d_k = A_0 + A_1(-1)^k + A_2 k + A_3(-1)^k k$ for $k > \deg P_0(s)$, and therefore it asymptotically solves a fourth order difference equation. As far as we know, even if some example of behaviour containing terms like $(-1)^k$ are known [387], this is the first time that we observe patterns with oscillations given by $k\,(-1)^k$. We conclude noting that the use of the algebraic entropy as integrability indicator is actually justified by the existence of finite order recurrence relations between the degrees d_k. Indeed the existence of such recurrence relations means that from a local property (the sequence of degrees) we may infer a global one (chaoticity/integrability/linearizability) [817].

2. Algebraic entropy for the non autonomous YdKN equation and its subcases.

In Section 2.4.7.5 we saw that the fluxes of all the generalized three point symmetries of the H^4 and H^6 equations are eventually related either to the YdKN (2.4.129) or to the non autonomous YdKN equation (2.4.198, 2.4.199).

It has been remarked that the non autonomous YdKN equation (2.4.198, 2.4.199) passes the necessary condition for the integrability which is only an indication of the integrability of such class of equation. In Section 2.4.7.5 we have shown that they are generalized symmetries of the H^4 and H^6 equations. Here we give a further evidence that the non autonomous YdKN might be an integrable DΔE based on the algebraic entropy test [82].

We recall briefly how to compute the algebraic entropy in the case of DΔEs of the form $du_n/dt = f_n\left(u_{n+1}, u_n, u_{n-1}\right)$ [208, 816]. First of all we assume that the equation is solvable for u_{n+1} uniquely. This is a condition on f_n. Then, starting from $n = 0$, we compute u_1 by substituting the initial conditions:

$$\left\{\frac{d^k u_{-1}}{dt^k}, \frac{d^k u_0}{dt^k}\right\}_{k=0}^{\infty}. \tag{C.1}$$

Knowing u_1 we can then calculate u_2 and so on. We define the degree of the iterate at the l-th step as the maximum between the degree of the numerator and of the numerator of u_l in the initial conditions (C.1). A great simplification in the explicit calculations is obtained if instead of a generic initial condition one parametrizes the curve of the initial condition rationally using the variable t:

$$u_{-1} = \frac{A_{-1}t + B_{-1}}{At + B}, \quad u_0 = \frac{A_0 t + B_0}{At + B}, \tag{C.2}$$

and then compute the degrees in t. Calculating N iterates, for a sufficiently large positive integer N, and constructing the generating function one can calculate the algebraic entropy without calculating the entire sequence. For more details on how the method is implemented see [336].

We look for the sequence of degrees of the iterate map for the non autonomous YdKN equation (2.4.198, 2.4.199) and its particular cases found in Section 2.4.6. We find for all the cases, except the symmetries of ${}_rH_1^\pi$, the symmetries (2.4.203, 2.4.205) for ${}_tH_1^\pi$ and the symmetries of the ${}_iD_2$ equations, the following values:

$$1, 1, 3, 7, 13, 21, 31, 43, 57, 73, 91, 111, 133, 157 \ldots \tag{C.3}$$

This sequence has the following generating function:

$$g(z) = \frac{1 - 2z + 3z^2}{(1-z)^2}, \tag{C.4}$$

which gives the following quadratic fit for the sequence (C.3):

$$d_l = l(l-1) + 1. \tag{C.5}$$

For the symmetry in the n direction (2.4.159a) of the equation ${}_rH_1^\pi$ we have the somehow different situation when the sequence growth is different according to the even or odd values of the m variable:

$$m = 2k \quad 1, 1, 3, 7, 10, 17, 23, 33, 42, 55, 67, 83, 98, 117 \ldots \tag{C.6a}$$

$$m = 2k+1 \quad 1, 1, 3, 4, 9, 13, 21, 28, 39, 49, 63, 76, 93, 109 \ldots. \tag{C.6b}$$

These sequences have the following generating functions and asymptotic fits:

$$m = 2k, \quad \begin{aligned} g(z) &= \frac{2z^5 - 3z^4 + 3z^3 + z^2 - z + 1}{(1-z)^3(z+1)}, \\ d_l &= \frac{3}{4}l^2 - l - \frac{5(-1)^l - 21}{8}, \end{aligned} \tag{C.7a}$$

$$m = 2k+1, \quad \begin{aligned} g(z) &= \frac{(z^2 + z + 1)(2z^2 - 2z + 1)}{(1-z)^3(z+1)}, \\ d_l &= \frac{3}{4}l^2 - \frac{3}{2}l - \frac{5(-1)^l - 19}{8}. \end{aligned} \tag{C.7b}$$

The symmetry in the m direction (2.4.200b) of the equation ${}_rH_1^\pi$ has the same behaviour by exchanging m with n in formulas (C.6-C.7). The n directional symmetry of the equation

${}_tH_1^\pi$ has almost the same growth for m odd as (C.6b, C.7b) however it is worthwhile to mention that the fit $d_l = \frac{3}{4}l^2 - \frac{5}{4}l + (-1)^l\frac{l}{4} + \frac{(-1)^l+15}{8}$ presents a term $l(-1)^l$, new in this kind of results. For m even we have the same growth as (C.3). The m directional symmetry of the equation ${}_tH_1^\pi$ has the same growth as the m even one of ${}_tH_1^\pi$ (C.6a, C.7a). For the symmetries (2.4.208) of the ${}_iD_2$ equations we have different growth according to the even or odd values of the m or n variables and similar sequences or slightly lower than in the case of equations H_1^ε, however always corresponding to a quadratic asymptotic fit.

This shows that the whole family of the non autonomous YdKN is integrable according to the algebraic entropy test.

For completeness let us just mention that the symmetries (2.4.206) of the ${}_iD_2$, $i = 1, 2, 3$ equations have a sequence growth of the same order than those considered above, i.e. quadratic growth and thus null entropy.

TABLE C.2. Sequences of growth, generating functions, analytic expression of the degrees and entropy for the trapezoidal $H^4$and H^6equations [reprinted from [336]].

Name	Degrees $\{ d_k \}$	Generating function $g(s)$	Degree fit d_k	Entropy η
L_0	1, 1, 1, 1, 1, 1, 1, 1, 1, 1, 1, 1, 1...	$\frac{1}{1-s}$	1	0
L_1	1, 2, 2, 5, 3, 8, 4, 11, 5, 14, 6, 17, 7...	$\frac{s^3+2s+1}{(s-1)^2(s+1)^2}$	$\frac{(-1)^k}{4}(-2k+1)+k+\frac{3}{4}$	0
L_2	1, 2, 4, 3, 7, 4, 10, 5, 13, 6, 16, 7, 19...	$\frac{-s^3+2s^2+2s+1}{(s-1)^2(s+1)^2}$	$\frac{(-1)^k}{4}(2k-1)+k+\frac{5}{4}$	0
L_3	1, 2, 2, 3, 3, 4, 4, 5, 5, 6, 6, 7, 7...	$\frac{-s^2+s+1}{s^3-s^2-s+1}$	$-\frac{(-1)^k}{4}+\frac{k}{2}+\frac{5}{4}$	0
L_4	1, 2, 4, 5, 7, 8, 10, 11, 13, 14, 16, 17, 19...	$\frac{s^2+s+1}{s^3-s^2-s+1}$	$\frac{(-1)^k}{4}+\frac{3k}{2}+\frac{3}{4}$	0
L_5	1, 2, 4, 6, 11, 10, 19, 14, 27, 18, 35, 22, 43...	$\frac{s^6+4s^4+2s^3+2s^2+2s+1}{(s-1)^2(s+1)^2}$	$(-1)^k\left(k-\frac{5}{2}\right)+3k-\frac{5}{2}$	0
L_6	1, 2, 4, 7, 8, 15, 12, 23, 16, 31, 20, 39, 24...	$\frac{3s^5+s^4+3s^3+2s^2+2s+1}{(s-1)^2(s+1)^2}$	$(-1)^k\left(-k+\frac{5}{2}\right)+3k-\frac{5}{2}$	0
L_7	1, 2, 4, 7, 11, 15, 19, 23, 27, 31, 35, 39, 43...	$\frac{s^4+s^3+s^2+1}{(s-1)^2}$	$4k-5$	0
L_8	1, 2, 4, 6, 8, 10, 12, 14, 16, 18, 20, 22, 24...	$\frac{s^2+1}{(s-1)^2}$	$2k$	0
L_9	1, 2, 2, 5, 3, 8, 4, 11, 5, 14, 6, 17, 7,...	$\frac{s^3+2s+1}{(s-1)^2(s+1)^2}$	$\frac{(-1)^k}{4}(-2k+1)+k+\frac{3}{4}$	0
L_{10}	1, 2, 3, 5, 5, 8, 7, 11, 9, 14, 11, 17, 13,...	$\frac{s^3+s^2+2s+1}{(s-1)^2(s+1)^2}$	$\frac{(-1)^k}{4}(-k+1)+\frac{5k}{4}+\frac{3}{4}$	0

Continued on next page

Table C.2 – *Continued from previous page*

Name	Degrees $\{d_k\}$	Generating function $g(s)$	Degree fit d_k	Entropy η
L_{11}	1, 2, 4, 5, 10, 8, 16, 11, 22, 14, 28, 17, 34,...	$\frac{3s^4+s^3+2s^2+2s+1}{(s-1)^2(s+1)^2}$	$\frac{(-1)^k}{4}(3k-5)+\frac{9k}{4}-\frac{3}{4}$	0
L_{12}	1, 2, 4, 5, 7, 8, 10, 11, 13, 14, 16, 17, 19,...	$\frac{s^2+s+1}{(s-1)^2(s+1)}$	$\frac{(-1)^k}{4}+\frac{3k}{2}+\frac{3}{4}$	0
L_{13}	1, 2, 4, 6, 11, 10, 18, 14, 25, 18, 32, 22, 39, ...	$\frac{4s^4+2s^3+2s^2+2s+1}{(s-1)^2(s+1)^2}$	$\frac{3(-1)^k}{4}(k-2)+\frac{11k}{4}-\frac{3}{2}$	0
L_{14}	1, 2, 3, 3, 6, 4, 9, 5, 12, 6, 15, 7, 18,...	$\frac{s^4-s^3+s^2+2s+1}{(s-1)^2(s+1)^2}$	$\frac{(-1)^k}{4}(2k-3)+k+\frac{3}{4}$	0
L_{15}	1, 1, 3, 3, 6, 5, 9, 7, 12, 9, 15, 11, 18,...	$\frac{s^4+s^3+s^2+s+1}{(s-1)^2(s+1)^2}$	$\frac{k}{4}\left((-1)^k+5\right)$	0
L_{16}	1, 2, 3, 2, 5, 2, 7, 2, 9, 2, 11, 2, 13,...	$\frac{-2s^3+s^2+2s+1}{(s-1)^2(s+1)^2}$	$\frac{(-1)^k}{2}(k-1)+\frac{k}{2}+\frac{3}{2}$	0
L_{17}	1, 1, 3, 3, 5, 5, 7, 7, 9, 9, 11, 11, 13, ...	$\frac{s^2+1}{(s-1)^2(s+1)}$	$\frac{(-1)^k}{2}+k+\frac{1}{2}$	0
L_{18}	1, 2, 4, 5, 11, 9, 19, 13, 27, 17, 35, 21, 43,...	$\frac{s^6+s^5+4s^4+s^3+2s^2+2s+1}{(s-1)^2(s+1)^2}$	$(-1)^k(k-2)+3k-3$	0
L_{19}	1, 2, 4, 6, 11, 10, 19, 14, 27, 18, 35, 22, 43,...	$\frac{s^6+4s^4+2s^3+2s^2+2s+1}{(s-1)^2(s+1)^2}$	$(-1)^k\left(k-\frac{5}{2}\right)+3k-\frac{5}{2}$	0
L_{20}	1, 1, 3, 2, 6, 3, 9, 4, 12, 5, 15, 6, 18,...	$\frac{s^4+s^2+s+1}{(s-1)^2(s+1)^2}$	$\frac{(-1)^k}{4}(2k-1)+k+\frac{1}{4}$	0

APPENDIX D

Translation from Russian of R I Yamilov, *On the classification of discrete equations*, reference [841].

Let us consider nonlinear DΔEs of the form

$$u_t = \varphi(u_1, u, u_{-1}) \tag{D.1}$$

satisfying the following conditions

$$-\frac{\partial\varphi}{\partial u_1} \Big/ \frac{\partial\varphi}{\partial u_{-1}} = y/y_{-1} \tag{D.2}$$

$$\partial_\varphi \ln y + 2\frac{\partial\varphi}{\partial u} = \Delta A \tag{D.3}$$

$$\partial_\varphi \ln \frac{\partial\varphi}{\partial u_1} = \Delta B \tag{D.4}$$

Conditions (D.2–D.4) mean that one can find functions y, A, B of a finite number of variables u, $u_{\pm 1}$, $u_{\pm 2}$, ... such that equations (D.2–D.4) are satisfied. The operator Δ is such that for any function $f = f(u_m, u_{m-1}, \dots, u_{m'})$

$$\Delta f = f - f_{-1}$$

where f_k ($k \in \mathbb{Z}$) is a function of the form

$$f_k = (u_{m+k}, \dots, u_{m'+k});$$

the operator ∂_φ is the derivative in the direction of the vector field φ in equation (D.1):

$$\partial_\varphi = \varphi_1 \frac{\partial}{\partial u_1} + \varphi\frac{\partial}{\partial u} + \varphi_{-1}\frac{\partial}{\partial u_{-1}}.$$

The conditions (D.2)–(D.4) are conditions on the vector field φ that are necessary for the existence of a nontrivial Lie-Bäcklund algebra and nontrivial conservation laws admitted by equation (D.1) (see Theorem D.I).

For continuous evolutionary equations the problem of studying equations with nontrivial Lie-Bäcklund algebras was set and solved by N.Kh. Ibragimov and A.B. Shabat (see [2], [3]).

Examples of equations satisfying (D.2)–(D.4) are

$$u_t = \left(u^3 + \alpha u + \beta\right)\left(\frac{1}{u_1 - u} + \frac{1}{u - u_1}\right), \tag{D.5}$$

$$u_t = \frac{u_1 u_{-1} + u_{-1}^2}{u_1 - u_{-1}}. \tag{D.6}$$

An element of the Lie-Bäcklund algebra of (D.5) is given by the function

$$(u^3+\alpha u+\beta)\Big[\frac{u_1^3+\alpha u_1+\beta}{(u_2-u_1)(u_1-u)^2}+\Big(u+\frac{u^3+\alpha u+\beta}{(u_1-u)(u-u_{-1})}\Big)\cdot$$
$$\cdot\Big(\frac{1}{u_1-u}+\frac{1}{u-u_{-1}}\Big)+\frac{u_{-1}^3+\alpha u_{-1}+\beta}{(u-u_{-1})(u_{-1}-u_{-2})}\Big],$$

and densities of its conservation law by

$$\ln\frac{u^3+\alpha u+\beta}{(u_1-u)^2},\qquad u+\frac{u^3+\alpha u+\beta}{(u_1-u)(u-u_{-1})}.$$

An element of the Lie-Bäcklund algebra of (D.6) is given by the function

$$\Big[\Big(\frac{u_1u_{-1}+u^2-1}{u_1-u_{-1}}\Big)^2+u^2-1\Big]\cdot\Big(\frac{1}{u_2-u}+\frac{1}{u-u_{-2}}\Big),$$

and densities of its conservation law by

$$\ln\frac{u^2+u_{-1}^2-1}{(u_1-u_{-1})^2},\qquad \frac{u^2+u_{-1}^2-1}{(u_1-u_{-1})(u-u_{-2})}.$$

1. Proof of the conditions (D.2–D.4). Let us deduce the necessary conditions (D.2–D.4). The equation

$$\partial_\varphi f-\frac{\partial\varphi}{\partial u_1}f_1-\frac{\partial\varphi}{\partial u}f-\frac{\partial\varphi}{\partial u_{-1}}f_{-1}=0 \tag{D.7}$$

is the defining equation for equation (D.1). A formal series $H=\sum_k h_k$ is a conservation law for equation (D.1) if there is a function h (density of the conservation law) such that $\partial_\varphi h=\Delta\psi$.

A list of evident properties of the operator Δ follows.

(1) $p+q=\Delta r\Rightarrow p+q_i=\Delta h$. (Proof: $p+q_1=p+q+\Delta q_1$ and induction).
(2) $p=\Delta q$, where $p=p(u_i,\dots,u_j)\Rightarrow \partial^2p/\partial u_i\partial u_j=0$.
(3) $p=\Delta q$, $p=p(u)\Rightarrow p\equiv 0$, $q=\alpha\in\mathbb{C}$.
(4) $p=\Delta q+\alpha$, $\alpha\in\mathbb{C}$, $p=p(u)\Leftrightarrow \delta p/\delta u=0$ where

$$\frac{\delta p}{\delta u}=\sum_i\frac{\partial p_i}{\partial u}$$

is the variational derivative of the function p.

It is convenient to study conservation laws through the adjoint defining equation

$$\partial_\varphi g+\frac{\partial\varphi_1}{\partial u}g_1+\frac{\partial\varphi}{\partial u}g+\frac{\partial\varphi_{-1}}{\partial u}g_{-1}=0 \tag{D.8}$$

With the help of properties (1) – (4) it is easy to prove that if h is the density of a conservation law, then the function $g=\delta h/\delta u$ is a solution of (D.8).

Theorem D.I *If* (D.7) *has a solution* $f=f(u_m,\dots,u_{m'})$ *with* $m\geq 2$ *and the adjoint defining equation has a solution* $g=g(u_N,\dots,u_{-N})$*, with* $N\geq 3$*, then the vector field* φ *satisfies conditions (1) – (4)*

PROOF. When $m \geq 1$, differentiating (D.7) with respect to u_{m+1} yields

$$\frac{\partial \varphi_m}{\partial u_{m+1}} \frac{\partial f}{\partial u_m} - \frac{\partial \varphi}{\partial u_1} \frac{\partial f_1}{\partial u_{m+1}} = 0.$$

It is easy to check that this equation can be rewritten in the form

$$0 = \Delta \left[\frac{\partial f_1}{\partial u_{m+1}} \left(\prod_{i=1}^{m} \frac{\partial \varphi_i}{\partial u_{i+1}} \right)^{-1} \right].$$

Now property (3) implies that

$$\frac{\partial f}{\partial u_m} = \alpha \prod_{i=0}^{m-1} \frac{\partial \varphi_i}{\partial u_{i+1}}, \qquad \alpha \in \mathbb{C}. \tag{D.9}$$

When $m \geq 2$, differentiating (D.7) with respect to u_m and dividing by $\partial f / \partial u_m$

$$\Rightarrow \partial_\varphi \ln \frac{\partial f}{\partial u_m} = \frac{\partial \varphi}{\partial u} - \frac{\partial \varphi_m}{\partial u_m} + \alpha^{-1} \Delta \left[\frac{\partial f_1}{\partial u_m} \left(\prod_{i=1}^{m-1} \frac{\partial \varphi_i}{\partial u_{i+1}} \right)^{-1} \right]$$

Using (D.9) this equation can be rewritten as

$$\sum_{i=1}^{m} \partial_\varphi \ln \frac{\partial \varphi_i}{\partial u_{i+1}} = \Delta E.$$

This is (D.4).

When $N > 1$, differentiating (D.8) with respect to u_{N+1} yields an equation for the partial derivative $\partial g / \partial u_N$ of a solution g

$$\frac{\partial \varphi_N}{\partial u_{N+1}} \frac{\partial g}{\partial u_N} + \frac{\partial \varphi_1}{\partial u} \frac{\partial g_1}{\partial u_{N+1}} = 0. \tag{D.10}$$

This equation can be written in the following form

$$\ln \left(-\frac{\partial \varphi_N}{\partial u_{N+1}} / \frac{\partial \varphi_1}{\partial u} \right) = \Delta \ln \frac{\partial g_1}{\partial u_{N+1}}.$$

This implies condition (D.2) by virtue of property (1) of Δ. With the help of this condition and relation (D.10) we obtain an equation for the partial derivative $\partial g / \partial u_N$ of a solution g

$$\frac{\partial g}{\partial u_N} = \beta y \prod_{i=1}^{N-1} \frac{\partial \varphi_i}{\partial u_{i+1}}.$$

Let us deduce now condition (D.3). Differentiate the left hand side of the adjoint defining equation (D.8) with respect to u_N $(N > 2)$ and divide the result by $\partial g / \partial u_N$, to obtain

$$\partial_\varphi \ln \frac{\partial g}{\partial u_N} = -\frac{\partial \varphi}{\partial u} - \frac{\partial \varphi_N}{\partial u_N} + \Delta \left[\frac{\partial \varphi_N}{\partial u_{N+1}} \frac{\partial g_1}{\partial u_N} \left(\frac{\partial g_1}{\partial u_{N+1}} \right)^{-1} \right].$$

Now using (I.7), condition (D.4) and property (1) we obtain condition (D.3). The Theorem D.I is proved. □

2. Nonlinear differential difference equations satisfying conditions (D.2–D.4). After a change of the form

(D.11) $$u = \sigma(v),\ t = \alpha\tau \quad (\alpha \in \mathbb{C})$$

equation (D.1) changes to

(D.12) $$v_\tau = \psi(v_1, v, v_{-1}), \quad \psi = \frac{\alpha}{\sigma'(v)}\varphi(\sigma(v_1), \sigma(v), \sigma(v_{-1})).$$

Equations (D.1) and (D.12) do not differ from the point of view of the presence of Lie-Bäcklund algebras and conservation laws.

If $f = f(u_m, u_{m-1}, \dots, u_{m'})$ is an element of the Lie-Bäcklund algebra of (D.1), then $(\alpha/\sigma'(v)) \cdot f(\sigma(v_m), \dots, \sigma(v_{m'}))$ is an element of the Lie-Bäcklund algebra of (D.12) similarly as equation $u_t = f$ is related by the change (D.11) to $v_\tau = (\alpha/\sigma'(v))f(\sigma(v_m), \dots)$. If $h = h(u_N, u_{N-1}, \dots, u_{N'})$ is the density of a conservation law of (D.1), then $h(\sigma(v_N), \dots, \sigma(v_{N'}))$ is a density of a conservation law of (D.12). It is clear that the vector field φ of equation (D.1) satisfies conditions (D.2–D.4) if and only if the vector field ψ of equation (D.12) satisfies the conditions $-(\partial\psi/\partial v_1)/(\partial\psi/\partial v_{-1}) = y'/y'_{-1}$, $\partial_{\psi \ln y'} + 2\partial\psi/\partial v = \Delta A'$, $\partial_\psi \ln(\partial\psi/\partial v_1) = \Delta B'$ where y', A' , B' some functions with a finite number of variables v , $v_{\pm 1}$, $v_{\pm 2}$, …

We will say that two equations related by the change (D.12) are equivalent.

Theorem D.II *Equations of the form* (D.1) *satisfying* (D.2), (D.4) *are equivalent to an equation of one of the types I-IV defined as*

- *Type I:* $u_t = \omega$.
- *Type II:* $u_t = \exp\omega + p(u)$.
- *Type III:* $u_t = \omega^{-1} + \sigma$.
- *Type IV:* $u_t = \tanh\omega + q(u)$.

Here ω *is a function of the variables* u_1, u, u_{-1} *of the special form*

(D.13) $$\omega = \frac{\partial}{\partial u}\Delta(z(u_1, u)).$$

Notice that for vector fields of types I–IV (as for any function of the form $\varphi = \phi(u, \omega)$) condition (D.3) can be satisfied setting

$$y = \frac{\partial^2 z}{\partial u_1 \partial u}.$$

PROOF. Condition (D.2) is a linear differential equation of first order over function φ that implies that function φ has the form

$$\varphi = \varphi(u, \omega)$$

where ω is a function of the form (D.13). Let us write an ordinary differential equation over function φ as a function of ω. Condition (D.4) and property (4) of the operator Δ imply that

$$\frac{\partial}{\partial u_3}\frac{\delta}{\delta u}\partial_\varphi \ln\frac{\partial\varphi}{\partial u_1} = 0.$$

Differentiating we obtain the equality

(D.14) $$\frac{\partial\varphi_2}{\partial u_3}\frac{\partial^2}{\partial u \partial u_2}\ln\frac{\partial\varphi_1}{\partial u_2} + \frac{\partial\varphi_1}{\partial u}\frac{\partial^2}{\partial u_1 \partial u_3}\ln\frac{\partial\varphi_2}{\partial u_3} = 0.$$

Using condition (D.2), we express in this equation the function $\partial\varphi_1/\partial u$ in terms of the functions $\partial\varphi_1/\partial u_2$, y_1, y we can write (D.14) in the form

$$0 = \Delta\left[\left(y_1\frac{\partial\varphi_2}{\partial u_3}\right)^{-1}\frac{\partial^2}{\partial u_1\partial u_3}\ln\frac{\partial\varphi_2}{\partial u_3}\right].$$

Thus, using property (3) of the operator Δ, we obtain the equation

$$\frac{\partial^2}{\partial u_1\partial u_{-1}}\ln\frac{\partial\varphi}{\partial u_1} = -\alpha y_{-1}\frac{\partial\varphi}{\partial u_1}, \quad (\alpha\in\mathbb{C}).$$

This equation, after dividing by yy_{-1}, becomes the ordinary differential equation

$$\left(\ln\phi_\omega\right)_{\omega\omega} = \alpha\phi_\omega. \tag{D.15}$$

Equation (D.15) has the following four types of solutions for the function $\phi = \phi(u,\omega)$:

$$\begin{aligned}
\phi &= R(u)\omega + P(u),\\
\phi &= \exp\left[R(u)\omega + P(u)\right] + Q(u),\\
\phi &= -2\alpha^{-1}\left[\omega + P(u)\right]^{-1} + Q(u)\\
\phi &= 2\alpha^{-1}R(u)\tanh\left[R(u)\omega + P(u)\right] + Q(u)
\end{aligned}$$

Notice that if $\omega = \omega(u_1,u,u_{-1})$ is a function of the form (D.13) then $\omega + f'(u)$ is also such a function, because $\omega + f'(u) = \frac{\partial}{\partial u}\Delta z + f'(u) = \frac{\partial}{\partial u}\Delta(z + f(u))$. Thus, any equation (D.1) with a vector field φ satisfying conditions (D.2) and (D.4) can be written in one of the following forms

$$u_t = r^2(u)\omega \tag{D.16}$$

$$u_t = r(u)\exp r(u)\omega + r(u)p(\sigma(u)) \tag{D.17}$$

$$u_t = 2\alpha^{-1}\omega^{-1} - 2\alpha^{-1}\varepsilon r(u) \quad (\varepsilon\in\mathbb{C}) \tag{D.18}$$

$$u_t = 2\alpha^{-1}r(u)\tanh r(u)\omega + 2\alpha^{-1}r(u)q(\sigma(u)) \tag{D.19}$$

where $\sigma(u)$ is a solution to the differential equation $\sigma'(u) = [r(u)]^{-1}$ and ω is a function of the form (D.13).

Performing in equations (D.16–D.19) a point transformation $v = f(u)$ we notice that the function ω is transformed onto another function of the form (D.13), multiplied by a function of one variable

$$\omega = \frac{\partial}{\partial u}\Delta z(u_1,u) = f'\circ f^{-1}(v)\frac{\partial}{\partial v}\Delta(f^{-1}(v_1), f^{-1}(v)).$$

This implies that equations (D.16–D.19) do not change in form under point transformations, and equations (D.1) are naturally divided in classes invariant under point transformations.

In particular, after the substitution $u\mapsto\sigma(u)$ (and for equations (D.3), (D.19) additionally after substituting $t\mapsto -2\alpha^{-1}t$, $t\mapsto 2\alpha^{-1}t$ respectively) equations (D.16–D.19) are transformed into equations of types I–IV respectively. □

3. List of non linear differential difference equations of type I satisfying conditions (D.2, D.4). Let us write a list of the nonlinear equations of type I satisfying conditions (D.3) and (D.4).

Theorem D.III *The vector field of a nonlinear equation of type I satisfies conditions* (D.3) *and* (D.4) *if and only if this equation is equivalent to one of the following equations*

(D.20) $$u_t = u(u_1 - u_{-1}),$$

(D.21) $$u_t = u^2(u_1 - u_{-1}),$$

(D.22) $$u_t = (1 - u^2)(u_1 - u_{-1}),$$

(D.23) $$u_t = f(u_1 + u) - f(u + u_{-1}), \quad f' = \alpha f^2 + \beta f + \gamma,$$

(D.24) $$u_t = f(u_1 - u) + f(u - u_{-1}), \quad f' = \alpha f^2 + \beta f + \gamma,$$

(D.25) $$u_t = \wp'(u)\left\{\left[\wp(u_1 + \theta) - \wp(u)\right]^{-1} + \left[\wp(u) - \wp(u_{-1} + \theta)\right]^{-1}\right\}$$

where $\wp$ *is an arbitrary Weierstrass function and* θ *its half-period.*

Besides Theorem D.III, it is possible to prove (see the proof of Theorem D.III) that an equation of type II has a vector field satisfying conditions (D.3) and (D.4) if and only if it is equivalent to equation

(D.26) $$u_t = \exp\left[f(u_1 - u) + f(u - u_{-1})\right] + \sigma, \quad f' = \alpha + \beta e^{-f} + \gamma e^{-2f}.$$

Further investigation of equations of types III-IV with non-trivial algebras of Lie-Bäcklund type and conservation laws leads to the study of an overdetermined system of a special type. Thus, for discrete equations of type III with $\sigma = 0$ the necessary conditions (D.3) and (D.4) are satisfied if and only if the corresponding function z (see formula (D.13)) satisfies the following systems of equations:

(D.27) $$\begin{aligned} z_{xx} &= \alpha(x)z_x^4 + \beta(x)z_x^3 + \gamma(x)z_x^2 + \delta(x)z_x + \sigma(x), \\ z_{yy} &= \alpha(y)z_y^4 + \beta(y)z_y^3 + \gamma(y)z_y^2 + \delta(y)z_y + \sigma(y). \end{aligned}$$

With equations of the type IV the related system is

$$\begin{aligned} z_{xx} &= \alpha(x)e^{4z_x} + \beta(x)e^{2z_x} + \gamma(x) + \delta(x)e^{-2z_x} + \sigma(x)e^{-4z_x}, \\ z_{yy} &= \alpha(y)e^{4z_y} + \beta(y)e^{2z_y} + \gamma(y) + \delta(y)e^{-2z_y} + \sigma(y)e^{-4z_y}. \end{aligned}$$

For equations (D.20–D.24) and also equation (D.26) it is possible to compute the simplest elements of the Lie-Bäcklund algebras and conservation laws. Formulas for the computation of elements of the algebras and densities of conservation laws of the discrete Korteweg-de Vries equation (D.20) can be found in reference [1]. In reference [5] we find other formulas as well as a full list of infinite Lie-Bäcklund algebras and an infinite set of conservation laws of equation (D.20). Following paper [5], it is not difficult to obtain the following result for equation (D.21): any element of the Lie-Bäcklund algebra of equation (D.21) is a linear combination of a finite set of elements of the algebra $f^{(i)}$:

$$\begin{aligned} f^{(1)} &= u^2(u_1 - u_{-1}), \\ h^{(i)} &: \ \Delta h^{(i)} = f^{(i)}/u, \\ f^{(i+1)} &= u^2\left[u_1(h_1^{(i)} + h^{(i)}) - u_{-1}(h_{-1}^{(i)} + h_{-2}^{(i)})\right]; \end{aligned}$$

the density of a any conservation law of (D.21) is a linear combination of a finite set of the densities $g^{(i)}$: $g^{(-2)} = \Delta G$ (G an arbitrary function), $g^{-1} = u^{-1}$, $g^{(0)} = \ln u$, $g^{(i)} = h^{(i)}$, $i = 1, 2, \ldots$.

Equation (D.25), due to the fact that θ is the half-period of $\wp$, is transformed, under the substitution $u \mapsto \wp(u)$ into an equation with rational functions in its vector field. When

$\theta = 0$ equation (D.25) is equivalent to equation (D.5). The equation

$$u_t = \tanh \omega$$
$$e^{2\omega} = -\frac{\wp(u_1) - \wp(u+\theta)}{\wp(u_1) - \wp(u-\theta)} \cdot \frac{\wp(u-\theta) - \wp(u_{-1})}{\wp(u+\theta) - \wp(u_{-1})}$$

where $\wp$ is an arbitrary Weierstrass function, θ is a regular point of $\wp$, $\wp' \neq 0$, is an example of an equation of type IV satisfying conditions (D.3) and (D.4). When $\wp(z) = 2^{-1}(1+\cos z)^{-1} + 1/12$, $\theta = \pi/2$, the substitution $u \mapsto \cos u$ transforms this equation into equation (D.6).

Equations (D.21–D.24) and (D.26) reduce to the discrete KdV equation (D.20). Equation (D.21) reduces to (D.20) with the substitution $u \mapsto u_1 u$ (i.e. if $u = u(n,t)$ is a solution of (D.21) then $v = v(n,t) = u_1 u$ is a solution of equation $v_t = v(v_1 - v_{-1})$). Equation (D.22) reduces to (D.20) through the substitution $u \mapsto (1+u_1)(1-u)$, equations (D.23), (D.24) through, respectively, substitutions $u \mapsto f(u_1+u)$, $u \mapsto f(u_1-u)$ to equation $u_t = (\alpha u^2 + \beta u + \gamma)(u_1 - u_{-1})$ and all such equations are equivalent to equations (D.20), (D.21) or (D.22). Equation (D.26) reduces to the equation $u_t = (\alpha u^2 + \beta u + \gamma)(u_1 - u_{-1})$ through the substitution $u \mapsto \exp f(u_1 - u)$.

Every conservation law of equation (D.20), using the previous substitutions, allows to construct a corresponding conservation law of equations (D.21–D.24), (D.26). Let equation (D.1) be transformed, through the non-point transformation $v = \psi(u_M, u_{M-1}, \dots, u_{M'})$, to equation

(D.28) $$v_t = \phi(v_1, v, v_{-1}), \quad \phi(\psi_1, \psi, \psi_{-1}) = \sum_k \frac{\partial \psi}{\partial u_k} \varphi_k,$$

and $h = h(v_N, v_{N-1}, \dots, v_{N'})$ be a density of a conservation law of equation (D.28). Then the function $h = h(\psi_N, \dots, \psi_{N'})$ is a density of a conservation law of equation (D.1).

The elements of the Lie-Bäcklund algebra of (D.21)–(D.24), (D.26) are more difficult to construct. For example, the equation

(D.29) $$u_t = \exp(u_1 - u) + \exp(u - u_{-1})$$

(see equations of the form (D.24)), substituting $v = \exp(u_1 - u)$, is reduced to the difference Korteweg-de Vries equation

(D.30) $$v_t = v(v_1 - v_{-1}).$$

In [5] it is shown that any element $f(v_m, v_{m-1}, \dots, v_{m'})$ of the Lie-Bäcklund algebra of (D.30) is such that $f = v\Delta g$, $g = g(v_m, v_{m-1}, \dots, v_{m'+1})$. From this follows that the function $g[\exp(u_m - u_{m-1}), \exp(u_{m-1} - u_{m-2}), \dots, \exp(u_{m'-1} - u_{m'})]$ is an element of the Lie-Bäcklund algebra of (D.29). Equation $u_t = g[\exp(u_m - u_{m-1}), \dots]$ under the substitution $v = \exp(u_1 - u)$ is converted into equation $v_t = f$. For equation (D.22) it is possible to first construct conservation laws through conservation laws of (D.20), and then use the following reasoning: if h is a density of a conservation law of equation $u_t = R(u)(u_1 - u_{-1})$, then $f = R(u)[R(u_1)g_1 - R(u_{-1})g_{-1}]$, where $g = \delta h / \delta u$, is an element of the Lie-Bäcklund algebra of this equation.

The difference Korteweg-de Vries equation (D.20) is at this time totally solved (as explained in [4]). In [4] the solution to equation (D.22) is found with the help of the solution to equation (D.20) using the transformation $u \mapsto (1+u_1)(1-u)$, that relates equation (D.22) to equation (D.20). Analogously, with the help of our indicated substitutions, it is possible to integrate any of the equations (D.21), (D.23), (D.24), (D.26).

Notice that one of the equations of the form (D.23) is equivalent to equation (D.21). We have highlighted equation (D.21) in order to show the relation between (D.21–D.24), (D.26) and (D.20). Besides, eqs. (D.20–D.22) exhaust the non-equivalent equations of the form (D.1) with polynomial vector field satisfying conditions (D.2) and (D.4).

PROOF. (Proof of Theorem D.III) Consider equations

$$u_t = \varphi(u_1, u, u_{-1}), \quad \varphi = \omega = \frac{\partial}{\partial u}\Delta z(u_1, u) \tag{D.31}$$

satisfying conditions (D.3), (D.4). Recall that the functions z and y from condition (D.2), are related by the formula $y = \partial^2 z / \partial u_1 \partial u$. It is clear that $y = \partial\varphi / \partial u_1$, so due to condition (D.4) $\partial_\varphi \ln y = \Delta B$ and condition (D.3) takes the form

$$\frac{\partial \varphi}{\partial u} = \frac{1}{2}\Delta(A - B). \tag{D.32}$$

Using property (4) of the operator Δ, we write the equality

$$\frac{\delta}{\delta u}\frac{\partial \varphi}{\partial u} = \frac{\partial^2 \varphi_1}{\partial u_1 \partial u} + \frac{\partial^2 \varphi}{\partial u^2} + \frac{\partial^2 \varphi_{-1}}{\partial u \partial u_{-1}} = 0.$$

Differentiating with respect to u_1 produces a PDE over the function y:

$$\frac{\partial^2 y}{\partial u^2} - \frac{\partial^2 y}{\partial u_1^2} = 0.$$

From here $y = F(u_1 + u) + G(u_1 - u)$, that is, the function φ has the form

$$\varphi = f(u_1 + u) - g(u_1 - u) - f(u + u_{-1}) - g(u - u_{-1}) + \tau(u).$$

Computing with this formula the derivative $\partial\varphi / \partial u$ and putting the result in formula (D.32) we can write a condition over function $\tau(u)$. This condition has the form $\tau'(u) = \Delta h$. Due to property (3) of Δ, the function $\tau(u)$ is constant, i.e. $\tau(u) = \alpha \in \mathbb{C}$. To economize letters let us denote the function $g - \frac{\alpha}{2}$ with the symbol g. Then the vector field φ of equation (D.31) necessarily has the form

$$\varphi = f(u_1 + u) - g(u_1 - u) - f(u + u_{-1}) - g(u - u_{-1}). \tag{D.33}$$

Let us use condition (D.4). Property (1) of the operator Δ permits to write that condition in the form

$$\varphi \frac{\partial}{\partial u} \ln \prod_i \frac{\partial \varphi_i}{\partial u_{i+1}} = \Delta h.$$

Thus as $\partial\varphi/\partial u_1 = f'(u_1 + u) - g'(u_1 - u)$ is a function of the variables u_1, u, the function $\varphi\,(\partial/\partial u) \ln \left[(\partial\varphi/\partial u_1)\, (\partial\varphi_{-1}/\partial u) \right]$ is a function of the variables u_1, u, u_{-1}. Due to property (2) of the operator Δ

$$\frac{\partial^2}{\partial u_1 \partial u_{-1}} \varphi \frac{\partial}{\partial u} \ln \frac{\partial \varphi}{\partial u_1}\frac{\partial \varphi_{-1}}{\partial u} = 0.$$

Developing the expression in the lhs, and taking into account that

$$\frac{\partial^2 \varphi}{\partial u_1 \partial u_{-1}} = 0,$$

we obtain

$$\frac{\partial \varphi}{\partial u_1}\frac{\partial^2}{\partial u_{-1}\partial u} \ln \frac{\partial \varphi_{-1}}{\partial u} + \frac{\partial \varphi}{\partial u_{-1}}\frac{\partial}{\partial u_1 \partial u} \ln \frac{\partial \varphi}{\partial u_1} = 0.$$

Observing that $\partial\varphi/\partial u_{-1} = -\partial\varphi_{-1}/\partial u$ (see (D.33)) and applying property (3) of the operator Δ (as it was done with equation (D.14)) we obtain

(D.34) $$\frac{\partial^2}{\partial u_1 \partial u} \ln \frac{\partial\varphi}{\partial u_1} = \sigma \frac{\partial\varphi}{\partial u_1} \qquad (\sigma \in \mathbb{C}).$$

It is useful from now on to recall that $\partial\varphi/\partial u_1 = f'(u_1+u) - g'(u_1-u)$. Introduce variables $\zeta = u_1 + u$, $\eta = u_1 - u$ and rewrite (D.34) as

(D.35) $$\left(\frac{\partial^2}{\partial\zeta^2} - \frac{\partial^2}{\partial\eta^2}\right) \ln y = \sigma y, \quad y = f'(\zeta) - g'(\eta).$$

From equation (D.35), knowing that $y_{\zeta\eta} = 0$, we obtain two equations

$$\frac{\partial}{\partial\zeta}\left(\frac{y_{\zeta\zeta\zeta}}{y_\zeta}\right) = 6\sigma y_\zeta, \quad \frac{\partial}{\partial\eta}\left(\frac{y_{\eta\eta\eta}}{y_\eta}\right) = -6\sigma y_\eta.$$

This means that the functions $f(\zeta)$ and $g(\eta)$ satisfy the following differential equations with constant coefficients

$$(f'')^2 = 2\sigma(f')^3 + c_1(f')^2 + c_2 f' + c_3,$$
$$(g'')^2 = 2\sigma(g')^3 + c_4(g')^2 + c_5 g' + c_6.$$

In order to obtain the remaining information about the coefficients c_1, c_2, ..., let us return to equation (D.35). If $f'(\zeta) \equiv 0$ condition (D.35) implies that

$$\frac{1}{2}c_5 g'(\eta) + c_6 = 0,$$

i.e. $c_5 = c_6 = 0$ (n.t. $g'(\eta) = c$ leads to $\varphi = -c(u_1 - u) - c(u - u_{-1})$ linear). If $g'(\eta) \equiv 0$ then from condition (D.35)

$$\frac{1}{2}c_2 f'(\zeta) + c_3 = 0,$$

i.e. $c_2 = c_3 = 0$. Finally if $f'(\zeta) \not\equiv 0$ and $g'(\eta) \not\equiv 0$ then

$$\left(c_1 - c_4\right) f'(\zeta)g'(\eta) - \frac{1}{2}\left(c_2 - c_5\right)\left(f'(\zeta) + g'(\eta)\right) + c_3 - c_6 = 0,$$

and thus $c_1 = c_4$, $c_2 = c_5$, $c_3 = c_6$, i.e. the functions f, g satisfy the same equation. Condition (D.35) is then completely satisfied.

Therefore, it only remains to list all equations of the forms

(D.36) $$u_t = g(u_1 - u) + g(u - u_{-1}), \quad \left(g''\right)^2 = 2\sigma\left(g'\right)^3 + \delta\left(g'\right)^2,$$

(D.37) $$u_t = f(u_1 + u) - f(u + u_{-1}) + \gamma, \quad \left(f''\right)^2 = 2\sigma\left(f'\right)^3 + \delta\left(f'\right)^2,$$

(D.38) $$u_t = f(u_1 + u) - g(u_1 - u) - f(u + u_{-1}) - g(u - u_{-1}),$$

where functions f' and g' satisfy the differential equation

(D.39) $$\left(h'\right)^2 = 2\sigma h^3 + \alpha h^2 + \beta h + \gamma,$$

for vector fields satisfying completely condition (D.4).

A function $h = h(x)$ satisfiying the equation $h''^2 = 2\sigma h'^3 + \delta h'^2$ is also a solution of the ordinary differential equation

(D.40) $$h' = \alpha h^2 + \beta h + \gamma$$

with certain constant coefficients α, β, γ. Let us obtain the equations of type (D.23). Let the function f from equation (D.37) be a solution of equation (D.40). From condition (D.4) we have

$$\Delta B = \partial_\varphi \ln \frac{\partial \varphi}{\partial u_1} = \frac{f''(u_1+u)}{f'(u_1+u)}\left(\partial_\varphi u_1 + \partial_\varphi u\right)$$
$$= \left[2\alpha f(u_1+u)+\beta\right]\left[f(u_2+u_1) - f(u+u_{-1}) + 2\gamma\right].$$

From this follows the next relation

$$\gamma\left[2\alpha f(u_1+u)+\beta\right] = \Delta G.$$

Using property (3) of the operator Δ and that function f must be a solution of equation (D.40), it is easy to show that for nonlinear equations of the form (D.37) the number γ must be zero. Now, condition (D.4) is satisfied completely. An equation (D.36) with a function g that is a solution of equation (D.40) satisfies condition (D.4).

It remains to look at equations of the form (D.38). For these equations, condition (D.4) implies additional conditions on functions f and g.

With this aim we introduce the function $F = (\delta/\delta u)\ln\left(\partial\varphi/\partial u_1\right)$ where φ is the vector field of equation (D.38). As the function $\ln\left(\partial\varphi/\partial u_1\right)$ is a density of a conservation law of equation (D.38), then the function F must satisfy (cf. (D.8))

$$\partial_\varphi F + \sum_i \frac{\partial \varphi_i}{\partial u} F_i = 0.$$

Differentiating this equation with respect to u_1 yields

$$\partial_\varphi \frac{\partial F}{\partial u_1} + \frac{\partial F}{\partial u}\frac{\partial \varphi}{\partial u_1} + \frac{\partial F}{\partial u_1}\frac{\partial \varphi_1}{\partial u_1} + \frac{\partial^2 \varphi}{\partial u_1 \partial u} F + \frac{\partial^2 \varphi_1}{\partial u_1 \partial u} F_1 + \frac{\partial \varphi}{\partial u}\frac{\partial F}{\partial u_1} + \frac{\partial \varphi}{\partial u_1}\frac{\partial F_1}{\partial u_1} = 0.$$

Using the identities

$$\frac{\partial \varphi_1}{\partial u} = -\frac{\partial \varphi}{\partial u_1}, \quad \frac{\partial F}{\partial u_1} = \frac{\partial^2 \ln(\frac{\partial\varphi}{\partial u_1})}{\partial u_1 \partial u} = \sigma \frac{\partial \varphi}{\partial u_1},$$

(see (D.34)) we rewrite the last relation in the form

$$\sigma\partial_\varphi \frac{\partial \varphi}{\partial u_1} + \frac{\partial \varphi}{\partial u_1}\left(\frac{\partial F}{\partial u} - \frac{\partial F_1}{\partial u_1}\right) + \sigma\frac{\partial \varphi}{\partial u_1}\left(\frac{\partial \varphi_1}{\partial u_1} + \frac{\partial \varphi}{\partial u}\right) + \frac{\partial^2 \varphi}{\partial u_1 \partial u} F - \frac{\partial^2 \varphi}{\partial u_1 \partial u_1} F_1 = 0.$$

Dividing both sides of this equality by $\partial\varphi/\partial u_1$

$$F\frac{\partial}{\partial u}\ln\frac{\partial \varphi}{\partial u_1} - F_1 \frac{\partial}{\partial u_1}\ln\frac{\partial \varphi}{\partial u_1} = \Delta P.$$

Recalling the definition of function F we rewrite this condition in the form

$$\left(\frac{\partial}{\partial u}\ln\frac{\partial \varphi}{\partial u_1}\right)^2 - \left(\frac{\partial}{\partial u_1}\ln\frac{\partial \varphi}{\partial u_1}\right)^2 = \Delta Q$$

Expressing the function φ in terms of the functions f and g we finally obtain a condition over f and g:

$$\text{(D.41)} \qquad \frac{f''(u_1+u)g''(u_1-u)}{\left[f'(u_1+u) - g'(u_1-u)\right]^2} = \Delta R.$$

From this it is not difficult to obtain the following condition that will facilitate further work: if equation (D.38) is nonlinear and functions g', g'' are finite in zero, then $g''(0) = 0$.

Thus, functions f', g' satisfy the equation

$$\left(h'\right)^2 = 2\sigma h^3 + \alpha h^2 + \beta h + \gamma$$

with the same coefficients. Suppose $\sigma = 0$. If $\alpha = \beta = 0$, $\gamma \neq 0$ condition (D.41) is not satisfied because on its lhs there is a function of only one variable (see property (3) of Δ).

If $\alpha = 0$, $\beta \neq 0$, using the consequences of condition (D.41) we obtain

$$f(x) = \frac{\beta}{12}(x+c)^3 + \gamma\beta^{-1}x + c_1, \quad g(x) = \frac{\beta}{12}x^3 + \gamma\beta^{-1}x + c_2,$$

where c_1, c_2, c are constants. After a shift and scaling of the variable u and a scaling of the time t, equation (D.38) takes the form

$$u_t = u\left(u_1^2 - u_{-1}^2\right) + \eta, \quad \eta \in \mathbb{C}.$$

Checking condition (D.4) proves that $\eta = 0$, and we obtain an equation equivalent to (D.20).

If $\alpha \neq 0$, $4\alpha\gamma = \beta^2$ condition (D.41) cannot be satisfied.

If $\alpha \neq 0$, $4\alpha\gamma \neq \beta^2$ using condition (D.41) we obtain the equation

$$u_t = \sinh u\left(\cosh u_1 - \cosh u_{-1}\right) + \eta, \quad \eta \in \mathbb{C}.$$

Condition (D.4) implies that $\eta = 0$. We obtain en equation equivalent to (D.22).

It remains to study the case $\sigma \neq 0$ in equation (D.39). There exist numbers $a, b \in \mathbb{C}$ such that $h = as + b$ with

$$s'^2 = 4s^3 - g_2 s - g_3, \quad g_2, g_3 \subset \mathbb{C}.$$

Thus functions f and g are of the form $f'(x) = a\wp(x + c_1) + b$, $g'(x) = a\wp(x + c_2) + b$, with $c_1, c_2 \in \mathbb{C}$, $\wp$ is any Weirstrass function with invariants g_2, g_3. From equation (D.41) (n.t. and the comment thereafter) follows the equality

$$\wp(2x + c_1 - c_2) = \wp(2x + c_1 + c_2)$$

i.e. c_2 is a half-period of the function $\wp$. Thus, after a point transformation $t \mapsto -\alpha^{-1}t$, $u \mapsto u + c_1 - c_2$ equation (D.38) becomes

$$u_t = \zeta(u_1 + u + \theta) - \zeta(u_1 - u + \theta) - \zeta(u + u_{-1} + \theta) - \zeta(u - u_{-1} + \theta) + \eta, \tag{D.42}$$

where $\eta \in \mathbb{C}$, $\theta = c_2$. Using the formula

$$\varphi = \eta + \zeta(2u) - \zeta(2u + 2\theta) - \frac{1}{2}\left(\frac{\partial}{\partial u}\ln\frac{\partial\varphi}{\partial u_1} - \frac{\partial}{\partial u}\ln\frac{\partial\varphi_{-1}}{\partial u}\right)$$

$$\partial_\varphi \ln\frac{\partial\varphi}{\partial u_1} = \varphi\frac{\partial}{\partial u}\ln\frac{\partial\varphi}{\partial u_1} + \varphi_1\frac{\partial}{\partial u_1}\ln\frac{\partial\varphi}{\partial u_1}$$

it is possible to show that equation (D.42) satisfies condition (D.4) in and only in the case when $\eta = \zeta(2u + 2\lambda) - \zeta(2u)$. With this value of η, it is not difficult to write Equation (D.42) in the form (D.25). Theorem 3 is proved. □

Annex. Currently, the author has succeeded in completing the work begun here on the classification of discrete equations of the form (D.1). The result is a complete list of nonlinear equations of the form (D.1) satisfying conditions (D.2–D.4). A central role in the proof is played by finding and investigating overdetermined (systems) of type(D.27). The proof given here of the theorem about equations of type I can be improved. It has been proved that equations (D.5) and (D.6) admit an infinite number of conservation laws. In the light of the last results, the most interesting equations of the form (D.1) satisfying conditions (D.2–D.4) seem to the author to be equation (D.5) and a generalization of equation (D.6). This material is under preparation for publication [843].

References:

[1] B. A. Dubrovin, V. B. Matveev, S. P. Novikov, *Non linear equations of Korteweg–de Vries type, finite-zone linear operators, and Abelian varieties*, Uspekhi Mat. Nauk, 1976, Volume 31, Issue 1(187), Pages 5–136 [English translation] Russian Mathematical Surveys, 1976, 31:1, 59–146.

[2] N. Kh. Ibragimov, A. B. Shabat, *Evolutionary equations with nontrivial Lie–Bäcklund group*, Funktsional. Anal. i Prilozhen., 14:1 (1980), 25–36 [English translation] Funct. Anal. Appl., 14:1 (1980), 19–28 [416].

[3] N. Kh. Ibragimov, A. B. Shabat, *Infinite Lie–Bäcklund algebras*, Funktsional. Anal. i Prilozhen., 14:4 (1980), 79–80 [English translation] Funct. Anal. Appl., 14:4 (1980), 313–315 [417].

[4] A. N. Leznov, M. V. Saveliev, V. G. Smirnov, *General solutions of the two-dimensional system of Volterra equations which realize the Bäcklund transformation for the Toda lattice*, TMF, 47:2 (1981), 216–224 [English translation] Theoret. and Math. Phys., 47:2 (1981), 417–422.

[5] R. I. Yamilov, *On conservation laws for the difference Korteweg-de Vries equation*, Din. Splosh. Sredy, 44 (1980), 164–173 , Russian Academy of Sciences - RAS (Rossiĭskaya Akademiya Nauk - RAN), Siberian Branch (Sibirskoe Otdelenie), Institute of Hydrodynamics named after M. A.
Lavrent'eva (Institut Gidrodinamiki Im. M. A. Lavrent'eva), Novosibirsk. [840]

APPENDIX E

No quad-graph equation can have a generalized symmetry given by the Narita-Itoh-Bogoyavlensky equation

Theorem *No quad-graph equation of the form (3.6.2, 3.6.3) can have a generalized symmetry of the form of the Narita-Itoh-Bogoyavlensky equation (3.6.38).*

PROOF. We use (3.6.20, 3.6.21) with $\ell = 2$ and $\ell' = -2$. Applying the operators T_1^{-1} and $-T_1$, we rewrite them in the form:

$$p^{(1)}_{1,0} - p^{(1)}_{-1,0} = \log \frac{u_{0,0}}{u_{0,1}}, \tag{E.1}$$

$$p^{(2)}_{1,0} - p^{(2)}_{-1,0} = \log \frac{u_{1,1}}{u_{1,0}}, \tag{E.2}$$

where $p^{(1)}_{0,0}$, $p^{(2)}_{0,0}$ are given by (3.6.27, 3.6.28). Studying (E.1, E.2), we will use in addition to (3.6.2) its equivalent form:

$$u_{-1,1} = f^{(-1,1)} = f^{(-1,1)}(u_{-1,0}, u_{0,0}, u_{0,1}).$$

The functions $p^{(j)}_{0,0}$, $j = 1, 2$ have the structure $p^{(j)}_{0,0} = P^{(j)}(u_{1,0}, u_{0,0}, u_{0,1})$. Therefore $p^{(j)}_{-1,0} = P^{(j)}(u_{0,0}, u_{-1,0}, \hat{f}_{0,0})$ and the right hand sides of (E.1, E.2) do not depend on $u_{2,0}$. The functions $p^{(j)}_{1,0} = P^{(j)}(u_{2,0}, u_{1,0}, f^{(-1,1)})$ depend on $u_{2,0}$, and from (E.1, E.2) we get: $\partial_{u_{2,0}} p^{(j)}_{1,0} = T_1 \partial_{u_{1,0}} p^{(j)}_{0,0} = 0$. Moreover, according to Proposition 5,

$$\partial_{u_{1,0}} p^{(j)}_{0,0} = 0, \qquad j = 1, 2. \tag{E.3}$$

From (E.3) with $j = 1$ we get $f^{(-1,1)}_{u_{1,0}u_{1,0}} = 0$, i.e.

$$f^{(-1,1)} = a_{0,0} u_{1,0} + b_{0,0} = A(u_{0,0}, u_{0,1}) u_{1,0} + B(u_{0,0}, u_{0,1}), \tag{E.4}$$

where $a_{0,0} \neq 0$ due to (3.6.3). Now $p^{(1)}_{0,0} = \log a_{0,0}$ and (E.1) is rewritten as:

$$\frac{a_{1,0}}{a_{-1,0}} = \frac{u_{0,0}}{u_{0,1}}. \tag{E.5}$$

Here only $a_{1,0}$ depends on $u_{1,0}$, and we get:

$$\frac{d a_{1,0}}{d u_{1,0}} = \partial_{u_{1,0}} a_{1,0} + a_{0,0} \partial_{u_{1,1}} a_{1,0} = 0.$$

Applying to it the shift operator T_1^{-1}, we get the more convenient form:

$$\partial_{u_{0,0}} a_{0,0} + a_{-1,0} \partial_{u_{0,1}} a_{0,0} = 0. \tag{E.6}$$

As $a_{-1,0} \neq 0$, only two cases are possible. The first one is when $\partial_{u_{0,0}} a_{0,0} = \partial_{u_{0,1}} a_{0,0} = 0$, i.e. $a_{0,0}$ is a constant. This is in contradiction with (E.5). So, $\partial_{u_{0,0}} a_{0,0} \neq 0$ and $\partial_{u_{0,1}} a_{0,0} \neq 0$.

From (E.3) with $j = 2$ we get

$$\frac{f^{(-1,1)}_{u_{0,0}u_{1,0}}}{f^{(-1,1)}_{u_{0,0}}} - \frac{f^{(-1,1)}_{u_{0,1}u_{1,0}}}{f^{(-1,1)}_{u_{0,1}}} = 0.$$

Using this equation together with (E.4, E.6), we get:

$$p^{(2)}_{0,0} = \log \frac{f^{(-1,1)}_{u_{0,0}u_{1,0}}}{f^{(-1,1)}_{u_{0,1}u_{1,0}}} = \log \frac{\partial_{u_{0,0}} a_{0,0}}{\partial_{u_{0,1}} a_{0,0}} = \log(-a_{-1,0}).$$

Applying T_1 we can rewrite (E.2) as:

$$\frac{a_{1,0}}{a_{-1,0}} = \frac{u_{2,1}}{u_{2,0}}. \tag{E.7}$$

Comparing (E.5, E.7) and using (E.4), we get

$$u_{2,1} = \frac{u_{0,0}}{u_{0,1}} u_{2,0} = a_{1,0}u_{2,0} + b_{1,0}.$$

As $a_{1,0}$, $b_{1,0}$ do not depend on $u_{2,0}$, we obtain from here $a_{1,0} = \frac{u_{0,0}}{u_{0,1}}$. Then from (E.5) we obtain $a_{-1,0} = 1$. These two last results are in contradiction, thus proving this theorem. □

Bibliography

[1] F. Kh. Abdullaev, B. B. Baizakov, S. A. Darmanyan, V. V. Konotop and M. Salerno, *Nonlinear excitations in arrays of Bose-Einstein condensates*, Phys. Rev. A **64** (2001), 043606 (10pp).

[2] N. H. Abel, *Méthode générale pour trouver des fonctions d'une seule quantité variable lorsqu'une propriété de ces fonctions est exprimé par une équation entre deux variables*, Mag. Naturvidenskab **1**/2 (1823), 1–10 (contained in *Oeuvres completes*, Vol. **I**, 1–10, Christiania, 1881).

[3] M. J. Ablowitz and P. A. Clarkson, *Solitons, nonlinear evolution equations and inverse scattering*, London Mathematical Society Lecture Note Series, vol. 149, Cambridge University Press, Cambridge, 1991, DOI 10.1017/CBO9780511623998. MR1149378

[4] M. J. Ablowitz, R. Halburd, and B. Herbst, *On the extension of the Painlevé property to difference equations*, Nonlinearity **13** (2000), no. 3, 889–905, DOI 10.1088/0951-7715/13/3/321. MR1759006

[5] M. J. Ablowitz, D. J. Kaup, A. C. Newell, and H. Segur, *The inverse scattering transform-Fourier analysis for nonlinear problems*, Studies in Appl. Math. **53** (1974), no. 4, 249–315, DOI 10.1002/sapm1974534249. MR450815

[6] M. J. Ablowitz and J. F. Ladik, *Nonlinear differential-difference equations*, J. Mathematical Phys. **16** (1975), 598–603, DOI 10.1063/1.522558. MR377223

[7] M. J. Ablowitz and J. F. Ladik, *A nonlinear difference scheme and inverse scattering*, Studies in Appl. Math. **55** (1976), no. 3, 213–229, DOI 10.1002/sapm1976553213. MR471341

[8] M. J. Ablowitz and J. F. Ladik, *Nonlinear differential-difference equations and Fourier analysis*, J. Mathematical Phys. **17** (1976), no. 6, 1011–1018, DOI 10.1063/1.523009. MR427867

[9] M. J. Ablowitz, B. Prinari, and A. D. Trubatch, *Discrete and continuous nonlinear Schrödinger systems*, London Mathematical Society Lecture Note Series, vol. 302, Cambridge University Press, Cambridge, 2004. MR2040621

[10] M. J. Ablowitz, A. Ramani, and H. Segur, *Nonlinear evolution equations and ordinary differential equations of Painlevé type*, Lett. Nuovo Cimento (2) **23** (1978), no. 9, 333–338. MR513779

[11] M. J. Ablowitz, A. Ramani and H. Segur, *A connection between nonlinear evolution equations and ordinary differential equations of P-type. I and II* J. Math. Phys. **21** (1980), 715–721 and 1006–1015.

[12] M. J. Ablowitz and H. Segur, *Solitons and the inverse scattering transform*, SIAM Studies in Applied Mathematics, vol. 4, Society for Industrial and Applied Mathematics (SIAM), Philadelphia, Pa., 1981. MR642018

[13] M. Abramowitz and I. A. Stegun (eds.), *Handbook of mathematical functions, with formulas, graphs, and mathematical tables*, Dover Publications, Inc., New York, 1966. MR0208797

[14] J. Aczél, *Lectures on functional equations and their applications*, Mathematics in Science and Engineering, Vol. 19, Academic Press, New York-London, 1966. Translated by Scripta Technica, Inc. Supplemented by the author. Edited by Hansjorg Oser. MR0208210

[15] J. Aczél, (ed.) *Functional equations: history, applications and theory*, Reidel, Amsterdam, 1984.

[16] V. È. Adler, *Nonlinear superposition principle for the Jordan NLS equation*, Phys. Lett. A **190** (1994), no. 1, 53–58, DOI 10.1016/0375-9601(94)90365-4. MR1283771

[17] V. È. Adler, *Bäcklund transformation for the Krichever-Novikov equation*, Internat. Math. Res. Notices **1** (1998), 1–4, DOI 10.1155/S1073792898000014. MR1601866

[18] V. È. Adler, *Legendre transformations on a triangular lattice* (Russian, with Russian summary), Funktsional. Anal. i Prilozhen. **34** (2000), no. 1, 1–11, 96, DOI 10.1007/BF02467062; English transl., Funct. Anal. Appl. **34** (2000), no. 1, 1–9. MR1747820

[19] V. È. Adler, *On the structure of the Bäcklund transformations for the relativistic lattices*, J. Nonlinear Math. Phys. **7** (2000), no. 1, 34–56, DOI 10.2991/jnmp.2000.7.1.4. MR1737252

[20] V. È. Adler, *Discrete equations on planar graphs*, J. Phys. A **34** (2001), no. 48, 10453–10460, DOI 10.1088/0305-4470/34/48/310. Symmetries and integrability of difference equations (Tokyo, 2000). MR1877466

[21] V. È. Adler, *Necessary integrability conditions for evolutionary lattice equations*, Theoret. and Math. Phys. **181** (2014), no. 2, 1367–1382, DOI 10.1007/s11232-014-0218-2. Russian version appears in Teoret. Mat. Fiz. **181** (2014), no. 2, 276–295. MR3344450

[22] V. È. Adler, A. I. Bobenko, and Yu. B. Suris, *Classification of integrable equations on quad-graphs. The consistency approach*, Comm. Math. Phys. **233** (2003), no. 3, 513–543, DOI 10.1007/s00220-002-0762-8. MR1962121

[23] V. È. Adler, A. I. Bobenko, and Yu. B. Suris, *Discrete nonlinear hyperbolic equations: classification of integrable cases* (Russian, with Russian summary), Funktsional. Anal. i Prilozhen. **43** (2009), no. 1, 3–21, DOI 10.1007/s10688-009-0002-5; English transl., Funct. Anal. Appl. **43** (2009), no. 1, 3–17. MR2503862

[24] V. È. Adler and A. B. Shabat, *On a class of Toda chains* (Russian, with English and Russian summaries), Teoret. Mat. Fiz. **111** (1997), no. 3, 323–334, DOI 10.1007/BF02634053; English transl., Theoret. and Math. Phys. **111** (1997), no. 3, 647–657. MR1472211

[25] V. È. Adler and A. B. Shabat, *Generalized Legendre transformations* (Russian, with English and Russian summaries), Teoret. Mat. Fiz. **112** (1997), no. 2, 179–194, DOI 10.1007/BF02634155; English transl., Theoret. and Math. Phys. **112** (1997), no. 2, 935–946 (1998). MR1483275

[26] V. È. Adler and A. B. Shabat, *First integrals of generalized Toda chains* (Russian, with Russian summary), Teoret. Mat. Fiz. **115** (1998), no. 3, 349–357, DOI 10.1007/BF02575487; English transl., Theoret. and Math. Phys. **115** (1998), no. 3, 639–646. MR1692410

[27] V. È. Adler, A. B. Shabat, and R. I. Yamilov, *The symmetry approach to the integrability problem* (Russian, with Russian summary), Teoret. Mat. Fiz. **125** (2000), no. 3, 355–424, DOI 10.1023/A:1026602012111; English transl., Theoret. and Math. Phys. **125** (2000), no. 3, 1603–1661. MR1839659

[28] V. È. Adler and S. Ya. Startsev, *On discrete analogues of the Liouville equation* (Russian, with Russian summary), Teoret. Mat. Fiz. **121** (1999), no. 2, 271–284, DOI 10.1007/BF02557219; English transl., Theoret. and Math. Phys. **121** (1999), no. 2, 1484–1495. MR1761918

[29] V. È. Adler and Yu. B. Suris, Q_4*: integrable master equation related to an elliptic curve*, Int. Math. Res. Not. **47** (2004), 2523–2553, DOI 10.1155/S107379280413273X. MR2077841

[30] V. È. Adler, S. I. Svinolupov, and R. I. Yamilov, *Multi-component Volterra and Toda type integrable equations*, Phys. Lett. A **254** (1999), no. 1-2, 24–36, DOI 10.1016/S0375-9601(99)00087-0. MR1688099

[31] V. È. Adler and A. P. Veselov, *Cauchy problem for integrable discrete equations on quad-graphs*, Acta Appl. Math. **84** (2004), no. 2, 237–262, DOI 10.1007/s10440-005-4557-8. MR2114898

[32] V. È. Adler, I. Yu. Cherdantsev, R. N. Garifullin, and et al., *In memory of Yamilov Ravil Islamovich*, Ufa Math. J. **12** (2020), no. 2, 119–120. MR4130349

[33] V. È. Adler and R. I. Yamilov, *Explicit auto-transformations of integrable chains*, J. Phys. A **27** (1994), no. 2, 477–492. MR1267427

[34] R. P. Agarwal, *Difference equations and inequalities: Theory, methods, and applications*, Monographs and Textbooks in Pure and Applied Mathematics, vol. 155, Marcel Dekker, Inc., New York, 1992. MR1155840

[35] V. Akcelik, B. Jaramaz, and O. Ghattas, *Nearly orthogonal two-dimensional grid generation with aspect ratio control*, J. Comput. Phys. **171** (2001), no. 2, 805–821, DOI 10.1006/jcph.2001.6811. MR1848734

[36] W. F. Ames, *Nonlinear partial differential equations in engineering*, Academic Press, New York-London, 1965. MR0210342

[37] W. F. Ames, R. L. Anderson, V. A. Dorodnitsyn, E. V. Ferapontov, R. K. Gazizov, N. H. Ibragimov, and S. R. Svirshchevskiĭ, *CRC handbook of Lie group analysis of differential equations. Vol. 1: Symmetries, exact solutions and conservation laws*, CRC Press, Boca Raton, FL, 1994. MR1278257

[38] S. C. Anco and G. Bluman, *Direct construction of conservation laws from field equations*, Phys. Rev. Lett. **78** (1997), no. 15, 2869–2873, DOI 10.1103/PhysRevLett.78.2869. MR1442661

[39] S. C. Anco and A. H. Kara, *Symmetry-invariant conservation laws of partial differential equations*, European J. Appl. Math. **29** (2018), no. 1, 78–117, DOI 10.1017/S0956792517000055. MR3742532

[40] I. M. Anderson, M. E. Fels, and P. J. Vassiliou, *Superposition formulas for exterior differential systems*, Adv. Math. **221** (2009), no. 6, 1910–1963, DOI 10.1016/j.aim.2009.03.010. MR2522831

[41] I. M. Anderson and N. Kamran, *The variational bicomplex for hyperbolic second-order scalar partial differential equations in the plane*, Duke Math. J. **87** (1997), no. 2, 265–319, DOI 10.1215/S0012-7094-97-08711-1. MR1443529

[42] R. L. Anderson, *A nonlinear superposition principle admitted by coupled Riccati equations of the projective type*, Lett. Math. Phys. **4** (1980), no. 1, 1–7, DOI 10.1007/BF00419796. MR565519

[43] R. L. Anderson, V. A. Baikov, R. K. Gazizov, W. Hereman, N. H. Ibragimov, F. M. Mahomed, S. V. Meleshko, M. C. Nucci, P. J. Olver, M. B. Sheftel′, A. V. Turbiner, and E. M. Vorob′ev, *CRC handbook of Lie group analysis of differential equations. Vol. 3: New trends in theoretical developments and computational methods; With 1 IBM-PC floppy disk (3.5 inch; HD)*, CRC Press, Boca Raton, FL, 1996. MR1383090

[44] R. L. Anderson, J. Harnad, and P. Winternitz, *Systems of ordinary differential equations with nonlinear superposition principles*, Phys. D **4** (1981/82), no. 2, 164–182, DOI 10.1016/0167-2789(82)90058-6. MR653771

[45] R. L. Anderson and N. H. Ibragimov, *Lie-Bäcklund transformations in applications*, SIAM Studies in Applied Mathematics, vol. 1, Society for Industrial and Applied Mathematics (SIAM), Philadelphia, Pa., 1979. MR520395

[46] D. V. Anosov and A. A. Bolibruch, *The Riemann-Hilbert problem*, Aspects of Mathematics, E22, Friedr. Vieweg & Sohn, Braunschweig, 1994, DOI 10.1007/978-3-322-92909-9. MR1276272

[47] V. I. Arnol′d, *Mathematical methods of classical mechanics*, Graduate Texts in Mathematics, vol. 60, Springer-Verlag, New York-Heidelberg, 1978. Translated from the Russian by K. Vogtmann and A. Weinstein. MR0690288

[48] D. J. Arrigo, *Symmetry analysis of differential equations: An introduction*, John Wiley & Sons, Inc., Hoboken, NJ, 2015. MR3328724

[49] D. J. Arrigo, P. Broadbridge, and J. M. Hill, *Nonclassical symmetry solutions and the methods of Bluman-Cole and Clarkson-Kruskal*, J. Math. Phys. **34** (1993), no. 10, 4692–4703, DOI 10.1063/1.530365. MR1235966

[50] D. J. Arrigo, J. M. Goard, and P. Broadbridge, *Nonclassical solutions are non-existent for the heat equation and rare for nonlinear diffusion*, J. Math. Anal. Appl. **202** (1996), no. 1, 259–279, DOI 10.1006/jmaa.1996.0316. MR1402601

[51] V. Artale and D. Levi, *Nonlinear internal waves in a one-dimensional channel* (English, with Italian summary), Nuovo Cimento C (1) **10** (1987), no. 1, 61–76, DOI 10.1007/BF02508697. MR985017

[52] V. Artale and D. Levi, *Nonlinear surface and internal waves in stratified shear flow*, Geoph. Astroph. Fluid Dyn. **54** (1990), 35–48.

[53] V. Artale, D. Levi, S. Marullo and R. Santoleri, *Analysis of nonlinear internal waves observed by Landsat thematic mapper*, J. Geophys. Res. **95** (1990), 16065–16073.

[54] U. M. Ascher and R. I. McLachlan, *Multisymplectic box schemes and the Korteweg-de Vries equation*, Appl. Numer. Math. **48** (2004), no. 3-4, 255–269, DOI 10.1016/j.apnum.2003.09.002. Workshop on Innovative Time Integrators for PDEs. MR2056917

[55] U. M. Ascher and L. R. Petzold, *Computer methods for ordinary differential equations and differential-algebraic equations*, Society for Industrial and Applied Mathematics (SIAM), Philadelphia, PA, 1998, DOI 10.1137/1.9781611971392. MR1638643

[56] A. Fratalocchi and G. Assanto, *Universal character of the discrete nonlinear Schrödinger equation*, Phys. Rev. A (3) **76** (2007), no. 4, 042108, 6, DOI 10.1103/PhysRevA.76.042108. MR2486290

[57] C. Athorne and A. Fordy, *Generalised KdV and MKdV equations associated with symmetric spaces*, J. Phys. A **20** (1987), no. 6, 1377–1386. MR893326

[58] C. Athorne and A. Fordy, *Integrable equations in* $(2+1)$ *dimensions associated with symmetric and homogeneous spaces*, J. Math. Phys. **28** (1987), no. 9, 2018–2024, DOI 10.1063/1.527463. MR904414

[59] C. Athorne and J. J. C. Nimmo, *On the Moutard transformation for integrable partial differential equations*, Inverse Problems **7** (1991), no. 6, 809–826. MR1140316

[60] D. Atkinson, *The spectral transform*, Internal report Groninghen University N. 155 (1979).

[61] J. Atkinson, *Bäcklund transformations for integrable lattice equations*, J. Phys. A **41** (2008), no. 13, 135202, 8, DOI 10.1088/1751-8113/41/13/135202. MR2451510

[62] J. Atkinson, *Linear quadrilateral lattice equations and multidimensional consistency*, J. Phys. A **42** (2009), no. 45, 454005, 7, DOI 10.1088/1751-8113/42/45/454005. MR2556647

[63] J. Atkinson, J. Hietarinta, and F. Nijhoff, *Seed and soliton solutions for Adler's lattice equation*, J. Phys. A **40** (2007), no. 1, F1–F8, DOI 10.1088/1751-8113/40/1/F01. MR2304925

[64] J. Atkinson, J. Hietarinta, and F. Nijhoff, *Soliton solutions for Q3*, J. Phys. A **41** (2008), no. 14, 142001, 11, DOI 10.1088/1751-8113/41/14/142001. MR2450374

[65] J. Atkinson and F. Nijhoff, *Solutions of Adler's lattice equation associated with 2-cycles of the Bäcklund transformation*, J. Nonlinear Math. Phys. **15** (2008), no. suppl. 3, 34–42, DOI 10.2991/jnmp.2008.15.s3.4. MR2452416

[66] J. Atkinson and F. Nijhoff, *A constructive approach to the soliton solutions of integrable quadrilateral lattice equations*, Comm. Math. Phys. **299** (2010), no. 2, 283–304, DOI 10.1007/s00220-010-1076-x. MR2679812

[67] F. Nijhoff and J. Atkinson, *Elliptic N-soliton solutions of ABS lattice equations*, Int. Math. Res. Not. IMRN **20** (2010), 3837–3895, DOI 10.1093/imrn/rnq010. MR2738345

[68] M. Audin, *Spinning tops: A course on integrable systems*, Cambridge Studies in Advanced Mathematics, vol. 51, Cambridge University Press, Cambridge, 1996. MR1409362

[69] O. Babelon, D. Bernard, and M. Talon, *Introduction to classical integrable systems*, Cambridge Monographs on Mathematical Physics, Cambridge University Press, Cambridge, 2003, DOI 10.1017/CBO9780511535024. MR1995460

[70] V. A. Baĭkov, R. K. Gazizov, and N. Kh. Ibragimov, *Approximate symmetries* (Russian), Mat. Sb. (N.S.) **136(178)** (1988), no. 4, 435–450, 590, DOI 10.1070/SM1989v064n02ABEH003318; English transl., Math. USSR-Sb. **64** (1989), no. 2, 427–441. MR965885

[71] M. Bakirova, V. Dorodnitsyn and R. Kozlov, *Invariant difference schemes for heat transfer equations with a source*, J. Phys. A: Math. Gen. **30**:23 (1997), 8139–8155.

[72] A. V. Bäcklund, *Ueber Flächentransformationen* (German), Math. Ann. **9** (1875), no. 3, 297–320, DOI 10.1007/BF01443337. MR1509862

[73] V. Bargmann, *On the connection between phase shifts and scattering potential*, Rev. Modern Physics **21** (1949), 488–493, DOI 10.1103/revmodphys.21.488. MR0032069

[74] H. Bateman, *Some recent researches on the motion of fluids*, Mon. Weather Rev. **43** (1915), 163–170.

[75] G. Baumann, *Symmetry analysis of differential equations with Mathematica®*, Springer-Verlag–TELOS, New York, 2000. With 1 CD-ROM (Windows, Macintosh and UNIX), DOI 10.1007/978-1-4612-2110-4. MR1753536

[76] R. J. Baxter, *Exactly solved models in statistical mechanics*, Academic Press, Inc. [Harcourt Brace Jovanovich, Publishers], London, 1982. MR690578

[77] J. Beckers, V. Hussin, and P. Winternitz, *Nonlinear equations with superposition formulas and the exceptional group G_2. I. Complex and real forms of $\mathfrak{g}_2$ and their maximal subalgebras*, J. Math. Phys. **27** (1986), no. 9, 2217–2227, DOI 10.1063/1.526993. MR854752

[78] J. Beckers, V. Hussin, and P. Winternitz, *Nonlinear equations with superposition formulas and the exceptional group G_2. II. Classification of the equations*, J. Math. Phys. **28** (1987), no. 3, 520–529, DOI 10.1063/1.527636. MR877223

[79] J. Beckers, L. Gagnon, V. Hussin, and P. Winternitz, *Superposition formulas for nonlinear superequations*, J. Math. Phys. **31** (1990), no. 10, 2528–2534, DOI 10.1063/1.528997. MR1072969

[80] V. Belinski and E. Verdaguer, *Gravitational solitons*, Cambridge Monographs on Mathematical Physics, Cambridge University Press, Cambridge, 2001, DOI 10.1017/CBO9780511535253. MR1839019

[81] M. P. Bellon, J.-M. Maillard, and C.-M. Viallet, *Rational mappings, arborescent iterations, and the symmetries of integrability*, Phys. Rev. Lett. **67** (1991), no. 11, 1373–1376, DOI 10.1103/PhysRevLett.67.1373. MR1123221

[82] M. P. Bellon and C.-M. Viallet, *Algebraic entropy*, Comm. Math. Phys. **204** (1999), no. 2, 425–437, DOI 10.1007/s002200050652. MR1704282

[83] E. D. Belokolos, A. I. Bobenko, V. Z. Enol'skii, A. R. Its and V. B. Matveev, *Algebro-geometric approach to nonlinear integrable equations*, Springer Berlin, Heidelberg, 1994.

[84] G. Berkeley, A. V. Mikhailov, and P. Xenitidis, *Darboux transformations with tetrahedral reduction group and related integrable systems*, J. Math. Phys. **57** (2016), no. 9, 092701, 15, DOI 10.1063/1.4962803. MR3549644

[85] J. Bertrand, *Théorème relatif au mouvement d'un point attiré vers un centre fixe*, C.R. Acad. Sci. Paris **77** (1873), 849–853.

[86] E. Bessel-Hagen, *Über die Erhaltungssätze der Elektrodynamik* (German), Math. Ann. **84** (1921), no. 3-4, 258–276, DOI 10.1007/BF01459410. MR1512036

[87] F. Beukers, J. A. Sanders, and J. P. Wang, *One symmetry does not imply integrability*, J. Differential Equations **146** (1998), no. 1, 251–260, DOI 10.1006/jdeq.1998.3426. MR1625751

[88] L. Bianchi, *Sulla trasformazione di Bäcklund per le superficie pseudosferiche*, Rend. Lincei Serie V **1** (1892), 3–12.

[89] A. Bihlo, *Invariant meshless discretization schemes*, J. Phys. A **46** (2013), no. 6, 062001, 12, DOI 10.1088/1751-8113/46/6/062001. MR3020279

[90] A. Bihlo, X. Coiteux-Roy, and P. Winternitz, *The Korteweg-de Vries equation and its symmetry-preserving discretization*, J. Phys. A **48** (2015), no. 5, 055201, 25, DOI 10.1088/1751-8113/48/5/055201. MR3300275

[91] A. Bihlo, J. Jackaman, and F. Valiquette, *On the development of symmetry-preserving finite element schemes for ordinary differential equations*, J. Comput. Dyn. **7** (2020), no. 2, 339–368, DOI 10.3934/jcd.2020014. MR4128814

[92] A. Bihlo, J. Jackaman, and F. Valiquette, *Invariant variational schemes for ordinary differential equations*, Stud. Appl. Math. **148** (2022), no. 3, 991–1020, DOI 10.1111/sapm.12466. MR4429078

[93] A. Bihlo and J.-C. Nave, *Invariant discretization schemes using evolution-projection techniques*, SIGMA Symmetry Integrability Geom. Methods Appl. **9** (2013), Paper 052, 23, DOI 10.3842/SIGMA.2013.052. MR3116188

[94] A. Bihlo and R. O. Popovych, *Invariant discretization schemes for the shallow-water equations*, SIAM J. Sci. Comput. **34** (2012), no. 6, B810–B839, DOI 10.1137/120861187. MR3023746

[95] A. Bihlo and F. Valiquette, *Symmetry-preserving numerical schemes*, Symmetries and integrability of difference equations, CRM Ser. Math. Phys., Springer, Cham, 2017, pp. 261–324. MR3677354

[96] A. Bihlo and F. Valiquette, *Symmetry-preserving finite element schemes: an introductory investigation*, SIAM J. Sci. Comput. **41** (2019), no. 5, A3300–A3325, DOI 10.1137/18M1177524. MR4021276

[97] G. W. Bluman and S. C. Anco, *Symmetry and integration methods for differential equations*, Applied Mathematical Sciences, vol. 154, Springer-Verlag, New York, 2002. MR1914342

[98] G. W. Bluman, A. F. Cheviakov, and S. C. Anco, *Applications of symmetry methods to partial differential equations*, Applied Mathematical Sciences, vol. 168, Springer, New York, 2010, DOI 10.1007/978-0-387-68028-6. MR2561770

[99] G. W. Bluman and J. D. Cole, *The general similarity solution of the heat equation*, J. Math. Mech. **18** (1968/69), 1025–1042. MR0293257

[100] G. W. Bluman and S. Kumei, *Symmetries and differential equations*, Applied Mathematical Sciences, vol. 81, Springer-Verlag, New York, 1989, DOI 10.1007/978-1-4757-4307-4. MR1006433

[101] G. W. Bluman and S. Kumei, *Symmetry-based algorithms to relate partial differential equations. I. Local symmetries*, European J. Appl. Math. **1** (1990), no. 3, 189–216, DOI 10.1017/S0956792500000176. MR1117349

[102] A. I. Bobenko and Yu. B. Suris, *Integrable systems on quad-graphs*, Int. Math. Res. Not. **11** (2002), 573–611, DOI 10.1155/S1073792802110075. MR1890049

[103] A. I. Bobenko and Yu. B. Suris, *Integrable noncommutative equations on quad-graphs. The consistency approach*, Lett. Math. Phys. **61** (2002), no. 3, 241–254, DOI 10.1023/A:1021249131979. MR1942362

[104] A. I. Bobenko and Yu. B. Suris, *Discrete differential geometry*: *Integrable structure*, Graduate Studies in Mathematics, vol. 98, American Mathematical Society, Providence, RI, 2008, DOI 10.1007/978-3-7643-8621-4. MR2467378

[105] O. I. Bogoyavlensky, *Integrable discretizations of the KdV equation*, Phys. Lett. A **134** (1988), no. 1, 34–38, DOI 10.1016/0375-9601(88)90542-7. MR972622

[106] O. I. Bogoyavlensky, *Overturning solitons* (Russian, with Russian summary), "Nauka", Moscow, 1991. Nelineĭ nye integriruemye uravneniya. [Nonlinear integrable equations]. MR1261186

[107] M. Boiti, M. Bruschi, F. Pempinelli, and B. Prinari, *A discrete Schrödinger spectral problem and associated evolution equations*, J. Phys. A **36** (2003), no. 1, 139–149, DOI 10.1088/0305-4470/36/1/309. MR1959017

[108] M. Boiti and F. Pempinelli, *Nonlinear Schrödinger equation, Bäcklund transformations and Painlevé transcendents* (English, with Italian and Russian summaries), Nuovo Cimento B (11) **59** (1980), no. 1, 40–58, DOI 10.1007/BF02739045. MR598413

[109] M. Boiti, F. Pempinelli, B. Prinari, and A. Spire, *An integrable discretization of KdV at large times*, Inverse Problems **17** (2001), no. 3, 515–526, DOI 10.1088/0266-5611/17/3/310. MR1843278

[110] A. V. Bocharov, '*DEliA: A system of exact analysis of differential equations using S. Lie approach*, report by Joint Venture OWIMEX Program Systems Institute of the USSR, Academy of Sciences, Pereslavl-Zalessky, USSR, 1989;
– –, *DEliA: project presentation*, SIGSAM Bull. **24** (1990), 37–38;
– –, *Will DEliA grow into an expert system?*, in Design and Implementation of Symbolic Computation Systems, Proceedings of the International Symposium, Capri, Italy, 1990, (A Miola ed.), Lecture Notes in Computer Science 429, Springer-Verlag, Berlin, 1990, 266–267.

[111] A. V. Bocharov, V. N. Chetverikov, S. V. Duzhin, N. G. Khor′kova, I. S. Krasil′shchik, A. V. Samokhin, Yu. N. Torkhov, A. M. Verbovetsky, and A. M. Vinogradov, *Symmetries and conservation laws for differential equations of mathematical physics*, Translations of Mathematical Monographs, vol. 182, American Mathematical Society, Providence, RI, 1999. Edited and with a preface by Krasil′shchik and Vinogradov; Translated from the 1997 Russian original by Verbovetsky [A. M. Verbovetskiĭ] and Krasil′shchik, DOI 10.1090/mmono/182. MR1670044

[112] R. Boll, *Classification of 3D consistent quad-equations*, J. Nonlinear Math. Phys. **18** (2011), no. 3, 337–365, DOI 10.1142/S1402925111001647. MR2846098

[113] R. Boll, *Corrigendum: Classification of 3D consistent quad-equations [MR2846098]*, J. Nonlinear Math. Phys. **19** (2012), no. 4, 1292001, 3, DOI 10.1142/S1402925112920015. MR3010833

[114] R. Boll, *Classification and Lagrangian structure of 3D consistent quad-equations*, Ph.D. Thesis, Technische Universität Berlin, 2012.

[115] T. C. Bountis, V. Papageorgiou, and P. Winternitz, *On the integrability of systems of nonlinear ordinary differential equations with superposition principles*, J. Math. Phys. **27** (1986), no. 5, 1215–1224, DOI 10.1063/1.527128. MR836021

[116] A. Bourlioux, C. Cyr-Gagnon, and P. Winternitz, *Difference schemes with point symmetries and their numerical tests*, J. Phys. A **39** (2006), no. 22, 6877–6896, DOI 10.1088/0305-4470/39/22/006. MR2233260

[117] A. Bourlioux, R. Rebelo, and P. Winternitz, *Symmetry preserving discretization of* SL(2, ℝ) *invariant equations*, J. Nonlinear Math. Phys. **15** (2008), no. suppl. 3, 362–372, DOI 10.2991/jnmp.2008.15.s3.35. MR2452447

[118] M. Boutin, *On orbit dimensions under a simultaneous Lie group action on n copies of a manifold*, J. Lie Theory **12** (2002), no. 1, 191–203. MR1885041

[119] T. Bridgman, W. Hereman, G. R. W. Quispel, and P. H. van der Kamp, *Symbolic computation of Lax pairs of partial difference equations using consistency around the cube*, Found. Comput. Math. **13** (2013), no. 4, 517–544, DOI 10.1007/s10208-012-9133-9. MR3085677

[120] T. J. Bridges, *A universal form for the emergence of the Korteweg-de Vries equation*, Proc. R. Soc. Lond. Ser. A Math. Phys. Eng. Sci. **469** (2013), no. 2153, 20120707, 15, DOI 10.1098/rspa.2012.0707. MR3031503

[121] M. Bruschi, D. Levi, and O. Ragnisco, *Discrete version of the modified Korteweg-de Vries equation with x-dependent coefficients* (English, with Italian summary), Nuovo Cimento A (11) **48** (1978), no. 2, 213–226, DOI 10.1007/BF02799676. MR520117

[122] M. Bruschi, D. Levi, and O. Ragnisco, *Toda lattice and generalised Wronskian technique*, J. Phys. A **13** (1980), no. 7, 2531–2533. MR578429

[123] M. Bruschi, S. V. Manakov, O. Ragnisco, and D. Levi, *The nonabelian Toda lattice-discrete analogue of the matrix Schrödinger spectral problem*, J. Math. Phys. **21** (1980), no. 12, 2749–2753, DOI 10.1063/1.524393. MR597590

[124] M. Bruschi and O. Ragnisco, *Existence of a Lax pair for any member of the class of nonlinear evolution equations associated to the matrix Schrödinger spectral problem*, Lett. Nuovo Cimento (2) **29** (1980), no. 10, 321–326. MR603212

[125] M. Bruschi and O. Ragnisco, *Extension of the Lax method to solve a class of nonlinear evolution equations with x-dependent coefficients associated to the matrix Schrödinger spectral problem*, Lett. Nuovo Cimento (2) **29** (1980), no. 10, 327–330. MR603213

[126] M. Bruschi and O. Ragnisco, *Bäcklund transformations and Lax technique*, Lett. Nuovo Cimento (2) **29** (1980), no. 10, 331–334. MR603214

[127] M. Bruschi and O. Ragnisco, *Nonlinear differential-difference matrix equations with n-dependent coefficients*, Lett. Nuovo Cimento (2) **31** (1981), no. 14, 492–496. MR631451

[128] M. Bruschi and O. Ragnisco, *Nonlinear differential-difference equations, associated Bäcklund transformations and Lax technique*, J. Phys. A **14** (1981), no. 5, 1075–1081. MR611973

[129] M. Bruschi, O. Ragnisco, and D. Levi, *Evolution equations associated with the discrete analog of the matrix Schrödinger spectral problem solvable by the inverse spectral transform*, J. Math. Phys. **22** (1981), no. 11, 2463–2471, DOI 10.1063/1.524805. MR640652

[130] C. J. Budd and G. J. Collins, *Symmetry based numerical methods for partial differential equations*, Numerical analysis 1997 (Dundee), Pitman Res. Notes Math. Ser., vol. 380, Longman, Harlow, 1998, pp. 16–36. MR1636919

[131] C. J. Budd and M. D. Piggott, *Geometric integration and its applications*, Handbook of numerical analysis, Vol. XI, Handb. Numer. Anal., XI, North-Holland, Amsterdam, 2003, pp. 35–139. MR2009771

[132] C. Budd and V. Dorodnitsyn, *Symmetry-adapted moving mesh schemes for the nonlinear Schrödinger equation*, J. Phys. A **34** (2001), no. 48, 10387–10400, DOI 10.1088/0305-4470/34/48/305. Symmetries and integrability of difference equations (Tokyo, 2000). MR1877461

[133] R. K. Bullough and P. Caudrey eds., *Solitons*, Springer, Berlin, 1980.

[134] J. M. Burgers, *A mathematical model illustrating the theory of turbulence*, Advances in Applied Mechanics, Academic Press, Inc., New York, N. Y., 1948, pp. 171–199. edited by Richard von Mises and Theodore von Kármán,. MR0027195

[135] J. M. Burgers, *The nonlinear diffusion equation: asymptotic solutions and statistical problems*, Springer, Dordrecht, 1974.

[136] S. Butler, M. Hay, *Simple identification of fake Lax pairs*, `arXiv:abs/1311.2406`.

[137] S. Butler, M. Hay, *Two definitions of fake Lax pairs*, AIP Conf. Proc. **1648** (2015), 180006, (5pp.)

[138] S. Butler and N. Joshi, *An inverse scattering transform for the lattice potential KdV equation*, Inverse Problems **26** (2010), no. 11, 115012, 28, DOI 10.1088/0266-5611/26/11/115012. MR2726180

[139] G. B. Byrnes, R. Sahadevan, and G. R. W. Quispel, *Factorizable Lie symmetries and the linearization of difference equations*, Nonlinearity **8** (1995), no. 3, 443–459. MR1331822

[140] F. Calogero, *Bäcklund transformations and functional relation for solutions of nonlinear partial differential equations solvable via the inverse scattering method*, Lett. Nuovo Cimento (2) **14** (1975), no. 15, 537–543. MR436188

[141] F. Calogero, *Why are certain nonlinear PDEs both widely applicable and integrable?*, What is integrability?, Springer Ser. Nonlinear Dynam., Springer, Berlin, 1991, pp. 1–62. MR1098335

[142] F. Calogero, *C-integrable nonlinear partial differential equations*, Important developments in soliton theory, Springer Ser. Nonlinear Dynam., Springer, Berlin, 1993, pp. 33–37. MR1280468

[143] F. Calogero, *Classical many-body problems amenable to exact treatments*, Lecture Notes in Physics. New Series m: Monographs, vol. 66, Springer-Verlag, Berlin, 2001. (Solvable and/or integrable and/or linearizable...) in one-, two- and three-dimensional space, DOI 10.1007/3-540-44730-X. MR1880249

[144] F. Calogero and A. Degasperis, *Inverse spectral problem for the one-dimensional Schrödinger equation with an additional linear potential*, Lett. Nuovo Cimento (2) **23** (1978), no. 4, 143–149. MR514526

[145] F. Calogero and A. Degasperis, *Solution by the spectral-transform method of a nonlinear evolution equation including as a special case the cylindrical KdV equation*, Lett. Nuovo Cimento (2) **23** (1978), no. 4, 150–154. MR514527

[146] F. Calogero and A. Degasperis, *Nonlinear evolution equations solvable by the inverse spectral transform associated with the matrix Schrödinger equation*, In Solitons (R K Bullough and P J Caudrey eds.), Springer, Berlin, Heidelberg, 1980.

[147] F. Calogero and A. Degasperis, *Spectral transform and solitons. Vol. I*, Lecture Notes in Computer Science, vol. 144, North-Holland Publishing Co., Amsterdam-New York, 1982. Tools to solve and investigate nonlinear evolution equations. MR680040

[148] F. Calogero and X.–D. Ji, *C-integrable nonlinear partial differential equations. I*, J. Math. Phys. **32** (1991), 875–87,
– –, *C-integrable nonlinear PDEs. II*, J. Math. Phys. **32** (1991), 2703–2717,
– –, *C-integrable nonlinear partial differential equations. III*, J. Math. Phys. **34** (1993), 5810–5831,
– –, *C-integrable nonlinear PDEs. IV*, J. Phys. A: Math. Gen. **29**:21 (1996), 6781–694.

[149] F. Calogero and M. C. Nucci, *Lax pairs galore*, J. Math. Phys. **32** (1991), no. 1, 72–74, DOI 10.1063/1.529096. MR1083088

[150] A. Campa, A. Giansanti, A. Tenenbaum, D. Levi, and O. Ragnisco, *Quasi-solitons on a diatomic chain at room temperature*, Phys. Rev. **48** (1993), 10168–10182.

[151] D. K. Campbell, P. Rosenau, and G. M. Zaslavsky eds., *Focus issue: the "Fermi–Pasta–Ulam" problem - the first 50 years*, Chaos **15**:1 (2005).

[152] R. Campoamor-Stursberg, M. A. Rodríguez, and P. Winternitz, *Symmetry preserving discretization of ordinary differential equations. Large symmetry groups and higher order equations*, J. Phys. A **49** (2016), no. 3, 035201, 21, DOI 10.1088/1751-8113/49/3/035201. MR3436668

[153] B. J. Cantwell, *Introduction to symmetry analysis*, Cambridge Texts in Applied Mathematics, Cambridge University Press, Cambridge, 2002. With 1 CD-ROM. MR1929892

[154] R. W. Carroll, *Topics in soliton theory*, North-Holland Mathematics Studies, vol. 167, North-Holland Publishing Co., Amsterdam, 1991. MR1140704

[155] É. Cartan, *Sur la structure des groupes des transformations finis et continus*, 2nd ed., Vuibert, Paris, 1933.

[156] K. M. Case, *On discrete inverse scattering problems. II*, J. Mathematical Phys. **14** (1973), 916–920, DOI 10.1063/1.1666417. MR332066

[157] K. M. Case, *The discrete inverse scattering problem in one dimension*, J. Mathematical Phys. **15** (1974), 143–146, DOI 10.1063/1.1666610. MR332023

[158] K. M. Case and S. C. Chiu, *The discrete version of the Marchenko equations in the inverse scattering problem*, J. Mathematical Phys. **14** (1973), 1643–1647, DOI 10.1063/1.1666237. MR332067

[159] K. M. Case and M. Kac, *A discrete version of the inverse scattering problem*, J. Mathematical Phys. **14** (1973), 594–603, DOI 10.1063/1.1666364. MR332065

[160] D. Catalano Ferraioli, *Nonlocal aspects of λ-symmetries and ODEs reduction*, J. Phys. A **40** (2007), no. 21, 5479–5489, DOI 10.1088/1751-8113/40/21/002. MR2343887

[161] D. Catalano Ferraioli and P. Morando, *Local and nonlocal solvable structures in the reduction of ODEs*, J. Phys. A **42** (2009), no. 3, 035210, 15, DOI 10.1088/1751-8113/42/3/035210. MR2525318

[162] E. Celledoni, A. Iserles, S. P. Nørsett, and B. Orel, *Complexity theory for Lie-group solvers*, J. Complexity **18** (2002), no. 1, 242–286, DOI 10.1006/jcom.2001.0615. MR1895086

[163] B. Champagne, W. Hereman, and P. Winternitz, *The computer calculation of Lie point symmetries of large systems of differential equations*, Comput. Phys. Comm. **66** (1991), no. 2-3, 319–340, DOI 10.1016/0010-4655(91)90080-5. MR1125408

[164] B. Champagne and P. Winternitz, *A MACSYMA program for calculating the symmetry group of a system of differential equations*, Preprint CRM-1278, Univ. de Montrèal, 1985.

[165] B. Champagne and P. Winternitz, *On the infinite-dimensional symmetry group of the Davey-Stewartson equations*, J. Math. Phys. **29** (1988), no. 1, 1–8, DOI 10.1063/1.528173. MR921757

[166] E. S. Cheb-Terrab, L. G. S. Duarte, and L. A. C. P. da Mota, *Computer algebra solving of second order ODEs using symmetry methods*, Comput. Phys. Comm. **108** (1998), no. 1, 90–114, DOI 10.1016/S0010-4655(97)00132-X. MR1606904

[167] H.-H. Chen and C.-S. Liu, *Bäcklund transformation solutions of the Toda lattice equation*, J. Math. Phys. **16** (1975), 1428–1430, DOI 10.1063/1.522688. MR386555

[168] W. Y. C. Chen, H. W. Galbraith, and J. D. Louck, *Angular momentum theory, umbral calculus, and combinatorics*, Comput. Math. Appl. **41** (2001), no. 9, 1199–1214, DOI 10.1016/S0898-1221(01)00091-8. Umbral calculus and its applications (Cortona, 1998). MR1828153

[169] I. Yu. Cherdantsev and R. I. Yamilov, *Master symmetries for differential-difference equations of the Volterra type*, Phys. D **87** (1995), no. 1-4, 140–144, DOI 10.1016/0167-2789(95)00167-3. The nonlinear Schrödinger equation (Chernogolovka, 1994). MR1361679

[170] I. Cherdantsev and R. Yamilov, *Local master symmetries of differential-difference equations*, Symmetries and integrability of difference equations (Estérel, PQ, 1994), CRM Proc. Lecture Notes, vol. 9, Amer. Math. Soc., Providence, RI, 1996, pp. 51–61, DOI 10.1090/crmp/009/06. MR1416826

[171] I. Cherednik, *Basic methods of soliton theory*, Advanced Series in Mathematical Physics, vol. 25, World Scientific Publishing Co., Inc., River Edge, NJ, 1996. Translated from the Russian by Takashi Takebe, DOI 10.1142/3084. MR1424040

[172] S. C. Chiu and J. F. Ladik, *Generating exactly soluble nonlinear discrete evolution equations by a generalized Wronskian technique*, J. Mathematical Phys. **18** (1977), no. 4, 690–700, DOI 10.1063/1.523328. MR428138

[173] D. V. Chudnovsky and G. V. Choodnovsky, *Pole expansions of nonlinear partial differential equations* (English, with Italian and Russian summaries), Nuovo Cimento B (11) **40** (1977), no. 2, 339–353, DOI 10.1007/BF02728217. MR448415

[174] D. V. Chudnovsky and G. V. Chudnovsky (eds.), *The Riemann problem, complete integrability and arithmetic applications*, Lecture Notes in Mathematics, vol. 925, Springer-Verlag, Berlin-New York, 1982. MR659853

[175] P. L. Christiansen and A. C. Scott eds., *Davydov's soliton revisited. Self-trapping of vibrational energy in proteins*, Springer, Boston, MA, 1990.

[176] M. T. Chu, *The generalized Toda flow, the* QR *algorithm and the center manifold theory*, SIAM J. Algebraic Discrete Methods **5** (1984), no. 2, 187–201, DOI 10.1137/0605020. MR745438

[177] O. A. Chubykalo, V. V. Konotop, L. Vázquez and V. E. Vekslerchik, *Some features of the repulsive discrete nonlinear Schrödinger equation*, Phys. Lett. **A169** (1992), 359–363.

[178] G. Cicogna, *"Weak" symmetries and adapted variables for differential equations*, Int. J. Geom. Methods Mod. Phys. **1** (2004), no. 1-2, 23–31, DOI 10.1142/S0219887804000058. MR2055287

[179] G. Cicogna, *Reduction of systems of first-order differential equations via* Λ*-symmetries*, Phys. Lett. A **372** (2008), no. 20, 3672–3677, DOI 10.1016/j.physleta.2008.02.041. MR2413867

[180] G. Cicogna and G. Gaeta, *Partial Lie-point symmetries of differential equations*, J. Phys. A **34** (2001), no. 3, 491–512, DOI 10.1088/0305-4470/34/3/312. MR1822788

[181] J. L. Cieśliński, *Lie symmetries as a tool to isolate integrable geometries*, Nonlinear evolution equations and dynamical systems (Baia Verde, 1991), World Sci. Publ., River Edge, NJ, 1992, pp. 260–268. MR1192584

[182] J. L. Cieśliński, *Group interpretation of the spectral parameter in the case of nonhomogeneous, nonlinear Schrödinger system*, J. Math. Phys. **34** (1993), no. 6, 2372–2384, DOI 10.1063/1.530122. MR1218993

[183] J. L. Cieśliński and A. Kobus, *Group interpretation of the spectral parameter. The case of isothermic surfaces*, J. Geom. Phys. **113** (2017), 28–37, DOI 10.1016/j.geomphys.2016.11.028. MR3603753

[184] A. C. Clairaut, *Sur l'integration ou la construction des equations différentielles du premier ordre*, Mem.Acad.r.Sci.Paris **2** (1740), 293–323.

[185] P. A. Clarkson (ed.), *Applications of analytic and geometric methods to nonlinear differential equations*, NATO Advanced Science Institutes Series C: Mathematical and Physical Sciences, vol. 413, Kluwer Academic Publishers Group, Dordrecht, 1993, DOI 10.1007/978-94-011-2082-1. MR1261642

[186] P. A. Clarkson and M. D. Kruskal, *New similarity reductions of the Boussinesq equation*, J. Math. Phys. **30** (1989), no. 10, 2201–2213, DOI 10.1063/1.528613. MR1016286

[187] P. A. Clarkson and F. W. Nijhoff (eds.), *Symmetries and integrability of difference equations*, London Mathematical Society Lecture Note Series, vol. 255, Cambridge University Press, Cambridge, 1999. Papers from the 2nd International Conference (SIDE II) held at the University of Kent at Canterbury, Canterbury, July 1–5, 1996, DOI 10.1017/CBO9780511569432. MR1705214

[188] A. M. Cohen and W. A. de Graaf, *Lie algebraic computation*, Comput. Phys. Comm. **97** (1996), no. 1-2, 53–62, DOI 10.1016/0010-4655(96)00021-5. MR1408293

[189] A. M. Cohen, W. A. de Graaf, and L. Rónyai, *Computations in finite-dimensional Lie algebras*, Discrete Math. Theor. Comput. Sci. **1** (1997), no. 1, 129–138. Lie computations (Marseille, 1994). MR1605035

[190] J. D. Cole, *On a quasi-linear parabolic equation occurring in aerodynamics*, Quart. Appl. Math. **9** (1951), 225–236, DOI 10.1090/qam/42889. MR42889

[191] F. A. Costabile, *Modern umbral calculus—an elementary introduction with applications to linear interpolation and operator approximation theory*, De Gruyter Studies in Mathematics, vol. 72, De Gruyter, Berlin, 2019. MR4321536

[192] R. Conte (ed.), *The Painlevé property: One century later*, CRM Series in Mathematical Physics, Springer-Verlag, New York, 1999, DOI 10.1007/978-1-4612-1532-5. MR1713574

[193] R. Conte and N. Boccara (eds.), *Partially integrable evolution equations in physics*, NATO Advanced Science Institutes Series C: Mathematical and Physical Sciences, vol. 310, Kluwer Academic Publishers Group, Dordrecht, 1990, DOI 10.1007/978-94-009-0591-7. MR1073547

[194] R. Conte and M. Musette, *A new method to test discrete Painlevé equations*, Phys. Lett. A **223** (1996), no. 6, 439–448, DOI 10.1016/S0375-9601(96)00783-9. MR1427884

[195] C. M. Cosgrove, *Higher-order Painlevé equations in the polynomial class. I. Bureau symbol* P2, Stud. Appl. Math. **104** (2000), no. 1, 1–65, DOI 10.1111/1467-9590.00130. MR1738749

[196] O. Costin and M. Kruskal, *Movable singular points of solutions of difference equations, their relation to solvability, and the investigation of superstable fixed points* (Russian, with Russian summary), Teoret. Mat. Fiz. **133** (2002), no. 2, 160–169, DOI 10.1023/A:1021134424352; English transl., Theoret. and Math. Phys. **133** (2002), no. 2, 1455–1462. MR2001530

[197] G. Darboux, *Sur les équations aux dérivées partielles du second ordre* (French), Ann. Sci. École Norm. Sup. **7** (1870), 163–173. MR1508571

[198] G. Darboux, *Leçons sur la théorie générale des surfaces et les applications géométriques du calcul infinitésimal. Quatrième partie*, Chelsea Publishing Co., Bronx, N.Y., 1972. Déformation infiniment petite et représentation sphérique; Réimpression de la première édition de 1896. MR0396214

[199] G. Darboux. *Leçons sur la théorie générale de surfaces, volume II*, Hermann, Paris, 1915.

[200] T. Dauxois, *Fermi, Pasta, Ulam and a misterious woman*, Physics Today **61** (2008), 55–57.

[201] D. David, N. Kamran, D. Levi, and P. Winternitz, *Subalgebras of loop algebras and symmetries of the Kadomtsev-Petviashvili equation*, Phys. Rev. Lett. **55** (1985), no. 20, 2111–2113, DOI 10.1103/PhysRevLett.55.2111. MR811845

[202] D. David, N. Kamran, D. Levi, and P. Winternitz, *Symmetry reduction for the Kadomtsev-Petviashvili equation using a loop algebra*, J. Math. Phys. **27** (1986), no. 5, 1225–1237, DOI 10.1063/1.527129. MR836022

[203] A. S. Davydov and N. Kislukha, *Solitary excitons in one-dimensional molecular chains*, Phys. Stat. Sol. B **59** (1973) 465–470.

[204] A. S. Davydov, *Solitons in molecular systems*, Phys. Scripta **20** (1979), no. 3-4, 387–394, DOI 10.1088/0031-8949/20/3-4/013. Special issue on solitons in physics. MR544482
– –, *Biology and Quantum Mechanics*, Pergamon Press, New York, 1982

[205] G. S. Deem, N. J. Zabusky and M. D. Kruskal, *Formation, propagation, and interaction of solitons in nonlinear dispersive media. This 16 mm film had three parts: (1) solution of the KdV equation with a sinusoidal initial condition; (2) solution of the modified-KdV equation with a compressive and rarefactive soliton initial condition; (3) solution of the nonlinear lattice with a high-frequency ("optical") initial condition.* A 16 mm film made at Bell Telephone Labs available at https://www.youtube.com / watch?v=D_TB57Bz5HA (1965).

[206] A. Degasperis, S. V. Manakov, and P. M. Santini, *Multiple-scale perturbation beyond the nonlinear Schroedinger equation. I*, Phys. D **100** (1997), no. 1-2, 187–211, DOI 10.1016/S0167-2789(96)00179-0. MR1436380

[207] A. Degasperis and M. Procesi, *Asymptotic integrability*, Symmetry and perturbation theory (Rome, 1998), World Sci. Publ., River Edge, NJ, 1999, pp. 23–37. MR1844104
- -, *Multiscale expansion and integrability of dispersive wave equations*, in Integrability, (A. V. Mikhailov ed.), Springer, Berlin, Heidelberg, 2009, 215–244.

[208] D. K. Demskoi and C.-M. Viallet, *Algebraic entropy for semi-discrete equations*, J. Phys. A **45** (2012), no. 35, 352001, 10, DOI 10.1088/1751-8113/45/35/352001. MR2969111

[209] G. Deng, G. Biondini, and S. Trillo, *Small dispersion limit of the Korteweg–de Vries equation with periodic initial conditions and analytical description of the Zabusky-Kruskal experiment*, Phys. D **333** (2016), 137–147, DOI 10.1016/j.physd.2016.03.003. MR3523496

[210] A. DiBucchianico and D. Loeb, *Umbral calculus*, Electr. J. Combin. **DS3** (2000) http://www.combinatorics.org.

[211] L. A. Dickey, *Soliton equations and Hamiltonian systems*, 2nd ed., Advanced Series in Mathematical Physics, vol. 26, World Scientific Publishing Co., Inc., River Edge, NJ, 2003, DOI 10.1142/5108. MR1964513

[212] A. Dimakis, F. Müller-Hoissen, and T. Striker, *Umbral calculus, discretization, and quantum mechanics on a lattice*, J. Phys. A **29** (1996), no. 21, 6861–6876, DOI 10.1088/0305-4470/29/21/017. MR1421904

[213] S. Dimas and D. Tsoubelis, *SYM: A new symmetry-finding package for Mathematica*, Proceedings of the 10th international conference in modern group analysis, University of Cyprus Press, 2004, 64–70.

[214] R. K. Dodd, *Generalised Bäcklund transformation for some non-linear partial differential-difference equations*, J. Phys. A **11** (1978), no. 1, 81–92. MR501057

[215] R. K. Dodd, J. C. Eilbeck, J. D. Gibbon, and H. C. Morris, *Solitons and nonlinear wave equations*, Academic Press, Inc. [Harcourt Brace Jovanovich, Publishers], London-New York, 1982. MR696935

[216] I. Dorfman, *Dirac structures and integrability of nonlinear evolution equations*, Nonlinear Science: Theory and Applications, John Wiley & Sons, Ltd., Chichester, 1993. MR1237398

[217] V. A. Dorodnitsyn, *Transformation groups in difference spaces* (Russian), Current problems in mathematics. Newest results, Vol. 34 (Russian), Itogi Nauki i Tekhniki, Akad. Nauk SSSR, Vsesoyuz. Inst. Nauchn. i Tekhn. Inform., Moscow, 1989, pp. 149–191, 195. Translated in J. Soviet Math. **55** (1991), no. 1, 1490–1517. MR1012326

[218] V. A. Dorodnitsyn, *Symmetry of finite difference equations*, in CRC Handbook of Lie Group Analysis of Differential Equations. Vol. I: Symmetries, exact solutions and conservation laws,(N H Ibragimov ed.), Chapman & Hall/CRC Press, Baton Rouge, 1993, 365–403.

[219] V. A. Dorodnitsyn, *A finite-difference analogue of Noether's theorem* (Russian, with Russian summary), Dokl. Akad. Nauk **328** (1993), no. 6, 678–682; English transl., Phys. Dokl. **38** (1993), no. 2, 66–68. MR1219157

[220] V. A. Dorodnitsyn, *Finite difference models entirely inheriting continuous symmetry of original differential equations*, Internat. J. Modern Phys. C **5** (1994), no. 4, 723–734, DOI 10.1142/S0129183194000830. MR1291538

[221] V. A. Dorodnitsyn, *Invariant discrete model for the Korteweg–de Vries equation*, Preprint CRM–2187, Université de Montréal, 1994.

[222] V. A. Dorodnitsyn, *Group properties of difference equations*, [in Russian] Fizmatlit, Moscow, 2001.

[223] V. A. Dorodnitsyn, *Applications of Lie groups to difference equations*, Differential and Integral Equations and Their Applications, vol. 8, CRC Press, Boca Raton, FL, 2011. MR2759751

[224] V. A. Dorodnitsyn and R. Kozlov, *A heat transfer with a source: the complete set of invariant difference schemes*, J. Nonlinear Math. Phys. **10** (2003), no. 1, 16–50, DOI 10.2991/jnmp.2003.10.1.3. MR1952864

[225] V. A. Dorodnitsyn, R. Kozlov, S. V. Meleshko, and P. Winternitz, *Lie group classification of first-order delay ordinary differential equations*, J. Phys. A **51** (2018), no. 20, 205202, 33, DOI 10.1088/1751-8121/aaba91. MR3803584

[226] V. A. Dorodnitsyn, R. Kozlov, S. V. Meleshko, and P. Winternitz, *Linear or linearizable first-order delay ordinary differential equations and their Lie point symmetries*, J. Phys. A **51** (2018), no. 20, 205203, 24, DOI 10.1088/1751-8121/aab3e9. MR3803585

[227] V. A. Dorodnitsyn, R. Kozlov, S. V. Meleshko and P. Winternitz, *Invariant solutions of delay differential equations*, AIP Conference Proceedings **1978**:47 (2018), 470091 (6pp).

[228] V. A. Dorodnitsyn, R. Kozlov, S. V. Meleshko and P. Winternitz, *Lie point symmetries of delay ordinary differential equations*, in Nonlinear Systems and Their Remarkable Mathematical Structures, Vol.2, (N Euler ed.), CRC Press, Boca Raton, 2020, 89-118.

[229] V. A. Dorodnitsyn, R. Kozlov, S. V. Meleshko, and P. Winternitz, *Second-order delay ordinary differential equations, their symmetries and application to a traffic problem*, J. Phys. A **54** (2021), no. 10, Paper No. 105204, 28, DOI 10.1088/1751-8121/abdc81. MR4232163

[230] V. A. Dorodnitsyn, R. Kozlov, and P. Winternitz, *Lie group classification of second-order ordinary difference equations*, J. Math. Phys. **41** (2000), no. 1, 480–504, DOI 10.1063/1.533142. MR1738593

[231] V. A. Dorodnitsyn, R. Kozlov, and P. Winternitz, *Continuous symmetries of Lagrangians and exact solutions of discrete equations*, J. Math. Phys. **45** (2004), no. 1, 336–359, DOI 10.1063/1.1625418. MR2026374

[232] V. A. Dorodnitsyn and P. Winternitz, *Lie point symmetry preserving discretizations for variable coefficient Korteweg-de Vries equations*, Nonlinear Dynam. **22** (2000), no. 1, 49–59, DOI 10.1023/A:1008365224018. Modern group analysis. MR1776184

[233] P. G. Drazin, *Solitons*, London Mathematical Society Lecture Note Series, vol. 85, Cambridge University Press, Cambridge, 1983, DOI 10.1017/CBO9780511662843. MR716135

[234] P. G. Drazin and R. S. Johnson, *Solitons: an introduction*, Cambridge Texts in Applied Mathematics, Cambridge University Press, Cambridge, 1989, DOI 10.1017/CBO9781139172059. MR985322

[235] B. A. Dubrovin, V. B. Matveev, and S. P. Novikov, *Nonlinear equations of Korteweg-de Vries type, finite-band linear operators and Abelian varieties* (Russian), Uspehi Mat. Nauk **31** (1976), no. 1(187), 55–136; English transl., Russ. Math. Surv. **31** (1976), no. 1, 59-146. MR0427869

[236] J. J. Duistermaat, *Discrete integrable systems: QRT maps and elliptic surfaces*, Springer Monographs in Mathematics, Springer, New York, 2010, DOI 10.1007/978-0-387-72923-7. MR2683025

[237] M. Dunajski, *Solitons, instantons, and twistors*, Oxford Graduate Texts in Mathematics, vol. 19, Oxford University Press, Oxford, 2010. MR2583958

[238] L. Eça, *2D orthogonal grid generation with boundary point distribution control*, J. Comput. Phys. **125** (1996), no. 2, 440–453, DOI 10.1006/jcph.1996.0106. MR1388157

[239] S. Elaydi, *An introduction to difference equations*, 3rd ed., Undergraduate Texts in Mathematics, Springer, New York, 2005. MR2128146

[240] A. M. Escobar-Ruiz, R. Linares, and P. Winternitz, *New infinite families of Nth-order superintegrable systems separating in Cartesian coordinates*, J. Phys. A **53** (2020), no. 44, 445203, 26, DOI 10.1088/1751-8121/abb341. MR4177084

[241] A. M. Escobar-Ruiz, J. C. López Vieyra, and P. Winternitz, *Fourth order superintegrable systems separating in polar coordinates. I. Exotic potentials*, J. Phys. A **50** (2017), no. 49, 495206, 35, DOI 10.1088/1751-8121/aa9203. MR3732726

[242] A. M. Escobar-Ruiz, P. Winternitz, and İ. Yurduşen, *General Nth-order superintegrable systems separating in polar coordinates*, J. Phys. A **51** (2018), no. 40, 40LT01, 12, DOI 10.1088/1751-8121/aadc23. MR3862455

[243] L. Euler. *Institutionum calculi integralis, volume III*, Imp. Acad. Imp. Scient., Petropoli, 1770.

[244] N. Euler and M. C. Nucci eds., *Nonlinear systems and their remarkable mathematical structures, Volume 2*, Chapman and Hall/CRC, Boca Raton, 2020.

[245] N. Euler (ed.), *Nonlinear systems and their remarkable mathematical structures. Vol. 1*, CRC Press, Boca Raton, FL, 2019, DOI 10.1201/9780429263743. MR3890066

[246] L. D. Faddeev, *Properties of the S-matrix of the one-dimensional Schrödinger equation*, English transl., AMS Transl. Ser. 2, **65**, AMS, Providence, RI., 1967, 139–166.

[247] L. D. Faddeev and L. A. Takhtajan, *Hamiltonian methods in the theory of solitons*, Reprint of the 1987 English edition, Classics in Mathematics, Springer, Berlin, 2007. Translated from the 1986 Russian original by Alexey G. Reyman. MR2348643

[248] G. Falqui and C.-M. Viallet, *Singularity, complexity, and quasi-integrability of rational mappings*, Comm. Math. Phys. **154** (1993), no. 1, 111–125. MR1220950

[249] E. Faou, *Geometric numerical integration and Schrödinger equations*, Zurich Lectures in Advanced Mathematics, European Mathematical Society (EMS), Zürich, 2012, DOI 10.4171/100. MR2895408

[250] Y. N. Fedorov and V. V. Kozlov, *A memoir on integrable systems*, Springer, Berlin Heidelberg, 2022.

[251] M. Fels and P. J. Olver, *Moving coframes. I. A practical algorithm*, Acta Appl. Math. **51** (1998), no. 2, 161–213, DOI 10.1023/A:1005878210297. MR1620769

[252] M. Fels and P. J. Olver, *Moving coframes. II. Regularization and theoretical foundations*, Acta Appl. Math. **55** (1999), no. 2, 127–208, DOI 10.1023/A:1006195823000. MR1681815

[253] E. Fermi, J. Pasta and S. Ulam, *Studies of nonlinear problems*, Los Alamos Rpt. LA1940 included in Fermi, Enrico Collected papers (Note e memorie). Vol. II: United States, 1939–1954, The University of Chicago Press, Chicago, Ill., 1965; Accademia Nazionale dei Lincei, Rome, 1955.

[254] E. V. Ferapontov, V. S. Novikov, and I. Roustemoglou, *On the classification of discrete Hirota-type equations in 3D*, Int. Math. Res. Not. IMRN **13** (2015), 4933–4974, DOI 10.1093/imrn/rnu086. MR3439096

[255] S. Flach and C. R. Willis, *Discrete breathers*, Phys. Rep. **295** (1998), no. 5, 181–264, DOI 10.1016/S0370-1573(97)00068-9. MR1607320

[256] H. Flaschka, *The Toda lattice. I. Existence of integrals*, Phys. Rev. B (3) **9** (1974), 1924–1925. MR408647

[257] H. Flaschka, *On the Toda lattice. II. Inverse-scattering solution*, Progr. Theoret. Phys. **51** (1974), 703–716, DOI 10.1143/PTP.51.703. MR408648

[258] R. Floreanini, J. Negro, L. M. Nieto, and L. Vinet, *Symmetries of the heat equation on the lattice*, Lett. Math. Phys. **36** (1996), no. 4, 351–355, DOI 10.1007/BF00714402. MR1384641

[259] R. Floreanini and L. Vinet, *Lie symmetries of finite-difference equations*, J. Math. Phys. **36** (1995), no. 12, 7024–7042, DOI 10.1063/1.531205. MR1359678

[260] V. Fock, *Theory of the hydrogen atom*, Phys. Z. **98** (1935), 145–154.

[261] A. S. Fokas, *A symmetry approach to exactly solvable evolution equations*, J. Math. Phys. **21** (1980), no. 6, 1318–1325, DOI 10.1063/1.524581. MR574692

[262] A. S. Fokas, *Symmetries and integrability*, Stud. Appl. Math. **77** (1987), no. 3, 253–299, DOI 10.1002/sapm1987773253. MR1002293

[263] A. S. Fokas, *On the integrability of linear and nonlinear partial differential equations*, J. Math. Phys. **41** (2000), no. 6, 4188–4237, DOI 10.1063/1.533339. MR1768651

[264] A. S. Fokas and B. Fuchssteiner, *The hierarchy of the Benjamin-Ono equation*, Phys. Lett. A **86** (1981), no. 6-7, 341–345, DOI 10.1016/0375-9601(81)90551-X. MR637024

[265] A. S. Fokas and P. M. Santini, *The recursion operator of the Kadomtsev-Petviashvili equation and the squared eigenfunctions of the Schrödinger operator*, Stud. Appl. Math. **75** (1986), no. 2, 179–185, DOI 10.1002/sapm1986752179. MR859179

[266] A. S. Fokas and P. M. Santini, *Conservation laws for integrable systems*, Symmetries and nonlinear phenomena (Paipa, 1988), CIF Ser., vol. 9, World Sci. Publ., Teaneck, NJ, 1988, pp. 7–21. MR1119851

[267] A. S. Fokas and P. M. Santini, *Recursion operators and bi-Hamiltonian structures in multidimensions. II*, Comm. Math. Phys. **116** (1988), no. 3, 449–474. MR937770

[268] A. S. Fokas and V. E. Zakharov (eds.), *Important developments in soliton theory*, Springer Series in Nonlinear Dynamics, Springer-Verlag, Berlin, 1993, DOI 10.1007/978-3-642-58045-1. MR1280466

[269] J. Ford, *The Fermi-Pasta-Ulam problem: paradox turns discovery*, Phys. Rep. **213** (1992), no. 5, 271–310, DOI 10.1016/0370-1573(92)90116-H. MR1170741

[270] A. P. Fordy and J. Gibbons, *Integrable nonlinear Klein-Gordon equations and Toda lattices*, Comm. Math. Phys. **77** (1980), no. 1, 21–30. MR588685

[271] A. P. Fordy and J. Gibbons, *Factorization of operators. I. Miura transformations*, J. Math. Phys. **21** (1980), no. 10, 2508–2510, DOI 10.1063/1.524357. MR586328

[272] A. P. Fordy and P. P. Kulish, *Nonlinear Schrödinger equations and simple Lie algebras*, Comm. Math. Phys. **89** (1983), no. 3, 427–443. MR709476

[273] A. R. Forsyth, *Theory of differential equations. 1. Exact equations and Pfaff's problem; 2, 3. Ordinary equations, not linear; 4. Ordinary linear equations; 5, 6. Partial differential equations*, Dover Publications, Inc., New York, 1959. Six volumes bound as three. MR0123757

[274] J. P. Françoise, G. L. Naber and T. S. Tsun, eds., *The encyclopedia of mathematical physics*, Academic Press, New York, 2006.

[275] I. L. Freire, *Note on Lie point symmetries of Burgers equations* (English, with English and Portuguese summaries), TEMA Tend. Mat. Apl. Comput. **11** (2010), no. 2, 151–157, DOI 10.5540/tema.2010.011.02.0151. MR2745988

[276] J. Friš, V. Mandrosov, Ya. A. Smorodinsky, M. Uhlíř, and P. Winternitz, *On higher symmetries in quantum mechanics*, Phys. Lett. **16** (1965), 354–356. MR187763

[277] P. Winternitz, Ya. A. Smorodinskiĭ, M. Uhlíř, and I. Friš, *Symmetry groups in classical and quantum mechanics*, Soviet J. Nuclear Phys. **4** (1967), 444–450. MR0211699

[278] W. I. Fushchich, N. I. Serov, and V. Ī. Chopik, *Conditional invariance and nonlinear heat equations* (Russian, with English summary), Dokl. Akad. Nauk Ukrain. SSR Ser. A **9** (1988), 17–21, 86. MR973028

[279] W. I. Fushchich and W. M. Shtelen, *On approximate symmetry and approximate solutions of the nonlinear wave equation with a small parameter*, J. Phys. A **22** (1989), no. 18, L887–L890. MR1015227

[280] W. I. Fushchich, W. M. Shtelen, and S. L. Slavutsky, *Reduction and exact solutions of the Navier-Stokes equations*, J. Phys. A **24** (1991), no. 5, 971–984. MR1118767

[281] W. I. Fushchich, W. M. Shtelen, and N. I. Serov, *Symmetry analysis and exact solutions of equations of nonlinear mathematical physics*, Mathematics and its Applications, vol. 246, Kluwer Academic Publishers Group, Dordrecht, 1993. Translated from the 1989 Russian original; Revised by the authors, DOI 10.1007/978-94-017-3198-0. MR1306731

[282] B. Fuchssteiner, *Application of hereditary symmetries to nonlinear evolution equations*, Nonlinear Anal. **3** (1979), no. 6, 849–862, DOI 10.1016/0362-546X(79)90052-X. MR548956

[283] B. Fuchssteiner, *The Lie algebra structure of nonlinear evolution equations admitting infinite-dimensional abelian symmetry groups*, Progr. Theoret. Phys. **65** (1981), no. 3, 861–876, DOI 10.1143/PTP.65.861. MR617703

[284] B. Fuchssteiner, *Mastersymmetries, higher order time-dependent symmetries and conserved densities of nonlinear evolution equations*, Progr. Theoret. Phys. **70** (1983), no. 6, 1508–1522, DOI 10.1143/PTP.70.1508. MR734645

[285] B. Fuchssteiner, *On the hierarchy of the Landau-Lifshitz equation*, Phys. D **13** (1984), no. 3, 387–394, DOI 10.1016/0167-2789(84)90139-8. MR776558

[286] B. Fuchssteiner and A. S. Fokas, *Symplectic structures, their Bäcklund transformations and hereditary symmetries*, Phys. D **4** (1981/82), no. 1, 47–66, DOI 10.1016/0167-2789(81)90004-X. MR636470

[287] B. Fuchssteiner and W.-X. Ma, *An approach to master symmetries of lattice equations*, Symmetries and integrability of difference equations (Canterbury, 1996), London Math. Soc. Lecture Note Ser., vol. 255, Cambridge Univ. Press, Cambridge, 1999, pp. 247–260, DOI 10.1017/CBO9780511569432.020. MR1705233

[288] G. Gaeta, *Nonlinear symmetries and nonlinear equations*, Mathematics and its Applications, vol. 299, Kluwer Academic Publishers Group, Dordrecht, 1994, DOI 10.1007/978-94-011-1018-1. MR1326373

[289] G. Gaeta, *Asymptotic symmetries and asymptotically symmetric solutions of partial differential equations*, J. Phys. A **27** (1994), no. 2, 437–451. MR1267424

[290] G. Gaeta, *A gauge-theoretic description of μ-prolongations, and μ-symmetries of differential equations*, J. Geom. Phys. **59** (2009), no. 4, 519–539, DOI 10.1016/j.geomphys.2009.01.004. MR2504472

[291] G. Gaeta, D. Levi, and R. Mancinelli, *Asymptotic symmetries of difference equations on a lattice*, J. Nonlinear Math. Phys. **12** (2005), no. suppl. 2, 137–146, DOI 10.2991/jnmp.2005.12.s2.9. MR2217099

[292] G. Gaeta and R. Mancinelli, *Asymptotic scaling in a model class of anomalous reaction-diffusion equations*, J. Nonlinear Math. Phys. **12** (2005), no. 4, 550–566, DOI 10.2991/jnmp.2005.12.4.9. MR2172005

[293] G. Gaeta and P. Morando, *On the geometry of lambda-symmetries and PDE reduction*, J. Phys. A **37** (2004), no. 27, 6955–6975, DOI 10.1088/0305-4470/37/27/007. MR2078325

[294] L. Gagnon, V. Hussin, and P. Winternitz, *Nonlinear equations with superposition formulas and exceptional group G_2. III. The superposition formulas*, J. Math. Phys. **29** (1988), no. 10, 2145–2155, DOI 10.1063/1.528141. MR962548

[295] O. Galor, *Discrete dynamical systems*, Springer-Verlag Berlin Heidelberg, 2007.

[296] R. Gambini and J. Pullin, *Loops, knots, gauge theories and quantum gravity*, Cambridge Monographs on Mathematical Physics, Cambridge University Press, Cambridge, 1996. With a foreword by Abhay Ashtekar, DOI 10.1017/CBO9780511524431. MR1439964

[297] R. Gambini and J. Pullin, *Consistent discretizations as a road to quantum gravity*, Approaches to Quantum Gravity Toward a New Understanding of Space, Time and Matter, (D Oriti ed.), Cambridge University Press, Cambridge, 2009, 378–392.

[298] F. Gantmacher, *On the classification of real simple Lie groups* (English, with Russian summary), Rec. Math. [Mat. Sbornik] N.S. **5 (47)** (1939), 217–250. MR0002141

[299] C. S. Gardner, *Korteweg-de Vries equation and generalizations. IV. The Korteweg-de Vries equation as a Hamiltonian system*, J. Mathematical Phys. **12** (1971), 1548–1551, DOI 10.1063/1.1665772. MR286402

[300] C. S. Gardner, J. M. Greene, M. D. Kruskal, and R. M. Miura, *Korteweg-deVries equation and generalization. VI. Methods for exact solution*, Comm. Pure Appl. Math. **27** (1974), 97–133, DOI 10.1002/cpa.3160270108. MR336122

[301] C. S. Gardner and G. K. Morikawa, *Similarity in the asymptotic behavior of collision-free hydromagnetic waves and water waves*, Courant Institute of Mathematical Sciences Report No. NYO 9082, May 1st, 1960 (unpublished).

[302] R. N. Garifullin, G. Gubbiotti, and R. I. Yamilov, *Integrable discrete autonomous quad-equations admitting, as generalized symmetries, known five-point differential-difference equations*, J. Nonlinear Math. Phys. **26** (2019), no. 3, 333–357, DOI 10.1080/14029251.2019.1613050. MR3948389

[303] R. N. Garifullin, E. V. Gudkova, and I. T. Habibullin, *Method for searching higher symmetries for quad-graph equations*, J. Phys. A **44** (2011), no. 32, 325202, 16, DOI 10.1088/1751-8113/44/32/325202. MR2821933

[304] R. N. Garifullin, A. V. Mikhailov, and R. I. Yamilov, *Discrete equation on a square lattice with a nonstandard structure of generalized symmetries*, Theoret. and Math. Phys. **180** (2014), no. 1, 765–780, DOI 10.1007/s11232-014-0178-6. Translation of Teoret. Mat. Fiz. **180** (2014), no. 1, 17–34. MR3344492

[305] R. N. Garifullin and R. I. Yamilov, *Generalized symmetry classification of discrete equations of a class depending on twelve parameters*, J. Phys. A **45** (2012), no. 34, 345205, 23, DOI 10.1088/1751-8113/45/34/345205. MR2960639

[306] R. N. Garifullin, R. I. Yamilov, *Integrable discrete non-autonomous quad-equations as Bäcklund autotransformations for known Volterra and Toda type semidiscrete equations*, J. Phys. Conf. Ser. **621** (2015), 012005 (18pp).

[307] R. N. Garifullin and R. I. Yamilov, *On integrability of a discrete analogue of Kaup-Kupershmidt equation*, Ufa Math. J. **9** (2017), no. 3, 158–164, DOI 10.13108/2017-9-3-158. MR3707512

[308] R. N. Garifullin and R. I. Yamilov, *Integrable modifications of the Ito-Narita-Bogoyavlensky equation*, SIGMA Symmetry Integrability Geom. Methods Appl. **15** (2019), Paper No. 062, 15 pages, DOI 10.3842/SIGMA.2019.062. MR3995889

[309] R. N. Garifullin and R. I. Yamilov, *On series of Darboux integrable discrete equations on square lattice*, Ufa Math. J. **11** (2019), no. 3, 99–108, DOI 10.1080/14029251.2019.1613050. MR4032394

[310] R. N. Garifullin and R. I. Yamilov, *Modified series of integrable discrete equations on a quadratic lattice with a nonstandard symmetry structure* (Russian, with Russian summary), Teoret. Mat. Fiz. **205** (2020), no. 1, 23–40, DOI 10.4213/tmf9899; English transl., Theoret. and Math. Phys. **205** (2020), no. 1, 1264–1278. MR4156832

[311] R. N. Garifullin, R. I. Yamilov, and D. Levi, *Classification of five-point differential-difference equations*, J. Phys. A **50** (2017), no. 12, 125201, 27, DOI 10.1088/1751-8121/aa5cc3. MR3622975

[312] R. N. Garifullin, R. I. Yamilov, and D. Levi, *Classification of five-point differential-difference equations II*, J. Phys. A **51** (2018), no. 6, 065204, 16, DOI 10.1088/1751-8121/aaa14e. MR3755587

[313] E. Gau, *Sur l'intégration des équations aux dérivées partielles du second ordre par la méthode de M. Darboux*, J. Math. Pures Appl. **7** (1911), 123–240.

[314] J.-P. Gazeau and P. Winternitz, *Symmetries of variable coefficient Korteweg-de Vries equations*, J. Math. Phys. **33** (1992), no. 12, 4087–4102, DOI 10.1063/1.529807. MR1191768

[315] I. M. Gel′fand and B. M. Levitan, *On the determination of a differential equation from its spectral function* (Russian), Izvestiya Akad. Nauk SSSR. Ser. Mat. **15** (1951), 309–360; English transl., AMS Transl. Ser. **1** (1955), no. 2, 253–305. MR0045281

[316] F. Gesztesy and H. Holden, *The Cole-Hopf and Miura transformations revisited*, Mathematical physics and stochastic analysis (Lisbon, 1998), World Sci. Publ., River Edge, NJ, 2000, pp. 198–214. MR1893107

[317] F. Gesztesy and H. Holden, *Soliton equations and their algebro-geometric solutions. Vol. I: (1+1)-dimensional continuous models*, Cambridge Studies in Advanced Mathematics, vol. 79, Cambridge University Press, Cambridge, 2003, DOI 10.1017/CBO9780511546723. MR1992536

[318] F. Gesztesy, H. Holden, J. Michor, and G. Teschl, *Soliton equations and their algebro-geometric solutions. Vol. II: (1+1)-dimensional discrete models*, Cambridge Studies in Advanced Mathematics, vol. 114, Cambridge University Press, Cambridge, 2008, DOI 10.1017/CBO9780511543203. MR2446594

[319] A. Giansanti, A. Campa, D. Levi, O. Ragnisco and A. Tenenbaum, *Vibrational properties and energy transport in acetanilide by molecular dinamics*, in Davydov's soliton revisited, (P L Christiansen and A C Scott eds.), Springer Science+Business Media, New York, 1990, 439-448.

[320] D. Gómez-Ullate, S. Lafortune, and P. Winternitz, *Symmetries of discrete dynamical systems involving two species*, J. Math. Phys. **40** (1999), no. 6, 2782–2804, DOI 10.1063/1.532728. MR1694230

[321] D. Gómez-Ullate, S. Lafortune, and P. Winternitz, *Symmetry classification of systems of differential-difference equations*, SIDE III—symmetries and integrability of difference equations (Sabaudia, 1998), CRM Proc. Lecture Notes, vol. 25, Amer. Math. Soc., Providence, RI, 2000, pp. 167–172, DOI 10.1090/crmp/025/16. MR1771719

[322] P. Winternitz, D. Gomez-Ullate, A. Iserles, D. Levi, P. J. Olver, R. Quispel, and P. Tempesta (eds.), *Group theory and numerical analysis*, CRM Proceedings & Lecture Notes, vol. 39, American Mathematical Society, Providence, RI, 2005. Papers from the workshop held in Montréal, QC, May 26–31, 2003, DOI 10.1090/crmp/039. MR2169291

[323] A. González-López, N. Kamran, and P. J. Olver, *Lie algebras of vector fields in the real plane*, Proc. London Math. Soc. (3) **64** (1992), no. 2, 339–368, DOI 10.1112/plms/s3-64.2.339. MR1143228

[324] E. Goursat. *Leçons sur l'intégration des équations aux dérivées partielles du second orde à deux variables indépendendantes, volume I-II*, Hermann, Paris, 1896.

[325] W. A. de Graaf, *An algorithm for the decomposition of semisimple Lie algebras*, Theoret. Comput. Sci. **187** (1997), no. 1-2, 117–122, DOI 10.1016/S0304-3975(97)00060-1. Computer algebra (Saint-Louis, 1996). MR1479978

[326] W. A. de Graaf, *Lie algebras: theory and algorithms*, North-Holland Mathematical Library, vol. 56, North-Holland Publishing Co., Amsterdam, 2000, DOI 10.1016/S0924-6509(00)80040-9. MR1743970

[327] B. Grammaticos, R. G. Halburd, A. Ramani, and C.-M. Viallet, *How to detect the integrability of discrete systems*, J. Phys. A **42** (2009), no. 45, 454002, 30, DOI 10.1088/1751-8113/42/45/454002. MR2556644

[328] B. Grammaticos, T. Tamizhmani and Y. Kosmann–Schwarzbach eds., *Discrete integrable systems*, Springer-Verlag Berlin/Heidelberg 2004.

[329] B. Grammaticos and R. Ramani, *Discrete Painlevé equations: a review*, in Discrete integrable systems, (B Grammaticos, T Tamizhmani and Y Kosmann-Schwarzbach eds.), Springer-Verlag Berlin/Heidelberg 2004, 245–321.

[330] B. Grammaticos, A. Ramani, *Toda equations as paradigms of integrable (continuous, discrete and ultradiscrete) systems*, in 40 years Anniversary of Toda lattice - history and perspective of nonlinear wave research (RIAM, Kyushu University, November 7-9, 2007), report N.11 (15pp).

[331] B. Grammaticos, A. Ramani, and V. Papageorgiou, *Do integrable mappings have the Painlevé property?*, Phys. Rev. Lett. **67** (1991), no. 14, 1825–1828, DOI 10.1103/PhysRevLett.67.1825. MR1125950

[332] B. Grammaticos, A. Ramani, and C. M. Viallet, *Solvable chaos*, Phys. Lett. A **336** (2005), no. 2-3, 152–158, DOI 10.1016/j.physleta.2005.01.026. MR2117728

[333] B. Grammaticos, T. Tamizhmani, A. Ramani, and K. M. Tamizhmani, *Growth and integrability in discrete systems*, J. Phys. A **34** (2001), no. 18, 3811–3821, DOI 10.1088/0305-4470/34/18/309. MR1840846

[334] T. J. Grant, *Bespoke finite difference schemes that preserve multiple conservation laws*, LMS J. Comput. Math. **18** (2015), no. 1, 372–403, DOI 10.1112/S1461157015000078. MR3349321

[335] C.-H. Gu Editor, *Soliton theory and its applications*, Springer-Verlag Berlin Heidelberg 1995.

[336] G. Gubbiotti, *Integrability properties of quad equations consistent on the cube*, PhD thesis, (http://www.matfis.uniroma3.it > TESI > gubbiotti), 2017.

[337] G. Gubbiotti, *Integrability of difference equations through algebraic entropy and generalized symmetries*, Symmetries and integrability of difference equations, CRM Ser. Math. Phys., Springer, Cham, 2017, pp. 75–151, DOI 10.1007/978-3-319-56666-5_3. MR3677348

[338] G. Gubbiotti and C. Scimiterna, *Reconstructing a lattice equation: a non-autonomous approach to the Hietarinta equation*, SIGMA Symmetry Integrability Geom. Methods Appl. **14** (2018), Paper No. 004, 21, DOI 10.3842/SIGMA.2018.004. MR3743752

[339] G. Gubbiotti, C. Scimiterna, and D. Levi, *Algebraic entropy, symmetries and linearization of quad equations consistent on the cube*, J. Nonlinear Math. Phys. **23** (2016), no. 4, 507–543, DOI 10.1080/14029251.2016.1237200. MR3557554

[340] G. Gubbiotti, C. Scimiterna, and D. Levi, *Linearizability and fake Lax pairs for nonlinear nonautonomous quad-graph equations subject to a consistency condition during passages around a cubic nucleus* (Russian, with Russian summary), Teoret. Mat. Fiz. **189** (2016), no. 1, 69–83, DOI 10.4213/tmf9079; English transl., Theoret. and Math. Phys. **189** (2016), no. 1, 1459–1471. MR3589022

[341] G. Gubbiotti, C. Scimiterna and D. Levi, *On partial differential and difference equations with symmetries depending on arbitrary functions*, Acta Polytech. **56** (2016), 193–201.

[342] G. Gubbiotti, C. Scimiterna, and D. Levi, *The non-autonomous YdKN equation and generalized symmetries of Boll equations*, J. Math. Phys. **58** (2017), no. 5, 053507, 18, DOI 10.1063/1.4982747. MR3648851

[343] G. Gubbiotti, C. Scimiterna, and D. Levi, *A two-periodic generalization of the Q_V equation*, J. Integrable Syst. **2** (2017), no. 1, xyx004, 13, DOI 10.1093/integr/xyx004. MR3691967

[344] G. Gubbiotti, C. Scimiterna, and R. I. Yamilov, *Darboux integrability of trapezoidal H^4 and H^6 families of lattice equations II: General solutions*, SIGMA Symmetry Integrability Geom. Methods Appl. **14** (2018), Paper No. 008, 51, DOI 10.3842/SIGMA.2018.008. MR3757534

[345] G. Gubbiotti and R. I. Yamilov, *Darboux integrability of trapezoidal H^4 and H^4 families of lattice equations I: first integrals*, J. Phys. A **50** (2017), no. 34, 345205, 26, DOI 10.1088/1751-8121/aa7fd9. MR3687574

[346] I. T. Habibullin, *Characteristic algebras of fully discrete hyperbolic type equations*, SIGMA Symmetry Integrability Geom. Methods Appl. **1** (2005), Paper 023, 9, DOI 10.3842/SIGMA.2005.023. MR2169846

[347] I. T. Habibullin, *Characteristic Lie rings, finitely-generated modules and integrability conditions for (2+1)-dimensional lattices*, Physica Scripta **87** (2013), 065005.

[348] I. T. Habibullin and M. Poptsova, *Classification of a subclass of two-dimensional lattices via characteristic Lie rings*, SIGMA Symmetry Integrability Geom. Methods Appl. **13** (2017), Paper No. 073, 26, DOI 10.3842/SIGMA.2017.073. MR3694020

[349] I. T. Habibullin, V. V. Sokolov, and R. I. Yamilov, *Multi-component integrable systems and nonassociative structures*, Nonlinear physics: theory and experiment (Lecce, 1995), World Sci. Publ., River Edge, NJ, 1996, pp. 139–168. MR1454861

[350] I. T. Habibullin, N. Zheltukhina, and A. Sakieva, *Discretization of hyperbolic type Darboux integrable equations preserving integrability*, J. Math. Phys. **52** (2011), no. 9, 093507, 12, DOI 10.1063/1.3628587. MR2867825

[351] F. A. Haggar, G. B. Byrnes and G. R. W. Quispel, *K-integrals and k-Lie symmetries in discrete dynamical systems*, Physica A **233** (1996), 379–394.

[352] E. Hairer, C. Lubich, and G. Wanner, *Geometric numerical integration: Structure-preserving algorithms for ordinary differential equations*, 2nd ed., Springer Series in Computational Mathematics, vol. 31, Springer-Verlag, Berlin, 2006. MR2221614

[353] R. G. Halburd, *Diophantine integrability*, J. Phys. A **38** (2005), no. 16, L263–L269, DOI 10.1088/0305-4470/38/16/L01. MR2131425

[354] R. G. Halburd and R. J. Korhonen, *Meromorphic solutions of difference equations, integrability and the discrete Painlevé equations*, J. Phys. A **40** (2007), no. 6, R1–R38, DOI 10.1088/1751-8113/40/6/R01. MR2343636

[355] J. Harnad and F. Balogh, *Tau functions and their applications*, Cambridge Monographs on Mathematical Physics, Cambridge University Press, Cambridge, 2021, DOI 10.1017/9781108610902. MR4411375

[356] J. Harnad, Y. Saint-Aubin, and S. Shnider, *Bäcklund transformations for nonlinear sigma models with values in Riemannian symmetric spaces*, Comm. Math. Phys. **92** (1984), no. 3, 329–367. MR734223

[357] J. Harnad, P. Winternitz, and R. L. Anderson, *Superposition principles for matrix Riccati equations*, J. Math. Phys. **24** (1983), no. 5, 1062–1072, DOI 10.1063/1.525831. MR702080

[358] J. Harnad, G. Sabidussi, and P. Winternitz (eds.), *Integrable systems: from classical to quantum*, CRM Proceedings & Lecture Notes, vol. 26, American Mathematical Society, Providence, RI, 2000, DOI 10.1090/crmp/026. MR1792130

[359] M. Hay, *Discrete Lax pairs, reductions and hierarchies*, PhD thesis, The University of Sidney, 2008 (https://core.ac.uk/download/41232257.pdf).

[360] M. Hay, *A completeness study on discrete,* 2×2 *Lax pairs*, J. Math. Phys. **50** (2009), no. 10, 103516, 29, DOI 10.1063/1.3177197. MR2572689

[361] M. C. Hay, *A completeness study on certain* 2×2 *Lax pairs including zero terms*, SIGMA Symmetry Integrability Geom. Methods Appl. **7** (2011), Paper 089, 12, DOI 10.3842/SIGMA.2011.089. MR2861187

[362] A. Head, *Lie and BigLie, programs to find Lie symmetries of differential equations*, Available online at http://www.cmst.csiro.au/LIE/LIE.htm, 1999.

[363] S. Helgason, *Differential geometry and symmetric spaces*, Pure and Applied Mathematics, Vol. XII, Academic Press, New York-London, 1962. MR0145455

[364] S. Helgason, *Analysis on Lie groups and homogeneous spaces*, Conference Board of the Mathematical Sciences Regional Conference Series in Mathematics, No. 14, American Mathematical Society, Providence, R.I., 1972. MR0316632

[365] S. Helgason, *Differential geometry, Lie groups, and symmetric spaces*, Pure and Applied Mathematics, vol. 80, Academic Press, Inc. [Harcourt Brace Jovanovich, Publishers], New York-London, 1978. MR514561

[366] W. Hereman, *Symbolic software for Lie symmetry analysis*, Lie Group Analysis of Differential Equations, vol. 3, (N H Ibragimov ed.), Chapman & Hall/CRC Press, Boca Raton, 1996, 367–413.

[367] R. Hernández Heredero and D. Levi, *The discrete nonlinear Schrödinger equation and its Lie symmetry reductions*, J. Nonlinear Math. Phys. **10** (2003), no. suppl. 2, 77–94, DOI 10.2991/jnmp.2003.10.s2.7. MR2062270

[368] R. Hernández Heredero, D. Levi, M. Petrera, and C. Scimiterna, *Multiscale expansion of the lattice potential KdV equation on functions of an infinite slow-varyness order*, J. Phys. A **40** (2007), no. 34, F831–F840, DOI 10.1088/1751-8113/40/34/F02. MR2371123

[369] R. Hernández Heredero, D. Levi, M. Petrera, and C. Scimiterna, *Multiscale expansion on the lattice and integrability of partial difference equations*, J. Phys. A **41** (2008), no. 31, 315208, 12, DOI 10.1088/1751-8113/41/31/315208. MR2425810

[370] R. Hernández Heredero, D. Levi, M. Petrera, and C. Scimiterna, *Multiscale expansion and integrability properties of the lattice potential KdV equation*, J. Nonlinear Math. Phys. **15** (2008), no. suppl. 3, 323–333, DOI 10.2991/jnmp.2008.15.s3.31. MR2452443

[371] R. Hernández Heredero, D. Levi, M. A. Rodríguez, and P. Winternitz, *Lie algebra contractions and symmetries of the Toda hierarchy*, J. Phys. A **33** (2000), no. 28, 5025–5040, DOI 10.1088/0305-4470/33/28/307. MR1779083

[372] R. Hernández Heredero, D. Levi, M. A. Rodríguez, and P. Winternitz, *Relation between Bäcklund transformations and higher continuous symmetries of the Toda equation*, J. Phys. A **34** (2001), no. 11, 2459–2465, DOI 10.1088/0305-4470/34/11/334. Kowalevski Workshop on Mathematical Methods of Regular Dynamics (Leeds, 2000). MR1831309

[373] R. Hernández Heredero, D. Levi, M. A. Rodríguez, and P. Winternitz, *Symmetries of differential difference equations and Lie algebra contractions*, Bäcklund and Darboux transformations. The geometry of solitons (Halifax, NS, 1999), CRM Proc. Lecture Notes, vol. 29, Amer. Math. Soc., Providence, RI, 2001, pp. 233–243, DOI 10.1090/crmp/029/20. MR1870921

[374] R. Hernández Heredero, D. Levi, and C. Scimiterna, *A discrete linearizability test based on multiscale analysis*, J. Phys. A **43** (2010), no. 50, 502002, 14, DOI 10.1088/1751-8113/43/50/502002. MR2740376

[375] R. Hernández Heredero, D. Levi, and C. Scimiterna, *Classification of discrete systems on a square lattice*, Theoret. and Math. Phys. **172** (2012), no. 2, 1097–1108, DOI 10.1007/s11232-012-0098-2. Russian version appears in Teoret. Mat. Fiz. **172** (2012), no. 2, 250–263. MR3170083

[376] R. Hernández Heredero, D. Levi, and P. Winternitz, *Symmetries of the discrete Burgers equation*, J. Phys. A **32** (1999), no. 14, 2685–2695, DOI 10.1088/0305-4470/32/14/009. MR1690237

[377] R. Hernández Heredero, D. Levi, and P. Winternitz, *Symmetry preserving discretization of the Burgers equation*, SIDE III—symmetries and integrability of difference equations (Sabaudia, 1998), CRM Proc. Lecture Notes, vol. 25, Amer. Math. Soc., Providence, RI, 2000, pp. 197–208, DOI 10.1090/crmp/025/19. MR1771722

[378] R. Hernández Heredero, D. Levi, and P. Winternitz, *Symmetries of the discrete nonlinear Schrödinger equation* (Russian, with Russian summary), Teoret. Mat. Fiz. **127** (2001), no. 3, 379–387, DOI 10.1023/A:1010439432232; English transl., Theoret. and Math. Phys. **127** (2001), no. 3, 729–737. MR1869959

[379] M. Hickman, W. Hereman, J. Larue, and Ü. Göktaş, *Scaling invariant Lax pairs of nonlinear evolution equations*, Appl. Anal. **91** (2012), no. 2, 381–402, DOI 10.1080/00036811.2011.629611. MR2876760

[380] J. Hietarinta, *Integrability tests for difference equations*, Proceedings of the Third World Congress of Nonlinear Analysts, Part 7 (Catania, 2000), Nonlinear Anal. **47** (2001), no. 7, 4641–4650, DOI 10.1016/S0362-546X(01)00577-6. MR1975858

[381] J. Hietarinta, *A new two-dimensional lattice model that is 'consistent around a cube'*, J. Phys. A **37** (2004), no. 6, L67–L73, DOI 10.1088/0305-4470/37/6/L01. MR2045902

[382] J. Hietarinta, *Searching for CAC-maps*, J. Nonlinear Math. Phys. **12** (2005), no. suppl. 2, 223–230, DOI 10.2991/jnmp.2005.12.s2.16. MR2217106

[383] J. Hietarinta, *Definitions and predictions of integrability for difference equations*, in Symmetries and integrability of difference equations, Summer school on symmetries and integrability of difference equations, Montréal 8-21 June 2008, (D Levi, P Olver, Z Thomova, P Winternitz eds.), Cambridge University Press, Cambridge, 2011, 83–114.

[384] J. Hietarinta, N. Joshi, and F. W. Nijhoff, *Discrete systems and integrability*, Cambridge Texts in Applied Mathematics, Cambridge University Press, Cambridge, 2016, DOI 10.1017/CBO9781107337411. MR3587455

[385] J. Hietarinta and C.-M. Viallet, *Singularity confinement and chaos in discrete systems*, Phys. Rev. Lett. **81** (1998), 325–328.

[386] J. Hietarinta and C.-M. Viallet, *Discrete Painlevé I and singularity confinement in projective space*, Chaos Solitons Fractals **11** (2000), no. 1-3, 29–32, DOI 10.1016/S0960-0779(98)00266-5. Integrability and chaos in discrete systems (Brussels, 1997). MR1729552

[387] J. Hietarinta and C.-M. Viallet, *Searching for integrable lattice maps using factorization*, J. Phys. A **40** (2007), no. 42, 12629–12643, DOI 10.1088/1751-8113/40/42/S09. MR2392894

[388] J. Hietarinta and C.-M. Viallet, *Weak Lax pairs for lattice equations*, Nonlinearity **25** (2012), no. 7, 1955–1966, DOI 10.1088/0951-7715/25/7/1955. MR2947929

[389] J. Hietarinta and D. Zhang, *Soliton solutions for ABS lattice equations. II. Casoratians and bilinearization*, J. Phys. A **42** (2009), no. 40, 404006, 30, DOI 10.1088/1751-8113/42/40/404006. MR2544270

[390] R. Hirota, *Nonlinear partial difference equations. I. A difference analogue of the Korteweg-de Vries equation*, J. Phys. Soc. Japan **43** (1977), 1424–1433.

[391] R. Hirota, *Nonlinear partial difference equations. II. Discrete-time Toda equation*, J. Phys. Soc. Japan **43** (1977), no. 6, 2074–2078, DOI 10.1143/JPSJ.43.2074. MR460935

[392] R. Hirota, *Nonlinear partial difference equations. III. Discrete sine-Gordon equation*, J. Phys. Soc. Japan **43** (1977), no. 6, 2079–2086, DOI 10.1143/JPSJ.43.2079. MR460936

[393] R. Hirota, *Nonlinear partial difference equations. IV. Bäcklund transformations for the discrete time Toda equation*, J. Phys. Soc. Japan **45** (1978), 321–332.

[394] R. Hirota, *Nonlinear partial difference equations V. Nonlinear equations reducible to linear equations*, J. Phys. Soc. Japan **46** (1979), 312–319.

[395] R. Hirota, *The direct method in soliton theory*, Cambridge Tracts in Mathematics, vol. 155, Cambridge University Press, Cambridge, 2004. Translated from the 1992 Japanese original and edited by Atsushi Nagai, Jon Nimmo and Claire Gilson; With a foreword by Jarmo Hietarinta and Nimmo, DOI 10.1017/CBO9780511543043. MR2085332

[396] R. Hirota and J. Satsuma, *A variety of nonlinear network equations generated from the Bäcklund transformation for the Toda lattice*, Progr. Theoret. Phys. Suppl. **59** (1976), 64–100, DOI 10.1143/PTPS.59.64. MR0452254

[397] R. Hirota and J. Satsuma, *A simple structure of superposition formula of the Bäcklund transformation*, J. Phys. Soc. Japan **45** (1978), 1741-1750.

[398] R. Hirota and S. Tsujimoto, *Conserved quantities of discrete Lotka-Volterra equations* (Japanese), Sūrikaisekikenkyūsho Kōkyūroku **868** (1994), 31–38. State of the art and perspectives in studies on nonlinear integrable systems (Japanese) (Kyoto, 1993). MR1330699

[399] R. Hirota and S. Tsujimoto, *Conserved quantities of a class of nonlinear difference-difference equations*, J. Phys. Soc. Japan **64** (1995), no. 9, 3125–3127, DOI 10.1143/JPSJ.64.3125. MR1358290

[400] E. Hopf, *The partial differential equation $u_t + uu_x = \mu u_{xx}$*, Comm. Pure Appl. Math. **3** (1950), 201–230, DOI 10.1002/cpa.3160030302. MR47234

[401] J. Hoppe, *Lectures on integrable systems*, Lecture Notes in Physics. New Series m: Monographs, vol. 10, Springer-Verlag, Berlin, 1992. MR1177167

[402] X.-B. Hu and J. Springael, *A Bäcklund transformation and nonlinear superposition formula for the Lotka-Volterra hierarchy*, ANZIAM J. **44** (2002), no. 1, 121–128, DOI 10.1017/S1446181100007975. Kruskal, 2000 (Adelaide). MR1919932

[403] X. Hu, Y. Li, and Y. Chen, *A direct algorithm of one-dimensional optimal system for the group invariant solutions*, J. Math. Phys. **56** (2015), no. 5, 053504, 17, DOI 10.1063/1.4921229. MR3391007

[404] Q. Huang and R. Zhdanov, *Realizations of the Witt and Virasoro algebras and integrable equations*, J. Nonlinear Math. Phys. **27** (2020), no. 1, 36–56, DOI 10.1080/14029251.2020.1683964. MR4023521

[405] P. E. Hydon, *Symmetry methods for differential equations*, Cambridge Texts in Applied Mathematics, Cambridge University Press, Cambridge, 2000. A beginner's guide, DOI 10.1017/CBO9780511623967. MR1741548

[406] P. E. Hydon, *Symmetries and first integrals of ordinary difference equations*, R. Soc. Lond. Proc. Ser. A Math. Phys. Eng. Sci. **456** (2000), no. 2004, 2835–2855, DOI 10.1098/rspa.2000.0643. MR1843844

[407] P. E. Hydon, *Conservation laws of partial difference equations with two independent variables*, J. Phys. A **34** (2001), no. 48, 10347–10355, DOI 10.1088/0305-4470/34/48/301. Symmetries and integrability of difference equations (Tokyo, 2000). MR1877457

[408] P. E. Hydon, *Difference equations by differential equation methods*, Cambridge Monographs on Applied and Computational Mathematics, vol. 27, Cambridge University Press, Cambridge, 2014, DOI 10.1017/CBO9781139016988. MR3467113

[409] P. E. Hydon and E. L. Mansfield, *A variational complex for difference equations*, Found. Comput. Math. **4** (2004), no. 2, 187–217, DOI 10.1007/s10208-002-0071-9. MR2049870

[410] P. E. Hydon and C.-M. Viallet, *Asymmetric integrable quad-graph equations*, Appl. Anal. **89** (2010), no. 4, 493–506, DOI 10.1080/00036810903329951. MR2647763

[411] E. Inonu and E. P. Wigner, *On the contraction of groups and their representations*, Proc. Nat. Acad. Sci. U.S.A. **39** (1953), 510–524, DOI 10.1073/pnas.39.6.510. MR55352

[412] N. H. Ibragimov, *Transformation groups applied to mathematical physics*, Mathematics and its Applications (Soviet Series), D. Reidel Publishing Co., Dordrecht, 1985. Translated from the Russian, DOI 10.1007/978-94-009-5243-0. MR785566

[413] N. H. Ibragimov ed., *CRC handbook of Lie group analysis of differential equations. Vol. II: Applications in engineering and physical sciences*, Chapman & Hall/CRC Press, Baton Rouge, 1994.

[414] N. H. Ibragimov, *A new conservation theorem*, J. Math. Anal. Appl. **333** (2007), no. 1, 311–328, DOI 10.1016/j.jmaa.2006.10.078. MR2323493

[415] N. H. Ibragimov, *Nonlinear self-adjointness and conservation laws*, J. Phys. A: Math. Theor. **44**:43 (2011), 432002 (9 pp).

[416] N. H. Ibragimov and A. B. Shabat, *Evolution equations with a nontrivial Lie-Bäcklund group* (Russian), Funktsional. Anal. i Prilozhen. **14** (1980), no. 1, 25–36, 96. MR565093

[417] N. H. Ibragimov and A. B. Shabat, *Infinite Lie-Bäcklund algebras* (Russian), Funktsional. Anal. i Prilozhen. **14** (1980), no. 4, 79–80; English transl., Funct. Anal. Appl. **14** (1980), no. 4, 313–315. MR595739

[418] E. L. Ince, *Ordinary Differential Equations*, Dover Publications, New York, 1944. MR0010757

[419] A. Iserles, *Brief introduction to Lie-group methods*, Collected lectures on the preservation of stability under discretization (Fort Collins, CO, 2001), SIAM, Philadelphia, PA, 2002, pp. 123–143. MR2026666

[420] Y. Itoh, *Integrals of a Lotka-Volterra system of odd number of variables*, Progr. Theoret. Phys. **78** (1987), no. 3, 507–510, DOI 10.1143/PTP.78.507. MR919145

[421] A. A. Izmest′ev, G. S. Pogosyan, A. N. Sissakian, and P. Winternitz, *Contractions of Lie algebras and separation of variables. The n-dimensional sphere*, J. Math. Phys. **40** (1999), no. 3, 1549–1573, DOI 10.1063/1.532820. MR1674634

[422] N. Jacobson, *Lie algebras*, Interscience Tracts in Pure and Applied Mathematics, No. 10, Interscience Publishers (a division of John Wiley & Sons, Inc.), New York-London, 1962. MR0143793

[423] X.-Y. Jia and N. Wang, *Geometric approach to Lie symmetry of discrete time Toda equation*, Chin. Phys. Lett. **26**:8 (2009), 080201.

[424] Z. Jiang, *Lie symmetries and their local determinacy for a class of differential-difference equations*, Phys. Lett. A **240** (1998), no. 3, 137–143, DOI 10.1016/S0375-9601(98)00099-1. MR1614411

[425] R. S. Johnson, *Water waves and Korteweg - de Vries equations*, J. Fluid Mech. **97** (1980), no. 4, 701–719, DOI 10.1017/S0022112080002765. MR573829

[426] C. Jordan, *Calculus of Finite Differences*, Hungarian Agent Eggenberger Book-Shop, Budapest, 1939. MR0000441

[427] N. Joshi and P. J. Vassiliou, *The existence of Lie symmetries for first-order analytic discrete dynamical systems*, J. Math. Anal. Appl. **195** (1995), no. 3, 872–887, DOI 10.1006/jmaa.1995.1393. MR1356647

[428] M. Juráš and I. M. Anderson, *Generalized Laplace invariants and the method of Darboux*, Duke Math. J. **89** (1997), no. 2, 351–375, DOI 10.1215/S0012-7094-97-08916-X. MR1460626

[429] V. G. Kac, *Highest weight representations of infinite-dimensional Lie algebras*, Proceedings of the International Congress of Mathematicians (Helsinki, 1978), Acad. Sci. Fennica, Helsinki, 1980, pp. 299–304. MR562619

[430] V. G. Kac, *Virasoro algebra*, in Encyclopedia of mathematics. URL : $http$: //encyclopediaofmath.org/index.php?title= Virasoro algebra&oldid=49148, EMS Press, Berlin, 2001.

[431] E. G. Kalnins, J. M. Kress, and W. Miller Jr., *Second-order superintegrable systems in conformally flat spaces. I. Two-dimensional classical structure theory*, J. Math. Phys. **46** (2005), no. 5, 053509, 28, DOI 10.1063/1.1897183. MR2143019

[432] E. G. Kalnins, J. M. Kress, and W. Miller Jr., *Second order superintegrable systems in conformally flat spaces. II. The classical two-dimensional Stäckel transform*, J. Math. Phys. **46** (2005), no. 5, 053510, 15, DOI 10.1063/1.1894985. MR2143027

[433] E. G. Kalnins, J. M. Kress, and W. Miller Jr., *Second order superintegrable systems in conformally flat spaces. III. Three-dimensional classical structure theory*, J. Math. Phys. **46** (2005), no. 10, 103507, 28, DOI 10.1063/1.2037567. MR2178607

[434] E. G. Kalnins, J. M. Kress, and W. Miller Jr., *Second order superintegrable systems in conformally flat spaces. IV. The classical 3D Stäckel transform and 3D classification theory*, J. Math. Phys. **47** (2006), no. 4, 043514, 26, DOI 10.1063/1.2191789. MR2226351

[435] E. G. Kalnins, J. M. Kress, and W. Miller Jr., *Second-order superintegrable systems in conformally flat spaces. V. Two- and three-dimensional quantum systems*, J. Math. Phys. **47** (2006), no. 9, 093501, 25, DOI 10.1063/1.2337849. MR2263655

[436] E. G. Kalnins, J. M. Kress, and W. Miller Jr., *Separation of variables and superintegrability*, IOP Expanding Physics, IOP Publishing, Bristol, 2018. The symmetry of solvable systems, DOI 10.1088/978-0-7503-1314-8. MR3942135

[437] M. Kanki, J. Mada, and T. Tokihiro, *Singularities of the discrete KdV equation and the Laurent property*, J. Phys. A **47** (2014), no. 6, 065201, 12, DOI 10.1088/1751-8113/47/6/065201. MR3159512

[438] I. Kaplansky, *The Virasoro algebra*, Comm. Math. Phys. **86** (1982), no. 1, 49–54. MR678001

[439] A. Kasman, *Glimpses of soliton theory: The algebra and geometry of nonlinear PDEs*, Student Mathematical Library, vol. 54, American Mathematical Society, Providence, RI, 2010, DOI 10.1090/stml/054. MR2731261

[440] D. J. Kaup, *On the inverse scattering problem for cubic eigenvalue problems of the class $\psi_{xxx} + 6Q\psi_x + 6R\psi = \lambda\psi$*, Stud. Appl. Math. **62** (1980), no. 3, 189–216, DOI 10.1002/sapm1980623189. MR571639

[441] W. G. Kelley and A. C. Peterson, *Difference equations: An introduction with applications*, Academic Press, Inc., Boston, MA, 1991. MR1142573

[442] P. H. M. Kersten and I. S. Krasil'shchik, *Symmetries and recursion operators for classical and supersymmetric differential equations*, Kluwer Acad. Publisher, Dordrecht, 2000.

[443] P. G. Kevrekidis, *The discrete nonlinear Schrödinger equation: Mathematical analysis, numerical computations and physical perspectives*, Springer Tracts in Modern Physics, vol. 232, Springer-Verlag, Berlin, 2009. Edited by Kevrekidis and with contributions by R. Carretero-González, A. R. Champneys, J. Cuevas, S. V. Dmitriev, D. J. Frantzeskakis, Y.-J. He, Q. Enam Hoq, A. Khare, K. J. H. Law, B. A. Malomed, T. R. O. Melvin, F. Palmero, M. A. Porter, V. M. Rothos, A. Stefanov and H. Susanto, DOI 10.1007/978-3-540-89199-4. MR2742565

[444] W. Killing, *Die Zusammensetzung der stetigen endlichen Transformationsgruppen* (German), Math. Ann. **36** (1890), no. 2, 161–189, DOI 10.1007/BF01207837. MR1510618

[445] E. Koelink, *Scattering theory with advanced topics in analysis*, TU, Deift, 2006.

[446] B. G. Konopel′chenko, *Nonlinear integrable equations: Recursion operators, group-theoretical and Hamiltonian structures of soliton equations*, Lecture Notes in Physics, vol. 270, Springer-Verlag, Berlin, 1987, DOI 10.1007/3-540-17567-9. MR882726

[447] B. G. Konopel′chenko, *Introduction to multidimensional integrable equations: The inverse spectral transform in $2+1$ dimensions*, Plenum Press, New York, 1992, DOI 10.1007/978-1-4899-1170-4. MR1219952

[448] D. J. Korteweg and G. de Vries, *On the change of form of long waves advancing in a rectangular canal, and on a new type of long stationary waves*, Philos. Mag. (5) **39** (1895), no. 240, 422–443, DOI 10.1080/14786449508620739. MR3363408

[449] A. M. Kosevich and A. S. Kovalev, *Vvedenie v nelineĭnuyu fizicheskuyu mekhaniku* (Russian), "Naukova Dumka", Kiev, 1989. With the collaboration of Yu. S. Kivshar. MR1027598

[450] Y. Kosmann-Schwarzbach, B. Grammaticos, and K. M. Tamizhmani (eds.), *Integrability of nonlinear systems*, Lecture Notes in Physics, vol. 495, Springer-Verlag, Berlin, 1997, DOI 10.1007/BFb0113690. MR1636289

[451] S. Konstantinou-Rizos, A. V. Mikhailov, and P. Xenitidis, *Reduction groups and related integrable difference systems of nonlinear Schrödinger type*, J. Math. Phys. **56** (2015), no. 8, 082701, 21, DOI 10.1063/1.4928048. MR3455334

[452] J. Krasil'shchik, A. Verbovetsky, and R. Vitolo, *The symbolic computation of integrability structures for partial differential equations*, Texts and Monographs in Symbolic Computation, Springer, Cham, 2017, DOI 10.1007/978-3-319-71655-8. MR3753682

[453] I. M. Krichever, *Elliptic analog of the Toda lattice*, Internat. Math. Res. Notices **8** (2000), 383–412, DOI 10.1155/S1073792800000222. MR1753690

[454] I. M. Krichever and S. P. Novikov, *Holomorphic bundles over algebraic curves, and nonlinear equations* (Russian), Uspekhi Mat. Nauk **35** (1980), no. 6(216), 47–68, 215; English transl., Russ. Math. Surv. **35** (1980), no. 6, 53–79. MR601756

[455] M. D. Kruskal, *Asymptotology in numerical computation: Progress and plans on the Fermi–Pasta–Ulam program*, in Proceedings IBM scientific computing symposium. large -scale problems in physics December 9–11, 1963, IBM, White Plans 1965, 43–59, discussion 59–62. In the Addendum with N. Zabusky introduce the word "soliton".

[456] M. D. Kruskal, N. Joshi, and R. Halburd, *Analytic and asymptotic methods for nonlinear singularity analysis: a review and extensions of tests for the Painlevé property*, Integrability of nonlinear systems (Pondicherry, 1996), Lecture Notes in Phys., vol. 495, Springer, Berlin, 1997, pp. 171–205, DOI 10.1007/BFb0113696. MR1636294

[457] M. D. Kruskal, R. M. Miura, C. S. Gardner, and N. J. Zabusky, *Korteweg-de Vries equation and generalizations. V. Uniqueness and nonexistence of polynomial conservation laws*, J. Mathematical Phys. **11** (1970), 952–960, DOI 10.1063/1.1665232. MR271527

[458] S. Kumei, *Invariance transformations, invariance group transformations, and invariance groups of the sine-Gordon equations*, J. Mathematical Phys. **16** (1975), no. 12, 2461–2468, DOI 10.1063/1.522487. MR385357

[459] S. Kumei and G. W. Bluman, *When nonlinear differential equations are equivalent to linear differential equations*, SIAM J. Appl. Math. **42** (1982), no. 5, 1157–1173, DOI 10.1137/0142079. MR673531

[460] B. A. Kuperschmidt, *Discrete Lax equations and differential-difference calculus* (English, with French summary), Astérisque **123** (1985), 212. MR785802

[461] B. A. Kupershmidt, *KP or mKP: Noncommutative mathematics of Lagrangian, Hamiltonian, and integrable systems*, Mathematical Surveys and Monographs, vol. 78, American Mathematical Society, Providence, RI, 2000, DOI 10.1090/surv/078. MR1752088

[462] E. A. Kuznetsov and G. E. Falkovich, *On the stability of a self-similar solution in the Burgers equation*, Phys. Lett. A **86** (1981), no. 4, 203–204, DOI 10.1016/0375-9601(81)90487-4. MR635735

[463] E. A. Kuznetsov and A. V. Mikhaĭlov, *Stability of stationary waves in nonlinear weakly dispersive media* (Russian, with English summary), Ž. Èksper. Teoret. Fiz. **67** (1974), no. 5, 1717–1727; English transl., Soviet Physics JETP **40** (1974), no. 5, 855–859. MR0387847

[464] S. Lafortune, P. Winternitz, and L. Martina, *Point symmetries of generalized Toda field theories*, J. Phys. A **33** (2000), no. 12, 2419–2435, DOI 10.1088/0305-4470/33/12/309. MR1751984

[465] S. Lafortune, S. Tremblay, and P. Winternitz, *Symmetry classification of diatomic molecular chains*, J. Math. Phys. **42** (2001), no. 11, 5341–5357, DOI 10.1063/1.1398583. MR1861346

[466] S. Lafortune and P. Winternitz, *Superposition formulas for pseudounitary matrix Riccati equations*, J. Math. Phys. **37** (1996), no. 3, 1539–1550, DOI 10.1063/1.531448. MR1377650

[467] G. L. Lamb Jr., *Elements of soliton theory*, Pure and Applied Mathematics, John Wiley & Sons, Inc., New York, 1980. MR591458

[468] J. S. W. Lamb and G. R. W. Quispel, *Cyclic reversing k-symmetry groups*, Nonlinearity **8** (1995), no. 6, 1005–1026. MR1363396

[469] P.-S. Laplace. *Recherches sur le calcul intégral aux différences partielles*, Mem.Acad.r.Sci.Paris, Paris, 1773 5–68.

[470] P. D. Lax, *Integrals of nonlinear equations of evolution and solitary waves*, Comm. Pure Appl. Math. **21** (1968), 467–490, DOI 10.1002/cpa.3160210503. MR235310

[471] D. Levi, *Nonlinear differential-difference equations as Bäcklund transformations*, J. Phys. A **14** (1981), no. 5, 1083–1098. MR611974

[472] D. Levi, *Toward a unification of the various techniques used to integrate nonlinear partial differential equations: Bäcklund and Darboux transformations vs. dressing method*, Rep. Math. Phys. **23** (1986), no. 1, 41–56, DOI 10.1016/0034-4877(86)90066-2. MR868630

[473] D. Levi, *Hierarchies of integrable equations obtained as nonisospectral (in x and t) deformations of the Schrödinger spectral problem*, Phys. Lett. A **119** (1987), no. 9, 453–456, DOI 10.1016/0375-9601(87)90414-2. MR879223

[474] D. Levi, *On a new Darboux transformation for the construction of exact solutions of the Schrödinger equation*, Inverse Problems **4** (1988), no. 1, 165–172. MR940796

[475] D. Levi, *Nonlinear internal solitary waves in straits with varying width and depth*, in Nonlinear topics in ocean physics, (A R Osborne ed.), North Holland, Amsterdam, 1991, 648-666.

[476] D. Levi, *Inhomogeneous Toda lattices and their symmetries*, in Group theory in physics: Proceedings of the international symposium held in honor of professor Marcos Moshinsky, (F A Seligman and K B Wolf eds.), AIP, New York, 1992, 97–108.

[477] D. Levi, *Symmetries and generalized symmetries for discrete dynamical systems*, Symmetries and integrability of difference equations (Canterbury, 1996), London Math. Soc. Lecture Note Ser., vol. 255, Cambridge Univ. Press, Cambridge, 1999, pp. 261–274, DOI 10.1017/CBO9780511569432.021. MR1705234

[478] D. Levi, *Lie symmetries for lattice equations*, in GROUP 24: Physical and mathematical aspects of symmetries: Proceedings of the 24th international colloquium on group theoretical methods in physics, Paris, 15-20 July 2002, (J–P Gazeau, R Kerner, J–P Antoine, S Métens and J–Y Thibon eds.), IOP, Bristol, 2003, 103–111.

[479] D. Levi, *Lie symmetries of differential and difference equations*, CCAST-WL Workshop Beijing, 2008.

[480] D. Levi and R. Benguria, *Bäcklund transformations and nonlinear differential difference equations*, Proc. Nat. Acad. Sci. U.S.A. **77** (1980), no. 9, 5025–5027, DOI 10.1073/pnas.77.9.5025. MR587276

[481] D. Levi and L. Martina, *Integrable hierarchies of nonlinear difference-difference equations and symmetries*, J. Phys. A **34** (2001), no. 48, 10357–10368, DOI 10.1088/0305-4470/34/48/302. Symmetries and integrability of difference equations (Tokyo, 2000). MR1877458

[482] D. Levi, L. Martina, and P. Winternitz, *Lie-point symmetries of the discrete Liouville equation*, J. Phys. A **48** (2015), no. 2, 025204, 18, DOI 10.1088/1751-8113/48/2/025204. MR3293063

[483] D. Levi, L. Martina, and P. Winternitz, *Structure preserving discretizations of the Liouville equation and their numerical tests*, SIGMA Symmetry Integrability Geom. Methods Appl. **11** (2015), Paper 080, 20, DOI 10.3842/SIGMA.2015.080. MR3403544

[484] D. Levi, L. Martina, and P. Winternitz, *The conformally invariant elliptic Liouville equation and its symmetry-preserving discretization* (Russian, with Russian summary), Teoret. Mat. Fiz. **196** (2018), no. 3, 419–433, DOI 10.4213/tmf9523; English transl., Theoret. and Math. Phys. **196** (2018), no. 3, 1307–1319. MR3849107

[485] D. Levi, J. Negro, and M. A. del Olmo, *Discrete derivatives and symmetries of difference equations*, J. Phys. A **34** (2001), no. 10, 2023–2030, DOI 10.1088/0305-4470/34/10/306. MR1828844

[486] D. Levi, J. Negro, and M. A. del Olmo, *Lie symmetries of difference equations*, Czechoslovak J. Phys. **51** (2001), no. 4, 341–348, DOI 10.1023/A:1017541506672. DI-CRM Workshop on Mathematical Physics (Prague, 2000). MR1848642

[487] D. Levi, J. Negro, and M. A. del Olmo, *Discrete q-derivatives and symmetries of q-difference equations*, J. Phys. A **37** (2004), no. 10, 3459–3473, DOI 10.1088/0305-4470/37/10/010. MR2039859

[488] D. Levi, M. C. Nucci, and M. A. Rodríguez, *λ-symmetries for the reduction of continuous and discrete equations*, Acta Appl. Math. **122** (2012), 311–321, DOI 10.1007/s10440-012-9745-8. MR2994001

[489] D. Levi and M. Petrera, *Continuous symmetries of the lattice potential KdV equation*, J. Phys. A **40** (2007), no. 15, 4141–4159, DOI 10.1088/1751-8113/40/15/006. MR2320058

[490] D. Levi, M. Petrera, and C. Scimiterna, *The lattice Schwarzian KdV equation and its symmetries*, J. Phys. A **40** (2007), no. 42, 12753–12761, DOI 10.1088/1751-8113/40/42/S18. MR2392903

[491] D. Levi, M. Petrera, and C. Scimiterna, *On the integrability of the discrete nonlinear Schrödinger equation*, Europhys. Lett. EPL **84** (2008), no. 1, Art. 10003, 4, DOI 10.1209/0295-5075/84/10003. MR2453187

[492] D. Levi, M. Petrera, C. Scimiterna, and R. I. Yamilov, *On Miura transformations and Volterra-type equations associated with the Adler-Bobenko-Suris equations*, SIGMA Symmetry Integrability Geom. Methods Appl. **4** (2008), Paper 077, 14, DOI 10.3842/SIGMA.2008.077. MR2470519

[493] D. Levi, L. Pilloni, and P. M. Santini, *Integrable three-dimensional lattices*, J. Phys. A **14** (1981), no. 7, 1567–1575. MR620593

[494] D. Levi and O. Ragnisco, *Extension of the spectral-transform method for solving nonlinear differential difference equations*, Lett. Nuovo Cimento (2) **22** (1978), no. 17, 691–696. MR507029

[495] D. Levi and O. Ragnisco, *Nonlinear differential-difference equations with N dependent coefficients: I*, J. Phys. A: Math. Gen. **12**:7 (1979), L157–L162.

[496] D. Levi and O. Ragnisco, *Nonlinear differential-difference equations with N-dependent coefficients. II*, J. Phys. A **12** (1979), no. 7, L163–L167. MR534245

[497] D. Levi and O. Ragnisco, *The inhomogeneous Toda lattice: its hierarchy and Darboux-Bäcklund transformations*, J. Phys. A **24** (1991), no. 8, 1729–1739. MR1117597

[498] D. Levi and O. Ragnisco (eds.), *SIDE III—symmetries and integrability of difference equations*, CRM Proceedings & Lecture Notes, vol. 25, American Mathematical Society, Providence, RI, 2000. Papers from the 3rd International Meeting held in Sabaudia, May 16–22, 1998, DOI 10.1090/crmp/025. MR1771703

[499] D. Levi and O. Ragnisco, *Dressing method and Bäcklund and Darboux transformations*, in Bäcklund & Darboux transformations: The geometry of soliton theory, (A. Coley, D. Levi, R. Milson, C. Rogers and P. Winternitz eds.), AMS, Providence, RI., 2001, 29–52.

[500] D. Levi, O. Ragnisco, and M. Bruschi, *Extension of the Zakharov-Shabat generalized inverse method to solve differential-difference and difference-difference equations* (English, with Italian and Russian summaries), Nuovo Cimento A (11) **58** (1980), no. 1, 56–66, DOI 10.1007/BF02730220. MR593274

[501] D. Levi, O. Ragnisco, and M. Bruschi, *Extension of the Zakharov-Shabat generalized inverse method to solve differential-difference and difference-difference equations* (English, with Italian and Russian summaries), Nuovo Cimento A (11) **58** (1980), no. 1, 56–66, DOI 10.1007/BF02730220. MR593274

[502] D. Levi, O. Ragnisco, and M. Bruschi, *Continuous and discrete matrix Burgers' hierarchies* (English, with Italian summary), Nuovo Cimento B (11) **74** (1983), no. 1, 33–51, DOI 10.1007/BF02721683. MR702965

[503] D. Levi, O. Ragnisco, and M. A. Rodríguez, *On nonisospectral flows, Painlevé equations and symmetries of differential and difference equations* (English, with English and Russian summaries), Teoret. Mat. Fiz. **93** (1992), no. 3, 473–480, DOI 10.1007/BF01016397; English transl., Theoret. and Math. Phys. **93** (1992), no. 3, 1409–1414 (1993). MR1233554

[504] D. Levi, O. Ragnisco, and A. Sym, *Bäcklund transformation vs. the dressing method*, Lett. Nuovo Cimento (2) **33** (1982), no. 13, 401–406. MR659008

[505] D. Levi, E. Ricca, Z. Thomova, and P. Winternitz, *Lie group analysis of a generalized Krichever-Novikov differential-difference equation*, J. Math. Phys. **55** (2014), no. 10, 103503, 12, DOI 10.1063/1.4896989. MR3390472

[506] D. Levi and M. A. Rodríguez, *Symmetry group of partial differential equations and of differential-difference equations: the Toda lattice versus the Korteweg-de Vries equation*, J. Phys. A **25** (1992), no. 15, L975–L979. MR1177544

[507] D. Levi and M. A. Rodríguez, *Lie symmetries for integrable evolution equations on the lattice*, J. Phys. A **32** (1999), no. 47, 8303–8316, DOI 10.1088/0305-4470/32/47/309. MR1733340

[508] D. Levi and M. A. Rodríguez, *Lie discrete symmetries of lattice equations*, J. Phys. A **37** (2004), no. 5, 1711–1725, DOI 10.1088/0305-4470/37/5/016. MR2043465

[509] D. Levi and M. A. Rodríguez, *Asymptotic symmetries and integrability: the KdV case*, Europhys. Lett. EPL **80** (2007), no. 6, Art. 60005, 4, DOI 10.1209/0295-5075/80/60005. MR2444011

[510] D. Levi and M. A. Rodríguez, *λ-symmetries for discrete equations*, J. Phys. A **43** (2010), no. 29, 292001, 9, DOI 10.1088/1751-8113/43/29/292001. MR2659779

[511] D. Levi and M. A. Rodríguez, *Construction of partial difference schemes: I. The Clairaut, Schwarz, Young theorem on the lattice*, J. Phys. A **46** (2013), no. 29, 295203, 19, DOI 10.1088/1751-8113/46/29/295203. MR3083452

[512] D. Levi and M. A. Rodríguez, *On the construction of partial difference schemes II: discrete variables and Schwarzian lattices*, Acta Polytech. **56**:3 (2016) 236–244.

[513] D. Levi and M. A. Rodríguez, *Yamilov's theorem for differential and difference equations*, Ufa Math. J. **13** (2021), no. 2, 152–159. MR4293325

[514] D. Levi, M. A. Rodríguez, and Z. Thomova, *Differential equations invariant under conditional symmetries*, J. Nonlinear Math. Phys. **26** (2019), no. 2, 281–293, DOI 10.1080/14029251.2019.1591731. MR3937142

[515] D. Levi, M. A. Rodríguez, and Z. Thomova, *The discretized Boussinesq equation and its conditional symmetry reduction*, J. Phys. A **53** (2020), no. 4, 045201, 22, DOI 10.1088/1751-8121/ab5b47. MR4054734

[516] D. Levi, M. A. Rodríguez, and Z. Thomova, *Conditional discretization of a generalized reaction-diffusion equation*, Quantum theory and symmetries, CRM Ser. Math. Phys., Springer, Cham, [2021] ©2021, pp. 149–156, DOI 10.1007/978-3-030-55777-5_14. MR4380836

[517] D. Levi and C. Scimiterna, *Integrability test of discrete nonlinear Schrödinger equations via multiscale reduction*, Appl. Anal. **89** (2010), no. 4, 507–527, DOI 10.1080/00036810903329969. MR2647764

[518] D. Levi and C. Scimiterna, *Linearizability of nonlinear equations on a quad-graph by a point, two points and generalized Hopf-Cole transformations*, SIGMA Symmetry Integrability Geom. Methods Appl. **7** (2011), Paper 079, 24, DOI 10.3842/SIGMA.2011.079. MR2861197

[519] D. Levi and C. Scimiterna, *Linearization through symmetries for discrete equations*, J. Phys. A **46** (2013), no. 32, 325204, 18, DOI 10.1088/1751-8113/46/32/325204. MR3085376

[520] D. Levi and C. Scimiterna, *Four points linearizable lattice schemes*, J. Geom. Symmetry Phys. **31** (2013), 93–104. MR3154685

[521] D. Levi, C. Scimiterna, Z. Thomova, and P. Winternitz, *Contact transformations for difference schemes*, J. Phys. A **45** (2012), no. 2, 022001, 9, DOI 10.1088/1751-8113/45/2/022001. MR2871398

[522] D. Levi, D. Sekera, and P. Winternitz, *Lie point symmetries and ODEs passing the Painlevé test*, J. Nonlinear Math. Phys. **25** (2018), no. 4, 604–617, DOI 10.1080/14029251.2018.1503435. MR3834406

[523] D. Levi and A. Sym, *Integrable systems describing surfaces of nonconstant curvature*, Phys. Lett. A **149** (1990), no. 7-8, 381–387, DOI 10.1016/0375-9601(90)90897-W. MR1075779

[524] D. Levi, A. Sym and G.-Z. Tu, *A working algorithm to isolate integrable surfaces in E3*, preprint 761, 10.10.1990 Physics Dep. Rome University "La Sapienza" and INFN, Rome Branch available in inspire-hep.net.

[525] D. Levi, P. Tempesta, and P. Winternitz, *Umbral calculus, difference equations and the discrete Schrödinger equation*, J. Math. Phys. **45** (2004), no. 11, 4077–4105, DOI 10.1063/1.1780612. MR2098119

[526] D. Levi, P. Tempesta and P. Winternitz, *Lorentz and Galilei invariance on lattices*, Phys. Rev. D **69** (2004), 105011 (6pp).

[527] D. Levi, Z. Thomova, and P. Winternitz, *Are there contact transformations for discrete equations?*, J. Phys. A **44** (2011), no. 26, 265201, 7, DOI 10.1088/1751-8113/44/26/265201. MR2801553

[528] D. Levi, S. Tremblay, and P. Winternitz, *Lie point symmetries of difference equations and lattices*, J. Phys. A **33** (2000), no. 47, 8507–8523, DOI 10.1088/0305-4470/33/47/313. MR1804124

[529] D. Levi, S. Tremblay, and P. Winternitz, *Continuous symmetries of equations on lattices*, Czechoslovak J. Phys. **51** (2001), no. 4, 349–356, DOI 10.1023/A:1017593523510. DI-CRM Workshop on Mathematical Physics (Prague, 2000). MR1848643

[530] D. Levi, S. Tremblay, and P. Winternitz, *Lie symmetries of multidimensional difference equations*, J. Phys. A **34** (2001), no. 44, 9507–9524, DOI 10.1088/0305-4470/34/44/311. MR1875206

[531] D. Levi and P. Winternitz, *Lie point symmetries of differential-difference equations*, Symmetries and integrability of difference equations (Estérel, PQ, 1994), CRM Proc. Lecture Notes, vol. 9, Amer. Math. Soc., Providence, RI, 1996, pp. 199–208, DOI 10.1090/crmp/009/19. MR1416839

[532] D. Levi, L. Vinet, and P. Winternitz, *Lie group formalism for difference equations*, J. Phys. A **30** (1997), no. 2, 633–649, DOI 10.1088/0305-4470/30/2/024. MR1442880

[533] D. Levi and P. Winternitz, *The cylindrical Kadomtsev-Petviashvili equation: its Kac-Moody-Virasoro algebra and relation to KP equation*, Phys. Lett. A **129** (1988), no. 3, 165–167, DOI 10.1016/0375-9601(88)90137-5. MR941073

[534] D. Levi and P. Winternitz, *Nonclassical symmetry reduction: example of the Boussinesq equation*, J. Phys. A **22** (1989), no. 15, 2915–2924. MR1007225

[535] D. Levi and P. Winternitz, *Continuous symmetries of discrete equations*, Phys. Lett. A **152** (1991), no. 7, 335–338, DOI 10.1016/0375-9601(91)90733-O. MR1092660

[536] D. Levi and P. Winternitz, *Symmetries of differential equations on a lattice. An example: the Toda lattice*, Modern group analysis: advanced analytical and computational methods in mathematical physics (Acireale, 1992), Kluwer Acad. Publ., Dordrecht, 1993, pp. 265–272. MR1259562

[537] D. Levi and P. Winternitz, *Dimensional reduction for equations involving discrete and continuous variables*, Applications of analytic and geometric methods to nonlinear differential equations (Exeter, 1992), NATO Adv. Sci. Inst. Ser. C: Math. Phys. Sci., vol. 413, Kluwer Acad. Publ., Dordrecht, 1993, pp. 405–413. MR1261681

[538] D. Levi and P. Winternitz, *Symmetries and conditional symmetries of differential-difference equations*, J. Math. Phys. **34** (1993), no. 8, 3713–3730, DOI 10.1063/1.530054. MR1230547

[539] D. Levi and P. Winternitz, *Symmetry analysis of discrete physical systems*, Nonlinear Physics, Theory and Experiment, (E Alfinito, M Boiti, L Martina and F Pempinelli eds.), World Scientific, Singapore, 1996, 199–216.

[540] D. Levi and P. Winternitz, *Lie point symmetries of differential-difference equations*, Symmetries and integrability of difference equations (Estérel, PQ, 1994), CRM Proc. Lecture Notes, vol. 9, Amer. Math. Soc., Providence, RI, 1996, pp. 199–208, DOI 10.1090/crmp/009/19. MR1416839

[541] D. Levi and P. Winternitz, *Symmetries of discrete physical systems*, Proceedings of the IV Wigner Symposium, (N M Atakishiyev, T H Seligman and K B Wolf eds.), World Scientific, Singapore, 1996, 382–391.

[542] D. Levi and P. Winternitz, *Symmetries of discrete dynamical systems*, J. Math. Phys. **37** (1996), no. 11, 5551–5576, DOI 10.1063/1.531722. MR1417160

[543] D. Levi and P. Winternitz, *Lie point symmetries and commuting flows for equations on lattices*, J. Phys. A **35** (2002), no. 9, 2249–2262, DOI 10.1088/0305-4470/35/9/314. MR1908722

[544] D. Levi and P. Winternitz, *Continuous symmetries of difference equations*, J. Phys. A **39** (2006), no. 2, R1–R63, DOI 10.1088/0305-4470/39/2/R01. MR2198959

[545] D. Levi and P. Winternitz, *Lie groups and numerical solutions of differential equations: Invariant discretization versus differential approximation*, Acta Polytech. **53**:5 (2013), 438–443.

[546] D. Levi, P. Winternitz, and R. I. Yamilov, *Lie point symmetries of differential-difference equations*, J. Phys. A **43** (2010), no. 29, 292002, 14, DOI 10.1088/1751-8113/43/29/292002. MR2659780

[547] D. Levi, P. Winternitz, and R. I. Yamilov, *Symmetries of the continuous and discrete Krichever-Novikov equation*, SIGMA Symmetry Integrability Geom. Methods Appl. **7** (2011), Paper 097, 16, DOI 10.3842/SIGMA.2011.097. MR2861179

[548] D. Levi and R. I. Yamilov, *Classification of evolutionary equations on the lattice The general theory*, arXiv:solv-int/9511006 v1 16 Nov 1995.

[549] D. Levi and R. I. Yamilov, *Conditions for the existence of higher symmetries of evolutionary equations on the lattice*, J. Math. Phys. **38** (1997), no. 12, 6648–6674, DOI 10.1063/1.532230. MR1483513

[550] D. Levi and R. I. Yamilov, *Dilation symmetries and equations on the lattice*, J. Phys. A **32** (1999), no. 47, 8317–8323, DOI 10.1088/0305-4470/32/47/310. MR1733341

[551] D. Levi and R. I. Yamilov, *Non-point integrable symmetries for equations on the lattice*, J. Phys. A **33** (2000), no. 26, 4809–4823, DOI 10.1088/0305-4470/33/26/307. MR1792901

[552] D. Levi and R. I. Yamilov, *Conditions for the existence of higher symmetries and nonlinear evolutionary equations on the lattice*, Algebraic methods in physics (Montréal, QC, 1997), CRM Ser. Math. Phys., Springer, New York, 2001, pp. 135–148. MR1847253

[553] D. Levi and R. I. Yamilov, *On the integrability of a new discrete nonlinear Schrödinger equation*, J. Phys. A **34** (2001), no. 41, L553–L562, DOI 10.1088/0305-4470/34/41/102. MR1876601

[554] D. Levi and R. I. Yamilov, *On a nonlinear integrable difference equation on the square*, Ufa Math. J. **1**:2 (2009), 101–105.

[555] D. Levi and R. I. Yamilov, *The generalized symmetry method for discrete equations*, J. Phys. A **42** (2009), no. 45, 454012, 18, DOI 10.1088/1751-8113/42/45/454012. MR2556654

[556] D. Levi and R. I. Yamilov, *Generalized symmetry integrability test for discrete equations on the square lattice*, J. Phys. A **44** (2011), no. 14, 145207, 22, DOI 10.1088/1751-8113/44/14/145207. MR2780422

[557] D. Levi and R. I. Yamilov, *Generalized Lie symmetries for difference equations*, in Symmetries and Integrability of Difference Equations, (D. Levi, P. Olver, Z. Thomova, P. Winternitz eds.), Cambridge University Press, Cambridge, 2011, 160–190.

[558] D. Levi and R. I. Yamilov, *Integrability test for discrete equations via generalized symmetries*, in Symposium in Memoriam of Marcos Moshinsky, (L. Benet, P. O. Hess, J. M. Torres and K. B. Wolf eds.), AIP Conf. Proc. **1323**:1 (2012), 203–214.

[559] E. E. Levi, *Sulla struttura dei gruppi finiti e continui*, Atti Accad.Sci. Torino **XL** (1905), 551–565 [in Italian].

[560] D. Lewis and P. J. Olver, *Geometric integration algorithms on homogeneous manifolds*, Found. Comput. Math. **2** (2002), no. 4, 363–392, DOI 10.1007/s102080010028. MR1930943

[561] A. N. Leznov, M. V. Savel′ev, and V. G. Smirnov, *General solutions for a two-dimensional system of Volterra equations realizing the Bäcklund transform for the Toda lattice* (Russian, with English summary), Teoret. Mat. Fiz. **47** (1981), no. 2, 216–224. MR626983

[562] A. N. Leznov, A. B. Shabat, and R. I. Yamilov, *Canonical transformations generated by shifts in nonlinear lattices*, Phys. Lett. A **174** (1993), no. 5-6, 397–402, DOI 10.1016/0375-9601(93)90197-8. MR1209078

[563] Y. S. Li and G. C. Zhu, *New set of symmetries of the integrable equations, Lie algebra and nonisospectral evolution equations. II. AKNS system*, J. Phys. A **19** (1986), no. 18, 3713–3725. MR869304

[564] S. Lie, *Vorlesungen uber Differentialgleichungen mit bekannten infinitesimalen Transformationen, bearbeitet und herausgegeben von Dr. G. Scheffers* , Teubner, Leipzig, 1891, reprint *Differentialgleichungen*, Chelsea, New York, 1967.

[565] S. Lie, *Vorlesungen über continuierliche Gruppen mit Geometrischen und anderen Anwendungen* (German), Chelsea Publishing Co., Bronx, N.Y., 1971. Bearbeitet und herausgegeben von Georg Scheffers; Nachdruck der Auflage des Jahres 1893. MR0392458

[566] S. Lie, *Klassifikation und Integration von gewohnlichen Differentialgleichunden zwischen x, y die eine Gruppe von Transformationen gestatten. I–IV. Gessamelte Abhandlungen*, Teubner, Leipzig, 1924 Vol. 5.

[567] J. Liouville, *Sur l'equation aux différences partielles* $\partial^2 \log \lambda/\partial u \partial v \pm \lambda/(aa^2) = 0$, J. Math. Pures Appl. **18** (1853), 71–72.

[568] Q. P. Liu, *Darboux transformations for supersymmetric Korteweg-de Vries equations*, Lett. Math. Phys. **35** (1995), no. 2, 115–122, DOI 10.1007/BF00750761. MR1347874

[569] O. Loos, *Jordan pairs*, Lecture Notes in Mathematics, Vol. 460, Springer-Verlag, Berlin-New York, 1975. MR0444721

[570] J. E. López-Sendino, J. Negro, M. A. del Olmo and E. Salgado, *Quantum mechanics and umbral calculus*, The 5th international symposium on quantum theory and symmetries, J. Phys.: Conf. Ser. **128** (2008), 012056 (15pp).

[571] J. E. López-Sendino, J. Negro and M. A. del Olmo, *Discrete Coulomb potential*, Phys. Atom. Nuclei **73** (2010), 384–390.

[572] W.-X. Ma and B. Fuchssteiner, *Algebraic structure of discrete zero curvature equations and master symmetries of discrete evolution equations*, J. Math. Phys. **40** (1999), no. 5, 2400–2418, DOI 10.1063/1.532872. MR1686569

[573] R. McLachlan and R. Quispel, *Six lectures on the geometric integration of ODEs*, Foundations of computational mathematics (Oxford, 1999), London Math. Soc. Lecture Note Ser., vol. 284, Cambridge Univ. Press, Cambridge, 2001, pp. 155–210. MR1836618

[574] S. Maeda, *Canonical structure and symmetries for discrete systems*, Math. Japon. **25** (1980), no. 4, 405–420. MR594539

[575] S. Maeda, *Extension of discrete Noether theorem*, Math. Japon. **26** (1981), no. 1, 85–90. MR613471

[576] S. Maeda, *Lagrangian formulation of discrete systems and concept of difference space*, Math. Japon. **27** (1982), no. 3, 345–356. MR663162

[577] S. Maeda, *The similarity method for difference equations*, IMA J. Appl. Math. **38** (1987), no. 2, 129–134, DOI 10.1093/imamat/38.2.129. MR983722

[578] S. Maeda, *Application of numerical schemes admitting symmetries*, Electron. Comm. Japan Part III Fund. Electron. Sci. **73** (1990), no. 9, 107–113 (1991), DOI 10.1002/ecjc.4430730912. MR1093281

[579] F. Magri, *A simple model of the integrable Hamiltonian equation*, J. Math. Phys. **19** (1978), no. 5, 1156–1162, DOI 10.1063/1.523777. MR488516

[580] F. M. Mahomed and P. G. L. Leach, *Lie algebras associated with scalar second-order ordinary differential equations*, J. Math. Phys. **30** (1989), no. 12, 2770–2777, DOI 10.1063/1.528511. MR1025217

[581] F. M. Mahomed and P. G. L. Leach, *The Lie algebra* $\mathrm{sl}(3, \mathbf{R})$ *and linearization*, Quaestiones Math. **12** (1989), no. 2, 121–139. MR998743

[582] J.-M. Maillet and F. Nijhoff, *Integrability for multidimensional lattice models*, Phys. Lett. B **224** (1989), no. 4, 389–396, DOI 10.1016/0370-2693(89)91466-4. MR1004628

[583] A. Makarov, J. Smorodinsky, Kh. Valiev, and P. Winternitz. *A systematic search for non-relativistic systems with dynamical symmetries*, Nuovo Cimento A **52** (1967), 1061–1084.

[584] V. G. Makhankov, *Soliton phenomenology*, Mathematics and its Applications (Soviet Series), vol. 33, Kluwer Academic Publishers Group, Dordrecht, 1990, DOI 10.1007/978-94-009-2217-4. MR1078628

[585] B. A. Malomed, *Soliton management in periodic systems*, Springer, New York, 2006. MR2683323

[586] S. V. Manakov, *Complete integrability and stochastization of discrete dynamical systems* (Russian, with English summary), Ž. Èksper. Teoret. Fiz. **67** (1974), no. 2, 543–555; English transl., Soviet Physics JETP **40** (1974), no. 2, 269–274 (1975). MR0389107

[587] E. L. Mansfield and P. E. Hydon, *On a variational complex for difference equations*, The geometrical study of differential equations (Washington, DC, 2000), Contemp. Math., vol. 285, Amer. Math. Soc., Providence, RI, 2001, pp. 121–129, DOI 10.1090/conm/285/04738. MR1874294

[588] N. Manton and P. Sutcliffe, *Topological solitons*, Cambridge Monographs on Mathematical Physics, Cambridge University Press, Cambridge, 2004, DOI 10.1017/CBO9780511617034. MR2068924

[589] V. A. Marčenko, *On reconstruction of the potential energy from phases of the scattered waves* (Russian), Dokl. Akad. Nauk SSSR (N.S.) **104** (1955), 695–698. MR0075402

[590] I. Marquette, *Superintegrability and higher order polynomial algebras*, J. Phys. A **43** (2010), no. 13, 135203, 15, DOI 10.1088/1751-8113/43/13/135203. MR2602100

[591] I. Marquette and C. Quesne, *New families of superintegrable systems from Hermite and Laguerre exceptional orthogonal polynomials*, J. Math. Phys. **54** (2013), no. 4, 042102, 16, DOI 10.1063/1.4798807. MR3088222

[592] I. Marquette, M. Sajedi, and P. Winternitz, *Fourth order superintegrable systems separating in Cartesian coordinates I. Exotic quantum potentials*, J. Phys. A **50** (2017), no. 31, 315201, 29, DOI 10.1088/1751-8121/aa7a67. MR3673481

[593] I. Marquette and P. Winternitz, *Higher order quantum superintegrability: a new "Painlevé conjecture"*, Integrability, supersymmetry and coherent states, CRM Ser. Math. Phys., Springer, Cham, 2019, pp. 103–131. MR3967787

[594] L. Martina, S. Lafortune, and P. Winternitz, *Point symmetries of generalized Toda field theories. II. Symmetry reduction*, J. Phys. A **33** (2000), no. 36, 6431–6446, DOI 10.1088/0305-4470/33/36/311. MR1789563

[595] L. Martina and P. Winternitz, *Analysis and applications of the symmetry group of the multidimensional three-wave resonant interaction problem* (English, with French summary), Ann. Physics **196** (1989), no. 2, 231–277, DOI 10.1016/0003-4916(89)90178-4. MR1027661

[596] M. Marvan, *On the spectral parameter problem*, Acta Appl. Math. **109** (2010), no. 1, 239–255, DOI 10.1007/s10440-009-9450-4. MR2579891

[597] S. Maxon, *Cylindrical and spherical solitons*, Rocky Mountain J. Math. **8** (1978), no. 1-2, 269–281, DOI 10.1216/RMJ-1978-8-1-269. MR479217

[598] K. Maruno, K. Kajiwara, S. Nakao, and M. Oikawa, *Bilinearization of discrete soliton equations and singularity confinement*, Phys. Lett. A **229** (1997), no. 3, 173–182, DOI 10.1016/S0375-9601(97)00171-0. MR1445501

[599] V. B. Matveev and M. A. Salle, *Darboux transformations and solitons*, Springer Series in Nonlinear Dynamics, Springer-Verlag, Berlin, 1991, DOI 10.1007/978-3-662-00922-2. MR1146435

[600] S. V. Meleshko, *Group classification of equations of two-dimensional motions of a gas* (Russian, with Russian summary), Prikl. Mat. Mekh. **58** (1994), no. 4, 56–62, DOI 10.1016/0021-8928(94)90138-4; English transl., J. Appl. Math. Mech. **58** (1994), no. 4, 629–635. MR1310992

[601] I. Merola, O. Ragnisco, and G. Z. Tu, *A novel hierarchy of integrable lattices*, Inverse Problems **10** (1994), no. 6, 1315–1334. MR1306807

[602] A. V. Mikhailov, *Integrability of a two-dimensional generalization of the Toda chain*, Pis'ma v Zh. Eksp. Teor. Fiz. **30**:7 (1979), 443-447 [in Russian]; English version in Sov. Phys. JETP Lett. **30**:7 (1979), 414–418.

[603] A. V. Mikhailov,*The reduction problem and the inverse scattering method*, Physica D **3** (1981), 73-117.

[604] A. V. Mikhailov (ed.), *Integrability*, Lecture Notes in Physics, vol. 767, Springer-Verlag, Berlin, 2009, DOI 10.1007/978-3-540-88111-7. MR2868499

[605] A. V. Mikhailov, *Formal diagonalisation of Lax-Darboux schemes* (English, with English and Russian summaries), Model. Anal. Inf. Sist. **22** (2015), no. 6, 795–817, DOI 10.18255/1818-1015-2015-6-795-817. MR3493712

[606] A. B. Shabat and A. V. Mikhailov, *Symmetries—test of integrability*, Important developments in soliton theory, Springer Ser. Nonlinear Dynam., Springer, Berlin, 1993, pp. 355–374. MR1280482

[607] A. V. Mikhailov, A. B. Shabat, and V. V. Sokolov, *The symmetry approach to classification of integrable equations*, What is integrability?, Springer Ser. Nonlinear Dynam., Springer, Berlin, 1991, pp. 115–184. MR1098338

[608] A. V. Mikhailov, A. B. Shabat, and R. I. Yamilov, *A symmetric approach to the classification of nonlinear equations. Complete lists of integrable systems* (Russian), Uspekhi Mat. Nauk **42** (1987), no. 4(256), 3–53; English transl., Russ. Math. Surv. **43** (1987), no. 4, 1–63. MR912060

[609] A. V. Mikhailov, A. B. Shabat, and R. I. Yamilov, *On an extension of the module of invertible transformations* (Russian), Dokl. Akad. Nauk SSSR **295** (1987), no. 2, 288–291; English transl., Soviet Math. Dokl. **36** (1988), no. 1, 60–63. MR902684

[610] A. V. Mikhailov, A. B. Shabat, and R. I. Yamilov, *Extension of the module of invertible transformations. Classification of integrable systems*, Comm. Math. Phys. **115** (1988), no. 1, 1–19. MR929143

[611] A. V. Mikhailov, J. P. Wang, and P. Xenitidis, *Recursion operators, conservation laws, and integrability conditions for difference equations*, Theoret. and Math. Phys. **167** (2011), no. 1, 421–443, DOI 10.1007/s11232-011-0033-y. Russian version appears in Teoret. Mat. Fiz. **167** (2011), no. 1, 23–49. MR3165750

[612] A. V. Mikhailov, J. P. Wang, and P. Xenitidis, *Cosymmetries and Nijenhuis recursion operators for difference equations*, Nonlinearity **24** (2011), no. 7, 2079–2097, DOI 10.1088/0951-7715/24/7/009. MR2805594

[613] A. V. Mikhailov and P. Xenitidis, *Second order integrability conditions for difference equations: an integrable equation*, Lett. Math. Phys. **104** (2014), no. 4, 431–450, DOI 10.1007/s11005-013-0668-8. MR3177991

[614] L. M. Milne-Thomson, *The Calculus of Finite Differences*, Macmillan & Co., Ltd., London, 1951. MR0043339

[615] W. Miller Jr., S. Post, and P. Winternitz, *Classical and quantum superintegrability with applications*, J. Phys. A **46** (2013), no. 42, 423001, 97, DOI 10.1088/1751-8113/46/42/423001. MR3119484

[616] W. Miller Jr. and A. V. Turbiner, *(Quasi)-exact-solvability on the sphere S^n*, J. Math. Phys. **56** (2015), no. 2, 023501, 14, DOI 10.1063/1.4906909. MR3390897

[617] R. M. Miura, *Korteweg-de Vries equation and generalizations. I. A remarkable explicit nonlinear transformation*, J. Mathematical Phys. **9** (1968), 1202–1204, DOI 10.1063/1.1664700. MR252825

[618] R. M. Miura, C. S. Gardner, and M. D. Kruskal, *Korteweg-de Vries equation and generalizations. II. Existence of conservation laws and constants of motion*, J. Mathematical Phys. **9** (1968), 1204–1209, DOI 10.1063/1.1664701. MR252826

[619] T. Miwa, M. Jimbo, and E. Date, *Solitons: Differential equations, symmetries and infinite-dimensional algebras*, Cambridge Tracts in Mathematics, vol. 135, Cambridge University Press, Cambridge, 2000. Translated from the 1993 Japanese original by Miles Reid. MR1736222

[620] R. V. Moody and J. Patera, *Discrete and continuous graded contractions of representations of Lie algebras*, J. Phys. A **24** (1991), no. 10, 2227–2257. MR1118531

[621] P. Moon and D. E. Spencer, *Field theory handbook: Including coordinate systems, differential equations and their solutions*, 2nd ed., Springer-Verlag, Berlin, 1988. MR947546

[622] B. Moritz and W. Schwalm, *Triangle lattice Green functions for vector fields*, J. Phys. A **34** (2001), no. 3, 589–602, DOI 10.1088/0305-4470/34/3/317. MR1822793

[623] L. Munteanu and S. Donescu, *Introduction to soliton theory: applications to mechanics*, Fundamental Theories of Physics, vol. 143, Kluwer Academic Publishers, Dordrecht, 2004. MR2085905

[624] C. Muriel and J. L. Romero, *New methods of reduction for ordinary differential equations*, IMA J. Appl. Math. **66** (2001), no. 2, 111–125, DOI 10.1093/imamat/66.2.111. MR1820174

[625] C. Muriel and J. L. Romero, *C^∞-symmetries and reduction of equations without Lie point symmetries*, J. Lie Theory **13** (2003), no. 1, 167–188. MR1958580

[626] K. Narita, *Solutions for the Mikhailov-Shabat-Yamilov difference-differential equations and generalized solutions for the Volterra and the Toda lattice equations*, Progr. Theoret. Phys. **99** (1998), no. 3, 337–348, DOI 10.1143/PTP.99.337. MR1633358

[627] K. Narita, *Soliton solution to extended Volterra equation*, J. Phys. Soc. Japan **51** (1982), no. 5, 1682–1685, DOI 10.1143/JPSJ.51.1682. MR668131

[628] E. Neher, *Jordan triple systems by the grid approach*, Lecture Notes in Mathematics, vol. 1280, Springer-Verlag, Berlin, 1987, DOI 10.1007/BFb0078217. MR911879

[629] N. N. Nehorošev, *Action-angle variables, and their generalizations* (Russian), Trudy Moskov. Mat. Obšč. **26** (1972), 181–198. MR0365629

[630] A. C. Newell, *Solitons in mathematics and physics*, CBMS-NSF Regional Conference Series in Applied Mathematics, vol. 48, Society for Industrial and Applied Mathematics (SIAM), Philadelphia, PA, 1985, DOI 10.1137/1.9781611970227. MR847245

[631] U. Niederer, *The maximal kinematical invariance group of the free Schrödinger equation*, Helv. Phys. Acta **45** (1972/73), no. 5, 802–810. MR400948

[632] F. W. Nijhoff, *On some "Schwarzian" equations and their discrete analogues*, Algebraic aspects of integrable systems, Progr. Nonlinear Differential Equations Appl., vol. 26, Birkhäuser Boston, Boston, MA, 1997, pp. 237–260. MR1418878

[633] F. W. Nijhoff, *Discrete Painlevé equations and symmetry reduction on the lattice*, Discrete integrable geometry and physics (Vienna, 1996), Oxford Lecture Ser. Math. Appl., vol. 16, Oxford Univ. Press, New York, 1999, pp. 209–234. MR1676599

[634] F. W. Nijhoff, *Lax pair for the Adler (lattice Krichever-Novikov) system*, Phys. Lett. A **297** (2002), no. 1-2, 49–58, DOI 10.1016/S0375-9601(02)00287-6. MR1912127

[635] F. W. Nijhoff, J. Atkinson, and J. Hietarinta, *Soliton solutions for ABS lattice equations. I. Cauchy matrix approach*, J. Phys. A **42** (2009), no. 40, 404005, 34, DOI 10.1088/1751-8113/42/40/404005. MR2544269

[636] F. W. Nijhoff and H. W. Capel, *The direct linearisation approach to hierarchies of integrable PDEs in $2+1$ dimensions. I. Lattice equations and the differential-difference hierarchies*, Inverse Problems **6** (1990), no. 4, 567–590. MR1065921

[637] F. W. Nijhoff and H. Capel, *The discrete Korteweg-de Vries equation*, Acta Appl. Math. **39** (1995), no. 1-3, 133–158, DOI 10.1007/BF00994631. KdV '95 (Amsterdam, 1995). MR1329559

[638] F. W. Nijhoff and V. G. Papageorgiou, *Similarity reductions of integrable lattices and discrete analogues of the Painlevé* II *equation*, Phys. Lett. A **153** (1991), no. 6-7, 337–344, DOI 10.1016/0375-9601(91)90955-8. MR1098879

[639] F. W. Nijhoff, V. G. Papageorgiou, and H. W. Capel, *Integrable time-discrete systems: lattices and mappings*, Quantum groups (Leningrad, 1990), Lecture Notes in Math., vol. 1510, Springer, Berlin, 1992, pp. 312–325, DOI 10.1007/BFb0101197. MR1183496

[640] F. W. Nijhoff, V. G. Papageorgiou, H. W. Capel, and G. R. W. Quispel, *The lattice Gel′fand-Dikiĭ hierarchy*, Inverse Problems **8** (1992), no. 4, 597–621. MR1178232

[641] F. W. Nijhoff, G. R. W. Quispel, and H. W. Capel, *Direct linearization of nonlinear difference-difference equations*, Phys. Lett. A **97** (1983), no. 4, 125–128, DOI 10.1016/0375-9601(83)90192-5. MR719638

[642] F. W. Nijhoff, A. Ramani, B. Grammaticos, and Y. Ohta, *On discrete Painlevé equations associated with the lattice KdV systems and the Painlevé VI equation*, Stud. Appl. Math. **106** (2001), no. 3, 261–314, DOI 10.1111/1467-9590.00167. MR1819383

[643] F. W. Nijhoff, Yu. B. Suris, and C.-M. Viallet, *Foreword [Proceedings of the Meeting on Symmetries and Integrability of Difference Equations (SIDE V)]*, J. Nonlinear Math. Phys. **10** (2003), no. suppl. 2, i–ii. Held at Giens, June 21–28, 2002. MR2062263

[644] F. W. Nijhoff and A. J. Walker, *The discrete and continuous Painlevé VI hierarchy and the Garnier systems*, Glasg. Math. J. **43A** (2001), 109–123, DOI 10.1017/S0017089501000106. Integrable systems: linear and nonlinear dynamics (Islay, 1999). MR1869690

[645] J. J. C. Nimmo and W. K. Schief, *Superposition principles associated with the Moutard transformation: an integrable discretization of a* $(2+1)$*-dimensional sine-Gordon system*, Proc. Roy. Soc. London Ser. A **453** (1997), no. 1957, 255–279, DOI 10.1098/rspa.1997.0015. MR1436830

[646] K. Nishinari and D. Takahashi, *Analytical properties of ultradiscrete Burgers equation and rule-184 cellular automaton*, J. Phys. A: Math. Gen. **31**:24 (1998), 5439–5450.

[647] E. Noether, *Invariante Variationsprobleme*, Nachr. Ges. Wiss. Goettingen, Math.-Phys. Kl. 2 (1918), 235–257; English version in Transp. Theory Stat. Phys. **1** (1971), 186–207.

[648] S. P. Novikov ed., *Integrable systems*, Selected Papers, Cambridge University Press, Cambridge, 1981.

[649] V. E. Zaharov, S. V. Manakov, S. P. Novikov, and L. P. Pitaevskiĭ, *Theory of Solitons* (Russian), "Nauka", Moscow, 1980. Metod obratnoĭ zadachi. [The method of the inverse problem]. MR573607

[650] M. C. Nucci and D. Levi, *λ-symmetries and Jacobi last multiplier*, Nonlinear Anal. Real World Appl. **14** (2013), no. 2, 1092–1101, DOI 10.1016/j.nonrwa.2012.09.001. MR2991136

[651] W. Oevel, *Rekursionsmechanismen für Symmetrien und Erhaltungssätze in integrablen Systemen* (German), Universität-Gesamthochschule-Paderborn, Fachbereich Mathematik-Informatik, Paderborn, 1984. Dissertation, Universität-Gesamthochschule Paderborn, Paderborn, 1984. MR843834

[652] W. Oevel and B. Fuchssteiner, *Explicit formulas for symmetries and conservation laws of the Kadomtsev-Petviashvili equation*, Phys. Lett. A **88** (1982), no. 7, 323–327, DOI 10.1016/0375-9601(82)90605-3. MR649207

[653] W. Oevel, B. Fuchssteiner, H. Zhang, and O. Ragnisco, *Mastersymmetries, angle variables, and recursion operator of the relativistic Toda lattice*, J. Math. Phys. **30** (1989), no. 11, 2664–2670, DOI 10.1063/1.528497. MR1019014

[654] W. Oevel, H. Zhang, and B. Fuchssteiner, *Mastersymmetries and multi-Hamiltonian formulations for some integrable lattice systems*, Progr. Theoret. Phys. **81** (1989), no. 2, 294–308, DOI 10.1143/PTP.81.294. MR997438

[655] Y. Ohta, K. M. Tamizhmani, B. Grammaticos, and A. Ramani, *Singularity confinement and algebraic entropy: the case of the discrete Painlevé equations*, Phys. Lett. A **262** (1999), no. 2-3, 152–157, DOI 10.1016/S0375-9601(99)00670-2. MR1729946

[656] M. A. del Olmo, M. A. Rodríguez, and P. Winternitz, *Simple subgroups of simple Lie groups and nonlinear differential equations with superposition principles*, J. Math. Phys. **27** (1986), no. 1, 14–23, DOI 10.1063/1.527381. MR816409

[657] M. A. del Olmo, M. A. Rodríguez, and P. Winternitz, *Superposition formulas for the rectangular matrix Riccati equations*, J. Math. Phys. **28** (1987), no. 3, 530–535, DOI 10.1063/1.527637. MR877224

[658] P. J. Olver, *Evolution equations possessing infinitely many symmetries*, J. Mathematical Phys. **18** (1977), no. 6, 1212–1215, DOI 10.1063/1.523393. MR521611

[659] P. J. Olver, *Applications of Lie groups to differential equations*, 2nd ed., Graduate Texts in Mathematics, vol. 107, Springer-Verlag, New York, 1993, DOI 10.1007/978-1-4612-4350-2. MR1240056

[660] P. J. Olver, *Equivalence, invariants, and symmetry*, Cambridge University Press, Cambridge, 1995, DOI 10.1017/CBO9780511609565. MR1337276

[661] P. J. Olver, *Moving frames*, J. Symbolic Comput. **36** (2003), no. 3-4, 501–512, DOI 10.1016/S0747-7171(03)00092-0. International Symposium on Symbolic and Algebraic Computation (ISSAC'2002) (Lille). MR2004039

[662] P. J. Olver and P. Rosenau, *Group-invariant solutions of differential equations*, SIAM J. Appl. Math. **47** (1987), no. 2, 263–278, DOI 10.1137/0147018. MR881350

[663] S. J. Orfanidis, *Discrete sine-Gordon equations*, Phys. Rev. D (3) **18** (1978), no. 10, 3822–3827, DOI 10.1103/PhysRevD.18.3822. MR515265

[664] A. Yu. Orlov and E. I. Schulman, *Additional symmetries for integrable equations and conformal algebra representation*, Lett. Math. Phys. **12** (1986), no. 3, 171–179, DOI 10.1007/BF00416506. MR865754

[665] A. Yu. Orlov and P. Winternitz, *Algebra of pseudodifferential operators and symmetries of equations in the Kadomtsev-Petviashvili hierarchy*, J. Math. Phys. **38** (1997), no. 9, 4644–4674, DOI 10.1063/1.532112. MR1468657

[666] L. V. Ovsjannikov, *Gruppovoĭ analiz differentsial′nykh uravneniĭ* (Russian), "Nauka", Moscow, 1978; English transl. in *Group analysis of differential equations*, Academic Press, New York, 1992. MR511921

[667] R. S. Palais, *The symmetries of solitons*, Bull. Amer. Math. Soc. (N.S.) **34** (1997), no. 4, 339–403, DOI 10.1090/S0273-0979-97-00732-5. MR1462745

[668] C. J. Papachristou, *Aspects of integrability of differential systems and fields: A mathematical primer for physicists*, SpringerBriefs in Physics, Springer, Cham, [2019] ©2019, DOI 10.1007/978-3-030-35002-4. MR3971595

[669] V. G. Papageorgiou, F. W. Nijhoff, and H. W. Capel, *Integrable mappings and nonlinear integrable lattice equations*, Phys. Lett. A **147** (1990), no. 2-3, 106–114, DOI 10.1016/0375-9601(90)90876-P. MR1063609

[670] V. G. Papageorgiou and A. G. Tongas, *Yang-Baxter maps and multi-field integrable lattice equations*, J. Phys. A **40** (2007), no. 42, 12677–12690, DOI 10.1088/1751-8113/40/42/S12. MR2392897

[671] A. D. Polyanin and V. F. Zaitsev, *Handbook of nonlinear partial differential equations*, Chapman & Hall/CRC, Boca Raton, FL, 2004. MR2042347

[672] L. Pauling, R. B. Corey, and H. R. Branson, *The structure of proteins: Two hydrogen-bonded helical configurations of the polypeptide chain*, PNAS **37**:4 (1951), 205–211.

[673] A. V. Penskoi and P. Winternitz, *Discrete matrix Riccati equations with superposition formulas*, J. Math. Anal. Appl. **294** (2004), no. 2, 533–547, DOI 10.1016/j.jmaa.2004.02.028. MR2061341

[674] P. L. Vu, *Inverse scattering problems and their application to nonlinear integrable equations*, CRC Press, Boca Raton, 2020.

[675] J. Praught and R. G. Smirnov, *Andrew Lenard: a mystery unraveled*, SIGMA Symmetry Integrability Geom. Methods Appl. **1** (2005), Paper 005, 7, DOI 10.3842/SIGMA.2005.005. MR2173592

[676] S. Pincherle and U. Amaldi, *Le operazioni distributive e le loro applicazioni all'analisi*, Zanichelli, Bologna, 1901.

[677] R. O. Popovych, V. M. Boyko, M. O. Nesterenko, and M. W. Lutfullin, *Realizations of real low-dimensional Lie algebras*, J. Phys. A **36** (2003), no. 26, 7337–7360, DOI 10.1088/0305-4470/36/26/309. MR2004893

[678] S. Post and P. Winternitz, *General Nth order integrals of motion in the Euclidean plane*, J. Phys. A **48** (2015), no. 40, 405201, 24, DOI 10.1088/1751-8113/48/40/405201. MR3405316

[679] E. Pucci and G. Saccomandi, *On the reduction methods for ordinary differential equations*, J. Phys. A **35** (2002), no. 29, 6145–6155, DOI 10.1088/0305-4470/35/29/314. MR1930776

[680] G. R. W. Quispel, H. W. Capel, V. G. Papageorgiou, and F. W. Nijhoff, *Integrable mappings derived from soliton equations*, Phys. A **173** (1991), no. 1-2, 243–266, DOI 10.1016/0378-4371(91)90258-E. MR1109691

[681] G. R. W. Quispel, H. W. Capel, and R. Sahadevan, *Continuous symmetries of differential-difference equations: the Kac-van Moerbeke equation and Painlevé reduction*, Phys. Lett. A **170** (1992), no. 5, 379–383, DOI 10.1016/0375-9601(92)90891-O. MR1190898

[682] G. R. W. Quispel, F. W. Nijhoff, H. W. Capel, and J. van der Linden, *Linear integral equations and nonlinear difference-difference equations*, Phys. A **125** (1984), no. 2-3, 344–380, DOI 10.1016/0378-4371(84)90059-1. MR761644

[683] G. R. W. Quispel and R. Sahadevan, *Lie symmetries and the integration of difference equations*, Phys. Lett. A **184** (1993), no. 1, 64–70, DOI 10.1016/0375-9601(93)90347-3. MR1254645

[684] O. Ragnisco, *Conservation laws for the whole class of nonlinear evolution equations associated to the matrix Schroedinger spectral problem*, Lett. Nuovo Cimento (2) **31** (1981), no. 18, 651–656. MR631460

[685] O. Ragnisco and G.-Z. Tu, *A new hierarchy of integrable discrete systems*, Preprint University of Rome, 1989.

[686] A. Ramani, A. S. Carstea, B. Grammaticos, and Y. Ohta, *On the autonomous limit of discrete Painlevé equations*, Phys. A **305** (2002), no. 3-4, 437–444, DOI 10.1016/S0378-4371(01)00619-7. MR1928121

[687] A. Ramani and B. Grammaticos, *Discrete Painlevé equations: coalescences, limits and degeneracies*, Phys. A **228** (1996), no. 1-4, 160–171, DOI 10.1016/0378-4371(95)00439-4. MR1399286

[688] A. Ramani and B. Grammaticos, *What is the discrete analogue of the Painlevé property?*, ANZIAM J. **44** (2002), no. 1, 21–32, DOI 10.1017/S1446181100007872. Kruskal, 2000 (Adelaide). MR1919922

[689] A. Ramani, B. Grammaticos, and J. Satsuma, *Bilinear discrete Painlevé equations*, J. Phys. A **28** (1995), no. 16, 4655–4665. MR1352181

[690] A. Ramani, B. Grammaticos, and K. M. Tamizhmani, *Painlevé analysis and singularity confinement: the ultimate conjecture*, J. Phys. A **26** (1993), no. 2, L53–L58. MR1210713

[691] A. Ramani, B. Grammaticos, J. Satsuma, and R. Willox, *On two (not so) new integrable partial difference equations*, J. Phys. A **42** (2009), no. 28, 282002, 6, DOI 10.1088/1751-8113/42/28/282002. MR2519724

[692] A. Ramani, B. Grammaticos, T. Tamizhmani, and K. M. Tamizhmani, *The road to the discrete analogue of the Painlevé property: Nevanlinna meets singularity confinement*: *Advances in difference equations, IV*, Comput. Math. Appl. **45** (2003), no. 6-9, 1001–1012, DOI 10.1016/S0898-1221(03)00076-2. MR2000573

[693] A. Ramani, N. Joshi, B. Grammaticos, and T. Tamizhmani, *Deconstructing an integrable lattice equation*, J. Phys. A **39** (2006), no. 8, L145–L149, DOI 10.1088/0305-4470/39/8/L01. MR2209299

[694] D. W. Rand, *PASCAL programs for identification of Lie algebras. I. RADICAL—a program to calculate the radical and nil radical of parameter-free and parameter-dependent Lie algebras*, Comput. Phys. Comm. **41** (1986), no. 1, 105–125, DOI 10.1016/0010-4655(86)90024-X. MR851302

- -, *PASCAL programs for identification of Lie algebras. III: Levi decomposition and canonical basis*, Comp. Phys. Comm. **46**:2 (1987), 311–322.

[695] D. W. Rand and P. Winternitz, *Nonlinear superposition principles: a new numerical method for solving matrix Riccati equations*, Comput. Phys. Comm. **33** (1984), no. 4, 305–328, DOI 10.1016/0010-4655(84)90136-X. MR770094

[696] D. W. Rand, P. Winternitz, and H. Zassenhaus, *PASCAL programs for the identification of Lie algebras. II. SPLIT—a program to decompose parameter-free and parameter-dependent Lie algebras into direct sums*, Comput. Phys. Comm. **46** (1987), no. 2, 297–309, DOI 10.1016/0010-4655(87)90006-3. MR905390

[697] D. Rand, P. Winternitz, and H. Zassenhaus, *On the identification of a Lie algebra given by its structure constants. I. Direct decompositions, Levi decompositions, and nilradicals*, Linear Algebra Appl. **109** (1988), 197–246, DOI 10.1016/0024-3795(88)90210-8. MR961578

[698] O. G. Rasin, *Infinitely many symmetries and conservation laws for quad-graph equations via the Gardner method*, J. Phys. A **43** (2010), no. 23, 235201, 11, DOI 10.1088/1751-8113/43/23/235201. MR2646677

[699] O. G. Rasin and P. E. Hydon, *Symmetries of integrable difference equations on the quad-graph*, Stud. Appl. Math. **119** (2007), no. 3, 253–269, DOI 10.1111/j.1467-9590.2007.00385.x. MR2349174

[700] O. G. Rasin and P. E. Hydon, *Conservation laws for integrable difference equations*, J. Phys. A **40** (2007), no. 42, 12763–12773, DOI 10.1088/1751-8113/40/42/S19. MR2392904

[701] R. Rebelo and F. Valiquette, *Symmetry preserving numerical schemes for partial differential equations and their numerical tests*, J. Difference Equ. Appl. **19** (2013), no. 5, 738–757, DOI 10.1080/10236198.2012.685470. MR3049052

[702] R. Rebelo and F. Valiquette, *Invariant discretization of partial differential equations admitting infinite-dimensional symmetry groups*, J. Difference Equ. Appl. **21** (2015), no. 4, 285–318, DOI 10.1080/10236198.2015.1007134. MR3326274

[703] R. Rebelo and P. Winternitz, *Invariant difference schemes and their application to* $\mathrm{sl}(2,\mathbb{R})$ *invariant ordinary differential equations*, J. Phys. A **42** (2009), no. 45, 454016, 10, DOI 10.1088/1751-8113/42/45/454016. MR2556658

[704] M. Reed and B. Simon, *Methods of modern mathematical physics. III*: *Scattering theory*, Academic Press [Harcourt Brace Jovanovich, Publishers], New York-London, 1979. MR529429

[705] G. J. Reid, *Algorithms for reducing a system of PDEs to standard form, determining the dimension of its solution space and calculating its Taylor series solution*, European J. Appl. Math. **2** (1991), no. 4, 293–318, DOI 10.1017/S0956792500000577. MR1142467

[706] G. J. Reid, *Finding abstract Lie symmetry algebras of differential equations without integrating determining equations*, European J. Appl. Math. **2** (1991), no. 4, 319–340, DOI 10.1017/S0956792500000589. MR1142468

[707] M. Remoissenet, *Waves called solitons*: *Concepts and experiments*, 3rd ed., Springer-Verlag, Berlin, 1999, DOI 10.1007/978-3-662-03790-4. MR1716486

[708] J. A. G. Roberts and Dinh T. Tran, *Algebraic entropy of (integrable) lattice equations and their reductions*, Nonlinearity **32** (2019), no. 2, 622–653, DOI 10.1088/1361-6544/aaecda. MR3903264

[709] T. M. Rocha Filho and A. Figueiredo, *[SADE] a Maple package for the symmetry analysis of differential equations*, Comp. Phys. Comm. **182**: 2 (2011), 46—476.

[710] M. A. Rodríguez and P. Winternitz, *Quantum superintegrability and exact solvability in n dimensions*, J. Math. Phys. **43** (2002), no. 3, 1309–1322, DOI 10.1063/1.1435077. MR1885008

[711] M. A. Rodríguez and P. Winternitz, *Lie symmetries and exact solutions of first-order difference schemes*, J. Phys. A **37** (2004), no. 23, 6129–6142, DOI 10.1088/0305-4470/37/23/011. MR2074627

[712] C. Rogers and W. F. Ames, *Nonlinear boundary value problems in science and engineering*, Mathematics in Science and Engineering, vol. 183, Academic Press, Inc., Boston, MA, 1989. MR1017035

[713] C. Rogers, W. K. Schief, and M. P. Stallybrass, *Initial/boundary value problems and Darboux-Levi-type transformations associated with a* $(2+1)$*-dimensional eigenfunction equation*, Internat. J. Non-Linear Mech. **30** (1995), no. 3, 223–233, DOI 10.1016/0020-7462(94)00045-C. MR1336911

[714] C. Rogers and W. K. Schief, *Bäcklund and Darboux transformations: Geometry and modern applications in soliton theory*, Cambridge Texts in Applied Mathematics, Cambridge University Press, Cambridge, 2002, DOI 10.1017/CBO9780511606359. MR1908706

[715] C. Rogers, W. K. Schief, and P. Winternitz, *Lie-theoretical generalization and discretization of the Pinney equation*, J. Math. Anal. Appl. **216** (1997), no. 1, 246–264, DOI 10.1006/jmaa.1997.5674. MR1487263

[716] C. Rogers and W. F. Shadwick, *Bäcklund transformations and their applications*, Mathematics in Science and Engineering, vol. 161, Academic Press, Inc. [Harcourt Brace Jovanovich, Publishers], New York-London, 1982. MR658491

[717] S. Roman, *The umbral calculus*, Pure and Applied Mathematics, vol. 111, Academic Press, Inc. [Harcourt Brace Jovanovich, Publishers], New York, 1984. MR741185

[718] S. Roman, *More on the umbral calculus, with emphasis on the q-umbral calculus*, J. Math. Anal. Appl. **107** (1985), no. 1, 222–254, DOI 10.1016/0022-247X(85)90367-1. MR786026

[719] G. C. Rota, *Finite operator calculus*, Academic Press, San Diego, 1975.

[720] A. Roy Chowdhury and G. Mahato, *A Darboux-Bäcklund transformation associated with a discrete nonlinear Schrödinger equation*, Lett. Math. Phys. **7** (1983), no. 4, 313–317, DOI 10.1007/BF00420181. MR719463

[721] S. N. M. Ruijsenaars, *Relativistic Toda systems*, Comm. Math. Phys. **133** (1990), no. 2, 217–247. MR1090424

[722] R. Sachs, *Asymptotic symmetries in gravitational theory*, Phys. Rev. (2) **128** (1962), 2851–2864. MR149909

[723] R. Sahadevan, *Nonlinear differential-difference and difference equations: integrability and exact solvability*, Comput. Math. Appl. **42** (2001), no. 3-5, 627–637, DOI 10.1016/S0898-1221(01)00182-1. Advances in difference equations, III. MR1838019

[724] R. Sahadevan, G. B. Byrnes, and G. R. W. Quispel, *Linearisation of difference equations using factorisable Lie symmetries*, Symmetries and integrability of difference equations (Estérel, PQ, 1994), CRM Proc. Lecture Notes, vol. 9, Amer. Math. Soc., Providence, RI, 1996, pp. 337–343, DOI 10.1090/crmp/009/30. MR1416850

[725] R. Sahadevan and S. Khousalya, *Similarity reduction of a* $(2+1)$ *Volterra system*, J. Phys. A **33** (2000), no. 19, L171–L176, DOI 10.1088/0305-4470/33/19/102. MR1769413

[726] R. Sahadevan and G. R. W. Quispel, *Lie symmetries and linearisation of the QRT mapping*, Phys. A **234** (1997), no. 3-4, 775–784, DOI 10.1016/S0378-4371(96)00321-4. MR1753356

[727] R. Sahadevan, O. G. Rasin, and P. E. Hydon, *Integrability conditions for nonautonomous quad-graph equations*, J. Math. Anal. Appl. **331** (2007), no. 1, 712–726, DOI 10.1016/j.jmaa.2006.09.026. MR2306034

[728] S. Saito, N. Saitoh, J. Yamamoto, and K. Yoshida, *A characterization of discrete time soliton equations*, J. Phys. Soc. Japan **70** (2001), no. 12, 3517–3523, DOI 10.1143/JPSJ.70.3517. MR1880642

[729] J. A. Sanders and J. P. Wang, *On the integrability of homogeneous scalar evolution equations*, J. Differential Equations **147** (1998), no. 2, 410–434, DOI 10.1006/jdeq.1998.3452. MR1633957

[730] J. A. Sanders and J. P. Wang, *Number theory and the symmetry classification of integrable systems*, Integrability, Lecture Notes in Phys., vol. 767, Springer, Berlin, 2009, pp. 89–118, DOI 10.1007/978-3-540-88111-7_3. MR2867547

[731] J. T. Sandevan, *Discrete dynamical systems. Theory and applications*, Oxford University Press, New York, 1990.

[732] P. M. Santini and A. S. Fokas, *Recursion operators and bi-Hamiltonian structures in multidimensions. I*, Comm. Math. Phys. **115** (1988), no. 3, 375–419. MR931668

[733] J. Satsuma J and T. Tokihiro eds., *SIDE IV–Symmetries and integrability of difference equations*, J. Phys. A: Math. Gen. **34**:48 (2001).

[734] J. Sherring, G. Prince and M. Jerie, *Dimsym - symmetry determination and linear differential equation package*, available in https://reduce-algebra.sourceforge.io/packages.php, 1996.

[735] W. K. Schief, *Generalized Darboux and Darboux-Levi transformations in 2 + 1-dimensions*, Ph.D. thesis, Loughborough University of Technology, UK, 1992.

[736] W. K. Schief, *A discrete Pinney equation*, Appl. Math. Lett. **10** (1997), no. 3, 13–15, DOI 10.1016/S0893-9659(97)00026-8. MR1381160

[737] W. K. Schief and C. Rogers, *Bäcklund transformations and the integrable discretisation of characteristic equations*, Chaos Solitons Fractals **11** (2000), no. 1-3, 107–113, DOI 10.1016/S0960-0779(98)00274-4. Integrability and chaos in discrete systems (Brussels, 1997). MR1729561

[738] P. C. Schuur, *Asymptotic analysis of soliton problems*, Lecture Notes in Mathematics, vol. 1232, Springer-Verlag, Berlin, 1986. An inverse scattering approach, DOI 10.1007/BFb0073054. MR874343

[739] W. Schwalm, B. Moritz, M. Giona, and M. Schwalm, *Vector difference calculus for physical lattice models*, Phys. Rev. E (3) **59** (1999), no. 1, 1217–1233, DOI 10.1103/PhysRevE.59.1217. MR1659511

[740] F. Schwarz, *A REDUCE package for determining Lie symmetries of ordinary and partial differential equations*, Comp. Phys. Comm. **27** (1982), 179–186.

[741] H. A. Schwarz, *Communication*, Arch. sci. phys. nat., **48** (1873), 38–44.

[742] C. Scimiterna, *Multiscale techniques for nonlinear difference equations*,PhD thesis 2009 (https://arcadia.sba.uniroma3.it/bitstream/2307/408/1/Tesi.pdf).

[743] C. Scimiterna and D. Levi, *C-integrability test for discrete equations via multiple scale expansions*, SIGMA Symmetry Integrability Geom. Methods Appl. **6** (2010), Paper 070, 17, DOI 10.3842/SIGMA.2010.070. MR2725013

[744] C. Scimiterna and D. Levi, *Integrability of differential-difference equations with discrete kinks*, Theoret. and Math. Phys. **167** (2011), no. 3, 826–842, DOI 10.1007/s11232-011-0066-2. Russian version appears in Teoret. Mat. Fiz. **167** (2011), no. 3, 496–513. MR3166388

[745] C. Scimiterna and D. Levi, *Three-point partial difference equations linearizable by local and nonlocal transformations*, J. Phys. A **46** (2013), no. 2, 025205, 13, DOI 10.1088/1751-8113/46/2/025205. MR3005898

[746] C. Scimiterna and D. Levi, *Classification of discrete equations linearizable by point transformation on a square lattice*, Front. Math. China **8** (2013), no. 5, 1067–1076, DOI 10.1007/s11464-013-0280-3. MR3091125

[747] A. C. Scott, *Davydov's soliton*, Phys. Rep. **217** (1992), 1–67.

[748] A. C. Scott (ed.), *Encyclopedia of nonlinear science*, Routledge, New York, 2005. MR2265413

[749] A. Sergyeyev, *New integrable* $(3 + 1)$*-dimensional systems and contact geometry*, Lett. Math. Phys. **108** (2018), no. 2, 359–376, DOI 10.1007/s11005-017-1013-4. MR3748368

[750] A. B. Shabat, *The infinite-dimensional dressing dynamical system*, Inverse Problems **8** (1992), no. 2, 303–308. MR1158181

[751] A. B. Shabat, *Dressing chains and lattices*, Proceedings of the Workshop on Nonlinearity, Integrability and All That: Twenty Years after NEEDS '79 (Gallipoli, 1999), World Sci. Publ., River Edge, NJ, 2000, pp. 331–342. MR1772998

[752] A. B. Shabat, *Universal models of soliton hierarchies* (Russian, with Russian summary), Teoret. Mat. Fiz. **136** (2003), no. 2, 197–208, DOI 10.1023/A:1025057719859; English transl., Theoret. and Math. Phys. **136** (2003), no. 2, 1066–1076. MR2025372

[753] A. B. Shabat and R. I. Yamilov, *Factorization of nonlinear equations of the Heisenberg model type*, Preprint Soviet Academy of Sciences, Ufa, 1987 [in Russian].

[754] A. B. Shabat and R. I. Yamilov, *Lattice representations of integrable systems*, Phys. Lett. A **130** (1988), no. 4-5, 271–275, DOI 10.1016/0375-9601(88)90608-1. MR949852

[755] A. B. Shabat and R. I. Yamilov, *Symmetries of nonlinear lattices* (Russian), Algebra i Analiz **2** (1990), no. 2, 183–208; English transl., Leningrad Math. J. **2** (1991), no. 2, 377–400. MR1062269

[756] A. B. Shabat and R. I. Yamilov, *To a transformation theory of two-dimensional integrable systems*, Phys. Lett. A **227** (1997), no. 1-2, 15–23, DOI 10.1016/S0375-9601(96)00922-X. MR1435916

[757] V. E. Zaharov and A. B. Shabat, *A scheme for integrating the nonlinear equations of mathematical physics by the method of the inverse scattering problem. I* (Russian), Funkcional. Anal. i Priložen. **8** (1974), no. 3, 43–53; English transl., Funct. Anal. Appl. **8** (1974), no. 3, 226–235. MR0481668

[758] I. P. Shestakov and V. V. Sokolov, *Multi-component generalizations of mKdV equation and nonassociative algebraic structures*, J. Algebra Appl. **20** (2021), no. 4, Paper No. 2150050, 24, DOI 10.1142/S021949882150050X. MR4251730

[759] S. Shnider and P. Winternitz, *Nonlinear equations with superposition principles and the theory of transitive primitive Lie algebras*, Lett. Math. Phys. **8** (1984), no. 1, 69–78, DOI 10.1007/BF00420043. MR731203

[760] S. Shnider and P. Winternitz, *Classification of systems of nonlinear ordinary differential equations with superposition principles*, J. Math. Phys. **25** (1984), no. 11, 3155–3165, DOI 10.1063/1.526085. MR761834

[761] L. Šnobl and P. Winternitz, *Classification and identification of Lie algebras*, CRM Monograph Series, vol. 33, American Mathematical Society, Providence, RI, 2014, DOI 10.1090/crmm/033. MR3184730

[762] V. V. Sokolov, *Symmetries of evolution equations* (Russian), Uspekhi Mat. Nauk **43** (1988), no. 5(263), 133–163, 240, DOI 10.1070/RM1988v043n05ABEH001927; English transl., Russian Math. Surveys **43** (1988), no. 5, 165–204. MR971467

[763] V. V. Sokolov, *Algebraic structures in integrability*, World Scientific Publishing Co. Pte. Ltd., Hackensack, NJ, [2020] ©2020. With a foreword by Victor Kac, DOI 10.1142/11809. MR4265175

[764] V. V. Sokolov and A. B. Shabat, *Classification of integrable evolution equations*, Mathematical physics reviews, Vol. 4, Soviet Sci. Rev. Sect. C: Math. Phys. Rev., vol. 4, Harwood Academic Publ., Chur, 1984, pp. 221–280. MR768941

[765] V. V. Sokolov and A. V. Zhiber, *On the Darboux integrable hyperbolic equations*, Phys. Lett. A **208** (1995), no. 4-6, 303–308, DOI 10.1016/0375-9601(95)00774-2. MR1363158

[766] M. Sorine and P. Winternitz, *Superposition laws for nonlinear equations arising in optimal control theory*, IEEE Transactions, AC **30** (1985), 266–272.

[767] S. Ya. Startsev, *On non-point invertible transformations of difference and differential-difference equations*, SIGMA Symmetry Integrability Geom. Methods Appl. **6** (2010), Paper 092, 14, DOI 10.3842/SIGMA.2010.092. MR2769923

[768] S. Ya. Startsev, *Darboux integrable discrete equations possessing an autonomous first-order integral*, J. Phys. A **47** (2014), no. 10, 105204, 16, DOI 10.1088/1751-8113/47/10/105204. MR3179736

[769] S. Ya. Startsev, *On Darboux non-integrability of Hietarinta equation*, Ufa Math. J. **13** (2021), no. 2, 160–169. MR4293326

[770] W.-H. Steeb, *Continuous symmetries, Lie algebras, differential equations and computer algebra*, World Scientific Publishing Co., Inc., River Edge, NJ, 1996, DOI 10.1142/3309. MR1626405

[771] H. Stephani, *Differential equations: Their solution using symmetries*, Cambridge University Press, Cambridge, 1989. MR1041800

[772] J. J. Stoker, *Water waves: The mathematical theory with applications*, Pure and Applied Mathematics, Vol. IV, Interscience Publishers, Inc., New York; Interscience Publishers Ltd., London, 1957. MR0103672

[773] O. Stormark, *Lie's structural approach to PDE systems*, Encyclopedia of Mathematics and its Applications, vol. 80, Cambridge University Press, Cambridge, 2000, DOI 10.1017/CBO9780511569456. MR1771254

[774] C. H. Su and C. S. Gardner, *Korteweg-de Vries equation and generalizations. III. Derivation of the Korteweg-de Vries equation and Burgers equation*, J. Mathematical Phys. **10** (1969), 536–539, DOI 10.1063/1.1664873. MR271526

[775] C. Sulem and P.-L. Sulem, *The nonlinear Schrödinger equation: Self-focusing and wave collapse*, Applied Mathematical Sciences, vol. 139, Springer-Verlag, New York, 1999. MR1696311

[776] Yu. B. Suris, *New integrable systems related to the relativistic Toda lattice*, J. Phys. A **30** (1997), no. 5, 1745–1761, DOI 10.1088/0305-4470/30/5/035. MR1450011

[777] Yu. B. Suris, *The problem of integrable discretization: Hamiltonian approach*, Progress in Mathematics, vol. 219, Birkhäuser Verlag, Basel, 2003, DOI 10.1007/978-3-0348-8016-9. MR1993935

[778] Yu. B. Suris, *Discrete time Toda systems*, J. Phys. A **51** (2018), no. 33, 333001, 64, DOI 10.1088/1751-8121/aacbdc. MR3829400

[779] S. I. Svinolupov, *On the analogues of the Burgers equation*, Phys. Lett. A **135** (1989), no. 1, 32–36, DOI 10.1016/0375-9601(89)90721-4. MR983341

[780] S. I. Svinolupov, *Jordan algebras and generalized Korteweg-de Vries equations* (Russian, with English summary), Teoret. Mat. Fiz. **87** (1991), no. 3, 391–403, DOI 10.1007/BF01017947; English transl., Theoret. and Math. Phys. **87** (1991), no. 3, 611–620. MR1129673

[781] S. I. Svinolupov, *Generalized Schrödinger equations and Jordan pairs*, Comm. Math. Phys. **143** (1992), no. 3, 559–575. MR1145599

[782] S. I. Svinolupov and V. V. Sokolov, *Factorization of evolution equations* (Russian, with Russian summary), Uspekhi Mat. Nauk **47** (1992), no. 3(285), 115–146, 207, DOI 10.1070/RM1992v047n03ABEH000895; English transl., Russian Math. Surveys **47** (1992), no. 3, 127–162. MR1185301

[783] S. I. Svinolupov and V. V. Sokolov, *Vector-matrix generalizations of classical integrable equations* (Russian, with English and Russian summaries), Teoret. Mat. Fiz. **100** (1994), no. 2, 214–218, DOI 10.1007/BF01016758; English transl., Theoret. and Math. Phys. **100** (1994), no. 2, 959–962 (1995). MR1311194

[784] S. I. Svinolupov, V. V. Sokolov, and R. I. Yamilov, *Bäcklund transformations for integrable evolution equations* (Russian), Dokl. Akad. Nauk SSSR **271** (1983), no. 4, 802–805. MR719823

[785] S. I. Svinolupov and R. I. Yamilov, *The multi-field Schrödinger lattices*, Phys. Lett. A **160** (1991), no. 6, 548–552, DOI 10.1016/0375-9601(91)91066-M. MR1141812

[786] S. I. Svinolupov and R. I. Yamilov, *Explicit self-transformations for multifield Schrödinger equations and Jordan generalizations of the Toda chain* (Russian, with English and Russian summaries), Teoret. Mat. Fiz. **98** (1994), no. 2, 207–219, DOI 10.1007/BF01015792; English transl., Theoret. and Math. Phys. **98** (1994), no. 2, 139–146. MR1291375

[787] *SymPy: Python library for symbolic mathematics*, SymPy Development Team, `http://www.sympy.org`, (2014)

[788] M. Tabor, *Chaos and integrability in nonlinear dynamics: An introduction*, A Wiley-Interscience Publication, John Wiley & Sons, Inc., New York, 1989. MR1007309

[789] P. Tamamidis and D. N. Assanis, *Generation of orthogonal grids with control of spacing*, J. Comput. Phys. **94** (1991), 437–453

[790] K. M. Tamizhmani, A. Ramani, B. Grammaticos, and Y. Ohta, *Integrability criteria for differential-difference systems: a comparison of singularity confinement and low-growth requirements*, J. Phys. A **32** (1999), no. 38, 6679–6685, DOI 10.1088/0305-4470/32/38/306. MR1733847

[791] P. Tempesta, P. Winternitz, J. Harnad, W. Miller Jr., G. Pogosyan, and M. Rodriguez (eds.), *Superintegrability in classical and quantum systems*, CRM Proceedings & Lecture Notes, vol. 37, American Mathematical Society, Providence, RI, 2004. Papers from the workshop held at the Université de Montréal, Montréal, QC, September 16–21, 2002, DOI 10.1090/crmp/037. MR2105429

[792] P. Tempesta, A. V. Turbiner, and P. Winternitz, *Exact solvability of superintegrable systems*, J. Math. Phys. **42** (2001), no. 9, 4248–4257, DOI 10.1063/1.1386927. MR1852545

[793] J. W. Thomas, *Numerical partial differential equations: finite difference methods*, Texts in Applied Mathematics, vol. 22, Springer-Verlag, New York, 1995, DOI 10.1007/978-1-4899-7278-1. MR1367964

[794] M. Toda, *Waves in nonlinear lattice*, Prog. Theor. Phys. Suppl. **45** (1970), 174–200.

[795] M. Toda, *Exact treatment of nonlinear lattice waves*, Prog. Theor. Phys. Suppl. **59** (1976), 1–35.

[796] M. Toda, *Theory of nonlinear lattices*, 2nd ed., Springer Series in Solid-State Sciences, vol. 20, Springer-Verlag, Berlin, 1989, DOI 10.1007/978-3-642-83219-2. MR971987

[797] M. Toda, *Nonlinear waves and solitons*, Mathematics and its Applications (Japanese Series), vol. 5, Kluwer Academic Publishers Group, Dordrecht; SCIPRESS, Tokyo, 1989. Translated from the Japanese. MR1029793

[798] M. Toda, *Vibration of a chain with nonlinear interaction*, J. Phys. Soc. Jpn. **22** (1967), 431–436.

[799] A. G. Tongas, D. Tsoubelis and V. G. Papageorgiou, *Symmetries and group invariant reductions of integrable partial difference equations*, in Proc. 10th Int. Conf. in Modern Group Analysis, (N H Ibragimov, C Sophocleous and P A Damianou eds.), University of Cyprus, 2004, 222–230.

[800] A. Tongas, D. Tsoubelis, and P. Xenitidis, *Affine linear and D_4 symmetric lattice equations: symmetry analysis and reductions*, J. Phys. A **40** (2007), no. 44, 13353–13384, DOI 10.1088/1751-8113/40/44/015. MR2384000

[801] S. Tremblay, B. Grammaticos, and A. Ramani, *Integrable lattice equations and their growth properties*, Phys. Lett. A **278** (2001), no. 6, 319–324, DOI 10.1016/S0375-9601(00)00806-9. MR1849830

[802] F. Tremblay, A. V. Turbiner, and P. Winternitz, *An infinite family of solvable and integrable quantum systems on a plane*, J. Phys. A **42** (2009), no. 24, 242001, 10, DOI 10.1088/1751-8113/42/24/242001. MR2515526

[803] T. Trogdon and S. Olver, *Riemann-Hilbert problems, their numerical solution, and the computation of nonlinear special functions*, Society for Industrial and Applied Mathematics (SIAM), Philadelphia, PA, 2016. MR3450072

[804] A. Turbiner and P. Winternitz, *Solutions of nonlinear differential and difference equations with superposition formulas*, Lett. Math. Phys. **50** (1999), no. 3, 189–201, DOI 10.1023/A:1007602926351. MR1762299

[805] M. G. Tzitzeica, *Sur une nouvelle classes de surfaces*, C. R. Acad. Sci. Paris **144** (1907), 1257–1259.

[806] M. G. Tzitzeica, *Sur une nouvelle classe de surfaces*, C. R. Acad. Sci. Paris **150** (1910), 955–956.

[807] C. U. Maheswari and R. Sahadevan, *On the conservation laws for nonlinear partial difference equations*, J. Phys. A **44** (2011), no. 27, 275203, 16, DOI 10.1088/1751-8113/44/27/275201. MR2805269

[808] F. Valiquette, *Discretizations preserving all Lie point symmetries of the Korteweg-de Vries equation*, Group Theoretical Methods in Physics: Proceedings of the XXV International Colloquium on Group Theoretical Methods in Physics, Cocoyoc, Mexico, 2–6 August, 2004, (G S Pogosyan, L E Vicent, and K B Wolf eds.), J. Phys. Conf. Ser. **185**, CRC Press, Baton Rouge, 2005, 539–544.

[809] F. Valiquette and P. Winternitz, *Discretization of partial differential equations preserving their physical symmetries*, J. Phys. A **38** (2005), no. 45, 9765–9783, DOI 10.1088/0305-4470/38/45/004. MR2198765

[810] P. H. van der Kamp, *Initial value problems for lattice equations*, J. Phys. A **42** (2009), no. 40, 404019, 16, DOI 10.1088/1751-8113/42/40/404019. MR2544283

[811] P. H. van der Kamp and J. A. Sanders, *On testing integrability*, J. Nonlinear Math. Phys. **8** (2001), no. 4, 561–574, DOI 10.2991/jnmp.2001.8.4.8. MR1862578

[812] A. P. Veselov and A. B. Shabat, *A dressing chain and the spectral theory of the Schrödinger operator* (Russian, with Russian summary), Funktsional. Anal. i Prilozhen. **27** (1993), no. 2, 1–21, 96, DOI 10.1007/BF01085979; English transl., Funct. Anal. Appl. **27** (1993), no. 2, 81–96. MR1251164

[813] E. Vessiot, *Sur les équations aux dérivées partielles du second ordre, $F(x, y, z, p, q, r, s, t) = 0$, intégrables par la méthode de Darboux* (French), J. Math. Pures Appl. (9) **21** (1942), 1–66. MR8867

[814] C.-M. Viallet, *Algebraic entropy for lattice equations*, (2006), `arXiv:math-ph/0609043v2`.

[815] C.-M. Viallet, *Integrable lattice maps: Q_V, a rational version of Q_4*, Glasg. Math. J. **51** (2009), no. A, 157–163, DOI 10.1017/S0017089508004874. MR2481234

[816] C.-M. Viallet, *Algebraic entropy for differential–delay equations*, (2014), `arXiv:1408.6161`.

[817] C.-M. Viallet, *On the algebraic structure of rational discrete dynamical systems*, J. Phys. A **48** (2015), no. 16, 16FT01, 21, DOI 10.1088/1751-8113/48/16/16FT01. MR3335696

[818] C.-M. Viallet, *Features of discrete integrability*, in Proceedings of the International Conference in Quantum Theory and Symmetries, Montréal 2019, (R MacKenze, M Paranjape, Z Thomova, P Winternitz and W Witczak-Kremp eds.), Springer, Berlin, 2021, 21–35.

[819] J. Villarroel, S. Chakravarty, and M. J. Ablowitz, *On a* $2+1$ *Volterra system*, Nonlinearity **9** (1996), no. 5, 1113–1128, DOI 10.1088/0951-7715/9/5/004. MR1416470

[820] A. M. Vinogradov ed., *Symmetries of partial differential equations conservation laws - applications - algorithms*, Springer, Dordrecht, 1989.

[821] V. Volterra, *Leçons sur la théorie mathématique de la lutte pour la vie* (French), Les Grands Classiques Gauthier-Villars. [Gauthier-Villars Great Classics], Éditions Jacques Gabay, Sceaux, 1990. Reprint of the 1931 original. MR1189803

[822] M. Wadati, *Transformation theories for nonlinear discrete systems*, Suppl. Progr. Theor. Phys. **59** (1976), 36–63.

[823] H. D. Wahlquist and F. B. Estabrook, *Bäcklund transformation for solutions of the Korteweg-de Vries equation*, Phys. Rev. Lett. **31** (1973), 1386–1390, DOI 10.1103/PhysRevLett.31.1386. MR415100

[824] A. J. Walker, *Similarity reductions and integrable lattice equations*, University of Leeds, Ph.D. thesis, 2001.Alain Robbe-Grillet

[825] J. Weiss, M. Tabor, and G. Carnevale, *The Painlevé property for partial differential equations*, J. Math. Phys. **24** (1983), no. 3, 522–526, DOI 10.1063/1.525721. MR692140

[826] R. M. Williams, *Discrete quantum gravity*, J. Phys.: Conf. Ser. **33** (2006), 38–48.

[827] P. Winternitz, *Nonlinear action of Lie groups and superposition principles for nonlinear differential equations*, Phys. A **114** (1982), no. 1-3, 105–113, DOI 10.1016/0378-4371(82)90268-0. MR678391

[828] P. Winternitz, *Comments on superposition rules for nonlinear coupled first-order differential equations*, J. Math. Phys. **25** (1984), no. 7, 2149–2150, DOI 10.1063/1.526425. MR748392

[829] P. Winternitz, *Kac-Moody-Virasoro symmetries of integrable nonlinear partial differential equations*, Symmetries and nonlinear phenomena (Paipa, 1988), CIF Ser., vol. 9, World Sci. Publ., Teaneck, NJ, 1988, pp. 358–375. MR1119865

[830] P. Winternitz, *Lie groups and solutions of nonlinear partial differential equations*, Integrable systems, quantum groups, and quantum field theories (Salamanca, 1992), NATO Adv. Sci. Inst. Ser. C: Math. Phys. Sci., vol. 409, Kluwer Acad. Publ., Dordrecht, 1993, pp. 429–495. MR1262843

[831] P. Winternitz, *Lie groups and difference equations*, Group22: Proceedings of the XXII International Colloquium in Group Theoretical Methods in Physics (Hobart, 1998), Int. Press, Cambridge, MA, 1999, pp. 113–124. MR1689901

[832] P. Winternitz, *Symmetries of discrete systems*, Discrete integrable systems, Lecture Notes in Phys., vol. 644, Springer, Berlin, 2004, pp. 185–243, DOI 10.1007/978-3-540-40357-9_6. MR2087742

[833] P. Winternitz, *Lie point symmetries of difference equations and symmetry preserving discretization of differential equations*, Symmetries and integrability of difference equations, (D Levi, P J Olver, Z Thomova and P Winternitz eds.), Cambridge University Press, Cambridge, 2011, 292–341.

[834] P. Winternitz and D. Levi, *Continuous symmetries of differential-difference equations*, Modern group theoretical methods in physics (Paris, 1995), Math. Phys. Stud., vol. 18, Kluwer Acad. Publ., Dordrecht, 1995, pp. 271–281. MR1361455

[835] G. B. Whitham, *Non-linear dispersive waves*, Proc. Roy. Soc. London Ser. A **283** (1965), 238–261, DOI 10.1098/rspa.1965.0019. MR176724

[836] T. Wolf, *The symbolic integration of exact PDEs*, J. Symbolic Comput. **30** (2000), no. 5, 619–629, DOI 10.1006/jsco.2000.1069. MR1797273

[837] P. D. Xenitidis, *Integrability and symmetries of difference equations: the Adler-Bobenko-Suris case*, Group analysis of differential equations and integrable systems, [s.n.], [s.l.], 2009, pp. 226–242. MR2759029

[838] P. D. Xenitidis, *Determining the symmetries of difference equations*, Proc. A. **474** (2018), no. 2219, 20180340, 20, DOI 10.1098/rspa.2018.0340. MR3895993

[839] P. D. Xenitidis and V. G. Papageorgiou, *Symmetries and integrability of discrete equations defined on a black-white lattice*, J. Phys. A **42** (2009), no. 45, 454025, 13, DOI 10.1088/1751-8113/42/45/454025. MR2556667

[840] R. I. Yamilov, *On conservation laws for the difference Korteweg-de Vries equation*, Din. Splosh. Sredy, **44** (1980), 164–173, Russian Academy of Sciences - RAS (Rossiĭskaya Akademiya Nauk - RAN), Siberian

Branch (Sibirskoe Otdelenie), Institute of Hydrodynamics named after M. A. Lavrent'eva (Institut Gidrodinamiki Im. M. A. Lavrent'eva), Novosibirsk [in Russian].

[841] R. I. Yamilov, *On the classification of discrete equations*, in Integrable systems, (A.B. Shabat ed.), Soviet Academy of Sciences, Ufa, 1982, 95–114 [in Russian].

[842] R. I. Yamilov, *Classification of discrete evolution equations*, Uspekhi Mat. Nauk **38**:6 (1983), 155–156 [in Russian].

[843] R. I. Yamilov, *Discrete equations of the form $du_n/dt = F(u_{n-1}, u_n, u_{n+1})$ ($n \in \mathbb{Z}$) with an infinite number of local conservation laws*, dissertation for Candidate of Science, Leningrad Branch of Steklov Mathematical Institute, Soviet Academy of Sciences, Leningrad, 1984 [in Russian].

[844] R. I. Yamilov, *Invertible substitutions of variables generated by Bäcklund transformations* (Russian, with English summary), Teoret. Mat. Fiz. **85** (1990), no. 3, 368–375, DOI 10.1007/BF01018403; English transl., Theoret. and Math. Phys. **85** (1990), no. 3, 1269–1275 (1991). MR1099132

[845] R. I. Yamilov, *Generalizations of the Toda lattice, and conservation laws*, Preprint Soviet Academy of Sciences, Ufa, 1989 (21pp) [in Russian]; English version in *Classification of Toda type scalar lattices*, in Proc. 8th Int. Workshop on Nonlinear Evolution Equations and Dynamical Systems NEEDS'92 (Dubna, Russia, 6–17 July 1992, (V. Makhankov, I. Puzynin and O. Pashaev eds.), World Scientific, Singapore, 1993, 423–431.

[846] R. I. Yamilov, *On the construction of Miura type transformations by others of this kind*, Phys. Lett. A **173** (1993), no. 1, 53–57, DOI 10.1016/0375-9601(93)90086-F. MR1201789

[847] R. I. Yamilov, *Construction scheme for discrete Miura transformations*, J. Phys. A **27** (1994), no. 20, 6839–6851. MR1306834

[848] R. I. Yamilov, *Symmetry approach to the classification from the standpoint of integrable differential difference equations. transformation theory*, Dissertation for Doctor of Science (Ufa: Russian Academy of Sciences) 2000 [in Russian].

[849] R. I. Yamilov, *Relativistic Toda chains and Schlesinger transformations* (Russian, with Russian summary), Teoret. Mat. Fiz. **139** (2004), no. 2, 209–224, DOI 10.1023/B:TAMP.0000026180.34362.c5; English transl., Theoret. and Math. Phys. **139** (2004), no. 2, 623–635. MR2084554

[850] R. I. Yamilov, *Symmetries as integrability criteria for differential difference equations*, J. Phys. A **39** (2006), no. 45, R541–R623, DOI 10.1088/0305-4470/39/45/R01. MR2275903

[851] R. I. Yamilov, *Integrability conditions for analogues of the relativistic Toda chain* (Russian, with Russian summary), Teoret. Mat. Fiz. **151** (2007), no. 1, 66–80, DOI 10.1007/s11232-007-0037-9; English transl., Theoret. and Math. Phys. **151** (2007), no. 1, 492–504. MR2347303

[852] R. I. Yamilov and D. Levi, *Integrability conditions for n and t dependent dynamical lattice equations*, J. Nonlinear Math. Phys. **11** (2004), no. 1, 75–101, DOI 10.2991/jnmp.2004.11.1.6. MR2031212

[853] J. Yang, *Nonlinear waves in integrable and nonintegrable systems*, Mathematical Modeling and Computation, vol. 16, Society for Industrial and Applied Mathematics (SIAM), Philadelphia, PA, 2010, DOI 10.1137/1.9780898719680. MR2721855

[854] Y. Yang, *Solitons in field theory and nonlinear analysis*, Springer Monographs in Mathematics, Springer-Verlag, New York, 2001, DOI 10.1007/978-1-4757-6548-9. MR1838682

[855] W. H. Young, *On the conditions for the reversibility of the order of partial differentiation*, Proc. Royal Soc. Edinburgh **29** (1908-09), 136–164.

[856] N. J. Zabusky, *A synergetic approach to problems of nonlinear dispersive wave propagation and interaction*, Proceedings of the Symposium on Nonlinear Partial Differential Equations, (W Ames ed.), Academic Press Inc., New York, 1967, 223–258.

[857] N. J. Zabusky and M. D. Kruskal, *Interaction of "solitons" in a collisionless plasma and the recurrence of the initial states*, Phys. Rev. Lett **15** (1965), 240–243.

[858] V. E. Zakharov (ed.), *What is integrability?*, Springer Series in Nonlinear Dynamics, Springer-Verlag, Berlin, 1991, DOI 10.1007/978-3-642-88703-1. MR1098334

[859] V. E. Zakharov and B. G. Konopel′chenko, *On the theory of recursion operator*, Comm. Math. Phys. **94** (1984), no. 4, 483–509. MR763748

[860] V. E. Zakharov and S. V. Manakov, *Soliton theory*, Sov. Sci. Rev. Sect. A: Physics Reviews **1** (1979), 133–190 [in Russian]; English version in Harwood Academic Publishers, London, 1979.

[861] V. E. Zakharov and A. V. Mikhailov, *Relativistically invariant two-dimensional models of field theory which are integrable by means of the inverse scattering problem method*; Russian transl., Soviet Phys. JETP **74** (1978), no. 6, 1953–1973. MR524247

[862] V. E. Zakharov and A. B. Shabat, *Exact theory of two-dimensional self-focusing and one-dimensional self-modulation of waves in nonlinear media* (Russian, with English summary), Ž. Èksper. Teoret. Fiz. **61** (1971), no. 1, 118–134; English transl., Soviet Physics JETP **34** (1972), no. 1, 62–69. MR0406174

[863] V. E. Zaharov and A. B. Shabat, *Integration of the nonlinear equations of mathematical physics by the method of the inverse scattering problem. II* (Russian), Funktsional. Anal. i Prilozhen. **13** (1979), no. 3, 13–22. MR545363

[864] Y. Zarmi, *The Bertrand theorem revisited*, Am. J. Phys. **70**:4 (2002), 446-449.

[865] Y. Zarmi, *Sine-Gordon equation in (1+2) and (1+3) dimensions: Existence and classification of traveling-wave solutions*, PLOS ONE **10**:5 (2015), e0124306 (21pp).

[866] H. Zhang, G. Z. Tu, W. Oevel, and B. Fuchssteiner, *Symmetries, conserved quantities, and hierarchies for some lattice systems with soliton structure*, J. Math. Phys. **32** (1991), no. 7, 1908–1918, DOI 10.1063/1.529205. MR1112724

[867] A. V. Zhiber, N. H. Ibragimov, and A. B. Shabat, *Liouville-type equations* (Russian), Dokl. Akad. Nauk SSSR **249** (1979), no. 1, 26–29; English transl., Sov. Math. Dokl. **20** (1979), 607–609. MR553975

[868] A. V. Zhiber, R. D. Murtazina, I. T. Habibullin, and A. B. Shabat, *Characteristic Lie rings and integrable models in mathematical physics*, Ufa Math. J. **4** (2012), no. 3, 17–85. MR3429920

[869] A. V. Zhiber and A. B. Shabat, *The Klein-Gordon equation with nontrivial group* (Russian), Dokl. Akad. Nauk SSSR **247** (1979), no. 5, 1103–1107; English transl., Sov. Phys. Dokl. **24** (1979), 607–609. MR550472

[870] A. V. Zhiber and A. B. Shabat, *Systems of equations $u_x = p(u, v)$, $v_y = q(u, v)$ that possess symmetries* (Russian), Dokl. Akad. Nauk SSSR **277** (1984), no. 1, 29–33; English transl., Sov. Math. Dokl. **30** (1984), 23–26. MR757064

[871] A. V. Zhiber and V. V. Sokolov, *A new example of a nonlinear hyperbolic equation possessing integrals* (Russian, with Russian summary), Teoret. Mat. Fiz. **120** (1999), no. 1, 20–26, DOI 10.1007/BF02557393; English transl., Theoret. and Math. Phys. **120** (1999), no. 1, 834–839. MR1737201

[872] A. V. Zhiber and V. V. Sokolov, *Exactly integrable hyperbolic equations of Liouville type* (Russian, with Russian summary), Uspekhi Mat. Nauk **56** (2001), no. 1(337), 63–106, DOI 10.1070/rm2001v056n01ABEH000357; English transl., Russian Math. Surveys **56** (2001), no. 1, 61–101. MR1845643

[873] A. V. Zhiber, V. V. Sokolov, and S. Ya. Startsev, *On nonlinear Darboux-integrable hyperbolic equations* (Russian), Dokl. Akad. Nauk **343** (1995), no. 6, 746–748; English transl., Doklady Math. **52** (1995), 128–130. MR1359404

Subject Index

Selected Published Titles in This Series

SELECTED PUBLISHED TITLES IN THIS SERIES

5 **Andrew Bruckner,** Differentiation of Real Functions, 1994

4 **David Ruelle,** Dynamical Zeta Functions for Piecewise Monotone Maps of the Interval

3 **V. Kumar Murty,** Introduction to Abelian Varieties, 1993

2 **M. Ya. Antimirov, A. A. Kolyshkin, and Remi Vaillancourt,** Applied Integral Transforms, 1993

1 **Dan Voiculescu, Alexandru Nica, and Kenneth J Dykema,** Free Random Variables, 1992